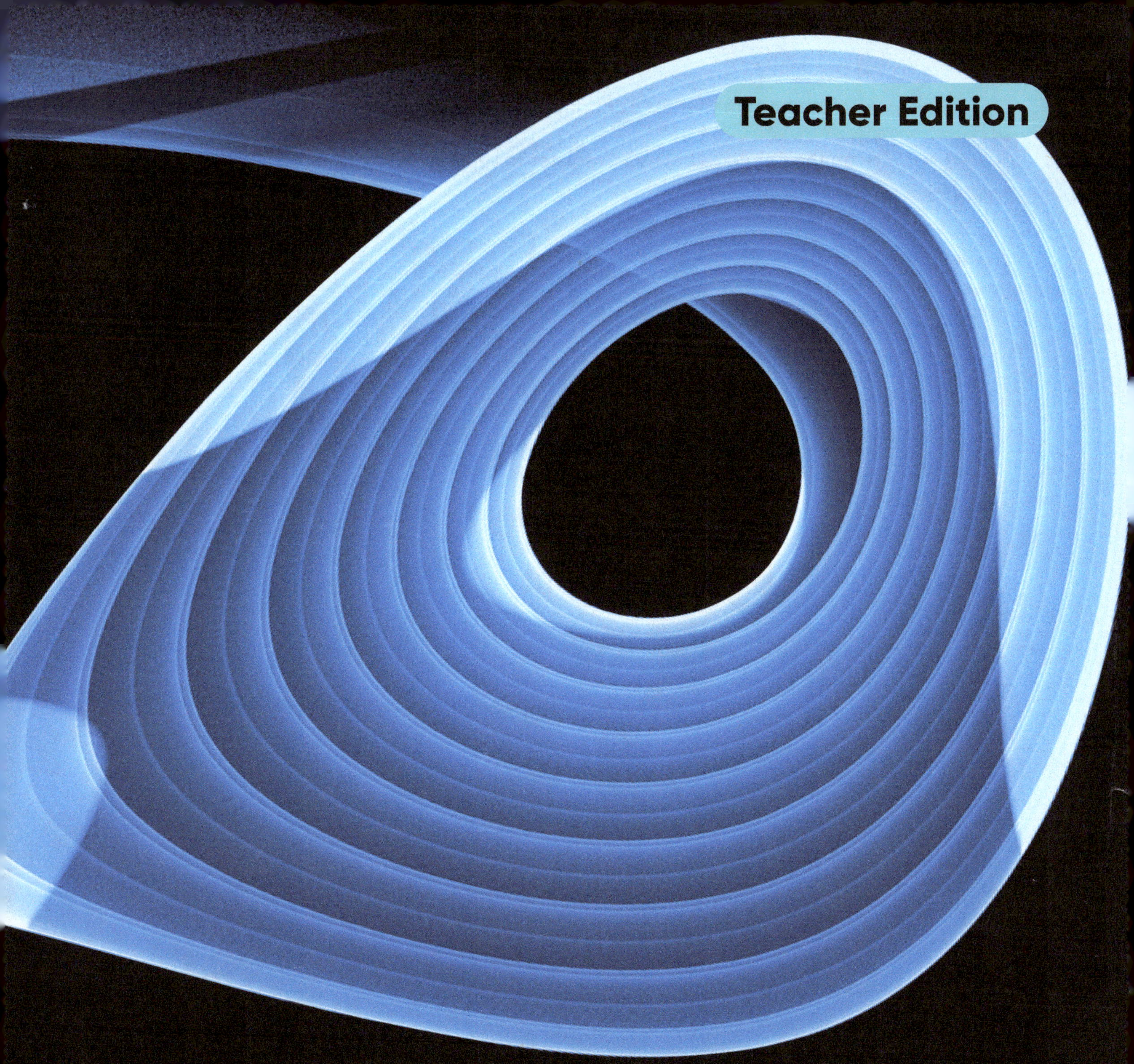

Teacher Edition

Grade 8

mathspace

2023 Virginia SOL

The front matter for our print textbooks will be ready soon.

Scan the QR code below to register for updates when it's available.

Mathspace Inc.
228 Park Ave S #15992
New York NY 10003-1502

For our full digital offering, visit mathspace.co

Mathspace Virginia: Grade 8 for 2023 SOLs - Print Teacher Edition

ISBN: 978-1-963022-17-9

Contents

1 Real Numbers

Big ideas

- Real numbers are either rational or irrational.
- The properties of real numbers can be applied to many types of expressions.
- Percents are useful for comparing a quantity to a whole amount.

Chapter outline

Square roots pop up in nature! The ratio between the spirals in seashells and pinecones can be closely estimated using square roots and Fibonacci numbers.

1. Real Numbers

Topic overview

Foundational knowledge

Evaluating standards proficiency

The skills book contains questions matched to individual standards. It can be used to measure proficiency for each.

Students should be proficient in these standards.

7.NS.2 — The student will reason and use multiple strategies to compare and order rational numbers.

7.NS.3 — The student will recognize and describe the relationship between square roots and perfect squares.

7.CE.2 — The student will solve problems, including those in context, involving proportional relationships.

Big ideas and essential understanding

Real numbers are either rational or irrational.
1.01 — A rational number is a real number that can be written as a ratio of two integers. An irrational number is a real number that cannot be written as a ratio of two integers.

1.02 — The square root of a number that is not a perfect square is always irrational.

The properties of real numbers can be applied to many types of expressions.
1.02 — The square root of a whole number that is not a perfect square always lies between two natural numbers.

1.03 — Comparing rational numbers in different forms can easily be done by converting them first to the same form.

1.04 — An increase of $n\%$ is the same as multiplying a given number by $(100 + n)\%$. A decrease of $n\%$ is the same as multiplying a given number by $(100 - n)\%$.

Percents are useful for comparing a quantity to a whole amount.
1.05 — There are many real-world applications of percents. Discounts are applications of percent decrease. Fees and markups are applications of percent increase.

1.06 — There are many real-world applications of percents. Sales tax, commissions, and tips are applications of percent increase.

Standards

8.NS.1 — The student will compare and order real numbers and determine the relationships between real numbers.

8.NS.1a — Estimate and identify the two consecutive natural numbers between which the positive square root of a given number lies and justify which natural number is the better approximation. Numbers are limited to natural numbers from 1 to 400.
1.02 Estimate square roots

8.NS.1b — Use rational approximations (to the nearest hundredth) of irrational numbers to compare, order, and locate values on a number line. Radicals may include both positive and negative square roots of values from 0 to 400 yielding an irrational number.
1.02 Estimate square roots
1.03 Compare and order real numbers

8.NS.1c — Use multiple strategies (e.g., benchmarks, number line, equivalency) to compare and order no more than five real numbers expressed as integers, fractions (proper or improper), decimals, mixed numbers, percents, numbers written in scientific notation, radicals, and π. Radicals may include both positive and negative square roots of values from 0 to 400. Ordering may be in ascending or descending order. Justify solutions orally, in writing or with a model.
1.03 Compare and order real numbers

8.NS.2 — The student will investigate and describe the relationship between the subsets of the real number system.

8.NS.2a — Describe and illustrate the relationships among the subsets of the real number system by using representations (e.g., graphic organizers, number lines). Subsets include rational numbers, irrational numbers, integers, whole numbers, and natural numbers.
1.01 The real number system

8.NS.2b — Classify and explain why a given number is a member of a particular subset or subsets of the real number system.
1.01 The real number system

8.NS.2c — Describe each subset of the set of real numbers and include examples and non-examples.
1.01 The real number system

8.CE.1 — The student will estimate and apply proportional reasoning and computational procedures to solve contextual problems.

8.CE.1a — Estimate and solve contextual problems that require the computation of one discount or markup and the resulting sale price.
1.05 Discounts, fees, and markups

8.CE.1b — Estimate and solve contextual problems that require the computation of the sales tax, tip and resulting total.
1.06 Sales tax, commissions and tips

8.CE.1c — Estimate and solve contextual problems that require the computation of the percent increase or decrease.
1.04 Percent increase and decrease
1.05 Discounts, fees, and markups
1.06 Sales tax, commissions and tips

Future connections

A2.EO.4 — The student will perform operations on complex numbers.

A.EO.1 — The student will represent verbal quantitative situations algebraically and evaluate these expressions for given replacement values of the variables.

A.EO.4 — The student will simplify and determine equivalent radical expressions involving square roots of whole numbers and cube roots of integers.

A.F.2 — The student will investigate, analyze, and compare characteristics of functions, including quadratic and exponential functions, and model quadratic and exponential relationships.

1.01 The real number system

Subtopic overview

Lesson narrative

In this lesson, students explore the real number system, including its various subsets: natural numbers, whole numbers, integers, rational numbers, and irrational numbers. The lesson provides a historical context and visual aids to help students understand how each subset fits within the larger set of real numbers. Students will classify numbers, identify their subsets, and solve problems involving these classifications. By the end of the lesson, students should be able to describe the relationships among the subsets of the real number system and accurately classify numbers into these subsets.

Learning objectives

Students: Page 4

> 🎓 **After this lesson, you will be able to...**
> - use visual representations to describe and illustrate the relationships among the subsets of the real number system.
> - classify a given number using a subset or subsets of the real number system.
> - explain why a given number is a member of a particular subset or subsets of the real number system.
> - describe each subset of the set of real numbers.
> - write examples and non-examples for each subset of the set of real numbers.

Key vocabulary

- decimal expansion
- density property
- integer
- irrational number
- natural number
- rational number
- real number system
- whole number

Essential understanding

A rational number is a real number that can be written as a ratio of two integers. An irrational number is a real number that cannot be written as a ratio of two integers.

Standards

This subtopic addresses the following Virginia 2023 Mathematics Standards of Learning standards.

Mathematical process goals

MPG2 — Mathematical Communication

Teachers can facilitate mathematical communication by encouraging students to express their understanding and reasoning about the real number system using mathematical language. For example, students may be asked to justify their classification of a number as rational or irrational, using appropriate mathematical terminology. Additionally, teachers can promote mathematical discussions in class where students can share their ideas about the relationships among the subsets of the real number system.

MPG3 — Mathematical Reasoning

Teachers can promote mathematical reasoning by guiding students to use logical reasoning to understand the relationships among the subsets of the real number system. For example, students can be asked to explain why an integer is also a rational number, and why a number can't be both rational and irrational. Teachers can also encourage students to use inductive reasoning to predict the classification of a number based on its properties.

MPG4 — Mathematical Connections

Teachers can help students make mathematical connections by relating the concept of the real number system to other mathematical concepts they have learned, such as rational numbers, square roots, and perfect squares. For instance, the concept of square roots can be connected to the classification of rational and irrational numbers. Teachers can also illustrate how the real number system applies to real-world contexts, such as measuring finite quantities or representing data.

MPG5 — Mathematical Representations

Teachers can help students create mathematical representations by encouraging students to organize the real number system using a graphic organizer. This will allow students to visualize and make connections among the subsets of numbers. These mathematical representations will combine written description with visual and symbolic representations of the subsets of the real number system.

Content standards

8.NS.2 — The student will investigate and describe the relationship between the subsets of the real number system.

8.NS.2a — Describe and illustrate the relationships among the subsets of the real number system by using representations (e.g., graphic organizers, number lines). Subsets include rational numbers, irrational numbers, integers, whole numbers, and natural numbers.

8.NS.2b — Classify and explain why a given number is a member of a particular subset or subsets of the real number system.

8.NS.2c — Describe each subset of the set of real numbers and include examples and non-examples.

Prior connections

7.NS.2 — The student will reason and use multiple strategies to compare and order rational numbers.

Future connections

A2.EO.4 — The student will perform operations on complex numbers.

Lesson Preparation

Suggested review

Depending on your students' level of prior knowledge, consider revisiting the following lessons:

Grade 7 — 1.01 Compare and order rational numbers

Tools

You may find these tools helpful:

- Frayer model graphic organizer

Lesson supports

The following supports may be useful for this lesson. More specific supports may appear throughout the lesson:

Graphic organizer of terms
Targeted instructional strategies

Have students create a graphic organizer, such as the Frayer model, of the types of numbers. Students should use one Frayer model graphic organizer per number type:

- Whole numbers
- Integers
- Rational numbers
- Irrational numbers
- Real numbers

The number type should be written in the center of the Frayer model. Provide students with some time to fill in what they can remember, then have them meet with three different partners to complete their templates.

Collect and display
English language learner support

As students are working, note how they describe the various types of numbers. Collect the different ways that students find to understand these concepts and display them in a common place for the students to access.

If students do not come up with alternative ways to word these concepts and are confused by them, suggest some of your own. For example:

- Natural numbers
  - Counting numbers
  - Whole numbers from 1 to infinity
- Whole numbers
  - Natural numbers and 0
  - Whole numbers from 0 to infinity
- Integers
  - Positive and negative whole numbers
  - Positive and negative numbers without fractions or decimals, including zero
- Rational numbers
  - Fractions
  - An integer divided by an integer
  - Can be written as decimals that end or repeat
- Irrational numbers
  - Numbers that cannot be written as fractions
  - Can be written as decimals that are infinite and do not have a repeating pattern

Visualizing the real number system
Student with disabilities support

Help students who have difficulty understanding the relationships between the classifications of numbers in the real number system by providing a print out of the visual below.

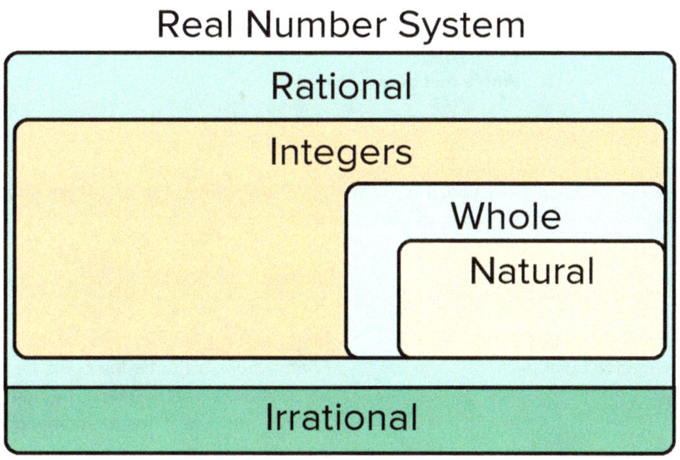

Real Number System

```
┌─────────────────────────────────────────────┐
│                  Rational                     │
│  ┌─────────────────────────────────────────┐ │
│  │               Integers                    │ │
│  │              ┌──────────────────────────┐ │ │
│  │              │           Whole           │ │ │
│  │              │  ┌──────────────────────┐ │ │ │
│  │              │  │       Natural         │ │ │ │
│  │              │  │                       │ │ │ │
│  │              │  └──────────────────────┘ │ │ │
│  │              └──────────────────────────┘ │ │
│  └─────────────────────────────────────────┘ │
│                                               │
├─────────────────────────────────────────────┤
│                 Irrational                    │
└─────────────────────────────────────────────┘
```

Encourage students fill in the diagram with their own examples of each of the types of numbers. With a physical copy of the real number system, students can draw on and highlight certain numbers to better understand the relationships between classifications.

Not knowing how to classify 0
Address student misconceptions

Students may only classify 0 as a whole number since that is the only place where it is specifically listed. Remind students that many numbers, including 0, can be classified multiple ways.

For students who do not list all the classifications, make them aware that any number that is a whole number is also an integer since integers are defined as positive and negative whole numbers. In addition, we can write $0 = \dfrac{0}{n}$ where $n \neq 0$, so it can also be classified as a rational number.

The real number system

Students are introduced to the real number system and the definitions of the classifications of numbers within this system. They will also identify numbers that belongs to each set, including how to classify a number based on its decimal expansion. The lesson discusses the density property and makes students aware that a number can belongs to multiple sets of numbers.

Examples

Students: Page 5–6

Example 1

Height above sea level is expressed as a positive quantity. Which set of numbers is the most appropriate for describing the position of a submarine relative to sea level?

A Integers   **B** Whole numbers

Create a strategy

If the height above sea level is a positive value, this implies that the sea level can be represented by zero. Anything below sea level can be represented by a negative value.

Apply the idea

The position of a submarine below sea level is negative. Whole numbers are counting number starting from 0, which do not include negative numbers.

The correct option is A.

Reflect and check

Other subsets of the real numbers could also be used to represent a submarine's position relative to sea level. For example, a submarine could be 4.5 meters below sea level, or −4.5 meters deep. The number −4.5 is a rational number because it has a terminating decimal.

Purpose

Check that students can apply number classifications to real-world examples of numbers.

Reflecting with students

Ask students to consider whether integers is the most appropriate classification or if a different set of numbers is more appropriate. The entire set of real numbers is the most appropriate, as the position of a submarine can be an integer, a rational number, or an irrational number.

Students: Page 6

Example 2

Using the diagram, complete the following statement:

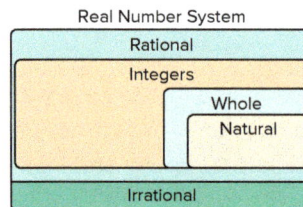

Real Number System

Rational
Integers
Whole
Natural

Irrational

A real number is either:

A a whole number or an irrational number.   **B** an integer or an irrational number.

C a rational number or an irrational number.   **D** an integer or a rational number.

Create a strategy

Use the diagram to consider how we can split the real numbers without excluding any numbers.

Apply the idea

- A: This is not correct because real numbers also include numbers that are not whole, such as fractions and negative integers.
- B: This statement is not correct because it excludes numbers that are not integers, like $\frac{1}{2}$ or $-\frac{3}{4}$.
- C: This is correct because real numbers are composed of exactly these two sets of numbers; every real number is either rational or irrational, covering all possibilities without overlap.
- D: This statement is not correct because it excludes irrational numbers such as π, which are also real numbers.

The rational numbers and irrational numbers together include all real numbers.

The correct option is C.

Purpose

Check that students understand that the real number system is composed of rational and irrational numbers.

Students: Page 7

Example 3

Real Number System

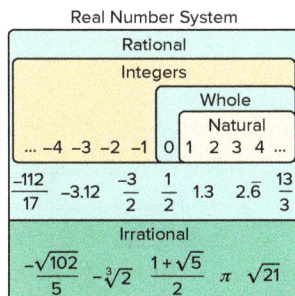

Consider that we can express $\sqrt{49}$ as $\sqrt{7^2}$.

Using the diagram, classify the number $\sqrt{49}$.

Select the three options that apply.

A $\sqrt{49}$ is an irrational number.　　　　B $\sqrt{49}$ is an integer.

C $\sqrt{49}$ is a rational number.　　　　D $\sqrt{49}$ is a whole number.

Create a strategy

Check to see if the square root can be simplified, then use the diagram to find the options that best apply.

Apply the idea

Since 49 is a perfect square, $\sqrt{49} = \sqrt{7^2} = 7$.

Irrational numbers cannot be expressed as fractions. However, $\sqrt{49}$ can be written as 7 or $\frac{7}{1}$, so it is not irrational.

Integers are numbers that can be written without a fractional component, such as 7, so $\sqrt{49}$ is an integer.

Rational numbers can be expressed as fractions, such as $\frac{7}{1}$, so $\sqrt{49}$ is rational.

Whole numbers are non-negative integers, such as 7, so $\sqrt{49}$ is a whole number.

The correct options are B, C, and D.

Purpose

Check that students know the definitions of the real number classifications and how to determine which set(s) a number belongs to.

Expected mistakes

Students may classify $\sqrt{49}$ as an irrational number purely based on the fact that it has a radical symbol.

Encourage students to consider whether $\sqrt{49} = \sqrt{7^2}$ can be simplified, and if it can, what the simplified value can be classified as.

Develop algorithmic thinking when classifying numbers
Targeted instructional strategies

use with Example 3

Encourage students to use algorithmic thinking by developing a general procedure for classifying any number in any form. Guide them to create a step-by-step algorithm that systematically checks the properties of a given number to determine all its applicable classifications. Have students test and refine their algorithm with a variety examples from the Worksheet to ensure it works correctly in each case. This approach helps students understand the hierarchy of number sets and enhances their logical reasoning skills. Here are a possible set of steps:

1. Simplify the number. If the number is an expression (e.g., $\sqrt{49}, \frac{14}{7}, 3^0$), simplify it to its simplest form.

2. Check if the number is a natural number (N), a positive integer greater than zero.

3. Check if the number is a whole number, a natural number or zero.

4. Check if the number is an integer (Z).

5. Check if the number is a rational number (Q).

   - It can be expressed as a fraction $\frac{a}{b}$, where a and b are integers and $b \neq 0$.
   - It could be a repeating, non-terminating decimal.

6. If the number hasn't fit any categories so far, then it is an irrational number, a non-terminating, non-repeating decimal, or a root that cannot be simplified to a rational number (e.g., $\sqrt{2}$). Confirm this classification.

7. Classify the number within the Real numbers (R), since all the above numbers are subsets of real numbers, confirm that the number is a real number.

Summarize all applicable classifications. List all the sets to which the number belongs based on the above assessments.

Students: Page 7

Example 4

Describe the decimal expansion of $\frac{7}{8}$.

Create a strategy

To find the decimal expansion of $\frac{7}{8}$, divide 7 by 8 using long division or a calculator.

Apply the idea

$$7 \div 8 = 0.875$$

The decimal expansion of $\frac{7}{8}$ is finite or terminating, and is therefore a rational number.

Reflect and check

The decimal expansion of $\frac{7}{8}$ is a terminating decimal because the denominator is a power of 2.

Fractions whose denominators are powers of 2, or products of powers of 2 and 5 (e.g., 2, 4, 5, 8, 10, 16, 20, ...), always yield terminating decimals.

Purpose

Show students how to convert a fraction into a decimal using division and help them understand why fractions with denominators that are powers of 2 or products of powers of 2 and 5 result in terminating decimals.

Reflecting with students

Encourage advanced learners to describe the decimal expansion by observation, without using long division or a calculator. Because the number is a ratio of two integers, we know the decimal expansion will either terminate or repeat. Next, they can notice that the denominator is a power of 2. Fractions whose denominators are powers of 2 always yield terminating decimals.

Students: Page 8

Example 5

Is $\sqrt{35}$ rational or irrational?

Create a strategy

Use a calculator to find and describe the decimal expansion of $\sqrt{35}$.

Apply the idea

The number $\sqrt{35}$ when squared equals 35.

This number cannot be written as the quotient of two integers.

Using a calculator, $\sqrt{35} = 5.9160797830996160\ldots$ It has a decimal expansion that is infinite or non-terminating and that does not have a repeating pattern.

$\sqrt{35}$ is an irrational number.

Reflect and check

A rational number can be expressed as a fraction $\frac{a}{b}$, where a and b are integers and b does not equal 0. However, the square root of a non-perfect square, like 35, does not simplify to a fraction of integers and therefore cannot be expressed as a precise ratio.

Purpose

Ensure students can distinguish between rational and irrational numbers, specifically understanding that the square root of a non-perfect square is an irrational number.

Reflecting with students

Ask students to consider what types of square root expressions will simplify to rational numbers and which will be irrational. Students should conclude that only perfect square radicands will simplify to rational numbers. 35 is not a perfect square, so $\sqrt{35}$ is irrational.

Students: Page 8

💡 Idea summary

The real numbers include rational numbers, irrational numbers, integers, and whole numbers as shown in the diagram below:

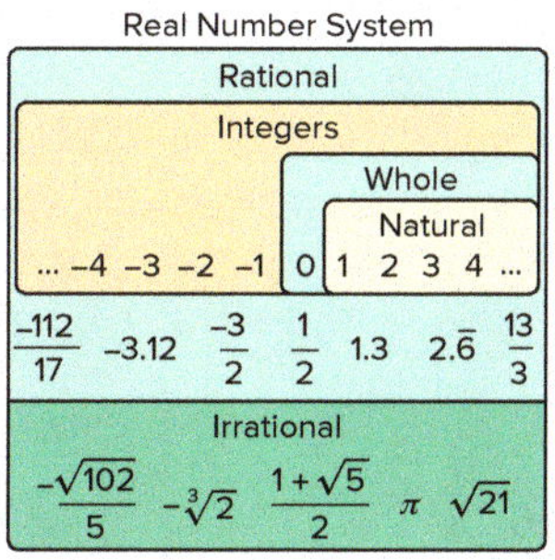

Real Number System

Rational

Integers

Whole

Natural

... −4 −3 −2 −1 0 1 2 3 4 ...

$\frac{-112}{17}$ −3.12 $\frac{-3}{2}$ $\frac{1}{2}$ 1.3 $2.\overline{6}$ $\frac{13}{3}$

Irrational

$-\frac{\sqrt{102}}{5}$ $-\sqrt[3]{2}$ $\frac{1+\sqrt{5}}{2}$ π $\sqrt{21}$

Practice

Students: Page 8–11

What do you remember?

1 Fill in the blanks with the words **real**, **irrational**, and **rational** to complete the sentences below.

A ☐ number is a number that can be written as the ratio of two integers. An ☐ number cannot be written as the ratio of two integers. Rational numbers and irrational numbers together form the set of ☐ numbers.

2 Which of the following statements is true about rational numbers?

    a All rational numbers can be expressed as terminating decimals.

    b Rational numbers include all real numbers.

    c Rational numbers include integers, fractions, and terminating or repeating decimals.

    d Rational numbers are always greater than zero.

3 Place the following numbers on a number line:

    a -4.25 **b** $\dfrac{1}{2}$ **c** $\sqrt{17}$ **d** $-\sqrt{9}$

4 Determine which of the following points on the number line could belong to each subset of the real numbers.

    a Whole   **b** Rational   **c** Irrational   **d** Integer

5 You are given a number line and the task to place the numbers $\dfrac{3}{4}$, $\sqrt{2}$, -5, and π on it.

    a Which of these numbers are rational, and which are irrational?

    b Explain how you would graph each rational number on the number line.

    c What are the challenges of graphing irrational numbers on the number line, and how might you approximate their placement?

6 Describe the decimal expansion of the following numbers as terminating, non-terminating and repeating, or non-terminating and non-repeating.

    a $-\dfrac{2}{7}$ **b** $\sqrt{2}$ **c** $\dfrac{40}{5}$ **d** $\dfrac{8}{15}$

    e $3\sqrt{5}$ **f** $1-\sqrt{7}$ **g** $\dfrac{1}{3}$ **h** π

Let's practice

7 Consider the following sets of numbers:

    a Identify the set of all integers.

    b Identify the set of whole numbers.

    c Identify the smallest whole number.

- $\{0, 1, 2, 3,...\}$
- $\{..., -3, -2, -1, 0, 1, 2, 3,...\}$
- $\{1, 2, 3,...\}$
- $\{..., -3, -2, -1, 0\}$

8 Classify the set of numbers containing $\left\{\dfrac{1}{6}, \dfrac{1}{3}, \dfrac{1}{2},...\right\}$

9 Consider the following diagram:

    a Fill in the blanks with the appropriate label: **Whole**, **Natural**, **Rational**, **Irrational**, and **Integers**.

    b Classify the following numbers:

        i 38   **ii** -42

        iii $\sqrt{6}$ **iv** 3.058

        v $\sqrt{81}$ **vi** $\dfrac{\pi}{3}$

       vii 0.5777 ...   **viii** $\dfrac{26}{45}$

Real Number System

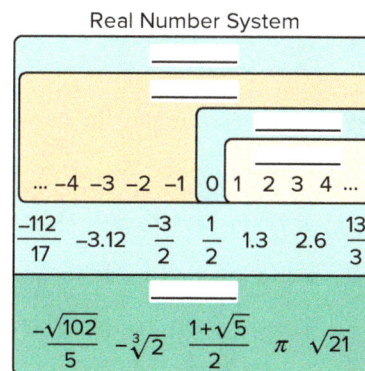

10 Determine whether each of the following statements is true or false:

    a $\sqrt{13}$ is a rational number.

    b Written in decimal form, irrational numbers have a repeating pattern of decimal digits.

    c A real number is either a rational or an irrational number.

11 Check all boxes that apply to the number.

	Natural	Whole	Integer	Rational	Irrational	Real
$\sqrt{49}$						
$\dfrac{1}{3}$						
0						
11						
-3						
$\sqrt{17}$						

12 Simplify each of the following numbers and state whether they are rational or irrational:

    a $\sqrt{16}$ **b** $\dfrac{2\pi}{8}$ **c** $-\dfrac{4\pi}{5\pi}$ **d** $\dfrac{0.4}{5}$

13 For each subset of the real numbers listed below, provide one example and one non-example. Explain why your choices are appropriate for each category.

    a Natural numbers   **b** Whole numbers   **c** Integers   **d** Rational numbers

14 Can a number be both rational and irrational? Explain your answer.

15 Every whole number is also a rational number but not every rational number is a whole number. Explain why this statement is true.

16 Consider the set of all prime numbers. Write down all subsets of the real number system that the set of prime numbers could be classified under.

17 Determine whether the statement is **always**, **sometimes**, or **never true**. Explain your reasoning.

    a A natural number is a whole number.   **b** An integer is a natural number.

    c A natural number is negative.   **d** A real number is an irrational number.

    e A rational number is a real number.   **f** A whole number is an irrational number.

18 Christa decides to express her favorite rational number as a decimal. Could the decimal be repeating, terminating or either?

19 Give a real-world example of an irrational number. Explain why it is irrational.

20 Akin states that 24.737337333... is a rational number because it has a repeating pattern. Is Akin correct? Explain your answer.

21 Jordan is working on a homework assignment about classifying numbers into the correct subsets of the real number system. He categorizes $\sqrt{16}$ as an irrational number, $\frac{\sqrt{9}}{2}$ as a rational number, and $-\frac{4}{2}$ as a whole number. Identify and explain any errors in Jordan's classifications.

22 State the set of numbers most appropriate to describe the following statements.

Choose from below options:
- Whole numbers
- Rational numbers
- Positive real numbers
- All real numbers
- Integers

a The populations of towns.

b The times of the runners (in seconds) in a 100 m sprint.

c Distance between Jupiter and Venus.

d The position of a submarine relative to sea level. (Note that height above sea level is expressed as a positive quantity.)

e The goal difference of a hockey team at the end of a season. (Note that the goal difference is the number of goals the team scored in the season minus the number of goals that were scored against them.)

f The cost (in dollars) of sending a text message.

Let's extend our thinking

23 **a** Is $\sqrt{4}$ a rational number?

b Is $\sqrt{3}$ a rational number?

c What must be true about all square roots that are rational?

24 For the following expressions, determine whether each results in an irrational number.

a $-\frac{5}{8} + \frac{3}{5}$ **b** $1 - \sqrt{5}$ **c** $\frac{1}{2} + \sqrt{2}$ **d** $3 \cdot \sqrt{49}$

25 Suppose $x = 2$, $y = \frac{2}{3}$, and $w = 2\sqrt{12}$. Determine whether each expression results in a rational number.

a $y \cdot w$ **b** $y - w$ **c** $x + y$ **d** $\frac{w}{y}$

26 Explain why the product of 3 and π is irrational.

27 Determine whether the following statement is true or false. Justify your reasoning.

The product of $\sqrt{144}$ and $\sqrt{684}$ is rational because one factor is rational.

28 **a** Name a rational number between $2\frac{1}{2}$ and 2.6 or explain why none exist.

b Name an integer between $2\frac{1}{2}$ and 2.6 or explain why none exist.

c Name an irrational number between $2\frac{1}{2}$ and 2.6 or explain why none exist.

d Use the examples above to determine whether the statement is **always**, **sometimes**, or **never true**. Explain your reasoning.

   i Two rational numbers have an integer in between them.

   ii Two rational numbers have an irrational number in between them.

   iii Two consecutive integers have an integer between them.

Answers

1.01 The real number system

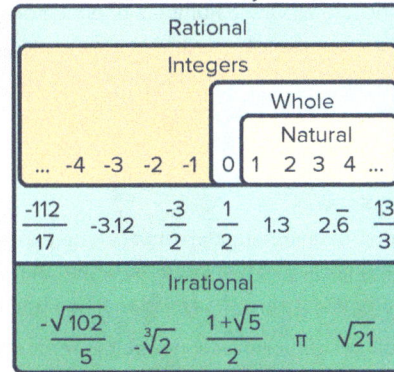

What do you remember?

1 A rational number is a number that can be written as the ratio of two integers. An irrational number cannot be written as the ratio of two integers. Rational numbers and irrational numbers together form the set of real numbers.

2 C

3 a

b

c

d

4 a iii   **b** i, ii, iii   **c** iv   **d** i, iii

5 a Rational Numbers: $\frac{3}{4}$ and −5 (because they can be expressed as fractions of integers). Irrational Numbers: $\sqrt{2}$ and π (because they cannot be exactly expressed as fractions of integers).

b To graph $\frac{3}{4}$, find the point three-fourths of the way from 0 to 1 on the number line. To graph −5, place a point on −5.

c Since irrational numbers cannot be precisely represented as fractions, their exact locations on the number line cannot be determined through simple fractional divisions. For example, $\sqrt{2}$ is known to be a little more than 1 but less than 2, so its approximate position can be placed between 1 and 2 on the number line. Similarly, π is approximately 3.14, so it can be placed a little after 3. These placements are estimates based on the known approximate values of these numbers.

6 a Non-terminating and repeating

b Non-terminating and non-repeating

c Terminating

d Non-terminating and repeating

e Non-terminating and non-repeating

f Non-terminating and non-repeating

g Non-terminating and repeating

h Non-terminating and non-repeating

Let's practice

7 a {... , −3, −2, −1, 0, 1, 2, 3, ...}

b {0, 1, 2, 3,...}   **c** 0

8 The set of rational numbers.

9 a

b i 38 is a rational, an integer and a whole number.

ii −42 is a rational number and an integer.

iii $\sqrt{6}$ is an irrational number

iv 3.058 is a rational number.

v $\sqrt{81}$ is a rational number, an integer, and a whole number.

vi $\frac{\pi}{3}$ is an irrational number.

vii 0.5777 ... is a rational number

viii $\frac{26}{45}$ ... is a rational number

10 a False   **b** False   **c** True

11

	Natural	Whole	Integer	Rational	Irrational	Real
$\sqrt{49}$	✓	✓	✓	✓		✓
$\frac{1}{3}$				✓		✓
0		✓	✓	✓		✓
11	✓	✓	✓	✓		✓
−3			✓	✓		✓
$\sqrt{17}$					✓	✓

12 a 4, Rational   **b** $\frac{\pi}{4}$, Rational

c $-\frac{4}{5}$, Rational   **d** 0.08, Rational

13 a Example: 5

Non-Example: −3

Natural numbers include counting numbers starting from 1. Therefore, 5 is a natural number, but −3 is not because natural numbers do not include negative values.

b Example: 0

Non-Example: −1

Whole numbers consist of all natural numbers plus zero. Hence, 0 is a whole number, while −1 is not included in the set of whole numbers as they do not contain negative numbers.

c Example: −10

Non-Example: 0.5

Integers include all whole numbers and their negative counterparts but do not include fractions or decimals. −10 is an integer, while 0.5 is not because it is a decimal.

d Example: $\frac{1}{2}$

Non-Example: $\sqrt{2}$

Rational numbers are numbers that can be expressed as a fraction or ratio of two integers, where the denominator is not zero. $\frac{1}{2}$ is a rational number, while $\sqrt{2}$ is irrational because it cannot be expressed as a fraction of two integers.

14 No. Irrational numbers can be defined by being not rational and so the two subsets of numbers are mutually exclusive.

15 Every rational number can be expressed in the form $\frac{p}{q}$ where p is an integer and q is a positive integer. A whole number can be expressed in this form where $q = 1$ and $p \geq 0$. That is to say, a whole number is a non-negative integer.

16 The set of prime numbers can be classified under the following subsets of the real number system: natural numbers, whole numbers, rational numbers, and real numbers.

17 **a** Always; The set of whole numbers includes the natural numbers.

 b Sometimes; The set of integers includes natural numbers, as well as their opposites and 0.

 c Never; The natural numbers are positive.

 d Sometimes; The set of real numbers includes irrational numbers, as well as rational numbers.

 e Always; The set of real numbers is made up of all rational and irrational numbers.

 f Never; Whole numbers can be written as the ratio of two integers, so they are not irrational.

18 Either.

19 A real-world example of an irrational number is the diagonal of a square whose sides are 1 unit in length. It's length would be $\sqrt{2}$. The square root of 2 is irrational because it cannot be expressed as a fraction or ratio of two integers. Its decimal representation is non-terminating and non-repeating, extending infinitely without a repeating pattern.

20 No. Although the decimal 24.737337333... may have a pattern, it cannot be expressed as either a repeating decimal or a terminating decimal, so the number cannot be rational.

21 Jordan made two errors in his classification. $\sqrt{16}$ is not an irrational number. This is incorrect because $\sqrt{16} = 4$. Therefore, $\sqrt{16}$ is a rational number, not irrational. Rational numbers include all numbers that can be expressed as a fraction or result from the division of two integers (where the denominator is not zero), as well as all integers and whole numbers. $-\frac{4}{2}$ is not a whole number. This classification is incorrect because when $-\frac{4}{2}$ is simplified, it equals −2, which is an integer. Integers include all whole numbers, their negative counterparts, and zero, but whole numbers do not include negatives. Therefore, $-\frac{4}{2}$ should be classified as an integer.

22 **a** Whole numbers

 b Rational numbers

 c Positive real numbers

 d All real numbers

 e Integers

 f Rational numbers

Let's extend our thinking

23 **a** Yes, $\sqrt{4}$ is rational because it equals 2, which can be expressed as a fraction $\frac{2}{1}$.

 b No, $\sqrt{3}$ is not rational. It cannot be expressed as a fraction of two integers. Its decimal representation is non-terminating and non-repeating.

 c For a square root to be rational, the number under the square root (the radicand) must be a perfect square. A perfect square is an integer that is the square of another integer. This ensures that the square root is an integer or a fraction that can be simplified to a ratio of two integers.

24 **a** No **b** Yes **c** Yes **d** No

25 **a** No **b** No **c** Yes **d** No

26 Since π is irrational, multiplying it by a rational number just results in the multiple of an irrational number.

27 False. $\sqrt{144} = 12$ and although this is rational, it is being multiplied by an irrational number. The product of a rational number and an irrational number is irrational because it cannot be written as the ratio of two integers.

28 **a** Answers vary. 2.55

 b None exist. 2 is too small, and 3 is too large. Since they are consecutive, no integer exists between the two numbers.

 c Answers vary. $\sqrt{6.5}$

 d **i** Sometimes.

  ii Always.

  iii Never.

1.02 Estimate square roots

Subtopic overview

Lesson narrative

In this lesson, students will learn to estimate square roots of non-perfect squares without using a calculator. They will explore how to approximate the values of irrational numbers by comparing them to the closest perfect squares. An interactive exploration helps students understand the relationship between the area of squares and their side lengths. Students will solve problems involving approximating square roots, such as estimating 75 and 95, by finding their positions on the number line. By the end of the lesson, students should be able to estimate and order irrational numbers accurately on a number line.

Learning objectives

Students: Page 12

After this lesson, you will be able to...

- identify the two consecutive natural numbers between which the positive square root of a given number lies.
- estimate and justify which natural number is the better approximation of a positive square root.
- compare, order, and locate values on a number line using rational approximations of square roots.

Key vocabulary

- perfect square
- square root (of a number)

Essential understanding

The square root of a whole number that is not a perfect square always lies between two natural numbers. The square root of a number that is not a perfect square is always irrational.

Standards

This subtopic addresses the following Virginia 2023 Mathematics Standards of Learning standards.

Mathematical process goals

MPG1 — Mathematical Problem Solving

Teachers can integrate this goal by presenting students with real-world problems that involve estimating square roots of non-perfect squares. For instance, teachers could create a problem scenario where students have to use square roots to calculate distances or areas. Students can then practice using the strategies they've learned to solve these problems, such as finding the two consecutive natural numbers between which the square root of a number lies.

MPG2 — Mathematical Communication

Teachers can integrate this process goal by having students utilize many forms of communication, including written, verbal, and pictorial representations. For example, students can explain to another student what strategies they used to determine the two whole numbers that the estimation of a non-perfect square falls between. Students should use appropriate vocabulary when discussing their strategies with classmates.

MPG3 — Mathematical Reasoning

Teachers can integrate this process goal by having students justify which natural number is the better approximation for the square root of a given number. This requires students to use logical reasoning to compare the given number with the midpoint of the two perfect squares. This process of reasoning can be facilitated through class discussions, group work, or individual written explanations.

MPG4 — Mathematical Connections

Teachers can integrate this process goal by encouraging students to make connections to the concept of positive square roots that they learned in grade 7. Students can use this knowledge to extend into the concept of negative squared numbers.

MPG5 — Mathematical Representations

Teachers can integrate this goal by encouraging students to represent their understanding of square roots in various ways. For example, students could use number lines to visually represent the square root of a number and its position between two consecutive natural numbers. Similarly, students could use tables to record the perfect squares and their roots, and then use this table to estimate the square root of non-perfect squares. This can help students to see that different representations can convey the same mathematical idea.

Content standards

8.NS.1 — The student will compare and order real numbers and determine the relationships between real numbers.

8.NS.1a — Estimate and identify the two consecutive natural numbers between which the positive square root of a given number lies and justify which natural number is the better approximation. Numbers are limited to natural numbers from 1 to 400.

8.NS.1b — Use rational approximations (to the nearest hundredth) of irrational numbers to compare, order, and locate values on a number line. Radicals may include both positive and negative square roots of values from 0 to 400 yielding an irrational number.

Prior connections

7.NS.3 — The student will recognize and describe the relationship between square roots and perfect squares.

Future connections

A.EO.1 — The student will represent verbal quantitative situations algebraically and evaluate these expressions for given replacement values of the variables.

A.EO.4 — The student will simplify and determine equivalent radical expressions involving square roots of whole numbers and cube roots of integers.

Lesson Preparation

Suggested review

Depending on your students' level of prior knowledge, consider revisiting the following lessons:

Grade 7 — 1.08 Square roots and perfect squares
Grade 8 — 1.01 The real number system

Tools

You may find these tools helpful:
- Unit tiles
- Blank number lines
- Step-by-step graphic organizer

Student lesson & teacher guide

Estimate square roots

In this lesson, students explore irrational numbers, specifically focusing on how to approximate the square roots of non-perfect squares without a calculator. Students are reminded that irrational numbers, like π and the square roots of numbers that are not perfect squares, have non-terminating, non-repeating decimal expansions.

▌ Exploration

Students: Page 12

> **▶ Interactive exploration**
> **Explore online to answer the questions**
>
> ⊕ **mathspace.co**

Use the interactive exploration in 1.02 to answer these questions.

1. Why is the side length of a square with an area of 25 units^2 equal to $\sqrt{25}$?
2. What do you notice about the areas of squares with whole number side lengths?
3. What do you notice about the side lengths of squares that have an area which is not a perfect square?
4. What are the closest perfect square area values that are smaller and larger than an area of 30 square units?
5. What does this tell us about the whole number side length values that $\sqrt{30}$ lies between? What are they?

Suggested student grouping: In pairs

The exploration focuses on understanding the relationship between the area of a square and its side length, particularly how changes in area affect side lengths, both for perfect and non-perfect squares. By manipulating a slider in a GeoGebra applet, students can explore how perfect squares can be used to estimate the values of square roots of non-perfect squares.

Ideal student responses

These ideal responses may differ from other correct student responses. Less formal responses can be connected with the more precise mathematical language presented here.

1. **Why is the side length of a square with an area of 25 units^2 equal to $\sqrt{25}$?**
   The side length of a square with an area of 25 square units is equal to $\sqrt{25}$ because the side length of a square is the distance along one edge of the square, and squaring that length (multiplying it by itself) gives the area. Since $5 \cdot 5 = 25$, the side length must be $\sqrt{25} = 5$ units.

2. **What do you notice about the areas of squares with whole number side lengths?**
   For squares with whole number side lengths, the areas are perfect squares.

3. **What do you notice about the side lengths of squares that have an area which is not a perfect square?**
   The side lengths are irrational numbers, which means they cannot be expressed as a simple fraction, and their decimal representation is non-repeating and non-terminating. This is because non-perfect square numbers do not have integer square roots.

4. **What are the closest perfect square area values that are smaller and larger than an area of 30 square units?**
   The closest perfect square area values smaller and larger than 30 square units are 25 and 36, respectively.

5. **What does this tell us about the whole number side length values that $\sqrt{30}$ lies between? What are they?**
   Since 25 and 36 are the perfect squares nearest to 30, and their square roots are 5 and 6, respectively, 30 must have a value between 5 and 6.

Purposeful questions

- What is the opposite of adding a number? What is the opposite of squaring a number?
- What types of numbers are 1, 4, 9, 16, 25, etc.?
- If the area is a perfect square number, what type of number is the side length?
- How does knowing the two perfect squares that a non-perfect square root falls between help you estimate its value?

Possible misunderstandings

- Students might not recognize that the areas of squares with whole number side lengths are perfect square numbers, and the side lengths of non-perfect square numbers are irrational numbers. Use guided questions to help them classify the numbers.

Concrete-Representational-Abstract (CRA) approach
Targeted instructional strategies

Concrete: Engage students with physical manipulatives to explore square roots. Provide square tiles or blocks and ask them to build squares with areas of perfect squares, such as 64 (an 8 × 8 square) and 81 (a 9 × 9 square). Then challenge them to consider squares with areas of non-perfect squares like 75 and 95. Since they can't form perfect squares with these areas, encourage them to estimate the side lengths by comparing them to the squares they've built.

For example, 75 must be closer to 9 than 8 since they only need 6 more squares to make a 9 × 9 grid, but they have 11 more squares than what is needed for an 8 × 8 grid and 6 < 11.

This hands-on activity helps students understand that the square root of a non-perfect square lies between two whole numbers.

Representational: Transition to visual representations by having students draw squares on graph paper. They can sketch squares with areas of perfect squares and label the side lengths. Then, ask them to draw squares representing non-perfect squares like 75 and 95, estimating the side lengths between the whole numbers they identified earlier. Encourage them to use shading or color to indicate the areas.

Abstract: Move to abstract reasoning by introducing numerical methods to estimate square roots. Teach students to identify the two consecutive whole numbers between which the square root of a non-perfect square lies. For example, since 64 < 75 < 81, they can conclude that $\sqrt{75}$ is between 8 and 9. Show them how to write this as an inequality: $8 < \sqrt{75} < 9$.

Guide them to make better approximations by considering how close the non-perfect square is to the nearest perfect squares. Explain how the differences between 75 and the nearest perfect squares, $81 - 75 = 6$ and $75 - 64 = 11$, relates to the tiles needed to make the next perfect square or the extra tiles they have from the previous perfect square. Since 75 is closer to 81, the square root is closer to 9. Have them practice placing these approximations on a number line and comparing irrational numbers using rational approximations.

Connecting the stages: Help students see the connections between the concrete, representational, and abstract stages. Discuss how building squares with tiles relates to their drawings, and how both relate to the numerical estimations they've made. Encourage them to reflect on how each stage deepened their understanding of square roots. Use questions to prompt thinking, such as, "How did building the squares help you estimate the square roots?" or "How do your drawings represent the ideas we explored with the tiles?" This will reinforce their learning and help them choose the best representation when solving problems.

Following the exploration, students are shown a numerical method for estimating the value of $\sqrt{75}$ and determining which natural number it is closest to so that they can estimate its location on a number line. Students also examine a visual method of estimating $\sqrt{75}$ using square tiles.

Examples

Students: Page 14

Example 1

Approximate $\sqrt{95}$ to the nearest tenth without using a calculator.

a   Identify the two natural numbers the square root lies between.

Create a strategy

To find the two natural numbers that $\sqrt{95}$ lies between, we need to find the two perfect squares that are just less than and just greater than 95. Then we will calculate their square roots.

Apply the idea

First, we identify the two closest perfect squares to 95, which are 81 and 100.

We can order these numbers using inequality signs:

$$81 < 95 < 100$$

We know this order will hold true if we take the square root of each term.

This means that

$$\sqrt{81} < \sqrt{95} < \sqrt{100}$$

and since 81 and 100 are perfect squares, we can simplify them to get

$$9 < \sqrt{95} < 10$$

This means that $\sqrt{95}$ is between the natural numbers 9 and 10.

Purpose

Show students how to estimate the value of a square root without a calculator by identifying the two whole numbers it lies between.

Expected mistakes

The student may incorrectly identify the nearest perfect squares. Instead of 81 and 100 , they may state that it is 64 or 121 . Remind students to work through all of the perfect squares to find the ones closest to the given number.

b Approximate $\sqrt{95}$ to the nearest tenth without using a calculator.

Create a strategy

We can use a grid to estimate the value of $\sqrt{95}$.

Apply the idea

We identified in part (a) that $\sqrt{95}$ lies between 9 and 10.

To determine whether $\sqrt{95}$ is closer to 9 or 10, we will look at how close 95 is to 9^2 and 10^2.

Create 9 × 9 and 10 × 10 grids, which represent 9^2 and 10^2, and place the smaller grid on top of the larger one.

The length of the total grid represents $\sqrt{100} = 10$ and the length of the blue grid represents $\sqrt{81} = 9$.

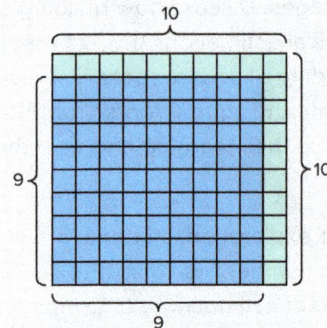

Mark the area up to 95 squares starting with the 81 blue squares. One square is equal to one unit. We cannot make a square with whole number side lengths using 95 smaller squares, so we can use the unmarked green squares to visually approximate how much closer the $\sqrt{95}$ is to 10 than to 9.

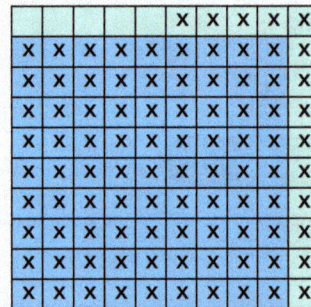

The blue squares represent the whole number of 9. The green squares represent a fraction, where the marked ones are the numerator and the total number of green squares is the denominator. There are 14 marked green squares and 19 total green squares, so the fraction part is $\frac{14}{19}$.

We then can combine the whole number, 9, and the fraction, $\frac{14}{19}$, to get our estimation for the $\sqrt{95}$.

$$\sqrt{95} \approx 9\frac{14}{19} \text{ or } 9.7$$

Reflect and check

We can use a calculator to square the fraction to see how close it is to 95.

$$\sqrt{95}^2 \approx \left(9\frac{14}{19}\right)^2$$

$$95 \approx 94.806$$

Our fractional estimate of $9\frac{14}{19}$ for the $\sqrt{95}$ is fairly close. It is only about 0.194 off from the exact number.

If we want to represent $\sqrt{95}$ using a square, its side lengths would be approximately 9.7 units.

Purpose

Show students how to use a visual grid-based method to approximate square roots without a calculator.

🎓 Advanced learners: Numerical methods for approximating square roots   use with Example 1
Targeted instructional strategies

Introduce students to an algorithm using numerical methods for approximating square roots without using visual aids. Using the provided method as an example, encourage students to find two perfect squares between which the target number lies and then use iterative testing of decimal values to narrow down the approximation.

From part (a), we found that $9 < \sqrt{95} < 10$. Next, we need to determine whether $\sqrt{95}$ is closer to 9 or 10.

$100 - 95 = 5$	The difference between the two largest squares
$95 - 81 = 14$	The difference between the two smallest squares

This shows us $\sqrt{95}$ is closer to 10. Since $\sqrt{95}$ is closer to 10, we can start squaring 9.5 and the tenths above it until we determine which tenths $\sqrt{95}$ lies between.

$$9.5^2 = 90.25$$
$$9.6^2 = 92.16$$
$$9.7^2 = 94.09$$
$$9.8^2 = 96.04$$

$94.09 < 95 < 96.04$	Compare the values
$\sqrt{94.09} < \sqrt{95} < \sqrt{96.04}$	Square root the numbers
$9.7 < \sqrt{95} < 9.8$	Evaluate the square roots of the perfect squares

Finally, we need to determine whether $\sqrt{95}$ is closer to 9.7 or 9.8.

$96.04 - 95 = 1.04$	The difference between the two largest squares
$95 - 94.09 = 0.91$	The difference between the two smallest values

This shows us $\sqrt{95}$ is closer to 9.7.

Therefore, $\sqrt{95} \approx 9.7$.

This approach deepens their understanding of exponents and roots by engaging them in strategic estimation and numerical reasoning. Encourage students to compare this method with the visual grid method to appreciate different strategies for approximating irrational numbers.

Additionally, prompt them to extend this exploration to higher-order roots, fostering connections to numerical analysis and algorithmic thinking. This strategy not only enhances computational skills but also promotes a deeper conceptual understanding of mathematical relationships.

🔊 Critique, correct, and clarify   use with Example 1
English language learner support

Before students share their responses, display the following incorrect statement:

"I know that $\sqrt{95}$ is closer to 8 because there are enough squares to make an 8×8 grid, but there are not enough squares to make a 9×9 grid."

Invite students to identify the error, critique the reasoning, and write a correct explanation. Invite one or two students to share their critiques and corrected explanations with the class.

Listen for and amplify the language students use to describe the difference between the squares. In particular, discuss how finding the "difference" or "subtracting $100 - 95$ and $95 - 81$" helps us determine the "distance between" or "closeness of" the values.

Example 2

Given the irrational numbers $\sqrt{2}$, $\sqrt{50}$, π, $\sqrt{29}$, $\sqrt{98}$.

a   Plot the irrational numbers on the following number line:

Create a strategy

Estimate the values of the numbers by determining which two perfect squares each number lies between. Then plot the estimations on the same number line.

Apply the idea

Estimate the value of $\sqrt{2}$ by comparing to the closest square numbers that are bigger and smaller.

$$\sqrt{1} < \sqrt{2} < \sqrt{4}$$
$$1 < \sqrt{2} < 2$$

The value of $\sqrt{2}$ is closer to $\sqrt{1}$ than $\sqrt{4}$ since $2 - 1 = 1$ and $4 - 2 = 2$.

So, we can place $\sqrt{2}$ between 1 and 2 but closer to 1 on the number line.

Estimate the value of $\sqrt{50}$ by comparing to the closest square numbers that are bigger and smaller.

$$\sqrt{49} < \sqrt{50} < \sqrt{64}$$
$$7 < \sqrt{50} < 8$$

The value of $\sqrt{50}$ is closer to $\sqrt{49}$ than $\sqrt{64}$ since $50 - 49 = 1$ and $64 - 50 = 14$. So, we can have $\sqrt{50}$ plotted between 7 and 8 but closer to 7.

Estimate the value of $\sqrt{29}$ by comparing to the closest square numbers that are bigger and smaller.

$$\sqrt{25} < \sqrt{29} < \sqrt{36}$$
$$5 < \sqrt{29} < 6$$

The value of $\sqrt{29}$ is closer to $\sqrt{25}$ than $\sqrt{36}$ since $29 - 25 = 4$ and $36 - 29 = 7$. So, we can place $\sqrt{29}$ between 5 and 6 but closer to 5.

Estimate the value of $\sqrt{98}$ by comparing to the closest square numbers that are bigger and smaller.

$$\sqrt{89} < \sqrt{98} < \sqrt{100}$$
$$9 < \sqrt{98} < 10$$

The value of $\sqrt{98}$ is closer to $\sqrt{100}$ than $\sqrt{81}$ since $100 - 98 = 2$ and $98 - 89 = 17$. So, we can place $\sqrt{98}$ between 9 and 10 but closer to 10.

Recall that π estimated to two decimals is 3.14.

Plotting the final point we get the number line:

Purpose

Show students how to estimate the value of irrational numbers and plot them on a number line.

Reflecting with students

Encourage students to consider the appropriate degree of precision when estimating and plotting irrational numbers on the number line. Guide them to refine their estimations based on how close the number is to each perfect square.

For example, instead of merely stating that $\sqrt{29}$ is between 5 and 6 , prompt students to recognize that since 29 is closer to 25 than to 36, $\sqrt{29}$ should be less than halfway between 5 and 6 . An imprecise response might be simply placing $\sqrt{29}$ halfway between 5 and 6 . Additionally, encourage students to verify the reasonableness of their estimations using technology.

Students: Page 17

b Arrange the irrational numbers from largest to smallest.

Create a strategy

In part (a), we plotted all the numbers on a number line. We can use the number line to order them from largest to smallest.

Apply the idea

Based on the positions of the numbers on the number line, going right to left, the order is:

$$\sqrt{98}, \sqrt{50}, \sqrt{29}, \pi, \sqrt{2}$$

Reflect and check

We can also determine the order of numbers from largest to smallest by squaring each of our numbers, confirming they are in the same order we found. We will estimate that $\pi = 3.14$.

$$\sqrt{98}^2, \sqrt{50}^2, \sqrt{29}^2, 3.14^2, \sqrt{2}^2$$

$$98, 50, 29, 9.8596, 2$$

The numbers are still in order from largest to smallest.

Purpose

Reinforce students' understanding of irrational numbers and how to compare them using a number line and squaring method.

🎓 Steps for estimating irrational numbers
Targeted instructional strategies

use with Example 2

Assist students with making a list of steps for estimating irrational, radical expressions or provide one for them. In particular, students can a step-by-step graphic organizer like the one shown.

Procedure:	
Steps	Details
Step 1:	
Step 2:	
Step 3:	
Step 4:	
Step 5:	

An example list of steps is shown below.
1. Determine the closest square numbers that are smaller and bigger than the number underneath the given square root.
2. Evaluate the square roots of those two square numbers to find the two integers the given expression lies between.
3. Find the difference between the radicand and each perfect square to find which value the square root expression is closest to.
4. Place the square root expression on the number line, closer to the value found in Step 3.

use with Example 2

Visualizing decimal place values
Student with disabilities support

Students may struggle to place decimal estimations on a number line. Take some time to review counting in tenths for both positive and negative values, and use a number line labeled in tenths as a visual aid.

If students continue to struggle with the decimal places on the number line, consider expanding the number line to only include the section between the two required integers. For example, use the number line below to label π.

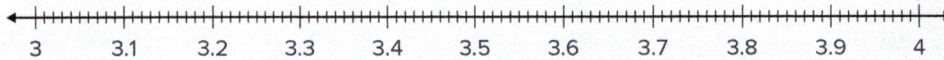

Students: Page 17

💡 **Idea summary**

We can estimate the values of irrational numbers and represent them on the number line.

To estimate square roots we can follow these steps:

- Determine the closest perfect squares that are bigger and smaller than the number underneath the root symbol.
- Find the square roots of the perfect squares to find what two whole numbers the given number lies between.

To determine which natural number the given square root lies closet to, we can find the differences between the perfect squares and the number underneath the square root symbol.

Practice

Students: Page 17–19

What do you remember?

1 What is a perfect square? Provide examples.

2 Evaluate:

   a $\sqrt{4}$ **b** $\sqrt{25}$ **c** $\sqrt{49}$ **d** $\sqrt{121}$

   e $\sqrt{169}$ **f** $\sqrt{225}$ **g** $\sqrt{400}$ **h** $\sqrt{625}$

3 Explain why the square root of 2 is considered an irrational number.

4 Which of the following square roots does not belong? Justify your answer.

   A $\sqrt{4}$ **B** $\sqrt{8}$ **C** $\sqrt{16}$ **D** $\sqrt{64}$

5 Consider the number 50.

   a What is the closest perfect square that is less than 50?

   b What is the closest perfect square that is greater than 50?

   c What is the closest integer that is less than $\sqrt{50}$?

   d What is the closest integer that is greater than $\sqrt{50}$?

Let's practice

6 Write a square root that has a value between 5 and 6.

7 Consider the number $\sqrt{22}$.

   a Complete the inequality with two consecutive perfect square numbers that 22 lies between:

$$\square < 22 < \square$$

   b Complete the inequality with two consecutive natural numbers that $\sqrt{22}$ lies between:

$$\square < \sqrt{22} < \square$$

8 Consider the number $\sqrt{59}$.

   a Complete the inequality with two consecutive perfect square numbers that 59 lies between:

$$\square < 59 < \square$$

   b Complete the inequality with two consecutive natural numbers that $\sqrt{59}$ lies between:

$$\square < \sqrt{59} < \square$$

9 Determine the two consecutive integers that will complete the following inequalities:

   a $\square < \sqrt{8} < \square$ **b** $\square < \sqrt{20} < \square$ **c** $\square < \sqrt{98} < \square$ **d** $\square < \sqrt{74} < \square$

   e $\square < -\sqrt{38} < \square$ **f** $\square < \sqrt{75} < \square$ **g** $\square < \sqrt{50} < \square$ **h** $\square < -\sqrt{140} < \square$

10 Which one of the following has a value between 10 and 11?

   A $\sqrt{143}$ **B** $\sqrt{102}$ **C** $\sqrt{90}$ **D** $\sqrt{167}$

SOL 11 Which statement best describes $\sqrt{60}$?

     A Exactly 7   **B** Exactly 30   **C** Between 7 and 8   **D** Between 8 and 9

SOL 12 Which pair of numbers are both between 8 and 9?

     A $\sqrt{30}$ and $\sqrt{42}$ **B** $\sqrt{65}$ and $\sqrt{90}$ **C** $\sqrt{57}$ and $\sqrt{72}$ **D** $\sqrt{72}$ and $\sqrt{80}$

13 State the integer that lies between $\sqrt{12}$ and $\sqrt{22}$.

14 Consider each of the following radical expressions.

     a State the integer that each of the following numbers is closet in value to:

     b Justify your answer.

     A $\sqrt{52}$ **B** $\sqrt{130}$ **C** $-\sqrt{67}$ **D** $\sqrt{221}$

15 For the following roots:

     i Evaluate, correct to two decimal places where necessary.

     ii State whether the solution is an exact value or an approximation.

     A $\sqrt{3}$ **B** $-\sqrt{16}$ **C** $\sqrt{60}$ **D** $\sqrt{63}$

16 Determine whether each number is a good estimate for each square root. Justify your answer.

     a Is 7 a good estimate for the square root of 50?

     b Is 4.2 a good estimate for the square root of 18?

     c Is 9 a good estimate for the square root of 99?

17 Approximate the following numbers to the nearest tenth without using a calculator.

     A $\sqrt{45}$ **B** $\sqrt{56}$ **C** $-\sqrt{72}$ **D** $\sqrt{83}$

18 Estimate the square root of 350 and explain your choice of approximation.

19 Plot the following on a number line:

     A $\sqrt{13}$ **B** $\sqrt{43}$ **C** $\sqrt{88}$ **D** $\sqrt{110}$

20 The radius of a circle with area A is approximately $\sqrt{\dfrac{A}{3}}$. The area of a circular mouse pad is 48 square inches. Estimate its radius.

21 What is the rational approximation of $-\sqrt{18}$ to the nearest tenth?

22 For each of the following pairs of numbers, select the number with the larger value:

     a $\sqrt{27}$ or $\sqrt{22}$ **b** $\sqrt{50}$ or $\sqrt{46}$ **c** $-\sqrt{9}$ and 3   **d** $-\sqrt{14}$ and -4

23 For each of the following pairs of numbers, select the number with the smaller value:

     a $-\sqrt{120}$ or -11 **b** $\sqrt{72}$ or 8   **c** $\sqrt{20}$ or 5   **d** $\sqrt{15}$ or 3

24 Which of the following could be represented by the dot on the number line?

     a $\sqrt{14}$ **b** $-\sqrt{6}$ **c** $\sqrt{34}$ **d** $\sqrt{45}$

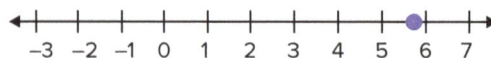

25 Place each of the following numbers on the same number line.

     a $-\sqrt{9}$ **b** $-\sqrt{7}$ **c** -3.2 **d** $-\sqrt{11}$

26 Find two numbers a and b that satisfy the diagram.

Let's extend our thinking

27 Bob has a square-shaped pool with an area of 59 m^2. What is the approximate length of each side of his pool to the nearest meter?

28 A toy block in the shape of a cube has a volume of 123 in^3. What is the approximate length of one side of the block to the nearest inch?

29 The distance (in nautical miles) you can see with a periscope is $1.19\sqrt{h}$, where h is the height of the periscope above the water. Can a periscope that is 8 feet above the water see twice as far as a periscope that is 4 feet above the water? Explain.

Use a calculator to find the distances.

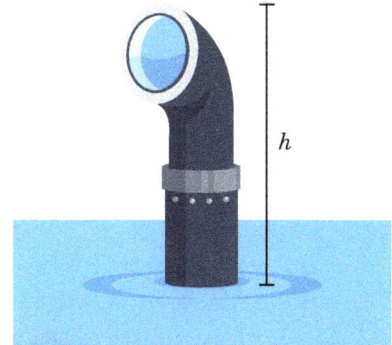

30 Sophia wants to create a square photo frame that has an area of 24 in^2.

   a What is the approximate length of each side of the photo frame to the nearest inch?

   b Find the approximate length of the material (to the nearest inch) she would need to create the frame.

31 Letisha is digging a hole for a pond in her yard. The hole is currently in the shape of a cube and has a volume of 54 ft^3.

   a What is the approximate length of one side of the hole?

   b How much wider (approximate to the nearest foot) do the sides need to be for the hole to be in the shape of a cube with a volume of 215 ft^3?

Answers

1.02 Estimate square roots

What do you remember?

1 A perfect square is a number raised to the power of two or can be obtained by multiplying a number by itself. Examples include:

- 4 is a perfect square because $2 \cdot 2 = 4$.
- 9 is a perfect square because $3 \cdot 3 = 9$.
- 16 is a perfect square because $4 \cdot 4 = 16$.

2   **a** 2   **b** 5   **c** 7   **d** 11

    e 13   **f** 15   **g** 20   **h** 25

3 The square root of $2 \left(\sqrt{2} \right)$ is considered an irrational number because it cannot be expressed as the quotient of two integers (as a fraction). It cannot be written in the form $\frac{a}{b}$, where a and b are integers, and $b \neq 0$.

4   **B** $\sqrt{8}$ does not belong. Only $\sqrt{8}$ is not a perfect square.

5   **a** 49   **b** 64   **c** 7   **d** 8

Let's practice

6 Any square root from $\sqrt{26}$ to $\sqrt{35}$

7   **a** $16 < 22 < 25$ **b** $4 < \sqrt{22} < 5$

8   **a** $49 < 59 < 64$ **b** $7 < \sqrt{59} < 8$

9   **a** $2 < \sqrt{8} < 3$ **b** $4 < \sqrt{20} < 5$

    c $9 < \sqrt{98} < 10$ **d** $8 < \sqrt{74} < 9$

    e $7 < -\sqrt{38} < -6$ **f** $8 < \sqrt{75} < 9$

    g $7 < \sqrt{38} < 8$ **h** $-12 < -\sqrt{140} < -11$

10   **B**

11   **C**

12   **D**

13   4

14   **a**   **i** 7

       ii $\sqrt{52}$ is between $\sqrt{49}$ and $\sqrt{64}$, but it is closer to $\sqrt{49}$.

    b   **i** 11

       ii $\sqrt{130}$ is between $\sqrt{121}$ and $\sqrt{144}$, but it is closer to $\sqrt{121}$.

    c   **i** −8

       ii $\sqrt{67}$ is between $\sqrt{64}$ and $\sqrt{81}$, but it is closer to $\sqrt{64}$. This means that $-\sqrt{67}$ is closer to −8.

d   **i** 15

    ii $\sqrt{221}$ is between $\sqrt{196}$ and $\sqrt{225}$, but it is closer to $\sqrt{225}$.

15   **a**   **i** 1.73   **ii** Approximation

    b   **i** −4   **ii** Exact value

    c   **i** 7.75   **ii** Approximation

    d   **i** 7.94   **ii** Approximation

16   **a** The square root of 49 is 7, and the square root of 64 is 8, so the square root of 50 should be slightly more than 7. Since 50 is very close to 49, 7 is a good estimate for the square root of 50.

    b The square root of 16 is 4, and the square root of 25 is 5. The number 118 falls between 16 and 25, so its square root should be between 4 and 5. Therefore, 4.2 is a reasonable estimate for the square root of 18, leaning towards the lower end of the range.

    c The square root of 81 is 9.

    The square root of 100 is 10. Since 99 is very close to 100, the square root of 99 should be very close to, but slightly less than, 10. The square root of 99 is further from the square root of 81 which is 9. Thus, 10 is a more reasonable estimate for the square root of 99.

17   **a** 6.7   **b** 7.5   **c** −8.5   **d** 9.1

18 The square root of 350 can be estimated by finding the square roots of numbers close to 350 that are perfect squares. The perfect squares closest to 350 are 324 (since $\sqrt{324} = 18$ and 361 (since $\sqrt{361} = 19$. Since 350 is closer to 324 than to 361, a reasonable estimate for $\sqrt{350}$ is slightly more than 18 but less than 19. Therefore, we can estimate $\sqrt{350} \approx 18.7$ considering it is nearer to the square root of 324.

19   **a**

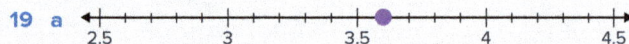

    b

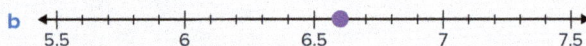

20   4 inches

21   −4.2

22   **a** $\sqrt{27}$ **b** $\sqrt{50}$ **c** 3   **d** $-\sqrt{14}$

23   **a** −11   **b** 8   **c** $\sqrt{20}$ **d** 3

24   **C**

25   **a**

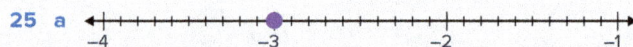

    b

26 For the number line to be accurate, $\sqrt{a}$ must be between 9 and $\sqrt{b}$, and $\sqrt{b}$ must be between $\sqrt{a}$ and 10. This implies that a must be a number whose square root is slightly more than 9, and b must be a number whose square root is slightly less than 10.

One possibility is $a = \sqrt{82}$ because $\sqrt{82} \approx 9.055$ and $b = 99$ because $\sqrt{99} \approx 9.95$.

Let's extend our thinking

27 8 m

28 5 in

29 Two find the distances each periscope can see, we'll use the formula provided:
$$\text{Distance} = 1.19\sqrt{h}$$

where h is the height of the periscope above the water.
Periscope with 8 ft height:
$$\text{Distance} = 1.19\sqrt{8}$$
$$\text{Distance} = 3.37$$

Periscope with 4 ft height:
$$\text{Distance} = 1.19\sqrt{4}$$
$$\text{Distance} = 2.38$$

The ratio of the distances is approximately: $\dfrac{3.37}{2.38} = 1.46$

The periscope that is 8 ft above the water does not see twice as far as the periscope that is 4 ft above the water. This is because the relationship between the height and distance is not linear but follows the square root function, resulting in diminishing gains as height increases.

30 a 5 in   b 20 in

31 a 4 ft   b 2 ft

1.03 Compare and order real numbers

Subtopic overview

Lesson narrative

In this lesson, students will learn how to compare and order real numbers, including rational and irrational numbers. The lesson begins by explaining the importance of converting numbers to a common form, such as decimal, to facilitate comparison. Students will practice ordering numbers in both ascending and descending order and use strategies to estimate and compare values. Through examples and exercises, students will develop their skills in arranging numbers on a number line and identifying the relative size of different types of real numbers. By the end of the lesson, students should be able to confidently compare and order any set of real numbers.

Learning objectives

Students: Page 20

After this lesson, you will be able to...
- compare up to five real numbers using various strategies.
- order up to five real numbers in ascending or descending order using various strategies.
- justify solutions orally, in writing or with a model.

Key vocabulary

- ascending order
- descending order

Essential understanding

Comparing rational numbers in different forms can easily be done by converting them first to the same form.

Standards

This subtopic addresses the following Virginia 2023 Mathematics Standards of Learning standards.

Mathematical Process Goals

MPG2 — Mathematical Communication

Teachers can integrate this process goal by having students utilize many forms of communication, including written, verbal, and pictorial representations. For example, students can explain to another student what strategies they used to compare and order a set of numbers in various forms. Students should use appropriate vocabulary when discussing their strategies with classmates.

MPG3 — Mathematical Reasoning

Teachers can help students develop mathematical reasoning by teaching them to use inductive and deductive reasoning skills when comparing and ordering real numbers. Students can be guided to justify steps in mathematical procedures, such as converting numbers to a common representation, and to use logical reasoning to analyze an argument and to determine whether conclusions are valid.

MPG4 — Mathematical Connections

Teachers can make mathematical connections by linking students' prior knowledge of comparing and ordering rational numbers to the new concept of comparing and ordering real numbers. Teachers can also relate the concept of the density property of real numbers to other mathematical concepts or real-world contexts, thus helping students see mathematics as an integrated field of study.

MPG5 — Mathematical Representations

Teachers can reach this goal by teaching students how to represent real numbers in various forms such as integers, fractions, decimals, mixed numbers, percents, numbers written in scientific notation, radicals, and π. Students can be guided to understand that these different representations can convey the same mathematical idea. Furthermore, teachers can show how to interpret these representations in real-world contexts, thus enabling students to see representation as both a process and a product.

Content standards

8.NS.1 — The student will compare and order real numbers and determine the relationships between real numbers.

8.NS.1b — Use rational approximations (to the nearest hundredth) of irrational numbers to compare, order, and locate values on a number line. Radicals may include both positive and negative square roots of values from 0 to 400 yielding an irrational number.

8.NS.1c — Use multiple strategies (e.g., benchmarks, number line, equivalency) to compare and order no more than five real numbers expressed as integers, fractions (proper or improper), decimals, mixed numbers, percents, numbers written in scientific notation, radicals, and π. Radicals may include both positive and negative square roots of values from 0 to 400. Ordering may be in ascending or descending order. Justify solutions orally, in writing or with a model.

Prior connections

7.NS.2 — The student will reason and use multiple strategies to compare and order rational numbers.

7.NS.3 — The student will recognize and describe the relationship between square roots and perfect squares.

Future connections

8.CE.1 — The student will estimate and apply proportional reasoning and computational procedures to solve contextual problems.

Lesson Preparation

Suggested review

Depending on your students' level of prior knowledge, consider revisiting the following lessons:

Grade 6 — 2.03 Convert between fractions, decimals, and percents
Grade 7 — 1.01 Compare and order rational numbers
Grade 8 — 1.01 The real number system
Grade 8 — 1.02 Estimate square roots

Tools

You may find these tools helpful:
- Blank number lines
- Sticky notes

Student lesson & teacher guide

Compare and order real numbers

The lesson explains how to compare and order real numbers, including irrational numbers, by converting them into a common form, like decimals. Students discover that expressing numbers in decimal form and plotting them on a number line makes it easier to determine which numbers are larger or smaller. Students review how to arrange numbers in ascending (smallest to largest) and descending (largest to smallest) order, noting that a number line naturally displays numbers in ascending order from left to right.

Examples

Students: Page 20

Example 1

For each of the following pairs of numbers, select the number with the smaller value.

a **A** 12   **B** 4π

Create a strategy

Find an approximate value for 4π and compare this to 12.

Apply the idea

Since $\pi \approx 3.14$ we know that π is greater than 3. So 4π must be greater than $4 \cdot 3 = 12$. 12 is smaller than 4π, so the correct answer is option A.

Purpose

Check that students can compare values rational and irrational values.

Expected mistakes

Students may forget to multiply π by 4 before comparing it to 12. Remind students that numbers must be in the same form to compare them.

Students: Page 21

b **A** π^2 **B** $3\sqrt{8}$

Create a strategy

Approximate the value of π^2 and compare it to an approximation of the square root of 8 multiplied by 3.

Apply the idea

We know that $\pi \approx 3.14$ so π^2 will be greater than 9 since $3^2 = 9$ and 3.14^2 must be larger.
To estimate the value of $\sqrt{8}$:

$$\sqrt{4} < \sqrt{8} < \sqrt{9} \qquad \text{Find the closest perfect squares}$$
$$2 < \sqrt{8} < 3 \qquad \text{Evaluate the square roots}$$

This means $\sqrt{8}$ will be less than 3. Multiplying a number less than 3 by 3 will result in a number less than 9. Since we know π^2 must be greater than 9, the correct answer is option B.

Reflect and check

We can check our answer with a calculator:

$$\pi^2 \approx 9.869604401089359\ldots$$
$$3\sqrt{8} \approx 8.48528137423857\ldots$$

Purpose

Demonstrate to students that they can use estimation and comparison skills to solve problems involving irrational numbers and square roots, even without a calculator.

Reflecting with students

To support students in developing mathematical precision, encourage them to actively consider the degree of precision needed for their estimations. Ask students whether estimating irrational numbers to the nearest whole number is sufficient for this problem, or if they should estimate them to the nearest tenths. Encourage them to justify why a whole-number estimate is suitable in this context, and discuss situations where more precise decimal approximations might be necessary.

Students: Page 21

C   A $-\sqrt{75}$ B $-\dfrac{17}{2}$

Create a strategy

Estimate the value of the square root by identifying the closest perfect squares and evaluating their roots. Convert the fraction into a decimal.

Apply the idea

Estimate the value of $\sqrt{75}$. The approximate value would then be multiplied by the negative coefficient.

$$\sqrt{64} < \sqrt{75} < \sqrt{81} \qquad \text{Find the closest perfect squares}$$
$$8 < \sqrt{75} < 9 \qquad \text{Evaluate the square roots}$$

See if $\sqrt{75}$ is closer to 8 or 9.

$$81 - 75 = 6 \qquad \text{Subtract the two largest values}$$
$$75 - 64 = 11 \qquad \text{Subtract the two smallest values}$$

This shows that $\sqrt{75}$ is closer to 9.

Therefore, $-\sqrt{75}$ must be somewhere between -8.5 and -9.

Converting $-\dfrac{17}{2}$ to decimal, we get -8.5.

So, $-\sqrt{75} < -8.5$.

The correct answer is option A.

Purpose

Show students how to estimate the value of square roots and fractions by converting them into more familiar forms.

Reflecting with students

Encourage advanced learners to estimate $\sqrt{75}$ to the nearest tenth without using a calculator. Since $\sqrt{75}$ is closer to 9, we can start squaring 8.5 and the tenths above it until we determine which tenths $\sqrt{75}$ lies between.

$$8.5^2 = 72.25$$
$$8.6^2 = 73.96$$
$$8.7^2 = 75.69$$
$$8.6 < \sqrt{75} < 8.7$$

Then, we need to determine whether $\sqrt{75}$ is closer to 8.6 or 8.7.

$$75.69 - 75 = 0.69 \qquad \text{The difference between the two largest squares}$$
$$75 - 73.69 = 1.04 \qquad \text{The difference between the two smallest values}$$

This shows us $\sqrt{75}$ is closer to 8.7. Therefore, $\sqrt{75} - 8.7$.

Stronger and clearer each time
English language learner support

use with Example 1

Begin by asking students to individually compare each pair of numbers in the question and write down their initial thoughts and reasoning. Encourage them to explain how they determined which number is smaller, even if they are unsure about the exact values. Remind students to use any mathematical vocabulary they know, such as "perfect square," "rational," or "irrational."

Next, have students pair up and share their explanations with a classmate. Encourage them to listen carefully to their partner and ask clarifying questions to understand each other's reasoning. Prompt them to focus on using precise mathematical language and to help each other express their ideas more clearly.

After the discussion, ask students to revise their initial explanations, incorporating any new vocabulary or ideas they gained from their partner. Then, have them switch to a new partner and repeat the process, further refining their explanations and language usage.

Finally, have students write a final, polished explanation for each comparison, using accurate mathematical terminology and clear reasoning. Encourage them to include terms like "square root," "perfect square," and "approximate value." This process helps students strengthen their understanding of the mathematical concepts while also developing their English language skills through repeated practice and peer feedback.

Students: Page 22

Example 2

Compare the numbers $-\frac{4}{3}$, $\frac{\pi}{3}$, -1.25, $\sqrt{3}$, and 130% and arrange them in ascending order.

Create a strategy

Convert the numbers into decimal form to compare and arrange them from least to greatest.

Apply the idea

$$-\frac{4}{3} = -1.\overline{3} \qquad \text{Convert to decimal}$$

The value of π is approximately 3.14, so:

$$\frac{\pi}{3} \approx 3.14 \div 3 \qquad \text{Divide by 3}$$

$$\approx 1.05 \qquad \text{Evaluate the division}$$

To estimate $\sqrt{3}$:

$$1 < 3 < 4 \qquad \text{Identify closest perfect squares}$$
$$\sqrt{1} < \sqrt{3} < \sqrt{4} \qquad \text{Square root the numbers}$$
$$1 < \sqrt{3} < 2 \qquad \text{Evaluate the square roots of the perfect squares}$$

Determining whether $\sqrt{3}$ is closer to 1 or 2:

$$4 - 3 = 1 \qquad \text{The difference between the two largest squares}$$
$$3 - 1 = 2 \qquad \text{The difference between the two smallest squares}$$

$\sqrt{3}$ is closer to 2, so it is somewhere between 1.5 and 2.

$$1.5 < \sqrt{3} < 2$$

Convert 130% to decimal:

$$\frac{130\%}{100} = 1.3$$

The list from smallest to largest is: $-\frac{4}{3}$, -1.25, $\frac{\pi}{3}$, 130%, $\sqrt{3}$.

Reflect and check

We can also plot the numbers in a number line which shows us our numbers from smallest to largest.

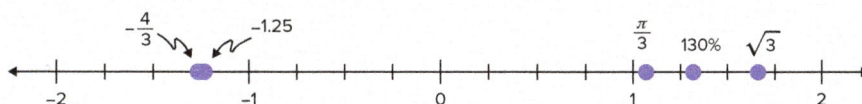

Purpose

Check that students can compare different types of real numbers by converting them into comparable forms.

Expected mistakes

Students might not consider the negative signs of $-\dfrac{4}{3}$ and -1.25 or they might say -1.25 is less than $-\dfrac{4}{3}$. It may be helpful to plot the values on a number line to create a visual for the comparison.

Scaffold decimal conversions and use number lines
Student with disabilities support   use with Example 2

Begin this example by scaffolding the process of converting each number into decimal form to support students who may struggle with conceptual processing and organization. Break down each conversion into step-by-step instructions and model them explicitly.

For example, show how to convert fractions to decimals by dividing the numerator by the denominator, and explain how to approximate irrational numbers like $\dfrac{\pi}{3}$ using a calculator or known approximate values, such as $\dfrac{\pi}{3} \approx \dfrac{3.14}{3} \approx 1.05$.

Once the numbers are converted, use a large number line that extends from -2 to 2. Have students plot each decimal value on the number line using sticky notes or markers. This hands-on activity helps students with visual-spatial processing difficulties by making abstract concepts more concrete and spatially relatable.

As students place each number, facilitate discussions about their relative positions and the reasoning behind the placement to deepen conceptual understanding and reinforce the organizational skills involved in arranging numbers in order.

Students may forget to convert values
Address student misconceptions   use with Example 2

Students may forget to convert values to a comparable form. For example, when given a decimal and percentage to compare, they may forget to change the percentage to a decimal or a decimal to a percentage to compare the values.

Remind students that they must convert values to the same form to decide which value is larger, smaller or equivalent.

Students: Page 22

> 💡 **Idea summary**
>
> In comparing and ordering real numbers, it is always helpful to convert all numbers you are comparing to the same form. Usually decimal form is most appropriate, especially when irrational numbers are involved.

Practice

Students: Page 23–25

What do you remember?

1 Express the following decimals as a:

   i Percentage   **ii** Simplified fraction

   a 0.98   **b** −0.43   **c** 0.075   **d** 0.185

   e −0.02   **f** −0.036   **g** 1.05   **h** −3.802

2 Convert the following fractions into percentages:

   a $\dfrac{3}{5}$ **b** $\dfrac{13}{20}$ **c** $-\dfrac{38}{25}$ **d** $\dfrac{1}{6}$

   e $-\dfrac{3}{3}$ **f** $\dfrac{1}{3}$ **g** $-\dfrac{2}{3}$ **h** $-\dfrac{7}{3}$

3 For each of the following expressions:

   i Rewrite the numbers as percentages.

   ii Fill in the ☐ with the inequality symbol, < or >, that would make the statement true.

   a −69% ☐ 0.63   **b** 71% ☐ 0.31   **c** −51% ☐ −0.67   **d** 0.052 ☐ 4.9%

   e $-\dfrac{38}{25}$ ☐ −131%   **f** $\dfrac{13}{20}$ ☐ 64%   **g** $-\dfrac{7}{8}$ ☐ 0.88   **h** 0.095 ☐ $-\dfrac{1}{10}$

4 State the larger value in each pair of numbers:

   a 0.91 or 82%   **b** −8.9 or −879%   **c** −45% or −0.31   **d** 37% or −0.61

   e 0.019 or 7.5%   **f** $-\dfrac{1}{4}$ or −76%   **g** $\dfrac{5}{6}$ or 0.3   **h** $\dfrac{91}{92}$ or 45%

5 Use a model or write an explanation to demonstrate how you know these numbers are in descending order:

$$1.5 \times 10^2, \frac{2}{5}, 25\%, -\frac{3}{4}.$$

6 Is it always necessary to convert numbers to a common format to order them correctly? Justify your answer.

7 Draw tick marks on the number line that would help you accurately place 6.4.

Let's practice

8 State whether the following statements are true or false:

   a $\dfrac{13}{20} > 64\%$ **b** $-\dfrac{38}{25} > -131\%$ **c** $\dfrac{15}{8} < \sqrt{2}$ **d** $-2\dfrac{3}{4} < -\sqrt{7}$

   e $4.5 < 2\sqrt{9}$ **f** $-3.25 < \sqrt{8}$ **g** $-\sqrt{15} > -3\sqrt{5}$ **h** $\sqrt{65} > -2\sqrt{16}$

9 State the smaller value in each pair of numbers:

   a −0.15 or $-\dfrac{1}{10}$ **b** 3.25 or $\dfrac{19}{5}$ **c** −4 or $-\sqrt{10}$ **d** $2\dfrac{1}{2}$ or $\sqrt{5}$

   e −4π or −12   **f** π^2 or 144   **g** $\sqrt{4}$ or $-\sqrt{\pi}$ **h** $\dfrac{\pi}{2}$ or $\sqrt{\pi}$

   i $-\sqrt{9}$ or $-\sqrt{25}$ **j** 5.65 or $\sqrt{33}$ **k** $2\sqrt{2}$ or $\sqrt{10}$ **l** $-\dfrac{5\pi}{2}$ or $-\sqrt{63}$

10 For each part, fill in the blank with <, =, or > to make the inequality true.

a $3^2 \; \square \; \sqrt{80}$

b $-7.3 \; \square \; -7\frac{3}{8}$

c $-3 \times 2.4 \; \square \; -\sqrt{50}$

d $\frac{2}{3} \times \pi \; \square \; \sqrt{5}$

e $4 \times 10^{-5} \; \square \; 3 \times 10^{-5}$

f $1 \times 10^{-2} \; \square \; 2 \times 10^{-1}$

g $1.2 \times 10^{-6} \; \square \; 1.1 \times 10^{-5}$

h $-4.75 \times 10^{-3} \; \square \; -5.25 \times 10^{-5}$

i $300 \; \square \; 3 \times 10^3$

j $-0.007 \; \square \; -6 \times 10^{-3}$

11 Which is greater: 0.3 or $\frac{1}{4}$? Justify your answer

12 Given the two numbers: $\frac{\pi}{2}$ and 2, come up with three different numbers based on the criteria below:

a Find a number that is smaller than both $\frac{\pi}{2}$ and 2.

b Find a number that is between $\frac{\pi}{2}$ and 2.

c Find a number that is larger than both $\frac{\pi}{2}$ and 2.

13 Place each letter on the corresponding point on the number line.

a $-\pi$ **b** $-\sqrt{8}$ **c** $5 - 2.85$ **d** 46%

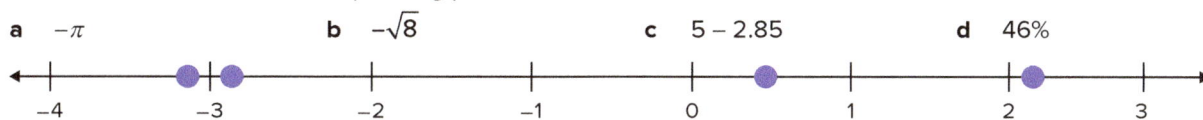

14 Order the following numbers of a number line.

a $5\frac{2}{3}$ **b** 2π **c** $-\sqrt{32}$ **d** -5

15 Erik claims he has correctly ordered the numbers 246%, 2.5×10^2 and 264 on the number line. Is he correct? Justify your thinking.

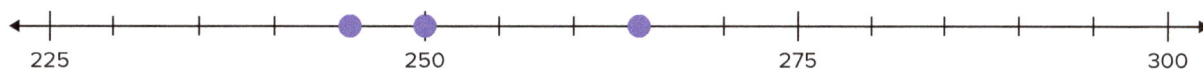

16 Which list of numbers is ordered from greatest to least?

a $46\%, -0.37, 2.1 \times 10^2, 6\frac{2}{5}$

b $2.1 \times 10^2, 46\%, -0.37, 6\frac{2}{5}$

c $46\%, 2.1 \times 10^2, 6\frac{2}{5}, -0.37$

d $2.1 \times 10^2, 6\frac{2}{5}, 46\%, -0.37$

17 Arrange each of the following sets of numbers in ascending order:

a $-0.89, \frac{3}{5}, 53\%$

b $-\frac{3}{3}, -\frac{1}{3}, -66\frac{2}{3}\%$

c $\sqrt{22}, 4.8, \frac{15}{11}, 3\sqrt{42}$

d $\sqrt{30}, \sqrt{3}, 4.3, \frac{11}{2}, \frac{19}{4}$

e $-2\sqrt{5}, \sqrt{3}, 1.1, \frac{3}{2}, -2.6$

f $-\sqrt{54}, -5\sqrt{6}, -3.1, -\frac{12}{17}, -\frac{21}{8}$

18 Arrange each of the following sets of numbers in descending order:

a $\frac{2}{3}, -33\frac{1}{3}\%, 100\%$

b $\frac{5}{8}, 63\frac{2}{5}\%, 0.7, \frac{13}{20}$

c $-2\sqrt{2}, -2\sqrt{3}, -3.6, -\frac{10}{3}$

d $-5.2, 3\sqrt{5}, \sqrt{35}, -\frac{32}{5}, \frac{23}{4}$

e $\sqrt{26}, 4, \sqrt{16}, \frac{58}{9}, 2\pi$

f $-6.5, -3\sqrt{6}, -\sqrt{85}, -\frac{96}{17}, -7.85$

19 Arrange the following numbers in ascending order.

a $7 \times 10^{-6}, 7 \times 10^{-5}, 7 \times 10^{-4}$

b $-6.2 \times 10^{-3}, -2.6 \times 10^{-3}, -1.5 \times 10^{-3}$

c $-1.4 \times 10^{-4}, 1.4 \times 10^4, -4.1 \times 10^{-4}$

d $-3.25 \times 10^{-6}, -4.72 \times 10^{-7}, 1.36 \times 10^{-9}$

Let's extend our thinking

20   For each of the following sets of numbers:

    i   State the largest value.   **ii**   State the smallest value.

    iii   State the value that is closest to 0.5.

    a -92%, $\dfrac{3}{4}$, $\dfrac{503}{1000}$, -0.1, 0.36, 60.1% **b** 71%, $-\dfrac{4}{6}$, $\dfrac{84}{1000}$, 0.7, -0.99, -50.8%

21   Determine whether 2π or $\sqrt{20} + 2$ is greater. Justify your answer.

22   Explain why $3 \cdot 0.\overline{33}$ is exactly the same as 1.

23   During a week of weather observations, a meteorologist recorded the daily temperatures in degrees Celsius.

    Day 1: Started at 15.5 °C and got $9\dfrac{1}{2}$ °C warmer.

    Day 2: Started at $26\dfrac{1}{8}$ °C and got 3 °C cooler.

    Day 3: Started at -2 °C and got $24\dfrac{5}{6}$ °C warmer.

    Order the ending daily temperatures from warmest to coolest, and explain which day was the coldest and which was the warmest.

24   A geography class is studying the heights of various mountains around the world. They record the following heights above sea level (in meters): 8848 (Mount Everest), 5.895×10^3 (Mount Kilimanjaro), $\pi \cdot 1000$ (a theoretical mountain called "Mount Pi"), and 4300 (Mount Meru).

    List the mountains from the lowest height to the highest height. Identify the shortest and the tallest mountain among them.

Answers

What do you remember?

1 a i 98% **ii** $\dfrac{49}{50}$

b i −43% **ii** $-\dfrac{43}{100}$

c i 7.5% **ii** $\dfrac{3}{40}$

d i 18.5% **ii** $\dfrac{37}{200}$

e i −2% **ii** $-\dfrac{1}{50}$

f i −3.6% **ii** $-\dfrac{9}{250}$

g i 105% **ii** $\dfrac{21}{20}$

h i −380.2% **ii** $-\dfrac{1901}{500}$

2 a 60% **b** 65% **c** −152% **d** $16\dfrac{2}{3}\%$

e −100% **f** $33\dfrac{1}{3}\%$ **g** −66.67% **g** −233.33%

3 a i −69% $\square$ 63% **ii** −69% < 0.63

b i 71% $\square$ 31% **ii** 71% > 0.31

c i −51% $\square$ −67% **ii** −51% > −0.67

d i 5.2% $\square$ 4.9% **ii** 0.052 > 4.9%

e i −152% $\square$ −131% **ii** $-\dfrac{38}{25} < -131\%$

f i 65% $\square$ 64% **ii** $\dfrac{13}{20} > 64\%$

g i −87.5% $\square$ 88% **ii** $-\dfrac{7}{8} < 0.88$

h i 9.5% $\square$ −10% **ii** $0.095 > -\dfrac{1}{10}$

4 a 0.91 **b** −879% **c** −0.31 **d** −37%

e 7.5% **f** $-\dfrac{1}{4}$ **g** $\dfrac{5}{6}$ **h** $\dfrac{91}{92}$

5 To demonstrate that these numbers are in descending order, we can convert each to a common format, such as decimals:

$1.5 \times 10^2 = 150$,

$\dfrac{2}{5} = 0.4$ (since $\dfrac{2}{5} \times 100 = 40\%$),

$25\% = 0.25$ (since $25\% = \dfrac{25}{100}$),

$-\dfrac{3}{4} = -0.75$ (since $-\dfrac{3}{4} \times 100 = -75\%$).

Therefore, 150 > 0.4 > 0.25 > −0.75, which shows that the numbers are in descending order.

6 No. Sometimes the order can be determined easily without a common format such as the set 9.1, $\dfrac{3}{4}$, and −π. It is clear since there is only one negative number that must besmallest. Proper fractions must be less than 1 so $\dfrac{3}{4}$ is in the middle and must be largest.

7

Let's practice

8 a True **b** False **c** False **d** True

e True **f** True **g** True **h** True

9 a −0.15 **b** 3.25 **c** −4 **d** $\sqrt{5}$

e −4π **f** π^2 **g** $-\sqrt{\pi}$ **h** $\dfrac{\pi}{2}$

i $-\sqrt{25}$ **j** 5.65 **k** $2\sqrt{2}$ **l** $-\sqrt{63}$

10 a $3^2 > \sqrt{80}$ **b** $-7.3 > -7\dfrac{3}{8}$

c $3 \times 2.4 < \sqrt{50}$ **d** $\dfrac{2}{3} \times \pi < \sqrt{5}$

e $4 \times 10^{-5} > 3 \times 10^{-5}$ **f** $1 \times 10^{-2} < 2 \times 10^{-1}$

g $1.2 \times 10^{-6} < 1.1 \times 10^{-5}$ **h** $-4.75 \times 10^{-3} < -5.25 \times 10{-5}$

i $300 < 3 \times 10^3$ **i** $-0.007 < -6 \times 10^{-3}$

11 0.3 is greater than $\dfrac{1}{4}$. To justify, convert $\dfrac{1}{4}$ to a decimal form, which is 0.25. Comparing 0.25 and 0.3 shows that 0.3 is larger.

12 a Answers Vary. One possible solution is. 1

b Answers Vary. One possible solution is 1.7

c Answers Vary. One possible solution is 3.

13

14

15 Erik is incorrect. 246 is not the same thing as 246%. It should be placed on the number line at 2.46.

16 D

17 a $-0.89, 53\%, \dfrac{3}{5}$ **b** $-66\dfrac{2}{3}\%, -\dfrac{3}{3}, -\dfrac{1}{3}$

c $\dfrac{15}{11}, \sqrt{22}, 4.8, 3\sqrt{42}$ **d** $\sqrt{3}, 4.3, \dfrac{19}{4}, \sqrt{30}, \dfrac{11}{2}$

e $-2\sqrt{5}, -2.6, 1.1, \dfrac{3}{2}, \sqrt{3}$

f $-5\sqrt{6}, -\sqrt{54}, -3.1, -\dfrac{21}{8}, -\dfrac{12}{17}$

18 a $100\%, \dfrac{2}{3}, -33\dfrac{1}{3}\%$

   b $0.7, \dfrac{13}{20}, 63\dfrac{2}{5}\%, \dfrac{5}{8}$

   c $-2\sqrt{2}, -\dfrac{10}{3}, -2\sqrt{3}, -3.6$

   d $3\sqrt{5}, \sqrt{35}, \dfrac{23}{4}, -5.2, -\dfrac{32}{5}$

   e $\dfrac{58}{9}, 2\pi, \sqrt{26}, \sqrt{16}, 4$

   f $-\dfrac{96}{17}, -6.5, -3\sqrt{6}, -7.85, -\sqrt{85}$

19 a $7 \times 10^{-6}, 7 \times 10^{-5}, 7 \times 10^{-4}$

   b $-6.2 \times 10^{-3}, -2.6 \times 10^{-3}, -1.5 \times 10^{-3}$

   c $-4.1 \times 10^{-4}, -1.4 \times 10^{-4}, 1.4 \times 10^{4}$

   d $-3.25 \times 10^{-6}, -4.72 \times 10^{-7}, 1.36 \times 10^{-9}$

Let's extend our thinking

20 a i $\dfrac{3}{4}$ ii -92% iii $\dfrac{503}{1000}$

   b i 71% ii -0.99 iii 0.7

21 The value of 2π is approximately 6.28, and the value of $\sqrt{20}+2$ is approximately 6.47. Therefore, $\sqrt{20}+2$ is greater than 2π.

22 The notation $0.\overline{33}$ represents an infinite repeating decimal, where the digit '3' repeat indefinitely. Mathematically, this can be expressed as $\dfrac{1}{3}$. Therefore, when we multiply $\dfrac{1}{3}$ by 3, the result is $3 \cdot \dfrac{1}{3} = 1$. This is because multiplying a number by its reciprocal always equals 1, illustrating why $3 \cdot 0.\overline{33} = 1$.

23 Ending daily temperatures ordered from warmest to coolest:

   Day 1: 25°C

   Day 2: 23.125°C

   Day 3: 22.833°C

   The warmest day at the end was **Day 1**, reaching 25°C after the temperature increase.

24 Listing the mountains from the lowest height to the highest height:

   Mount Pi: 3141.59 meters

   Mount Meru: 4300 meters

   Mount Kilimanjaro: 5895 meters

   Mount Everest: 8848 meters

   The shortest mountain among them is Mount Pi with a height of 3141.59 meters, and the tallest mountain is Mount Everest with a height of 8848 meters.

1.04 Percent increase and decrease

Subtopic overview

Lesson narrative

In this lesson, students will learn how to calculate percent increase and decrease. The lesson begins with a basic introduction to the concepts, demonstrating how to find the new amount after a percentage change. Students will practice converting percentages to fractions and perform calculations to determine the final values after changes. The lesson includes an exploration where students estimate new values and verify their calculations through contextual problems. By solving these problems, students develop strategies to estimate and calculate exact percentage changes. By the end of the lesson, students should be proficient in calculating both percent increases and decreases in different contexts.

Learning objectives

Students: Page 26

🎓 **After this lesson, you will be able to...**
- estimate the solution to real-world problems involving percent increase or percent decrease.
- solve real-world problems involving percent increase or percent decrease.

Key vocabulary

- percent change

Essential understanding

An increase of $n\%$ is the same as multiplying a given number by $(100 + n)\%$. A decrease of $n\%$ is the same as multiplying a given number by $(100 - n)\%$.

Standards

This subtopic addresses the following Virginia 2023 Mathematics Standards of Learning standards.

Mathematical process goals

MPG1 — Mathematical Problem Solving

Teachers can assist students in applying mathematical concepts and skills to solve problems involving percent increase and decrease. Teachers can create problems from real-world data, such as changes in populations or prices, and guide students in applying appropriate strategies to solve these problems. Teachers can encourage students to develop a repertoire of skills and strategies for solving these problems, such as sing the formula for percent increase or decrease and using benchmark percentages when necessary.

MPG2 — Mathematical Communication

Teachers can encourage students to communicate their thinking and reasoning as they work through problems involving percent increase and decrease. Students can be encouraged to use the language of mathematics, including specialized vocabulary and symbolic notation, to express their ideas with precision. Teachers can prompt students to use mathematical terms, such as 'percent increase', 'percent decrease', 'original value', and 'change in value', when explaining their reasoning. Utilizing the specialized vocabulary of mathematics helps students express their ideas with precision and accuracy. Teachers can facilitate mathematical discussions, allowing students to clarify their thinking and deepen their understanding of the concepts being studied.

MPG3 — Mathematical reasoning

Teachers can integrate this process goal by supporting students through problem solving with contextual situations. Strategies for teachers to use can include: analyzing contextual situations, evaluating evidence, collecting data and drawing accurate and appropriate conclusions.

MPG4 — Mathematical Connections

Teachers can help students make connections between their prior knowledge of estimating and determining percentages and their new understanding of percent increase and decrease. Teachers can link these concepts to real-world situations, such as changes in population or prices, to help students see mathematics as an integrated field of study.

Content standards

8.CE.1 — The student will estimate and apply proportional reasoning and computational procedures to solve contextual problems.

8.CE.1c — Estimate and solve contextual problems that require the computation of the percent increase or decrease.

Prior connections

7.CE.2 — The student will solve problems, including those in context, involving proportional relationships.

Future connections

A.F.2 — The student will investigate, analyze, and compare characteristics of functions, including quadratic and exponential functions, and model quadratic and exponential relationships.

Lesson Preparation

Suggested review

Depending on your students' level of prior knowledge, consider revisiting the following lessons:

Grade 7 — 2.04 Percentages of whole numbers

Tools

You may find these tools helpful:
- Scientific calculator

Percent increase and decrease

The lesson begins by explaining that percentage increases and decreases can be found by calculating the percentage of the original amount, then adding or subtracting it from the original amount. Then students engage in an exploration about finding the percent increase or decrease in a single step.

> **Interactive exploration**
> Explore online to answer the questions
>
> 🌐 **mathspace.co**

Use the interactive exploration in 1.04 to answer these questions.

1. Make the whole amount 60. Drag the slider left to decrease by 40%. What is 60 reduced by 40%?
2. What percent multiplier is used to reduce by 40%?
3. Repeat the steps and decrease other amounts by different percents. How do you find a single multiplier when decreasing by a percent?
4. Make the whole amount 80. Drag the slider right to increase by 30%. What is 80 increased by 30%?
5. What percent multiplier is used to increase by 30%?
6. Repeat the steps and increase other amounts by different percents. How do you find a single multiplier when increasing by a percent?

Suggested student grouping: Small groups

percentage. The applet facilitates experiments with different amounts and percentage changes, using a slider to dynamically adjust values and observe the effects of applying a single multiplier for both increases and decreases.

Ideal student responses

These ideal responses may differ from other correct student responses. Less formal responses can be connected with the more precise mathematical language presented here.

1. **Make the whole amount 60. Drag the slider left to decrease by 40%. What is 60 reduced by 40%?**
    60 reduced by 40% is 36.

2. **What percent multiplier is used to reduce by 40%?**
    The percent multiplier is 0.6 or 60% of the original number (because 100% − 40% = 60%).

3. **Repeat the steps and decrease other amounts by different percents. How do you find a single multiplier when decreasing by a percent?**
    To find the single multiplier for any decrease, subtract the percentage decrease from 100% and convert it to decimal form. For example, to decrease by 25%, the multiplier would be 100% − 25% = 75%, which as a decimal is 0.75.

4. **Make the whole amount 80. Drag the slider right to increase by 30%. What is 80 increased by 30%?**
    80 increased by 30% is 104.

5. **What percent multiplier is used to increase by 30%?**
    The percent multiplier is 1.3 or 130% of the original number (because 100% + 30% = 130%).

6. **Repeat the steps and increase other amounts by different percents. How do you find a single multiplier when increasing by a percent?**
    To find the single multiplier for any increase, add the percentage increase to 100% and convert it to decimal form. For example, to increase by 20%, the multiplier would be 100% + 20% = 120% which as a decimal is 1.20.

Purposeful questions

- If you decrease an amount by 40%, what percentage remains? How can we use this to find the remaining amount in one step?
- What percentage represents a whole? What is that percentage as a decimal?
- Can you identify a single multiplier that represents a 25% increase? Explain why it works.

Possible misunderstandings

- Students may not understand how the percentage decrease relates to the percent multiplier. Help them see how the percent multiplier is simply the percentage of the original that remains after the decrease.

The lesson summarizes what students discovered in the exploration - that the amount after a percent increase or decrease has been applied can be found by multiplying the original amount by a single percentage.

Examples

Students: Page 27

Example 1

Elena and Mikee currently work at the same company and each make $70 000 in a year. They both get promoted and received a raise. Mikee received a 7% raise, and Elena received a 9% raise.

a Estimate Elena's new salary.

Create a strategy

Use a benchmark of a 10% raise because it is close to her actual raise of 9%.

Apply the idea

First we need to find the estimate of Elena's raise.

$$\text{Estimated raise} = 70\,000 \cdot 10\% \quad \text{Multiply the current salary by 10\%}$$

$$= 70\,000 \cdot \frac{10}{100} \quad \text{Convert the percent to a fraction}$$

$$= 7000 \quad \text{Evaluate}$$

To find Elena's estimated new salary, we add the estimated raise to her original salary.

$$\text{Estimated new salary} = 70\,000 + 7000 \quad \text{Add the current salary and the estimated raise}$$

$$= \$77\,000 \quad \text{Evaluate the addition}$$

Elena's estimated new annual salary is approximately $77 000.

Purpose

Show students how to use benchmark percentages to estimate the result of a percent increase in a real-world context.

Reflecting with students

Ask students to describe whether Elena's actual new salary will be more or less than this estimate. Students should recognize that 10% is higher than the actal percent increase, so this estimate will be higher than Elena's actual new salary. Thus, Elena's new salary is less than $77000.

Students: Page 27

b Estimate Mikee's new salary.

Create a strategy

Estimate using percentages close to his raise of 7%. We can quickly approximate 5% raise and a 10% raise. Halfway between them would be a 7.5% raise which is very close to 7%.

Apply the idea

Find a 5% raise.

$$5\% \text{ raise} = 70\,000 \cdot 5\%$$ Multiply the current salary by 5%

$$= 70\,000 \cdot \frac{5}{100}$$ Convert the percent to a fraction

$$= 3500$$ Evaluate

Find a 10% raise.

$$10\% \text{ raise} = 70\,000 \cdot 10\%$$ Multiply by 10%

$$= 70\,000 \cdot \frac{10}{100}$$ Convert the percent to a fraction

$$= 7000$$ Evaluate

Now we will find the amount that is halfway between $3500 and $7000.

$$\text{Average} = \frac{(3500 + 7000)}{2}$$ Add the amounts and divide the sum by 2

$$= \frac{(10\,500)}{2}$$ Evaluate the addition

$$= 5250$$ Evaluate the division

We have found that a 7.5% raise is $5250. We can add that to the original salary to estimate the new salary after a 7% raise.

$$\text{New Salary} = 70\,000 + 5250$$ Add the estimated raise to the original salary

$$= \$75\,250$$ Evaluate

Mikee's estimated new annual salary is $75 250.

Purpose

Show students how to estimate a salary increase using percentages and basic algebraic operations.

Students: Page 28

c   Calculate Elena and Mikee's new salaries exactly.

Create a strategy

Calculate the exact raise of 9% for Elena and 7% for Mikee. Multiply their current salary by their raise percentage and then add that to their current salary. This will give them each their new annual salary.

Apply the idea

First we need to find the 9% raise for Elena.

$$\text{Raise} = 70\,000 \cdot 9\%$$ Multiply the current salary by 9%

$$= 70\,000 \cdot \frac{9}{100}$$ Convert the percent to a fraction

$$= 6300$$ Evaluate

To find Elena's new annual salary, we add the raise to her original salary.

$$\text{New annual salary} = 70\,000 + 6300$$ Add the current salary and the raise

$$= \$76\,300$$ Evaluate the addition

Elena's new annual salary is $76 300.

Second we need to find the 7% raise for Mikee.

$$\text{Raise} = 70\,000 \cdot 7\%$$ Multiply the current salary by 7%

$$= 70\,000 \cdot \frac{7}{100}$$ Convert the percent to a fraction

$$= 4900$$ Evaluate

To find Mikee's new annual salary, we add the raise to his original salary.

$$\text{New annual salary} = 70\,000 + 4900$$ Add the current salary and the raise

$$= \$74\,900$$ Evaluate the addition

Mikee's new annual salary is $74 900.

Reflect and check

Recall that we estimated Elena's salary to be $77 000 and her actual salary is $76 300. That's a difference of $77 000 – $76 300, so we only overestimated by $700.

We estimated Mikee's salary to be $75 250 and his actual salary is $74 900. That's a difference of $75 250 – $74 900, so we only overestimated by $350.

Our estimates were very close to the exact values.

Purpose

Show students how to calculate new salaries after a given percentage increase, reinforcing their understanding of percentage calculations in real-world contexts.

Expected mistakes

Students might not add the amount each salary increased by to the original salaries. Show students that they found the amount by which the salaries increased by, but they have not yet found Elena and Mikee's new salaries.

Calculating exact amounts instead of estimating
Address student misconceptions

use with Example 1

Students might calculate the exact amounts of Elena and Mikee's new salaries instead of providing estimates as the questions ask. They may compute the precise increase, resulting in an exact figure like $76,300, rather than using estimation techniques.

Encourage students to pay close attention to keywords in the problem, such as "estimate," to understand that an approximate answer is appropriate. Introduce estimation strategies like using benchmark percentages that are easier to compute mentally, such as using 10% instead of 9%. Show how rounding percentages and amounts simplifies calculations while still providing a reasonable estimate.

Students: Page 29

Example 2

A bag of biscuits weighs 120 kg. The weight of the bag decreases by 48%.

a   Estimate the new weight of the bag.

Create a strategy

Use a 50% decrease as an estimation for the 48% decrease.

Apply the idea

50% is the same as $\frac{1}{2}$ so we just need to find half of the 120 kg bag and that is the new weight.

$$120 \cdot \frac{1}{2} = 60$$

The estimated new weight of the bag is approximately 60 kg.

Purpose

Show students how they can use estimation to simplify calculations involving percentages, particularly in real-world applications such as estimating the weight of items.

Reflecting with students

To promote attention to precision, encourage students to always include units with their answers. When estimating the new weight of the bag, they should state the result as " 60 kg." An imprecise response might simply state "The new weight is 60," which is unclear about what that number represents. By consistently including units, students demonstrate a precise understanding of the quantities involved.

b   Find the new weight of the bag.

Create a strategy

We can calculate the new weight by determining what is 48% of 120 kg and then subtracting that amount from 120 kg.

Apply the idea

$$\text{New Weight} = 120 - (120 \cdot 48\%) \qquad \text{Subtract 48\% of the original weight}$$

$$= 120 - \left(120 \cdot \frac{48}{100}\right) \qquad \text{Rewrite the percent as a fraction}$$

$$= 120 - 57.6 \qquad \text{Evaluate the multiplication}$$

$$= 62.4 \text{ kg} \qquad \text{Evaluate the subtraction}$$

The new weight of the bag is 62.4 kg.

Reflect and check

Decreasing by 48% is the same as multiplying by a 52% because 100% – 48% = 52%. This means that 52% of the weight will remain after the decrease.

We can also calculate the new weight by multiplying the original weight by 52%.

$$\text{New Weight} = 120 \cdot 52\% \qquad \text{Multiply current weight by 52\%}$$

$$= 120 \cdot \frac{52}{100} \qquad \text{Rewrite the percent as a fraction}$$

$$= 62.4 \text{ kg} \qquad \text{Evaluate the multiplication}$$

This confirms the new weight of the bag is 62.4 kg.

Purpose

Show students how to calculate a percentage decrease and apply it to practical situations, like determining the new weight of an object.

Reflecting with students

Encourage students to evaluate the reasonableness of their answer. For instance, since a 48% decrease is nearly half of the original weight, an answer close to 60 kg is reasonable. Ask guiding questions like, "Does your answer seem reasonable compared to the original weight?" or "What would a 50% decrease look like?" This practice not only helps students catch mistakes but also fosters a habit of thoughtful reflection, leading to more precise and accurate work.

Three reads   use with Example 2
English language learner support

Advise students to read through the instructions a few times, focusing on gathering different information each time in order to build up understanding of what the question is asking.

On the first read, students should aim to identify the scenario presented in the question. Ask students, "What do you think is happening in this question?" or "Can you explain what this question is about?"

On the second read, students should aim to interpret the problem by answering questions like, "What is the question asking you to find?" and "What information should be included in the answer?"

On the third read, students should look for important information in the instructions. In this question, the important information includes:

- The original weight of the bag is 120 kilograms.
- The weight decreases by 48%.
- The first part is just an estimate while the second part asks for an exact amount. The answer should be in kilograms.

Students can be prompted by framing these as questions like "Will the answer be a weight or a percentage? What are the units used in this situation?" or "Should the new weight of the bag be more or less than the original weight?"

> 💡 **Idea summary**
>
> To increase x by $y\%$, we can calculate: $x \cdot (100 + y)\,\%$
>
> To decrease x by $y\%$, we can calculate: $x \cdot (100 - y)\,\%$

Calculate percent change

Students explore the concept of percent change and how it is calculated. The lesson provides the formulas for calculating the percent increase or decrease and include an example of each.

🎓 Advanced learners: Generalizing formulas for percentage change
Targeted instructional strategies

Encourage students to derive the general formulas for calculating percentage increase and decrease on their own. Begin by presenting a variety of problems with different original amounts and percentage changes. Ask students to identify patterns in how the new amounts are calculated from the original amounts and the percentages. Guide them to express these patterns algebraically, leading them to formulate the general equations:

$$\text{Percent Increase} = \frac{\text{Increase}}{\text{Original Amount}} \cdot 100\%$$

$$\text{Percent Decrease} = \frac{\text{Decrease}}{\text{Original Amount}} \cdot 100\%$$

By deriving these formulas themselves, students deepen their understanding of the relationships between percentages, fractions, and decimal multipliers. This exploratory approach fosters critical thinking and allows students to generalize their learning to any percentage change context.

Example 3

Jenna used to earn $950 per month. Her new monthly salary is $1235 per month. Determine the percentage increase of Jenna's salary.

Create a strategy

To calculate the percentage increase, we need to find the difference between Jenna's new salary and her starting salary. Then, we divide this difference by her starting salary and convert it to a percentage.

Apply the idea

Starting salary = $950 and New salary = $1235

Increase in salary = 1235 − 950	Subtract the starting salary from the new salary
= 285	Evaluate the subtraction
Percentage increase = $\left(\frac{285}{950}\right) \cdot 100\%$	Divide the increase by the previous salary and multiply by 100%
= 30%	Evaluate

Jenna's salary increased by 30%.

Reflect and check

We can check our work by adding Jenna's starting salary to 30% of her starting salary to see if it equals her new salary.

New salary = $950 + ($950 · 30%)   Multiply the starting salary by 30%

= $950 + $285   Add the starting salary and amount increased

= $1235   Evaluate

This confirms that Jenna's salary increased by 30%.

Purpose

Show students how to calculate the percentage increase of a value, an important skill for understanding changes in quantities in a variety of real-world situations.

Expected mistakes

Students might divide the increase by the new salary rather than the original salary. Help students understand that we want to find how much the salary increased from her original salary, so the percent increase is a proportion of the original written as a percentage.

Template for percent increase and decrease
Student with disabilities support

use with Example 3

For students who struggle with organization and multistep problems, provide students with a template for solving percent increase and decrease problems to help organize their thoughts and calculations. Write the steps as a guide, like the ones shown, with space for students to complete the necessary steps.

1. Identify the original salary and the new salary.
2. Calculate the amount of the increase.
3. Divide the increase by the original salary.
4. Convert the result to a percentage.

For additional support, provide fill-in-the-blank templates for the steps of algebraic work.

$$\text{Increase in salary} = \square \qquad \text{Subtract the starting salary from the new salary}$$

$$= \square \qquad \text{Evaluate the subtraction}$$

$$\text{Percentage increase} = \left(\frac{\square}{\square}\right) \times \square \qquad \text{Divide the increase by the original salary and convert to a percent}$$

$$= \square \times \square \qquad \text{Evaluate the division}$$

$$= \square \qquad \text{Evaluate the multiplication}$$

Encourage students to fill in each section with the appropriate numbers from the problem. This structured approach helps students focus on one step at a time, reducing feelings of being overwhelmed and enabling them to understand and solve percentage increase questions more effectively.

Example 4

Last year, a local bookstore sold an average of 300 books per month. This year, the average monthly sales dropped to 216 books. Determine the percentage decrease in the bookstore's monthly book sales.

Create a strategy

To find the percent decrease, subtract this year's average monthly sales from last year's, then divide the result by last year's average and convert to a percent.

Apply the idea

Last year's sales = 300 and This year's sales = 216

$$\text{Decrease in sales} = 300 - 216 \qquad \text{Subtract this year's sales from last year's sales}$$
$$= 84 \qquad \text{Evaluate the subtraction}$$
$$\text{Percentage decrease} = \left(\frac{84}{300}\right) \cdot 100\% \qquad \text{Divide the decrease by the last year sales and multiply by 100\%}$$
$$= 28\% \qquad \text{Evaluate}$$

The percentage decrease in the bookstore's monthly book sales is 28%.

Reflect and check

We can check our work by subtracting 28% of last year's sales of 300 books from last year's total amount.

$$\text{New book sales} = 300 - (300 \cdot 28\%) \qquad \text{Subtract 28\% of the last year's sales from last year's sales}$$
$$= 300 - 84 \qquad \text{Evaluate the multiplication}$$
$$= 216 \qquad \text{Evaluate the subtraction}$$

This confirms that the book store had a 28% decrease in sales.

Purpose

Students demonstate that they can calculate a percentage decrease in a real-world scenario and interpret the result.

Reflecting with students

Ask students how they would know whether the percentage was an increase or a decrease if the problem instead said, "Determine the percentage change in the bookstore's monthly book sales." Students should notice that the sales decreased, so the percentage change would be a percent decrease.

Concrete-Representational-Abstract (CRA) approach
Targeted instructional strategies

use with Example 4

Concrete: Begin by engaging students with physical manipulatives to represent the bookstore's monthly book sales. Since handling 300 individual items might be impractical, have students write the books as a ratio (300 : 216) and simplify the ratio $\frac{300}{3} : \frac{216}{3} = 100 : 72$). Provide students with 100 counters (like blocks or chips) to represent last year's sales groups and 72 counters for this year's sales groups. Have students arrange the 100 counters on a 10 × 10 grid to visualize the total sales from last year. Then, have them remove 28 counters to represent the drop to 72 groups this year. This hands-on activity helps students concretely see that there was a decrease of 28 groups out of 100, which will relate to a 28% decrease.

Representational: Help students transition from the physical manipulatives to visual representations. Encourage them to draw a hundreds grid, which is a 10 × 10 square divided into 100 smaller squares. Have them shade all 100 squares to represent last year's sales of 100 groups. Then, ask them to erase or cross out 28 squares to represent the decrease, leaving 72 squares shaded for this year's sales. This visual shows the decrease in sales as the unshaded portion of the grid. Discuss how this visual helps them understand that 28 out of 100 groups were lost, which directly translates to a 28% decrease. In other words, each square represents 1% of the total groups, so the 28 unshaded squares indicate a 28% decrease.

Abstract: Assist students in connecting their visual representations to abstract mathematical calculations.

Explain that the decrease they observed can be written as a fraction $\frac{28}{100}$, which equals 28%. Show them how this fraction comes from the difference between the original 100 groups and the current 72 groups. Reinforce that the manipulatives and drawings they've used represent the numbers in the calculation $100 - 72 = 28$, and this difference divided by the original amount $\left(\frac{28}{100}\right)$ gives the percentage decrease.

Teach students to perform the mathematical calculations using numbers and symbols. Guide them through calculating the percentage decrease using the simplified group numbers:

1. Find the decrease in groups:

$$\text{Decrease} = \text{Original groups} - \text{New groups} = 100 - 72 = 28$$

2. Calculate the percentage decrease:

$$\text{Percentage decrease} = \left(\frac{\text{Decrease}}{\text{Original groups}}\right) \times 100\% = \left(\frac{28}{100}\right) \times 100\% = 28\%$$

Alternatively, relate this back to the original book numbers:

1. Find the decrease in books:

$$\text{Decrease} = 300 - 216 = 84 \text{ books}$$

2. Calculate the percentage decrease:

$$\text{Percentage decrease} = \left(\frac{84}{300}\right) \times 100\% = 28\%$$

Emphasize that both methods yield the same percentage decrease because grouping the books into sets of 3 maintains the proportion between the numbers.

Students: Page 31

> ### 💡 Idea summary
>
> To find the increase as a percent of the original quantity, write it as a fraction and multiply by 100%:
>
> $$\text{Percent Increase} = \frac{\text{Increase}}{\text{Original quantity}} \cdot 100\%$$
>
> To find the decrease as a percent of the original quantity, write it as a fraction and multiply by 100%:
>
> $$\text{Percent Decrease} = \frac{\text{Decrease}}{\text{Original quantity}} \cdot 100\%$$

Practice

Students: Page 31–34

What do you remember?

1 How do you know whether a percent of change is a percent of increase or a percent of decrease?

2 Which of the following multipliers represent a percentage increase? Select all that apply.

    A 103%   **B** 65%   **C** 1.09   **D** 0.99

3 Which of the following multipliers represent a percentage decrease? Select all that apply.

    A 84%   **B** 102%   **C** 1.4   **D** 0.79

4 Here is a grid of 100 blue squares. Each small square represents 1% of the total:

    a Choose the image that represents 3% more than the original grid:

        A   B   C   D

    b How many squares were added to the original grid to increase it by 3%.

    c Fill in the blanks to complete an expression that represents the 3% increase in squares:
        ☐% + ☐% = 103% = 1.03

    d Choose the image that represents 6% less than the original grid:

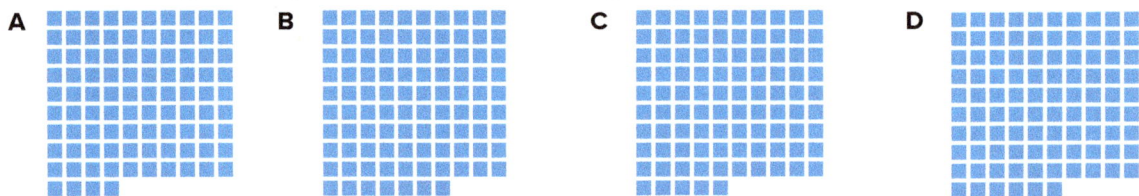

        A   B   C   D

    e How many squares were subtracted from the original grid to decrease it by 6%.

    f Fill in the blanks to complete an expression that represents the 6% decrease in squares:
        ☐% − ☐% = 94% = 0.94

5 What percentage should a quantity be multiplied by in order to increase the quantity by:

    a 10%   **b** 20.7%

6 What percentage should a quantity be multiplied by in order to decrease the quantity by:

   a 22%   **b** 3.55%

7 Consider increasing 1300 by 40%.

   a Find 10% of 1300.

   b How could you use the percentage you just found to calculate 40% of 1300?

   c Find 40% of 1300.

   d Add your answer in (c) to the original amount of 1300.

   e Calculate 140% of 1300.

   f Are your answers to (d) and (e) the same?

8 Consider decreasing 1500 by 15%.

   a Find 10% of 1500.

   b Find 20% of 1500.

   c Subtract your answer in part (b) from the original amount of 1500.

   d Calculate 80% of 1500.

   e Are your answers to parts (c) and (d) the same?

Let's practice

9 Hiro wants to decrease 120 by 40%.

   a Calculate 40% of 120.

   b Should Hiro add or subtract the amount from part (a) from 120? Explain your answer.

   c Decrease 120 by 40%.

   d Solve the proportion for x: $\dfrac{60}{100} = \dfrac{x}{120}$

   e How could you use a proportion to solve for a percent decrease?

10 Mark collected 20 eggs from his chicked yesterday. If he collected 15% more eggs today, which proportion could be solved to find the number of eggs he collected today?

   A $\dfrac{12}{100} = \dfrac{20}{x}$ **B** $\dfrac{112}{100} = \dfrac{20}{x}$ **C** $\dfrac{112}{100} = \dfrac{x}{20}$ **D** $\dfrac{100}{20} = \dfrac{x}{12}$

11 A shelter starts off the weekend with 50 dogs to adopt. If 25% of them were adopted, which letter in the diagram could represent the remaining number of dogs?

12 A recipe originally calls for 480 grams of flour. Libby is increasing the recipe by 16%. Which is the best estimate for the amount of flour she will need?

   A 72 grams   **B** 555 grams   **C** 580 grams   **D** 600 grams

13 Last year the 8th grade had 350 students. This year the 8th grade had 307 students. Which is the best estimate of the percent change?

   A 9% decrease   **B** 14% increase   **C** 12% decrease   **D** 12% increase

14 Joseph currently earns $50 000 in a year. He got promoted and received an 8% raise.

    a What benchmark percent could help you estimate Joseph's new annual salary?

    b Using your percent from part (a), estimate Joseph's new annual salary.

    c Calculate Joseph's new annual salary.

15 Abed wants to increase 1500 by 30%.

    a Calculate 30% of 1500 and add the answer to 1500.

    b Calculate 130% of 1500.

    c Explain why the answers to parts (a) and (b) are the same.

16 For each of the following:

    i State whether the value increased or decreased.

    ii Calculate the percentage change rounded to two decimal places.

    a 82 changed to 38   **b** 33 changed to 47

    c $63 changed to $98   **d** 122 minutes changed to 109 minutes

    e 79 m changed to 157 m   **f** 141 g changed to 142 g

    g 199 km changed to 92 km   **h** 135 students changed to 85 students

17 A bag of rice weighs 220 lb. When some rice is poured out, the weight of the bag decreases by 40%. Find the new weight of the bag of rice.

18 A bag of potatoes weighs 110 lb. When more potatoes are added, the weight of the bag increases by 35% find the new weight of the bag of potatoes.

19 A car company increased the number of miles that could be driven with 20 gallons of gas from 450 miles to 567 miles. What is closest to the percent increase in the amount of miles that can be driven?

    A 20%   **B** 26%   **C** 30%   **D** 36%

20 Adrian previously wage was $1250 and his new wage is $1700. Determine the percentage increase of Adrian's wage.

21 The price of a phone was reduced by 25% and then again by 25%. Find the overall percentage decrease.

22 During the editing phase when making a movie, the director ordered that the movie be cut down to 124 minutes, which was a 38% decrease from the original duration. What was the movie's original duration length?

23 Describe and correct the error in finding the percent increase from 16 to 42.

$$\times \frac{42 - 16}{42} \approx 0.62 = 62\%$$

24 Explain why a change from 30 to 60 is a 100% increase, but a change from 60 to 30 is a 50% decrease.

25 The table shows population data for a community.

    a What is the percent of change from 2014 to 2019?

    b Use this percent of change to predict the population in 2024.

Year	Population
2014	124, 000
2019	138, 000

Let's extend our thinking

26 The Run4Fun charity race is increasing in popularity. A year ago 60 000 people registered to run, and this number is expected to increase by 6% this year and then by another 8% next year. How many people are expected to run next year?

27 Bob has invested heavily in a certain company. One day it was reported that the company's factory was damaged by a fire and its share price fell by 40%. However, when it was revealed the next day that the company had insurance and wouldn't be too badly affected, the shares recovered by 40%.

   a Express the share price after the original announcement as a percentage of the previous day's price.

   b Express the share price after the insurance announcement as a percentage of the price after the original fire announcement.

   c Calculate the final share price as a percentage of the price before either announcement.

   d Now, evaluate the percentage change over the two days.

   e Is this change an overall increase or decrease? Explain.

28 The length of a rope is 120 meters. The length is increased by 15% and then the result is decreased by 15%. What is the new length of the rope, rounded to the nearest meter?

29 Kevin earns a salary of $1500 per week for a 45-hour week. His weekly salary is increased by 8% and his number of hours worked is decreased by 12%. Calculate his new hourly salary.

30 If a discount of a% is applied to an amount and then an increase of b% is applied, write an expression for the single multiplier that is applied to the amount.

Answers

1.04 Percent increase and decrease

What do you remember?

1 A percent increase will be over 100%. A percent decrease will be under 100%.

2 A, C

3 A, D

4
a B
b 3 more squares were added.
c 100% + 3% = 103% = 1.03
d A
e 6 squares were taken away.
f 100% − 6% = 94% = 0.94

5 **a** 110%   **b** 120.7%

6 **a** 78%   **b** 96.45%

7
a 130
b 40% of 1300 should be 4 times as much as 10%
c 520   **d** 1820
e 1820   **f** Yes

8
a 150   **b** 300
c 1200   **d** 1200
e Yes

Let's practice

9
a 48
b Subtract, because he wants to lower the value.
c 72
d $x = 72$
e To find a percent decrease using a proportion, set up a fraction where the numerator is the amount of decrease and the denominator is the original value. Then, set this fraction equal to another fraction where the numerator is the unknown percent (as a decimal) and the denominator is 100.

10 C

11 C

12 B

13 C

14 **a** 10%   **b** $55 000
c $54 000

15
a 1950
b 1950
c Yes it is equivalent since 100% of 150 is 150.

16
a **i** Decrease   **ii** 53.66%
b **i** Increase   **ii** 42.42%
c **i** Increase   **ii** 55.56%
d **i** Decrease   **ii** 10.66%
e **i** Increase   **ii** 98.73%
f **i** Increase   **ii** 0.71%
g **i** Decrease   **ii** 53.77%
h **i** Decrease   **ii** 37.04%

17 132 lb

18 148.5 lb

19 B

20 36%

21 43.75%

22 200 minutes

23 The error in the calculation is using the new value (42) as the denominator instead of theoriginal value (16). The correct percent increase is calculated by dividing the increase invalue (26 − 18) by the original value (16), and then multiplying by 100 to get the percentage.

24 A change from 30 to 60 is a 100% increase because the amount increased (30) is equal to theoriginal amount (30), which is a 100% of the original. However, a change from 60 to 30 is a 50% decrease because the amount decreased (30) is half of the original amount (60), whichis a 50% of the original.

25
a The percent change from 2007 to 2013 is approximately 11.3%.
b To predict the population in 2024 using the percent change, we apply the percent increase to the 2019 population. If we assume the same rate of change, we can expect the 2024 population to be 138 000 + (138 000 · 11.3%). This predicts a population of approximately 153 594.

Let's extend our thinking

26 **a** 68 688 people

27 **a** 60%   **b** 140%   **c** 84%   **d** −16%
e Decrease. Positive values indicate a percentage increase while negative values indicate a percentage decrease, since the percentage change is negative it means an overall decrease.

28 117 m

29 $40.91

30 Multiplier is $\left(1 - \dfrac{a}{100}\right) \cdot \left(1 + \dfrac{b}{100}\right)$

1.05 Discounts, fees, and markups

Subtopic overview

Lesson narrative

In this lesson, students will explore the concepts of discounts, fees, and markups, which involve percent increases and decreases. The lesson begins by explaining the definitions of a discount, a markup, and a fee (a percent increase for a service). Students will learn the general method for calculating these percentages. Students will engage in solving practical examples, such as calculating the discounted price of an item, determining the new price of a product after a markup, and finding the overall cost including a service fee. They will apply strategies such as converting percentages to fractions, setting up proportions, and using the percent increase and decrease formulas. By the end of the lesson, students should be proficient in calculating and interpreting discounts, markups, and fees, and understand their impact on the overall price of goods and services.

Learning objectives

Students: Page 35

After this lesson, you will be able to...
- estimate the solution to real-world problems involving a single discount or markup.
- solve real-world problems involving a single discount or markup.

Key vocabulary

- discount
- fee
- markup
- percent decrease
- percent increase
- regular price
- sale price

Essential understanding

There are many real-world applications of percents. Discounts are applications of percent decrease. Fees and markups are applications of percent increase.

Standards

This subtopic addresses the following Virginia 2023 Mathematics Standards of Learning standards.

Mathematical process goals

MPG1 — Mathematical Problem Solving

Teachers can integrate this goal into their instruction by posing real-world problems involving discounts, fees, and markups. This allows students to apply their mathematical skills and knowledge, such as calculating percentages, to solve these problems. For example, teachers can ask students to calculate the final price of an item after a discount or markup is applied, or to determine the additional cost of a fee on a base price.

MPG2 — Mathematical Communication

Teachers can encourage students to communicate their thinking and reasoning as they work through problems involving discouns, fees, and markups. Students can be encouraged to use the language of mathematics, including specialized vocabulary and symbolic notation, to express their ideas with precision.

MPG4 — Mathematical Connections

Teachers can integrate this goal into instruction by drawing connections between the concept of percentages and its application in calculating discounts, fees, and markups. Furthermore, teachers can relate these mathematical concepts to real-world situations like shopping, thereby reinforcing the practical relevance of the mathematics being studied.

Content standards

8.CE.1 — The student will estimate and apply proportional reasoning and computational procedures to solve contextual problems.

8.CE.1a — Estimate and solve contextual problems that require the computation of one discount or markup and the resulting sale price.

8.CE.1c — Estimate and solve contextual problems that require the computation of the percent increase or decrease.

Prior connections

7.CE.2 — The student will solve problems, including those in context, involving proportional relationships.

Future connections

A.F.2 — The student will investigate, analyze, and compare characteristics of functions, including quadratic and exponential functions, and model quadratic and exponential relationships.

Lesson Preparation

Suggested review

Depending on your students' level of prior knowledge, consider revisiting the following lessons:

Grade 7 — 2.04 Percentages of whole numbers

Tools

You may find these tools helpful:
- Scientific calculator

Discounts, fees, and markups

Students discover that a discount is the result of a percent decrease, and fees and markups are the result of a percent increase. They learn that the final price can be calculated in the same way they calculated percent increase and decreases from the previous lesson. They review the definitions of regular price and sale price so that they know which values are given in the questions and which values they need to solve for.

Examples

The following supports may be useful for the examples in this section.

Consistent use of problem identification charts
Student with disabilities support

Introduce students to a standardized problem identification chart that they can use for every problem involving discounts, markups, and fees. This chart should have columns such as "Original price," "Percent increase/decrease," "Amount of increase/decrease," and "New price." Teach students to think algorithmically and first read the problem carefully, then identify what information was given and what information is needed, then fill in the chart accordingly to find the desired quantity.

Below the table, students can describe how to find each piece of information. For example:
- Amount of increase/decrease, given percent and original price

  1. Change the percent to a decimal or fraction
  2. Multiply the value from step 1 by the original price

- New price, given original price and percent

  - For an increase, add the amount of increase to the original price
  - For a decrease, subtract the amount of decrease from the original price

- Percent increase/decrease, given new price and original price

  1. Subtract the original price from the new price to find the amount of increase/decrease
  2. Divide the amount of increase/decrease by the original price
  3. Multiply the result by 100%

     - If positive, then it is increase
     - If negative, then it is decrease

- Original price, given percent and new price

  1. Subtract the percent increase/decrease from 100%
  2. Convert the percent from step 1 to a decimal or fraction
  3. Divide the amount of increase/decrease by the value from step 2

By consistently using this method, students can clearly see whether they need to find the amount of the increase/decrease, the percentage involved, or the new price after the change. This strategy supports students with organizational difficulties and helps reduce confusion by breaking down the problem into manageable parts. It also aids in memory retention as students repeatedly practice identifying and categorizing the essential elements of each problem.

This approach can also build up computational thinking skills by helping students break down problems and identify which approach is best.

Example 1

Steph is going to buy a hat that is marked at 75% off. The original price is $36.

a What is the value of the discount in dollars?

Create a strategy

To find the discount amount, multiply the original price by the percent discount. Remember that percent means 'divided by 100'.

Apply the idea

$$\text{Discount} = 36 \cdot \frac{75}{100}$$ Multiply the original price by the percent discount in fraction form

$$= \frac{2700}{100}$$ Evaluate the multiplication

$$= 27$$ Evaluate the division

The discount is $27.

Purpose

Students demonstrate that they can find the amount of discount given a percent and the original cost.

Expected mistakes

Students may forget units in their answers. Remind students that when solving problems in real-world scenarios, the answer should include any relevant units. This helps them maintain mathematical precision and clearly communicate their results.

Students: Page 35

b What is the price that Steph will pay for the hat?

Create a strategy

To find the discounted price, subtract the discount from the original price.

Apply the idea

$$\text{Discounted price} = \$36 - \$27$$ Subtract the discount from the original price

$$= \$9$$ Evaluate

The discounted price is $9.

Reflect and check

We have decreased the amount by 75%, which means that Steph is paying the remaining 25%, or 100% – percent decrease.

We could have calculated the discounted price by:

$$\text{Discounted price} = 36 \cdot \frac{25}{100}$$ Rewrite 25% as a fraction and multiply by the original price

$$= \frac{900}{100}$$ Evaluate the multiplication

$$= 9$$ Evaluate the division

We get the same answer. The discounted price is $9.

Purpose

Students demonstrate that they can use the amount of discount to calculate sale price by subtracting the amount of discount from the original cost.

Reflecting with students

Have students to consider the reasonableness of their answers. Prompt them with questions such as, "Does the price reflect a 75% discount?" or "Does this price seem to be three-quarters of the original price?"

Students: Page 36–37

Example 2

A watch that normally costs $75 is marked up by 20%. What is the new price of the watch?

Create a strategy

Remember that markup involves percent increase. Increasing by 20% is the same as multiplying by 100% + 20% or 120%.

Apply the idea

$$\text{Markup price} = \$75 \cdot \frac{120}{100} \qquad \text{Convert 120\% to a fraction and multiply to the original price}$$

$$= \frac{9000}{100} \qquad \text{Evaluate the multiplication}$$

$$= 90 \qquad \text{Evaluate the division}$$

The markup price or new amount of the watch is $90.

Reflect and check

We could have also found the new price by translating the statement into a proportion. To set up the proportion, we set the percent as a fraction equal to the new price over the original price.

$$\frac{\text{percent}}{100} = \frac{\text{part}}{\text{whole}} \qquad \text{Set up a proportion}$$

$$\frac{120}{100} = \frac{x}{75} \qquad \text{Substitute 120 for the percent, } x \text{ for the part, and 75 for the whole.}$$

$$100x = 9000 \qquad \text{Cross multiply}$$

$$\frac{100x}{100} = \frac{9000}{100} \qquad \text{Evaluate the division}$$

$$x = 90 \qquad \text{The markup price is \$90}$$

Purpose

Show students how to calculate a new price after a percentage markup, emphasizing the importance of understanding percentage increase in real-world situations like shopping and sales.

🎓 Advanced learners: Use a parallel task to extend learning   use with Example 2
Targeted instructional strategies

To extend learning for advanced students, introduce a parallel task where they calculate the original price given the marked-up price, instead of calculating the new price. Present the problem as, "A 20% markup is applied to a watch, and then it is sold for $90. What was the original price of the watch?"

This requires students to reverse the percentage increase process, encouraging them to think critically about inverse operations. Students should recognize that the new price is found by multiplying the original price by 120%, so the original price can be found by dividing the new price by 120%.By working through this task, students can deepen their understanding of the relationship between percentage increases and inverse operations.

Example 3

An artist was hired to paint a portrait that will cost $4000. The contractor also includes a service fee of $1500 for the overall cost of the contract.

What percent of the cost of the portrait is the artist's fee?

Create a strategy

Find the percent rate as a ratio of the fee and the cost of the house.

Apply the idea

$\dfrac{\text{percent}}{100} = \dfrac{\text{part}}{\text{whole}}$	Set up a proportion
$\dfrac{x}{100} = \dfrac{1500}{4000}$	Substitute 1500 for the part, 4000 for the whole, and x for the percent
$\dfrac{x}{100} = \dfrac{37.5}{100}$	Divide the numerator and denominator by 40
$x = 37.5$	Multiply both sides by 100

The artist charges a percent fee of 37.5%.

Reflect and check

Notice that the fee is an increase in the price of service. This means we could have used the percent increase formula to find the percent amount of the fee.

$\text{Percent Increase} = \dfrac{\text{Increase}}{\text{Original Amount}} \cdot 100\%$	Write the percent increase formula
$= \dfrac{1500}{4000} \cdot 100\%$	Substitute known amounts
$= 0.375 \cdot 100\%$	Evaluate the division
$= 37.5\%$	Evaluate the multiplication

Purpose

Show students how to calculate the percentage of a part relative to a whole using proportions, reinforcing the concept of percentages in real-world contexts.

Expected mistakes

Students might think they need to convert the value to a percent, not recognizing that they set up the proportion so that x already represents a percent. Remind them that "percent" translates to "out of 100", so $\dfrac{x}{100}$ is the same as $x\%$.

Example 4

A TV normally sells for $1792.94, but is currently on sale.

In each of the following scenarios, calculate the percent discount correct to two decimal places.

a   The TV is discounted by $149.50.

Create a strategy

A discount is a percent decrease so we can use the percent decrease formula:

$$\text{Percent Decrease} = \frac{\text{Decrease}}{\text{Original Amount}} \cdot 100\%$$

Apply the idea

The discount amount is $149.50 and the regular price is $1792.94.

$$\text{Percent Discount} = \frac{149.50}{1792.94} \cdot 100\% \qquad \text{Substitute the given values into the formula}$$

$$= 8.34\% \qquad \text{Evaluate}$$

Purpose

Show students how to calculate the percent discount given the original price and the discount amount. This helps students understand the concept of percent decrease in a real-world context.

Reflecting with students

Remind students to consider the reasonableness of their answers by estimating the percent discount before performing detailed calculations. Since $149.50 is less than 10% of $1792.94, an answer around 8% makes sense. By comparing their calculated answer to this estimate, they can verify its reasonableness and ensure they've performed the calculation correctly.

Students: Page 38

b   The TV is on sale for $1428.74.

Create a strategy

Calculate the discount amount and then use the percent decrease formula:

$$\text{Percent Decrease} = \frac{\text{Decrease}}{\text{Original Amount}} \cdot 100\%$$

Apply the idea

The regular price is $1792.94.

$$\text{Discount Amount} = 1792.94 - 1428.74 \qquad \text{Subtract 1428.74 from 1792.94}$$

$$= \$364.20 \qquad \text{Evaluate}$$

$$\text{Percent Discount} = \frac{364.20}{1792.94} \cdot 100\% \qquad \text{Substitute the given values into the formula}$$

$$= 20.31\% \qquad \text{Evaluate}$$

Purpose

Show students how to calculate the percent decrease in price given the original and sale price of an item, reinforcing the application of the percent decrease formula.

Expected mistakes

Students might misread the problem and think that the given price is the amount of the discount. Have them re-read the problem carefully and ask them whether the price is a sale price or a discount.

Three reads

English language learner support

Advise students to read through the instructions a few times, focusing on gathering different information each time in order to build up understanding of what the question is asking.

On the first read, students should aim to identify the scenario presented in the question. Ask students, "What do you think is happening in this question?" or "Can you explain what this question is about?"

On the second read, students should aim to interpret the problem by answering questions like, "What is the question asking you to find?" and "What information should be included in the answer?"

On the third read, students should look for important information in the instructions. In this question, the important information includes:

- The original price of the TV is $1792.94.
- For part (a), the discount is $149.50.
- For part (b), the discounted price is $1st428.74.

Students can be prompted by framing these as questions like "Is the amount given a discount or the new price?" or "What is the original price?"

Students: Page 38

> **Idea summary**
>
> Discounts are examples of percent decreases. We subtract the amount of discount from the original price to find the sale price. We can also multiply the original price by 100% − percent decrease to get the sale price.
>
> Fees and markups are examples of percent increases. We add the fee or the markup to the original price to find the full price. We can also multiply the original cost by 100% + percent increase to find the full price.

Practice

Students: Page 38–42

What do you remember?

1   Fill in the blank to write each percent as a ratio.

   a $24\% = \dfrac{\square}{100}$ b $136\% = \dfrac{\square}{100}$ c $36\% = \dfrac{\square}{100}$ d $224\% = \dfrac{\square}{100}$

2   Select the term that is best used to describe the situation:

   a   A shopkeeper buys an item for $35 and adds 10% to the price.

   b   A company charges a certain amount proportional to the total.

   c   A shopkeeper reduces the price of an item by 15% during a sale.

   - Discount
   - Fee
   - Mark-up
   - Selling price
   - Cost price

3 Find the percent increase in price after the mark-ups are applied:

    a   The new price of a wallet after a mark-up is calculated by multiplying the original price by 110%.

    b   The new price of a hat after a mark-up is calculated by multiplying the original price by 121%.

4 Find the percent decrease in price after the discounts are applied:

    a   The new price of a phone after a discount is calculated by multiplying the original price by 95%.

    b   The new price of a shirt after a discount is calculated by multiplying the original price by 77%.

5 Which would you rather pay? Explain your reasoning.

    a   9% tax on a discounted price or 9% tax on the original price

    b   20% markup on a $20 shirt or $20 markup on a $20 shirt

6 Nadya gets a discount of $35 from a cabinet originally priced at $100. What is the discounted price?

Let's practice

7 Sienna runs an artisanal craft shop and decides to mark up all handmade products by 150%.

Use a model to estimate the retail price of the crafts. Then use a calculator to find the selling price.

    a   The retail price of a handcraft vase is $75.

$0 $75   Selling price

    b   The retail price of a handcraft scarf is $30.

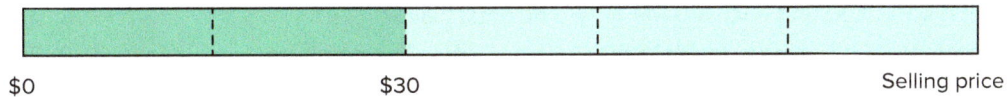

$0 $30   Selling price

    c   The retail price of a painted picture frame is $40.

$0 $40   Selling price

8 A clothing store is having a sale on all items. If a shirt originally costs $50 and is now being sold for $35.50, estimate the percent discount on the shirt and explain your strategy.

9 Steph is going to buy a dress that is marked as 25% off. The original price was $36.

    a   Find the discount amount.

    b   Find the discounted price that Steph will pay for the dress.

10 Duncan is going to buy a sweatshirt that is marked up by 25%. The original price was $30.

    a   Find the marked up amount.

    b   Find the marked up price that Duncan will pay for the sweatshirt.

11 Calculate the sale price of an item which is sold at:

    a   A discount of 18% to its regular price of $100.   **b**   A discount of $95.64 to its regular price of $546.59.

12 A bicycle is on sale for 60% off the original price. Which of the methods can you use to find the sale price? Which method do you prefer? Explain.

    a   Multiply $37.45 by 0.6.   **b**   Multiply $37.45 by 0.4.

    c   Multiply $37.45 by 0.6, then add to $37.45.   **d**   Multiply $37.45 by 0.6, then subtract from $37.45.

13 Ace purchased a bicycle that was on sale for 30% off. The original price of the bicycle was $880. What was the total amount of money Ace saved?

    A $616   **B** $377   **C** $352   **D** $264

14 Very Cheap Auto car company is having a sale so that car prices are discounted by 10%. If the original price of a car was $12 500, find the discounted price of the car.

15 A tennis racket marked at a price of $90 is advertized to be selling at 45% off the marked price. Find the discounted price.

16 Calculate the marked up price of an item:

    a   Priced at $100 marked up by 20%.   **b**   Priced at $85.25 marked up by 12%.

17 A brush set priced at $140 is marked up by 45%. What is the marked up price of the brush set?

18 A store is having a 13% off sale. Calculate the original price of a discounted item that is on sale for $3723.75.

19 A laptop priced at $120 is marked up to $156. What is the percent mark up?

20 Sally earns $610 per week. Her weekly work related expenses are $185 on travel and $86 on lunch. What percent of her weekly income goes towards work related expenses?

21 A computer normally sells for $3605.67, but is currently on sale. In each of the following scenarios, calculate the percent discount. Round each answer to two decimal places.

    a   The computer is discounted by $325.80.   **b**   The computer is on sale for $3222.37.

22 A contractor was hired to build a house which will cost $300 000. The contractor also includes a percent fee of $45 000 for the overall cost of the contract.

What percent of the cost of the house is the contractor's percent fee?

23 Three different stores are offering the same model of headphones at a discount.

    a Create a tape diagram to represent each deal.

    b Determine which store offers the best deal.

       i Regular price: $60   **ii** Regular price: $64   **iii** Regular price: $68

30% OFF   **40% OFF**   **up to 50% OFF**

Let's extend our thinking

24 A flute priced at $600 is marked up by 10% and then marked up again by 40%.

    a Instead of calculating 2 separate mark-ups, Elizabeth claims that she can calculate the final price in a single step using the expression 600 · 150%, because 100 + 10 + 40 = 150.

      Is Elizabeth correct? Explain your answer.

    b Instead of calculating 2 separate mark-ups, Judy claims that she can calculate the final price in a single step using the expression 600 · 110% · 140%.

      Is Judy correct? Explain your answer.

    c Find the final price of the flute.

25 Alex is planning to buy a new laptop for school. The laptop's original price is $800, but it is currently on sale with a 20% discount. Additionally, Alex wants to buy a protective case and an external hard drive for the laptop. The case costs $50 with no discount, and the external hard drive has a 10% markup from its original price of $100. There is also a flat shipping fee of $15 for the entire order. Alex has saved $750 in his bank account for this purchase.

    a Calculate the sale price of the laptop after applying the discount.

    b Determine the final price of the external hard drive after applying the markup.

    c Calculate the total cost of Alex's purchase, including the laptop, case, external hard drive, and shipping fee.

    d Based on the total cost, determine if Alex has enough money saved for the purchase. If not, calculate how much more Alex needs.

26 The price of a $234 heater is discounted by 14% and later marked up by 14%. What is the final sale price?

27 Hiro wants to buy a washing machine and has two brands to choose. Which brand gives Hiro greater discount?

    • Brand A: Originall sells at $1300 with a discount of 20%.

    • Brand B: Originall sells at $850 with a discount of 35%.

28 A dining table selling for $1000 is discounted at 5% and then again at 6%.

    a What is the price of the dining table after the first discount is applied?

    b What is the price of the dining table after the second discount is applied?

    c The percent discount is changed by adding the previous percents (5% + 6%) to get 11%. Find the new discounted price.

    d Is a single discount of 11% equal to successive discounts of 5% and 6%?

29 Which offer gives the greater discount? Explain.

    a Offer 1: A single discount of 13%   **b** Offer 1: A single discount of 16%

      Offer 2: Successive discounts of 6% and 11%   Offer 2: Successive discounts of 5% and 11%

30 Jane wants to purchase 10 plates and has three offers to choose. Each plate cost $4.50 Which offer gives her a greater discount? Explain.

- Offer 1: Buy 3 plates for $13.50 and get two plates free.
- Offer 2: Buy 4 plates for only $12 and get one plate free.
- Offer 3: A discount of 40% for each plate.

31 A washing machine originally priced at $80 is discounted by 15%. A few months later, the washing machine is marked up from its new price by 45%.

What is the final price of the washing machine after applying both the discount and the mark up? Give your answer in dollars, rounded to the nearest cent.

32 SmartCash is a mobile application that lets you instantly pay bills, buy load, send money and shop. One can make a cash-in transaction over the counter of a convenience store. A charge of 2% cash-in fee is to be deducted from the total amount in excess of $1500 that was cashed-in over the counter.

If Sally made a cash-in transaction of $2000 over the counter, how much money will reflect in her account?

Answers

1.05 Discounts, fees, and markups

What do you remember?

1 **a** $24\% = \dfrac{24}{100}$ **b** $136\% = \dfrac{136}{100}$

  c $36\% = \dfrac{36}{100}$ **d** $224\% = \dfrac{224}{100}$

2 **a** Mark-up **b** Fee **c** Discount

3 **a** Increased by 10%
  b Increased by 21%

4 **a** Decreased by 5% **b** Decreased by 23%

5 **a** You would rather pay a 9% tax on a discounted price than on the original price. This is because the discounted price would be lower than the original price, and applying the same tax rate to a lower base amount results in less tax paid overall. The savings from the discount reduce the amount on which the tax is calculated, leading to lower total cost.
  b You would rather pay a 20% markup on a $20 shirt than a $20 markup on the same shirt. A 20% markup on a $20 shirt results in a $4 increase ($24 total), whereas a flat $20 markup would double the price to $40. Thus, the percentage markup leads to a lower overall cost than the flat dollar amount markup.

Let's practice

6 $65

7 **a** $187.5 **b** $75 **c** $100

8 To estimate the percent discount, first determine the amount of the discount by subtracting the sale price from the original price. Then, use friendly percents to approximate the discount.
Discount Amount: $50 − $35.50 = $14.50
10% of $50 is $5, 30% of $50 is $15. This is close to $14.50, so the percent discount is approximately 30%.

9 **a** $9 **b** $27

10 **a** $7.50 **b** $37.50

11 **a** $82.00 **b** $450.95

12 Both methods b and d can be used to find the sale price. Answers vary on preference. Method b tends to be more efficient since it is a single step.

13 **D**

14 $11250

15 $49.50

16 **a** $120.00 **b** $95.48

17 $203.00

18 $4280.17

19 30%

20 44.43%

21 **a** 9.04% **b** 10.63%

22 15%

23 **i** **i**
  ii No

  ii **i**
  ii No

  iii **i**
  ii Yes

Let's extend our thinking

24 **a** No, because we must add each percent to 100% and then multiply them. Elizabeth added 10% to 100%, but she did not add 40% to 100%.
  b Yes, because she first added both percents to 100% and then multiplied.
  c $924

25 **a** $800 − (0.20 \cdot 800) = \$640$
  b $100 + (0.10 \cdot 100) = \$110$
  c $640 + $110 + $50 = $800
  d Alex does not have enough saved. He needs to save $50 more.

26 $229.41

27 Offer 2

28 $950 **b** $893.00 **c** $890 **d** No

29 **a** Offer 2. Consider that the price of an item is $100. A single discount of 13% would cost $87 while a succesive discounts of 6% and 11% would cost only $83.66.
  b Offer 1. Consider that the price of an item is $100. A single discount of 16% would cost only $84 while a succesive discounts of 5% and 11% would cost $84.55.

30 Offer 2. Offer 1 and Offer 3 would cost Jane $27 in buying 10 plates while Offer 2 would only cost $24 which is cheaper among the other two offers.

31 $98.6

32 1990

1.06 Sales tax, commissions and tips

Subtopic overview

Lesson narrative

In this lesson, students will learn to calculate sales tax, tips, and commissions by applying their knowledge of percentages. The lesson begins with an explanation of tipping customs and practices, including how to calculate tips as a percentage of the bill before tax. Students will work through examples that require them to convert percentages to decimals and fractions, and then apply these conversions to find the tip amounts. Next, the lesson addresses sales tax, explaining its application and calculation on purchases. Students will solve problems involving varying sales tax rates and determine the total cost of items after tax. They will practice converting percentages to fractions and decimals, and perform multiplications to find the additional cost due to tax. Finally, the lesson covers commissions, detailing how they are calculated as a percentage of sales above a certain threshold. Students will engage with examples that involve determining commission amounts and adding these to base salaries to find total earnings. By the end of the lesson, students should be proficient in calculating tips, sales tax, and commissions, and understand their real-world applications.

Learning objectives

Students: Page 43

> **After this lesson, you will be able to...**
> * estimate the solution to real-world problems involving sales tax, commission, and tips.
> * solve real-world problems involving sales tax, commission, and tips.

Key vocabulary

* comission
* sales tax
* tip

Essential understanding

There are many real-world applications of percents. Sales tax, commissions, and tips are applications of percent increase.

Standards

This subtopic addresses the following Virginia 2023 Mathematics Standards of Learning standards.

Mathematical process goals

MPG1 — Mathematical Problem Solving

Teachers can integrate this goal into their instruction by guiding students through the process of calculating sales tax, commissions, and tips. Teachers can demonstrate how to apply the mathematical concept of percentages to solve these real-world problems. They can provide students with a variety of practice problems and encourage them to apply their newly acquired skills to determine the total cost of an item including tax, or the amount of commissions or tips earned.

MPG2 — Mathematical Communication

Teachers can encourage students to communicate their mathematical thinking through discussions and written work. This can be achieved by having students explain their strategies for calculating sales tax, commissions, and tips, and justify their methods and answers. Teachers can also introduce and use specialized mathematical vocabulary related to percentages, tax, and commissions in classroom discussions and instructions to enhance students' mathematical language and communication skills.

MPG4 — Mathematical Connections

Teachers can make connections to students' prior knowledge of percentages and apply it to the new context of calculating sales tax, commissions, and tips. They can also link these concepts to real-life situations like shopping, budgeting, and working in sales or service industries, thus making mathematical connections to other disciplines and real-world contexts.

Content standards

8.CE.1 — The student will estimate and apply proportional reasoning and computational procedures to solve contextual problems.

8.CE.1b — Estimate and solve contextual problems that require the computation of the sales tax, tip and resulting total.

8.CE.1c — Estimate and solve contextual problems that require the computation of the percent increase or decrease.

Prior connections

7.CE.2 — The student will solve problems, including those in context, involving proportional relationships.

Future connections

A.F.2 — The student will investigate, analyze, and compare characteristics of functions, including quadratic and exponential functions, and model quadratic and exponential relationships.

Lesson Preparation

Suggested review

Depending on your students' level of prior knowledge, consider revisiting the following lessons:

Grade 7 — 2.04 Percentages of whole numbers
Grade 8 — 1.04 Percent increase and decrease

Tools

You may find these tools helpful:

- Scientific calculator

Tips

Students learn that a tip is an additional percentage of the bill, given as a sign of appreciation for good service, typically calculated before tax. This increase is a percent increase, calculated in the same that they have calculated percent increase in previous lessons.

Examples

Students: Page 43

Example 1

Anasofia and her daughter have finished her meal at a restaurant and received exceptional service. She will be leaving a 25% tip. Help her calculate the tip if the meal, before sales tax, came to $42.30.

Create a strategy

To find the tip amount, we need to calculate 25% of $42.30. We have to convert the percentage into a fraction and multiply by the amount of the bill.

Apply the idea

$\text{Tip} = 42.30 \cdot 25\%$ Multiply 42.30 by 25%

$= 43.25 \cdot \dfrac{25}{100}$ Convert the percentage to a fraction

$= \$10.58$ Evaluate and round

Purpose

Students demonstrate that they can find the amount of tip given a percent and a cost.

Expected mistakes

Students might find the total cost of the meal rather than finding the amount of the tip alone. Point out that the problem only asks for the tip, which is 25% of the bill, so the answer should be less than the price of the meal.

Consistent use of problem identification charts

Student with disabilities support

use with Example 1

Use the standardized problem identification chart from the previous lesson with this example and the other examples in this lesson. Teach students to think algorithmically and first read the problem carefully, then identify what information was given and what information is needed, then fill in the chart accordingly to find the desired quantity.

Original price	Percent increase	Amount of increase	New price
42.30	25%	×	−

Since the problem states the original price and percent increase and asks for the amount of the tip, they can follow these steps to find the answer.

1. Change the percent to a decimal or fraction
2. Multiply the value from step 1 by the original price

> ### 💡 Idea summary
>
> A **tip** is an amount of money left for the staff, in addition to paying the bill, as a sign that we appreciate good service. It is calculated as a percentage of the bill, typically before tax is applied.

Sales tax

This lesson explains that in most of the U.S., sales tax is added to the marked price of goods or services at the time of purchase, while in some countries it is included in the price. Students learn that tax is a percent increase on a service or item. Once the amount of tax is calculated, a final price can be found by adding the tax to the original cost.

Students: Page 44

Example 2

The sales tax on clothing in Florida is 6%.

a   Calculate how much tax you need to pay on a jacket if it costs $38.20 before tax.

Create a strategy

To calculate how much tax will be charged, we need to calculate 6% of $38.20. We have to convert the percentage into a fraction and multiply by the cost of the jacket.

Apply the idea

$$\text{Tax} = 38.20 \cdot 6\% \qquad \text{Multiply 38.20 by 6\%}$$
$$= 38.20 \cdot \frac{6}{100} \qquad \text{Convert the percentage to a fraction}$$
$$= \frac{229.2}{100} \qquad \text{Evaluate the multiplication}$$
$$= 2.29 \qquad \text{Evaluate the division and round}$$

The amount of tax needed to pay is $2.29.

Purpose

Students demonstrate that they can find the amount of tax on an item given a percent and an original cost.

Reflecting with students

Encourage students to consider the appropriate degree of precision when calculating the sales tax on the jacket. Since we're dealing with money, it's important for them to express their final answer in dollars and cents, rounding to the nearest cent.

Guide them to see that while the exact calculation might yield $2.292, the tax should be reported as $2.29 because we cannot pay fractions of a cent. Point out that an imprecise response like $2.3 lacks the necessary precision for financial transactions, while overly precise answers like $2.292 are not practical.

Students: Page 44

b Find the total amount to pay for the jacket.

Create a strategy

Add the cost of the jacket and the sales tax.

Apply the idea

$$\begin{aligned} \text{Total cost} &= \$40.60 + \$2.84 && \text{Add the total cost and the tax} \\ &= \$43.44 && \text{Evaluate} \end{aligned}$$

The total amount to pay is $43.44.

Reflect and check

We can apply the concept of percent increase to find the total cost of the jacket including tax.

Remember that adding 7% to $40.60 is the same as multiplying $40.60 by 100% + 7% or 107%.

$$\begin{aligned} \text{Total cost} &= 40.60 \cdot \frac{107}{100} && \text{Convert the percent to fraction} \\ &= \frac{4344.2}{100} && \text{Evaluate the multiplication} \\ &= 43.44 && \text{Evaluate the division} \end{aligned}$$

The total amount to pay is $43.44.

Purpose

Students demonstrate that they can find the total cost of an item after adding a tax amount.

Students: Page 45

Example 3

Jaleel is at the grocery store. He added up the total of the items in his cart using the prices seen on the shelves. His current total, or the subtotal, is $123.50. When he checks out, the total is $133.38.

a How much did Jaleel pay in sales tax?

Create a strategy

We have to find the difference in the two prices by subtracting the subtotal from the total at the checkout.

Apply the idea

$$\begin{aligned} \text{Cost of tax} &= \$133.38 - \$123.50 && \text{Subtract the subtotal from the cost at checkout} \\ &= 9.88 && \text{Evaluate} \end{aligned}$$

The cost of the tax is $9.88.

Purpose

Students demonstrate that they can find the amount of commission if an employee sells over a certain value.

b What is the sales tax percentage?

Create a strategy

We know that the total amount paid is equal to the subtotal times the tax percentage. We can substitute the price from check out and the price he calculated and solve for the missing percentage.

Apply the idea

$$\$9.88 = 123.50 \cdot \frac{x}{100}$$ Substitute the tax cost and the subtotal

$$\frac{\$9.88}{123.50} = \frac{\$123.50}{\$123.50} \cdot \frac{x}{100}$$ Divide both sides by \$123.50

$$0.08 = \frac{x}{100}$$ Evaluate the division

$$0.08 \cdot 100 = \left(\frac{x}{100}\right) \cdot 100$$ Multiply both sides by 100

$$8 = x$$ Evaluate

The tax percentage is 8%

Purpose

Students demonstrate that they can calculate the percentage increase when given the increase amount and original price.

Expected mistakes

Students might find the percentage of the total price with the tax amount, rather than finding the percentage of the pre-tax amount. Remind students that tax is applied to the subtotal, not to the final amount, which already includes the tax amount.

Advanced learners: Exploration of varying sales tax rates and their impact
use with Example 3

Targeted instructional strategies

Encourage students to investigate how different sales tax percentages affect the total cost of Jaleel's purchase. Have them calculate the sales tax and final total if the sales tax were, say, 5%, 7.5%, or 10% instead of 8%. This allows students to see how changes in the tax rate impact the overall amount paid, deepening their understanding of proportional relationships.

Additionally, prompt students to consider why sales tax rates might differ across various regions and how that affects consumers and businesses. This exploration not only reinforces their computational skills but also connects mathematical concepts to real-world economic considerations, fostering critical thinking and real-life applicability.

Students: Page 45

💡 Idea summary

Sales tax is an amount to be paid in addition to the marked price of goods or the stated price of a service, and is added on to the marked price, at the point of sale.

Commission

Students learn that a commission is a percentage of an employee's total sales that an employee earns as payment. They discover that an employee's salary might completely consist of commission, or the commission can be an addition to their regular salary, in which case it is called a retainer.

Students: Page 46

Example 4

Alton works in a jewelry store and is paid a commission of 5.5% on the value of jewelry sold in excess of $1500, in addition to his monthly salary of $3000. Calculate his monthly income if he sells $3200 worth of jewelry this month.

Create a strategy

Calculate the amount that commissions are earned on, then multiply by the percentage to find the commission.

Apply the idea

Amount to earn commission $= 3200 - 1500$	Subtract 1500 from sales of 3200
$= 1700$	Evaluate
Commission $= 1700 \cdot \dfrac{5.5}{100}$	Convert 5.5% to fraction and multiply to amount earned
$= \dfrac{9350}{100}$	Evaluate the multiplication
$= 93.5$	Evaluate the division

Alton gets a commission of $93.50.

To find his total income, we have to add the amount of his commission to his monthly salary.

Total income $= 3000 + 93.50$	Add his salary to his commission
$= \$3093.50$	Evaluate the addition

Alton's total monthly income is $3093.50

Purpose

Students demonstrate that they can solve contextual problems that require the computation of a commission and resulting total pay.

Expected mistakes

Since the commission only applies to values over a certain amount, students may not subtract the cost from the amount needed to receive the commission. Help them understand that "in excess of " means Alton only receives commission on the amount of sales over $1500.

Reflecting with students

Considering an employee may sell multiple items in a day, ask students whether it would make more sense for commission to be taken on each item individually or as a total at the end of the day. Would there be a difference in the commission earned? Would this difference work in favor of the employee or the employer or both?

Three reads
English language learner support

use with Example 4

Advise students to read through the instructions a few times, focusing on gathering different information each time in order to build up understanding of what the question is asking.

On the first read, students should aim to identify the scenario presented in the question. Ask students, "What do you think is happening in this question?" or "Can you explain what this question is about?"

On the second read, students should aim to interpret the problem by answering questions like, "What is the question asking you to find?" and "What information should be included in the answer?"

On the third read, students should look for important information in the instructions. In this question, the important information includes:

- Alton receives a commission of 5.5% on sales over $1500.
- Alton receives a monthly salary of $3000.
- This month, Alton sold $3200 worth of jewelry.

Students can be prompted by framing these as questions like "On how much of this month's sales will Alton receive commission on?" or "What is the total amount that Alton will make in salary plus commission?"

Students: Page 46

> **💡 Idea summary**
>
> **Commission** applies the concept of percentage increase, and is paid to an employee based on how much they sell.

Practice

Students: Page 46–49

What do you remember?

1 Determine whether each situation represents sales tax, commission, or tips.

  a A car dealer gets 10% of the price of each car he sells.

  b Mark pays his waiter 20% of the price of his meal.

  c A shop is required to charge an additional 6.8% on each sale.

2 Determine each percent.

  a $30 is what percent of $120

  b $1.50 is what percent of $10

  c $6.50 is what percent of $20

  d $26.10 is what percent of $18

3 Jason just got the bill for dinner for $35.47.

  a If he wants to leave a 22% tip, what should he multiply the bill total by to determine the amount of the tip?

  b If he wants to leave a 22% tip, what should he multiply the bill total by to find the total cost of the meal?

4 Calculate the sales tax of each bill.

  a Sales tax: 9%
    Bill before tax: $42.30

  b Sales tax: 6%
    Bill before tax: $16.50

5 Erin earns a 9.8% commission on each sale. If he makes a sale for $1 850, what is his approximate commission?

  A $120   **B** $135   **C** $160   **D** $185

6 Calculate the amount of sales tax that will be charged to each purchase:

    a A meal in Alabama, which costs $37.48 before tax, where the sales tax is 4%.

    b A dress in Indiana, which costs $40.60 before tax, where the sales tax is 7%.

    c A prescribed drug in Rhode Island, which costs $11.40 before tax, where the sales tax on prescription drugs is 7%.

7 Calculate the 5.3% tax applicable to each purchase:

    a Sheilah buys a pair of shoes priced $80

    b Zion purchases a pair of jeans marked $39.50

    c Kira buys a dress that costs $35 before tax

    d Mariko wants a purse worth $77.80

8 James is paid a commission of 2% of the value of the products he sells. Calculate his commission if his sales for the month are $261 000.

9 For each of the following, determine the amount of the tip:

    a Sharon has finished her meal at a restaurant which came to $36.60 but received bad service. She will only be leaving a 10% tip.

    b David is paying for a meal which came to $173.70. He received great service, so he is giving a 20% tip.

    c Vanessa wants to leave a 15% tip on a bill of $64.54.

10 Han is paid a commission of 3% of the value of trucks sold. Estimate his commission if he sells 2 trucks with a total value of $68 000.

11 Sean's dinner bill total, including tax, was $47.52. If Sean plans to tip 21% of this amount, estimate the value of the tip.

12 A restaurant includes an 18% service charge on the bill when a table has 6 or more people.

    a Calculate the minimuim tip that a waiter will get for serving a table of 6 people that had a bill of $238.51.

    b Assuming the table decided not to tip extra, find the total amount of their bill.

13 The sales tax on nonprescription drugs in South Dakota is 4%.

    a Calculate how much tax you need to pay if some nonprescription drugs cost $237.50 before tax.

    b Find the total cost after applying the tax.

14 You go to a stylist for a haircut. The cost of the haircut is $32.50.

    a If local tax is 8%, find the total tax paid on the haircut.

    b Find the amount of a 15% tip for the stylist, calculated on the pre-tax total.

    c Find the total paid for the haircut.

15 Complete the following sales receipt.

SALES RECEIPT

Description	Amount
White vinegar	$ 6.51
Olive oil	$ 5.85
Sugar	$ 8.29
Rice milk	$ 7.21
Cheese	$ 4.00
Soy milk	$ 6.23
Subtotal	
Tax 4.5%	
Total purchase price	

16 Calculate the sales price, including 4% tax, for an item with an original price of $20.

17 The state of Orgeon has no sales tax. For each situation, determine the total paid with tip.
    a Thomas wants to leave a 20% tip on a bill of $116.72.
    b Anya wants to leave a 24% tip on a bill of $84.60.

18 The pre-tax total from a trip to the convenience store was $27.30. If state tax is 6.5%, estimate the total of the bill including tax and explain your strategy.

19 Garry was charged $77 for consultation and $17 for herbal supplements from a medical professional. The totals do not include the 10% tax payable for this service. If 10% tax is only applied to the consultation fee, what is the total bill that Garry must pay?

20 Two friends, Jamie and Taylor, go out to dinner at a restaurant. When the bill comes, they agree to leave a 15% tip. The pre-tax amount of the meal is $40, and the amount including tax is $43.20. Jamie and Taylor calculate the tip and total bill individually.

Jamie's calculation:

    $40 \cdot 0.15 = \$6$ Amount of tip

    $43.20 + 6 = \$49.20$ Total bill

Taylor's calculation:

    $43.20 \cdot 0.15 = \$6.48$ Amount of tip

    $43.20 + 6.48 = \$49.68$ Total bill

Explain why Jamie and Taylor came up with different totals.

21 Marta earns a $354 commission for total sales of $5900. What is her percent commission?

22 Ben is paid a commission based on his weekly sales. In one week, he sells $7800 worth of lighting. His total pay for the week was $390.

Find the commission rate Ben is paid, as a percentage of his sales.

23 Samantha and Cam went to a restaurant for lunch. The total bill including tip was $28.44. They left $5.82 for a tip. What percent tip did Samantha and Cam leave?

24 Complete each sales receipt by filling in the blanks. Note that the tip percentage is applied to the pre-tax total.

a

Uber	
Tip Fare	$ 24.60
Tax (8%)	$ _____
Total (with just tax)	$ _____
Tip (18%)	$ _____
Total	$ _____

b

Cafe'	
Tip Fare	$ 14.54
Tax (____ %)	$ 1.31
Total (with just tax)	$ 15.85
Tip (____ %)	$ 3.05
Total	$ 18.90

Let's extend our thinking

25 Sally is paid 7% commission on the first $2600 of goods sold and 1% on any value thereafter. Sally sells goods to the value of $15 400.
    a Calculate the commission earned on the first $2600 of goods sold.
    b Calculate the commission earned on the amount of goods sold in excess of $2600.
    c Find Sally's total commission earned.

26   Jemimah is dining at a restaurant and wants to calculate the total bill, including a 6% sales tax and a 20% tip, in one step. The pre-tax amount of the meal is $50.

Determine what single percentage Jemimah should multiply the pre-tax meal amount by to calculate the total bill, including both sales tax and tip, in one step. Explain your reasoning.

27   Jeremiah is working at a restaurant and has been an excellent waiter. He always receives tips for his great service. During the first three hours of work, he already received three tips from the following persons:

- Lesley left a 20% tip from her $145 bill.
- Neil left a 25% tip from his $185.25 bill.
- Hiro left a 22% tip from his $155.75 bill.

If Jeremiah earns $6 per hour, find his total earnings in the first three hours, including the tips. Round your answer to the nearest dollar.

28   Calculate the price of the following items before 5.3% tax was added:

a   The sales price of a dress, including tax, is $40.

b   The price of an iPad including tax, is $238.

29   A shopping bill costs $109 including 5.3% tax. This bill includes fruits and vegetables amounting to $62. If fruits and vegetables do not incur 5.3% tax, calculate:

a   The cost of the other items on the bill.

b   The cost of the other items prior to adding 5.3% tax.

30   If a commission of 20% is paid on all sales, find the total value of sales if the gross income is $238.

Answers

1.06 Sales tax, commissions and tips

What do you remember?

1 a Commission

   b Tip

   c Sales tax

2 a 25%   b 15%   c 32.5%   d 145%

3 a 0.22   b 1.22

4 a $3.81   b $0.99

5 D

Let's practice

6 a $1.50   b $2.84   c $0.80

7 a $4.24   b $2.09   c $1.86   d $4.12

8 $5220

9 a $3.66   b $34.74   c $9.68

10 $2040

11 $10

12 a $42.93   b $281.44

13 a $9.50   b $247

14 $2.60   b $4.88   c $39.98

15

SALES RECEIPT

DESCRIPTION	AMOUNT
White vinegar	$ 6.51
Olive oil	$ 5.85
Sugar	$ 8.29
Rice milk	$ 7.21
Cheese	$ 4.00
Soy milk	$ 6.23
Subtotal	$ 38.09
Tax 4.5%	$ 1.71
Total purchase price	$ 39.80

16 $20.80

17 a $140.06   b $104.90

18 To estimate the total, first approximate the tax by finding 1% of 30 and 0.5% of 30. Then, multiply the 1% of 30 by 6. Finally, add the tax to the original total.

Answers may vary. 1% of $30 is approximately $0.3 and half of this is 0.15. 0.3 times 6 is $1.8. Therefore, the tax is about $1.8 + 0.15 = 1.95. So the bill including tax is approximately $30.

19 $101.70

20 Jamie calculated the tip based solely on the pre-tax meal amount, which did not include the additional cost of sales tax. Taylor calculated the tip based on the total bill amount, including the sales tax. Because Taylor's base amount for calculating the tip was higher due to the inclusion of sales tax, Taylor's tip amount and total bill is slightly larger than Jamie's.

21 6%

22 5%

23 25.73%

24 a

Uber	
Tip Fare	$ 24.60
Tax (8%)	$ 1.97
Total (with just tax)	$ 26.57
Tip (18%)	$ 4.43
Total	$ 31.00

b

Cafe'	
Tip Fare	$ 14.54
Tax (9%)	$ 1.31
Total (with just tax)	$ 15.85
Tip (21%)	$ 3.05
Total	$ 18.90

Let's extend our thinking

25 a $182   b $128   c $310

26 Jeremiah should multiply the meal amount by 26% or 0.26. When calculating the tip and tax separately, we would multiply each percent by the meal amount and add the results to find the total:

$$50\,(0.20) + 50\,(0.06)$$

However, we can combine these into a single calculation by the distributive property:

$$50\big(0.20\big) + 50\big(0.06\big) = 50\big(0.20 + 0.06\big)$$
$$= 50\big(0.26\big)$$

This shows that we can multiply the meal total by 26% to find the total bill with tax and tip in one step.

27 $128.00

28 a $37.99   b $226.02

29 a $47   b $44.63

30 $1190

Topic 1 Assessment: Real Numbers

1 Which of the following numbers is the best estimation of $\sqrt{82}$?

    A   41   **B**   9   **C**   8   **D**   10

2 Place the following numbers on the number line and list them in ascending order:

$$\sqrt{3}, -\sqrt{7}, -\frac{28}{10}, -0.1, -\pi$$

3 State whether each of the following numbers are rational or irrational:

    a $\frac{7}{10}$ **b** $\sqrt{15}$ **c**   8   **d** $-0.\overline{3}$

4 Determine the two consecutive integers that will complete the following inequalities:

    a $\square < \sqrt{21} < \square$ **b** $\square < \sqrt{58} < \square$ **c** $\square < \sqrt{83} < \square$ **d** $\square < \sqrt{312} < \square$

5 For each of the following pairs of numbers, select the number with the larger value:

    a 3×10^{-1} or $\frac{1}{3}$ **b** $\sqrt{47}$ or 6.1   **C** -5 or $-\sqrt{53}$ **d** $\frac{2}{5}$ or 50%

6 Identify all categories that each number represents. Choose from the options:

- Whole numbers
- Rational numbers
- Integer
- Irrational number
- Natural number

    a -3 **b** $\frac{\pi}{4}$ **C** $\frac{1}{3}$ **d**   0.25

7 Two local grocery stores are running sales on birthday cakes this week. Grocery Store A normally sells cakes for $35.00 and is running a 19% discount. Grocery Store B normally sells cakes for $30.00 and is running a 11% discount. Which store is the better buy?

8 Calculate the final costs of the following products with the given discount or markups. Round to two decimals if necessary.

    a   Bananas: $1.25   **b**   Tuna: $2.35   **c**   Ice cream: $5.99   **d**   Game: $45.00

        Discount: 10%   Markup: 8%   Markup: 20%   Discount: 23%

9 Marco went out to dinner with his family. The total before tip or sales taxs $64.32.

    a   Sales tax is 7%. How much is the bill after tax?

    b   Marco is going to leave a tip that is 20% of the total before tax. What is the total amount of money he should pay with both tax and a tip?

10 A town's population increased from 5000 to 5400 in a year. Calculate the percentage increase.

11 Elisa and her friend, Irvane, go out to dinner. The bill total before tip was $49.72. Irvane plans to pay the bill and leave a 20% tip. They do the math and get a total of $50.81. Is this total correct? If not, find the value of the tip they actually left, and the total they should have paid.

12 For the following roots:

    a Evaluate, correct to two decimal places where necessary.

      i $\sqrt{2}$ **ii** $\sqrt{20}$ **iii** $\sqrt{78}$ **iv** $\sqrt{127}$

    b Order the numbers from least to greatest.

13 Fill in the blanks with the set of numbers that fit in the diagram. Give at least 2 examples of each of these types of numbers.

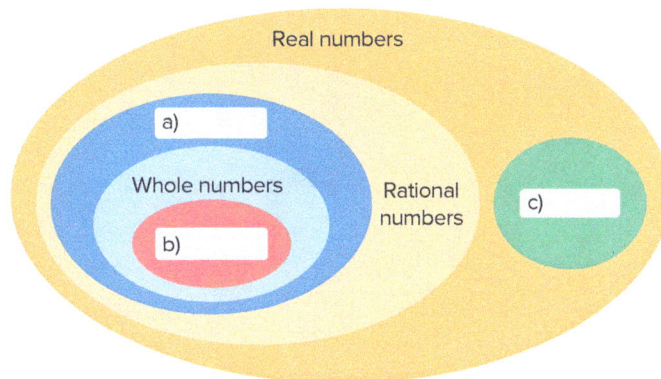

14 List an example and a non-example of a number in each subset of the real numbers:

    a Natural number   **b** Rational number   **c** Whole number

SOL **15** Which statement best describes $\sqrt{60}$?

    A Exactly 7   **B** Exactly 30

    C Between 7 and 8   **D** Between 29 and 31

16 Patrick caught 23 insects for his science experiment last week and 9 insects this week. Which value is closest to the percentage change in insects caught?

    A 61% increase   **B** 61% decrease   **C** 64% increase   **D** 16% decrease

17 Marta wanted to buy a shirt that was normaly $19.99. The store is running a 15% sale. Estimate the cost by choosing which is closest to the sale price of the shirt.

    A $14.99   **B** $16.50   **C** $17.00   **D** $15.50

Performance task

18 The period of a pendulum is the time (in seconds) it takes the pendulum to swing back and forth once. The period T is often estimated to be 10% greather than $\sqrt{L}$ where L is the length of the pendulum (in feet).

    a Without using a calculator, about how long does it take a 4 foot pendulum to swing back and forth once?

    b Without using a calculator, about how long does it take a 11 foot pendulum to swing back and forth once?

    c Some pendulums need more exact measurements to keep accurate time. We can use the exact formula $T = 2\pi\sqrt{\dfrac{L}{g}}$, where g is the acceleration due to gravity which is 32 feet per second squared. Will the time a pendulum takes to complete one swing back and forth ever be rational? Justify your thinking.

    d Design a pendulum that takes approximately 20 seconds for one swing back and forth.

Answers

Topic 1 Assessment: Real numbers

1 **B**

   8.NS.1a

2

   $$-\pi, -\frac{28}{10}, -\sqrt{7}, -0.1, \sqrt{3}$$

   8.NS.1b, 8.NS.1c

3 a Rational   b Irrational
   c Rational   d Rational

   8.NS.2b

4 a $4 < \sqrt{21} < 5$ b $7 < \sqrt{58} < 8$
   c $9 < \sqrt{83} < 10$ d $17 < \sqrt{312} < 18$

   8.NS.1a

5 a $\frac{1}{3}$ b $\sqrt{47}$ c –5   d 0.5

   8.NS.1c

6 a Rational number, Integer   c Rational number
   b Irational number   d Rational number

   8.NS.2b

7 Grocery Store B is the better buy.

   8.CE.1a

8 a $1.13   b $2.54
   c $7.19   d $34.65

   8.CE.1a

9 a $68.82   b $81.68

   8.CE.1b

10 a 8%

   8.CE.1c

11 They are incorrect. They only left a $1.09 tip. Their total should have been $59.66.

   8.CE.1b

12 a i 1.41   ii 4.47   iii 8.83   iv 11.27
   b $\sqrt{2}, \sqrt{20}, \sqrt{78}, \sqrt{127}$

   8.NS.1b, 8.NS.1c

13 a Integers: –5, –1 are possible examples.
   b Natural numbers: 1, 5 are possible examples.
   c Irrational Numbers: $\sqrt{2}, \pi$ are possible examples.

   8.NS.2a, 8.NS.2c

14 a Example: 5   b Example: $\frac{1}{2}$
       Non-example: 3.2   Non-example: $\sqrt{19}$

   c Example: 15
      Non-example: 0.1

   8.NS.2c

15 C

   8.NS.1a

16 B

   8.CE.1c

17 C

   8.CE.1a

Performance Task

18 a For $L = 4$ feet:

   $$\sqrt{4} = 2$$

   So, T is 10% greater than 2:

   $T = 2 + 0.1 \times 2 = 2 + 0.2 = 2.2$ seconds

   b For $L = 11$ feet:

   $$\sqrt{11} \approx 3.32$$

   So, T is 10% greater than 3.32:

   $T = 3.32 + 0.1 \times 3.32 \approx 3.32 + 0.332 \approx 3.652$ seconds

   c The time T involves the term 2π, which is irrational, and $\sqrt{\dfrac{L}{g}}$, which can be rational or irrational depending on the values of L and g. However, because π is always irrational, 2π times any number will also be irrational. Therefore, the time will always be irrational.

   d Design a pendulum that takes approximately 20 seconds for one swing back and forth. Using the exact formula $T = 2\pi\sqrt{\dfrac{L}{g}}$ and solving for L:

   $$20 = 2\pi\sqrt{\frac{L}{32}}$$

   $$10 = \pi\sqrt{\frac{L}{32}}$$

   $$\frac{10}{\pi} = \sqrt{\frac{L}{32}}$$

   Squaring both sides:

   $$\left(\frac{10}{\pi}\right)^2 = \frac{L}{32}$$

   $$L = 32\left(\frac{10}{\pi}\right)^2 \approx 32 \times \frac{100}{9.87} \approx 324.6$$

   So, the length L should be approximately 324.6 feet to achieve a period of 20 seconds.

   8.NS.1b, 8.NS.2b, MP1, MP4

2 Expressions, Equations, & Inequalities

Big ideas

- Expressions are the building blocks of algebra. They can be used to represent and interpret real-world situations.
- An equals sign indicates an equivalent relationship between two expressions.
- A standard algorithm can be followed to solve a wide range of equations. This algorithm is reliable and useful in a variety of situations, but there is often a more efficient method that can be used based on the structure of the equation.
- A solution set is the collection of all values that make an equation or inequality true.

Chapter outline

Astronauts on the moon weigh just $\frac{1}{6}$ of their Earth weight! Equations help us figure out how much things weigh in space.

2. Expressions, Equations, & Inequalities

Topic Overview

Foundational knowledge

Evaluating standards proficiency

The skills book contains questions matched to individual standards. It can be used to measure proficiency for each.

7.PFA.2 — The student will simplify numerical expressions, simplify and generate equivalent algebraic expressions in one variable, and evaluate algebraic expressions for given replacement values of the variables.

7.PFA.3 — The student will write and solve two-step linear equations in one variable, including problems in context, that require the solution of a two-step linear equation in one variable.

7.PFA.4 — The student will write and solve one- and two-step linear inequalities in one variable, including problems in context, that require the solution of a one- and two-step linear inequality in one variable.

Big ideas and essential understanding

Expressions are the building blocks of algebra. They can be used to represent and interpret real-world situations.
2.01 — The structure of an expression can reveal important details about the situation it represents.

2.02 — The structure of an expression can reveal important details about the situation it represents.

2.03 — The structure of an equation can reveal important details about the situation it represents.

An equals sign indicates an equivalent relationship between two expressions.
2.04 — The properties of equality allow an equation to be manipulated without changing the equivalence of the expressions on either side.

2.04 — The value(s) of the variables that make every equation in a system true make up its solution set.

A standard algorithm can be followed to solve a wide range of equations. This algorithm is reliable and useful in a variety of situations, but there is often a more efficient method that can be used based on the structure of the equation.
2.04 — A standard algorithm can be used to solve any linear equation accurately and efficiently.

A solution set is the collection of all values that make an equation or inequality true.

2.05 — An inequality is a mathematical sentence that compares two expressions and uses the symbols $>$, $<$, $\geq$, or $\leq$.

2.05 — The value(s) of the variables that make an equation or inequality true make up its solution set.

2.06 — The value(s) of the variables that make an equation or inequality true make up its solution set. Inequalities have an infinite number of solutions so their solution sets are often represented on a number line.

Standards

8.PFA.1 — The student will represent, simplify, and generate equivalent algebraic expressions in one variable.

8.PFA.1a — Represent algebraic expressions using concrete manipulatives or pictorial representations (e.g., colored chips, algebra tiles), including expressions that apply the distributive property.
2.01 Represent algebraic expressions
2.02 Simplify expressions and distributive property

8.PFA.1b — Simplify and generate equivalent algebraic expressions in one variable by applying the order of operations and properties of real numbers. Expressions may need to be expanded (using the distributive property) or require combining like terms to simplify. Expressions will include only linear and numeric terms. Coefficients and numeric terms may be rational.
2.01 Represent algebraic expressions
2.02 Simplify expressions and distributive property

8.PFA.4 — The student will write and solve multistep linear equations in one variable, including problems in context that require the solution of a multistep linear equation in one variable.

8.PFA.4a — Represent and solve multistep linear equations in one variable with the variable on one or both sides of the equation (up to four steps) using a variety of concrete materials and pictorial representations.
2.03 Write and represent multistep equations
2.04 Solve multistep equations

8.PFA.4b — Apply properties of real numbers and properties of equality to solve multistep linear equations in one variable (up to four steps). Coefficients and numeric terms will be rational. Equations may contain expressions that need to be expanded (using the distributive property) or require combining like terms to solve.
2.04 Solve multistep equations

8.PFA.4c — Write a multistep linear equation in one variable to represent a verbal situation, including those in context.
2.03 Write and represent multistep equations
2.04 Solve multistep equations

8.PFA.4d — Create a verbal situation in context given a multistep linear equation in one variable.
2.03 Write and represent multistep equations

8.PFA.4e — Solve problems in context that require the solution of a multistep linear equation.
2.04 Solve multistep equations

8.PFA.4f — Interpret algebraic solutions in context to linear equations in one variable.
2.03 Write and represent multistep equations
2.04 Solve multistep equations

8.PFA.4g — Confirm algebraic solutions to linear equations in one variable.
2.03 Write and represent multistep equations
2.04 Solve multistep equations

8.PFA.5 — The student will write and solve multistep linear inequalities in one variable, including problems in context that require the solution of a multistep linear inequality in one variable.

8.PFA.5a — Apply properties of real numbers and properties of inequality to solve multistep linear inequalities (up to four steps) in one variable with the variable on one or both sides of the inequality. Coefficients and numeric terms will be rational. Inequalities may contain expressions that need to be expanded (using the distributive property) or require combining like terms to solve.
2.06 Solve multistep inequalities

8.PFA.5b — Represent solutions to inequalities algebraically and graphically using a number line.
2.05 Write and represent multistep inequalities
2.06 Solve multistep inequalities

8.PFA.5c — Write multistep linear inequalities in one variable to represent a verbal situation, including those in context.
2.05 Write and represent multistep inequalities
2.06 Solve multistep inequalities

8.PFA.5d — Create a verbal situation in context given a multistep linear inequality in one variable.
2.05 Write and represent multistep inequalities

8.PFA.5e — Solve problems in context that require the solution of a multistep linear inequality in one variable.
2.06 Solve multistep inequalities

8.PFA.5f — Identify a numerical value(s) that is part of the solution set of a given inequality.
2.05 Write and represent multistep inequalities
2.06 Solve multistep inequalities

8.PFA.5g — Interpret algebraic solutions in context to linear inequalities in one variable.
2.06 Solve multistep inequalities

Future connections

A.EO.1 — The student will represent verbal quantitative situations algebraically and evaluate these expressions for given replacement values of the variables.

A.EO.2 — The student will perform operations on and factor polynomial expressions in one variable.

A.F.1 — The student will investigate, analyze, and compare linear functions algebraically and graphically, and model linear relationships.

A.EI.1 — The student will represent, solve, explain, and interpret the solution to multistep linear equations and inequalities in one variable and literal equations for a specified variable.

A.EI.2 — The student will represent, solve, explain, and interpret the solution to a system of two linear equations, a linear inequality in two variables, or a system of two linear inequalities in two variables.

A2.EI.2 — The student will represent, solve, and interpret the solution to quadratic equations in one variable over the set of complex numbers and solve quadratic inequalities in one variable.

2.01 Represent algebraic expressions

Subtopic overview

Lesson narrative

In this lesson, students will learn to represent algebraic expressions using variables, coefficients, and constants. They will explore how to model real-world situations with expressions, identifying and interpreting terms and coefficients. The lesson includes an exploration where students create expressions to represent costs of school supplies and interpret their meanings. Examples provided guide students through identifying terms, coefficients, constants, and creating equivalent expressions using algebra tiles. By the end of the lesson, students should understand how to identify terms and coefficients, construct and simplify algebraic expressions, and represent them visually using algebra tiles.

Learning objectives

Students: Page 52

After this lesson, you will be able to...
- interpret and identify parts of an expression, such as terms, coefficients and constants, in a real-world context.
- write algebraic expressions to represent real-world situations.
- use manipulatives or pictorial representations to represent algebraic expressions and generate equivalent algebraic expressions.

Key vocabulary

- coefficient
- term (of an expression)
- constante
- variable
- expression

Essential understanding

The structure of an expression can reveal important details about the situation it represents.

Standards

This subtopic addresses the following Virginia 2023 Mathematics Standards of Learning standards.

Mathematical Process Goals

MPG2 — Mathematical Communication

Teachers can enhance mathematical communication by encouraging students to use appropriate terminology such as "variables", "constants", "algebraic expressions" and "manipulatives" in their discussions and explanations. They can provide opportunities for students to share their thought processes, and to articulate how they used concrete manipulatives or pictorial representations to represent variables and constants in algebraic expressions. This utilization of mathematical language not only clarifies students' understanding but also reinforces the correct use of these terms.

MPG4 — Mathematical Connections

Teachers can create mathematical connections by linking the lesson's content to students' prior knowledge of variables and algebraic expressions. For instance, they can remind students that a variable is a symbol used to represent an unknown number, and an algebraic expression is a combination of variables, numbers, and operations. By making these connections, teachers can help students see the relevance of previous learnings to new concepts.

MPG5 — Mathematical Representations

Teachers can promote understanding of mathematical representations by guiding students through the process of representing algebraic expressions using concrete manipulatives or pictorial representations. They can also provide examples of different algebraic expressions and demonstrate how these can be represented using colored chips or algebra tiles. By doing so, teachers can help students understand that different representations can convey the same mathematical idea.

Content standards

8.PFA.1 — The student will represent, simplify, and generate equivalent algebraic expressions in one variable.

8.PFA.1a — Represent algebraic expressions using concrete manipulatives or pictorial representations (e.g., colored chips, algebra tiles), including expressions that apply the distributive property.

8.PFA.1b — Simplify and generate equivalent algebraic expressions in one variable by applying the order of operations and properties of real numbers. Expressions may need to be expanded (using the distributive property) or require combining like terms to simplify. Expressions will include only linear and numeric terms. Coefficients and numeric terms may be rational.

Prior connections

7.PFA.2 — The student will simplify numerical expressions, simplify and generate equivalent algebraic expressions in one variable, and evaluate algebraic expressions for given replacement values of the variables.

Future connections

A.EO.1 — The student will represent verbal quantitative situations algebraically and evaluate these expressions for given replacement values of the variables.

A.EO.2 — The student will perform operations on and factor polynomial expressions in one variable.

A.F.1 — The student will investigate, analyze, and compare linear functions algebraically and graphically, and model linear relationships.

Lesson Preparation

Suggested review

Depending on your students' level of prior knowledge, consider revisiting the following lessons:

Grade 6 — 6.01 Algebraic expressions
Grade 7 — 3.03 Represent algebraic expressions

Lesson supports

The following supports may be useful for this lesson. More specific supports may appear throughout the lesson:

Anatomy of an algebraic term
Targeted instructional strategies

Break down individual terms into their basic components so that students can understand them on a fundamental level. For any given term, show that it consists of:

- The coefficient
- The variable(s)

Each coefficient has a positive or negative sign, even if the sign is not shown (such as the positive sign of the leading coefficient).
Breaking down a term visually can assist with this explanation.

$$\underset{\text{Coefficient}}{\underset{\displaystyle 4}{\underbrace{}}}\ \underset{\text{Variables}}{\underbrace{}}$$

$$\overset{\text{Sign}}{\overbrace{}}\, -4\,x^2 y$$

This strategy can also be used to help explain less obvious cases like:

$$x \rightarrow \overset{\text{Sign}}{\overbrace{}}\, \underset{\text{Coefficient}}{+1}\ \underset{\text{Variable}}{x}$$

Clarifications about algebraic terms
Address student misconceptions

Common misconceptions when first learning about algebraic expressions are:

- Not including the negative sign as part of the coefficient
- Equating the absence of a coefficient to mean that the coefficient is 0 (consider that x has a coefficient of 1, not 0)
- Assuming that all symbols must be variables (consider that π is not a variable)

Having students highlight and label key words that indicate operations and order in the problem will help ensure students are making sense of the problem as they work.

Student lesson & teacher guide

Translate algebraic expressions

Students are reminded of important vocabulary that arises when working with algebraic expressions (such as expression, variable, coefficient, constant, and term) before engaging in an exploration to uncover the meaning of the parts of an expression.

Collect and display
English language learner support

As students are working, note how students describe the concepts of "expression," "variable," "coefficient," and "term". Collect the different ways that students find to understand these concepts and display them in a common place for the students to access.

If students do not come up with alternative ways to word these concepts and are confused by them, suggest some of your own. For example:
- Variable
  - A letter that stands for a number we don't know
  - Something that can change
  - An unknown in the problem
- Coefficient
  - The number that tells how many times to take the variable
  - The amount of the variable we have
  - The number in front of the letter
- Term
  - A piece of the expression
  - A number and a variable multiplied together
  - Parts separated by plus or minus signs
- Expression
  - A math sentence without an equals sign
  - Numbers and letters combined
  - A way to show a calculation

Encourage students to refer to this display as they work through problems, helping them connect their own words to the formal mathematical vocabulary. This will support their understanding of key terms and aid them in expressing their mathematical thinking more precisely.

Exploration

Students: Page 53

Exploration

In order to write an expression that can be used to model the total cost of buying new school supplies, Mr. Okware defines the variables:

Let x represent the cost of a folder, y represent the cost of a calculator, and z represent the cost of a pencil pack.

1. What could these expressions represent in this context?
   - $x + y$
   - $x + y + z$
   - $5y$
   - $2x + 10z$
   - $x + 3y + 4z$
   - $4(4x + y + 2z)$

2. In this context, what do the coefficients describe?

3. What expressions could we write that wouldn't make sense in this context?

Suggested student grouping: In pairs

Students are given variables that represent different school item costs. Then, they are tasked with interpreting given expressions with the respect to the scenario. The aim is for students to realize that algebraic expressions represent specific real-world contexts, and some algebraic expressions do not make sense in a given context.

Ideal student responses

These ideal responses may differ from other correct student responses. Less formal responses can be connected with the more precise mathematical language presented here.

1. **What could the following expressions represent in this context?**

$x + y$	cost of one folder and one calculator
$x + y + z$	cost of one folder, one calculator, and one pencil pack
$5y$	cost of five calculators
$2x + 10z$	cost of two folders and ten pencil packs
$x + 3y + 4z$	cost of one folder, three calculators, and four pencil packs
$4(4x + y + 2z)$	four times the cost of four folders, one calculator, and two pencil packs

2. **In this context, what do the coefficients describe?**
   The coefficients describe the quantity of each item.

3. **What expressions could we write that wouldn't make sense in this context?**
   Expressions that don't have integer coefficients would not make sense because it is not possible to purchase only part of an item. Expressions that include multiple variables in a single term, for example $4xy$, would also not make sense because the product of two item costs does not represent anything meaningful.

Purposeful questions

- What does each variable represent? If we add them together, what does the sum represent?
- If x is the cost of one folder, then what does $2x$ represent?
- If we divide the cost of a calculator by 2, what might that represent in context?
- Would it make sense to multiply the cost of two items together?

Possible misunderstandings

- Students may misinterpret the question and think that the variables represent the quantities of each item. Point out that the variables represent the cost of one item. In this case, the coefficients would be the quantity of items that needs to be purchased.

Students should develop the understanding that each part of an algebraic expression has a meaning and can provide information about the context it represents.

Examples

The following supports may be useful for the examples in this section.

Choose unambiguous variables
Student with disabilities support

When using multiple variables in a question, make sure that they are visually distinct. Combinations to avoid include:

- b and d
- m and n
- p and q
- u, v, and w

Variables that look like numbers or symbols should also be avoided whenever possible. This will also depend on handwriting, but the most common examples are:

- *o* and 0
- *l* and 1
- *z* and 2
- *s* and 5
- *g* and 9
- *t* and +

Students: Page 53

Example 1

For the algebraic expression $5x + 7y - 12x + 21$:

a Determine the number of terms.

Create a strategy	**Apply the idea**
Terms are separated by plus or minus signs in the expression.	The algebraic expression $5x + 7y - 12x + 21$ contains four terms: $5x$, $7y$, $-12x$, and 21.

Purpose
Show students how to identify the number of terms in an algebraic expression

Expected mistakes
Students might count each coefficient and variable as a separate term. Remind them that terms are separated by plus (+) or minus (−) signs, so a variable and its coefficient together are one term.

Students: Page 53

b Identify the coefficient of the first term.

Create a strategy	**Apply the idea**
The coefficient of a term is the number that is multiplied by the variable in the term.	The first term is $5x$, so the coefficient of the first term is 5.

Purpose
Show students that they can identify and differentiate between coefficients, constants, and terms within an algebraic expression.

Students: Page 53

c Identify the constant term.

Create a strategy	**Apply the idea**
The constant term in an algebraic expression is the term that does not contain any variable.	In the expression $5x + 7y - 12x + 21$, the constant term is 21.

Purpose
Show students how to identify constant terms in an algebraic expression and understand their role in the overall expression.

    d   Determine if the expression contains like terms.

Create a strategy

Like terms in an expression are terms that contain the same variable raised to the same power.

Apply the idea

The expression $5x + 7y - 12x + 21$ contains like terms, $5x$ and $12x$, since they both contain the variable x raised to the first power.

Purpose

Students demonstrate that they can identify and analyze like terms in an algebraic expression

Reflecting with students

Challenge students to combine the like terms and simplify the expression.

Example 2

A courier service charges a fixed fee of $30 plus $0.75 per kilogram for delivering packages. Write an algebraic expression for the total delivery cost of a package weighing m kilograms.

Create a strategy

The total cost includes a fixed fee and a variable fee based on the weight of the package.

Apply the idea

For a package weighing m kilograms, the total delivery cost can be represented by the algebraic expression $30 + 0.75m$. This expression combines the fixed fee of $30 with the variable fee of $0.75 per kilogram.

Purpose

Show students how to use algebra to represent real-world situations.

Advanced learners: Investigate cost variations of different package weights

use with Example 2

Targeted instructional strategies

Encourage students to explore the algebraic expression $30 + 0.75m$ by substituting various values for m to calculate the total delivery cost for different package weights. Have them create a table listing different weights, such as 0 kg, 5 kg, 10 kg, 20 kg, and even non-integer values like 7.5 kg and compute the corresponding costs. This hands-on exploration helps students see how changes in the variable m affect the total cost.

Invite them to plot these values on a graph with package weight on one axis and total cost on the other. Discuss how the cost increases by $0.75 for each additional kilogram. Encourage them to observe the consistent pattern: as the weight increases, the total cost increases in a steady way. By analyzing this pattern, students can make predictions about the cost for any given weight, reinforcing their understanding of variables and how they can represent real-world relationships.

This informal exploration allows students to intuitively grasp how the delivery cost depends on the package weight, even if they have not formally studied linear functions yet.

Example 3

Write an algebraic expression for the phrase "eight more than the quotient of 9 and x".

Create a strategy

Translate the terms into mathematical symbols and operations.

Apply the idea

The phrase "eight more than" indicates that we need to add 8.

The "quotient of 9 and x" indicates division, which we can write as a fraction, with 9 as the numerator and x as the denominator.

We can combine the whole description into a single expression:

$$\frac{9}{x} + 8$$

Purpose

Students demonstrate that they can translate verbal descriptions into algebraic expressions

Expected mistakes

The student may accidentally write the division as $\frac{x}{9}$. Remind students that when we are writing division that is verablly written as "the quotient of", the first term is the numerator.

Example 4

The perimeter of a rectangle can be expressed by $2(l + w)$. Explain what each of the factors represents.

Create a strategy

First, we need to identify each of the factors. One factor is 2 and the other is $(l + w)$.

We know that the perimeter of an object is the distance around the outside edges. A rectangle has 4 sides, but 2 are called lengths and 2 are called widths.

Apply the idea

$$\text{Perimeter} = l + w + l + w$$

We can see from the perimeter formula that there are 2 groups of $l + w$.

This shows that the factor $l + w$ represents adding the length and width, and the 2 means we added the length and width twice.

Reflect and check

Another way to represent the perimeter of a rectangle is $2l + 2w$.

Purpose

Demonstrate to students that they can interpret and translate algebraic expressions in the context of geometric problems.

Reflecting with students

Encourage students to substitute different values for the length and width of rectangles to see how the expression always results in the perimeter of the rectangle.

> ### 💡 Idea summary
> Expressions can be used to represent mathematical relationships. In an expression, sums often represent totals and coefficients and factors represent multiplication. When interpreting an expression in context, we can use the units to help understand the meaning.

Represent algebraic expressions

This lesson demonstrates how algebra tiles can visually represent algebraic expressions. The tiles represent variables (e.g., x), positive units (+1), and negative units (−1). By rearranging visual models, they are shown how equivalent expressions can be depicted using algebra tiles.

Example 5

Write an equivalent algebraic expression for the following:

$$\boxed{+x} \quad \boxed{+x} \qquad \boxed{+1}\,\boxed{+1}\,\boxed{+1}\,\boxed{+1}\,\boxed{+1}$$

Create a strategy

Count the number of $\boxed{+x}$ tiles. Then count the number of $\boxed{+1}$ tiles.

There are many ways to write expressions that are algebraically equivalent by rearranging the terms and combining like terms, but for simplicity, we'll directly reflect the layout shown by the tiles.

Apply the idea

From the image, we observe two x tiles and five unit tiles. To express this algebraically we can write:

$$x + x + 1 + 1 + 1 + 1 + 1$$

Another way to write the expression is:

$$2x + 5$$

Purpose

Show students how to translate a visual representation of algebraic expressions using algebra tiles into algebraic expression.

Reflecting with students

Ask students to describe how the expression would be different if each of the +1 tiles were -1 tiles instead. Then, ask them how the expression would change if we doubled the amount of +x tiles shown.

Example 6

Represent the expression $-3x - 5 + 8$ using algebra tiles.

Create a strategy

We can use a negative variable tile and positive and negative unit tiles to represent the given expression.

Apply the idea

Reflect and check

We could also use pictorial models to represent the expression.

The box represents the variable, and the strawberries represent the units. The 3 boxes and 5 strawberries are grayed-out to show they have been removed or "subtracted".

Purpose

Challenge students to use tangible representations like algebra tiles or pictorial models to visualize and understand algebraic expressions.

Expected mistakes

The student may neglect the negative sign on the first coefficient and represent the equation in tiles using 3 as the coefficient and placing three $+x$ tiles.

Real-life context to promote understanding of abstract expressions   use with Example 6
Student with disabilities support

To support students who struggle with abstract concepts, introduce real-life contexts to make the expression $-3x - 5 + 8$ more meaningful. Present a scenario such as points in a game, where x represents a round in the game. Then, $-3x$ means it costs 3 coins to play each round, -5 could represent the cost of personalizing their character in the game, and $+8$ could be a bonus for signing up to play.

Encourage students to create a visual timeline or number line illustrating these changes. Use arrows pointing down for a loss of coins (negative values) and arrows pointing up for a gain in coins (positive values). This helps students visualize the cumulative effect of the coin changes.

By relating negative and positive terms to tangible experiences, students can better grasp why certain terms are negative or positive and how they combine. This contextual approach enhances language processing skills by linking mathematical expressions to everyday situations, making the abstract concepts more concrete and understandable.

Practice

What do you remember?

1 Match the terms with their definitions:

a Constant

b Variable

c Coefficient

d Algebraic term

i Symbol that represents an unknown number

ii A purely numeric term in an algebraic expression

iii The value that indicates how many of a variable in a term

iv Term including a variable

2 For the algebraic expressions:

i Determine the number of terms.

iii Identify the constant term.

ii Identify the numerical coefficient of the first term.

iv Determine if the expression contains like terms.

a $2x + 4$

b $7y + 3 + 5x$

c $3x + 2y - 8x + 9$

d $8p + 5$

3 How can you use algebra tiles to represent algebraic expressions?

4 Represent the following expressions using algebra tiles.

a $4x - 2$

b $2x + 3 + 4x$

Let's practice

5 Represent the following expressions using algebra tiles.

a $-4x - 3 + 2$

b $-2(2x - 3)$

6 Write a word statement for each expression:

a $2(b + 9)$

b $x(3 - y)$

c $4(n + 1) - 11$

d $\dfrac{3(m + 7)}{10}$

7 Each cyclist in a cycling touring group carries the items shown.

a Write an expression to represent the total weight carried by the group. Let x represent the number of cyclists.

b Explain what the expression means in context.

3.4 lbs   8.52 lbs   6.48 lbs

8 A shipping company charges a flat fee of $25 plus $0.50 per kilogram for shipping packages.

   a Write an algebraic expression for the total cost of shipping a package that weighs w kilograms.

   b Draw a pictorial model of your expression.

9 The representatives of 9 countries are attending a global summit, and each country is able to send $3r - 4$ representatives. Write an expression for the total number of representatives attending the summit.

10 For the following expressions:

   i   Represent using algebra tiles.

   ii Use the algebra tiles to write an equivalent expression.

   a $4(x + 8) - 2$ **b** $5 + 3(x + 4)$

11 Refer to the models A and B.

   a Which expression is **best** represented by model A? Explain your choice.

   A $4x + 4$ **B** $4x - 4c$

   C $2(2x + 2)$ **D** $-4x - 4$

Model A

   b Which expression is **best** represented by model B? Explain your choice.

   A $4x + 4$ **B** $4x - 4c$

   C $2(2x + 2)$ **D** $-4x - 4$

Model B

12 These tiles represent the expression $10 - 4x + 2 - 3x$.

Which expression is equivalent to $10 - 4x + 2 - 3x$? Use the algebra tiles to explain your thinking.

   $x + 12$ $-7x + 12$ $7x + 12$ $6x - 1$

Let's extend our thinking

13 For the expression $-\dfrac{1}{4}x + 2 - \dfrac{1}{2}x + 3$:

   a Draw a model of the expression using something other than algebra tiles.

   b Use your model to write an equivalent expression.

14 How can manipulatives help you to simplify $-2(3p + 5 - 4)$?

15 The length of a rectangle is 5 units more than half of its width:

   a What is its length as an algebraic expression?

   b Find an expression for its area. Assume its width to be w.

   c Draw a model of your expression.

Answers

2.01 Represent algebraic expressions

What do you remember?

1 a ii b i c iii d iv

2 a i 2 ii 2 iii 4 iv No
 b i 3 ii 7 iii 3 iv No
 c i 4 ii 3 iii 9 iv Yes
 d i 2 ii 8 iii 5 iv No

3 Algebra tiles are manipulative tools used to represent algebraic expressions, where square and rectangular tiles correspond to variables, constants, and their operations, providing a visual for to grasp abstract algebraic concepts.

4 a

x	x	x	x	-1
				-1

b

		1				
x	x	1	x	x	x	x
		1				

Let's practice

5 a

$-x$		
$-x$	-1	1
$-x$	-1	1
$-x$	-1	1

b

$-x$		
$-x$	1	1
$-x$	1	1
$-x$	1	1

6 a The sum of b and 9, multiplied by 2.
 b y subtracted from 3, then multiplied by x.
 c Eleven subtracted from the product of 4 and the sum of n and 1.
 d 3 lots of the sum of m and 7 divided by 10.

7 a $3.4x + 8.52x + 6.48x$
 b The total weight of items in the cycling touring group is expressed as $3.4x + 8.52x + 6.48x$, where (x) represents the number of cyclists.

8 a $25 + 0.5w$
 b

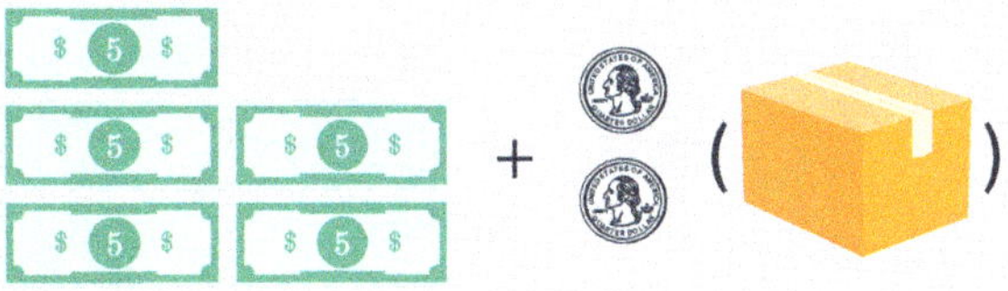

9 $9(3r - 4)$ representatives

10 a i

	+1	+1	+1	+1	+1	+1
$+x$	+1	+1	+1	+1	+1	+1
$+x$	+1	+1	+1	+1	+1	+1
$+x$	+1	+1	+1	+1	+1	+1
$+x$	+1	+1	+1	+1	+1	+1

 ii $4x + 30$

 b i

+1	+1	+1	+1	+1
$+x$	+1	+1	+1	+1
$+x$	+1	+1	+1	+1
$+x$	+1	+1	+1	+1
$+x$	+1	+1	+1	+1

 ii $3x + 17$

11 a **A**
 Model A shows four x and four 1 tiles, which can be represented by $4x + 4$.
 b **C**
 Model B shows 2 groups of two x and two 1 tiles, which can be represented by $2(2x + 2)$.

12 $-7x + 12$
We can combine like terms as shown in the algebra tiles, which would give us $-7x + 12$.

13 a

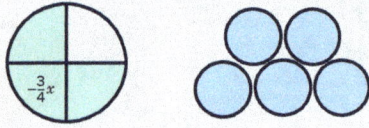

b

$$-\frac{3}{4}+5$$

14 Manipulatives, such as tiles or blocks, can aid in simplifying $-2(3p + 5 - 4)$ by visually representing and combining the terms step by step, making the distributive property and combining like terms more tangible for understanding.

15 a Length $= \frac{1}{2}w + 5$

b Area $= \left(\frac{1}{2}w + 5\right) \times w$

c

2.02 Simplify expressions and distributive property

Subtopic overview

Lesson narrative

In this lesson, students will build on their prior knowledge of the distributive property in creating equivalent expressions to expand algebraic expressions with integer coefficients. The lesson begins with an exploration using an applet which models distributive property. Students will use the structure of the expressions to recognize patterns and create generalizable formulas and help them evaluate products more efficiently using distributive property. Students will also justify whether an expression is equivalent to a given expression by using properties of operations. By the end of the lesson, students will be able to use distributive property as one of the strategies to add, subtract, and expand algebraic expressions containing parenthesis. This skill will also prepare students to factor algebraic expressions with integers and apply distributive properties with rational numbers.

Learning objectives

Students: Page 60

After this lesson, you will be able to...
- apply the order of operations and properties of real numbers to simplify algebraic expressions.
- generate equivalent expressions using the distributive property.
- represent algebraic expressions using concrete manipulatives.

Key vocabulary

- distributive property
- simplify (an expression)

Essential understanding

The structure of an expression can reveal important details about the situation it represents.

Standards

This subtopic addresses the following Virginia 2023 Mathematics Standards of Learning standards.

Mathematical Process Goals

MPG1 — Mathematical Problem Solving

Teachers can integrate this goal into their instruction by presenting students with real-life problems that require the use of the distributive property to solve. For example, teachers can pose a problem about calculating the cost of multiple items with the same price, prompting students to apply the distributive property to find the total cost. Teachers can also encourage students to create their own problems from real-world data and situations, which they can then solve using the distributive property.

MPG4 — Mathematical Connections

Teachers can establish mathematical connections by linking the distributive property to students' prior knowledge of equivalent algebraic expressions. For instance, showing how the distributive property can be visually demonstrated using colored chips or algebra tiles can relate to previous lessons on concrete manipulatives and pictorial representations. Additionally, connecting the application of the distributive property to real-world scenarios helps students see the relevance of what they are learning.

MPG5 — Mathematical Representations

Teachers can incorporate this goal by showing students different methods of representing the distributive property. This can include algebraic expressions, visual models using colored chips or algebra tiles, and real-life examples. By making connections among these different representations, students can understand that the same mathematical idea can be conveyed in different ways, making representation both a process and a product.

Content standards

8.PFA.1 — The student will represent, simplify, and generate equivalent algebraic expressions in one variable.

8.PFA.1a — Represent algebraic expressions using concrete manipulatives or pictorial representations (e.g., colored chips, algebra tiles), including expressions that apply the distributive property.

8.PFA.1b — Simplify and generate equivalent algebraic expressions in one variable by applying the order of operations and properties of real numbers. Expressions may need to be expanded (using the distributive property) or require combining like terms to simplify. Expressions will include only linear and numeric terms. Coefficients and numeric terms may be rational.

Prior connections

7.PFA.2 — The student will simplify numerical expressions, simplify and generate equivalent algebraic expressions in one variable, and evaluate algebraic expressions for given replacement values of the variables.

Future connections

8.PFA.4 — The student will write and solve multistep linear equations in one variable, including problems in context that require the solution of a multistep linear equation in one variable.

A.EO.1 — The student will represent verbal quantitative situations algebraically and evaluate these expressions for given replacement values of the variables.

A.EO.2 — The student will perform operations on and factor polynomial expressions in one variable.

Lesson Preparation

Suggested review

Depending on your students' level of prior knowledge, consider revisiting the following lessons:

Grade 7 — 3.01 Order of operations with integers
Grade 7 — 3.05 Equivalent algebraic expressions

Student lesson & teacher guide

Order of operations and properties of real numbers

Students recall the order of operations and the properties of real numbers, which are used to simplify algebraic expressions.

Collect and display
English language learner support

As students are working, note how students describe the different properties of real numbers. Collect the different ways that students find to understand these concepts and display them in a common place for the students to access.

If students do not come up with alternative ways to word these concepts and are confused by them, suggest some of your own. For example:
- Commutative property
  - When the numbers change positions
  - When the numbers move (or commute)
  - Swapping the terms does not change the result
- Associative property
  - When the parentheses are around different numbers
  - When the numbers associate (are grouped) in different ways
  - Grouping different terms does not change the result
- Identity property
  - When the identity of a number doesn't change
  - Adding zero keeps the number the same
  - Multiplying by one keeps the number the same
- Inverse property
  - Undoing the addition or multiplication
  - Subtracting by the same number to make zere
  - Dividing by the same number to make one

Encourage students to refer to this collective display as they work through problems and explain their reasoning. By anchoring their informal language to precise mathematical vocabulary, you support students in connecting their understanding of the concepts with the correct terminology, enhancing both their mathematical reasoning and language development.

Use physical manipulatives to represent properties of real numbers
Student with disabilities support

Use manipulatives like colored blocks, counters, or algebra tiles to visually represent numbers and their properties. This hands-on approach helps students concretely understand abstract concepts.

Ways to represent the properties with different physical manipulatives include:

- *Commutative property*: To show $4 \cdot 6 = 6 \cdot 4$, create 4 groups of 6 blocks and 6 groups of 4 blocks. Show both arrangements have 24 blocks.
- *Associative property*: To show $(2 + 3) + 4 = 2 + (3 + 4)$, group 2 red blocks and 3 blue blocks, then add 4 green blocks. Rearrange to show both groupings sum to 9.
- *Distributive property*: To show $2(3 + 4)$, use 2 sets of blocks grouped into 3 and 4 blocks. Rearrange to show the sum of the groups is equal to the product of 2 times 7.
- *Identity property*: To show $5 \cdot 1 = 5$, use 5 blocks in one group to show that multiplying by 1 keeps the number unchanged.
- *Inverse property*: To show $6 + (-6) = 0$, Pair 6 red blocks with 6 blue blocks to represent positive and negative values, showing they cancel out to zero.

The properties do not apply to subtraction and division
Address student misconceptions

Students might assume that since these properties are the properties of real numbers, that they apply to subtraction and division as well. However, we cannot generalize the rules for these operations. Show examples of when the properties do not apply to subtraction or division:

- Commutative property: $5 - 3 \neq 3 - 5$ or $6 \div 3 \neq 3 \div 6$
- Associative property: $(6 - 3) - 2 \neq 6 - (3 - 2)$ or $(8 \div 4) \div 2 \neq 8 \div (4 \div 2)$

Examples

Students: Page 61

Example 1

Consider the expression $2x + \dfrac{1}{8} - \dfrac{1}{4} - 7x$.

Complete the following work with properties or statements as reasoning in each step:

1. $2x + \dfrac{1}{8} - \dfrac{1}{4} - 7x = 2x + \dfrac{1}{8} + \left(-\dfrac{1}{4}\right) + (-7x)$ Inverse property of addition

2. $\quad = 2x + (-7x) + \dfrac{1}{8} + \left(-\dfrac{1}{4}\right)$ ☐

3. $\quad = 2x + (-7x) + \dfrac{1}{8} + \left(-\dfrac{1}{4} \cdot \dfrac{2}{2}\right)$ ☐

4. $\quad = 2x + (-7x) + \dfrac{1}{8} + \left(-\dfrac{2}{8}\right)$ ☐

5. $\quad = -5x - \dfrac{1}{8}$ ☐

Create a strategy

The reason for step 1 has been stated. For steps 2, 3, 4, and 5, find what has changed from the previous step and state the property that allows us to make that change.

Apply the idea

For step 2, the positions of the numbers are changed.

$$2x + \frac{1}{8} + \left(-\frac{1}{4}\right) + (-7x) = 2x + (-7x) + \frac{1}{8} + \left(-\frac{1}{4}\right)$$ Commutative Property of Addition

For step 3, $-\frac{1}{4}$ is multiplied by $\frac{2}{2}$, which equals 1. This shows a number being multiplied by 1.

$$2x + (-7x) + \frac{1}{8} + \left(-\frac{1}{4}\right) = 2x + (-7x) + \frac{1}{8} + \left(-\frac{1}{4} \cdot \frac{2}{2}\right)$$ Identity Property of Multiplication

For step 4, we performed the multiplication of $-\frac{1}{4}$ and $\frac{2}{2}$.

$$2x + (-7x) + \frac{1}{8} + \left(-\frac{1}{4} \cdot \frac{2}{2}\right) = 2x + (-7x) + \frac{1}{8} + \left(-\frac{2}{8}\right)$$ Evaluate the multiplication

For step 5, like terms were combined.

$$2x + (-7x) + \frac{1}{8} + \left(-\frac{2}{8}\right) = -5x - \frac{1}{8}$$ Combine like terms

The complete work is:

1	$2x + \frac{1}{8} - \frac{1}{4} - 7x = 2x + \frac{1}{8} + \left(-\frac{1}{4}\right) + (-7x)$	Inverse property of addition
2	$= 2x + (-7x) + \frac{1}{8} + \left(-\frac{1}{4}\right)$	Commutative Property of Addition
3	$= 2x + (-7x) + \frac{1}{8} + \left(-\frac{1}{4} \cdot \frac{2}{2}\right)$	Identity Property of Multiplication
4	$= 2x + (-7x) + \frac{1}{8} + \left(-\frac{2}{8}\right)$	Evaluate the multiplication
5	$= -5x - \frac{1}{8}$	Combine Like terms

Purpose

Show students how to use properties of real numbers to simplify expressions.

Reflecting with students

Review the properties used in each step. Discuss the importance of using these properties correctly to simplify the expression. Ask students to discuss different methods they could use to simplify the expression, such as applying the properties in a different order.

Students: Page 61

> 💡 **Idea summary**
>
> Simplifying an expression involves rewriting it in its most basic form without changing the value of the expression.
>
> We use the order of operations and the properties of real numbers to simplify expressions.

The distributive property with algebraic terms

Exploration

Students: Page 62

> **Interactive exploration**
> Explore online to answer the questions
>
> ⊕ **mathspace.co**

Use the interactive exploration in 2.02 to answer these questions.

1. What patterns exist between the number of +xs, +1s and the expression in the last column?
2. If someone says that $2(3x + 4) = 6x + 3$, should we agree or disagree? Why?

Suggested student grouping: Small groups

Students will manipulate an applet to explore the distributive property with algebraic terms. The aim of the exploration is for students to realize that an expression with a number in front of parentheses represents groups the terms inside the parentheses, but the tiles in total are the result of multiplying the number to each term inside the parentheses.

Ideal student responses

These ideal responses may differ from other correct student responses. Less formal responses can be connected with the more precise mathematical language presented here.

1. **What relationships exist between the terms in the algebraic expression and the total number of +x and +1 tiles?**

The number of +x tiles is equal to the product of the number in front of the parentheses and the coefficient of x. The number of +1 tiles is equal to the product of the number in front of the parentheses and the constant term.

2. **If someone says that $2(3x + 4) = 6x + 3$, should we agree or disagree? Why?**

No. There should be 2 groups of +4, which would result in +8, not +3. The correct equation should be $2(3x+4)=6x+8$.

Purposeful questions

- What does the number in front of the parentheses represent in terms of the algebra tiles?
- How does changing the number in front of the parentheses affect the number of +x and +1 tiles?
- Do you notice any relationships between the coefficients and constants in the original expression and the coefficient and constant of the total number of tiles?

Possible misunderstandings

- Students may not see the relationship if they do not write down the expression represented by the algebra tiles. Encourage students to only look at the algebra tiles, ignoring the given expression, and write a simplified expression for those tiles.

Following the exploration, students recall that a value outside a set of parentheses indicates the number of groups of the terms inside the parentheses. A visual representation of algebra tiles is shown to support this. Then, students discover that the expression can be simplified by multiplying the value to both terms inside the parentheses, and this is known as the distributive property.

Example 2

Consider the expression $4(t + 6)$. Use a model to simplify this expression.

Create a strategy

We can use an area model to represent this expression. The dimensions of this rectangle are 4 and $(t + 6)$ so its area is $4(t + 6)$.

Apply the idea

The area model will look like this:

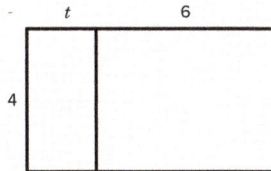

We can find the total area of the large rectangle by adding the area of each of the smaller rectangles.

$$\text{Area Large Rectangle} = \text{Area Small Rectangle}_1 + \text{Area Small Rectangle}_2$$

$$4(t + 6) = 4 \cdot t + 4 \cdot 6$$

$$= 4t + 24$$

Reflect and check

We can also represent the expression using algebra tiles. The expression $4(t + 6)$ can be represented as 4 groups of $(t + 6)$.

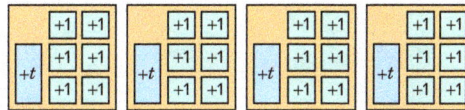

or 4 groups of t and 4 groups of 6.

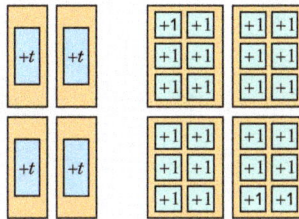

Purpose

Students demonstrate that they can represent algebraic expressions involving the distributive property using a pictorial representation.

🎓 **Advanced learners: Create an alternative real-world context**   use with Example 2
Targeted instructional strategies

Invite students to create real-world scenarios, other than area, that can be represented by the expression. Encourage them to think creatively about situations involving repeated groups, such as calculating total cost, combining quantities, or accumulating units over time. For example, they might imagine a context where 4 schools bring in cans for a food drive. Each school has t students and 6 teachers, and they need to find the total number of cans collected.

By creating their own contexts, students can see how the distributive property applies in various situations beyond geometry. Have them share and explain their scenarios with the class, highlighting how the expression models their specific situation. This approach helps students make meaningful connections between algebraic expressions and real-life applications, deepening their understanding of the distributive property.

Example 3

Simplify the expressions using the distributive property.

a $-2(3x - 1)$

Create a strategy

Use the distributive property $a(b + c) = a \cdot b + a \cdot c$.

Apply the idea

$-2(3x - 1) = (-2) \cdot 3x + (-2) \cdot (-1)$ Distributive property

$= -6x + 2$ Evaluate the multiplication

Purpose

Students demonstrate that they can apply the distributive property when the coefficient is negative.

Expected mistakes

Students might not distribute the negative sign when multiplying -2 to the second term. Encourage students to write each product separately, and remind them that a negative multiplied by a negative is a positive.

---

b $-(5 - s)$

Create a strategy

Use the distributive property $a(b + c) = a \cdot b + a \cdot c$.
The negative sign outside the parentheses represents multiplying by -1.

Apply the idea

$-(5 - s) = -1(5 - s)$ Rewrite using -1

$= -1 \cdot 5 - (-1) \cdot s$ Distributive property

$= -5 - (-s)$ Evaluate the multiplication

$= -5 + s$ Simplify the adjacent signs

Purpose

Demonstrate to students how the distributive property applies to expressions with a negative sign outside the parentheses, which is equivalent to -1.

Expected mistakes

Students might only distribute the negative to the first term, and forget to distribute it to the second term. Encourage them to write the negative as -1, so they can clearly see what they need to multiply it to both terms inside the parentheses.

---

c $-6.1(3.5y + 4.5)$

Create a strategy

Use the distributive property.

Apply the idea

$-6.1(3.5y + 4.5) = (-6.1) \cdot 3.5y + (-6.1) \cdot 4.5$ Distributive property

$= -21.35y + (-27.45)$ Evaluate the multiplication

$= -21.35y - 27.45$ Inverse property of addition

Reflect and check

We can check our work by substituting a value for y and confirming that the original expression and the simplified expression yield the same result.

Let's let $y = 1$ and see if both expressions evaluate to the same number.

$$-6.1(3.5\,(1) + 4.5) = -21.35(1) - 27.45$$
$$-6.1(3.5 + 4.5) = -21.35 - 27.45$$
$$-6.1(8) = -48.8$$
$$-48.8 = -48.8$$

Since both expressions evaluate to -48.8 when $y = 1$, we can be confident that these two expressions are equivalent.

Purpose

Challenge students to apply the distributive property to expressions with rational numbers.

Students: Page 65

d $\frac{1}{4}(x - 4) - \frac{1}{5}x$

Create a strategy

Use the distributive property to rewrite the expression in parentheses. Then combine like terms and make sure all fractions are fully simplified.

Apply the idea

$$\frac{1}{4}(x - 4) - \frac{1}{5}x = \frac{1}{4}x - \frac{1}{4}\cdot 4 - \frac{1}{5}x \qquad \text{Distributive property}$$

$$= \frac{1}{4}x - 1 - \frac{1}{5}x \qquad \text{Multiplicative inverse}$$

$$= \frac{1}{4}x - \frac{1}{5}x - 1 \qquad \text{Commutative property of addition}$$

$$= \left(\frac{1}{4}\cdot\frac{1}{5}\right)x - \left(\frac{1}{5}\cdot\frac{4}{4}\right)x - 1 \qquad \text{Multiplicative identity}$$

$$= \frac{5}{20}x - \frac{4}{20}x - 1 \qquad \text{Evaluate the multiplication}$$

$$= \frac{1}{20}x - 1 \qquad \text{Combine like terms}$$

Reflect and check

$\frac{1}{20}x - 1$ can also be written as $\frac{x}{20} - 1$. Always check if fractions are fully simplified.

Purpose

Show students how to apply the distributive property with a fraction and simply the expression when it contains an additional term.

Reflecting with students

Encourage students to check their result by substituting a value, such as 1, into the original and simplified expressions. Both substitutions should result in the same value.

Algorithmic thinking with a consistent procedure
Student with disabilities support

To help students who struggle with organization, introduce algorithmic thinking by providing a consistent, step-by-step procedure for applying the distributive property. Explain that algorithmic thinking involves creating a clear sequence of instructions to solve a problem efficiently. Begin by outlining a specific algorithm for distributing the term outside the parentheses to each term inside. Encourage students to follow these steps:

1. Rewrite the expression inside the parentheses with a box and a multiplication sign directly before each term.
2. Write the value outside the parentheses in each box.
3. Evaluate the multiplication.
4. Combine like terms if necessary.

Using the expression $-2(3x - 1)$ as an example, the steps of the solution are shown below:

1. $\square \cdot 3x + \square \cdot -1$
2. $-2 \cdot 3x + -2 \cdot -1$
3. $-6x + 2$

Provide students with practice problems that include different combinations of positive and negative signs, and encourage them to consistently apply this algorithm to ensure accuracy. Emphasize that using algorithmic thinking helps automate the distribution process, reduces errors, and improves their ability to simplify algebraic expressions efficiently.

Students: Page 65–66

Example 4

A student incorrectly used the distributive property and wrote $7(4x + 3) = 28x + 3$.

Which one of the following is the best explanation to help the student correct their error?

A They have multiplied $4x$ and 7 rather than adding them.

B They have forgotten to multiply the second part of the sum, 3, by the number outside the brackets, 7.

C They have added $4x$ and 7 rather than multiplying them.

D They have multiplied the wrong term in the sum by 7. They should multiply 3, instead of $4x$, by 7.

Create a strategy

Use the distributive property $a(b + c) = a \cdot b + a \cdot c$.

Apply the idea

The correct option is B, because they have forgotten to multiply the second part of the sum, 7, by the number outside the brackets, 3.

Purpose
Show students how to identify and correct errors in a worked example of the distributive property.

Expected mistakes
Students may not understand how to verbalize the mistake and choose an answer that contains portions of their thinking, such as answer choice D, which mentions multiplying by 3. They may not read the entire answer choice or know how to connect their thinking to the words.

When working with error analysis, suggest to students that a helpful exercise can be to work through the problem correctly first, then compare the options to their own work.

Encourage students to correct the mistake. Have them discuss in pairs what the student could do to avoid making the same mistake in the future.

Students: Page 66

> ### 💡 Idea summary
>
> **Distributive property:**
>
> For all numbers a, b, and c,
>
> $$a(b + c) = a \cdot b + a \cdot c$$
>
> and
>
> $$a(b - c) = a \cdot b - a \cdot c$$

Practice

Students: Page 66–68

What do you remember?

1 Explain how to use the distributive property to simplify $7(4x + 3)$.

2 In the equation $2(s + 9) = 2s + 9$, the distributive property has been incorrectly applied. Write the correct right hand side of this equation.

3 Explain how to use the distributive property to simplify $\frac{1}{4}(4x + 8)$.

4 In the equation $3.5(2y + 6) = 7y + 6$, the distributive property has been incorrectly applied. Write the correct right hand side of this equation.

5 These tiles represent the expression $2(-4x + 1)$.

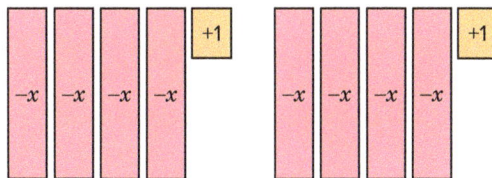

Which expression is equivalent to $2(-4x + 1)$?

A $-8x + 1$	**B** $-8x + 2$	**C** $8x + 1$	**D** $8x + 2$

Let's practice

6 For each expression:

    i Create a representation with algebra tiles.   **ii** Simplify the expression.

    a $4(t + 6)$ **b** $3(6 - r)$

7 Simplify each expression:

a $(r + 3) \cdot 5$ b $-9(r + 7)$ c $-6(n - 4)$ d $-(5 - a)$

e $-9(5 + 4c)$ f $-8(-9n - 5)$

8 Simplify each expression:

a $\frac{1}{6}(t + 6)$ b $\frac{1}{3}(6 + 12r)$ c $\left(\frac{s}{2} + 3\right) \cdot \frac{2}{3}$ d $-4\left(\frac{k}{2} + \frac{1}{8}\right)$

e $0.5(2x + 3)$ f $2.4(9x + 1.5)$ g $1.6(3.5 + 12.5x)$ h $-4.4(1.35 - 3.95x)$

9 Simplify each expression:

a $5 + 3(x + 4)$ b $8(3x + 4) - 6x$ c $6(y + 8) - 5y - 3$ d $9x - 2 - 3(x + 4)$

10 Simplify each expression:

a $7(y + 9) + 5(y + 6)$ b $-7(x + 4) + 8(x + 9)$ c $6.2(3p - 3) - (4p - 5.1)$ d $4.5(2x + 4) - 0.5(6x - 2)$

e $\frac{1}{2}(x + 2) + \frac{2}{3}(x + 9)$ f $\frac{3}{4}(12x + 8) - \frac{3}{5}(-20 - 35x)$

11 These tiles represent the expression $2(7x - 3) - 5x$.

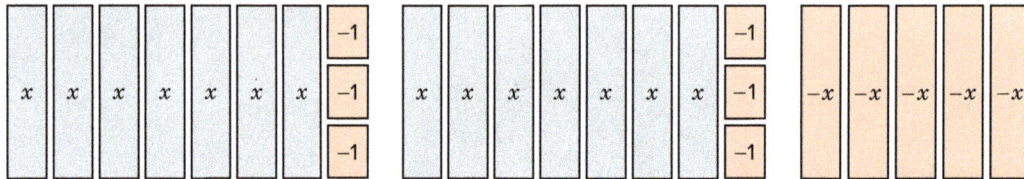

a Write the expression represented by the algebra tiles.

b Simplify the expression.

12 Complete the statements so they are true:

a $3(2x + \square) = \square x + 12$ b $\square(\square s + 5) = -8s + 10$ c $\square(3g - \square) = -6g + 14$ d $5(\square p + \square) = 15p - 30$

Let's extend our thinking

13 A solar energy company offers a range of solar panels, each with a power rating of $m + 14$ megawatts, where m is the panel model number.

a Write an expression in terms of m for the total amount of power produced by 9 solar panels.

b Simplify the expression.

14 The representatives of 8 countries are attending a global summit, and each country is able to send $3u + 4$ representatives.

a Write an expression for the total number of representatives attending the summit.

b Simplify the expression.

15 Consider the rectangle shown:

a Determine the areas of the two smaller rectangles, A and B.

b Write an expression for the area of the larger rectangle using your answer in part (a).

c Write an expression for the area of the larger rectangle using the given base length, $m + n$.

d Relate your answer in parts (b) and (c) using an equation.

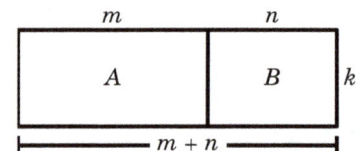

16 Complete the statements:

    a $3.5(2.4x + \square) = \square x + 33.6$

    b $\square(\square s + 4.16) = -12.5s + 10.4$

    c $\square(13.5g - \square) = -75.6g + 454.72$

    d $3.05(\square p - \square q) = 93.33p - 274.866q$

17 Let the length of the rectangle below be $L = \dfrac{5}{6}y + \dfrac{19}{4}$ and the width be $W = \dfrac{1}{6}y - \dfrac{23}{8}$.

    a   Write the perimeter of the figure in terms of y.

    b   Write an expression that could be used to find the area of the figure.

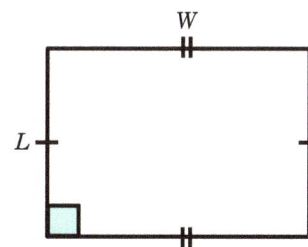

Answers

2.02 Simplify expressions and distributive property

What do you remember?

1 Distribute the 7 to the first term in parentheses $4x$ to get $28x$. Then distribute the 7 to the second term in parentheses to get 21. Because the 7 and 3 are both positive, we can keep the addition symbol resulting in the expression $28x + 21$.

2 $2s + 18$

3 Distribute the $\frac{1}{4}$ to the first term in parentheses $4x$ to get x. Then distribute the $\frac{1}{4}$ to the second term in parentheses 8 to get 2. Because the terms are both positive, we can keep the addition symbol resulting in the expression $x + 2$.

4 $7y + 21$

5 **B**

6 a i
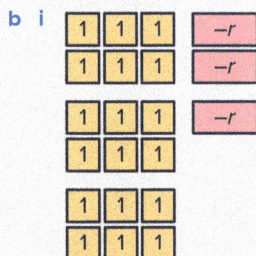

   ii $4t + 24$

   b i

   ii $18 - 3r$

Let's practice

7 a $5r + 15$ b $-9r - 63$

   c $-6n + 24$ d $-5 + a$

   e $-45 - 36c$ f $72n + 40$

8 a $\frac{1}{6}t + 1$ b $2r + 4r$

   c $\frac{s}{3} + 2$ d $-2k - \frac{1}{2}$

   e $x + 1.5$ f $21.6x + 3.6$

   g $5.6 + 20x$ h $-5.94 + 17.38x$

9 a $3x + 17$ b $18x + 32$

   c $y + 45$ d $6x - 14$

10 a $12y + 93$ b $x + 44$

   c $14.6p - 13.5$ d $6x + 19$

   e $\frac{7x}{6} + 7$ f $30x + 18$

11 a $14x - 6 - 5x$ b $9x - 6$

12 a $3(2x + 4) = 6x + 12$ b $2(-4s + 5) = -8s + 10$

   c $-2(3g - 7) = -6g + 14$ d $5(3p + (-6)) = 15p - 30$

Let's extend our thinking

13 a $9(m + 14)$ megawatts   b $(9m + 126)$

14 a $8(3u + 4)$ representatives   b $(24u + 32)$

15 a Area of $A = mk$
       Area of $B = nk$

   b $mk + nk$

   c $k(m + n)$

   d $mk + nk = k(m + n)$

16 a $3.5(2.4x + 9.6) = 8.4x + 33.6$

   b $2.5(-5s + 4.16) = -12.5s + 10.4$

   c $-5.6(13.5g - 81.2) = -75.6g + 454.72$

   d $3.05(30.6p - 90.12q) = 93.33p - 274.866q$

17 a $2y + \frac{15}{4}$ b $\left(\frac{5}{6}y + \frac{19}{4}\right)\cdot\left(\frac{1}{6}y - \frac{23}{8}\right)$

2.03 Write and represent multistep equations

Subtopic overview

Lesson narrative

In this lesson, students will learn to write and represent multistep equations. They will explore how to create equations from real-world scenarios, identify key phrases indicating mathematical operations, and use algebra tiles for visual representation. The lesson includes several examples and practice problems where students convert verbal descriptions into algebraic equations, such as calculating costs of school supplies or planning a garden border. By the end of the lesson, students should be able to write and represent linear equations, interpret solutions in context, and create verbal scenarios from given equations.

Learning objectives

Students: Page 69

After this lesson, you will be able to...
- represent multistep linear equations using a variety of concrete materials and pictorial representations.
- write a multistep linear equation in one variable to represent a verbal situation.
- create a verbal situation in context given a multistep linear equation in one variable.
- interpret given algebraic solutions to linear equations in context.
- confirm algebraic solutions to linear equations in one variable.

Key vocabulary

- equation

Essential understanding

The structure of an equation can reveal important details about the situation it represents.

Standards

This subtopic addresses the following Virginia 2023 Mathematics Standards of Learning standards.

Mathematical process goals

MPG3 — Mathematical Reasoning

Teachers can help students develop mathematical reasoning by guiding them through the process of analyzing multi-step equations and determining the validity of their solutions. This can be achieved by engaging students in inductive and deductive reasoning exercises related to the topic. For instance, teachers can provide students with incorrectly solved equations and ask them to identify the mistakes and correct them.

MPG4 — Mathematical Connections

Teachers can help students make mathematical connections by linking their prior knowledge of two-step linear equations to the concept of multi-step equations. They can also discuss how multi-step equations are applicable in other disciplines and real-world situations. For example, teachers can illustrate how multi-step equations can be used in financial planning or physics problems, thus connecting mathematics to other fields of study.

MPG5 — Mathematical Representations

Teachers can integrate mathematical representations into their lessons by using concrete materials and pictorial representations, such as algebra tiles or colored chips, to represent multistep equations. They can also guide students to represent these equations in different forms table of values, graphs, or word problems. This helps students see that different representations can express the same mathematical relationships and deepens their understanding of multi-step equations.

Content standards

8.PFA.4 — The student will write and solve multistep linear equations in one variable, including problems in context that require the solution of a multistep linear equation in one variable.

8.PFA.4a — Represent and solve multistep linear equations in one variable with the variable on one or both sides of the equation (up to four steps) using a variety of concrete materials and pictorial representations.

8.PFA.4c — Write a multistep linear equation in one variable to represent a verbal situation, including those in context.

8.PFA.4d — Create a verbal situation in context given a multistep linear equation in one variable.

8.PFA.4f — Interpret algebraic solutions in context to linear equations in one variable.

8.PFA.4g — Confirm algebraic solutions to linear equations in one variable.

Prior connections

7.PFA.3 — The student will write and solve two-step linear equations in one variable, including problems in context, that require the solution of a two-step linear equation in one variable.

Future connections

A.EI.1 — The student will represent, solve, explain, and interpret the solution to multistep linear equations and inequalities in one variable and literal equations for a specified variable.

8.PFA.5 — The student will write and solve multistep linear inequalities in one variable, including problems in context that require the solution of a multistep linear inequality in one variable.

A.EI.1 — The student will represent, solve, explain, and interpret the solution to multistep linear equations and inequalities in one variable and literal equations for a specified variable.

Lesson Preparation

Suggested review

Depending on your students' level of prior knowledge, consider revisiting the following lessons:

Grade 7-3.03 — Represent algebraic expressions
Grade 8-2.01 — Represent algebraic expressions

Student lesson & teacher guide

Write and represent multi-step equations

Students explore how to represent real-life scenarios as equations using pictorial models or algebra tiles.

The lesson emphasizes understanding how verbal phrases translate to mathematical expressions and how order affects the interpretation of terms. Students learn to verify solutions to equations using substitution to ensure accuracy.

Examples

Students: Page 70

Example 1

Write each situation as an algebraic equation.

a   Four less than twice a number equals the same number increased by six.

Create a strategy

Represent the 'number' with a variable. 'Less' indicates subtraction. 'Twice' means multiply by 2, and 'increased' indicates addition.

Apply the idea

Choosing the variable x as our variable and the term *equals* separating our expressions, we can write the equation:

$$2x - 4 = x + 6$$

Reflect and check

An equation represents a balance between two expressions. On one side, we start with twice a number, reduce it by four, and on the other, we simply add six to the same number. This following scale helps visualize the equation.

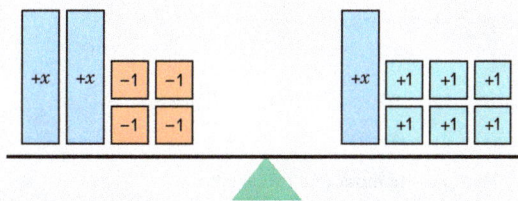

Purpose
Show students how to translate a word problem into an algebraic equation.

Expected mistakes
Students may misinterpret "four less than twice a number" and translate each word directly, which would result in $4 - 2x$. Remind students that "four less than" signifies that 4 is being subtracted from something, so 4 should be written after the subtraction symbol.

b The product of six and the sum of a number and three is the difference of the number and twelve.

Create a strategy

Use a variable to represent the 'number'. 'Product' indicates multiplication, 'sum' means addition, and 'difference' means subtraction.

Apply the idea

Notice, we are finding the product of a sum. So, we need to multiply the entire sum $x + 3$ by 6. This will require us to use parentheses: $6(x + 3)$.

Choosing the variable x as our variable and the term *is* separating our expressions, we write the equation:

$$6(x + 3) = x - 12$$

Reflect and check

To better understand our equation $6(x + 3) = x - 12$, let's visualize it using algebra tiles.

On one side of the equation, we have $6(x + 3)$, which represents six groups of $x + 3$. This visually shows the repeated addition that is caused by multiplication.

On the other side of the equation, we have $x - 12$, represented by one x tile and twelve negative unit tiles.

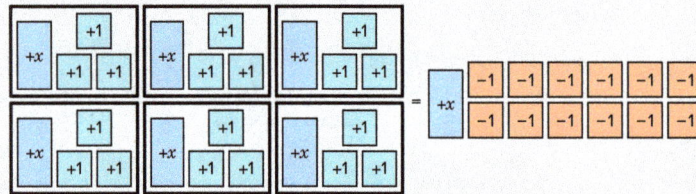

Purpose

Demonstrate to students how to translate a word problem into an algebraic equation by identifying key terms and using appropriate mathematical operations.

c The quotient of four more than triple a number and two is nine.

Create a strategy

Use a variable to represent the 'number'. 'Quotient' indicates division, 'triple' means multiplied by 3, and 'more than' indicates addition.

Apply the idea

The word *and* acts as the separator between the numerator and denominator for the fraction representing the quotient. This tells us that 'four more than triple a number' is the numerator and 'two' is the denominator.

Using the word *is* to separate the expressions, we have the equation

$$\frac{3x + 4}{2} = 9$$

Reflect and check

We can visualize the equation using algebra tiles. On the left are three x tiles and four 1 tiles, half grayed-out to represent dividing by 2. This is equal to nine 1 tiles on the right.

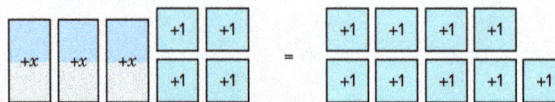

Purpose

Show students how to translate a word problem into an algebraic equation, using keywords to identify mathematical operations and relationships.

🎓 Highlight key words to aid in translating word problems
Student with disabilities support

use with Example 1

Encourage students to break down complex word problems into smaller, manageable parts by highlighting or underlining key terms and phrases that indicate mathematical operations, such as "increased by," "twice," "product of," and "quotient of." Students can use the table from the lesson as a reference.

Addition	Subtraction	Multiplication	Division
plus	minus	times	divided by
the sum of	the difference of	the product of	the quotient of
increased by	decreased by	multiplied by	half
total	fewer than	each/per	split
more than	less than	twice	equally shared
added to	subtracted from	double	

Offer consistent strategies, such as:

- Place parentheses around the phrases on each side of "equals" or "is"
- Highlight "a number" and place an x above it
- Underline "sum of," "difference of," "quotient of," or "product of" and circle the two following terms

The examples in this problem may be marked up as shown:

$$2 \cdot x - 4 = x + 6$$

a (Four less than twice a number) equals (the same number increased by six)

$$6(x + 3) = x - 12$$

b (The product of six and the sum of a number and three) is (the difference of the number and twelve)

$$(3 \cdot x + 4) \div 2 = 9$$

c (The quotient of four more than triple a number and two) is (nine)

This approach helps students with language processing difficulties systematically translate verbal descriptions into algebraic expressions.

Example 2

Carrie is going shopping to buy school supplies. She plans to buy 6 notebooks, 2 packages of pens, and 1 three-ring binder. The total cost will be $30.

She remembers that the binder costs twice as much as a notebook, and a pack of pens costs the same as six notebooks.

a Write an algebraic equation to represent the situation.

Create a strategy

Let's assign a variable to represent the cost of one notebook. Since the binder costs twice as much as a notebook and a pack of pens costs the same as six notebooks, we can write the costs of the binder and pens in terms of the notebook's cost. Finally, we'll use the total amount Carrie can spend to set up our equation.

Apply the idea

We know the sum of everything Carrie buys is $30, so:

6 notebooks + 2 pens + 1 binder = $30

Let n represent the cost of one notebook in dollars. So, the cost of 6 notebooks is $6n$.

The cost of one pack of pens is the same as the cost of 6 notebooks or $6n$, so the cost of 2 packs of pens is $2 \cdot 6n$.

The cost of one binder twice (or two times) the cost of a notebook, n, so the cost of a binder is $2n$.

Adding these up and setting them equal to $30, we get the equation:

$$6n + 2 \cdot 6n + 2n = 30$$

Purpose

Show students how to translate a real-world situation into an algebraic equation by identifying the relationships between different quantities and representing these relationships with variables and operations.

b Draw a pictorial model to represent the equation you wrote in part (a).

Create a strategy

To create a pictorial model of the equation, we will use visual representations for each component of the shopping list: notebooks, packs of pens, and the binder.

Apply the idea

Let's use a circle to represent a notebook, a square for a pack of pens, and a triangle for the binder.

Since the binder costs twice as much as a notebook, we'll place two circles inside the triangle to show this relationship. Similarly, a pack of pens, costing the same as six notebooks, will have six circles within a square.

Carrie plans to buy 6 notebooks (6 circles), 2 packs of pens (2 squares, each containing 6 circles), and 1 binder (1 triangle containing 2 circles). The total cost represented by these symbols must equal the total cost of $30.

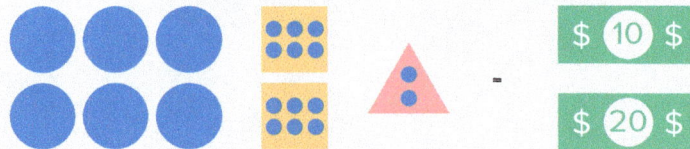

Purpose

Show students how to translate an algebraic equation into a visual representation, reinforcing their understanding of the relationships between variables in the equation.

Expected mistakes

Students might draw 20 circles rather than differentiating between notebooks, pens, and a binder. Use the picture given in this example to help them see how the rectangle and triangle show that there is a difference between the items, but the circles inside each shape should the equivalence in prices.

Students: Page 73

c Carrie solves the equations using the information she knows. Her value for n is 1.5. Verify and interpret the solution.

Create a strategy

To verify Carrie's solution, we'll substitute her value for n into our equation and see if the result matches the total cost of $30.

Apply the idea

Substituting $x = 1.5$ into the equation:

$$6(1.5) + 2 \cdot 6(1.5) + 2(1.5) = 30$$
$$9 + 18 + 3 = 30$$
$$30 = 30$$

This shows that the value of 1.5 for x makes the equation true, so it is the solution.

Purpose

Students that they can verify and interpret a solution to a real-world problem.

Reflecting with students

Challenge advanced learners, or any student who is ready, to determine the cost of a pack of pens and a binder. Since a notebook is $1.50, a binder is $2 \cdot \$1.50 = \$3$, and a pack of pens is $6 \cdot \$1.50 = \$9$.

Students: Page 73

Example 3

Write a real world scenario that could be represented by the equation $2x + 2(x + 4) = 23$.

Create a strategy

To create a real-world scenario for the given equation, we'll think of situations where two quantities, both related to the same unit (x), add up to a total.

Apply the idea

Imagine Jamie is planning a garden border that has two parallel sides of equal length, x, and two additional sides that are each 4 ft longer than the shorter sides. Jamie wants to enclose the garden with a small fence.

The equation $2x + 2(x + 4) = 23$ represents the total length of fencing needed to surround the garden. Here, $2x$ is the combined length of the shorter parallel sides, and $2(x + 4)$ is the combined length of the longer sides. The total length of the fencing required is 23 ft.

Purpose

Demonstrate to students how to translate a mathematical equation into a real-world scenario.

Reflecting with students

Encourage students to attend to mathematical precision by clearly defining variables and consistently using appropriate units throughout the task. When guiding them to create a verbal situation for the equation, emphasize the importance of specifying what each term represents, such as money in dollars or length in feet. Encourage students to organize their thoughts clearly and verify that their scenario accurately represents the given equation.

Stronger and clearer each time
English language learner support

Invite students to individually write their own real-world situation that could be modeled by the given equation. Encourage them to describe what the variable, coefficients, and constant represent in their context.

If students struggle to think of scenarios, prompt them with questions such as:

- Can you think of a situation where two related quantities are being added together, and you know the total? What if one quantity is a certain amount more than the other?
- What are some situations where a calculation is repeated (e.g., buying two of something or measuring two sides)?
- Notice that both terms on the left side are doubled. Can you think of something that happens twice, and then twice again, but there is more the second time?

Then, have students share their situations with a partner. Encourage them to listen carefully, ask questions, and provide constructive feedback to help clarify and strengthen each other's explanations.

After the first round, have students meet with a new partner to share their revised situations and receive further feedback. Encourage them to refine their language, making their descriptions more precise and detailed, and to incorporate any new vocabulary or ideas they have learned.

Finally, ask students to revise their original written situations, incorporating the feedback and new ideas from their discussions. By iterating through this process, students will strengthen both their understanding of the mathematical concepts and their ability to express them in English.

Students: Page 73

Idea summary

Key phrases for operations in equations:

Addition	Subtraction	Multiplication	Division
plus	minus	times	divided by
the sum of	the difference of	the product of	the quotient of
increased by	decreased by	multiplied by	half
total	fewer than	each/per	split
more than	less than	twice	equally shared
added to	subtracted from	double	

Key phrases for equality and additional notation needed in an equation:

Equal	Parentheses
is/are	quantity
equals	times the sum
amounts to	times the difference
totals	

Practice

Students: Page 74–77

What do you remember?

1 Emma took 5 violin lessons last month. For each lesson, she memorized 3 major scales and x minor scales. Select all the expressions that represent how many scales Emma memorized last month.

  A $3 + 5x$ **B** $5(3 + x)$ **C** $15 + 5x$ **D** $5 + 3x$

2 Ariana and her family went berry picking at a farm. She filled one large basket with 15 ripe strawberries to make strawberry jam. She also filled 5 small baskets with b strawberries each to give to her friends as presents. Select all the expressions that represent how many strawberries Ariana picked in all.

  A $15 + b + b + b + b + b$ **B** $20 + b$ **C** $15 + 5b$ **D** $75b$

3 Match the phrases on the left to the equations on the right.

  a A number tripled is −9

  b A number divided by -3 is −9

  c 3 more than a number is −9

  d 3 minus a number is −9

  e A third of a number is −9

  f 3 less than a number is −9

  i $3 - n = -9$

  ii $3n = -9$

  iii $n \div (-3) = -9$

  iv $n - 3 = -9$

  v $n + 3 = -9$

  vi $\dfrac{1}{3}n = -9$

4 Which set of algebra tiles represents the equation $5x + 9 = 24$?

5 James says the solution to the equation $\dfrac{2x}{5} + 3 = 7$ is $x = 10$. Fiona says it is $x = 20$.

  a How can you find out who is correct without actually solving the equation?

  b Who is correct?

6 Create a real world-situation that could be represented by the equation: $20 + 10x = 80$.

Let's practice

7 Write the following statements as equations, where x represents the number:

  a Five less than 2 times a number equals the number increased by 7.

  b Twice a number increased by 15 totals the difference of the number and 1.

  c 8 times a number increased by the product of 6 and the sum of the number and 11 is 25.

  d Seven times a number is equal to $\dfrac{25}{3}$ more than twice the number.

  e The quotient of 9 and the difference of 2 and a number is the number.

  f The quotient of a number and −3 is $-\dfrac{4}{3}$.

8 Jeff modeled a cube out of construction paper from the net shown. The cube he started has a side length of x inches. Jeff wants to increase the 4 vertical edges that represent the height by 5 inches to create a new rectangular prism.

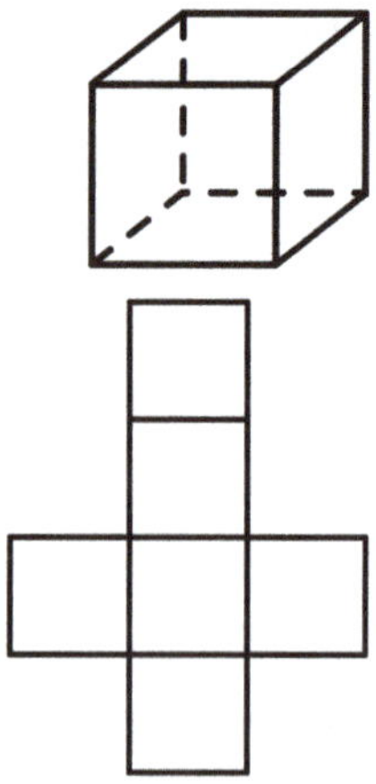

He wants to glue a decorative string along the edges of the new rectangular prism, and has 80 inches of string to use. Jeff writes the equation $8x + 5(x + 4) = 80$ to solve for how long the sides need to be to best use the string. He realizes he made a mistake.

Describe his mistake and write the correct equation.

9 Jamie had x pencils in her pencil case. She finds one more, and then shares all her pencils equally with her friend, Teddy. Her sister then gives her 5 more pencils. Jamie now has 7 pencils left in her pencil case.

    a Explain why the equation that can represent this situation is given by:

$$\frac{x + 1}{2} + 5 = 7$$

    b Draw a pictorial model of the equation.

    c Jamie claims that she started with 8 pencils. Is this correct?

10 Consider the scale:

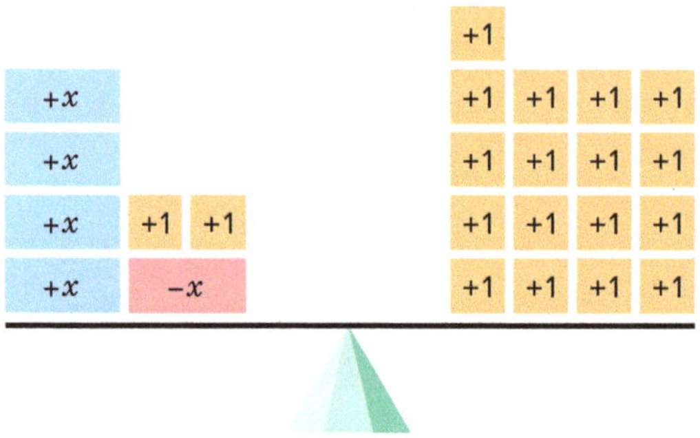

    a Write an equation to represent the scale.

    b Identify whether the scale would remain balanced after the following:

      i Take away a +1 tile from the left side of the scale and add a +1 tile to the right side of the scale.

      ii Take away a +1 tile from the left side of the scale and take away a +1 tile from the right side of the scale.

      iii Add a +1 tile to the left side of the scale and add a +1 tile to the right side of the scale.

      iv Add a +1 tile to the left side of the scale and take away a +1 tile from the right side of the scale.

11 Which set of algebra tiles represents the equation $3(x + 2) = -9$?

    A

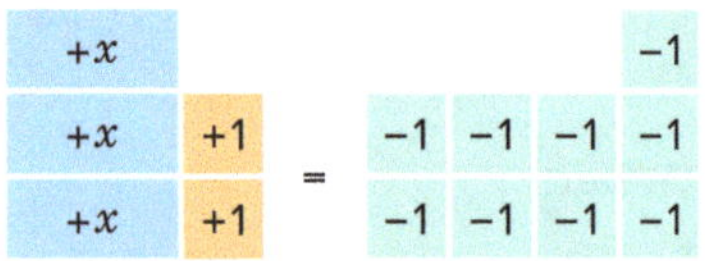

    B

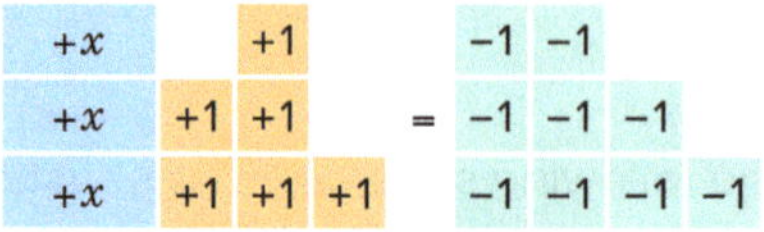

    C

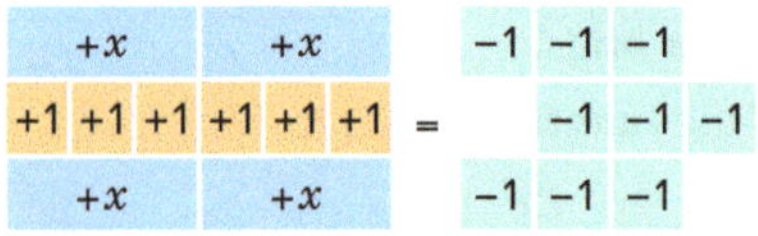

    D

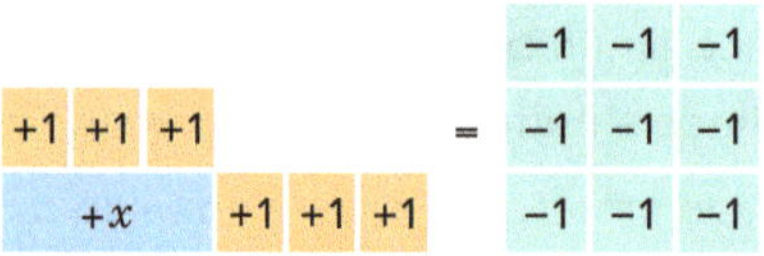

12 Write an equation that says the length of the green line is equal to the length of the black line.

13 Write an equation using parentheses to show that the length of the green line is equal to the length of the black line.

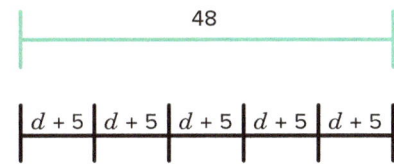

14 Jared has 3 groups of $x + 5$ video games, while Jenny has 2 groups of $3x - 2$ video games. Together they have 47 video games.

    a Write an equation for this situation.

    b After writing and solving their equation, the solution is $x = 4$. Jared says that with this solution, he has 7 more video games than Jenny. Is he correct?

15 Sharon is designing a garden and knows that the length, l, will be 8.5 feet more than the width. The garden will have a fence surrounding the perimeter, which totals 92 feet.

    a Given that perimeter is $2l + 2w$, write an equation to find l, the length of her garden.

    b When Sharon looked back at her design notebook for her flower beds, she found the equation $5x = 60$ with no description nearby. Write a possible scenario for Sharon's garden design that this equation could represent.

16 Identify the correct verbal situation for the equation $7(4x + 2) = 11x - 5$.

    A A number is multiplied by 4 and then 2 is added. Then the result is multiplied by 7. This is equal to 5 times the number minus 11.

    B A number is multiplied by 4 and then 2 is added. Then the result is multiplied by 7. This is equal to 11 times the number minus 5.

    C A number is multiplied by 7 and then 2 is added. Then the result is multiplied by 4. This is equal to 11 times the number minus 5.

    D A number is multiplied by 7 and then 2 is added. Then the result is multiplied by 4. This is equal to 5 times the number minus 11.

17 Create a verbal situation in context for each of the equations:

    a $8x + 9 = 3x - 17$ **b** $\dfrac{2x - 3}{5} = 4$ **c** $3(x + 7) = 12 - x$

18 Hazel is on a roadtrip and she can model the distance from her home, in miles, using the equation $D = 2x + 5$, where x is the number of hours she has been driving. Hazel substitutes $D = 19$ into the equation to get $19 = 2x + 5$. Solving this equation, she gets the solution $x = 7$.

What is the real-world interpretation of this solution?

19 For each equation, determine if the given value is a solution to the equation. Explain your reasoning.

    a $4(5m - 1) + 5m = 20 - (4m - 5)$ for $m = 1$ **b** $5 - \dfrac{x}{3} = \dfrac{4(x + 1)}{6}$ for $x = 7$

20 You are organizing a charity event to collect toy donations for a local animal shelter. You would like to collect 50 toys total so each animal can have a toy.

    • You have received 3 boxes of donations. Each box contains the same number of toys.

    • A pet store also donated 15 toys.

    • You receive another donation of 4 boxes of toys, each holding the same number of toys as the original boxes.

    • To be sure you have enough you donate 5 more toys.

Draw a pictorial model and write an equation to represent this situation.

Let's extend our thinking

21 Two taxi companies in town claim to have the best rates for their customers. Taxi Company A charges a flat rate of $10 plus $2 for each mile driven. Taxi Company B charges a $5 flat rate plus $4 for each mile driven. These companies claim there is only one mileage where the rides would cost the same amount. Find this distance.

22 The lifespan of a mouse is 3 years less than a hedgehog. The lifespan of a deer is 2 times that of the mouse. Let x represent the age of the hedgehog. If the total lifespan of these three animals is 11 years, calculate the lifespan for the hedgehog.

23 Sarah has three times as many books as Tom. If she gives away 15 books to Tom, they will have the same number of books. How many books does each person have initially? Let x represent the number of books Tom has initially.

24 Consider the shape in the diagram:

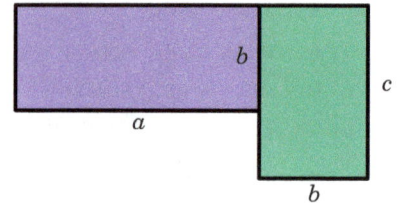

If A represents the total area of the shape, rewrite the following sentence as an equation:

a "The total area of the shape is equal to the product of a and b, added to the product of b and c".

b Rewrite the total area found in part (a) as a factor of b.

Answers

2.03 Write and represent multistep equations

What do you remember?

1 **B, C**

2 **A, C**

3 **a** ii. $3n = -9$ **b** iii. $n \div (3) = -9$

    c v. $n + 3 = -9$ **d** i. $3 - n = -9$

    e vi. $\frac{1}{3}n = -9$ **f** iv. $n - 3 = -9$

4 **D**

5 **a** Substitute the value for x into the equation.

    b James is correct.

6 Answers vary.

A person pays an initial fee of $20 to rent a bicycle and an additional $10 for each hour ($x$) they use the bicycle. After a certain number of hours, the total cost of renting the bicycle reaches $80.

Let's practice

7 **a** $2x - 5 = x + 7$ **b** $2x + 15 = x - 1$

    c $8x + 6(x + 11) = 25$ **d** $7x = 2x + \frac{25}{3}$

    e $\frac{9}{2-x} = x$ **f** $\frac{x}{-3} = -\frac{4}{3}$

8 Jeff writes an equation that represents 5 edges increasing by 4 inches instead of 4 edges increasing by 5 inches. The correct equation is $8x + 4(x + 5) = 80$.

9 **a** The $x + 1$ represents the original number of pencils x plus the one she found. This is then divided by 2 because she gave half the pencils away. The $+5$ represents the 5 pencils her sister gave her. The 7 on the right is the number of pencils she had left at the end.

    b

    c No

10 **a** $4x - x + 2 = 17$

    b **i** No   **ii** Yes   **iii** Yes   **iv** No

11 **B**

12 $m + 9m = 55m$

13 $5(d + 5) = 48$

14 **a** $3(x + 5) + 2(3x - 2) = 47$ **b** Yes

15 **a** $2l + 2(l - 8.5) = 92$

    b A possible scenario could be Sharon wanting to arrange 60 flower plants into 5 rows, and wanting to determine how many plants will be in each row.

16 **B**

17 **a** Imagine you're organizing a fundraiser bake sale at your school. You start with 8 trays, each containing (x) cookies, and you add 9 more cookies to make them more appealing. Meanwhile, you also prepared 3 trays containing (x) brownies. However, you and your friends end up eating 17 brownies before the sale. After sorting out the trays, you realize that the total number of cookies you have is the same as the total number of brownies.

    b You have 2 boxes (x) of toys and plan to donate them equally among 5 animal shelters. You found 3 broken toys and removed them from the box. After organizing the toys, you found out that each animal shelter would get 4 toys.

    c You're planning to bake cupcakes for 3 of your friends. Each friend will receive (x) cupcakes you'll be able to make within a day, along with an additional 7 cupcakes you prepared beforehand. You plan to add edible beads on the total number of cupcakes, but you only have 12, and you need to buy more based on the number of (x) cupcakes you'll make.

18 $x = 7$ is the number hours that Hazel has been driving if she is 19 miles away from her home.

19 **a** When $m = 1$, both sides of the equation are equal to 21. Therefore, $m = 1$ is a solution.

    b When $x = 7$, the left side of the equation is $\frac{8}{3}$, but the right side of the equation is $\frac{16}{3}$. Since the two sides are not equal, $x = 7$ is not a solution.

20

$3x + 15 + 4x + 5 = 50$

Let's extend our thinking

21 2.5 miles

22 $x = 15$

The lifespan of the mouse is $x - 3$, the lifespan of the hedgehog is x, and the lifespan of the deer is $2(x - 3)$.

The total of the three animals' lifespans is 11 years, so $x - 3 + x + 2(x - 3) = 11$.

Expanding and simplifying, $x - 3 + x + 2x - 6 = 11$. So, $4x - 9 = 11$. Adding 9 to both sides gives $4x = 20$ and dividing both sides by 4 we have $x = 5$.

23 Let x represent the number of books Tom has initially. Then, Sarah has $3x$ books. If Sarah gives away 15 books to Tom, they will have the same number of books, so:

$$3x - 15 = x + 15$$

Solving for x, we get:

$$2x = 30$$
$$x = 15$$

So, Tom initially has 15 books, and Sarah initially has $3 \cdot 15 = 45$ books.

24 a $A = ab + bc$ b $A = b(a + c)$

2.04 Solve multistep equations

Subtopic overview

Lesson narrative

In this lesson, students will learn to solve multistep linear equations. They will practice using both visual models and algebraic methods to solve equations with variables on both sides. The lesson includes examples and problems where students apply the distributive property, combine like terms, and use properties of equality to isolate and solve for the variable. By the end of the lesson, students should be able to solve complex linear equations involving multiple steps and understand the reasoning behind each step.

Learning objectives

Students: Page 78

After this lesson, you will be able to...
- solve multistep linear equations using a variety of concrete materials and pictorial representations.
- use properties of real numbers and properties of equality to justify steps in solving multistep equations.
- write and solve a multistep linear equation that represents a verbal or contextual situation.
- interpret algebraic solutions to linear equations in context.
- confirm algebraic solutions to linear equations.

Key vocabulary

- inverse operations
- properties of equality
- properties of real numbers
- zero pair

Essential understanding

The properties of equality allow an equation to be manipulated without changing the equivalence of the expressions on either side. The value(s) of the variables that make every equation in a system true make up its solution set. A standard algorithm can be used to solve any linear equation accurately and efficiently.

Standards

This subtopic addresses the following Virginia 2023 Mathematics Standards of Learning standards.

Mathematical process goals

MPG1 — Mathematical Problem Solving

Teachers can integrate this goal by guiding students through the process of solving multi-step equations in one variable. They can provide real-world problems and lead students to apply appropriate strategies to solve them. The process of problem-solving can be further enhanced by using do/undo tables to help students visualize how the process of solving an equation involves undoing all operations.

MPG3 — Mathematical Reasoning

Teachers can achieve this goal by helping students understand and apply properties of real numbers and properties of equality during the process of solving multi-step equations. They can also guide students in using inductive and deductive reasoning skills to make, test, and evaluate mathematical statements and justify steps in mathematical procedures.

MPG2 — Mathematical Communication

Teacher can encourage students to notice structure as they solve problems. Students should be able to describe strategies they use to solve based on the structure of the equation. Students can practice the think-aloud process while working in pairs or share their written ideas with a partner. One student could be the "driver" explaining each step while the other students writes down the steps to solve an equation

MPG5 — Mathematical Representations

Teachers can guide students in representing and interpreting multi-step equations in various forms, such as algebraic expressions and verbal situations. They can provide examples of word problems and guide students through the process of translating the verbal situation into a multistep equation and solving it. This would help students understand that different representations can convey the same mathematical idea and that representation is both a process and a product.

Content standards

8.PFA.4 — The student will write and solve multistep linear equations in one variable, including problems in context that require the solution of a multistep linear equation in one variable.

8.PFA.4a — Represent and solve multistep linear equations in one variable with the variable on one both sides of the equation (up to four steps) using a variety of concrete materials and pictorial representations.

8.PFA.4b — Apply properties of real numbers and properties of equality to solve multistep linear equations in one variable (up to four steps). Coefficients and numeric terms will be rational. Equations may contain expressions that need to be expanded (using the distributive property) or require combining like terms to solve.

8.PFA.4c — Write a multistep linear equation in one variable to represent a verbal situation, including those in context.

8.PFA.4e — Solve problems in context that require the solution of a multistep linear equation.

8.PFA.4f — Interpret algebraic solutions in context to linear equations in one variable.

8.PFA.4g — Confirm algebraic solutions to linear equations in one variable.

Prior connections

7.PFA.3 — The student will write and solve two-step linear equations in one variable, including problems in context, that require the solution of a two-step linear equation in one variable.

Future connections

A.EI.1 — The student will represent, solve, explain, and interpret the solution to multistep linear equations and inequalities in one variable and literal equations for a specified variable.

8.PFA.5 — The student will write and solve multistep linear inequalities in one variable, including problems in context that require the solution of a multistep linear inequality in one variable.

A.EI.1 — The student will represent, solve, explain, and interpret the solution to multistep linear equations and inequalities in one variable and literal equations for a specified variable.

Lesson Preparation

Suggested review

Depending on your students' level of prior knowledge, consider revisiting the following lessons:

Grade 6 — 6.03 One-step equations with addition and subtraction

Grade 6 — 6.03 One-step equations with multiplication and division

Grade 7 — 4.02 Solve two-step equations

Grade 8 — 2.02 Simplify expressions and distributive property

Grade 8 — 2.03 Write and represent multistep equations

Student lesson & teacher guide

Solve equations using visual models

Students explore the process of solving equations using a balance scale model. They discover the role of equivalence in maintaining balance on the scale and use the method of creating and eliminating zero pairs to isolate variables.

Examples

Students: Page 80

Example 1

Consider the equation

$$4(x - 3) + 8 = 12$$

a Represent the equation using algebra tiles.

Create a strategy

To represent this equation with algebra tiles, represent each part of the equation as a different type of tile. You will need variable tiles and positive and negative unit tiles.

Apply the idea

Let's start with the $4(x - 3)$ part of the expression on the left hand side of the equation. The parentheses tell us that we have 4 groups of $x - 3$. To create $x - 3$ we can use one x tile and three -1 tiles. Then we will create 4 groups of this to represent $4(x - 3)$.

The additional 8 unit tiles on the left side represent the +8.

For the right side, use 12 unit tiles to represent the number 12.

Purpose

Show students how to represent algebraic equations visually using algebra tiles, to enhance their understanding of algebraic manipulation.

Expected mistakes

Students might simplify the left side of the equation before representing it with algebra tiles. Clarify that the question wants them to represent the original equation, before simplification, with algebra tiles. Provide assistance with representing the original equation, as needed.

Students: Page 80–81

b Use the algebra tiles to solve the equation.

Create a strategy

Remove any zero pairs, then add tiles as needed to create and remove additional zero pairs until only one x tile remains on the left side of the equation.

Apply the idea

Identify the zero pairs on the left side of the equation. One +1 tile and one −1 tile create a zero pair. There are 8 zero pairs on the left side of the equation.

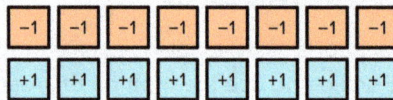

Removing the 8 zero pairs, we have:

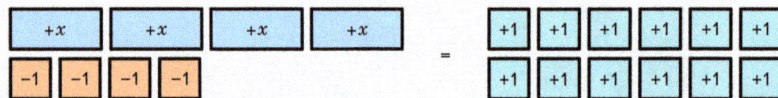

To be able to remove the −1 tiles from the left side, we need to add four +1 tiles. To keep the balance we will do the same on the right side of the equation.

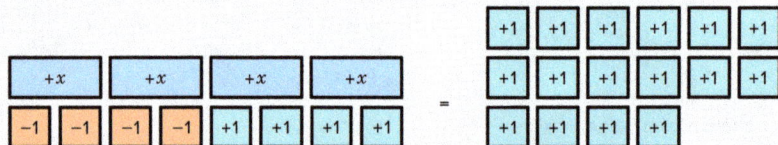

Removing the zero pairs we created on the left, we are left with four x tiles on the left and 16 unit tiles on the right.

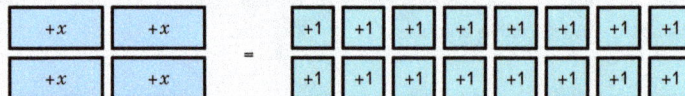

Let's see if we can divide the 16 unit tiles on the right into 4 equal sized groups so that one group represents one x tile.

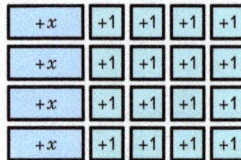

Removing 3 groups from each side will keep the equation balanced and leave us with one x tile on the left and 4 unit tiles on the right. So $x = 4$.

Reflect and check

After finding the solution, substitute it into the original equation to verify it.

$$4(x - 3) + 8 = 12 \qquad \text{Write the equation}$$
$$4(4 - 3) + 8 = 12 \qquad \text{Substitute } x = 4$$
$$4(1) + 8 = 12 \qquad \text{Evaluate the subtraction}$$
$$4 + 8 = 12 \qquad \text{Evaluate the multiplication}$$
$$12 = 12 \qquad \text{Evaluate the addition}$$

This confirms that $x = 4$ is the solution to the equation.

Purpose

Show students that solving algebraic equations can be visualized using algebra tiles, which allows them to see the concept of balance and zero pairs.

Concrete-Representational-Abstract (CRA) approach   use with Example 1
Targeted instructional strategies

Concrete: Engage students with physical manipulatives such as algebra tiles to represent the equation $4(x - 3) + 8 = 12$. Provide them with x tiles to represent the variable and unit tiles to represent positive and negative ones. Instruct students to create four groups of $(x - 3)$ using the tiles, which means they will have four x tiles and twelve -1 unit tiles on the left side. Then, have them add eight $+1$ unit tiles to the left side to represent the $+8$. On the right side, students should place twelve $+1$ unit tiles to represent the 12. Encourage students to physically manipulate the tiles to simplify and solve the equation by making zero pairs and balancing both sides. Remind students that we can show division by creating an equal number of equal-sized groups on each side, then determining how many tiles are in one group on each side.

Representational: Transition students from manipulatives to drawing representations of the algebra tiles. Ask them to sketch the tiles they used, drawing rectangles for x tiles and small squares for unit tiles, with shading or signs to indicate positive and negative values. Have them illustrate the four groups of $(x - 3)$, the additional $+8$ units on the left side, and the 12 units on the right side. Guide them in simplifying the drawings by crossing out zero pairs (a positive and a negative unit) and grouping like terms visually. This helps students see the steps they took with the manipulatives in a pictorial form.

Abstract: Move students to solving the equation using abstract symbols and algebraic notation. Write the original equation $4(x - 3) + 8 = 12$ on the board. Demonstrate how to apply the distributive property to expand $4(x - 3)$ into $4x - 12$. Then, combine like terms to simplify the equation to $4x - 4 = 12$. Guide students through the steps of adding 4 to both sides to get $4x = 16$ and then dividing both sides by 4 to find $x = 4$. Emphasize the connections between these symbolic steps and the earlier concrete and representational activities.

Making zero pairs rather than groups for division   use with Example 1
Address student misconceptions

Students often struggle with understanding division in equations and may mistakenly try to eliminate x tiles by adding or subtracting them to make zero pairs, just as they do with unit tiles. To address this misconception, explicitly teach students how to represent division using algebra tiles or pictorial models. Emphasize that division in this context involves partitioning the total number of units into equal parts to find the value of the variable.

When solving an equation like $4x = 16$, use algebra tiles to represent both sides: place four x tiles to represent $4x$ and sixteen unit tiles to represent 16. Explain that since we have four groups of x, we need to find the value of one x. Demonstrate how to equally divide the sixteen unit tiles among the four x tiles. Arrange the unit tiles into four groups of four, showing that each x corresponds to four units, so $x = 4$.

Example 2

A local community library is organizing a book drive. Jordan collects 5 boxes of books and 2 individual books. A local bookstore donates 3 boxes of books and 12 individual books to the drive.

Each box contains an equal number of books, and Jordan and the bookstore both donated the same total number of books.

a   Write an equation to represent this situation.

Create a strategy

Represent the unknown value as a variable. In this case the unknown is the number of books in each box so we can use the variable b.

The equation should represent how the total number of books collected by Jordan and the bookstore are equal.

Apply the idea

Jordan donates 5 boxes of books. We can represent this algebraically as $5b$ (or 5 times the number of books in a box).

Jordan also donates 2 more individual books which we can show as 2. So the total number of books Jordan donates is $5b + 2$.

The bookstore donates 3 boxes of books. We can represent this algebraically as $3b$ (or 3 times the number of books in a box).

The bookstore also donates 12 more individual books which we can show as 12, for a total of $3b + 12$.

Since Jordan and the bookstore each donate the same number of books, we can set each donation amount equal to each other.

This gives us the equation $5b + 2 = 3b + 12$.

Purpose

Show students how to represent real-world situations involving equal quantities using algebra. This example challenges students to use variables to represent unknown quantities and form algebraic equations that accurately depict real-world scenarios.

Expected mistakes

Students might think there should be two variables in the equation; one for the box of books and one for the individual books. Point out that the number of individual books is not variable (or is constant), so only the box of books requires a variable.

Students: Page 82

b   Represent the equation using a pictorial model.

Create a strategy

Use an image of a box to represent the number of boxes and an image of a book to represent each book.

Apply the idea

On the left, draw 5 boxes to represent the boxes of books and a separate pile of books to represent the 2 individual books donated by Jordan. On the right, draw 3 boxes to represent the boxes of books and a separate pile of books to represent the 12 individual books donated by the bookstore.

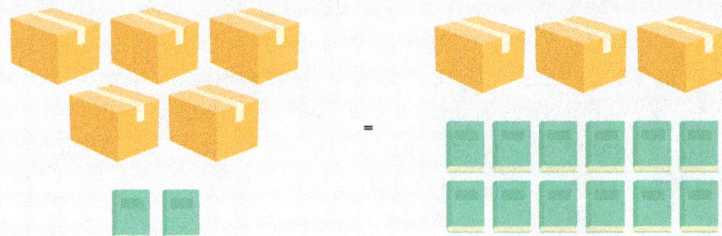

Purpose

Show students how to represent and solve an equation using a pictorial model.

Students: Page 82–84

c   Use the pictorial model to solve the equation.

Create a strategy

We need to remove the 2 books from the left side of the equation and the 3 boxes on the right side of the equation while still keeping it balanced. Then we need to isolate a single box on the left side of the equation.

Apply the idea

Pictorial model of the equation from part (b).

We first want to subtract 2 books from each side of the equation. Removing a book is represented by a grayed out book.

Now we can see that 5 boxes of books is equal to 3 boxes and 10 books. We can represent this as the equation $5b = 3b + 10$.

We want to subtract 3 boxes from each side of the equation. Removing a box is represented by a grayed out box.

Now we see that 2 boxes of books contain a total of 10 books. We can represent this as the equation $2b = 10$.

If we separate the books into two groups of the same size, each containing 5 books, we can find how many books will go in one box. Assuming the same number of books is in each box, one box contains 5 books. This relationship can be expressed with the equation $b = 5$.

Purpose

Show students how to use a pictorial model to solve real life problems that can be represented by equations.

Reflecting with students

Emphasize the importance of using appropriate units throughout the task. Remind students to clearly define b as the number of books per box and to include units in their answers. An imprecise response might be "$b = 5$", while a precise response would be "Each box contains 5 books."

Additionally, encourage students to verify their final solution for reasonableness by relating it back to the context of the problem. After finding that $b = 5$, prompt them to consider if it makes sense that each box contains 5 books. Have them substitute $b = 5$ back into the original equation to check if both sides are equal, confirming the solution is correct.

Advanced learners: Remove scaffolding to encourage independent problem-solving

use with Example 2

Targeted instructional strategies

For your advanced learners, consider removing the step-by-step scaffolding provided in the lesson and challenge students to solve the problem independently using a pictorial model. Present the scenario and ask students to use a visual representation to determine how many books are in each box.

Encourage students to think critically about how to depict the relationships between the quantities using an equation and a drawing or diagram. This approach allows students to apply their understanding of algebraic concepts and develop their own strategies for solving the equation. By engaging in this independent problem-solving activity, students can deepen their comprehension and enhance their ability to represent and solve complex problems visually.

Three reads

use with Example 2

English language learner support

Advise students to read through the instructions a few times, focusing on gathering different information each time in order to build up understanding of what the question is asking.

On the first read, students should aim to identify the scenario presented in the question. Ask students, "What do you think is happening in this question?" or "Can you explain what this question is about?"

On the second read, students should aim to interpret the problem by answering questions like, "What is the question asking you to find?" and "What information should be included in the answer?"

On the third read, students should look for important information in the instructions. In this question, the important information includes:

- Jordan has 5 boxes of books and 2 individual books.
- The local bookstore donates 3 boxes of books and 12 individual books.
- Jordan and the local bookstore donated the same number of books.

Students can be prompted by framing these as questions like "What might be the variable in this situation?" or "Which two quantities or expressions are equal?"

> 💡 **Idea summary**
>
> We can use visual models to represent and solve equations pictorially. We can use **zero pairs** and apply the same changes to both sides to maintain the balance or equality of the equation.

Solve equations algebraically

The lesson reminds students of the properties of equality and the properties of real numbers, explaining that inverse operations and these properties can be used to algebraically solve equations with variables on both sides.

Example 3

Solve the following equations:

a $5(2y + 2) + 3(4y - 5) + 5y = 45$

Create a strategy

Use the distributive property and properties of equality to remove the parentheses and isolate the variable.

Apply the idea

$5(2y + 2) + 3(4y - 5) + 5y = 45$	Given equation
$5 \cdot 2y + 5 \cdot 2 + 3 \cdot 4y - 3 \cdot 5 + 5y = 45$	Distributive property
$10y + 10 + 12y - 15 + 5y = 45$	Evaluate the multiplication
$10y + 12y + 5y + 10 - 15 = 45$	Associative property of addition
$27y - 5 = 45$	Combine like terms
$27y - 5 + 5 = 45 + 5$	Addition property of equality
$27y = 50$	Combine like terms
$\dfrac{27y}{27} = \dfrac{50}{27}$	Division property of equality
$y = \dfrac{50}{27}$	Evaluate the division

Reflect and check

We don't always show the steps for all of the properties we are using; notice the use of the additive inverse, multiplicative inverse, and multiplicative identity in between steps. For instance, when we have:

$5(2y + 2) + 3(4y - 5) + 5y = 45$	Given equation
$27y - 5 + 5 = 45 + 5$	Addition property of equality
$27y + 0 = 45 + 5$	Additive inverse
$27y = 45 + 5$	Additive identity
$27y = 50$	Evaluate the addition
$\dfrac{27y}{27} = \dfrac{50}{27}$	Division property of equality
$1 \cdot y = \dfrac{50}{27}$	Multiplicative inverse
$y = \dfrac{50}{27}$	Multiplicative identity

Purpose

Show students how to apply the distributive property and properties of equality to solve complex equations with multiple terms.

Reflecting with students

Remind students to check their answers by substituting them back into the original equation.

$5(2y + 2) + 3(4y - 5) + 5y = 45$ is a true statement, so substituting $y = \frac{50}{27}$ should result in 45 on both sides of the equation.

Students: Page 86

b $-\frac{3}{2}\left(x + \frac{5}{4}\right) + \frac{11}{3} = -\frac{125}{6}$

Create a strategy

Use the distributive property and properties of equality to remove the parentheses and isolate the variable.

Apply the idea

$-\frac{3}{2}\left(x + \frac{5}{4}\right) + \frac{11}{3} = -\frac{125}{6}$	Given equation
$-\frac{3}{2} \cdot x + -\frac{3}{2} \cdot \frac{5}{4} + \frac{11}{3} = -\frac{125}{6}$	Distributive property
$-\frac{3}{2}x - \frac{15}{8} + \frac{11}{3} = -\frac{125}{6}$	Evaluate the multiplication
$-\frac{3}{2}x + \frac{43}{24} = -\frac{125}{6}$	Combine like terms
$-\frac{3}{2}x + \frac{43}{24} - \frac{43}{24} = -\frac{125}{6} - \frac{43}{24}$	Subtraction property of equality
$-\frac{3}{2}x = -\frac{181}{8}$	Combine like terms
$-\frac{3}{2}x \cdot -\frac{2}{3} = -\frac{181}{8} \cdot -\frac{2}{3}$	Multiplication property of equality
$x = \frac{181}{12}$	Evaluate the multiplication

Reflect and check

We can check our work by substituting the solution, $x = \frac{181}{12}$, into the original equation and simplifying.

$-\frac{3}{2}\left(\frac{181}{12}\right) - \frac{15}{8} + \frac{11}{3} = -\frac{125}{6}$	Substitute $x = \frac{181}{12}$
$\frac{181}{8} - \frac{15}{8} + \frac{11}{3} = -\frac{125}{6}$	Evaluate the multiplication
$-\frac{125}{6} = -\frac{125}{6}$	Evaluate the subtraction and addition

Since the left and right side of the equation is equal, we can confirm our solution of $x = \frac{181}{12}$

Purpose

Demonstrate to students how to use the distributive property and properties of equality to solve a complex equation involving fractions and parentheses.

Reflecting with students

Teach students to solve the equation using an alternative method by using the multiplication property of equality. After applying the distributive property:

$$-\frac{3}{2}x - \frac{15}{8} + \frac{11}{3} = -\frac{125}{6}$$

Instruct students to multiply both sides of the equation by the lowest common multiple of the denominators.

$$24\left[-\frac{3}{2}x - \frac{15}{8} + \frac{11}{3}\right] = 24\left[-\frac{125}{6}\right]$$

$$-36x - 45 + 88 = -500$$

Students can then solve for x and simplify the result to find the solution.

c $9x + 40 - 7x = 4x$

Create a strategy

Rearrange the equation so that all of the x terms are on the same side.

Apply the idea

$9x + 40 - 7x = 4x$	Given equation
$9x + 40 - 7x - 4x = 4x - 4x$	Subtraction property of equality
$9x + 40 - 7x - 4x = 0$	Additive inverse
$-2x + 40 = 0$	Combine like terms
$-2x + 40 - 40 = 0 - 40$	Subtraction property of equality
$-2x = -40$	Combine like terms
$\dfrac{-2x}{-2} = \dfrac{-40}{-2}$	Division property of equality
$x = 20$	Evaluate the division

Reflect and check

Another way we could have solved is by moving the variable terms to the right side of the equation:

$9x + 40 - 7x = 4x$	Given equation
$9x + 40 - 7x + 7x = 4x + 7x$	Addition property of equality
$9x + 40 = 11x$	Combine like terms
$9x - 9x + 40 = 11x - 9x$	Subtraction property of equality
$40 = 2x$	Combine like terms
$\dfrac{40}{2} = \dfrac{2x}{2}$	Division property of equality
$20 = x$	Evaluate the division
$x = 20$	Symmetric property of equality

Even though we used different steps, we still relied on the properties of equality and real numbers and ended up with $x = 20$.

Purpose

Show students how to approach solving a linear equation with variables on both sides by using the properties of equality to isolate the variable.

Students: Page 87

d $12.6x + 28.25 = 1.1x + 85.75$

Create a strategy

The coefficent for x will be positive if we rearrange the equation by moving the x values to the left side since $12.6 > 1.1$.

Apply the idea

$12.6x + 28.25 = 1.1x + 85.75$	Given equation
$12.6x - 1.1x + 28.25 = 1.1x - 1.1x + 85.75$	Subtraction property of equality
$11.5x + 28.25 = 85.75$	Combine like terms
$11.5x + 28.25 - 28.25 = 85.75 - 28.25$	Subtraction property of equality
$11.5x = 57.5$	Combine like terms
$\dfrac{11.5x}{11.5} = \dfrac{57.5}{11.5}$	Division property of equality
$x = 5$	Evaluate the division

Purpose

Show students how to solve an algebraic equation containing rational numbers and with variables on both sides.

Students: Page 87–88

e $2(x - 3) = x - 3$.

Create a strategy

Use the distributive property first and then modify the equation so that all of the x terms are on the same side.

Apply the idea

$2(x - 3) = x - 3$	Given equation
$2 \cdot x - 2 \cdot 3 = x - 3$	Distributive property
$2x - 6 = x - 3$	Evaluate the multiplication
$2x - 6 + 6 = x - 3 + 6$	Addition property of equality
$2x = x - 3 + 6$	Additive inverse
$2x = x + 3$	Evaluate the subtraction
$2x - x = x - x + 3$	Subtraction property of equality
$2x - x = 3$	Additive inverse
$x = 3$	Combine like terms

Reflect and check

Substitute the solution, $x = 3$, back into the original equation to verify its correctness.

$2(3 - 3) = 3 - 3$	Substitute $x = 3$
$6 - 6 = 3 - 3$	Evaluate the multipication
$0 = 0$	Simplify

Since the left and right side of the equation is equal, we can confirm our solution of $x = 3$

Purpose

Show students how to solve equations with parentheses and variables on both sides by applying the distributive property and the principles of equality.

Expected mistakes

Students might think there is no solution, especially if they try to divide both sides by $x - 3$. Make them aware that dividing by $x - 3$ is only a valid algebraic operation if $x \neq 3$. Students can then substitute $x = 3$ to see that this makes the equation true.

Develop algorithmic thinking with multistep equations
Targeted instructional strategies

use with Example 3

Encourage students to think algorithmically by having them work in pairs or small groups to create a set of steps that could be used to solve any multistep equations algebraically. A possible set of questions could include:

- $5(2y + 2) + 3(4y - 5) + 5y = 45$, example 3a
- $-\dfrac{3}{2}\left(x + \dfrac{5}{4}\right) + \dfrac{11}{3} = -\dfrac{125}{6}$, example 3b
- $9x + 40 - 7x = 4x$, example 3c
- $12.6x + 28.25 = 1.1x + 85.75$, example 3d
- $2(x - 3) = 2x - 3$, a variation of example 3e

Have one student work on solving equations and have another student take notes, asking questions about the other student's thinking process. Possible questions include:

- Why did you do that operation first?
- What is the purpose of that step?
- How can you be sure that is the correct answer?
- Could you have solved that in fewer steps?

A possible algorithm students come up with could be:

1. Distribute any parentheses
2. Combine like terms on the same side of the equation
3. Get the terms with variables on the same side of the equation
4. Add or subtract constants so they are on the opposite side of the equation as the variable term
5. Multiply or divide to isolate the variable
6. State the solution, that there is no solution, or an infinite number of solutions.
7. Verify by substituting into the original equation

Scaffold by providing the reasoning for each step in the solution
Student with disabilities support

use with Example 3

For students who have difficulty remembering the procedure for solving equations and organizing their work, provide the reasoning for each step as a scaffold. For example:

$5(2y + 2) + 3(4y - 5) + 5y = 45$	Given equation
$\square = \square$	Apply the distributive property to remove the parentheses
$\square = 45$	Evaluate the multiplication
$\square = 45$	Group together like terms on the left side
$\square = 45$	Combine like terms on the left side
$\square = \square$	Add the constant term to both sides
$\square = \square$	Divide both sides by the coefficient of y

Encourage students to fill in the blanks with the appropriate expressions and calculations. This scaffolded approach supports organization by providing a clear framework for each step, helping students focus on one operation at a time. It reduces cognitive load on memory and attention by explicitly stating the reasoning behind each step, which can be particularly beneficial for students who struggle with processing multistep procedures. By using this template, students can systematically work through complex equations and build confidence in their problem-solving abilities.

2.04 Solve multistep equations   149
mathspace.co

> 💡 **Idea summary**
>
> When solving multistep equations involving parentheses:
> - Use the distributive property to clear parentheses.
> - Combine like terms to simplify each side.
> - Add or subtract to get the variable term on one side of the equals sign.
> - Multiply or divide to isolate the variable and solve the equation.
>
> To solve an equation with variables on both sides, we can rearrange the equation so that all of the variables are on the same side of the equation and all of the constants are on the other side of the equation.

Practice

Students: Page 88–92

What do you remember?

1 Simplify the expression using the distributive property:

a $4(p - 8)$ **b** $\frac{3}{4}(n + 12)$

2 Match the properties with the statements that demonstrate them.

a	Associative property	**i**	$4.8(3.1 + 7.2) = 4.8(3.1) + 4.8(7.2)$
b	Commutative property	**ii**	$-9.5 \cdot (1.2 \cdot 3.5) = (-9.5 \cdot 1.2) \cdot 3.5$
c	Inverse property	**iii**	$17.6 + (-17.6) = 0$
d	Identity property	**iv**	$155.9 + 10.44 = 10.44 + 155.9$
e	Distributive property	**v**	$5.66 + 0 = 5.66$

3 To move all the variables to the left hand side of the equation, what should be done to both sides of the equation $9x = 7 + 8x$?

4 Consider the equation $1.5(n + 0.5) = 2$ for $n = 0.5$.

     a Find the value of the left-hand side of the equation when $n = 0.5$.

     b Consider your answers from parts (a) and (b).

     c Is $n = 0.5$ a solution of $1.5(n + 0.5) = 2$?

5 Consider the equation $9(n + 5) = 4n + 50$ for $n = 1$.

     a Find the value of the left-hand side of the equation when $n = 1$.

     b Find the value of the right-hand side of the equation when $n = 1$.

     c Consider your answers from parts (a) and (b).

        Is $n = 1$ a solution of $9(n + 5) = 4n + 50$?

6 Determine whether the given value of the variable is a solution to the equation.

     a $4x - 2 = 18$ where $x = 5$ **b** $5x + 12 = 31$ where $x = 4$

     c $3p - 9 = 7p - 1$ where $p = 0$ **d** $-\frac{x}{24} + 7 = 5x$ where $x = -3$

7 Using n to represent the unknown number, form an equation that represents the relations.

    a Three more than three times a number is equal to six more than four times the number.

    b Six more than two times a number is equal to five more than two times the number.

    c Two more than four times a number is equal to two added to quadruple the number.

Let's practice

8 For $3(x + 1) - 4 = 5$:

    a Represent the equation using algebra tiles.

    b Use the algebra tiles to solve the equation.

9 For $2(3x - 2) = 6 + x$:

    a Represent the equation using algebra tiles.

    b Use the algebra tiles to solve the equation.

10 Haruko is collecting postcards and currently has four packs, each containing (p) cards. She buys 4 more postcards, and her brother gives her another 2 packs. Upon counting, she now has 22 postcards in total.

    a Draw a pictorial model of the equation.

    b Use the pictorial model to find how many postcards are in each pack.

11 Trechelle is participating in a toy drive with her local community center. She and her friend Graciela have collected the same number of stuffed animal toys for the drive. Trechelle has filled 7 boxes and has another 3 toys that will not fit in the boxes. Graciela has filled 2 boxes and has 13 toys that will not fit in the boxes.

If each box holds the same number of toys, the equation to represent this situation is:

$$7b + 3 = 2b + 13$$

    a Represent the equation using pictorial representation.

    b Use the pictorial representation to find how many toys are in each box.

12 For each of the following statements:

    i Write the statement as an equation in which x represents the number.

    ii Solve the equation for x, justifying each step using properties of equality.

    a 5 times a number increased by ten is the product of three and that same number plus 12.

    b A number is decreased by 5. The result of that is then doubled and equals 5 times that same number increased by 23.

    c One third of a number increased by 5 is the product of 6 and that number.

    d Twice a number increased by 5 and then 2 is three times that number decreased by 8.

13 Solve for the variable:

    a $-3(4 - m) = -12$ **b** $0.7(1.2 + 0.4g) = 2.1$

    c $\frac{1}{2}(8c + 6) = 5$ **d** $6.4 = -4.2(1.7 + 3.4k)$

    e $-5.26(g + 4.32) + 9.78 = -42.67$ **f** $-\frac{3}{8}\left(n + \frac{7}{12}\right) + \frac{2}{3} = -\frac{5}{12}$

    g $-2\frac{6}{7}\left(t - 2\frac{3}{4}\right) - 7\frac{5}{7} = -32\frac{9}{11}$ **h** $40 = 10\left(\frac{f}{6} - 1\right)$

    i $-4\left(12 - \frac{u}{3}\right) = -64$ **j** $\frac{1}{4}\left(\frac{d}{3} - 9\right) = 5$

    k $-3\left(\frac{2}{3}j + 5\right) = 12$ **l** $-7 = -2\left(\frac{3b}{5} + 11\right)$

14 Solve for the variable:

    a $5c - 3 = -4c + 33$ **b** $10b + 8 = -8b - 2$

    c $\dfrac{1}{4}y = 4 + \dfrac{1}{2}y$ **d** $\dfrac{2m}{3} = \dfrac{m}{2} - \dfrac{1}{4}$

    e $2.8r + 3.9 = 9r - 7.2$ **f** $5(2x + 5) = 2x + 13$

    g $4(3s - 2) = 6s + 16$ **h** $0.15 - 0.2w = 0.3(w + 3)$

    i $\dfrac{2}{5}(r - 2) = \dfrac{3}{20}(r - 4)$ **j** $0.4(0.3y - 1) = 2 + 0.75y$

    k $0.4a + 7.2 - 1.8a + 4.1 = 9.7 - 0.8a + 5.2$ **l** $5(2m + 1) - 3(3m - 1) = 2(x + 1)$

15 Determine whether the given value of the variable is a solution to the equation.

    a $p = 0$;   **b** $x = 2$;

        $3p - 9 + 4p = 7$ $3(x - 4) + 2(3x - 5) = -4$

SOL **16**   What is the solution to $\dfrac{m + 6}{3} - 4 = 10$?

    A   6   **B**   11   **C**   14   **D**   36

17 Consider the equation $\dfrac{6x + 90}{5} = 36$.

    a   Velma started to solve the equation as follows:

$$\frac{6x + 90}{5} = 36$$

$$6x + 90 = 180$$

        Determine the property that justifies Velma's first step of work.

    b   Velma continued to solve the equation as shown:

$$6x + 90 = 180$$

$$6x = 90$$

        Determine the property that justifies this step of work.

    c   Velma finished solving the equation as follows:

$$6x = 90$$

$$x = 15$$

        Determine the property that justifies Velma's final step of work.

18 Solve for the variable:

    a $\dfrac{3x + 6}{2} = 15$ **b** $\dfrac{x + 2}{3} - 25 = -22$ **c** $\dfrac{-3(x - 3)}{4} = 0$ **d** $\dfrac{9(x + 10)}{2} = -6$

    e $\dfrac{4x - 40}{8} = 3x$ **f** $\dfrac{4x + 84}{8} = -3x$ **g** $\dfrac{-6x + 54}{4} = 3x$ **h** $\dfrac{3x + 6}{7} = x + 2$

19 Ines wants to put a fence around her rectangular yard. The length of her yard is 5 feet more than the width. The entire perimeter of her yard is 85 feet.

     Ines set up an equation, solved it and got a solution of –12. Without solving the equation yourself, explain if –12 is a reasonable solution.

20 Richie attempted to solve the equation

$9(4x - 6) + 6(x + 4) - 4 = 18.$

His work is shown:

a What was his mistake?

b Solve the equation correctly.

$$9(4x - 6) + 6(x + 4) - 4 = 18$$

$$36x - 54 + 6x + 24 - 4 = 18$$

$$30x - 54 = 18$$

$$30x = 18 + 54$$

$$30x = 72$$

$$x = \frac{12}{5}$$

21 Solve for the variable:

a $4(5x + 1) + 3(5x - 5) = 0$

b $28 = 7(7 + x) - 5(3x + 6) - 2x$

c $5(2x - 9) + \frac{1}{4}(-16x - 8) = 1$

d $5.5y - 3.2(2.5y - 4.6) = 2.3$

22 For each of the following equations:

i Solve each equation.

ii Justify the steps using properties of equality and real numbers.

iii Verify the solution.

a $24 - 6x = 14 + 4x$ **b** $16 + 4x = 12x$ **c** $3(n - 3) = 2(n + 6)$ **d** $5 + 30x = 4(3 + 4x)$

23 Rob owns a coffee shop and is looking at finding a new coffee distributor for his beans. Distributor A sells their beans for $5 a pound, plus a flat $10 shipping fee. Distributor B sells their beans for $2 a pound, plus $1 per pound for shipping, plus a $40 processing fee.

a Write an equation to represent what amount of pounds, p, would be a breakeven point for the two companies.

b Solve your equation.

c What is the meaning of the solution?

24 Create a verbal situation in context for each of the equations:

a $5x - 14 = 2x + 3$ **b** $\frac{4x + 7}{4} = 3$ **c** $2(x - 12) = 9 + 4x$

Let's extend our thinking

25 The length of a rectangle is $2x + 5$ and the width is x. The perimeter of the rectangle is 22 which is equal to twice the sum of the length and the width.:

a Write an equation for the perimeter of the rectangle in terms of x.

b Find the value of x.

c Interpret your solution in the context of the given scenario.

26 A number is multiplied by 5 and then 2 is added. Then the result is multiplied by 6. This is equal to 10 times the number minus 8.

a Write an equation for this problem.   **b** Solve the equation to find the number.

27 A rectangle with a height of $3x + 7$ ft and a width of $4x$ ft has the same perimeter as a square with side length $2x + 9$ ft.

Find the value of x.

28 Complete the equation so that $x = -7$ is the only solution.

$2(x - 12) + 9 = 3x - \square$

29 Hyun is a college student planning to buy his dream bike, which costs \$3500. To earn enough money for the bike, he has taken a part-time job at a local store that pays \$20 per hour. He started working on January 15 and plans to earn enough before spring break begins on March 20. Use the graph to determine how many hours he will have to work to earn enough to afford his dream bike.

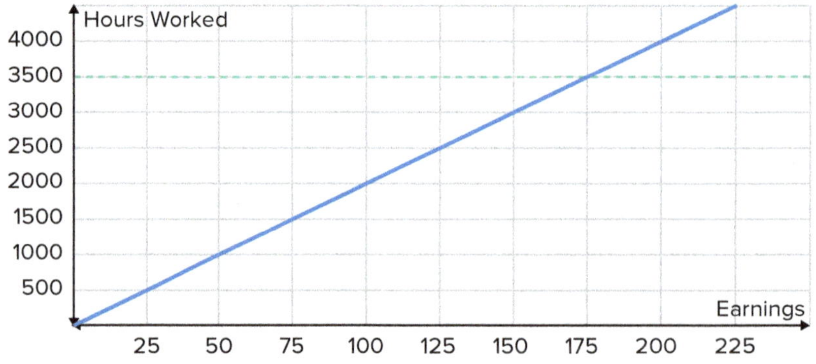

30 Suppose the Allmans want to rent a convertible for the day. They have a choice of two rental companies.

- A one-day rental at Nifty Car Rental costs \$30 plus 60 cents per mile.
- A one-day rental at Shazam Car Rental costs \$55 but only charges 35 cents per mile.

a If the Allmans drive 225 miles, which company would be a better deal? Explain.

b If the Allmans only have \$80 to spend, which company would be a better deal? Explain.

c At what number of miles will the two companies cost the same?

d Which car rental company should the Allman's choose and why?

31 Solve the equation $13(2x - 8) = -26$ in two different ways.

a Method 1: start by using the distributive property.

b Method 2: start by dividing both sides by 13.

c Was one method easier than the other? Explain your answer.

Answers

2.04 Solve multistep equations

What do you remember?

1   **a** $4p - 32$ **b** $\frac{3}{4}n + 9$

2   **a**   ii. $-9.5 \cdot (1.2 \cdot 3.5) = (-9.5 \cdot 1.2) \cdot 3.5$

     b   iv. $155.9 + 10.44 = 10.44 + 155.9$

     c   iii. $17.6 + (-17.6) = 0$

     d   v. $5.66 + 0 = 5.66$

     e   i. $4.8\,(3.1 + 7.2) = 4.8\,(3.1) + 4.8\,(7.2)$

3   Subtract $8x$ or add $-8x$

4   **a**   1.5   **b**   No

5   **a**   54   **b**   54   **c**   Yes

6   **a**   Yes   **b**   No   **c**   No   **d**   No

7   **a** $3n + 3 = 4n + 6$ **b** $2n + 6 = 2n + 5$

     c $4n + 2 = 4n + 2$

Let's practice

8   **a**

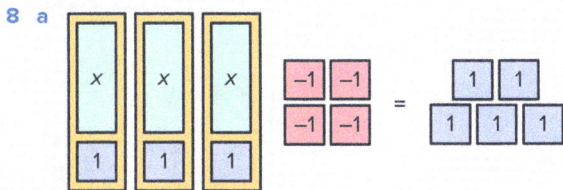

     b $x = 2$

9   **a**

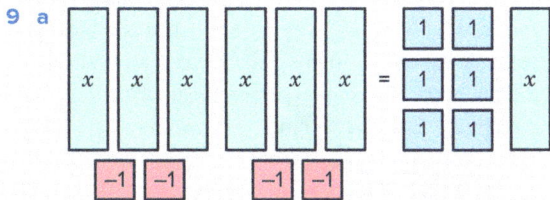

     b $x = 2$

10   **a**

     b $p = 3$ postcards

11   **a**

     b $b = 2$ toys

---

12   **a**   **i** $5x + 10 = 3\,(x + 12)$

ii	
$5x + 10 = 3\,(x + 12)$	Given equation
$5x + 10 = 3x + 36$	Distributive property of equality
$5x + 10 - 10 = 3x + 36 - 10$	Subtraction property of equality
$5x - 3x = 3x + 26 - 3x$	Subtraction property of equality
$\dfrac{2x}{2} = \dfrac{26}{2}$	Division property of equality
$x = 13$	Solution

   b   **i** $2(x - 5) = 5x + 23$

ii	
$2(x - 5) = 5x + 23$	Given equation
$2x - 10 = 5x + 23$	Distributive property of equality
$2x - 10 + 10 = 5x + 23 + 10$	Addition property of equality
$2x - 5x = 5x + 33 - 5x$	Subtraction property of equality
$\dfrac{-3x}{-3} = \dfrac{33}{-3}$	Division property of equality
$x = -11$	Solution

   c   **i** $\frac{1}{3}x + 5 = 6x$

ii	
$\frac{1}{3}x + 5 = 6x$	Given equation
$3 \cdot \left(\frac{1}{3}x + 5\right) = (6x) \cdot 3$	Multiplication property of equality
$x + 15 - x = 18x - x$	Subtraction property of equality
$\dfrac{15}{17} = \dfrac{17x}{17}$	Division property of equality
$\dfrac{15}{17} = x$	Solution

   d   **i** $2x + 5 + 2 = 3(x - 8)$

ii	
$2x + 5 + 2 = 3(x - 8)$	Given equation
$2x + 5 + 2 = 3x - 24$	Distributive property of equality
$2x + 5 + 2 + 24 = 3x - 24 + 24$	Addition property of equality
$2x + 31 - 2x = 3x - 2x$	Subtraction property of equality
$31 = x$	Solution

13   **a** $m = 0$ **b** $g = 4.5$ **c** $c = \frac{1}{2}$ **d** $k = -0.95$

     e $g = 5.65$ **f** $n = \frac{83}{36}$ **g** $t = 11\frac{59}{110}$ **h** $f = 30$

     i $u = -12$ **j** $d = 87$ **k** $j = -13.5$ **l** $b = -12.5$

14 a $c = 4$ **b** $b = -\dfrac{5}{9}$ **c** $y = -16$

   d $m = -1.5$ **e** $r = 1.79$ **f** $x = -\dfrac{3}{2}$

   g $s = 4$ **h** $w = -1.5$ **i** $r = \dfrac{4}{5}$ or $r = 0.8$

   j $y \approx -3.81$ **k** $a = -6$ **l** $m = 6$

15 a No   **b** Yes

16 D

17 a Multiplication property of equality

   b Addition property of equality or Subtraction Property of equality

   c Multiplication property of equality or Division property of equality

18 a $x = 8$ **b** $x = 7$ **c** $x = 3$ **d** $x = -\dfrac{34}{5}$

   e $x = -2$ **f** $x = -3$ **g** $x = 3$ **h** $x = -2$

19 A negative solution such as −12 for the width is not a reasonable solution in this context, as it would not make sense in for the yard to have a negative measurement.

20 a In combining the variables on the left side of the equation, he mistakenly subtracted $6x$ from $36x$ instead of adding $6x$ and $36x$.

   b $x = \dfrac{26}{21}$

21 a $x = \dfrac{11}{35}$ **b** $x = -0.9$ **c** $x = 8$ **d** $y = 4.97$

22 a i $x = 1$

ii		
$24 - 6x - 4x = 14 + 4x - 4x$	Subtraction property of equality
$24 - 10x = 14$	Simplification
$24 - 24 - 10x = 14 - 24$	Subtraction property of equality
$-10x = -10$	Simplification
$x = 1$	Division property of equality

iii	
$24 - 6(1) = 14 + 4(1)$	Substitute $x = 1$
$18 = 18$	Evaluate

   b i $x = 1$

ii	
$16 + 4x - 4x = 12x - 4x$	Subtraction property of equality
$16 = 8x$	Simplification
$2 = x$	Division property of equality

iii	
$16 + 4(2) = 12(2)$	Substitute $x = 2$
$24 = 24$	Evaluate

c i $n = 21$

ii	
$3n - 9 = 2n + 12$	Distributive property of multiplication
$3n - 2n - 9 = 2n - 2n + 12$	Subtraction property of equality
$n - 9 = 12$	Simplification
$n - 9 + 9 = 12 + 9$	Addition property of equality
$n = 21$	Simplification

iii	
$3(21 - 3) = 2(21 + 6)$	Substitute $n = 21$
$3(18) = 2(27)$	Evaluate
$54 = 54$	Evaluate

d i $n = 21$

ii	
$5 + 30x = 12 + 16x$	Distributive property of multiplication
$5 + 30x - 16x = 12 + 16x - 16x$	Subtraction property of equality
$5 + 14x = 12$	Simplification
$5 - 5 + 14x = 12 - 5$	Subtraction property of equality
$14x = 7$	Simplification
$x = \dfrac{7}{14} = 0.5$	Division property of equality

iii	
$5 + 30(0.5) = 4[3 + 4(0.5)]$	Substitute $x = 0.5$
$5 + 15 = 4(5)$	Evaluate
$20 = 20$	Evaluate

23 a let p be the number of pounds of beans sold:
$$5p + 10 = 2p + p + 40$$

   b $p = 15$

   c The breakeven point is 15 pounds.

24 a Imagine you're organizing a fundraiser bake sale at your school. You start with 5 trays, each containing (x) cookies. However, you noticed 14 cookies that are slightly burnt and remove them. Meanwhile, you also prepared 2 trays containing (x) brownies and added 3 more brownies to make them more appealing. After sorting out the trays, you realize that the total number of cookies you have is the same as the total number of brownies.

   b You have 4 boxes (x) of toys and plan to donate them equally among 4 animal shelters. You add 7 more toys to ensure you have enough. After organizing the toys, you found out that each animal shelter would get 3 toys.

   c You're planning to bake cupcakes for 2 of your friends. Each friend will receive (x) cupcakes you'll be able to make within a day, less 12 cupcakes you'll keep for yourself. You plan to add edible beads on the total number of cupcakes your friends would receive, but you only have 9, and you need to buy 4 more based on the number of (x) cupcakes you'll make.

Let's extend our thinking

25 a Example answer: $22 = 2(2x + 5 + x)$ or $2(2x + 5 + x) = 22$

b $x = 2$

c The width of the rectangle is 2 units. The length of the rectangle is $2x + 5$, so when $x = 2$, the length becomes $2(2) + 5 = 9$ units. Therefore, the rectangle has a length of 9 units and a width of 2 units.

26 a $6(5x + 2) = 10x - 8$ **b** -1

27 $x = \dfrac{11}{3}$

28 $2(x - 12) + 9 = 3x - 8$

29 Hyun needs to work for 175 hours to earn $3500 and afford his dream bike.

30 a Shazam Car Rental would be a better deal because it would cost $133.75 compared to Nifty Car Rental's $165 for 225 miles.

b With an 80 budget, Nifty Car Rental allows them to drive approximately 83.33 miles, while Shazam Car Rental offers about 71.43 miles. Therefore, Nifty Car Rental offers more mileage for their budget and is the better deal.

c The two companies will cost the same when the Allmans drive exactly 100 miles.

d The choice depends on their needs: If they plan to drive more than 100 miles, Shazam Car Rental is more cost-effective. If they have a strict budget and will travel less than 100 miles and want to maximize how far they can go, Nifty Car Rental is the better option due to the higher mileage it offers within that budget.

31 a $13(2x - 8) = -26$

$26x - 104 = -26$

$26x = 78$

$x = 3$

b $13(2x - 8) = -26$

$2x - 8 = -2$

$2x = 6$

$x = 3$

c Answers will vary

Sample answer 1: Method 1 is easier because it uses less division.

Sample answer 2: Method 2 is easier because it makes the numbers smaller.

2.05 Write and represent multistep inequalities

Subtopic overview

Lesson narrative

In this lesson, students will learn to write and represent multistep inequalities. They will explore how to model real-world situations with inequalities, identify key phrases indicating inequality symbols, and solve for variable values that satisfy the inequalities. The lesson includes practice problems where students write inequalities from verbal descriptions, solve them, and graph the solution sets on number lines. By the end of the lesson, students should be able to construct and interpret multistep inequalities and represent their solutions graphically.

Learning objectives

Students: Page 93

> 🎓 **After this lesson, you will be able to...**
> - write multistep linear inequalities in one variable to represent a verbal or contextual situation.
> - create a verbal situation in context given a multistep linear inequality in one variable.
> - identify whether a numerical value is part of the solution set of an inequality.

Key vocabulary

- linear inequality
- solution set

- non-viable solution
- viable solution

- solution (to an inequality)

Essential understanding

An inequality is a mathematical sentence that compares two expressions and uses the symbols $>$, $<$, $\leq$, or $\geq$. The value(s) of the variables that make an equation or inequality true make up its solution set.

Standards

This subtopic addresses the following Virginia 2023 Mathematics Standards of Learning standards.

Mathematical process goals

MPG2 — Mathematical Communication

Teachers can emphasize mathematical communication by instructing students on the specific vocabulary and phrases used in writing and representing inequalities. They can teach students how to use words such as 'greater than', 'less than', 'at most', 'at least', and how these terms translate into inequality symbols. Teachers can also encourage students to explain their thought process, either verbally or in writing, when translating real-world problems into multi-step inequalities. This way, students can effectively communicate their mathematical understanding and reasoning.

MPG4 — Mathematical Connections

Teachers can make mathematical connections by relating the concept of writing and representing multi-step inequalities to real-world situations. By using real-world examples that can be represented by multi-step inequalities, students can see how math is used in everyday life and other disciplines.

Content standards

8.PFA.5 — The student will write and solve multistep linear inequalities in one variable, including problems in context that require the solution of a multistep linear inequality in one variable.

8.PFA.5b — Represent solutions to inequalities algebraically and graphically using a number line.

8.PFA.5c — Write multistep linear inequalities in one variable to represent a verbal situation, including those in context.

8.PFA.5d — Create a verbal situation in context given a multistep linear inequality in one variable.

8.PFA.5f — Identify a numerical value(s) that is part of the solution set of a given inequality.

Prior connections

7.PFA.4 — The student will write and solve one- and two-step linear inequalities in one variable, including problems in context, that require the solution of a one- and two-step linear inequality in one variable.

8.PFA.4 — The student will write and solve multistep linear equations in one variable, including problems in context that require the solution of a multistep linear equation in one variable.

Future connections

A.EI.1 — The student will represent, solve, explain, and interpret the solution to multistep linear equations and inequalities in one variable and literal equations for a specified variable.

A.EI.2 — The student will represent, solve, explain, and interpret the solution to a system of two linear equations, a linear inequality in two variables, or a system of two linear inequalities in two variables.

A2.EI.2 — The student will represent, solve, and interpret the solution to quadratic equations in one variable over the set of complex numbers and solve quadratic inequalities in one variable.

Lesson Preparation

Suggested review

Depending on your students' level of prior knowledge, consider revisiting the following lessons:

Grade 6 — 6.05 Write inequality statements

Grade 7 — 4.04 Write and represent inequalities

Grade 7 — 4.07 Solve problems with inequalities

Student lesson & teacher guide

Write and represent multistep inequalities

This lesson reviews writing inequalities, graphing inequalities on a number line, and identifying solutions to inequalities from previous grades. Students are introduced to the terms "linear inequality," "viable solutions," and "non-viable solutions." The lesson summarizes specific keywords and phrases that indicate which inequality symbol should be used to represent a contextual statement or scenario.

Examples

Students: Page 94

Example 1

Consider the inequality:

$$k \geq -25$$

a Represent the inequality on a number line.

Create a strategy

To represent the solution to the inequality on a number line, first, identify the critical value, which is –25. Then, decide how to show that all values equal to and greater than this are included in the solution set.

Apply the idea

Draw a horizontal line to represent the number line. Mark the point –25 on this line.

Use a closed (filled) circle to indicate that –25 is included in the solution set (since the inequality includes equal to), and draw an arrow extending to the right from –25, showing that all values greater than –25 are included.

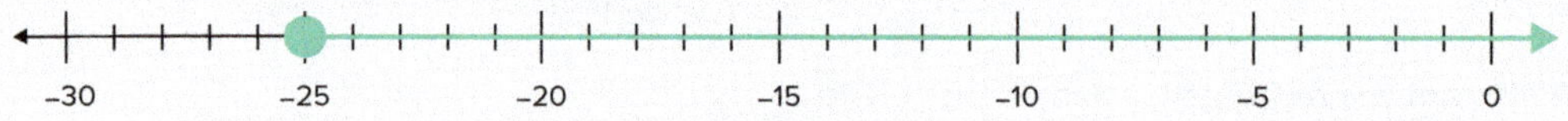

Purpose

Show students how to represent the solution of an inequality on a number line, highlighting the importance of accurately indicating the critical value and the direction of the solution set.

Expected mistakes

Students may neglect to fill in the circle at −25 . Remind students that the ≥ symbol is "greater than *or equal to*" which corresponds to a filled or closed circle. If the variable cannot be equal to the value (strictly greater than), an open or unfilled circle is used.

Students: Page 95

b Name 3 values that are in the solution set.

Create a strategy

Since the inequality $k \geq -25$ includes all values greater than or equal to −25, we can choose any three values greater than or equal to this. We can use the number line from part (a) to help us.

Apply the idea

Some of the values in the solution set are −25, 0, and 10.

We can see that they are all included in the shaded section of the number line:

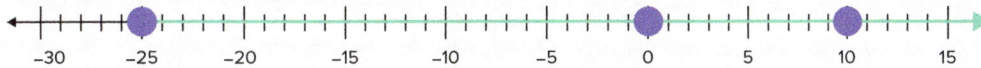

We can also see that when substituted for k in $k \geq -25$ they make the inequality true.

$$-25 \geq -25 \qquad \text{Reads '−25 is greater than or equal to −25'}$$
$$0 \geq -25 \qquad \text{Reads '0 is greater than or equal to −25'}$$
$$10 \geq -25 \qquad \text{Reads '10 is greater than or equal to −25'}$$

Purpose

Show students that they can determine values that satisfy an inequality by choosing any values that are within the range of the solution set, and verifying these values by substituting them back into the inequality.

Ray direction and inquality symbol
Student with disabilities support

use with Example 1

Associate the inequality sign to the direction of the ray on the number line. When the variable is on the left side and the number is on the right side of the inequality sign, such as $k \geq -25$, the inequality symbol ≥ looks like the arrowhead of a ray that points to the right direction.

The graph of $k \geq -25$ is:

The graph of $k \leq -25$ will have a ray with an arrowhead pointing towards the left direction:

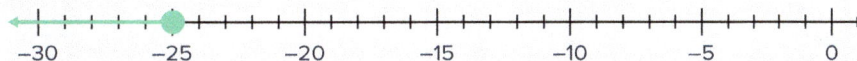

Students can use this method to determine the shaded region, then check the reasonableness of their solution region by substituting values on that side of the number line to ensure they satisfy the inequality.

Example 2

Sasha is drawing a pentagon-shaped house design and wants to make sure the combined length of two sides representing the roof are at least twice as long as the sum of the remaining sides.

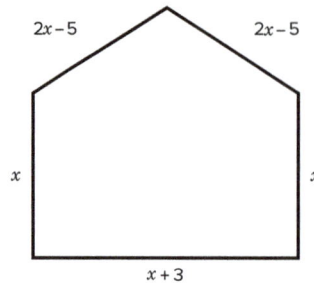

a Write two different inequalities that Sasha could use to represent this situation.

Create a strategy

The two expressions representing the roof are each $2x - 5$, the word *twice* will multiply their sum by 2, the remaining three expressions will be added, and *at least* represents ≥. Possible inequalities can show different levels of simplification or rearranging of variables.

Apply the idea

Possible inequalities representing the situation could include:

- $2(2x - 5) + 2(2x - 5) \geq x + x + (x + 3)$
- $2(2x - 5 + 2x - 5) \geq 2x + x + 3$
- $2(4x - 10) \geq 3x + 3$
- $x + x + x + 3 \leq 2(4x - 10)$

There are many other possibilities, and these represent just a few.

Purpose

Show students how to write an inequality that represents a real-world situation.

Reflecting with students

Challenge advanced learners to write 5 different inequalities, where one of the inequalities represents the solution, $x \geq \dfrac{23}{5}$.

b Is 4 a possible value for x?

Create a strategy

Choose one of the inequalities written in part (a) and substitute $x = 4$.

Apply the idea

$$2(2x - 5) + 2(2x - 5) \geq x + x + (x + 3) \qquad \text{Write the inequality}$$
$$2((2 \cdot 4) - 5) + 2((2 \cdot 4) - 5) \geq 4 + 4 + (4 + 3) \qquad \text{Substitute } x = 4$$
$$12 \geq 15 \qquad \text{Evaluate}$$

Since 12 is not greater than or equal to 15, $x = 4$ does not make the inequality true. Thus, $x = 4$ is not a possible value for x to satisfy the given inequality.

Purpose

Ensure students can test a given value in an inequality to determine whether it's a possible solution.

Three reads
English language learner support

Advise students to read through the instructions a few times, focusing on gathering different information each time in order to build up understanding of what the question is asking.

On the first read, students should aim to identify the scenario presented in the question. Ask students, "What do you think is happening in this question?" or "Can you explain what this question is about?"

On the second read, students should aim to interpret the problem by answering questions like, "What is the question asking you to find?" and "What information should be included in the answer?"

On the third read, students should look for important information in the instructions. In this question, the important information includes:

- The two sides representing the roof are $2x - 5$ and $2x - 5$, so the combined length is $2x - 5 + 2x - 5$.
- Twice the combined length of the sides of the roof is $2(2x - 5 + 2x - 5)$.
- The sum of the remaining sides is $x + x + x + 3$.
- "At least" means twice the combined length of the sides of the roof is greater than or equal to the sum of the remaining sides.

Students can be prompted by framing these as questions like "What are the two sides of the roof?" or "Which sides are the 'remaining sides'?"

Students: Page 96

Example 3

Calandra charges \$37.72 to style hair, as well as an additional \$6 per foil. Pauline would like the total cost for her styling to be no more than \$95.86.

a   Write an inequality that represents the number of foils Pauline could get.

Create a strategy

Pauline has *no more than* \$95.86 to spend. "No more than" means "less than or equal to."

Apply the idea

We can write an inequality in words that represents the cost to style Pauline's hair:

$$\text{Cost of styling} + \text{Cost per foil} \cdot \text{Number of foils} \le \text{Total Pauline can spend}$$

Translating that into an algebraic expression we get: $37.72 + 6N \le 95.86$ where N represents the number of foils.

Purpose
Demonstrate to students how to translate a real-life situation involving cost and budget into an algebraic inequality.

b The solution for the inequality is $N \leq 9.69$. Determine whether $N = -2$ is a viable solution to the inequality in the context of the question.

Create a strategy

Keep in mind it is not realistic to get part of a foil or a negative number of foils.

Apply the idea

Pauline can get a maximum of 9 foils and a minimum of 0 foils, so while –2 is mathematically part of the solution set for the inequality $N \leq 9.69$, it is not a viable solution in this context.

Reflect and check

Unlike a value that is not in a solution set of an inequality, this is an example of a solution that was mathematically valid and part of the original solution set, but when considering the context, we have found that it is non-viable.

Purpose

Show students that it is important to consider the context of a problem when interpreting the solution to an inequality. Even though a value may be mathematically part of the solution set, it may not be a viable solution in the real-world context of the problem.

Expected mistakes

Students might say it is viable since it satisfies the inequality, not considering the context. Remind students that, when an inequality represents a real-world situation, the value must make sense in context too.

Reflecting with students

Extend this problem by asking students to determine all the values in the solution set. In this context, the only viable solutions are $N = \{0,1,2,3,4,5,6,7,8,9\}$.

Example 4

Given the inequality $3x + 5 - 6x > 8$, create a verbal situation that can be represented by this inequality.

Create a strategy

Think of a scenario where the result needs to exceed a certain value.

Apply the idea

Imagine you are saving money for a new video game. You start with $5. Each week, you can save $3 from your allowance, but you also spend $6 on other expenses. To determine when you can afford the game, which costs more than $8, you use the inequality $3x + 5 - 6x > 8$, where x represents the number of weeks.

Purpose

Show students how to translate a mathematical inequality into a real-life situation, reinforcing the concept that algebra is a tool for solving real-world problems.

Reflecting with students

Encourage students to attend to mathematical precision by clearly defining variables and consistently using appropriate units throughout the task. When guiding them to create a verbal situation for the inequality, emphasize the importance of specifying what each term represents, such as money in dollars or time in weeks. Encourage students to organize their thoughts clearly and verify that their scenario accurately represents the given inequality.

Idea summary

The following key terms can be used to indicate operations and quantities in real-world situations:

Addition	Subtraction	Multiplication	Division	Parentheses
plus	minus	times	divided by	quantity
the sum of	the difference of	the product of	the quotient of	times the sum
increased by	decreased by	multiplied by	half	times the difference
total	fewer than	each/per	split	
more than	less than	twice	equally shared	
added to	subtracted from	double		

Phrases that describe inequalities include

>	<	≥	≤
greater than	less than	greater than or equal to	less than or equal to
more than	is below	at least	at most
over	under	minimum	maximum
		no less than	no more than

Because inequalities have infinite solutions, inequalities used to represent real-world situations often include solutions that are unreasonable in context and therefore non-viable.

Practice

Students: Page 97–99

What do you remember?

1 Match the inequality to its verbal statement:
  a There must be fewer than 20 students in each class.
  b In order to watch the movie, participants must be over 20 years old.
  c Raul has at least 20 friends in his neighborhood.
  d Mrs. Stephens assigns at most 20 problems for homework every night.
  e Alex can download no more than 20 songs in a week.
  f In order for a swimming class to be held, there must be a minimum of 20 students.

  i $x \leq 20$
  ii $x \geq 20$
  iii $x > 20$
  iv $x < 20$

2 Match the following inequalities with their verbal representations:
  a 10 times the sum of a number and 3 is greater than 70
  b 9 less than 4 times a number is greater than 5 times a number minus 7
  c 3 more than 10 times a number is less than 70
  d The difference of 9 and 4 times a number is greater than 5 times a number minus 7

  i $10x + 3 < 70$
  ii $10(x + 3) > 70$
  iii $9 - 4n > 5n - 7$
  iv $4n - 9 > 5n - 7$

3 Consider the inequality $3 - 3x < 2x - 4$. State whether or not the following values of x satisfy the inequality:
  a $x = \frac{12}{5}$ **b** $x = 1$ **c** $x = \frac{7}{5}$ **d** $x = 2$

4 Consider the inequality $6x - 5 \leq 4x + 3$. State whether or not the following values of x satisfy the inequality:

    a $x = -4$ **b** $x = 0$ **c** $x = 4$ **d** $x = 10$

5 If $x = 9$, is each inequality true or false?

    a $x + 3 > 12$ **b** $2x \geq 49$ **c** $x - 20 < 0$ **d** $4 - 2x > -25$

6 For each of these inequalites:

    i   Describe the range of values that satisfy each inequality.

    ii   Draw a number line representing inequality.

    a $x > 17$ **b** $x < -17$ **c** $x \geq -17$ **d** $x \leq -17$

7 Write the inequality shown on each number line:

    a

    b

    c

    d

Let's practice

8 Cora knows that the perimeter of the equilateral triangle shown is a minimum of double the perimeter of the rectangle shown.

She writes the inequality $54 \geq 2(2x + x)$ and realizes she made a mistake. What mistake did she make?

9 Sophia originally has a budget for art supplies of $32, but will be receiving $3p + 1$ more dollars from each of her 2 family members this evening for her birthday, so she looks again at her options at the store. She sees at least $52 worth of items she would like to buy.

    a   Write an inequality that shows how much money she will need to receive from each family member.

    b   She receives $p = 4$ dollars from each family member at her birthday. Is that enough to buy everything she wants?

10 Sarah is gathering signatures on a petition to save the orangutans. If she gets more signatures than she did yesterday, she will be featured on the "Save the Orangutans" website. Her online petitition shows that yesterday she had 3 groups of $x + 20$ people plus an additional x people sign.

So far today, she has received 9 groups of x people signing the petition. Write an inequality representing this.

11 Given the inequality $2(x + 5) < 24 + x$, create a verbal situation that can be represented by this inequality.

Let's extend our thinking

12 The table shows the times of six swimmers in a 1500 m race. Only swimmers who are a maximum of 1 minute slower than the average can qualify for the race.

Which swimmers qualify for the race?

Swimmer	Time (minutes)
James	21
Harry	24
Lachlan	20
Adam	27
Glen	31
Kenneth	23

13 Patricia is shopping for a new printer, and is comparing the price of two suitable models. The first costs $280 and has ink cartridges that cost $40, while the second costs $430 but has ink cartridges that only cost $20.

a Patricia wants to know how many ink cartridges she would need to buy before the second model is cheaper. If x represents the number of ink cartridges bought, write the inequality that describes the situation.

b Patricia wants to compare the better option to a third option that can be described by the expression $35x + 375$. Write a description of what this expression could represent.

14 When breeding certain types of fish, it is recommended that the number of female fish is more than triple the number of male fish.

a If the number of females is represented by f and the number of males is represented by m, write an inequality that represents the recommended relationship between f and m.

b Find a number of female and male fish that would satisfy the recommendation.

15 Ryan wants to save up enough money so that he can buy a new sports equipment set, which costs $40.00. Ryan has $22.10 that he saved from his birthday. In order to make more money, he plans to wash neighbors' windows for $2 per window.

a Let w be the number of windows that Ryan washes. Solve for w, correct to two decimal places.

b State whether the following statements are correct. Explain your thinking.

i Ryan must wash more than 9 windows to be able to afford the equipment.

ii Ryan must wash at least 8 windows to be able to afford the equipment.

iii If Ryan washed 8 windows, and 95% of another window, he could afford the equipment.

iv The number of windows Ryan must wash to be able to afford the equipment must be greater than or equal to 9.

Answers

2.05 Write and represent multistep inequalities

What do you remember?

1 **a** iv   **b** iii   **c** ii   **d** i
   e i   **f** ii

2 **a** ii   **b** iv   **c** i   **d** iii

3 **a** Yes   **b** No   **c** No   **d** Yes

4 **a** Yes   **b** Yes   **c** Yes   **d** No

5 **a** False   **b** False   **c** True   **d** True

6 **a** **i** The range of values that satisfy this inequality includes all real numbers greater than 17. There is no upper limit, so the range extends indefinitely to the right on the number line.

     ii

   b **i** The range of values that satisfy this inequality includes all real numbers less than −17. There is no lower limit, so the range extends indefinitely to the left on the number line.

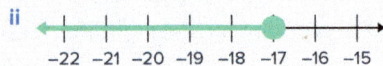

     ii

   c **i** The range of values that satisfy this inequality includes all real numbers less than −17. There is no lower limit, so the range extends indefinitely to the left on the number line.

     ii

   d **i** The range of values that satisfy this inequality includes all real numbers less than −17. There is no lower limit, so the range extends indefinitely to the left on the number line.

     ii

7 **a** $x \geq 3$ **b** $x < -2$ **c** $x > -1$ **d** $x \leq 2$

Let's practice

8 She did not double the perimeter of the rectangle.

9 **a** $32 + 2(3p + 1) \geq 52$ **b** Yes

10 $9x > 4x + 60$

11 Example answer:

Imagine a small business that starts each month with a budget of $24 for office supplies. Each pack of supplies they order costs $2 and they already have 5 packs they can order without exceeding their monthly budget. The inequality representing this situation is $x < 14$, where x represents the number of additional packs of supplies they can order.

Let's extend our thinking

12 James, Harry, Lachlan, and Kenneth

13 **a** $20x + 430 < 40x + 280$

   b This expression could represent a printer that costs $375 with ink cartridges that cost $35 each.

14 **a** $f > 3m$

   b Example answer: 14 female fish and 4 male fish.

15 **a** $w \geq 8.95$

   b **i** No, washing exactly 9 windows would generate $18, which with the $22.10 he already has is enough money to afford the equipment.

     ii No, washing 8 windows would only generate $16 which would give him a total of $38.10, which is not enough money to afford the equipment.

     iii No, he gets paid for washing the whole window, not parts.

     iv Yes, if Ryan washed 9 windows he would earn enough to buy the equipment. If he washed more than 9 windows, he would have more than enough money.

2.06 Solve multistep inequalities

Subtopic overview

Lesson narrative

In this lesson, students will learn to solve multistep inequalities involving one variable. They will practice using properties of inequalities, such as addition, subtraction, multiplication, and division, to isolate the variable and find the solution set. The lesson includes examples where students solve inequalities algebraically and represent solutions on number lines. Students will also apply their knowledge to real-world scenarios, interpreting solutions in context. By the end of the lesson, students should be able to solve and graph multistep inequalities and understand their practical applications.

Learning objectives

Students: Page 100

After this lesson, you will be able to...
- apply properties of real numbers and properties of inequality to solve multistep linear inequalities.
- represent solutions to inequalities algebraically and graphically.
- write and solve multistep linear inequalities in one variable representing a verbal or contextual situation.
- interpret algebraic solutions to linear inequalities in context.

Key vocabulary

- solution set

Essential understanding

The value(s) of the variables that make an equation or inequality true make up its solution set. Inequalities have an infinite number of solutions so their solution sets are often represented on a number line.

Standards

This subtopic addresses the following Virginia 2023 Mathematics Standards of Learning standards.

Mathematical process goals

MPG1 — Mathematical Problem Solving

Teachers can encourage students to apply their mathematical knowledge and skills in solving multi-step inequalities. They can do this by presenting students with problem situations and guiding them through the process of identifying the correct mathematical procedure to solve them. For example, teachers can present a real-world problem that can be represented by a multi-step inequality, and guide students in solving it using the properties of real numbers and inequalities.

MPG3 — Mathematical Reasoning

Teachers can promote mathematical reasoning by guiding students through the process of solving multi-step inequalities using both inductive and deductive reasoning. For example, teachers can explain the importance of reversing the inequality sign when multiplying or dividing both sides of an inequality by a negative number and ask students to reason why this is necessary. Teachers can also encourage students to use logical reasoning to evaluate their solutions and determine their validity.

MPG4 — Mathematical Connections

Teachers can make mathematical connections by linking the process of solving multi-step inequalities to students' prior knowledge of solving one- and two-step inequalities and multi-step equations. They can also relate this process to real-world situations, helping students see the practical applications of the mathematical concepts and skills they're learning.

Content standards

8.PFA.5 — The student will write and solve multistep linear inequalities in one variable, including problems in context that require the solution of a multistep linear inequality in one variable.

8.PFA.5a — Apply properties of real numbers and properties of inequality to solve multistep linear inequalities (up to four steps) in one variable with the variable on one or both sides of the inequality. Coefficients and numeric terms will be rational. Inequalities may contain expressions that need to be expanded (using the distributive property) or require combining like terms to solve.

8.PFA.5b — Represent solutions to inequalities algebraically and graphically using a number line.

8.PFA.5c — Write multistep linear inequalities in one variable to represent a verbal situation, including those in context.

8.PFA.5e — Solve problems in context that require the solution of a multistep linear inequality in one variable.

8.PFA.5f — Identify a numerical value(s) that is part of the solution set of a given inequality.

8.PFA.5g — Interpret algebraic solutions in context to linear inequalities in one variable.

Prior connections

7.PFA.4 — The student will write and solve one- and two-step linear inequalities in one variable, including problems in context, that require the solution of a one- and two-step linear inequality in one variable.

8.PFA.4 — The student will write and solve multistep linear equations in one variable, including problems in context that require the solution of a multistep linear equation in one variable.

Future connections

A.EI.1 — The student will represent, solve, explain, and interpret the solution to multistep linear equations and inequalities in one variable and literal equations for a specified variable.

A.EI.2 — The student will represent, solve, explain, and interpret the solution to a system of two linear equations, a linear inequality in two variables, or a system of two linear inequalities in two variables.

A2.EI.2 — The student will represent, solve, and interpret the solution to quadratic equations in one variable over the set of complex numbers and solve quadratic inequalities in one variable.

Lesson Preparation

Suggested review

Depending on your students' level of prior knowledge, consider revisiting the following lessons:

Grade 6 — 6.06 Solutions to inequalities
Grade 7 — 4.05 Solve one-step inequalities
Grade 7 — 4.06 Solve two-step inequalities

Tools

You may find these tools helpful:
- Number line
- Balance scale
- Blocks

Lesson supports

The following supports may be useful for this lesson. More specific supports may appear throughout the lesson:

Solve inequalities as if they were equations
Targeted instructional strategies

Let students know that the solving methods for inequalities are the same as those for solving equations, with the only difference being the inequality symbol, which changes direction when multiplying or dividing both sides by a negative value.

For students who have difficulty adapting to the inequality aspect of the questions, rewrite the inequality as an equation for them to solve, and have them write out each step of work. Then go through the steps to check if the inequality symbol changes direction at any point.

For example, for the inequality

$$5 - 4x < 21$$

we can start by solving the equation

$$5 - 4x = 21$$

$$-4x = 16 \quad \text{Subtraction property of inequality}$$

$$x = -4 \quad \text{Division property of inequality}$$

Looking back at the steps of work, we can see that we divided by a negative value in one step, so the inequality symbol will change direction.

Since the starting inequality uses the symbol $<$, the symbol for the solution will be $>$.

So the solution to the inequality will be $x > -4$.

It should also be noted that, when taking this approach, using the symmetric property of equality will also change the direction of the inequality symbol.

Student lesson & teacher guide

Multistep inequalities

Students are reminded of the properties of inequality, which can be applied to solve inequalities. Students review what the solution set of an inequality looks like algebraically and graphically. The lesson also reminds them to flip an inequality when multiplying or dividing by a negative number.

Concrete-Representational-Abstract (CRA) approach
Targeted instructional strategies

Concrete: Begin by using physical objects to model multistep linear inequalities. Use a real balance scale to represent the inequality, where one side is heavier than the other to show the unequal relationship. To represent the variable x, use a box with blocks inside. Place the box on one side of the scale and add or remove blocks (such as cubes or small weights), but do not allow students to see the number of blocks inside to make the concept of an unknown quantity more tangible. Also, place blocks on the scale pans to represent constants.

For example, to model the inequality $x + 2 > 5$, place the box with four blocks inside and two blocks on one side, and five blocks on the other side of the scale. Engage students by physically adding or removing blocks from the box and the scale pans, demonstrating how the balance tips and illustrating the properties of inequalities through these manipulations.

Representational: Transition to drawings that represent the physical models. Have students sketch the balance scale with the box and blocks. Use a rectangle to depict the box (the variable x) and small circles or squares for the blocks (constants). Encourage students to draw the steps they take to solve the inequality, such as adding or removing blocks from both sides of the scale. This visual representation helps students connect the concrete actions to mathematical concepts.

Abstract: Move on to solving inequalities using mathematical symbols and numbers. Teach students to write the inequality using algebraic notation, such as $x + 2 > 5$. Guide them through the steps of solving the inequality algebraically: subtract 2 from both sides to isolate x, resulting in $x > 3$. Emphasize the importance of maintaining the direction of the inequality sign (unless multiplying or dividing by a negative number).

Connecting the stages: Throughout the lesson, connect each stage to help students understand the relationships. After using the physical models, discuss how the drawings represent the same concepts. When transitioning to abstract symbols, explain how each algebraic step mirrors the actions taken with the manipulatives and drawings. For example, removing two blocks from both sides of the scale corresponds to subtracting 2 from both sides of the inequality, but one side of the scale was always heavier than the other.

Examples

Students: Page 100

Example 1

Consider the inequality $\dfrac{-8-3x}{2} \leq 5$

a Solve the inequality.

Create a strategy

We want to isolate x on one side of the inequality and a number on the other.

Apply the idea

$\dfrac{-8-3x}{2} \leq 5$	Original inequality
$-8-3x \leq 10$	Multiplication property of inequality
$-3x \leq 18$	Addition property of inequality
$x \geq -6$	Division property of inequality

Reflect and check

Solving an inequality is similar to solving an equation. However, we need to reverse the direction of the inequality when multiplying or dividing by a negative number.

Purpose

Show students how to solve an inequality using the properties of inequality.

Expected mistakes

In the final divison, the student may forget to reverse the inequality symbol and state that the solution is $x \leq -6$. Remind students that the inequality symbol must be reversed when multiplying or dividing by a negative number.

Students: Page 101

b Plot the inequality on a number line.

Apply the idea

Plot the solution set of the inequality $x \geq -6$. Note that since we include -6 the point should be filled.

Reflect and check

What if the solution was $x > -6$?

Endpoints included in the solution are filled points.

Endpoints not included in the solution are unfilled points.

Purpose

Show students how to represent an inequality on a number line.

Expected mistakes

Students might use an unfilled circle at -6 rather than a filled circle. Ask students whether -6 would make the inequality true. It does, so it should be included in the solution set by a filled circle.

c Is $x = 3$ in the solution set for the inequality?

Create a strategy

We can determine if $x = 3$ is in the solution set by using the number line or algebraically substituting the solution into the inequality.

Apply the idea

$$\frac{-8 - 3x}{2} \leq 5 \qquad \text{Original inequality}$$

$$\frac{-8 - 3(3)}{2} \leq 5 \qquad \text{Substitute } x = 3$$

$$\frac{-8 - 9}{2} \leq 5 \qquad \text{Evaluate the multiplication}$$

$$\frac{-17}{2} \leq 5 \qquad \text{Evaluate the subtraction}$$

$$-8.5 \leq 5 \qquad \text{Evaluate the division}$$

Since $x = 3$ leads to a true statement, we can confirm that $x = 3$ is in the solution set for the inequality.

Reflect and check

By using the number line, we can see that the point $x = 3$ is in the solution set of the inequality because it is in the shaded area.

Any points that are not in the shaded area on the number line are not in the solution set.

Purpose

Show students how to evaluate whether a solution is viable or nonviable by substituting it into the original inequality.

🎓 **Use code to check solutions to inequalities**   use with Example 1
Targeted instructional strategies

Incorporate code by showing students how to write a simple Python program to check if a given value is a solution to an inequality. Guide them in coding the inequality and using conditional statements to verify the solution. This activity enhances their understanding of inequalities while introducing basic programming skills. Encourage students to test different values and inequalities, reinforcing the concept through immediate feedback from the code.

An example of Python code is shown:

```
1  #Check if x = 3 is a solution of the inequality: (-8 - 3x)/2 <= 5

2  x=3

3  left_side = (-8 - 3 * x) / 2

4  right_side = 5

5  if left_side <= right_side:

6    print(f"x = 3 is a solution since {left_side} <= {right_side}")

7  else

8    print(f"x = 3 is not a solution since {left_side} > {right_side}")
```

This code calculates the left side of the inequality with the chosen value of x, compares it to the right side, and prints out whether $x = 3$ satisfies the inequality.

Provide a number line for students to use as they complete problems to help them determine which side of the number line to shade in relation to the critical value. Students can mark integer solutions on the number line that satisfy the inequality, then join those points to determine the shaded region.

For example, $0 \geq -6$ and $4 \geq -6$. We can plot these solutions as points on the number line.

Both of the solutions lie to the right of -6, so that is the region that must be shaded.

Students: Page 102

Example 2

Arlene charges $42.75 for a pet grooming session, plus an additional $5 for each special treatment. Clarisse wants the total cost for her pet's grooming to be no more than $102.80.

a Write an inequality that represents the number of special treatments Clarisse could get for her pet.

Create a strategy

Clarisse wants the total cost to be *no more than* $102.80. "No more than" means she could spend exactly that amount or less than that amount but not anything greater. This is "less than or equal to."

Apply the idea

We can write an inequality in words that represents the cost for a pet grooming session:

cost of grooming + cost per treatment · number of treatments ≤ total Clarisse can spend

Translating that into an algebraic expression, we get: $42.75 + 5s \leq 102.80$ where s represents the number of special treatments.

Purpose

Show students how to write an inequality that represents a real-world problem.

Reflecting with students

Encourage advanced learners to explore how the problem changes if Clarisse wants to find the *minimum* number of special treatments she can afford, such as wanting to spend more than $50. Ask them to set up a new inequality that represents this scenario, considering the lowest total cost. This exercise will require students to reverse their usual approach and consider how many treatments she can get, starting from a base price upwards.

b How many special treatments could Clarisse get for her pet and still afford the pet grooming?

Create a strategy

Solve the inequality and then write the solution set.

Apply the idea

$$42.75 + 5s \le 102.80 \quad \text{Original inequality}$$
$$5s \le 60.05 \quad \text{Subtract 42.75 from both sides}$$
$$s \le 12.01 \quad \text{Divide both sides by 5}$$

According to the solution, Clarisse could get 12.01 special treatments for her pet or fewer. However, since she can't get a partial treatment, a more realistic solution is that she can get 12 special treatments or less.

se

nts demonstrate that they can solve an inequality and interpret the solution to solve the real-world

m.

ting with students

students consistently use appropriate units and symbols throughout their work. Guide students to

er the context of the problem when interpreting their solutions; for example, while the inequality $s \le 12.01$

ematically correct, it's not practical to have a fraction of a treatment, so they should conclude that

e can get up to 12 special treatments.

ht the difference between precise responses—which adjust mathematical answers to fit the real-

ontext—and imprecise ones that might, for instance, suggest getting 12.01 treatments or ignore the

icality of negative numbers of treatments.

c Is $s = -2$ a viable solution?

Create a strategy

When considering negative solutions to a contextual problem we need to decide whether a negative value is possible in the context.

Apply the idea

Keep in mind it is not realistic to get part of a special treatment or a negative number of special treatments. Clarisse can get a maximum of 12 special treatments for her pet, so –2 is mathematically part of the solution set for the inequality $s \le 12.01$, but it is not a viable solution in this context because the smallest number of treatments she could get is 0.

Reflect and check

Unlike a value that is not in a solution set of an inequality, this is an example of a solution that was mathematically valid and part of the original solution set, but when considering the context, we have found that it is non-viable.

dents that not all mathematical solutions are viable in a given context.

Critique, correct, and clarify
English language learner support

Provide students with an incorrect solution to the problem. For example:

"Since Clarisse wants to spend no more than $102.80, we use the $\geq$ symbol.

$$42.75 + 5s \geq 102.80$$

$$5s \geq 60.05$$

$$s \geq 12.01$$

Therefore, Clarisse can get at least 12 special treatments."

Ask students to critique this solution by identifying any errors or misunderstandings and correct the solution.

Guide students to clarify that "no more than" means the total cost should be less than or equal to $102.80, so the correct inequality should use $\leq$ instead of $\geq$. Additionally, since Clarisse cannot get a fraction of a treatment, she can get up to 12 special treatments.

This activity helps students deepen their understanding of inequality symbols, accurately interpret phrases like "no more than," and communicate their mathematical reasoning more clearly. It also reinforces the importance of considering the context when determining viable solutions.

Students: Page 102

💡 **Idea summary**

Just like the properties of equality, the properties of inequality can justify how we solve inequalities.

The multiplication and division properties of inequality change the meaning of an inequality when multiplying or dividing by a negative number, meaning we have to reverse the inequality symbol when applying the property.

Because inequalities have infinitely many solutions, inequalities used to represent real-world situations often include solutions that are unreasonable in context and therefore non-viable.

Practice

Students: Page 103–105

What do you remember?

1 Plot the following inequalities on a number line:

    a $x > -9$ **b** $x \leq 15$

2 Describe the range of values that satisfy each inequality.

    a $x \geq 29$ **b** $x < -29$

3 Which of the following inequalities represents the solution for x in $10 \leq 6 - 4x$?

    A $x \leq -4$ **B** $x \geq -1$ **C** $x \leq 1$ **D** $x \geq 4$

4 Which of the following number lines represent the solution for $4x - 7 < 5$?

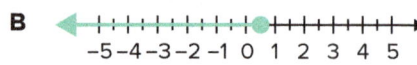

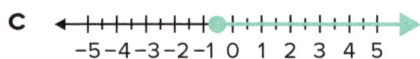

A ◄—|+|+|+|+|+|+|+|●|+|+|►
$-5\,-4\,-3\,-2\,-1\ 0\ 1\ 2\ 3\ 4\ 5$

B ◄—|+|+|+|+|+|●|+|+|+|+|+|►
$-5\,-4\,-3\,-2\,-1\ 0\ 1\ 2\ 3\ 4\ 5$

C ◄—|+|+|+|+|+|●|+|+|+|+|+|►
$-5\,-4\,-3\,-2\,-1\ 0\ 1\ 2\ 3\ 4\ 5$

D ◄—|+|+|+|+|+|+|+|+|●|+|+|►
$-5\,-4\,-3\,-2\,-1\ 0\ 1\ 2\ 3\ 4\ 5$

5 Consider the inequality $21 + 7x > 21$:

   a Solve the inequality.

   b State whether the following values of x satisfy the inequality:

      i $x = 0$ **ii** $x = 1$ **iii** $x = 14$ **iv** $x = 7$

6 Rochelle tried to solve the following inequality but made a mistake in her work:

$$\text{Step 0:}\quad -4 - 2x > 10$$

$$\text{Step 1:}\quad -2x > 14$$

$$\text{Step 2:}\quad x > -7$$

Determine which step is incorrect and explain the error.

7 Which of the following inequalities represents the solution for r in "5 more than $2r$ is less than 39"?

  A $r > 17$ **B** $r < 17$ **C** $r > 22$ **D** $r < 22$

8 Write each of the following relations as an inequality using mathematical symbols. Solve your inequality.

   a The sum of 3 groups of p, and 9, is less than 24.

   b Half of x is no more than five.

   c The product of negative four and x is at most three.

Let's practice

9 Consider the inequality: $5(x + 3) \le 35$.

   a Solve for x.

   b State whether each of the following is part of the solution set:

      i $x = -4$ **ii** $x = 8$ **iii** $x = 4$ **iv** $x = 0$

(SOL) 10 Graph the solution to this inequality:

$$-\frac{2}{3}(9x + 6) \le 0$$

◄—|+|+|+|+|+|+|+|+|+|+|+|+|+|+|►
$-3\quad -2\quad -1\quad 0\quad 1\quad 2\quad 3$

11 For the following inequalities:

   i Solve for x.   **ii** Plot the solutions on a number line.

   a $3x - 7 < 8$ **b** $4 < 6x - 2$ **c** $-6x - 7 \le 5$ **d** $2(x - 3) < -16$

   e $1.5x + 8 > 12.5$ **f** $\frac{4}{5}x + 2 \le -1.2$ **g** $2 - 3.6x < 20.9$ **h** $2\left(\frac{2}{3}x - 1\right) \ge 8$

   i $3\left(\frac{4}{7}w + 2\right) - \frac{2}{3} \le 6$ **j** $\frac{5.2 - 8.4x}{5} + 7 \le -1.2$

SOL **12** Which graph best represents the solution to $\frac{10}{12} \geq \frac{1}{3}x + \frac{1}{4}$?

A

B

C

D

13 Percy tried to plot the solution to the inequality $2x + 20 + 2x + 4 \geq -8$ on a number line, however, his answer is incorrect.

Identify the errors and explain how to rectify them.

14 Consider the situation: "3 less than 3 groups of p is no more than 24".
   a Write the relation as an inequality.
   b Solve the inequality.
   c Find the largest value p can take.

15 Consider the inequality: $4 - 2x < 3x - 2$.
   a Solve for x.
   b State whether each of the following is part of the solution set:
   i $x = \frac{11}{5}$ **ii** $x = 1$ **iii** $x = \frac{6}{5}$ **iv** $x = 2$

16 Solve the following inequalities and justify each step using properties of equality:
   a $9 - 5x > -2x$ **b** $3n - 1 > 5n - 13$ **c** $2x \leq 4(x + 2)$ **d** $\frac{2x + 8}{-3} > x + 6$
   e $\frac{2}{5}x - 8 \leq 9 - \frac{3}{5}x$ **f** $2x + \frac{3}{2} \leq -15 + x$ **g** $0.8(2x - 3) \leq -4(0.2x - 1.5)$ **h** $\frac{x + 2}{3} + 7 \leq 2x - 4$
   i $2(3x - 5) \leq x + 10$ **j** $1.2x - 3.6 - 0.6x \geq 18 - 0.9x$

17 James is saving up to buy a laptop that is selling for $550. He has $410 in his bank account and expects a nice sum of money for his birthday next month.
   a Write the inequality that models the situation.
   b Plot the solution to the inequality on a number line.
   c If James gets $135 for his birthday, can he afford the laptop?

18 To get a grade of C in an accounting course, Uther must obtain an average score of at least 75 over his four exams. So far he has taken the first three exams and achieved scores of 68, 60, and 86.
   a Write the inequality that models Uther getting at least a C in the course.
   b Solve the inequality you wrote in part (a).
   c Describe the solution regarding Uther's score.

19 Skye was given $72 for a birthday present. This present, along with earnings from a Saturday job, is being set aside for a mountain bike. The job pays $5.00 per hour, and the bike costs $379.

   a Set up and solve an inequality to find the minimum number of hours that Skye needs to work to be able to buy the mountain bike. Round your answer to one decimal place.

   b If Skye can only work her job for a whole number of hours, find the minimum number of hours she must work to afford her bike.

20 The length of a rectangle is $3x + 4$ units and the width is x units. The perimeter of the rectangle needs to be at least 32 units.

   a Write an inequality for the perimeter of the rectangle in terms of x.

   b Solve the inequality your wrote in part (a).

21 A tennis team plans to buy shirts for its team and supporters. Ace Graphics charges a $75 setup fee plus $7 per shirt. Grant Designs has charges $10.50 per shirt and has a one-time fee of $9.

   a Write an inequality do determine how many shirts the team needs to order for Ace Graphics to be the better option.

   b Solve the inequality your wrote in part (a).

   c Explain your answer in the context of the problem.

22 For an evening of babysitting, you charge $2 plus $5 per hour.

   a If you are babysitting eight evenings and you want to work the same number of hours each evening, write an inequality to determine how many hours you must work each evening to earn at least $121.

   b Can you reach your goal if you babysit for 15 hours in total for the eight evenings?

Let's extend our thinking

23 At a sport clubhouse the coach wants to rope off a rectangular area that is adjacent to the building. He uses the length of the building as one side of the area, which measures 26 m. He has at most 42 m of rope available to use.

   If the width of the roped area is W, form an inequality and solve for the range of possible widths.

24 A particular cargo plane can carry a maximum weight of 4750 lbs on board. If the average weight of a crate of food is 75 lbs and the weight of the crew is 80 lbs, then the number of crates on board the plane at any one time must satisfy $75c + 80 \leq 4750$, where c represents the number of crates.

   a There are 58 crates already on board and 8 more at the hangar waiting to be loaded. Find the combined weight of everything on board if all of the new crates are loaded.

   b Can all of the new crates safely be loaded on board?

Answers

What do you remember?

1 a

b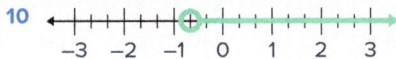

2 a x can be equal to 29 or any number greater than 29.

b x can be equal to any number less than -29, not including -29.

3 B

4 D

5 a $x > 0$

b i No   **ii** Yes   **iii** Yes   **iv** Yes

6 Step 2 is incorrect. At this step she divided both sides by a negative number, and so she should have reversed the inequality symbol.

7 B

8 a $3p + 9 < 24$

$p < 5$

b $\dfrac{x}{2} \le 5$

$x \le 10$

c $-4x \le 3$

$x \ge \dfrac{3}{4}$

Let's practice

9 a $x \le 4$

b i Yes   **ii** No   **iii** Yes   **iv** Yes

10

11 a i $x < 5$

ii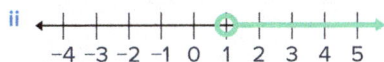

b i $x > 1$

ii

c i $x \ge -2$

ii

d i $x < -5$

ii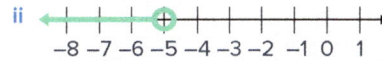

e i $x > 3$

ii

f i $x \le -4$

ii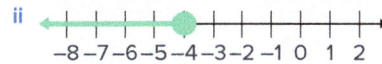

g i $x > -5.25$

ii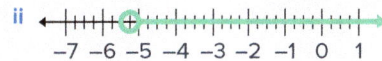

h i $x \ge 7.5$

ii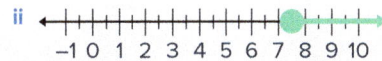

i i $x \le \dfrac{7}{18}$

ii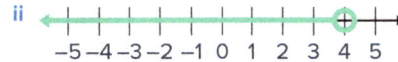

j i $x \ge 5.5$

ii

12 A

13 The dot at point -8 should be filled.

14 a $3p - 3 \le 24$ **b** $p \le 9$ **c** $p = 9$

15 a $x > \dfrac{6}{5}$

b i Yes   **ii** No   **iii** No   **iv** Yes

16 a

$9 - 5x > -2x$	Given
$9 - 5x + 5x > -2x + 5x$	Addition property of inequality
$\dfrac{9}{3} > \dfrac{3x}{3}$	Division property of inequality
$x < 3$	Solution

b

$3n - 1 > 5n - 13$	Given
$3n - 1 + 1 > 5n - 13 + 1$	Addition property of inequality
$3n - 5n > 5n - 5n - 12$	Subtraction property of inequality
$\dfrac{-2n}{-2} < \dfrac{-12}{-2}$	Division property of inequality
$n < 6$	Solution

c

$2x \le 4(x+2)$	Given
$2x \le 4x + 8$	Distributive property
$2x - 4x \le 4x - 4x + 8$	Subtraction property of inequality
$\dfrac{-2x}{-2} \le \dfrac{8}{-2}$	Division property of inequality
$x \ge -4$	Solution

d

$\dfrac{2x+8}{-3} > x + 6$	Given
$-3 \cdot \dfrac{2x+8}{-3} < (x+6) \cdot -3$	Multiplication property of inequality
$2x + 8 < -3x - 18$	Distributive property
$2x + 8 - 8 < -3x - 18 - 8$	Subtraction property of inequality
$2x + 3x < -3x + 3x - 26$	Addition property of inequality
$\dfrac{5x}{5} < \dfrac{-26}{5}$	Division property of inequality
$x < \dfrac{-26}{5}$	Solution

e

$\dfrac{2}{5}x - 8 \le 9 - \dfrac{3}{5}x$	Given
$\dfrac{2}{5}x + \dfrac{3}{5}x - 8 \le 9 - \dfrac{3}{5}x + \dfrac{3}{5}x$	Addition property of inequality
$x - 8 + 8 \le 9 + 8$	Subtraction property of inequality
$x \le 17$	Solution

f

$2x + \dfrac{3}{2} \le -15 + x$	Given
$2 \cdot \left(2x + \dfrac{3}{2}\right) \le (-15 + x) \cdot 2$	Multiplication property of inequality
$4x + 3 \le -30 + 2x$	Distributive property
$4x + 3 - 3 \le -30 - 3 + 2x$	Subtraction property of inequality
$4x - 2x \le -33 + 2x - 2x$	Subtraction property of inequality
$\dfrac{2x}{2} \le \dfrac{-33}{2}$	Division property of inequality
$x \le \dfrac{-33}{2}$	Solution

g

$0.8(2x - 3) \le -4(0.2x - 1.5)$	Given
$1.6x - 2.4 \le -0.8x + 6$	Distributive property
$1.6x - 2.4 + 2.4 \le -0.8x + 6 + 2.4$	Addition property of inequality
$1.6x + 0.8x \le -0.8x + 0.8x + 8.4$	Addition property of inequality
$\dfrac{2.4x}{2.4} \le \dfrac{8.4}{2.4}$	Division property of inequality
$x \le 3.5$	Solution

h

$\dfrac{x+2}{3} + 7 \le 2x - 4$	Given
$\dfrac{x+2}{3} + 7 - 7 \le 2x - 4 - 7$	Subtraction property of inequality
$3 \cdot \dfrac{x+2}{3} \le (2x - 11) \cdot 3$	Multiplication property of inequality
$x + 2 \le 6x - 33$	Distributive property
$x + 2 - 2 \le 6x - 33 - 2$	Subtraction property of inequality
$x - 6x \le 6x - 6x - 35$	Subtraction property of inequality
$\dfrac{-5x}{-5} \le \dfrac{-35}{-5}$	Division property of inequality
$x \ge 7$	Solution

i

$2(3x - 5) \le x + 10$	Given
$6x - 10 \le x + 10$	Distributive property
$6x - 10 + 10 \le x + 10 + 10$	Addition property of inequality
$6x - x \le x - x + 20$	Subtraction property of inequality
$\dfrac{5x}{5} \le \dfrac{20}{5}$	Division property
$x \le 4$	Solution

j

$1.2x - 3.6 - 0.6x \ge 18 - 0.9x$	Given
$1.2x - 3.6 + 3.6 - 0.6x \ge 18 + 3.6 - 0.9x$	Addition property of inequality
$1.2x - 0.6x + 0.9x \ge 21.6 - 0.9x + 0.9x$	Addition property of inequality
$\dfrac{1.5x}{1.5} \ge \dfrac{21.6}{1.5}$	Division property of inequality
$x \ge 14.4$	Solution

17 **a** $x + 410 \ge 550$

b

c No. He needs at least \$140 to afford the laptop.

18 **a** $\dfrac{x + 68 + 60 + 86}{4} \ge 75$

b $x \ge 86$

c To get an overall grade of C Uther must score at least 86 on the last exam.

19 **a** $h \ge 61.4$ **b** 62

20 **a** Example answer. $2(3x + 4) + 2x \ge 32$

b $x \ge 3$

21 **a** Example answer. $7x + 75 < 10.50x + 9$

b $18.857 < x$

c The team must buy 19 shirts or more for Ace Graphics to be the better option.

22 a $8(2 + 5x) \geq 121$

b Solving the inequality from part (a) gives you an answer of $x \geq 2.625$. If you multiply 2.625 by eight, you find that you must work at least 21 hours to reach your goal, so working 15 hours will not be enough hours.

Let's extend our thinking

23 $W \leq 8$

24 a 5030 lbs

b No

Topic 2 Assessment: Expressions, Equations, & Inequalities

1 Select all expressions that are equivalent to $5x-2$.

    A $3x+2x-4+2$

    B $6x-x-3$

    C $2-4x+3x$

    D $7x-3-5+x$

2 Rewrite and simplify the expressions.

    a $-3(2y-8)$

    b $5a-3a+6$

    c $\frac{2}{3}(21-6x)+10x$

    d $2(6g+3)-6(2g-1)$

SOL   **3**   Sasha believes that the model below represents the expression $3(2x+3)-(3x+2)$. Elyssa believes the model represents the expression $3(2x+3)-2(x-1)$. Who is correct? Justify your answer.

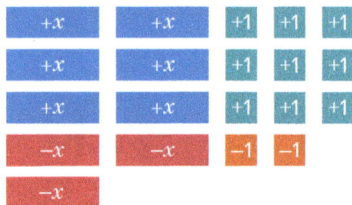

4 Solve the equation $\dfrac{3x+9}{4}=6$ and justify each step.

5 Solve the equations:

    a $11-4x=39$

    b $\dfrac{3}{4}a-5=\dfrac{5}{2}a+2$

    c $2b+7(b-4)=3b-10$

SOL   **6**   Nico was asked to solve the equation $\dfrac{n+3}{2}-5=6$. He believes the solution to the equation is $n=19$. Explain how Nico could confirm that his solution is correct.

7 Match the property that justifies each numbered step in solving $135+5(4x+9)=5x$. Note: Not all choices will be used.

    1 $135+20x+45=5x$ **a**   Subtraction property of equality

    2 $20x+135+45=5x$ **b**   Division property of equality

    3 $135+45=-15x$ **c**   Commutative property of addition

    4 $\dfrac{180}{-15}=\dfrac{-15x}{-15}$ **d**   Addition property of equality

    5   Solution: $x=-12$ **e**   Distributive property

SOL   **8**   Find the value of x in the model.

9   Katie has $41 and saves $7.50 per week. James has $32 and saves $8.25 per week.

    a   Write an equation that would determine the number of weeks, w, for the two to save the same amount of money.

    b   Solve the equation for w.

    c   Write a sentence showing how the solution fits the context of the situation, including the amount of money earned.

10   Which verbal situation describes the equation $2x + 2(x - 5) = 45$?

    A   The new bakery is taking inventory of supplies. A bag of flour contains 45 cups, and a batch of cookies needs 5 cups of flour less than a 2 -tier wedding cake recipe. The bakery wants to make two batches of cookies and a wedding cake and wants to see if one bag of flour is enough.

    B   A rectangle with a perimeter of 45 inches has a length that is 5 inches less than the width, x.

    C   A family of four kids are comparing the size of their feet in inches. Two kids have the same size feet, and the other two have a foot length that is 5 inches more than their siblings. The total length of their feet is 45 inches.

11   For the following inequalities:

    i   Solve for x.   **ii**   Plot the solutions on a number line.

    a $12x - 6 > 2(5x + 3)$ **b** $4(-3x - 5) \geq 28$

    c $\dfrac{6x + 15}{-7} \leq 3$ **d** $-7.5x + 8 > 24 - 3.5x$

12   Identify each value of x that satisfies $10x - 8 > 8x + 4$.

    A   4   **B**   6   **C**   10   **D**   15

    E   −8   **F**   −5   **G**   −2

13   Fill out the table below that uses the given properties of inequality to solve the given inequality.

	Step
Given	$2(x - 6) - 2x \leq 4x + 16$
Distributive property	
Commutative property of addition	
Distributive property (combine like terms)	
Addition property of equality	
Multiplication property of equality	

14   The Rocky Ridge High Marching Band is selling candy bars for a fundraiser. Riley and Allie want to sell at least 55 bars for the fundraiser. Allie currently has sold twice as many bars as Riley, and plans to sell 4 more bars later that day.

    a   Write an inequality to represent this situation, where r represents the number of bars Riley has sold so far.

    b   If Riley sold 16 bars so far, is this enough to reach their goal? Justify your answer.

    c   Assuming this trend continues, what is the minimum number of bars each will need to sell in order to reach the goal?

    d   Suppose their goal changes from 55 to $2r + 6$. What could this new expression represent in context of the problem?

Performance task

15 Doctors recommend staying within a target heart rate range when training for an athletic activity or sport to optimize performance. Your doctor gives you an assignment to do before the starting the season so that you will know whether you are exerting yourself too hard or not enough.

a You can find your resting heart rate by counting your pulse for one minute immediately after you wakes up for 7 days in a row. You find your average resting heart rate to be 61 over 7 days and remember that the last day's pulse was 55 and all the others were the same. Find your pulse on all the other days. Justify your answer by showing your steps.

b Maximum heart, m, rate can be calculated by subtracting your age, a, from 220 . Write an equation that she can use to find m.

c You can find your "heart rate reserve", h, by subtracting your average resting heart rate, r, from your maximum heart rate, m. Write an equation that she can use to find h.

d The Karvonen Formula calculates a person's target heart rate for different intensity levels of exercise. You can find your target heart rate, t, by adding your resting heart rate to the product of her heart rate reserve and her percent of intensity i (in decimal form). Write the Karvonen Formula.

e Your friend uses something called the Karvonen Formula to find the upper and lower limit of her target heart rate range. The upper limit of her target heart rate range is found using an intensity level of 80%. The lower limit of her target heart rate range is found using an intensity level of 60%. Write an equation for her upper limit, u, and her lower limit, l.

f When running a scrimmage with your friends, you found your heart rate to be 165 . Is this within the ideal range of the Kavonen formula?

Answers

Topic 2 Assessment: Expressions, Equations, & Inequalities

1 **A**

8.PFA.1b

2 **a** $-6y+24$ **b** $2a+6$
c $6x+14$ **d** 12

8.PFA.1b

3 Sasha is correct. Her initial expression becomes $6x+9-3x-2$ after distributing. This is represented by the 6 positive x terms, 9 positive 1 terms, 3 negative x terms, and 2 negative 1 terms. Elyssa's expression distributes to become $6x+9-2x+2$, which is missing a negative x term, and shows the incorrect sign for the 2 .

8.PFA.1a

4
$$\frac{3x+9}{4}=6 \quad \text{Given equation}$$
$$3x+9=24 \quad \text{Multiplication property of equality}$$
$$3x=15 \quad \text{Subtraction property of equality}$$
$$x=5 \quad \text{Division property of equality}$$

8.PFA.4b

5 **a** $x=-7$ **b** $a=-4$ **c** $b=3$

8.PFA.4b

6 Nico could confirm his solution by substituting 16 for n and evaluating to see if the two sides of the equation are balanced. If the left side of the equation is 6 after evaluating, then the solution is correct.

8.PFA.4g

7 1 e   2 c   3 a   4 b   5 d

8 $x=-3$

8.PFA.4a, 8.PFA.4b

9 **a** $41+7.5w=32+8.25w$
b $w=12$
c This solution means that after 12 weeks, Katie and James will have both earned $\$131$.

8.PFA.4c, 8.PFA.4e, 8.PFA.4f

10 **B**

8.PFA.4d

11 **a** **i** $x>6$ **ii**
b **i** $x\le-4$ **ii**
c **i** $x\ge-6$ **ii**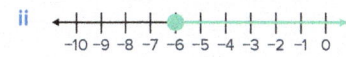
d **i** $x<-4$ **ii**

8.PFA.5a, 8.PFA.5b

12 **C, D**

8.PFA.5f

13

	Step
Given	$2(x-6)-2x\le4x+16$
Distributive property	$2x-12-2x\le4x+16$
Commutative property of addition	$2x-2x-12\le4x+16$
Distributive property (combine like terms)	$-12\le4x+16$
Addition property of equality	$-28\le4x$
Multiplication property of equality	$-7\le x$

8.PFA.5a

14 **a** $r+2r+4\ge55$
b No, this is not enough for the goal. If Riley sold 16 bars, this means that Allie sold $2(16)+4$, or 36 bars. This only totals 52 bars, which falls short of the goal of 55 bars.
c Riley will need to sell 17 bars, and Allie will need to sell 38 bars.
d (Answers may vary)The expression $2r+6$ could represent a student who has sold six more than double the amount of bars Riley has currently sold. The inequality would now represent Riley and Allie wanting to sell at least as many bars as this person.

8.PFA.5c, 8.PFA.5d, 8.PFA.5e, 8.PFA.5f, 8.PFA.5g

15 a Your average resting heart rate over 7 days is 61 bpm. The last day's pulse was 55 bpm, and all the others were the same. Let x be the pulse on the other 6 days. Then:

$$\frac{6x + 55}{7} = 61$$

Simplifying this equation:

$6x + 55 = 427$

$6x = 372$

$x = 62$

Therefore, your pulse on the other 6 days was 62 bpm.

b $m = 220 - a$

c $h = m - r$

d The Karvonen Formula to find the target heart rate t is:

$t = r + h \cdot i$

e The equations for the upper limit u and the lower limit l of the target heart rate range are:

$u = r + h \cdot 0.8$

$l = r + h \cdot 0.6$

f Answers may vary based on student's age. For a 13 year old student, we have:

$$h = (220 - 13) - 61 = 146$$
$$u = 61 + 0.8(146) = 177.8$$
$$u = 61 + 0.6(146) = 148.6$$

Since 165 is in this range, your heart rate is in an optimal range.

8.PFA.1b, 8.PFA.4a, 8.PFA.4c, 8.PFA.4e, MP1

③ Relations & Functions

Big ideas

- There are many ways to represent a function (equation, table, graph, written description, etc.). The way a function is represented can affect what conclusions can be made.
- Functions provide a representation for how related quantities vary. This makes functions a good way to represent many real-world situations.
- A family of functions is defined by a unique set of characteristics shared by all functions that belong to that family. These characteristics give insight into the types of real-world situations that a function models.

Chapter outline

Did you know? A vending machine is a perfect example of a function. You press one button (input), and you get exactly one snack (output)—no surprises!

3. Relations & Functions

Topic Overview

Foundational knowledge

Evaluating standards proficiency

The skills book contains questions matched to individual standards. It can be used to measure proficiency for each.

6.MG.3 — The student will describe the characteristics of the coordinate plane and graph ordered pairs.

7.PFA.1 — The student will investigate and analyze proportional relationships between two quantities using verbal descriptions, tables, equations in $y = mx$ form, and graphs, including problems in context.

Big ideas and essential understanding

There are many ways to represent a function (equation, table, graph, written description, etc.). The way a function is represented can affect what conclusions can be made.

3.01 — A relation is a relationship between two quantities that can be represented in a variety of ways including a table of values, ordered pairs, or points plotted in the coordinate plane.

3.02 — A relation is a function when each input only has one output. This can be determined algebraically or graphically.

Functions provide a representation for how related quantities vary. This makes functions a good way to represent many real-world situations.

3.03 — The domain is all the possible inputs or x-values of a function. The range is all the possible outputs or y-values of the function.

3.04 — When two variables are related, the dependent variable changes based on the value of the independent variable. Because of this the independent variable is the input and the dependent variable is the output.

A family of functions is defined by a unique set of characteristics shared by all functions that belong to that family. These characteristics give insight into the types of real-world situations that a function models.

3.05 — Linear functions have a constant rate of change, often referred to as slope, and can be represented by the equation $y = mx + b$.

3.06 — Linear functions have a constant rate of change, often referred to as slope, and can be represented by the equation $y = mx + b$.

3.07 — Linear functions have a constant rate of change, this determines the real-world situations that linear functions can model.

Standards

8.PFA.2 — The student will determine whether a given relation is a function and determine the domain and range of a function.

8.PFA.2a — Determine whether a relation, represented by a set of ordered pairs, a table, or a graph of discrete points is a function. Sets are limited to no more than 10 ordered pairs.
3.02 Identify functions

8.PFA.2b — Identify the domain and range of a function represented as a set of ordered pairs, a table, or a graph of discrete points.
3.03 Domain and range

8.PFA.3 — The student will represent and solve problems, including those in context, by using linear functions and analyzing their key characteristics (the value of the y-intercept b) and the coordinates of the ordered pairs in graphs will be limited to integers).

8.PFA.3a — Determine how adding a constant (b) to the equation of a proportional relationship $y = mx$ will translate the line on a graph.
3.06 Slope-intercept form

8.PFA.3b — Describe key characteristics of linear functions including slope (m), y-intercept (b), and independent and dependent variables.
3.04 Independent and dependent variables
3.05 Characteristics of linear functions
3.06 Slope-intercept form
3.07 Problem solving with linear functions

8.PFA.3c — Graph a linear function given a table, equation, or a situation in context.
3.05 Characteristics of linear functions
3.06 Slope-intercept form
3.07 Problem solving with linear functions

8.PFA.3d — Create a table of values for a linear function given a graph, equation in the form of $y = mx + b$, or context.
3.05 Characteristics of linear functions
3.06 Slope-intercept form
3.07 Problem solving with linear functions

8.PFA.3e — Write an equation of a linear function in the form $y = mx + b$, given a graph, table, or a situation in context.
3.06 Slope-intercept form
3.07 Problem solving with linear functions

8.PFA.3f — Create a context for a linear function given a graph, table, or equation in the form $y = mx + b$.
3.06 Slope-intercept form
3.07 Problem solving with linear functions

Future connections

A.F.2 — The student will investigate, analyze, and compare characteristics of functions, including quadratic and exponential functions, and model quadratic and exponential relationships.

A.F.1 — The student will investigate, analyze, and compare linear functions algebraically and graphically, and model linear relationships.

Continuous Assessment

⚑ Measure standards proficiency with check-ins

Before starting a new topic, it's a great time to go online and have students complete a Skills Check-in to measure their readiness for the topic.

3.01 Review: Plot points and represent relations

Subtopic overview

Lesson narrative

In this lesson, students will review how to plot points on a coordinate plane and represent relations. They will begin by learning how coordinates describe a point's position relative to the origin, with positive coordinates indicating directions up or right, and negative coordinates indicating directions down or left. Students will practice plotting points by following the horizontal and vertical values of given ordered pairs. The lesson also introduces the concept of relations, defined as sets of ordered pairs representing relationships between two variables. Students will explore different ways to represent relations, including input-output tables and graphs in the coordinate plane. They will practice writing and plotting relations using ordered pairs, and understand how these pairs can be visualized on a graph. By the end of the lesson, students should be proficient in plotting points and representing relations using various methods.

Learning objectives

Students: Page 108

After this lesson, you will be able to...
- define relation.
- use a table of values, ordered pairs, or points plotted on a coordinate plane to represent a given relation between two sets.
- convert between different representations of a relation.
- find the unknown elements in a relation given its rule.

Key vocabulary

- coordinate plane
- origin
- x-axis
- coordinates
- plot (a point)
- y-axis
- table of values
- quadrant
- ordered pair
- relation

Essential understanding

A relation is a relationship between two quantities that can be represented in a variety of ways including a table of values, ordered pairs, or points plotted in the coordinate plane.

Standards

This subtopic addresses the following Virginia 2023 Mathematics Standards of Learning standards.

Mathematical process goals

MPG4 — Mathematical Connections

Teachers can help students relate the concept of relations and ordered pairs to the broader topic of the coordinate plane. For instance, while explaining relations, teachers can remind students of the relationship between coordinates and distances from each

MPG5 — Mathematical Representations

Teachers can meet this goal by guiding students to represent mathematical ideas in multiple ways. For instance, when students are learning to graph ordered pairs in the four quadrants and on the axes of a coordinate plane, teachers can also show how these pairs can be represented in a table or as algebraic expressions. This approach helps students understand that the same mathematical idea can be represented in different ways.

Prior connections

6.MG.3 — The student will describe the characteristics of the coordinate plane and graph ordered pairs.

Future connections

8.PFA.2 — The student will determine whether a given relation is a function and determine the domain and range of a function.

Lesson Preparation

Suggested review

Depending on your students' level of prior knowledge, consider revisiting the following lessons:

Grade 6 — 4.05 Integers in the coordinate plane

Tools

You may find these tools helpful:
- Highlighter

Student lesson & teacher guide

Plot points on the coordinate plane

Students recall that a pair of coordinates describes a point's position from the origin on the coordinate plane and how each coordinate can be used to plot points. Students also review how to write the coordinates of a point that is plotted on a coordinate plane.

Exploration

Students: Page 109

> **Interactive exploration**
> Explore online to answer the questions
>
> ⊕ **mathspace.co**

Use the interactive exploration in 3.01 to answer these questions.

1. How does the first coordinate relate to the location on the graph?
2. How does the second coordinate relate to the location on the graph?
3. How do the signs of the coordinates relate to the quadrants on the graph?

Suggested student grouping: Individual

Students will practice plotting points on a graph using a GeoGebra applet. They will be given a set of coordinates, and they will have to plot these points accurately on the graph. This will help them understand the relationship between the coordinates and their location on the graph.

Ideal student responses

These ideal responses may differ from other correct student responses. Less formal responses can be connected with the more precise mathematical language presented here.

1. **How does the first coordinate relate to the location on the graph?**

   The first coordinate, also known as the x-coordinate, determines the horizontal position of the point on the graph. If the x-coordinate is positive, the point is to the right of the y-axis, and if it is negative, the point is to the left.

2. **How does the second coordinate relate to the location on the graph?**

   The second coordinate, or the y-coordinate, determines the vertical position of the point on the graph. If the y-coordinate is positive, the point is above the x-axis, and if it's negative, the point is below the x-axis.

3. **How do the signs of the coordinates relate to the quadrants on the graph?**

   The signs of the coordinates help determine which quadrant the point is in. If both coordinates are positive, the point is in the first quadrant. If the x-coordinate is negative and the y-coordinate is positive, the point is in the second quadrant. If both coordinates are negative, the point is in the third quadrant. Lastly, if the x-coordinate is positive and the y-coordinate is negative, the point is in the fourth quadrant.

Purposeful questions

- If the first coordinate is positive, on which side of the y-axis will it lie? If it is negative?
- If the second coordinate is positive, on which side of the x-axis will it lie? If it is negative?
- Which section is quadrant 1? Which is quadrant 2 ? Quadrants 3 and 4 ?
- What would be the coordinates of a point that lies on the x-axis? What about the y-axis?

Possible misunderstandings

- Students might confuse the order of the coordinates, thinking that the y-coordinate comes before the x-coordinate. Remind them that x comes before y in the alphabet, and the same is true with coordinate pairs.
- Students might not understand that the origin (0, 0) is the point where the x-axis, and y-axis, intersect, and it's not in any of the four quadrants.

Concrete-Representational-Abstract (CRA) approach
Targeted instructional strategies

Concrete: Begin by engaging students with a large, physical coordinate grid on the floor or wall. Use tape or chalk to create the x-axis and y-axis, including both positive and negative directions. Provide students with markers or cards labeled with ordered pairs. Have students physically move along the axes to plot points based on given ordered pairs. For example, for the ordered pair (3, –2), a student would move 3 units to the right and 2 units down from the origin. This hands-on activity helps students understand how coordinates describe positions relative to the origin.

Representational: After the physical plotting, discuss with students how their movements can be represented on paper. Provide students with graph paper and have them draw the x-axis and y-axis, labeling positive and negative directions. Guide them in plotting the same ordered pairs from the concrete activity onto their paper grids. Show them how the steps they took correspond to movements along the axes on a drawn grid. Encourage them to draw arrows showing the movement from the origin to each point, mirroring their physical movements. Introduce input-output tables where students list the x-values and corresponding y-values of the ordered pairs.

Abstract: Discuss with students how the graphs and tables they have created represent mathematical relationships. Introduce the concept of relations as sets of ordered pairs using mathematical notation. Teach students how to write ordered pairs in the form (x, y) and understand that these represent precise locations on the coordinate plane. Work with students to convert between input-output tables, ordered pairs, and graphs. Practice finding unknown elements in a relation by applying given rules or formulas. Encourage students to describe relationships using mathematical language and symbols, reinforcing their abstract understanding of relations.

Examples

Students: Page 109

Example 1

Consider the points A and B.

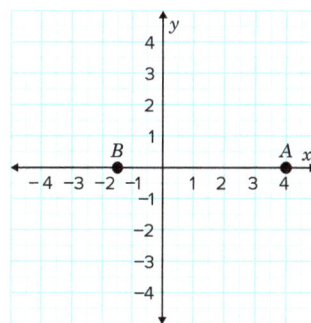

a Write the coordinates of the plotted points.

Create a strategy

Count the spaces the points lie from the origin and consider the direction from the origin.

Apply the idea

Point A lies 4 units to the right of the origin. Point B lies 1.5 units to the left of the origin.

The coordinates are: $A(4, 0)$ and $B(–1.5, 0)$.

Purpose

Students demonstrate that they can identify the coordinates of points in a Cartesian plane by considering the direction and distance from the origin.

Expected mistakes

Students might not write zero as the y-coordinate, instead writing only the x-coordinate. Show them the values on the y-axis and help them see how it is a vertical number line. In the middle of that number line lies $y = 0$.

Students: Page 109

b   Which axis do points A and B lie on?

Create a strategy

The horizontal axis is called the x-axis and the vertical axis is called the y-axis.

Apply the idea

Both points lie on the horizontal axis or x-axis.

Purpose

Students show that they can identify the axis on which a point lies in a coordinate plane.

Students: Page 110

Example 2

Consider the point $(6, -8)$.

a   Plot the point on the coordinate plane.

Create a strategy

The first value of the ordered pair tells us how to move along the x-axis, while the second value tells us how to move along the y-axis.

Apply the idea

The x-coordinate of 6 tells us to move 6 spaces to the right of the origin, and the y-coordinate of -8 tells us to move 8 spaces down from the origin.

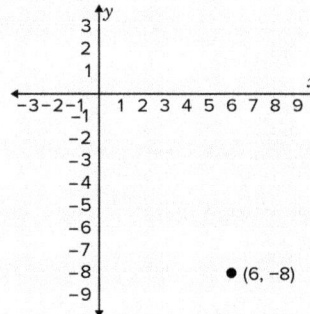

Purpose

Show students that they can plot points on a coordinate plane by understanding the role of x and y coordinates in an ordered pair.

b In which quadrant does the point (6, −8) lie?

Create a strategy

The coordinate plane is divided into four quadrants by the x-axis and y-axis. Refer to the image:

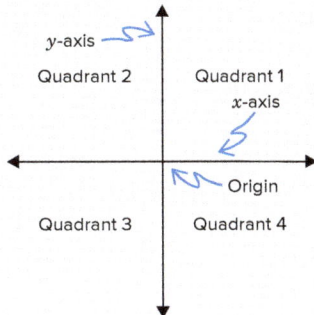

Apply the idea

Looking at the plotted point on the coordinate plane, we can say that (6, −8) lies in quadrant 4.

Purpose

Students show that they can identify the quadrant in which a point lies on a coordinate plane.

Reflecting with students

Invite students to share with the class the things that help them remember how the quadrants are labeled. For example, some students might remember that quadrant 1 is the top right corner, and the subsequent quadrants are labeled counterclockwise from quadrant 1. Other students might use the signs of the x and y-values to remember the quadrant numbers.

Graphing support
Student with disabilities support

use with Example 2

Students who struggle with fine-motor skills may have a hard time counting to plot the points on the graph. Provide students with a larger grid if needed and encourage them to count using the tip of their pencil from square to square.

Idea summary

The **coordinate plane** is made up of a horizontal and a vertical axes and has 4 sections called **quadrants**.

- The **horizontal axis** is called the x-axis and runs from left the right.
- The **vertical axis** is called the y-axis and runs up and down.
- The two lines meet at the **origin**, which has coordinates (0,0).

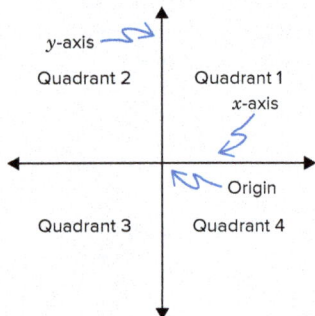

A pair of coordinates describes a point's position away from the origin. A negative coordinate indicates a direction of left or down from the origin. A positive coordinate indicates a direction of right or up from the origin.

To name coordinates, we write the horizontal value, then the vertical value that a point is away from the origin. Points are written as ordered pairs (x, y).

Relations

Students will be introduced to the concept of relations and explore different ways that relations can be represented.

Examples

Example 3

Write the relation {(2, 2), (4, 4), (6, 3), (7, 5)} in the table.

x	2	4	6	7
y				

Create a strategy

Write the second coordinate of each ordered pair in the y row, below the x-value it corresponds to.

Apply the idea

x	2	4	6	7
y	2	4	3	5

Purpose

Check that students can express a relation given as a set of ordered pairs in a table of values.

Students might struggle to identify x and y-values in representations of relations, making it difficult for them to convert between the representations. With a set and a table of values specifically, use color coding to distinguish between x and y-values.

For example, you might write x-values in blue and y-values in yellow, as shown.

$$\{(2, 2), (4, 4), (6, 3), (7, 5)\}$$

x	2	4	6	7
y				

Additionally, you can provide extra practice with matching x and y-values using more tables and ordered pairs. This can help build confidence and reinforce the concept.

Students: Page 112

Example 4

Consider the relation: $\{(-9, -5), (-5, -10), (-5, -4), (-3, 7), (-2, -4), (-1, 1)\}$. Represent the relation on the coordinate plane.

Create a strategy

The first value of each ordered pair tells us how to move along the x-axis, while the second value tells us how to move along the y-axis.

Apply the idea

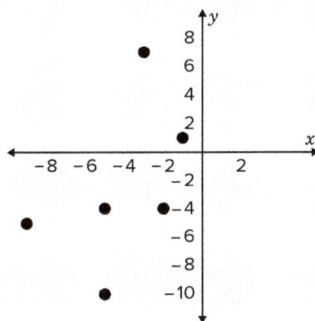

Purpose

Check that students can express a relation given as a set of ordered pairs on the coordinate plane.

Reflecting with students

In 8th grade, students are expected to interpret relations represented as a set of ordered pairs, a table, or a graph. Use this example to help them make connections between all three representations by asking students to also create a table of values.

x	−9	−5	−5	−3	−2	−1
y	−5	−10	−4	7	−4	1

Encourage students to explore how the relation can be represented in different forms by comparing the set of ordered pairs and the graph on the coordinate plane. Begin by having students plot the given ordered pairs on graph paper. Then, ask them to create a table of values for the relation.

Facilitate a class discussion where students share their representations with each other. Ask guiding questions such as, "How can we identify the x and y-values from each representation?" and "How does the graph help you understand the relation between the x and y-values?" Prompt students to observe any patterns or relationships they see in the data.

Highlight the connections between the different representations. For example, point out how each ordered pair corresponds to a point on the graph and a row in the table. Emphasize mathematical vocabulary by encouraging students to use terms like "ordered pair," "coordinate," "relation," "table of values," "x-axis," and "y-axis" during the discussion.

Students: Page 112

Example 5

A relation is defined as follows: $y = -4$ if x is positive and $y = 4$ if x is 0 or negative.

a Complete the table.

x	-4	-3	-2	-1	0	1	2	3	4
y									

Create a strategy

For each positive x-value: $y = -4$, otherwise $y = 4$.

Apply the idea

When $x = 1, 2, 3, 4$: $y = -4$.

When $x = 0, -1, -2, -3, -4$: $y = 4$.

x	-4	-3	-2	-1	0	1	2	3	4
y	4	4	4	4	4	-4	-4	-4	-4

Purpose

Check that students can express a relation given using specific criteria as a table of values.

Expected mistakes

Students might get confused by a relation represented as a description or may misread the criteria. It may be helpful for them to label the table with the criteria before adding the values.

For example, drawing a line starting at the positive values in the table (starting at 1) could help students visually see where to start writing -4.

x	-4	-3	-2	-1	0	1	2	3	4
y									

x is 0 or negative x is positive

b Plot the points on the coordinate plane.

Create a strategy

Plot the points from the completed table in part (a).

Apply the idea

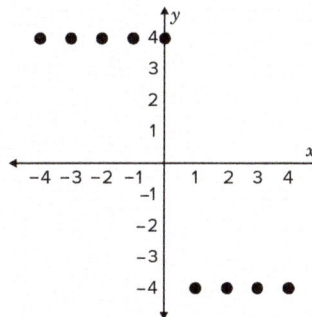

Purpose

Check that students can graph ordered pairs using a given table of values.

Reflecting with students

Encourage advanced learners or all students to create their own relations between two sets of data. Invite them to define a unique rule or pattern that connects two variables, and represent it using an input-output table, ordered pairs, and a graph on the coordinate plane.

Prompt students to analyze how changes in their rule affect the relation's representation across different formats. Additionally, have students exchange their relations with classmates to interpret and represent each other's rules, fostering collaborative learning and critical thinking.

Students: Page 113

> 💡 **Idea summary**
>
> A **relation** is a set of ordered pairs which represent a relationship.
>
> We can express the same relation in several different ways: as a set of ordered pairs, an input-output table, a graph on the coordinate plane, or as an equation in terms of x and/or y that describes a graph.

Practice

Students: Page 113–117

What do you remember?

1   Consider the mapping shown:

   a   Complete the table.

x	−1	0	1	2
y				

   b   State the inputs.

   c   State the outputs.

   d   If the input is −1, what is the corresponding output?

2   Represent the relationship of the inputs and outputs as a set of ordered pairs.

Input	Output
1	7
3	5
6	2
9	9

3   Explain what a relation is in your own words.

4   Evaluate if the following statements are true or false based on your understanding of the coordinate plane and relations. If the statement is false, correct the statement to make it true.

   a   The horizontal axis on the coordinate plane is called the y-axis.

   b   The point (0, 0) is referred to as the origin on the coordinate plane.

   c   A point with a positive x-coordinate and a negative y-coordinate lies in the second quadrant.

   d   The ordered pair (4, −3) indicates a point that is 4 units away from the y-axis and −3 units from the x-axis.

   e   When graphing the point (2, 5), you move 2 units up and 5 units to the right from the origin.

   f   Points with the same y-coordinate lie on a line parallel to the x-axis.

   g   A relation is a set of points where each x-coordinate is unique.

5   Match the following parts of the coordinate plane to the letter on the graph:

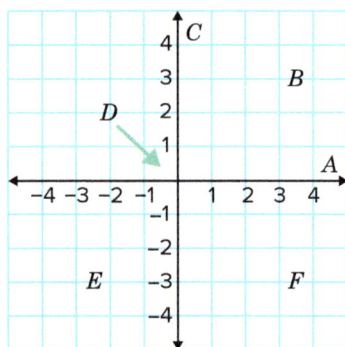

   a   The x-axis

   b   The y-axis

   c   The origin

   d   The first quadrant

   e   The third quadrant

   f   The fourth quadrant

6 Identify the following for the points labeled on the graph.

   i The ordered pair

   ii The quadrant or axis the point lies on

    a Point A

    b Point B

    c Point C

    d Point D

Let's practice

7 Given the table of values, write the list of ordered pairs:

a

x	3	6	8	10
y	4	6	2	8

b

x	7	3	5	7	12
y	7	2	5	3	8

c

x	−4	−1	3	6	9
y	9	5	7	8	3

d

x	$\frac{1}{5}$	$\frac{1}{4}$	$\frac{1}{3}$	$\frac{1}{2}$
y	3	5	6	8

8 Express each relation in a table of values:

   a {(2, 2), (4, 4), (6, 3), (2, 5)}

   b {(3, 9), (6, 2), (8, 6), (10, 3)}

   c {(1, 2), (4, 9), (7, 1), (10, 3)}

   d {(3, 4), (6, 6), (8, 2), (10, 8)}

9 For each relation represented by a table, represent the relation on a coordinate plane:

a

x	1	2	3	4	5
y	6	12	18	24	30

b

x	1	2	3	4	5
y	5	1	−3	−7	−11

c

x	1	2	7	9	13
y	−3	−3	−3	−3	−3

d

x	5	10	15	20	25
y	15	30	45	60	75

10 For each relation represented by a graph, represent the relation in a table of values:

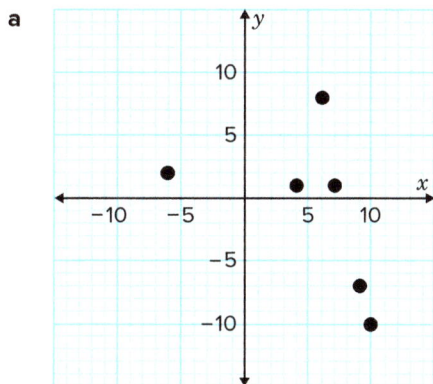

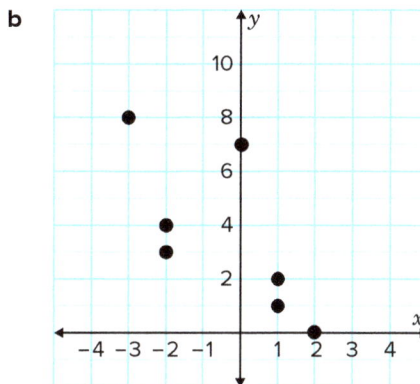

a

b

c

d

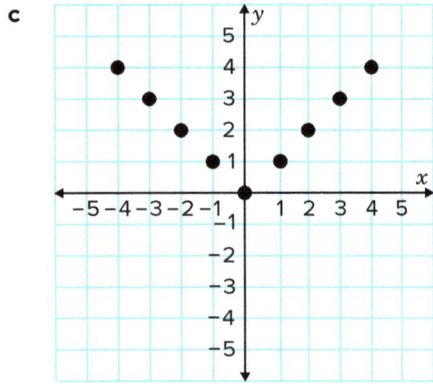

11 Express each relation as a series of points on the coordinate plane:

   a {(1, 2), (3, 5), (7, 8), (8, 11)}

   b {(2, 3), (5, 6), (9, 11), (14, 16)}

   c {(5, 5), (10, 9), (11, 12), (13, 17)}

   d {(2, 5), (7, −3), (5, 2), (−4, −9)}

12 Express each relation as a list of points:

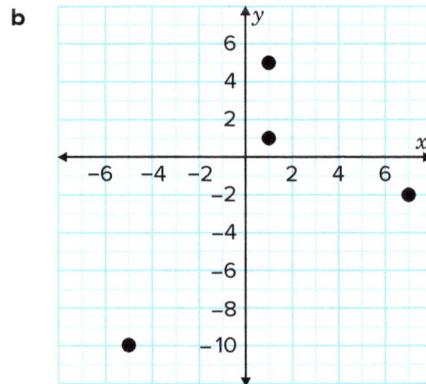

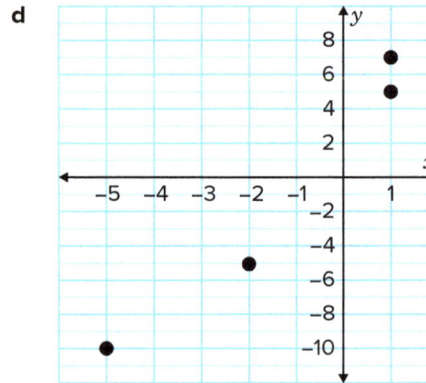

a

b

c

d

13 A relation is defined as follows: $y = 7$ if x is positive and $y = -7$ if x is 0 or negative.

   a Complete the table for this relation.

x	−4	−3	−2	−1	0	1	2	3	4
y									

   b Plot the points on a coordinate plane.

14 A relation's rule is the output value is one less than twice the input value.

    a Complete the table for this relation.

x	−4	−3	−2	−1	0	1	2	3	4
y									

    b Plot the points on a coordinate plane.

15 Given the relation's rule $y = \frac{1}{2}x - 3$, find the missing values for the incomplete ordered pairs.

    a (4, □)   **b** (□, 5)   **c** (2, □)   **d** (□, −1)

16 Explain how to find the missing value in this table:

a	4	5	6	7	8
b	6	10	14		22

17 Gwen sells bananas in bunches of 4.

    a Complete the table:

    b Describe the relationship between the number of bunches and the number of bananas in words.

Bunches (x)	1	2	3	4
Bananas (y)				

    c Create a graph for the number of bananas, y, in terms of the number of bunches, x.

18 Tarek buys some decks of playing cards that contain 52 cards each.

    a Complete the table:

    b Describe the relationship between the number of decks and the number of cards in words.

Decks (d)	1	2	3	4
Cards (c)				

    c Write a rule of the relation for the number of cards, c, in terms of the number of decks, d.

Let's extend our thinking

19 Plot the following points on the coordinate plane and label them accordingly.

$A(-3, 2)$ $B(1, 2.5)$ $C(-3, -3)$ $D\left(3, \frac{1}{2}\right)$ $E\left(-\frac{3}{4}, 3\right)$

20 A bookstore is offering two free notebooks for every five notebooks purchased. The relationship between the total cost and the number of notebooks is shown in the table.

    a Based on the table, how much will you pay for one notebook?

    b Based on the table, how much will you pay for 5 notebooks?

    c Using this relation, how much will one pay for 10 notebooks?

Number of Notebooks	Total Cost (dollars)
1	4
5	20
10	40
15	60
20	80

21 The graph depicts the growth of a plant in centimeters over time in days.

    a Express the relation as a set of ordered pairs.

    b If the plant continues to grow at a constant rate, how tall might the plant be after 6 days?

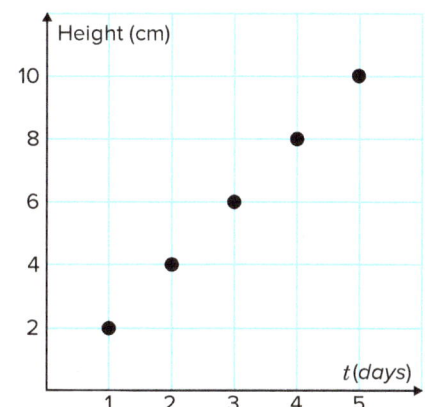

22 A partial graph of the $y = \frac{1}{3}x + 2$ relation is provided. Complete the graph by determining and plotting at least 3 additional points that satisfy the given relation.

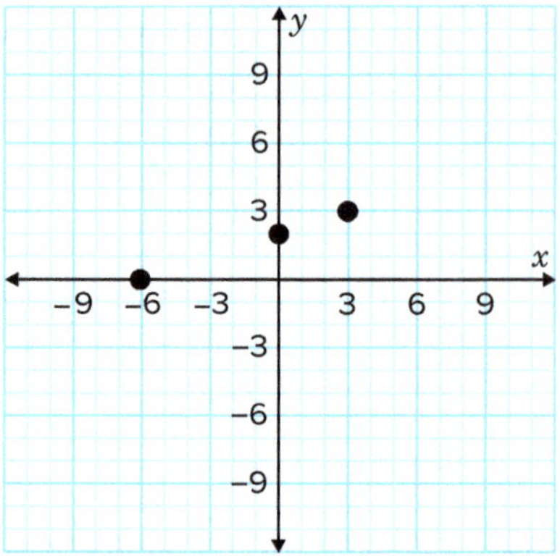

23 Huda opens a bank account and deposits $300. At the end of each week, she adds $10 to her account.

    a Complete the table which shows the balance of Huda's account over the first four weeks:

Week (W)	0	1	2	3	4
Account total (A)	$300	$310			

    b Write the proportional relationship for Huda's account total, A, in terms of the number of weeks W, for which she has been adding to her account.

    c Find the amount of money in Huda's account after twelve weeks.

24 Let the height of a potted plant be y cm. As the plant grows, the height increases according to the equation $y = 3t + 6$, where t is the elapsed time in weeks.

    a Complete the table of values:

Time (t weeks)	0	1	2	3
Height of plant (y cm)				

    b Plot the points in the table on a coordinate plane.

    c Is the relationship proportional? Explain your choice.

Answers

What do you remember?

1 a

x	−1	0	1	2
y	2	0	2	4

  b −1, 0, 1, 2   **c** 0, 2, 4   **d** 2

2 {(1, 7), (3, 5), (6, 2), (9, 9)}

3 A relation is a set of ordered pairs, where each pair consists of an input and an output.

4 a False, the horizontal axis is called the x-axis.

  b True

  c False, it would like in the fourth quadrant.

  d True

  e False, you move 2 units right and 5 units up from the origin.

  f True

  g False, a relation can have the same x-coordinate with different y-coordinates.

5 a A   **b C**   **c D**   **d B**

  e E   **f F**

6 a i (3, −4)   **ii** Quadrant 4

  b i (−3, 0)   **ii** x-axis

  c i (−2, 3)   **ii** Quadrant 2

  d i (0, 4)   **ii** y-axis

Let's practice

7 a {(3, 4), (6, 6), (8, 2), (10, 8)}

  b {(7, 7), (3, 2), (5, 5), (7, 3), (12, 8)}

  c {(−4, 9), (−1, 5), (3, 7), (6, 8), (9, 3)}

  d $\left\{ \left(\frac{1}{5}, 3\right), \left(\frac{1}{4}, 5\right), \left(\frac{1}{3}, 6\right), \left(\frac{1}{2}, 8\right) \right\}$

8 a

x	y
2	2
4	4
6	3
2	5

b

x	y
3	9
6	2
8	6
10	3

c

x	y
1	2
4	9
7	1
10	3

d

x	y
3	4
6	6
8	2
10	8

9 a

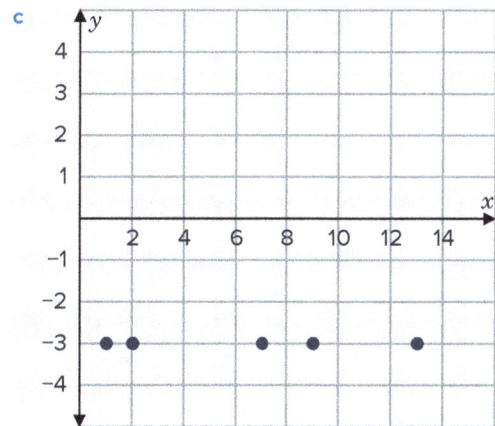

  b

  c

  d

10 a

x	−6	4	7	7	9	10
y	2	1	1	8	−7	−10

b

x	−3	−2	−2	0	1	1	2
y	8	4	3	7	2	1	0

c

x	−4	−3	−2	−1	0	1	2	3	4
y	4	3	2	1	0	1	2	3	4

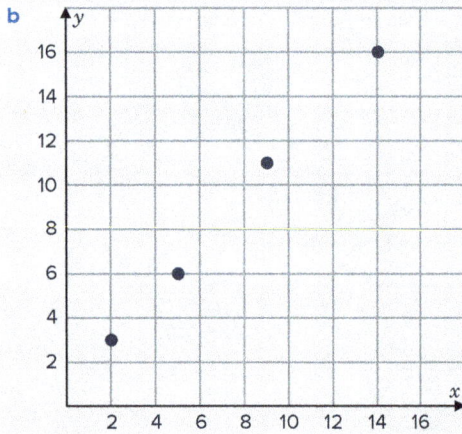

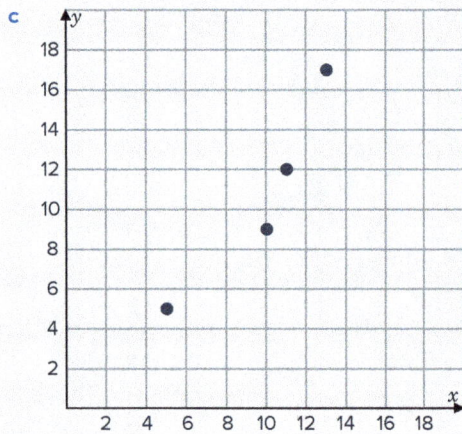

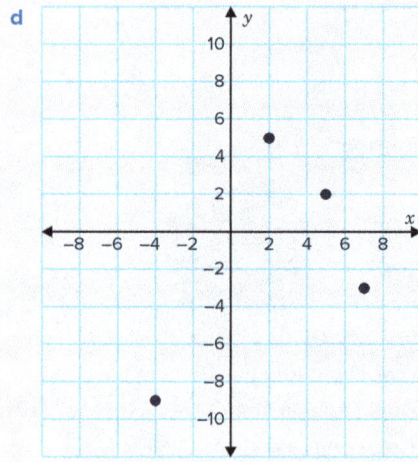

d

x	−4	−3	−2	−1	0	1	2	3	4
y	−4	3	−2	1	0	1	−2	3	−4

11 a

b

c

d

12 a $\{(2,5),(2,7),(−3,−4),(−9,13)\}$

b $\{(1,5),(1,1),(7,−2),(−5,−10)\}$

c $\{(1,5),(7,−2),(−5,−10),(13,−13)\}$

d $\{(1,5),(1,7),(−2,−5),(−5,−10)\}$

13 a

x	−4	−3	−2	−1	0	1	2	3	4
y	−7	−7	−7	−7	−7	7	7	7	7

b

14 a

x	−4	−3	−2	−1	0	1	2	3	4
y	−9	−7	−5	−3	−1	1	3	5	7

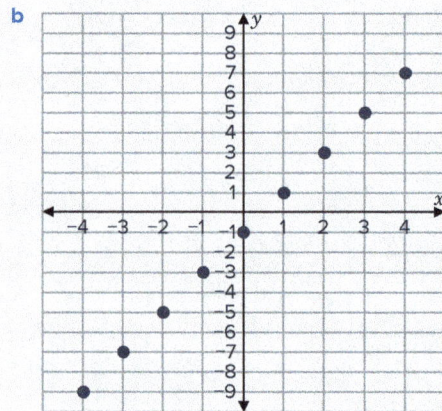

b

15 a $y = \frac{1}{2} \times 4 - 3 = 2 - 3 = -1$, so the pair is (4, −1)

b $5 = \frac{1}{2}x - 3$, solving for x gives $x = 16$, so the pair is (16, 5)

c $y = \frac{1}{2} \times 2 - 3 = 1 - 3 = -2$, so the pair is (2, −2)

d $-1 = \frac{1}{2}x - 3$, solving for x gives $x = 4$, so the pair is (4, −1)

16 Answers may vary.

Method 1: continue the pattern of adding 4 to the b values and the missing number is 14 + 4 = 18.

Method 2: Find the rule for the table $b = 4a - 10$ and use it when $a = 7$ to find $b = 4 \times 7 - 10 = 18$.

17 a

Bunches(x)	1	2	3	4
Bananas(y)	4	8	12	16

b The number of bananas is equal to four times the number of bunches.

c

18 a

Decks (d)	1	2	3	4
Cards (c)	52	104	156	208

b The number of cards is equal to fifty-two times the number of decks.

c $c = 52d$

Let's extend our thinking

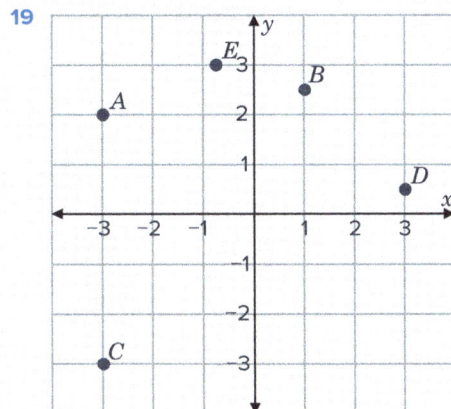

19

20 a $4   **b** $20   **c** $40

21 a {(1, 2), (2, 4), (3, 6), (4, 8), (5, 10)}

b 12 cm

22

23 a

Week (W)	0	1	2	3	4
Account total (A)	$300	$310	$320	$330	$340

b $A = 300 + 10W$

c $420

24 a

Time (t weeks)	0	1	2	3
Height of plant (y cm)	6	9	12	15

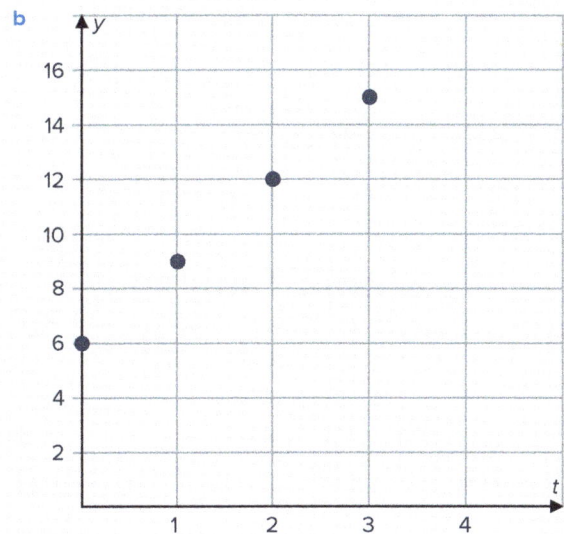

b

c No, since the ratio changes with different values of time, the relationship is not proportional.

3.02 Identify functions

Subtopic overview

Lesson narrative

In this lesson, students will learn to identify functions and distinguish them from general relations. The lesson begins with the definition of a function as a special type of relation where each input corresponds to exactly one output. Students explore this concept through examples, such as the relationship between the number of boba teas ordered and their cost, emphasizing the unique output for each input. The lesson continues by examining various representations of functions, including tables, ordered pairs, and graphs. Students will use the vertical line test to determine whether a graph represents a function, which involves checking if a vertical line intersects the graph at more than one point. Through examples and guided practice, students will classify relationships as functions or not by analyzing scenarios like the radius and circumference of a circle, the amount of time studied and test scores, and temperature readings at different times. By the end of the lesson, students should understand how to identify functions and apply the vertical line test to graphical representations.

Learning objectives

Students: Page 118

> 🎓 **After this lesson, you will be able to...**
> - determine whether a given relation is a function from a table, a set of ordered pairs or a graph of discrete points.
> - define a function.

Key vocabulary

- function
- relation

Essential understanding

A relation is a function when each input only has one output. This can be determined algebraically or graphically.

Standards

This subtopic addresses the following Virginia 2023 Mathematics Standards of Learning standards.

Mathematical process goals

MPG3 — Mathematical Reasoning

Teachers can promote mathematical reasoning by asking questions that require students to justify their conclusions about whether a relation is a function. For example, when a student determines a relation is not a function, the teacher could ask the student to explain why this is the case using the properties of functions.

MPG4 — Mathematical Connections

Teachers can show students how the concept of functions is related to other mathematical concepts they have learned, such as algebraic expressions and equations. For instance, the teacher can make a connection between the concept of a function and the students' prior knowledge of the coordinate plane by explaining how functions can be represented graphically. Furthermore, real-world examples can be used to illustrate the practical applications of functions in different fields, thus connecting mathematics to other disciplines and everyday life.

MPG5 — Mathematical Representations

Teachers can instruct students how to represent functions using various methods such as ordered pairs, tables of values, and graphs. By practicing these different representations, students will understand that mathematical ideas can be expressed in multiple ways. For example, a function can be represented as a set of ordered pairs, a table of values, or a graph. Teachers can also guide students to connect these different representations to understand that they all represent the same mathematical idea.

Content standards

8.PFA.2 — The student will determine whether a given relation is a function and determine the domain and range of a function.

8.PFA.2a — Determine whether a relation, represented by a set of ordered pairs, a table, or a graph of discrete points is a function. Sets are limited to no more than 10 ordered pairs.

Prior connections

6.MG.3 — The student will describe the characteristics of the coordinate plane and graph ordered pairs.

Future connections

A.F.2 — The student will investigate, analyze, and compare characteristics of functions, including quadratic and exponential functions, and model quadratic and exponential relationships.

Lesson Preparation

Suggested review

Depending on your students' level of prior knowledge, consider revisiting the following lessons:

Grade 8 — 3.01 Review: Plotting points and representing relations

Rich task

Facilitate this engaging activity so students can activate and apply their knowledge

Function Machines

Tools

You may find these tools helpful:

- Graphing calculator

Lesson supports

The following supports may be useful for this lesson. More specific supports may appear throughout the lesson:

Examples and non-examples of functions
Student with disabilities support

Provide students with examples of relations that are and are not functions. Be sure to include examples represented as tables, lists and graphs. In each example, clearly explain why the relations are or are not functions. Below are a few examples that could be used:

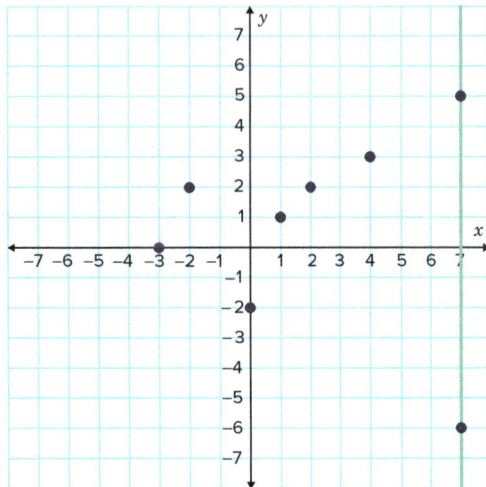

Relation that is not a function

Relation that is a function

x	−2	−1	−1	0	1	2
y	−5	−4	0	1	2	−3

Relation that is not a function

x	5	8	10	13	15	16
y	−2	1	1	−2	5	6

Relation that is a function

Relation that is not a function:

$$\{(1, 5), (2, -3), (2, 4), (3, 1), (4, -1), (5, 7)\}$$

Relation that is a function:

$$\{(-1, -3), (2, -3), (0, 2), (-3, 0), (1, -1), (4, -1)\}$$

All functions are relations, but not vice versa
Address student misconceptions

Students might think a set of ordered pairs, table, or graph is either a relation or a function, and cannot be both. Address this misconception by explaining to students that:

1. A relation is a relationship between a set of input values and output values.
2. All tables, graphs, equations, and sets of ordered pairs are relations.
3. Some relations are functions.
4. All functions are relations.

Student lesson & teacher guide

Functions

Students will connect what they've learned about relations to learning about the characteristics of functions. Functions are relations where the x-values are all unique. They discover that functions can be represented as a set of ordered pairs, a table of values, or a graph of discrete points.

Examples

Students: Page 120

Example 1

Determine whether each situation represents a function.

a The radius of a circle and its circumference.

Create a strategy

A relation is a function if each input matches to one output.

Apply the idea

The circumference of a circle, given the radius, is given by $2\pi r$. A circle with a specific radius (input) will not have more than one circumference (output).

The relationship of a circle's radius and its circumference is a function.

Purpose

Show students that they can determine whether a relationship between two measurements constitutes a function by examining whether each input corresponds to one and only one output.

Reflecting with students

Encourage students to pair up with a classmate and share their reasoning for why this is a function. Encourage them to use precise mathematical language in their explanations.

Students: Page 120

b Time studied by students in a class and grade on a test.

Create a strategy

A relation represents a function if each input matches with only one output. A relation is not a function if an input can pair with more than one output.

Apply the idea

Two students can study the same amount of time (input) but get different scores on a test (outputs).

This situation does not represent a function.

Purpose

Show students that not all relationships between variables represent a function.

Students: Page 120

c Age of a town's population compared to height.

Apply the idea

Two people of the same age (input) could have different heights (output). This relation is not a function.

Reflect and check

A relation that is not a function can become a function with additional constraints. If we limited the relation to one person's age and their height, the relation would become a function.

Purpose

Show students that a relation between two variables does not necessarily constitute a function, and that additional constraints may be required to form a function.

Reflecting with students

Ask students how old they are and point out that many of them are the same age. Then, ask them if they all are the same height. If age is the input and height is the output, then there are multiple outputs that correspond to a single input.

Students: Page 120

d   The temperature reading on a home's thermostat at a particular time of day.

Apply the idea

Each moment in time will have a single temperature reading, so the relation is a function.

Reflect and check

Even though the thermostat may register the same temperature at multiple times during the day, this is still a function because each time is associated with a single temperature.

Purpose

Show students that they can apply their understanding of the definition of a function to determine whether a given relation is a function or not.

Students: Page 120

Example 2

Determine if each relation is a function.

a

x	−8	−7	−6	−3	2	7	9	9	10
y	8	13	−18	−16	−15	−2	−4	11	−9

Create a strategy

The relation is a function if each unique x-value pairs with one y-value.

Apply the idea

We see from the table that the x-value of 9 is paired with two different output values.

x	−8	−7	−6	−3	2	7	9	9	10
y	8	13	−18	−16	−15	−2	−4	11	−9

So, the points do not represent a function.

Purpose

Show students how to determine if a relation is a function using a table of values.

Expected mistakes

Students might assume that each column in the table represents a unique input. Consider having them graph the relation to help them better visualize why this relation is not a function.

b

Create a strategy

A relation is a function if each element in the domain only maps to one element in the range.

Apply the idea

The ordered pairs represented by the graph are

$$\{(-1, -3), (2, -3), (0, 2), (-3, 0), (1, -1), (4, -1)\}$$

Since all of the domain values are different, the relation is a function.

Purpose

Challenge students o determine whether a relation is a function by examining a graph.

Expected mistakes

Students might think that the relation is not a function because there are repeated y-values. Remind students that a function has one *output* for every *input*, meaning that every x-value should correspond to only one y-value. It is fine for different inputs to map to the same output.

Students: Page 121

c $\{(-1, 1), (3, 3), (2, -1), (7, 1)\}$

Apply the idea

Each x-coordinate is paired with only one y-coordinate. This means that the set of ordered pairs is a function.

Purpose

Show students how to determine if a set of ordered pairs forms a function by applying the definition of a function.

Identifying functions with algorithmic thinking   use with Example 2
Targeted instructional strategies

Encourage your students to apply algorithmic thinking by developing a systematic method to determine whether a set of ordered pairs represents a function. Guide them to create a set of steps that would work for a set of points. After they've devised their algorithm, have them test it using the provided data to justify whether the relation is a function. This approach helps students formalize their reasoning and understand the criteria for a function.

1   List all the x-values from the table or points.

2   For each x-value, go through the list of ordered pairs and check if for that particular x-value more than one y-value associated with it.

3   Repeat for all x-values.

4   If every x-value has exactly one y-value, conclude that the relation is a function; otherwise, it is not.

Note that if the same ordered pair appears more than once, it does not count as an x-value having more than one y-value associated with it.

Students may think that a relation is not a function if multiple inputs are associated with a single output. They might mistakenly believe that for a relation to be a function, each output must correspond to only one input, misapplying the concept that functions have unique outputs for each input.

To address this misconception, clarify that the definition of a function requires each input to have exactly one output, but allows for multiple inputs to share the same output. Encourage students to create their own examples of functions where different inputs yield the same output, and discuss why these still satisfy the definition of a function.

Students: Page 121–122

Example 3

Oprah makes scarves to sell at the market. It costs her $2 to produce each one, and she sells them for $5.

a Complete the graph of the points representing the relation between the number of scarves she manages to sell and her total profit for when 1, 2, 3, 4 and 5 scarves are sold. The first point has been plotted for you.

Create a strategy

The total profit can be found using the formula:

$$\text{Total profit} = \text{Total revenue} - \text{Total cost}$$

Apply the idea

Total revenue = Number of scarves sold · $5

Total cost = Number of scarves sold · $2

Profit for 1 scarf = $1 \cdot 5 - 1 \cdot 2$	Substitute the number of scarves
= $3	Evaluate
Profit for 2 scarves = $2 \cdot 5 - 2 \cdot 2$	Substitute the number of scarves
= $6	Evaluate
Profit for 3 scarves = $3 \cdot 5 - 3 \cdot 2$	Substitute the number of scarves
= $9	Evaluate
Profit for 4 scarves = $4 \cdot 5 - 4 \cdot 2$	Substitute the number of scarves
= $12	Evaluate
Profit for 5 scarves = $5 \cdot 5 - 5 \cdot 2$	Substitute the number of scarves
= $15	Evaluate

Plot the pairs of values found.

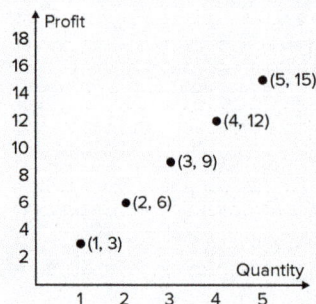

Purpose

Check that students can express a relation on a graph using the context of a real-world problem.

Expected mistakes

Students may use the cost of making each quantity of scarves as the y-values, rather than the profit. Make students aware that the profit (y-values) is the difference between the cost of making the scarves and what she sells them for.

Reflecting with students

Encourage students to consistently include units in their calculations and final answers to promote mathematical precision. When students calculate the profit for selling scarves, remind them to express their answers with the dollar sign, such as "$9" instead of just "9". This clarifies that the number represents dollars, not just an abstract figure.

Similarly, ask students to label the axes on their graphs with appropriate units on the x-axis and y-axis to ensure the graph is easily understood.

Students: Page 122

b Is this relation a function?

Create a strategy

This relation is a function if for every quantity sold, there is exactly one total profit.

Apply the idea

Checking each pair of values in the graph, each quantity sold is associated with only one total profit, so this relation does represent a function.

Purpose

Check that students can identify if a relation shown on a graph represents a function.

Students: Page 122

> 💡 **Idea summary**
>
> The relation is a function if each x-value input, or domain, is paired with exactly one y-value output, or range.

The vertical line test

This section begins with an exploration of how the vertical line test can be used to identify functions.

🎓 **Concrete-Representational-Abstract (CRA) approach**
Targeted instructional strategies

Concrete: Begin by engaging students with physical manipulatives to explore the concept of functions. Use input-output machines made from boxes or folders, where students place number cards (inputs) into the machine and receive corresponding output cards based on a specific rule. For example, set up a "machine" that adds 2 to any input number. Students will see that each input has exactly one output. You might also use string or yarn to connect input objects to output objects on a bulletin board, physically demonstrating the one-to-one relationship. This helps students understand that in a function, each input is linked to only one output.

Representational: Transition to visual representations by having students create tables of values and plot points on a coordinate grid. Encourage them to record the input and output pairs from the concrete activities into tables. Then, ask students to plot these ordered pairs on graph paper to create a visual graph of the relation. Introduce the vertical line test by demonstrating how vertical lines interact with the plotted points. Explain that if any vertical line passes through more than one point on the graph, the relation is not a function. Use this visual tool to help students understand how to identify functions from graphs of discrete points.

Abstract: Shift to working with abstract representations by having students identify functions from tables of values or sets of ordered pairs without visual aids. Present tables and sets of points and ask students to determine if they represent functions by analyzing the inputs and outputs. Teach students to check whether each input value corresponds to exactly one output value. Provide examples of relations where inputs have multiple outputs and discuss why these are not functions. Encourage students to apply the definition of a function to these abstract representations, reinforcing their understanding without the need for physical manipulatives or graphs.

Exploration

Students: Page 122

> **Interactive exploration**
> Explore online to answer the questions
>
> ⊕ **mathspace.co**

Use the interactive exploration in 3.02 to answer these questions.

1. What similarities did you notice in the graphs that were labeled as functions?
2. How did the vertical line help you determine which graphs represented functions?
3. Would a horizontal line be useful in determining if a relation is a function?

Suggested student grouping: In pairs

Students will use a GeoGebra applet to explore whether various relations are functions. The aim of the exploration is for students to discover that a relation is a function if the vertical line passes through the function at only one point.

Ideal student responses

These ideal responses may differ from other correct student responses. Less formal responses can be connected with the more precise mathematical language presented here.

1. **What similarities did you notice in the graphs that were labeled as functions?**
   In the graphs that were labeled as functions, the vertical line crossed the graph at only one point regardless of where it was positioned.

2. **How did the vertical line help you determine which graphs represented functions?**
   The vertical line helped to determine the graphs that represented functions by highlighting where there was only one output for every input. If a vertical line crossed the graph at more than one point, then there was more than one output for an input.

3. **Would a horizontal line be useful in determining if a relation is a function?**
   A horizontal line would not be useful because a function can have the same output (*y*-value) for different inputs (*x*-values).

Purposeful questions
- What is the definition of a function?
- How does this vertical line relate to the definition of a function?
- What do you notice about the relations that are not functions? How does the vertical line help show that it is not a function?

Possible misunderstandings
- For graphs with discrete domains, students might not see the points turn orange when the line intersects them. This could cause them to misunderstand the vertical line test. Have them drag the slider very slowly and stop when the line intersects the points to show them how the color of the points change.

Students will learn how to use the vertical line test to determine if the graph of a relation is a function or not. They will examine a graph where a vertical line passes through two points, showing the relation is not a function, and a graph where any vertical lines only crosses the graph at a single point, verifying it is a function.

Examples

The following supports may be useful for the examples in this section.

Stronger and clearer each time
English language learner support

Use this routine to help students clarify their reasoning as to why a relation is or is not a function. Give students a few minutes to individually justify their answers to Examples 4 and 5. Challenge them to write explanations that do not use the vertical line test as their only justification.

Next, provide time for students to meet in pairs to discuss their responses. Each pair will have 2–3 minutes to critique their reasoning, and students should meet with 2–3 different partners. In these pairs, students should ask their partner clarifying questions such as:

- Does it pass or fail the vertical line test? How do you know?
- Why can we use the vertical line test to say this is a function?
- How does the vertical line test verify that this is not a function?

After meeting with a few different partners, give students a few minutes to adjust and refine their initial justifications. Then, invite students to share their responses with the class. Highlight responses that use the definition of a function as justification.

Students: Page 123

Example 4

Determine whether or not the graph describes a function:

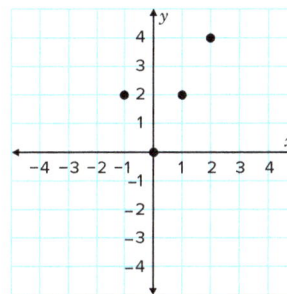

Create a strategy

Draw a vertical line through the points in the graph and check if it only crosses one point at a time.

Apply the idea

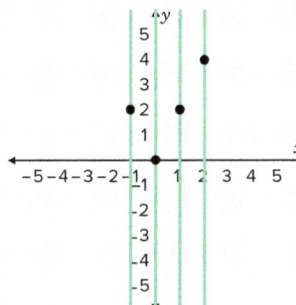

Each vertical line passes through one point at a time, so the graph is a function.

Purpose

Check that students can use the vertical line test to determine if a graph can be described as a function or a relation.

Expected mistakes

Students may misunderstand the concept of the vertical line test. They may need help connecting that the vertical line test checks for x-values that have more than one output. If the vertical line passes through multiple points, it would show an x-value that has more than one output.

Reflecting with students

Challenge advanced learners to explain their answer without using the vertical line test as their justification. Refer them back to the definition of a function if they need help getting started.

An example explanation could be, "This function is defined by four discrete points. Each point has a distinct x-value, which means that there is only one output for each of the four distinct inputs. Therefore, this graph represents a function."

Students: Page 124

Example 5

Use the vertical line test to justify whether the table shown is a function.

x	−2	4	−1	2	−3	−2	1
y	1	5	−4	3	5	−1	4

Create a strategy

Graph the points on the coordinate plane and determine if a vertical line passes through more than one point of the relation.

Apply the idea

After graphing the points and using a vertical line to check for points that share an x-value, we see

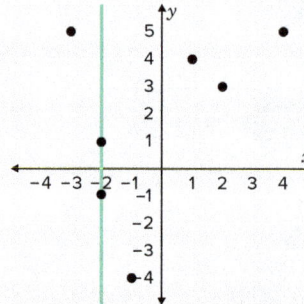

Since the vertical line passes through both (−2, 1) and (−2, −1), the relation is not a function.

Purpose

Show students how to use the vertical line test to determine if a relation is a function.

Idea summary

While all functions are relations, not all relations are functions.

The vertical line test for functions:

When looking at a graph, if you can draw a vertical line anywhere so that it crosses the graph of the relation in more than one place, then it is not a function.

Practice

Students: Page 124–129

What do you remember?

1   Express this relation in a table of values.

$$\left\{\left(-3, \frac{1}{2}\right), (-2, -0.5), (-1, 0), \left(0, -\frac{2}{3}\right), (1, 1.5), \left(2, -\frac{5}{4}\right), (3, 2.75), (4, -3.5)\right\}$$

2   Express this relation as a series of points on the coordinate plane.

$$\{(-3, 2), (-1, 2), (0, 0), (2, -1), (3, -1), (3, 1), (4, 3), (6, -2)\}$$

3   Express the relation shown on the coordinate plane in a table of values.

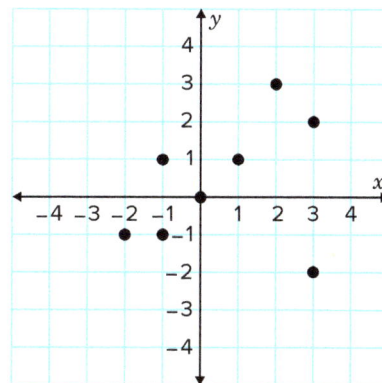

4 Determine which statement(s) are true. Select all that apply.

    A All relations are functions.   **B** Some relations are functions.

    C All functions are relations.   **D** No functions are relations.

    E No relations are functions.

5 Classify the following statements as describing a relation or a function:

    a A graph where, for some value of x, there exists two or more different values of y.

    b Substituting a certain value of x into the rule gives only one value of y.

    c An imaginary vertical line passes through more than one point on the graph.

Let's practice

6 Classify the sets of points as a relation or a function:

    a $\{(1, 5), (1, 1), (7, -2), (-5, -10)\}$ **b** $\{(1, 5), (7, -2), (-5, -10), (13, -13)\}$

    c $\{(1, 5), (1, 7), (-2, -5), (-5, -10)\}$ **d** $\{(-1, -9), (0, 0), (1, 9), (2, 18)\}$

    e $\{(-2, 4), (-1, 1), (2, 4), (6, 36)\}$ **f** $\left\{(0, 0), \left(1, \frac{1}{3}\right), \left(2, \frac{2}{3}\right), (3, 1)\right\}$

7 The pairs of values in the table represent a relation between x and y. Determine whether they represent a function:

    a

x	−4	−3	−2	−1	0	1	2	3	4
y	−4	−3	−2	−1	0	−1	−2	−3	−4

    b

x	0	1	4	8	9	12	16	18	20
y	0	1	2	$2\sqrt{2}$	3	$2\sqrt{3}$	4	$3\sqrt{2}$	$2\sqrt{5}$

    c

x	−9	−7	−6	−5	−3	−2	3	5	10
y	10	10	10	10	10	10	10	10	10

    d

x	−11	−8	−5	1	5	7	10	10	13
y	11	14	−15	−12	−14	−3	−8	13	−9

8 Determine whether each graph is a relation or a function:

    a

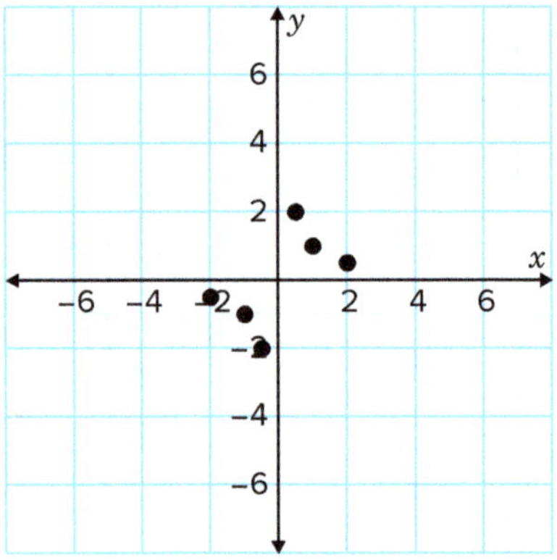

    b

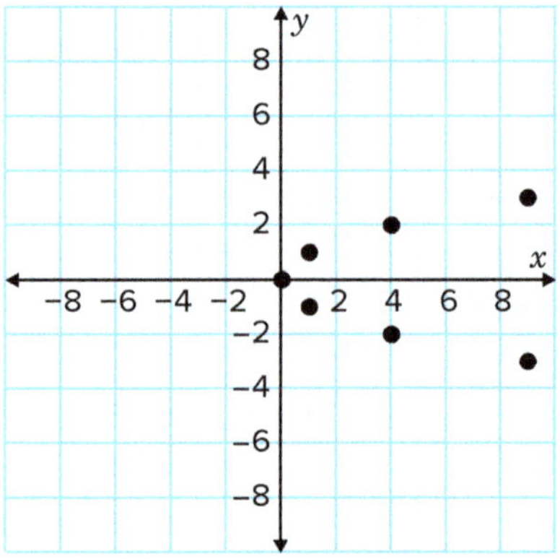

c

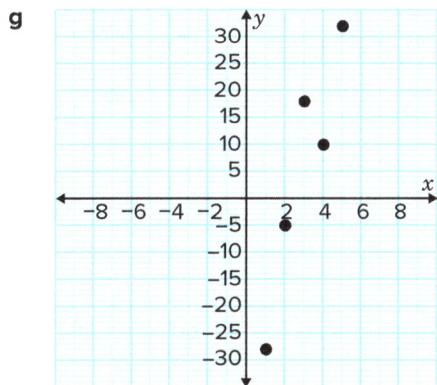

d

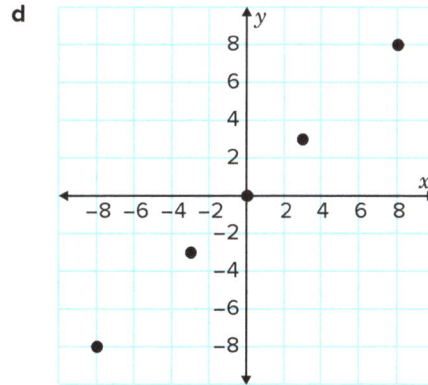

e

f

g

h

9 Select which of the following is a function.

A {(2, 5), (7, −3), (5, 2), (−4, −9)}

B {(2, 5), (2, 7), (−3, −4), (−9, 13)}

C

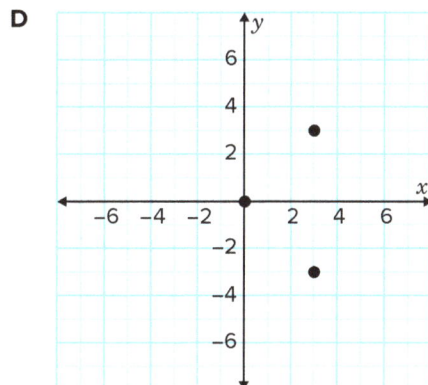

D

10 Consider:

x	−4	−3	0	−1	0	1	2	3	4
y	4	3	2	1	0	1	2	3	4

a Plot the points on a coordinate plane.   **b** Do they represent a function?

11 A relation is defined as:

$y = 1$ if x is positive and $y = -1$ if x is 0 or negative.

a Complete the table for this relation:

x	−4	−3	−2	−1	0	1	2	3	4
y									

b Plot the points on a coordinate plane.

c Do these values represent a function?

12 Quick Connect offers internet service at $1.50 per hour plus a service fee of $0.50 for access.

a Complete the table.

b Is this relation a function?

Internet usage (hours)	Total cost (dollars)
1	
2	
3	
4	
5	

13 A particular internet service provider charges $40 a month on their 20 GB plan and charges $10 for each additional 10 GB used.

a Complete the table.

b Is this relation a function?

Total GB used	Total charge (dollars)
10	
20	
30	
40	
50	

14 Consider the ordered pairs:

$$\{(-9, -5), (-5, -10), (-5, -4), (-3, 7), (-2, -4), (-1, 1)\}$$

a Plot the ordered pairs on a coordinate plane.

b Which ordered pair would need to be removed from the set so that the remaining ordered pairs represent a function?

15 Consider the set of ordered pairs where the x-values are prime numbers less than 10 and the corresponding y-values are the remainder when x is divided by 3.

Which of the following statements is true about the set?

A It is a function because each input x has exactly one output y.

B It is not a function because some inputs x have more than one output y.

C It is not a function because it consists of ordered pairs.

D It is not a function because not all prime numbers are included.

16 Which table of values is a function? Explain your reasoning.

Table 1:

Input	Output
A	1
B	2
C	3
D	4
E	1

Table 2:

Input	Output
1	A
2	B
3	C
4	D
1	E

17 Which set of ordered pairs is a function? Explain your reasoning.
- $\{(-1, 2), (0, 3), (1, 4), (2, 5), (3, 6)\}$
- $\{(2, -1), (2, 0), (2, 1), (2, 2), (2, 3)\}$

Let's extend our thinking

18 Explain how we can identify functions from a set of points, a table, and a graph. Use the given examples to support your answer.

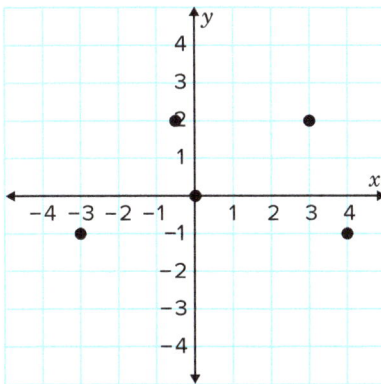

x	-4	-4	-2	-1	0
y	7	11	15	19	23

$\{(4, 2), (-7, -11), (-4, 2), (0, 2), (4, 2)\}$

19 The sets of points represent a relation between x and y. Find a value of k so that the relation does not represent a function:

   a $\{(8, 6), (5, k), (4, 9), (2, 1)\}$ **b** $\{(k, 3), (9, 8), (1, 4), (5, 7)\}$ **c** $\{(9, k), (6, 2), (5, 3), (1, 4)\}$

   d $\{(2, 9), (7, 6), (k, 4), (3, 1)\}$ **e** $\{(k, 5), (8, 9), (1, 4), (3, 2)\}$ **f** $\{(6, 9), (k, 4), (2, 8), (7, 3)\}$

   g $\{(k, 6), (2, 8), (7, 4), (5, 9)\}$ **h** $\{(5, k), (1, 7), (2, 9), (3, 8)\}$

20 Given the relations represented by the tables below, complete the table to make it a function.

a

x	-1.5	-0.5	0.5	1.5
y		7		11

b

x				
y	-5	0	5	10

c

x	1	2	3	4
y	2		2	

d

x	-3	-3	-3	-3
y	4		4	

21 Create a graph on a coordinate plane of a function with at least five points.

22 Daniel bakes cookies to sell at the local fair. It costs him \$2.50 to bake a dozen, and he sells them for \$5.00 per dozen.

   a   Consider when 1, 2, 3, 4, and 5 dozen cookies are sold. Plot the points representing the relation between the number of dozens sold and his total profit.

   b   Determine whether the relation is a function. Explain your reasoning.

23 A snack vending machine in a community center dispenses different types of snacks like chips, candy bars, gum, and cookies. The following table shows the selected snack type and the dispensed item.

Selected	chips	candy bar	gum	chips	candy bar	cookies
Dispensed	chips	gum	gum	chips	candy bar	cookies

    a   Is this machine operating correctly?

    b   Is this relation a function? Explain your reasoning.

24 You have been given the task to manage the school's snack bar during a sports event. The bar sells only one kind of snack at a fixed price, and you need to keep track of sales. Create a set of ordered pairs that represent the number of snacks sold and the total money made, assuming each snack costs $2.00.

    a   Create your own set of ordered pairs for the snacks sold and the total revenue. Ensure your set of ordered pairs represents a function.

    b   Explain why the relation you created is a function.

Answers

What do you remember?

1

x	y
−3	$\frac{1}{2}$
−2	−0.5
−1	0
0	$-\frac{2}{3}$
1	1.5
2	$-\frac{5}{4}$
3	2.75
4	−3.5

2

3

x	y
−2	−1
−1	1
−1	−1
0	0
1	1
2	3
3	−2
3	2

4 **B** and **C**

5 **a** Relation   **b** Function   **c** Relation

Let's practice

6 **a** Relation   **b** Function   **c** Relation
    d Function   **e** Function   **f** Function

7 **a** Yes   **b** Yes   **c** Yes   **d** No

8 **a** Function   **b** Relation   **c** Function
    d Function   **e** Function   **f** Relation
    g Function   **h** Function

9 **A**

10 **a**

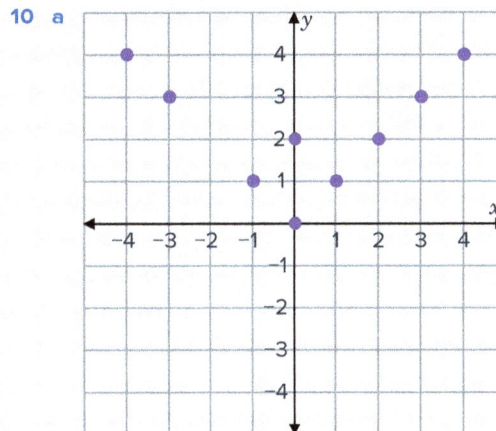

   b No

11 **a**

x	−4	−3	−2	−1	0	1	2	3	4
y	−1	−1	−1	−1	−1	1	1	1	1

   b

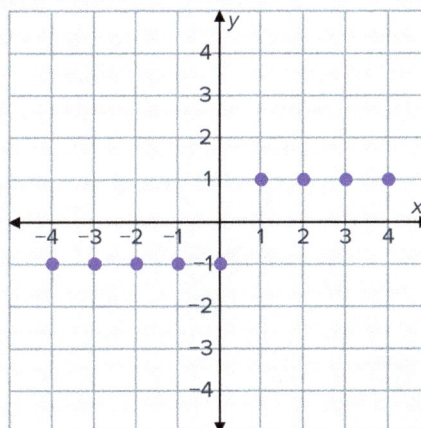

   c Yes

12 a

Internet usage (hours)	Total cost (dollars)
1	2.00
2	3.50
3	5.00
4	6.50
5	8.00

b Yes, because each amount of usage time has exactly one cost associated with it.

13 a

Total GB used	Total charge (dollars)
10	40
20	40
30	50
40	60
50	70

b Yes

14 a

b $(-5,-10)$ or $(-5,-4)$

15 A

16 The first table because each input from the set $\{(A, B, C, D, E)\}$ is associated with exactly one output. The second table is not a function because the input value 1 is associated with two output values A and E.

17 The first set is a function because each input has a unique output. The second set is not a function because the input 2 is associated with multiple outputs.

Let's extend our thinking

18 When identifying a function from a table you scan the graph from left to right to see if any points overlap vertically. If points overlap vertically they are a single x-value with two different y-values. In a table of values or a list of ordered pairs you look to see if there are any x-values going to two different y-values.

19 a No such value of k exists.

b 5 or 9 or 1

c No such value of k exists.

d 3 or 2 or 7

e 8 or 1 or 3

f 6 or 2 or 7

g 2 or 7 or 5

h No such value of k exists.

20 a Answers may vary.

x	y
−1.5	5
−0.5	7
0.5	9
1.5	11

b Answers may vary.

x	y
1	−5
2	0
3	5
4	10

c Answers may vary.

x	y
1	2
2	2
3	2
4	2

d Answers may vary.

x	y
−3	4
−3	4
−3	4
−3	4

21 Answers may vary

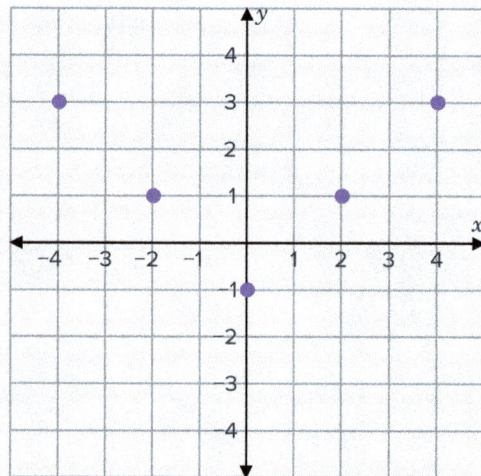

22 a

Profit (dollars) vs Dozens of cookies

b Yes, because each amount of dozens sold has exactly one profit value associated with it, which satisfies the definition of a function.

23 a No, because the selected item does not always match the dispensed item.

b No. For a relation to be considered a function, each input should be associated with exactly one output. In this case, the same selected snack type sometimes leads to different dispensed items, so it does not meet the definition of a function.

24 a Answers may vary. Example of ordered pairs could be: (1, 2.00) (2, 4.00), (3, 6.00), (4, 8.00), (5, 10.00), ... (10, 20.00)

b The relation is a function because each input (number of snacks sold) has exactly one unique output (total money made), which satisfies the definition of a function.

3.03 Domain and range

Subtopic overview

Lesson narrative

In this lesson, students will explore the concepts of domain and range, which describe the set of all possible input (x-values) and output (y-values) for a given relation, respectively. The lesson defines the domain as the set of all independent variable values and the range as the set of all dependent variable values. Students will learn to identify and list these values from various representations, such as ordered pairs, tables, and graphs. The lesson provides several examples for practice. Students will determine the domain and range for specific sets of ordered pairs and graphical representations. They will also analyze whether a given relation is a function by checking if each input corresponds to only one output. Through these exercises, students will apply strategies to list unique values in the domain and range and use the vertical line test to verify if a relation is a function. By the end of the lesson, students should be able to confidently identify the domain and range of relations and determine if these relations are functions.

Learning objectives

Students: Page 130

🎓 **After this lesson, you will be able to...**
- identify the domain of a function represented as a set of ordered pairs, a table, or a graph of discrete points.
- identify the range of a function represented as a set of ordered pairs, a table, or a graph of discrete points.

Key vocabulary

- discrete
- domain
- range

Essential understanding

The domain is all the possible inputs or x-values of a function. The range is all the possible outputs or y-values of the function.

Standards

This subtopic addresses the following Virginia 2023 Mathematics Standards of Learning standards.

Mathematical Process Goals

MPG1 — Mathematical Problem Solving

Teachers can integrate this goal into their instruction by providing students with practice problems that require them to identify the domain and range of functions represented in different ways. Teachers can encourage students to apply problem-solving strategies to these problems, such as analyzing the problem, devising a plan, carrying out the plan, and checking the solution. By solving these problems, students can develop their problem-solving skills and apply their understanding of domain and range in various contexts.

MPG4 — Mathematical Connections

Teachers can integrate this goal by linking the concepts of domain and range to students' prior knowledge and to real-world contexts. For example, teachers can show students how their prior knowledge of ordered pairs, tables of values, and graphs is connected to identifying the domain and range of functions. They can also use real-world examples, such as the number of hours worked and the amount earned, to illustrate how functions, and specifically the concepts of domain and range, are used in everyday life.

MPG5 — Mathematical Representations

Teachers can demonstrate this by teaching students how to represent the domain and range of functions in different ways, such as ordered pairs, tables of values, and graphs of discrete points. They can guide students to understand that these different representations can convey the same mathematical idea: the relationships between independent and dependent variables in functions. Teachers can also help students connect these mathematical representations to realworld contexts, illustrating the process and product of mathematical representation.

Content standards

8.PFA.2 — The student will determine whether a given relation is a function and determine the domain and range of a function.

8.PFA.2b — Identify the domain and range of a function represented as a set of ordered pairs, a table, or a graph of discrete points.

Prior connections

7.PFA.1 — The student will investigate and analyze proportional relationships between two quantities using verbal descriptions, tables, equations in $y = mx$ form, and graphs, including problems in context.

Future connections

A.F.2 — The student will investigate, analyze, and compare characteristics of functions, including quadratic and exponential functions, and model quadratic and exponential relationships.

Lesson Preparation

Suggested review

Depending on your students' level of prior knowledge, consider revisiting the following lessons:

Grade 8 — 3.01 Review: Plot points and represent relations
Grade 8 — 3.02 Identify functions

Student lesson & teacher guide

Domain and range

Students will learn the definitions of domain and range and discover how to identify the domain and range of a relation expressed as a set of ordered pairs, a table of values, and a graph.

Examples

Students: Page 130

Example 1

Consider the relation {(1, 2), (5, 3), (2, 7), (5, −1)}. State the domain and range.

Create a strategy

Look at the x-coordinates to determine the domain, then look at the y-coordinates for the range.

Apply the idea

The domain is {1, 2, 5} and the range is {−1, 2, 3, 7}.

Reflect and check

Notice that the value of 5 was only included once in the domain. That is because we are only concerned with what all of the possible x values are and not how many times they showed up.

This relation is not considered a function because the input 5 corresponds to two different outputs, 3 and −1. In a function, each input can correspond to only one output.

Purpose

Ensure that students can find the domain and range of function written as a set of ordered pairs.

Expected mistakes

Students may write the y-values as the domain and the x-values as the range.

Reflecting with students

Note that it is common to write the domain and range values in ascending order, but it is not required. Remind students that they only need to write domain and range values once, even if they show up multiple times.

Students: Page 131

Example 2

Consider the relation in the table.

a   What is the domain of the relation?

x	1.5	6	3	8.2	2
y	3	2.8	7	1	2.4

Create a strategy

Write all the x-values of the relation.

Apply the idea

Domain = {1.5, 6, 3, 8.2, 2}

Purpose

Check that students can find the domain of a relation expressed as a table of values.

Reflecting with students

Encourage students to use precise mathematical notation when stating the domain of a relation. Instead of simply listing the x-values like "1.5, 6, 3, 8.2, 2," guide them to present the domain using proper mathematical notation. Emphasize the importance of using curly braces to denote a set and explicitly labeling it as the domain.

Students: Page 131

---

b   What is the range of the relation?

Create a strategy

Write all the y-values of the relation.

Apply the idea

Range = {3, 2.8, 7, 1, 2.4}

Purpose

Check that students can find the range of a relation expressed as a table of values.

Reflecting with students

Teach advanced learners how to write the domain and range using proper set notation. In set notation, the curly brackets denote a set, and the symbol $\in$ can be read as "belongs to." In proper set notation, the domain and range can be written as:

$$x \in \{1.5, 6.3, 8.2, 2\}$$
$$y \in \{3.2, 8.7, 1.2, 4\}$$

This notation helps us see that x can only be one of the values listed in the first set, and y can only be one of the values listed in the second set.

Students: Page 131

c   Is this relation a function?

Create a strategy

A relation is a function if every x-value in the domain has only one corresponding y-value in the range.

Apply the idea

The x-value in the domain has only one corresponding y-value in the range. So, this relation is a function.

Purpose

Check that students can use the domain of a relation to determine if it is also a function.

Reflecting with students

Ask students to give you a point that could be added to the list of ordered pairs that would make this a relation but not a function.

Critique, correct, and clarify
English language learner support

use with Example 2

Provide students with an incorrect answer about the domain or range of the function. For example, present them with a written response like: "The range of this relation is {1.5, 2, 3, 6, 8.2}." Ask students to read this statement carefully and identify any errors or misunderstandings in the reasoning.

Encourage students to discuss in pairs or small groups what is incorrect about this identification. Ensure that students realize that the range should consist of the y-values of the relation, not the x-values, and invite them to correct the statement.

Students: Page 131

Example 3

Consider the relation on the graph.

a   What is the domain of the relation?

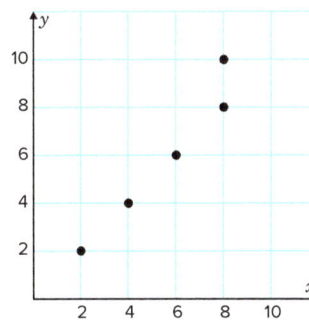

3.03 Domain and range   **235**

mathspace.co

Create a strategy

Write all the unique x-values of the points.

Apply the idea

First, identify the coordinates of the points in the graph:

Then write the unique x-values:

$$Domain = \{2, 4, 6, 8\}$$

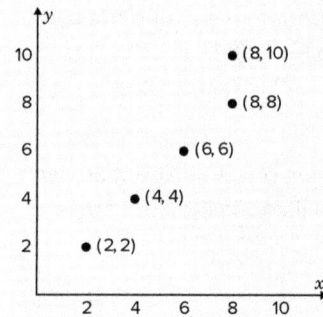

Purpose

Check that students can find the domain of a relation expressed on the coordinate plane.

Expected mistakes

Students might not consider the scale of each axis and instead assume that the points have consecutive, integer coordinates. Remind them of the importance of carefully reading the scale of each axis.

Students: Page 132

b   What is the range of the relation?

Create a strategy

Write all the unique y-values of the points from the graph in part (a).

Apply the idea

$$Range = \{2, 4, 6, 8, 10\}$$

Purpose

Check that students can find the range of a relation expressed on the coordinate plane.

Students: Page 132

c   Is this relation a function?

Create a strategy

A relation is a function if every x-value in the domain has only one corresponding y-value in the range.

Apply the idea

The x-value of 8 in the domain has two corresponding y-values of 8 and 10 in the range. So, this relation is not a function.

Reflect and check

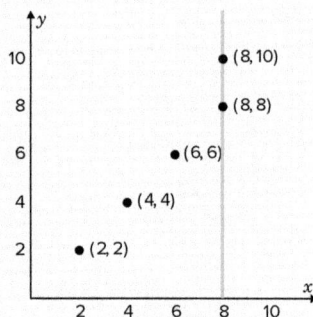

By applying the vertical line test to the graph, we can see that a vertical line drawn at $x = 8$ would intersect the graph at two points. This confirms that the relation is not a function since there's an x-value that is paired with more than one y-value.

Purpose

Check that students can use the domain of a relation to determine if it is also a function.

Reflecting with students

Ask students to give you a point that could be taken away from the graph to make it a function.

Identifying domain and range from different representations   use with Example 3

Student with disabilities support

Some students may find it challenging to identify the domain and range directly from a graph, especially if they have difficulties with visual-spatial processing. Encourage students to convert the points on the graph into a table of values. Guide them to list the ordered pairs in a two-row table, with one row for the x-values (domain) and another for the y-values (range). Providing a table template can support their organizational skills and make the task less daunting.

This approach allows them to work with numerical data in a format that may be more accessible and familiar to them. As they fill in the table, students can more easily identify the unique domain and range values by examining the entries. This strategy helps students make connections between different representations of relations and reinforces their understanding of the underlying concepts.

Students: Page 132

> **Idea summary**
>
> **Domain** - a relation's inputs, or x-values
>
> **Range** - a relation's outputs, or y-values
>
> The domain and range of a discrete relation are typically written in ascending order without repeating values.

Practice

Students: Page 132–137

What do you remember?

1 Identify the following mathematical terms.

    a A set of ordered pairs which represent a relationship

    b A special type of relation, where each input only has one output

    c All of the possible x values of a relation

    d All of the possible y values of a relation

2 Consider the following representations of a function:

   i   Identify the input value(s).

   ii Identify the output value(s).

a

x	y
−5	1.5
0	−1
7	−6

b

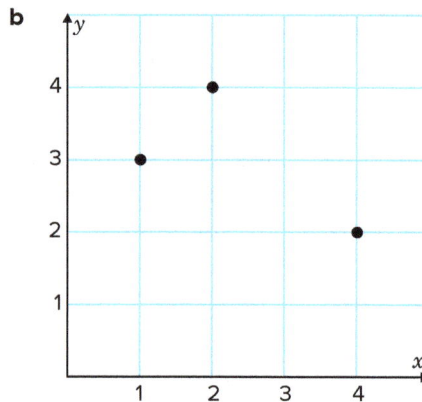

c $\left\{\left(\frac{4}{5}, 0\right), \left(0, 2\frac{1}{2}\right), \left(-2, -\frac{1}{5}\right)\right\}$

3 Consider the function represented by the following set of ordered pairs:

$\{(0, -3), (2, 0), (13, 2)\}$

Based on the set of ordered pairs, answer the following questions:

   a What is the input value associated with the output value of 2?

      A −3   **B** 0   **C** 2   **D** 13

   b Fill in the blank with the correct output value when the input value is 0:

      $0 \rightarrow \square$

   c Identify the complete range of the function.

      A $\{0, 2, 13\}$ **B** $\{-3, 0, 2\}$ **C** $\{0, 2, 13\}$ **D** $\{-3, 0, 13\}$

4 Consider the function represented by the graph:

   a What is the input value associated with the output value of 0?

      A −3   **B** 0   **C** 2   **D** 3

   b Fill in the blank with the correct output value when the input value is 3:

      $3 \rightarrow \square$

   c Identify the complete domain of the function.

      A $\{-4, 0, 3\}$ **B** $\{-2, 1, 2\}$

      C $\{-4, -2, 2\}$ **D** $\{-4, 2, 3\}$

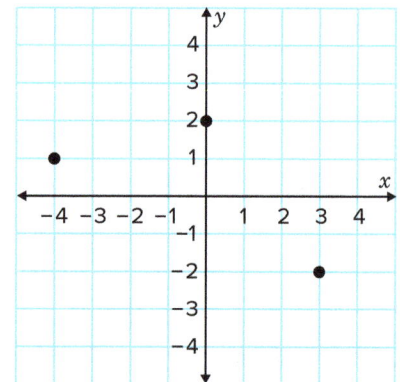

5 Consider the table of values of a function:

   a Write the ordered pairs represented by the table.

   b List the domain of the points.

x	−3	2	0
y	1	−2	3

      A $\{-3, 0, 2\}$ **B** $\{1, -2, 3\}$ **C** $\{-3, 2, 3\}$ **D** $\{-2, 0, 1\}$

   c List the range of the points.

      A $\{-3, 0, 2\}$ **B** $\{1, -2, 3\}$ **C** $\{-3, 2, 3\}$ **D** $\{-2, 1, 3\}$

6 The table lists values of a function f at selected points. Complete the domain and range for the function f given the table.

   a Domain: $\{-4, -1, 0, \square, 5\}$

   b Range: $\{-4, \square, 0, 1, 3\}$

x	−4	−1	0	2	5
$f(x)$	0	3	−2	1	−4

7 Consider the ordered pairs of the function:

$$A(-2, 4) \qquad B(1, -1) \qquad C(3, 2)$$

    a Fill in the blanks to complete the domain of the points.

        Domain: {☐, ☐, ☐}

    b Fill in the blanks to complete the range of the points.

        Range: {☐, ☐, ☐}

Let's practice

8 For each of the following relations:

    i Find the domain.

    ii Find the range.

    iii Determine whether or not the relation is a function.

    a {(8, 5), (1, 3), (3, 4), (9, 1), (2, 9)}   **b** {(2, 3), (3, 8), (6, 1), (8, 7), (3, 2)}

    c {(4, 4), (8, 9), (2, 8), (7, 9), (3, 1)}   **d** {(16, 14), (8, 28), (−15, 14), (−20, −28)}

    e $\left\{(7, 0.5), \left(-5, \frac{3}{4}\right), (5, 0.75), \left(7, \frac{1}{2}\right)\right\}$ **f** $\left\{\left(\frac{2}{3}, 12\right), \left(-\frac{1}{2}, 0\right), \left(0, \frac{1}{2}\right), \left(\frac{2}{3}, -12\right)\right\}$

9 For each of the following relations:

    i Find the domain.

    ii Find the range.

    iii Determine whether or not the relation is a function.

    a

x	1	6	3	8	2
y	3	2	7	1	2

    b

x	7	7	8	5	3
y	1	9	3	2	6

    c

x	2	6	9	1	5
y	6	5	1	8	3

    d

x	2	2	2	2	2
y	−1	1	3	5	7

    e

x	−10	−20	−30	−20	−10
y	−5	−4	−3	−2	−1

    f

x	−4	−1.5	0	1.5	4
y	0.75	1.5	3.25	4	4.75

10 For each of the following relations:

    i Find the domain.

    ii Find the range.

    iii Determine whether or not the relation is a function.

    a

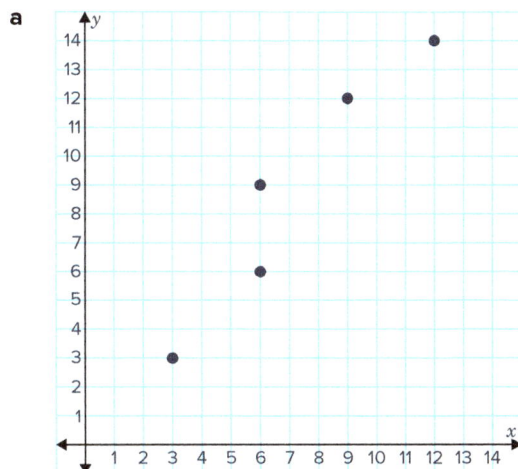

    b

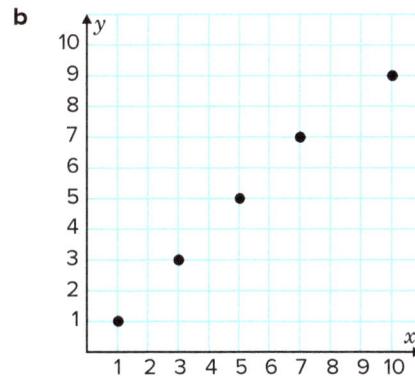

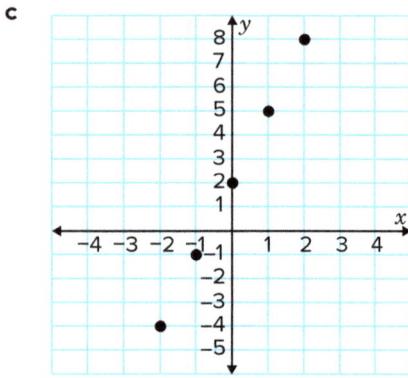

c

d

e

f

SOL 11 Which set has a domain of $\{2, 3\}$ and a range of $\{5, 8\}$?

  A $\{(2, 8), (3, 5), (5, 8)\}$ **B** $\{(3, 5), (2, 5), (8, 3)\}$ **C** $\{(2, 8), (3, 5)\}$ **D** $\{(5, 2), (8, 3)\}$

SOL 12 For the following sets of ordered pairs, select the one that has a domain of $\{-3, 0, 4\}$ and a range of $\{-1, 2, 5\}$.

  A $\{(-3, -1), (0, 2), (4, 5)\}$ **B** $\{(-3, 5), (0, -1), (4, 2)\}$ **C** $\{(0, -1), (4, 5), (-3, 2)\}$ **D** $\{(4, -1), (-3, 5), (0, 2)\}$

13 Given the domain of $\{-2, 0, 3\}$ and a range of $\{-1, 1, 4\}$, determine whether each of the following relations have the given domain and range.

  a $\{(-2, 0), (3, -1), (1, 4)\}$ **b**

x	−2	0	3
y	−1	1	4

  c

14 Given the domain $\{-4, 0, 5\}$ and the range $\{-2, 3, 6\}$, complete the table to show a relation that matches the given domain and range.

x	−4		5
$f(x)$		3	

15 Given the domain {−3, 1, 4} and the range {−2, 0, 3}, plot the missing point on the coordinate plane that completes the relation with the given domain and range.

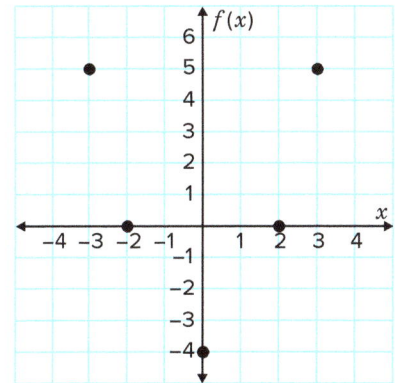

16 Given the domain {−1, 2, 4} and the range {3, −2, 0}, complete the list of ordered pairs to match the given domain and range.

$$\{(-1, 3), (2, \square), (\square, 0)\}$$

17 Consider the list of ordered pairs: {(1, 4), (−2, −3), (3, 1), (0, −5), (−4, 2)}

Fill in the blanks to complete the domain and range.

a Domain: {□, −2, 0, 1, 3}

b Range: {□, −3, 1, 2, 4}

18 Consider the graph of a function f. Based on the graph, complete the domain and range.

a Domain: {−3, −2, □, 2, 3}

b Range: {−4, 0, □}

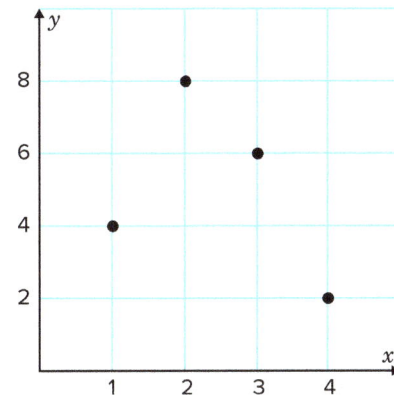

Let's extend our thinking

19 Consider the following representations of Function A and Function B.

Function A:

x	1	2	3	4
y	2	4	6	8

Function B:

a Identify the domain and range of Function A.

b Identify the domain and range of Function B.

c Compare and contrast the domain and range of Function A and Function B.

20 At an indoor ski facility, the temperature is set to 18° F at the opening time of 9 a.m. At 10 a.m., the temperature is immediately brought down to 5° F and left for 3 hours before immediately taking it down again to –10° F where it stays until the close time of 7 p.m. Let x be the number of hours since opening and y be the temperature of the ski facility.

    a State the domain of the function.

    b Kate entered the ski facility at 2:30 p.m. What was the temperature inside the facility?

    c State the range of the function.

21 The graph of a function is shown.

    a State the range of the function.

    b A ball is thrown from an apartment window in a high-rise building. The height of the ball above ground over time can be modeled by the function shown in the graph, where the ball is thrown at $x = 0$.

        i State the range of the function for this context using words.

        ii State the range of the function for this context mathematically.

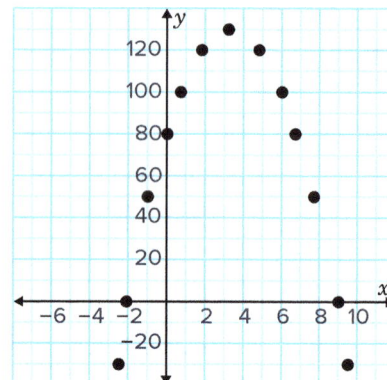

What do you remember?

1 **a** Relation   **b** Function

     c Domain   **d** Range

2 **a** **i** Input = −5, 0, 7   **ii** Output = −6, −1, 1.5

     b **i** Input = 1, 2, 4   **ii** Output = 2, 3, 4

     c **i** Input = $-2, 0, \frac{4}{5}$ **ii** Output = $-\frac{1}{5}, 0, 2\frac{1}{2}$

3 **a** **D**   **b** $0 \rightarrow -3$ **c** **B**

4 **a** **C**   **b** $3 \rightarrow -2$ **c** **A**

5 **a** A: (−3, 1), B: (2, −2), C: (0, 3)   **b** **A**

     c **B**

6 **a** Domain: {−4, −1, 0, 2, 5}   **b** Range: {−4, −2, 0, 1, 3}

7 **a** Domain: {−2, 1, 3}   **b** Range: {−1, 2, 4}

Let's practice

8 **a** **i** Domain = 8, 1, 3, 9, 2

     ii Range = 5, 3, 4, 1, 9

     iii Function

   b **i** Domain = 2, 3, 6, 8

     ii Range = 3, 8, 1, 7, 2

     iii Not a function

   c **i** Domain = 4, 8, 2, 7, 3

     ii Range = 4, 8, 9, 1

     iii Function

   d **i** Domain = 16, 8, −15, −20

     ii Range = 14, 28, 14, −28

     iii Function

   e **i** Domain = −5, 5, 7

     ii Range 0.5, 0.75 or Range = $\frac{1}{2}, \frac{3}{4}$

     iii Function

   f **i** Domain = $\frac{1}{2}, 0, \frac{2}{3}$

     ii Range = $-12, 0, \frac{1}{2}, 12$

     iii Not a function

9 **a** **i** Domain = 1, 6, 3, 8, 2

     ii Range = 3, 2, 7, 1

     iii Function

   b **i** Domain = 7, 8, 5, 3

     ii Range = 1, 9, 3, 2, 6

     iii Not a function

   c **i** Domain = 2, 6, 9, 1, 5

     ii Range = 6, 5, 1, 8, 3

     iii Function

   d **i** Domain = 2

     ii Range = −1, 1, 3, 5, 7

     iii Not a function

   e **i** Domain = −30, −20, −10

     ii Range = −5, −4, −3, −2, −1

     iii Not a function

   f **i** Domain = −4, −1.5, 0, 1.5, 4

     ii Range = 0.75, 1.5, 3.25, 4, 4.75

     iii Function

10 **a** **i** Domain = 3, 6, 9, 12

     ii Range = 3, 6, 9, 12, 14

     iii Not a function

   b **i** Domain = 1, 3, 5, 7, 10

     ii Range = 1, 3, 5, 7, 9

     iii Function

   c **i** Domain = −2, −1, 0, 1, 2

     ii Range = −4, −1, 2, 5, 8

     iii Function

   d **i** Domain = 0, 1, 4, 9

     ii Range = 0, 2, −2, 4, −4, 6, −6

     iii Function

   e **i** Domain = −8, 0, 8

     ii Range = −8, 0, 8

     iii No

   f **i** Domain = 0, 4, 9

     ii Range = 0, 4, 6

     iii Yes

11 **C**

12 **A**

13 **a** No   **b** Yes   **c** No

14 Answers may vary.

x	$f(x)$
−4	5
0	3
5	−2

15

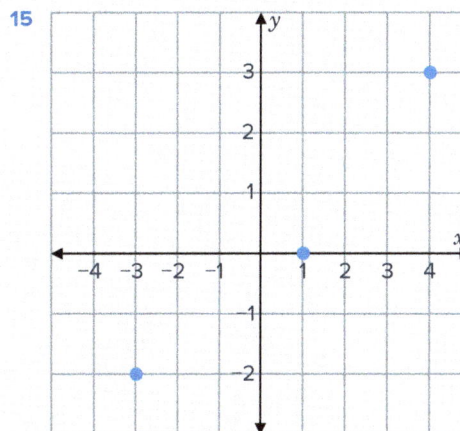

16 $\{(-1, 3), (2, -2), (4, 0)\}$

17 a Domain: $\{-4, -2, 0, 1, 3\}$

    b Range: $\{-5, -3, 1, 2, 4\}$

18 a Domain: $\{-3, -2, 0, 2, 3\}$

    b Range: $\{-4, 0, 5\}$

Let's extend our thinking

19 a The domain of Function A is $\{1, 2, 3, 4\}$ and the range is $\{2, 4, 6, 8\}$.

    b The domain of Function B is $\{1, 2, 3, 4\}$ and the range is $\{2, 4, 6, 8\}$.

    c Both Function A and Function B have the same domain $\{1, 2, 3, 4\}$ and the same range $\{2, 4, 6, 8\}$. These functions are represented in different ways, but have the same domain and range.

20 a All rational numbers from 0 to 10.

    b $-10°F$

    c $\{-10, 5, 18\}$

21 a $\{-30, 0, 50, 80, 100, 120, 130\}$

    b i A reasonable range for this context would include all y-values greater than or equal to 0 and all y-values less than or equal to 130. The ball won't fall below ground level, but should include all y-values in between the points on the graph up to 130.

       ii $y \geq 0$ and $y \leq 130$.

3.04 Independent and dependent variables

Subtopic overview

Lesson narrative

In this lesson, students will learn about independent and dependent variables in the context of bivariate data. The lesson begins by defining bivariate data as numerical data involving two variables organized into pairs. It explains how the independent variable is the one that can be controlled or chosen freely, while the dependent variable changes in response to the independent variable. Students will explore examples to identify these variables, such as the relationship between time spent lifting weights and the number of pull-ups a person can do, with time being the independent variable and pull-ups being the dependent variable. The lesson includes activities where students analyze scenarios like the effect of the season on the number of people at the beach and the relationship between months and savings. They will plot data points on a graph, placing the independent variable on the horizontal axis and the dependent variable on the vertical axis. Through these exercises, students will develop a clear understanding of how to identify and differentiate between independent and dependent variables, and how to represent their relationships graphically. By the end of the lesson, students should be proficient in determining which variable influences the other and in using graphs to illustrate these relationships.

Learning objectives

Students: Page 138

After this lesson, you will be able to...
- identify the independent and dependent variables given a graph, table, or a situation in context.
- describe the independent and dependent variables of a function.

Key vocabulary

- bivariate data
- input

- dependent variable
- output

- independent variable

Essential understanding

When two variables are related, the dependent variable changes based on the value of the independent variable. Because of this the independent variable is the input and the dependent variable is the output.

Standards

This subtopic addresses the following Virginia 2023 Mathematics Standards of Learning standards.

Mathematical Process Goals

MPG3 — Mathematical Reasoning

Teachers can integrate mathematical reasoning into their instruction by guiding students through the process of deducing the relationship between independent and dependent variables in various types of functions. Teachers can also help students use logical reasoning to analyze and justify their conclusions about the domain and range of a function. Encouraging students to explain their reasoning when identifying variables in practice problems can further develop their mathematical reasoning skills.

MPG4 — Mathematical Connections

Teachers can create mathematical connections by linking the concept of independent and dependent variables to other mathematical topics and real-world situations. For example, teachers can relate the concept of functions to students' prior knowledge of algebraic expressions. Additionally, teachers can use examples from real-world situations to illustrate the practical applications of these concepts, thereby establishing a connection between mathematics and the real world.

MPG5 — Mathematical Representations

Teachers can integrate the goal of mathematical representations by demonstrating the use of various methods such as ordered pairs, tables, graphs, and equations to represent independent and dependent variables. Teachers can guide students to understand that different representations can convey the same mathematical idea and encourage them to make connections between these representations. Moreover, teachers can show how to interpret these representations in real-world contexts, thus helping students see representations as both a process and a product.

Content standards

8.PFA.3 — The student will represent and solve problems, including those in context, by using linear functions and analyzing their key characteristics (the value of the y-intercept b) and the coordinates of the ordered pairs in graphs will be limited to integers).

8.PFA.3b — Describe key characteristics of linear functions including slope (m), y-intercept (b), and independent and dependent variables.

Prior connections

7.PFA.1 — The student will investigate and analyze proportional relationships between two quantities using verbal descriptions, tables, equations in $y = mx$ form, and graphs, including problems in context.

8.PFA.2 — The student will determine whether a given relation is a function and determine the domain and range of a function.

Future connections

A.F.1 — The student will investigate, analyze, and compare linear functions algebraically and graphically, and model linear relationships.

Lesson Preparation

Suggested review

Depending on your students' level of prior knowledge, consider revisiting the following lessons:

Grade 8 — 3.03 Domain and range

Student lesson & teacher guide

Independent and dependent variables

Students will be introduced to the definition of bivariate data and explore the relationship between independent and dependent variables when analyzing data. They learn that the independent variable affects the dependent variable, and the independent variable appears on the horizontal axis when plotted on a graph.

Examples

Students: Page 138–139

Example 1

Consider the following variables:
- Number of people at the beach
- Season

a   Which of the following statements makes sense?

   A   The season affects the number of people at the beach.

   B   The number of people at the beach affects the season.

Create a strategy

Determine which variable has an effect on the other.

Apply the idea

During the summer season, more people tend to go to the beach. There are typically fewer people at the beach during other seasons, and it is much llikely that people will visit during winter. Therefore, the correct answer is option A.

Purpose

Check that students can identify how one variable is related to another.

Students: Page 139

b   Which is the dependent variable and which is the independent variable?

Create a strategy

Recall that the independent variable is not affected by the other variable. In contrast, the dependent variable is affected or changed by the other variable.

Apply the idea

Based on the previous problem, we can see that the independent variable is the season and the dependent variable is the number of people at the beach, since the season affects the number of people at the beach, but changing the number of people at the beach will not change the season.

Purpose

Check that students can accurately identify the dependent and independent variable within the context of a real-world problem.

Expected mistakes

Students might confuse independent and dependent variables if they do not understand the definitions or language involved. Help them collect phrases that indicate which variable is independent of the other, such as the variable that "affects," "influences," or "changes" the other.

Reflecting with students

Encourage advanced learners or all students to create their own examples where one variable influences another, such as how the amount of sunlight affects plant growth, or how the speed of a car affects its stopping distance. Have students share their examples and explain the reasoning behind identifying the independent and dependent variables. This practice fosters critical thinking as they analyze different relationships, and it helps them recognize the widespread applicability of independent and dependent variables in everyday life.

🎓 Use structure to identify variables
Targeted instructional strategies

use with Example 1

This example provides a helpful structure, or scaffolding, for future problems. Encourage students to consider a process like:

1. Given the two variables fill in this sentence frame the two possible ways.

    The ☐ impacts the ☐.

    The ☐ impacts the ☐.

2. Determine which sentence makes more sense.

3. For the chosen sentence that makes more sense, the first box represents the independent variable and the second box represents the dependent variable.

Identifying a set of steps that can be followed can help build confidence with computational thinking.

📶 Discussion supports
English language learner support

use with Example 1

Initiate a class discussion focusing on identifying the independent and dependent variables in this scenario. Provide students with sentence stems to support their participation and help them express their ideas using mathematical language. Examples of sentence stems include:

*   I think the independent variable is ☐ because...
*   The dependent variable is ☐ because it depends on...
*   The ☐ affects the ☐ because....
*   An example of how ☐ impacts ☐ is...

Encourage students to use these stems as they share their thoughts in pairs or small groups before discussing as a whole class. As students contribute, model the use of key vocabulary such as "independent variable," "dependent variable," "affects," and "depends on." This strategy supports English language learners by providing linguistic structures to frame their ideas, making it easier for them to engage in the discussion and develop a deeper understanding of the relationship between the variables.

Students: Page 139

Example 2

The linear graph shows the relationship between Lesley's savings (in dollars) over a few months.

a   Which variable is the dependent variable?

    A   Number of months   B   Savings

Create a strategy

The dependent variable is placed on the vertical axis and is affected by the independent variable.

Apply the idea

The amount of savings is determined by number of months and is on the vertical axis, making it the dependent variable. So, the correct answer is B.

Purpose

Check that students can analyze a graph with a real-world context and determine which variable is dependent.

Reflecting with students

Students can check the reasonableness of their answer by asking themselves if the savings depends on how many months have passed by or if the number of months that have passed depends on the amount of savings Lesley has.

Let students know that, in general, if one of the variables represents a measurement of time (such as hours, weeks, months, etc), that variable will be in the independent variable. This is because we cannot control time, so it cannot be changed by other factors.

Students: Page 139

b   Which variable is the independent variable?

    A   Number of months   B   Savings

Create a strategy

An independent variable is a variable that stands alone and is not changed by the other variables you are measuring.

Apply the idea

From the previous problem, we know that the amount of savings is the dependent variable, so this means the number of months is the independent variable. The correct answer is A.

Purpose

Check that students can analyze a graph with a real-world context and determine which variable is independent.

Clarify independent and dependent variables with relatable scenarios   use with Example 2
Student with disabilities support

To support students who struggle with conceptual processing, relate the concept of independent and dependent variables to real-life situations that are meaningful to them. Begin by discussing everyday examples, such as how the amount of time spent studying (independent variable) affects test scores (dependent variable). Encourage students to share their own examples, fostering a deeper connection to the material.

Then, link these examples back to the graph in the problem by illustrating how Lesley's savings depend on the number of months. Use guided questioning to help students see the cause-and-effect relationship between the variables. For example, ask:

- As the number of months increases, what happens to Lesley's savings?
- Can the amount of money in Lesley's savings account change the months of the year? Why or why not?
- Which variable do we have control over: the number of months or the amount of savings?
- Can you give another example where one thing depends on the amount of time that has passed?

By connecting abstract concepts to familiar experiences and guiding students through thoughtful questions, students can better grasp the idea of independent and dependent variables and apply this understanding to interpret graphs.

> ### 💡 Idea summary
> **Independent variable** - the **input;** it can be changed freely and does not depend on any other variables.
>
> **Dependent variable** - the **output;** it changes as a result of the independent variable and *depends* on the value of the independent variable
>
> We plot the data points with the value of the independent variable on the horizontal (x) axis and the value of the dependent variable on the vertical (y) axis.

Practice

What do you remember?

1 Match the following terms with their correct definitions or concepts:

   a Independent Variable   **b** Dependent Variable

   c Proportional Relationship   **d** Slope

   e y-intercept   **f** Domain

   g Range

   i The set of all possible output values of a function.

   ii A measure of the steepness of a line, represented as the ratio of the vertical change to the horizontal change between two points on the line.

   iii The variable that is being affected when one variable is experimented or changed.

   iv The set of all possible input values to a function.

   v The point where the graph of a function crosses the y-axis.

   vi Graphically represented on the coordinate plane as a straight line.

   vii The variable that is manipulated or changed to see the effect on the other variable.

2 In your own words, explain why it is helpful to know the difference between independent and dependent variables.

3 Compare and contrast independent and dependent variables.

4 Which of the following best describes the connection between domain, input, independent variable, and range, output, dependent variable?

   A Domain and input are unrelated to independent variable while range, output, and dependent variable have no direct connection.

   B Independent and dependent variables are only applicable in scientific experiments, not in mathematical functions where domain and range are used.

   C Range and domain determine the shape of the graph, whereas independent and dependent variables are used to label the axes.

   D Domain and range correspond to the independent and dependent variables of a function, respectively, with the domain representing all possible input values and the range representing all possible output values.

5 For each pair of variables, fill in the statement: ⬚ affects ⬚.

    a Temperature (°F), Distance from the equator (miles)

    b Time of travel (minutes), Distance covered (yards)

    c Time spent working, Wages earned

    d Time spent to finish a novel, Number of pages

    e Number of pets, Time spent caring for pets (hours)

    f Time spent training (hours), Athletic performance

Let's practice

6 Consider the following pair of variables:

    • Ticket sales

    • Revenue from a show

    i State the dependent variable.   **ii** State the independent variable.

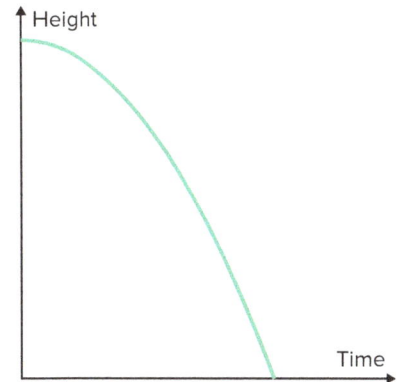

7 Consider the bivariate data for the number of ice cream cones sold and temperature outside (°F). The temperature affects the number of ice cream cones sold.

    a State the independent variable.   **b** State the dependent variable.

8 The graph shows the height of a ball after it is dropped off the side of a building:

    a Which variable is the dependent variable?

    b Which variable is the independent variable?

9 For the following sets of axes, state whether the variables have been placed in the correct position:

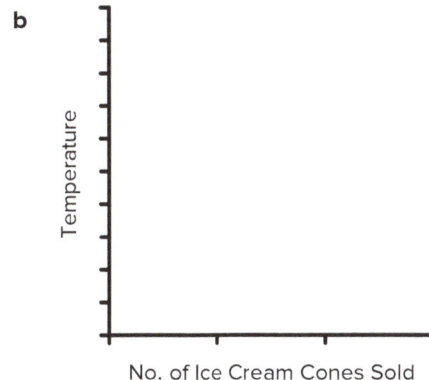

c

y-axis: Fitness level, x-axis: Time spent exercising

d

y-axis: Time spent using mobile phone, x-axis: Battery health

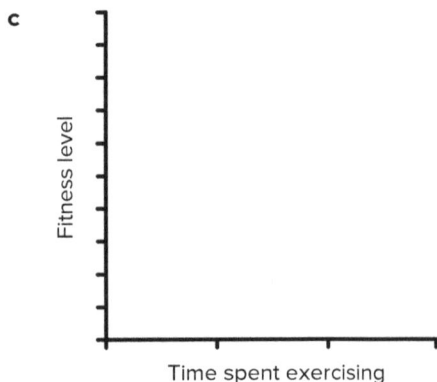

10 Analyze the following tables. Use the context given by the labels to guide your answer. Identify the:

    i Independent Variable   **ii** Dependent Variable

a

Age (years)	Maximum heart rate (beats per minute)
15	205
20	200
25	195
30	190

b

Hours studied	Test score (%)
1	65
2	75
3	85
4	95

c

Number of batteries	Flashlight runtime (hours)
1	2
2	4
3	6
4	8

d

Daily water intake (liters)	Level of hydration
1	Low
2	Moderate
3	High
4	Very high

11 Consider the data for the number of hours spent in studying Statistics and the scores after the test. The scores of the students in a class are affected by the number of hours they spent studying. What is the dependent variable?

12 A study was performed to find the relationship between the number of people in attendance at various musical performances and the number of years of training the performer had.

    a Identify the dependent variable.   **b** Identify the independent variable.

13 A kite can rise to a height of 161 m when the wind is blowing at 6 km per hour and to a height of 322 m when the wind is blowing at 12 km per hour.

    a Identify the dependent variable.   **b** Identify the independent variable.

14 In a study on houses, it was found that an independent variable was the amount of money spent (in dollars) on renovations.

    a Identify the dependent variable.   **b** Identify the independent variable.

15 In a study on basketball players, it was found that the dependent variable is the number of three-point shots scored.

    a Identify the dependent variable.   **b** Identify the independent variable.

16 A car rental company charges $25 per day to rent a car. The total cost depends on the number of days you rent the car.

    a What is the independent variable?   **b** What is the dependent variable?

    c Write a function to represent the situation.

17 Mrs. Robinson is ordering custom t-shirts for a family reunion. Each t-shirt costs $15.50.

    a The final cost depends on the number of ⬚.

    b In this scenario, what is the input? What is the output?

    c Write a function to represent the situation.

    d If Mrs. Robinson orders 20 t-shirts, the total cost is $310. How would you write this using function notation?

18 A student is analyzing the relationship between the number of chapters read in a book (independent variable) and the total number of pages read (dependent variable). They have created the following table to show this relationship:

Chapters read	Total pages read
1	30
2	60
3	90
5	150
4	110

Review the table and describe any errors found based on the concept of a proportional relationship. Explain how you would correct the table and describe the relationship between the independent and dependent variables.

Let's extend our thinking

19 The table shows the number of books read per year by six students and the mark they received in their yearly math exam:

Number of books read per year	5	15	20	30	40	50
Mark in yearly math exam	50	65	70	80	90	100

    a Which variable is the dependent variable?

    b Which variable is the independent variable?

    c Describe the relationship between the variables.

20 The graph shows the relationship between sea temperature and the amount of healthy coral:

    a Which variable is the dependent variable?

    b Which variable is the independent variable?

    c Describe the relationship between the variables based on the trend of the graph.

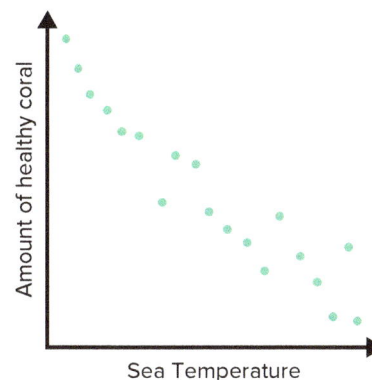

21 A lemonade stand sells lemonade for $2 per cup. The total earnings can be represented by the equation $f(x) = 2x$, where x is the number of cups sold.

    a Which variable is independent?

    b Which variable is dependent?

    c Explain the role of function notation in explaining the independent and dependent variables in this context.

22 Imagine a local artist charges a flat rate to create a custom piece of art regardless of the time it takes. The artist charges $200 for a custom piece. The total earnings depend on the number of pieces commissioned. If the roles of the independent and dependent variables were reversed, how would this situation change?

Answers

3.04 Independent and dependent variables

What do you remember?

1 a vii   **b** iii   **c** vi   **d** ii

  e v   **f** iv   **g** i

2 Answers may vary.

Understanding the difference between independent and dependent variables helps to establish the relationship between two variables in a relation, function, or an equation. The independent variable is the one that you can change or control in an experiment or scenario; it's the cause. The dependent variable is the one that changes in response to the independent variable; it's the effect.

3 Answers may vary.

Independent variables are the variables that are manipulated or changed in an experiment or equation to observe their effect on the dependent variables. They are the cause in a cause-and-effect relationship. In contrast, dependent variables are the variables that are tested and measured in an experiment or equation; they are the effect in a cause-and-effect relationship. While independent variables are controlled by the experimenter or the conditions of the experiment, dependent variables change in response to the independent variables. In a graph of a function, the independent variable is typically represented on the x-axis, whereas the dependent variable is represented on the y-axis.

4 D

5 a Distance from the equator affects temperature.

  b Distance covered affects time of travel or time of travel affects distance covered.

  c Time spent working affects wages earned.

  d Number of pages affects time spent to finish a novel.

  e Number of pets affects time spent caring for pets (hours).

  f Time spent training (hours) affects athletic performance.

Let's practice

6 i Revenue made from a show

  ii Number of ticket sales

7 a Temperature outside

  b Number of ice cream cones sold

8 a Height of the ball

  b Amount of time passed

9 a No   **b** No   **c** Yes   **d** No

10 a i Age   **ii** Maximum Heart Rate

  b i Hourse Studied   **ii** Test Score

  c i Number of Batteries   **ii** Flashlight Runtime

  d i Daily Water Intake   **ii** Level of Hydration

11 Scores of the students

12 a Dependent variable - Number of people at the performance

  b Independent variable - Number of years of training the performer had

13 a Dependent variable - height of the kite

  b Independent variable - speed of the wind

14 a Dependent variable - The amount of money the house sells for

  b Independent variable - The amount of money put into the renovations

15 a Dependent variable - the number of three-point shots scored

  b Independent variable - Time spent training (hours)

16 a The independent variable is the number of days the car is rented.

  b The dependent variable is the total cost of renting the car.

  c The function can be represented as $c(d) = 25d$, where c is the total cost and d is the number of days the car is rented.

17 a t-shirts she orders.

  b The input is the number of t-shirts ordered, and the output is the total cost.

  c The function can be represented as $T(n) = 15.50n$, where T is the total cost and n is the number of t-shirts ordered.

  d Using function notation, this can be written as $T(20) = 310$.

18 There is an inconsistency in the table: the total number of pages read does not increase proportionally with the number of chapters read. Given the pattern in the first three rows, reading one additional chapter corresponds to reading 30 more pages. Therefore, the total pages read for '4' chapters should be '120' instead of '110'. To correct the table, adjust the total pages read for '4' chapters to '120'. This would maintain the proportional relationship where the independent variable 'Chapters Read' directly affects the dependent variable 'Total Pages Read'.

19 **a** Mark in yearly math exam

   b Number of books read per year

   c The more books you read, the higher your yearly math exam mark will be.

20 **a** Amount of healthy coral

   b Sea temperature

   c • The amount of healthy coral decreses as the sea temperature increases.

   • The amount of healthy coral increses as the sea temperature decreases.

21 **a** The independent variable is the number of cups sold (x).

   b The dependent variable is the total earnings $(f(x))$, as it depends on the number of cups sold.

   c The function notation $f(x) = 2x$ shows that for each cup sold, the earnings increase by $2, highlighting the direct relationship between the cups sold and the earnings.

3.05 Characteristics of linear functions

Subtopic overview

Lesson narrative

In this lesson, students will explore the characteristics of linear functions, focusing on multiple representations including equations, tables, and graphs. The lesson begins with an introduction to key concepts such as slope and intercepts. Students will learn that the slope represents the rate of change in a linear function, indicating how the y-values change relative to the x-values. The slope can be positive, negative, zero, or undefined, and is calculated as the ratio of the vertical change to the horizontal change between points on a line. The lesson includes examples where students interpret tables of values to identify the slope and use it to predict future values. For instance, they will analyze the growth of a plant over several weeks, determining that the plant grows by one inch every two weeks based on the given data. Students will practice plotting points from a table onto a coordinate plane, drawing the corresponding line, and identifying key points such as the y-intercept. Additionally, students will work with linear equations, such as ($y = 3x + 1$), to complete tables of values and plot graphs. They will use the equation to find specific y-values for given x-values, plot these points, and draw the line representing the function. By the end of the lesson, students will understand how to identify and interpret the slope and intercepts of linear functions, and how these characteristics are represented in different formats. This foundational knowledge will enable them to analyze and predict the behavior of linear relationships effectively.

Learning objectives

Students: Page 144

After this lesson, you will be able to...

- graph a linear function given a table.
- create a table of values for a linear function given a graph, equation in the form $y = mx + b$, or context.
- describe the slope and y-intercept of a linear function given a table or graph.

Key vocabulary

- slope

- y-intercept

Essential understanding

Linear functions have a constant rate of change, often referred to as slope, and can be represented by the equation $y = mx + b$.

Standards

This subtopic addresses the following Virginia 2023 Mathematics Standards of Learning standards.

Mathematical Process Goals

MPG2 — Mathematical Communication

Teachers can promote mathematical communication by encouraging students to articulate their understanding and reasoning when determining the slope and y-intercept of linear functions. For instance, in a class discussion or in written responses, students can be asked to describe the process they used to determine the slope or y-intercept from a table, graph, or equation, and to explain the meaning of these in context. This emphasizes the importance of mathematical language in expressing mathematical ideas precisely and effectively.

MPG3 — Mathematical Reasoning

Teachers can foster mathematical reasoning by guiding students to apply deductive reasoning in working with linear functions. For instance, students can be asked to justify why the slope determined from a table, graph, or equation represents the rate of change in the dependent variable compared to the independent variable. They can also be asked to reason out why the y-intercept is the value of the dependent variable when the independent variable is zero. This practice will help students to make and validate mathematical statements.

MPG5 — Mathematical Representations

Teachers can incorporate this goal by teaching students to represent linear functions using different methods. They can guide students to understand that a linear function can be represented as a graph, a table, or an equation, and that these different representations can convey the same idea. For example, the teacher can demonstrate how to graph a linear function given a table of values, and how to create a table of values for a linear function given a graph or equation. By interpreting these representations in the context of real-world problems, students will understand that representation is both a process and a product.

Content standards

8.PFA.3 — The student will represent and solve problems, including those in context, by using linear functions and analyzing their key characteristics (the value of the y-intercept b) and the coordinates of the ordered pairs in graphs will be limited to integers).

8.PFA.3b — Describe key characteristics of linear functions including slope (m), y-intercept (b), and independent and dependent variables.

8.PFA.3c — Graph a linear function given a table, equation, or a situation in context.

8.PFA.3d — Create a table of values for a linear function given a graph, equation in the form of $y = mx + b$, or context.

Prior connections

7.PFA.1 — The student will investigate and analyze proportional relationships between two quantities using verbal descriptions, tables, equations in $y = mx$ form, and graphs, including problems in context.

8.PFA.2 — The student will determine whether a given relation is a function and determine the domain and range of a function.

Future connections

A.F.1 — The student will investigate, analyze, and compare linear functions algebraically and graphically, and model linear relationships.

Lesson Preparation

Suggested review

Depending on your students' level of prior knowledge, consider revisiting the following lessons:

Grade 7 — 2.05 Unit rate and slope

Grade 7 — 2.06 Graph proportional relationships

Grade 7 — 2.07 Compare representations of proportional relationships

Tools

You may find these tools helpful:
- Cardboard
- Counters or chips
- Colored pencils
- Grid paper

Student lesson & teacher guide

Characteristics of linear functions

This lesson explores the representation and interpretation of linear functions, particularly through their graphical and tabular forms. Students will learn how to determine key characteristics of linear functions, such as the slope and the y-intercept, by examining data in tables and on graphs.

Concrete-Representational-Abstract (CRA)
Targeted instructional strategies

Concrete: Begin by engaging students with physical manipulatives to explore the concept of slope. Use items like ramps made from boards or cardboard to represent different inclines. Provide coordinate grid mats and have students place physical objects, like counters or tiles, at specific points to form a line. Encourage them to measure the vertical change (rise) and horizontal change (run) between the top and bottom of the ramps or between two points on the grid mats using rulers or measuring tapes.

Representational: Next, transition to drawing representations of what they explored concretely. Have students sketch the ramps they used, labeling the rise and run for each. Use graph paper to plot the points from the concrete activity and draw lines connecting them. Encourage students to draw right triangles beneath the lines to visually represent the rise over run. Provide diagrams of coordinate planes with lines, and have students highlight and label the rise and run on each line. Using images of graphs, help students see how the steepness of a line relates to its slope.

Abstract: Finally, introduce the symbolic representation of slope as $\dfrac{\text{Change in } y}{\text{Change in } x}$. Explain how this ratio calculates the steepness between two points on a line. Guide students through calculating slope from pairs of coordinates they plotted earlier. Practice finding the slope from tables by determining the change in y-values over the change in x-values. Work with the equation $y = mx + b$ to show how the slope, m, represents the constant rate of change in a linear function.

Connecting the Stages: Throughout the lesson, help students make connections between all three stages. When working with the slope ratio, remind them of the rise and run they measured with the manipulatives. Ask them to compare their drawings to the calculations they perform, highlighting how the numbers correspond to the measurements in their sketches.

Compare and connect

English language learner support

Encourage students to compare and connect different representations of linear functions: equations, tables, graphs, and real-world contexts. Begin by presenting the linear function in multiple forms that all describe the same relationship.

For example, these four representations could describe the same information:
- Verbal description: Annabelle is growing a tomato plant for the school garden. When she first planted it, the plant was 1 cm tall. She continues to track the growth and notices that it is growing 2 cm each week.
- Algebraically: $y = 2x + 1$
- Numerically in a table:

x	0	1	2	3	4	5
y	1	3	5	7	9	11

Graphically

Ask students to work in pairs to identify the slope and y-intercept in each representation. Prompt them with questions like, "In which representation is it easiest to identify the slope?" and "How does the rate of change appear in the context versus the equation?" Encourage students to discuss how each representation conveys the same information in different ways and to explain their reasoning.

Facilitate a class discussion where students share their findings and make connections between the representations. Highlight how the slope represents the rate of change in all forms—for instance, the plant grows 2 cm each week (slope) and was initially 1 cm tall (y-intercept). Emphasize the importance of understanding these connections to deepen their comprehension of linear functions.

Examples

Students: Page 146

Example 1

Consider the equation $y = 3x + 1$.

a Complete the table of values shown:

x	−1	0	1	2
y				

Create a strategy

Substitute each value from the table into the given equation.

Apply the idea

For $x = -1$:

$$y = 3 \cdot (-1) + 1 \qquad \text{Substitute } -1 \text{ for } x$$
$$= -2 \qquad \text{Evaluate}$$

Similarly, if we substitute the other values of x, ($x = 0$, $x = 1$, $x = 2$), into $y = 3x + 1$, we get:

x	−1	0	1	2
y	−2	1	4	7

Purpose

Students will demonstrate that they can complete a table of values using a given linear equation.

Ask students to consider whether we could use other x-values to create the table of values. Explain that choosing x-values that are small and easy to work with can be helpful when creating a table and minimize calculation errors, but we could use any x-values to build a table.

Students: Page 146

b   Plot the points from the table of values.

Create a strategy

For an ordered pair (a, b) from the given table of values found in part (a), identify where $x = a$ along the x-axis and $y = b$ along the y-axis.

Apply the idea

Based on the table of values, the ordered pairs of points to be plotted on the coordinate plane are $(-1, -2)$, $(0, 1)$, $(1, 4)$ and $(2, 7)$.

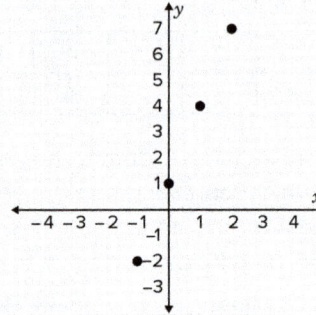

Purpose

Students will demonstrate that they can plot the points in a table of values on a coordinate plane.

Students: Page 147

c   Draw the graph of $y = 3x + 1$.

Create a strategy

Use the plotted points on the coordinate plane from part (b).

Apply the idea

The equation $y = 3x + 1$ must pass through each of the plotted points.

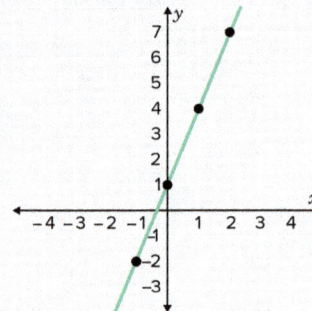

Reflect and check

Both the graph and the table from part (a) show the y-intercept at $(0, 1)$. If 0 had not been included in the table, the graph or equation $y = 3x + 1$ could be used to determine the y-intercept.

Purpose

Check that students can draw a graph of a linear equation in slope-intercept form using points created in a table of values.

Expected mistakes

Students may plot the points, but forget to connect the points to create a line. Have a conversation with student about when they should include the line in the graph.

For linear functions specifically, x could represent any value, not only integer values. That is why we connect the specific points that we find with a line.

Students: Page 147–148

Example 2

The graph of a linear function is shown.

a Determine the slope.

Create a strategy

Find points graphed on the line and draw slope triangles using these points. The slope is the ratio of $\dfrac{\text{vertical change}}{\text{horizontal change}}$.

Apply the idea

We will use slope triangles to help determine the vertical and horizontal change.

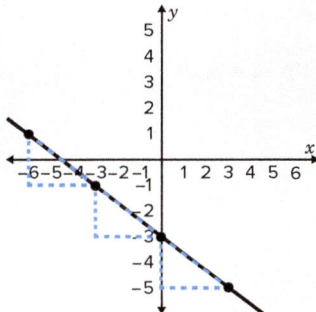

To get from one point to the next point further to the right, we move 2 units down and 3 units right.

The ratio of vertical change to horizontal change is $\dfrac{-2}{3}$ or $-\dfrac{2}{3}$.

The slope is $-\dfrac{2}{3}$.

Reflect and check

Making a table of values from the graph can also be used to find the slope.

x	–6	–3	0	3
y	1	–1	–3	–5

From the table, we see the y-values change by –2 and the x-values change by +3. The slope is the ratio of the change in y to the change in x, so the slope is $-\dfrac{2}{3}$.

Purpose

Students will understand how to determine the slope of a linear function from a graph.

Reflecting with students

Encourage students to verify their final solutions for reasonableness. After calculating the slope, have them compare their result to the visual representation of the line. For example, if they calculate a slope of $-\frac{2}{3}$, they should confirm that the line indeed slopes downward from left to right, matching a negative slope.

Prompt them to consider the steepness of the line; does the ratio of vertical change to horizontal change they found accurately reflect how steep the line appears on the graph? By routinely verifying solutions for reasonableness, students deepen their understanding and improve their mathematical precision.

Students: Page 148

b Identify the y-intercept.

Create a strategy

The y-intercept is the point $(0, y)$ where the line crosses the y-axis.

Apply the idea

The line crosses the y-axis at the point $(0, -3)$.

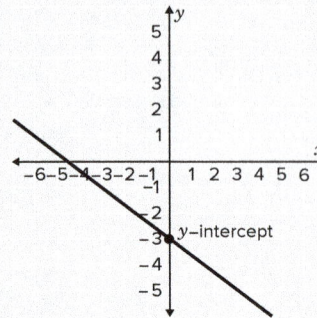

The y-intercept is $(0, -3)$.

Reflect and check

The y-intercept is found in a table by finding the y-value that goes with the x-value of 0.

x	−6	−3	0	3
y	1	−1	−3	−5

The y-value that goes with the x-value of 0 is −3, so the y-intercept is −3 or written as an ordered pair $(0, -3)$.

Purpose

Show students how to identify the y-intercept from a graph and a table of values.

Identifying points with integer coordinates and color-coding slope triangles

use with Example 2

Student with disabilities support

Guide students in identifying points on the line that have integer coordinates, which makes calculating the rise and run more straightforward. Supply manipulatives such as grid paper and colored pencils, so students can physically draw the rise and run between these specific points on the line.

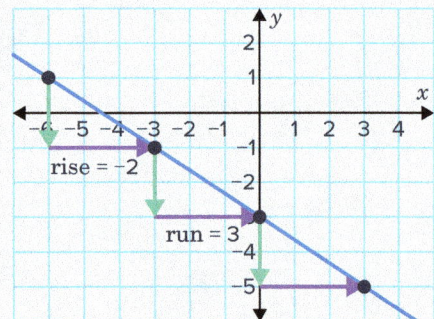

Encourage students to color-code the vertical and horizontal sides of the slope triangles, using one color for the rise and another for the run, to help them visually distinguish these components. This visual differentiation aids in understanding that the slope is the ratio of the vertical change to the horizontal change between two points on the line.

Including an image or demonstration of a slope triangle on the graph, with points labeled and the sides highlighted and labeled, can serve as a helpful reference for students.

Counting rise and run in the wrong direction
Address student misconceptions

Students may incorrectly calculate the "rise" and "run" when determining the slope by counting from the lower point to the upper point, regardless of the location of the points, or by measuring the horizontal change from right to left instead of from left to right. This can lead to incorrect signs for the slope.

Encourage students to always calculate the slope by moving from the left point to right point on the graph, aligning with the positive direction on the x-axis. Remind them that the "run" is the horizontal change from the left point to the right point, and the "rise" is the vertical change corresponding to that movement. Use slope triangles drawn from the left point to the right point to visually reinforce the correct direction.

Students: Page 149

Example 3

Consider the table of values.

a   Identify the coordinates of the y-intercept.

x	–1	0	1	2
y	–5	–2	1	4

Create a strategy

We can use the fact that the y-intercept is the point where the x-coordinate is always 0.

Apply the idea

Based on the table, when the value of x is 0, the value of y is –2. This means that the y-intercept is (0, –2).

Reflect and check

By graphing the points from the table, the y-intercept is represented by a point on the y-axis. The pair (0, –2) from the original table shows as a point on the y-axis, which makes it the y-intercept.

Purpose

Show students how to identify the y-intercept from a table of values and understand its graphical representation.

Expected mistakes

Students might say the y-intercept is $y = -2$, rather than stating the coordinates. Remind students that the y-intercept is a point on the graph, and the question asks for the coordinates of that point, which are written as an ordered pair.

b Determine the slope.

Create a strategy

Find the change in y-values, change in x-values, then write as the ratio of $\dfrac{\text{change in } y}{\text{change in } x}$.

Apply the idea

The y-values $\{-5, -2, 1, 4\}$ change by $+3$.

The x-values $\{-1, 0, 1, 2\}$ change by $+1$.

The ratio of the change in y to the change in x is $\dfrac{3}{1}$. This ratio should be simplified as a reduced improper fraction or integer, if possible, before being written as a slope. The slope is 3.

Reflect and check

The slope can also be found on a graph by writing and simplifying the ratio of vertical change to horizontal change between points from left to right.

The vertical change is $+3$, the horizontal change is $+1$, so the ratio would be $\dfrac{3}{1}$. This would simplify to a slope of 3.

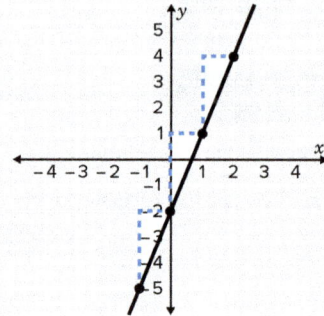

Purpose

Show students how to calculate the slope of a line using changes in x and y-values from a table.

Example 4

On a summer day, the temperature starts at 80° in the morning and rises 2° every hour for several hours. Use this information to complete the table of values.

Hours (x)	0	2	4	6	8
Temperature (y)					

Create a strategy

Use the initial value of x and y with the unit rate to determine missing values. The table's x-values do not change by 1, so the change in y will need to be adjusted to match the change in x.

Apply the idea

The initial value of the table will be (0, 80) because the starting temperature is 80°.

Notice that we are told the temperature rises by 2° every 1 hour but the table is numbered in 2 hour increments. To find the temperature at 2 hours, we will start a 80° and increase by 2° twice.

$$80 = 80 + 0 \qquad \text{0 hours}$$
$$82 = 80 + 2 \qquad \text{1 hour}$$
$$84 = 80 + 2 + 2 \qquad \text{2 hours}$$

We see that for every 2 hours, the temperature increases by 4°. To complete the table, each previous y-value will be increased by 4.

$$80 = 80 + 0 \qquad \text{0 hours}$$
$$84 = 80 + 4 \qquad \text{2 hours}$$
$$88 = 84 + 4 \qquad \text{4 hours}$$
$$92 = 88 + 4 \qquad \text{6 hours}$$
$$96 = 92 + 4 \qquad \text{8 hours}$$

Hours (x)	0	2	4	6	8
Temperature (y)	80	84	88	92	96

Purpose

Demonstrate to students how to use a given initial value and rate of change to complete a table of values representing a real-world situation, and help them understand the importance of context when interpreting the rate of change.

Expected mistakes

Students might ignore the x-values in the table only add 2 to each of the outputs. Point out that the temperature rises by 2° each hour, not every 2 hours.

Reflecting with students

Challenge advanced learners to take the given information and derive a mathematical equation that models the relationship between time and temperature. Start by asking them to identify and define the independent and dependent variables. Then, students will need to identify the starting temperature and the slope and use this information to build the equation.

After expressing this relationship algebraically as $y = 2x + 80$, where y is the temperature and x is the number of hours, students should verify that their equation is accurate by substituting in x-values. Encourage them to use this equation to predict temperatures at times not listed in the table, thus extending their learning. By deriving the equation themselves, students engage in higher-order thinking and see how patterns in data can be generalized mathematically.

Use spreadsheets to solve problems use with Example 4
Targeted instructional strategies

To incorporate coding into this lesson using spreadsheets, start by having your students set up a table in Excel or Google Sheets to calculate the temperatures at the specified hours. Show them how to input the starting temperature and the rate of temperature increase into separate cells. Then, guide them to use formulas that reference these cells to compute the temperature for each hour. This approach allows students to see how changing the initial values affects the results, reinforcing their understanding of linear relationships and demonstrating how spreadsheets can automate repetitive calculations.

1. Set Up the table headers:

	A	B	C	D	E
1	Hours (x)	Temperature (y)			
2					
3					
4					
5					
6					

2. Input the Starting Values:

	A	B	C
1	Hours (x)	Temperature (y)	
2	0	80	
3			
4			
5			
6			

Advanced option with dynamic values:

	A	B	C	D	E
1	Hours (x)	Temperature (y)		Starting temp:	80
2				Temp increase:	2
3					
4					
5					
6					

3. Create the Formula to Calculate Temperature:

	A	B	C
1	Hours (x)	Temperature (y)	
2	0	80	
3	=A2+2	=B2+2*A3	
4			
5			
6			

Advanced option with dynamic values:

	A	B	C	D	E
1	Hours (x)	Temperature (y)		Starting temp:	80
2	0	=$E$2		Temp increase:	2
3	=A2+2	=B2+$E$3*A3			
4					
5					
6					

4. Drag the Formula to Other Cells:

	A	B	C
1	Hours (x)	Temperature (y)	
2	0	80	
3	2	84	
4	4	88	
5	6	92	
6	8	96	

Advanced option with dynamic values:

	A	B	C	D	E
1	Hours (x)	Temperature (y)		Starting temp:	80
2	0	80		Temp increase:	2
3	2	84			
4	4	88			
5	6	92			
6	8	96			

5. Review the Results:

   Ensure the results are reasonable by spot checking one manually.

💡 **Idea summary**

Linear functions can be represented as equations, tables of values and graphs.

The **slope** of a linear function represents

$$\frac{\text{change in } y}{\text{change in } x} = \frac{\text{vertical change}}{\text{horizontal change}}$$

The y-**intercept** of a linear function in a table is the ordered pair $(0, y)$ and the point on a graph where the function crosses the y-axis.

Practice

What do you remember?

1 Match the following terms with their definitions:

a	Linear Function	**i**	A relationship with a constant rate of change whose graph is a straight line
b	Slope	**ii**	The rate of change between any two points on a line
c	y-intercept	**iii**	The point where the line crosses the y-axis
d	Independent Variable	**iv**	The variable in a function with a value that is chosen
e	Dependent Variable	**v**	The variable in a function whose value is determined by the value chosen for the other variable

2 For the following linear relationships, create a table of values:

a $y = 3x$ **b** $y = \dfrac{1}{2}x$ **c** $y = -4x$ **d** $y = -\dfrac{2}{3}x$

3 For the following linear relationships, identify the following characteristics:

i Slope   **ii** y-intercept

a

x	2	4	6	8
y	10	20	30	40

b

x	1	2	3	4
y	5	10	15	20

c

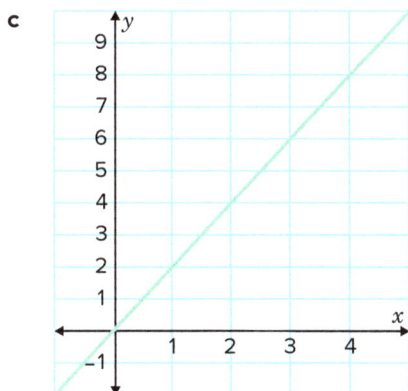

d

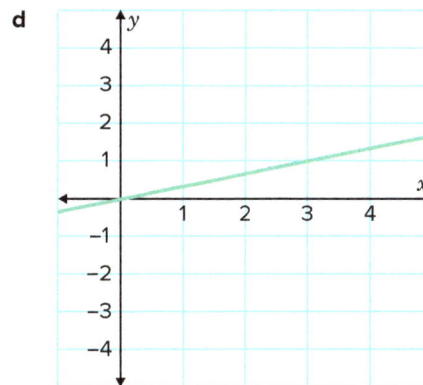

4 For each table of values:

   i Plot the points on a coordinate plane.

   ii State whether the graph of the equation is linear. Explain your reasoning.

a

x	−1	0	1	2
y	−3	−6	−12	−24

b

x	−2	−1	0	1
y	−8	−4	0	4

c

x	0	1	2	3
y	−4.5	−3	−1.5	0

d

x	−3	−2	−1	0
y	1	0	1	4

e

x	−5	0	5	10
y	−1	0	1	2

f

x	−4	0	4	8
y	−0.5	0	0.5	0.25

5 Consider this line, where Point $A(4, 0)$ and Point $B(0, 16)$ both lie on the line:

   a Find the slope of the line.

   b As x increases, is the value of y increasing or decreasing?

Let's practice

6 For each table of values:

   i State the value of m, the slope.

   ii State the value of b, the y-intercept.

a

x	1	2	3	4
y	2	4	6	8

b

x	0	1	2	3
y	−3	0	3	6

c

x	−9	−6	−3	0
y	−11	−15	−19	−23

d

x	−2	0	2	4
y	8	4	0	−4

7 For each graph:

   i Determine the slope, m, of the line.

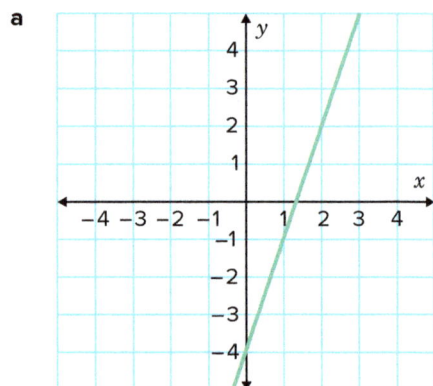

   ii Determine the y-intercept, b, of the line.

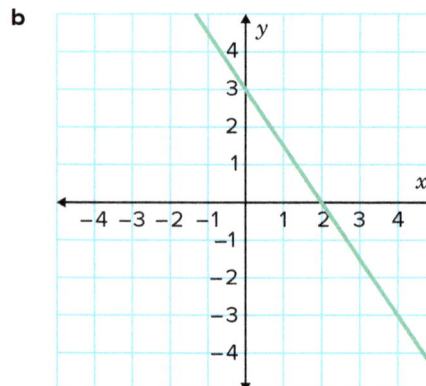

a

b

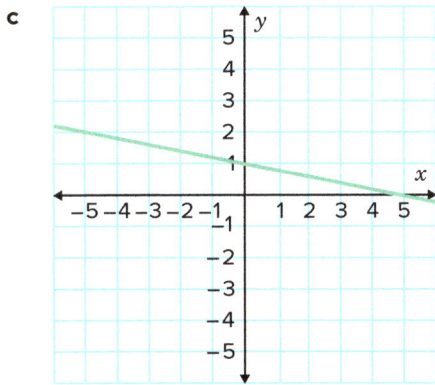

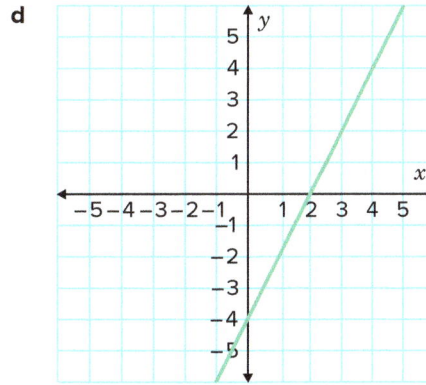

8 For each table of values:

   i Find the slope, m.   **ii** Find the y-intercept, b.   **iii** Complete the table of values.

a

x	0	1	2	3	4	24
y	0	4	8	12	16	

b

x	0	1	2	3	4	21
y	9	14	19	24	29	

c

x	0	1	2	3	4	25
y	−23	−21	−19	−17	−15	

d

x	0	1	2	3	4	70
y	27	22	17	12	7	

9 For each graph, complete the given table of values:

a

x	−1	0	1	2
y				

b

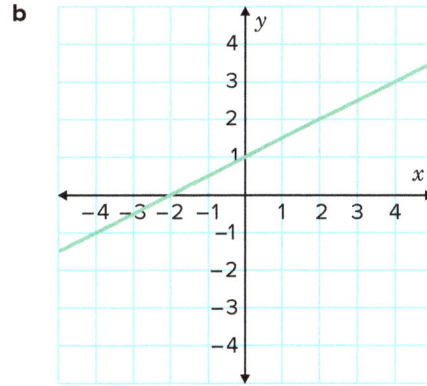

x	−4	−2	0	4
y				

10 For each equation, complete the given table of values:

a $y = 5x + 6$

x	−10	−5	0	5
y				

b $y = 3x + 2$

x	0	1	2	3	4
y					

11 Consider the equation $y = -\dfrac{x}{4} + 5$.

a Complete the table of values:

x	−8	−4	0	4
y				

b Plot the points from the table on a coordinate plane.

c Sketch the graph of $y = -\dfrac{x}{4} + 5$.

d Is the graph of the equation linear?

12 For each equation and table of values:

i Complete the given table of values.

ii Plot the points in the table of values and sketch the graph of the equation.

iii Is the graph of the equation linear?

a $y = -4x$

x	−1	0	1	2
y				

b $y = 3x^2$

x	−1	0	1	2
y				

c $y = -\dfrac{x}{5}$

x	−10	−5	0	5
y				

d $y = \dfrac{x}{2} + 5$

x	−4	−2	0	2
y				

e $y = -2x + 1$

x	−1	0	1	2
y				

f $y = -\dfrac{x}{4} + 5$

x	−8	−4	0	4
y				

g $y = \dfrac{2}{x}$

x	−6	−4	−2	−1
y				

h $y = 3x + 4$

x	−2	0	4	6
y				

13 A race car starts the race with 60 gallons of fuel. From there, it uses fuel at a rate of 1 gallon per minute. Complete the table of values.

Number of minutes passed (x)	0	5	10	15	20	60
Amount of fuel left in tank (y)						

14 Neil is running a 30-mile ultramarathon at a speed of 5 miles per hour.

   a Based on the graph, is the function increasing or decreasing?

   b Is the slope positive or negative?

   c Find Neil's initial distance from the finish line.

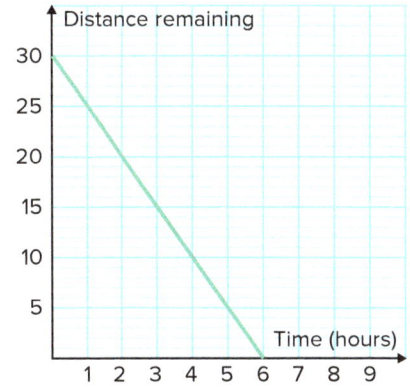

Let's extend our thinking

15 The table shows the water level of a well that is being emptied at a constant rate with a pump, where the time is measured in minutes and the water level in feet.

Time (x)	0	3	7	10
Water level (y)	63	56.7	48.3	42

   a Is the function increasing or decreasing?

   b Explain the meaning of the slope in this context.

   c Explain the meaning of the y-intercept in this context.

   d Find the water level after 18 minutes.

16 A diver starts at the surface of the water and begins to descend below the surface at a constant rate. The graph shows the depth below the surface of the water over the first 4 minutes.

   a Find the rate of change in ft/min.

   b Determine the depth of the diver after 6 minutes.

   c Find how long the diver takes to reach 11.41 ft beneath the surface.

17 For each representation of a linear function:

   i Find the slope, m.

   ii State the value of b, the y-intercept.

   iii Compare and contrast how slope and y-intercepts are represented in each function.

   a

x	1	2	3	4
y	3	5	7	9

   b

   c $y = -\dfrac{1}{2}x + 3$

Answers

What do you remember?

1 a i   **b** ii   **c** iii   **d** iv
   e v

2 a

x	0	1	2	3
y	0	3	6	9

b

x	0	2	4	6
y	0	1	2	3

c

x	0	1	2	3
y	0	−4	−8	−16

d

x	0	3	6	9
y	0	−2	−4	−6

3 a i $m = 5$ ii $b = 0$
   b i $m = 5$ ii $b = 0$
   c i $m = 2$ ii $b = 0$
   d i $m = \dfrac{1}{3}$ ii $b = 0$

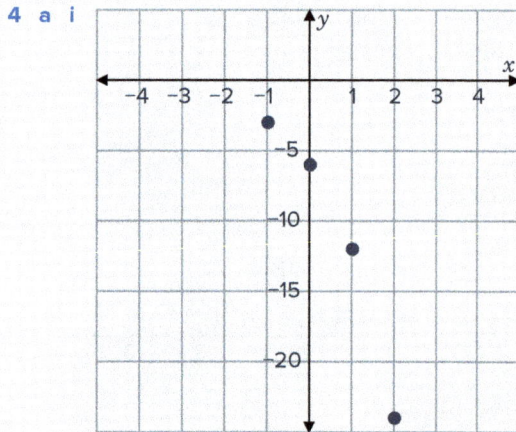

4 a i

   ii No it is not linear because the output values do not have a constant rate of change.

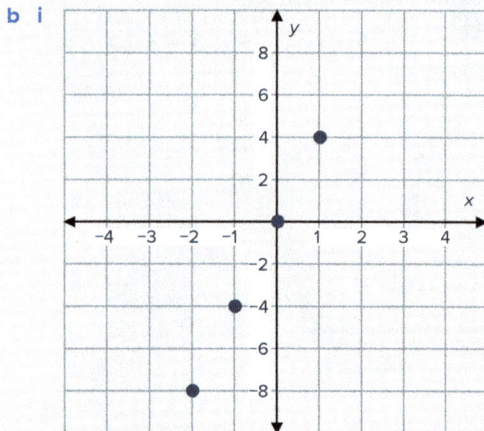

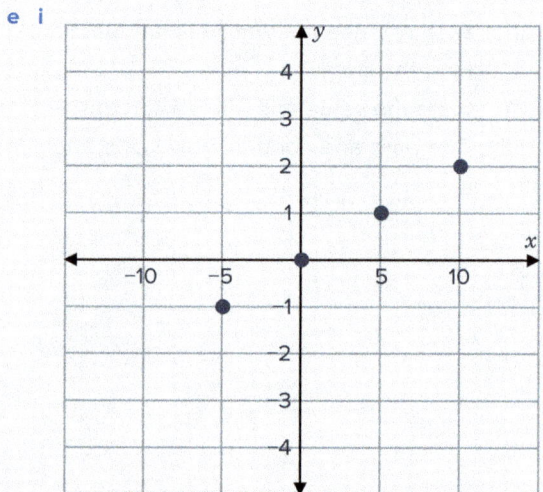

b i

ii Yes, because the output values have a constant rate of change of 4.

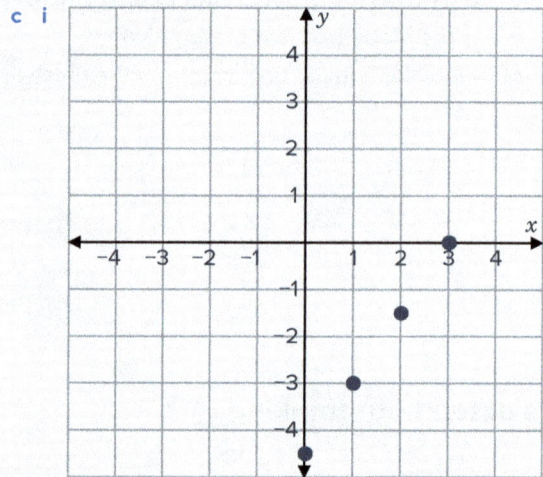

c i

ii Yes, because the output values have a constant rate of change of 1.5.

d i

ii No it is not linear because the output values do not have a constant rate of change.

e i

ii Yes, because the output values have a constant rate of change of 1 for every 5 units.

f **i**

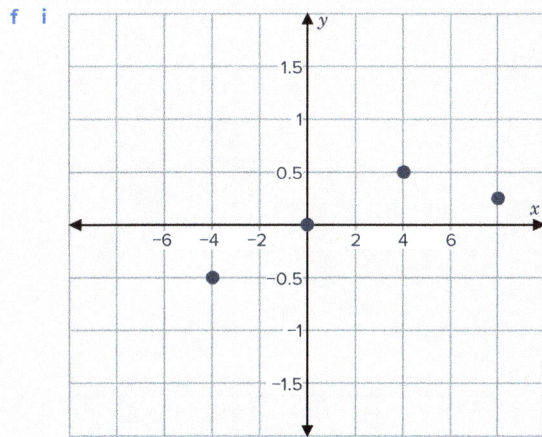

ii No it is not linear because the output values do not have a constant rate of change.

5 **a** −4   **b** Decreasing

Let's practice

6 **a** **i** $m = 2$ **ii** $b = 0$

   b **i** $m = 3$ **ii** $b = -3$

   c **i** $m = -\dfrac{4}{3}$ **ii** $b = -23$

   d **i** $m = -2$ **ii** $b = 4$

7 **a** **i** $m = 3$ **ii** $b = -4$

   b **i** $m = -\dfrac{3}{2}$ **ii** $b = 3$

   c **i** $-\dfrac{1}{5}$ **ii** 1

   d **i** 2   **ii** −4

   e **i** 0   **ii** 3

   f **i** $\dfrac{2}{5}$ **ii** 2

8 **a** **i** $m = 4$ **ii** $b = 0$ **iii** $y = 4x$

     iv

x	0	1	2	3	4	24
y	0	4	8	12	16	96

   b **i** $m = 5$ **ii** $b = 9$ **iii** $y = 5x + 9$

     iv

x	0	1	2	3	4	21
y	9	14	19	24	29	114

   c **i** $m = 2$ **ii** $b = -23$ **iii** $y = 2x - 23$

     iv

x	0	1	2	3	4	25
y	−23	−21	−19	−17	−15	27

   d **i** $m = -5$ **ii** $b = 27$ **iii** $y = -5x + 27$

     iv

x	0	1	2	3	4	70
y	27	22	17	12	7	−323

9 **a**

x	−1	0	1	2
y	5	2	−1	−4

   b

x	−4	−2	0	4
y	−2	0	1	3

10 **a**

x	−10	−5	0	5
y	−44	−19	6	31

   b

x	0	1	2	3	4
y	2	5	8	11	14

11 **a**

x	−8	−4	0	4
y	7	6	5	4

   b

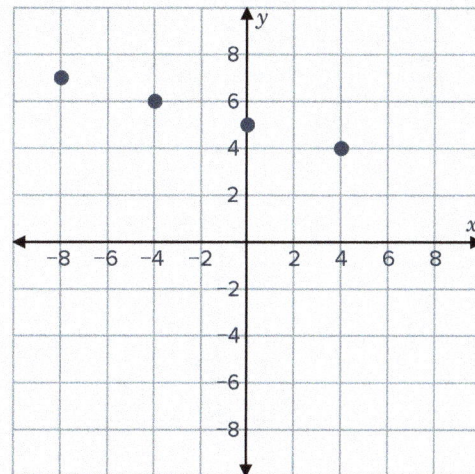

   c

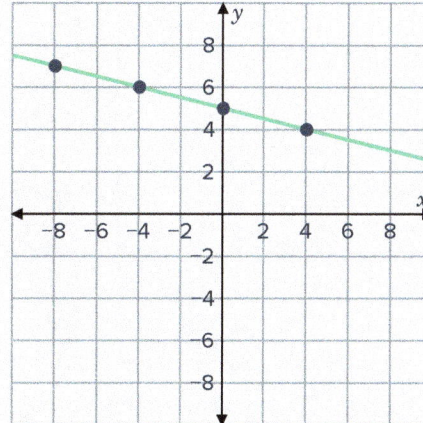

   d Yes

12 a i

x	–1	0	1	2
y	4	0	–4	–8

ii

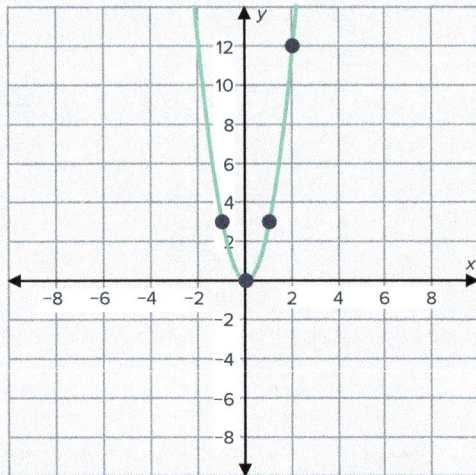

iii Yes

b i

x	–1	0	1	2
y	3	0	3	12

ii

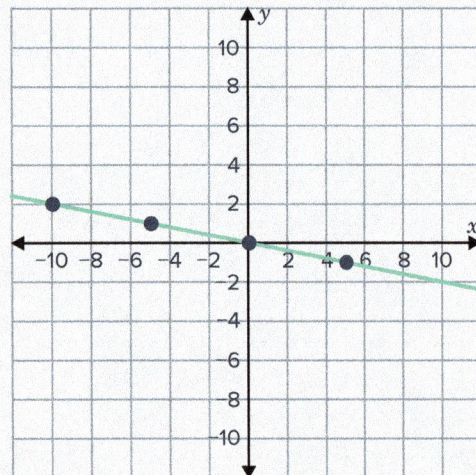

iii No

c i

x	–10	–5	0	5
y	2	1	0	–1

ii

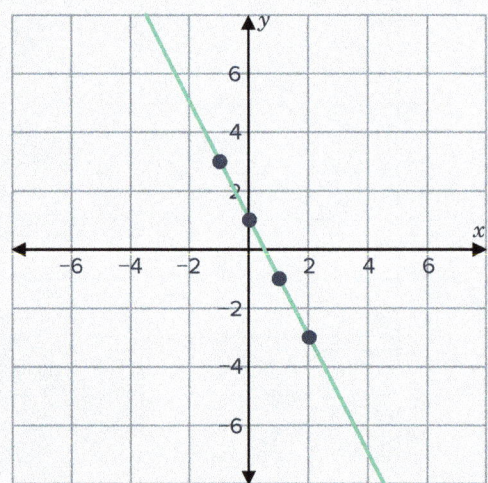

iii Yes

d i

x	–4	–2	0	2
y	3	4	5	6

ii

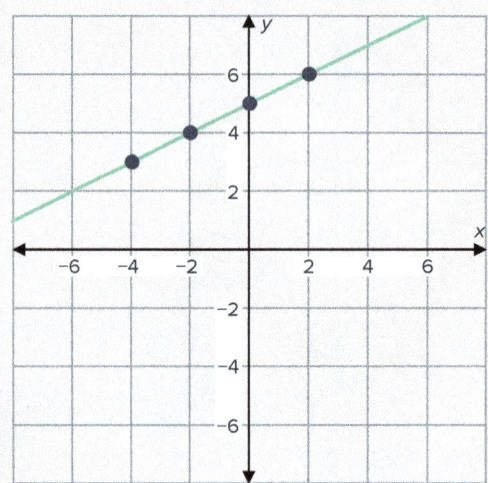

iii Yes

e i

x	–1	0	1	2
y	3	1	–1	–3

ii

iii Yes

f i

x	–8	–4	0	4
y	7	6	5	4

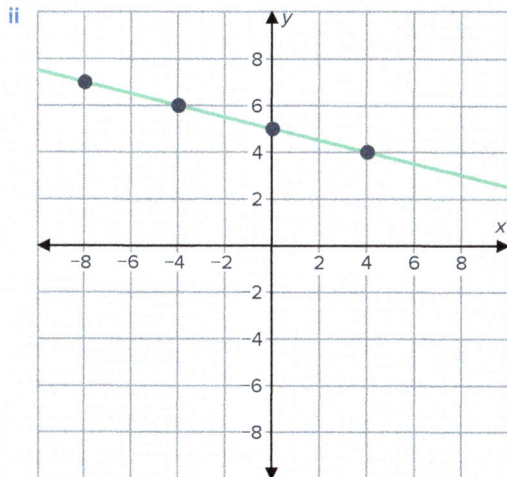

ii

iii Yes

g i

x	-6	-4	-2	-1
y	$-\dfrac{1}{3}$	$-\dfrac{1}{2}$	-1	-2

ii

iii No

h i

x	-2	0	4	6
y	-2	4	16	22

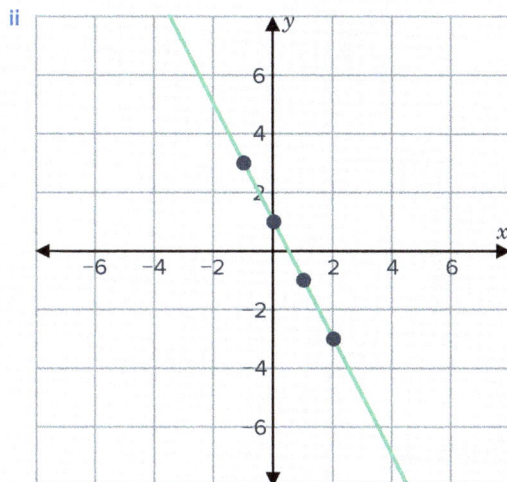

ii

iii Yes

13

Number of minutes passed (x)	0	5	10	15	20	60
Amount of fuel left in tank (y)	60	55	50	45	40	0

14 a Decreasing

b Negative

c 30 miles

Let's extend our thinking

15 a Decreasing

b The amount of water being pumped/emptied per minute.

c The initial water level.

d 25.2 ft

16 a 2.5 ft/min   **b** 15 ft   **c** $x = 4.564$

17 a i $m = 2$ **ii** $b = 1$

iii The slope of a table is calculated by finding the changing in y values divided by the change in x values to find the constant rate of change. This is similar to a graph where the constant rate of change can either be counted out or calculated. The y-intercept needs to be found by either identifying or using the constant rate of change to find the corresponding y-value when $x = 0$.

b i $m = 2$ **ii** $b = -1$

iii The slope of a graph is calculated by finding the changing in y values divided by the change in x values to find the constant rate of change, similar to the table of values. The y-intercept needs to be found by either identifying the y-value of the point where the function crosses the y-axis, which is a process similar to identifying the constant of the equation in the form $y = mx + b$.

c i $m = -\dfrac{1}{2}$ **ii** $b = 3$

iii The slope of an equation is represented by the coefficient of x in the form $y = mx + b$. The y-intercept is the constant of the equation in the form $y = mx + b$. While it is simple to identify the values, it is still necessary to interpret these values for the given function. For example, the slope is negative, therefore the function is decreasing, and the function has a positive y-intercept so the function begins above the x-axis.

3.06 Slope-intercept form

Subtopic overview

Lesson narrative

In this lesson, students will learn about the slope-intercept form of a linear equation. The lesson focuses on how to use this form to identify the characteristics of a linear function and graph it. Students will explore how the slope (m) affects the steepness and direction of the line, while the y-intercept (b) indicates where the line crosses the y-axis. The lesson includes an exploration where students will experiment with different slopes and y-intercepts to see how changes in these values affect the graph of the line. This hands-on activity allows students to manipulate the slope and y-intercept in real-time using graphing tools or software, fostering a deeper understanding of the relationship between the equation and its graphical representation. Students will practice extracting the slope and y-intercept from a given equation and use these values to create a table of points, which can then be plotted on a coordinate plane. They will also practice shifting lines vertically by altering the y-intercept and observe how these changes affect the graph. Additionally, the lesson includes strategies for graphing lines directly from the slope-intercept form. By the end of the lesson, students should be proficient in writing, interpreting, and graphing linear equations in slope-intercept form, enabling them to analyze and understand linear relationships effectively.

Learning objectives

Students: Page 156

> 🎓 **After this lesson, you will be able to...**
> - describe how adding a constant to the equation of a proportional relationship $y = mx$ will translate the line on a graph.
> - identify the constant that was added to the translated graph of a proportional relationship $y = mx$.
> - identify and describe the slope and y-intercept of a linear function given an equation in slope-intercept form.
> - graph a linear function given an equation in slope-intercept form.
> - write an equation of a linear function in the form $y = mx + b$, given a graph or table.

Key vocabulary

- slope-intercept form
- translation

Essential understanding

Linear functions have a constant rate of change, often referred to as slope, and can be represented by the equation $y = mx + b$.

Standards

This subtopic addresses the following Virginia 2023 Mathematics Standards of Learning standards.

Mathematical Process Goals

MPG1 — Mathematical Problem Solving

Teachers can offer a variety of problems where students need to identify the slope and y-intercept from different linear equations. They can also provide challenges where students have to write the equation from a given slope and y-intercept or from a graph. The practice problems will allow students to apply different problem-solving strategies and understand which works best in different scenarios.

MPG4 — Mathematical Connections

Teachers can emphasize the connection between the slope-intercept form of a linear function and proportional relationships. By discussing how adding a constant (b) to the equation of a proportional relationship translates the line on a graph, students can see the integration and application of different mathematical concepts. This helps them see mathematics as a coherent subject.

MPG5 — Mathematical Representations

Teachers can guide students to represent linear functions using a variety of methods, including algebraic equations, tables of values, and graphs. For instance, students can be asked to create a table of values and a graph using the slope and y-intercept identified from an equation. Teachers can also discuss how different representations — equations, tables, and graphs — communicate the same idea about a linear function. Additionally, teachers can introduce the concept of interpreting the slope and y-intercept in the context of real-world situations. Students can be guided to understand that the slope can represent a rate of change and the y-intercept a starting value in a given context. Through these activities, students can understand that representation is both a process and a product in mathematics.

Content standards

8.PFA.3 — The student will represent and solve problems, including those in context, by using linear functions and analyzing their key characteristics (the value of the y-intercept) and the coordinates of the ordered pairs in graphs will be limited to integers).

8.PFA.3a — Determine how adding a constant (b) to the equation of a proportional relationship $y = mx$ will translate the line on a graph.

8.PFA.3b — Describe key characteristics of linear functions including slope (m), y-intercept (b), and independent and dependent variables.

8.PFA.3c — Graph a linear function given a table, equation, or a situation in context.

8.PFA.3d — Create a table of values for a linear function given a graph, equation in the form of $y = mx + b$, or context.

8.PFA.3e — Write an equation of a linear function in the form $y = mx + b$, given a graph, table, or a situation in context.

8.PFA.3f — Create a context for a linear function given a graph, table, or equation in the form $y = mx + b$.

Prior connections

7.PFA.1 — The student will investigate and analyze proportional relationships between two quantities using verbal descriptions, tables, equations in $y = mx$ form, and graphs, including problems in context.

8.PFA.2 — The student will determine whether a given relation is a function and determine the domain and range of a function.

Future connections

A.F.2 — The student will investigate, analyze, and compare characteristics of functions, including quadratic and exponential functions, and model quadratic and exponential relationships.

A.F.1 — The student will investigate, analyze, and compare linear functions algebraically and graphically, and model linear relationships.

Lesson Preparation

Suggested review

Depending on your students' level of prior knowledge, consider revisiting the following lessons:

Grade 7 — 2.05 Unit rate and slope
Grade 7 — 2.06 Graph proportional relationships
Grade 8 — 3.05 Characteristics of linear functions

Tools

You may find these tools helpful:
- Graph paper

Student lesson & teacher guide

Slope-intercept form

Exploration

Students: Page 156

> **Interactive exploration**
> Explore online to answer the questions
>
> ⊕ mathspace.co

Use the interactive exploration in 3.06 to answer these questions.

1. How do different values of m affect the graph?
2. How do different values of b affect the graph?

Suggested student grouping: Individual

Students interactively manipulate the parameters m and b of the linear equation $y = mx + b$ using sliders and observe how each changes the key features of the line. The aim of the exploration is for students to realize that m affects the slope or steepness of the line, while b changes the y-intercept and translates the line vertically.

Ideal student responses

These ideal responses may differ from other correct student responses. Less formal responses can be connected with the more precise mathematical language presented here.

1. **How do different values of m affect the graph?**

   The value of m controls the steepness and direction of the line. If m is positive, the line slopes upward from left to right; if m is negative, the line slopes downward. The greater the magnitude of m, the steeper the line. When $m = 1$, the line increases at a 45-degree angle, indicating a uniform rise of one unit vertically for each unit moved horizontally.

2. **How do different values of b affect the graph?**

   Changing b shifts the line up or down without affecting its slope. It also determines where the line crosses the y-axis. A positive b results in a positive y-intercept, while a negative b results in a negative y-intercept. When $b = 0$, the line passes through the origin.

Purposeful questions
- As the value of m increases, how does the steepness of the line change?
- What happens when you make m negative?
- How does the line shift when b is increased or decreased?
- Does the value of b affect the line's steepness or direction?

Possible misunderstandings
- Students might confuse the effects of the slope m and the y-intercept, b. They might think changing b affects the steepness of the line or that changing m shifts the line up and down. Systematically choose different values of m and b to help students see exactly what happens as each value changes.

Collect and display
English language learner support

As students experiment with different slopes and y-intercepts using graphing tools, listen for the words and phrases they use to describe how the graphs change. Collect their language on a visible chart or board that all students can access.

If students do not come up with varied ways to express these concepts, provide some examples such as:
- Slope (m)
  - Steeper when m is bigger
  - Flatter when m is small
  - Goes up to the right for positive m
  - Goes down to the right for negative m
- y-intercept (b):
  - Crosses the y-axis at b
  - Shifts up when b increases
  - Shifts down when b decreases
  - Starts at point (0, b)

Encourage students to refer to this chart during discussions and when explaining their observations. By continually updating and referencing this collective display, you support students in building their mathematical vocabulary and deepen their understanding of how changing m and b affects the graph of a linear equation.

This lesson explores how manipulating the equation of a linear equation in slope-intercept form, $y = mx + b$, affects the slope and y-intercept of the corresponding graph. Students discover that the value of m affects the steepness of the line, and the value of b affects the y-intercept. When adding a constant to $y = mx$ the aranh is translated unwards and subtractinc a constant tranclates the cranh downward. Students learn that the slope and y-intercept can be used to graph a line represented by an equation in slope-intercept form.

Examples

Students: Page 157

Example 1

Consider the equation $y = -4x + 5$.

a State the slope and y-intercept of the equation.

Create a strategy

We can use the general slope-intercept form: $y = mx + b$, where m is the slope and b is the y-intercept.

Apply the idea

The equation $y = -4x + 5$ is in the form $y = mx + b$:

$$m = -4 \qquad \text{Identify the slope}$$
$$b = 5 \qquad \text{Identify the } y\text{-intercept}$$

Purpose

Students demonstrate that they can identify the slope and y-intercept from an equation in slopeintercept form.

Reflecting with students

Encourage students to attend to precision by using exact mathematical terms and symbols when identifying the slope and y-intercept. Remind them that in the slope-intercept form, m represents the slope and b represents the y-intercept. Emphasize the importance of correctly labeling these values by stating $m = -4$ and $b = 5$.

Students: Page 157–158

b Complete the table of values for the given equation:

x	–1	0	1	2
y				

Create a strategy

Substitute each x-value from the table into the given equation and evaluate to find y.

Apply the idea

For $x = -1$:

$$y = -4 \cdot (-1) + 5 \qquad \text{Substitute } -1 \text{ for } x$$
$$= 9 \qquad \text{Evaluate}$$

Similarly, if we substitute the other values of x, $(x = 0, x = 1, x = 2)$, into $y = -4x + 5$, we get:

x	–1	0	1	2
y	9	5	1	–3

Purpose

Students demonstrate they can identify the corresponding outputs of specific inputs for a given equation.

Example 2

Consider the following graph of a line:

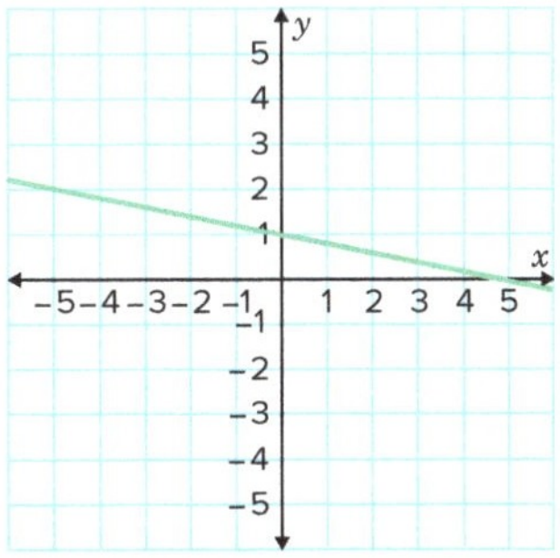

a What is the slope of the line shown in the graph?

Create a strategy

The slope is the ratio of vertical change to horizontal change or $\dfrac{\text{change in } y}{\text{change in } x}$.

Choose two points on the graph and draw a slope triangle to help find the vertical and horizontal change.

Apply the idea

Choosing the points (–5, 2) and (0, 1), we see that to get from (–5, 2) to (0, 1), we need to move 1 unit down and 5 units to the right.

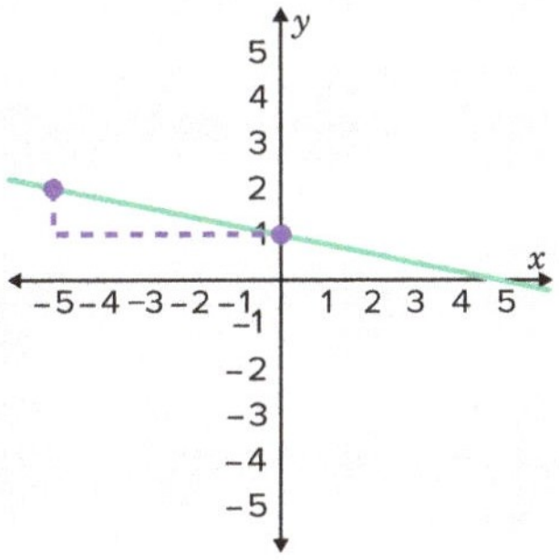

The vertical change is –1 and the horizontal change is +5.

The slope is $\dfrac{-1}{5}$ or $-\dfrac{1}{5}$.

Purpose

Show students how to determine the slope of a line from a graph by selecting two points and calculating the ratio of the vertical change to the horizontal change.

Reflecting with students

Challenge students to use the slope to complete the following table of values:

x	0	1	2	3	4	5
y	1					0

Students: Page 159

b What is the y-value of the y-intercept of the line shown in the graph?

Create a strategy

The y-intercept is the point where the line intersects the y-axis.

Apply the idea

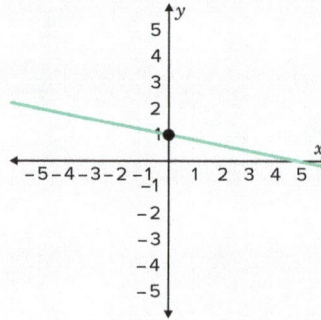

Looking at the graph, the line intersects the y-axis at point (0, 1). Thus, the y-value of the y-intercept is $y = 1$.

Purpose

Show students how to identify the y-intercept from a graph and understand its significance in the equation of the line.

Students: Page 159

c Write the equation of the line in slope-intercept form.

Create a strategy

Substitute the values of the slope and y-intercept in the slope-intercept form of the equation of a line.

Apply the idea

$y = mx + b$ — Slope-intercept form

$y = -\frac{1}{5}x + 1$ — Substitute the values of the slope and y-intercept

Purpose

Demonstrate to students how to convert given information about a line's slope and y-intercept into the slope-intercept form of the line's equation.

Expected mistakes

Students might write the equation as $y = -\frac{1}{5} + 1$, forgetting to include the variable x. Ask students to check their answer by substituting in $x = -5, 0, 5$, which may help them realize their mistake.

Approach unscaffolded problems
Targeted instructional strategies

Ask students how they would have approached the problem if it had just been:

Consider the following graph of a line:

Write the equation of the line in slope-intercept form.

Encourage them to recognize that the parts of this example begin to decompose the problem into smaller more manageable parts. From there, have them work with a partner to write out a set of steps for finding the equation of a line from a graph by adding more detail, including how they can check their answer.

Students: Page 159–160

Example 3

Graph the line $y = 3x + 2$ using its slope and y-intercept.

Create a strategy

Determine the slope, y-intercept, and a starting point using the slope-intercept form of the line.

Apply the idea

The slope is 3, and the y-intercept is 2.

This means that one of the points on $y = 3x + 2$ is (0, 2). We can plot this point and move across 1 and up 3 to get the next point:

Then we can draw a line through these two points:

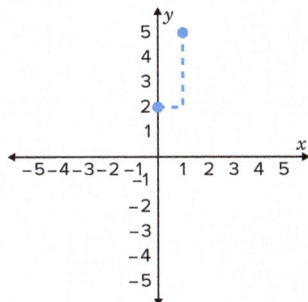

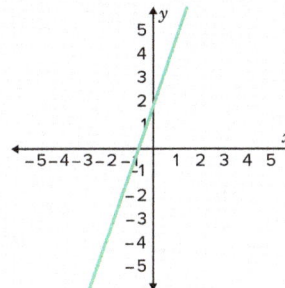

Reflect and check

A table can also be used to graph the line by substituting chosen values of x into the equation $y = 3x + 2$ and determining the corresponding y-values before graphing.

For the x-values

x	-2	-1	0	1
y				

$3(-2) + 2 = -4$ Substituting -2

$3(-1) + 2 = -1$ Substituting -1

$3(0) + 2 = 2$ Substituting 0

$3(1) + 2 = 5$ Substituting 1

x	-2	-1	0	1
y	-4	-1	2	5

This creates a table of values that can be graphed as ordered pairs. This would create the same line graphed by the slope and y-intercept.

Purpose

Show students how to graph a line using its slope and y-intercept, and how to use this information to determine other points on the line.

Advanced learners: Analyzing alternative graphing methods
Targeted instructional strategies

use with Example 3

Present advanced learners with the following example of a student's method for graphing the line $y = 3x + 2$.

"I started by plotting the y-intercept at (0, 2). Then, I moved down 3 units and left 1 unit to plot another point on the line. Then, I connected the points to graph the line."

Invite students to analyze this explanation and determine if it is correct. The aim is for students to see that moving down 3 and left 1 corresponds to a rise of -3 and a run of -1, which simplifies to a positive slope since $\frac{-3}{-1} = 3$. This realization reinforces the concept that different movements can represent the same slope. By validating the student's method, advanced learners deepen their understanding of slope as a ratio and enhance their analytical skills by justifying why the approach works.

Scaffolded checklist for graphing equations in slope-intercept form
Student with disabilities support

use with Example 3

To help students write equations of lines in slope-intercept form with increasingly complex information, provide a scaffolded set of steps for students to use to determine how to use the given information to write the slope-intercept form of an equation.

1. Find the slope, m.
   - *Given slope:* Skip to step 2.
   - *Given two points (x_1, y_1) and (x_2, y_2):* use slope formula $\frac{y_2 - y_1}{x_2 - x_1}$ and simplify.
   - *Given a graph:* choose two points on the graph, use $\frac{\text{rise}}{\text{run}}$ and simplify.

2. Find the y-intercept, b.
   - *Given the y-intercept:* Skip to step 3.
   - *Given a point:* Substitute values of m and (x, y) into $y = mx + b$ and solve for b.
   - *Given two points:* Choose either (x_1, y_1) or (x_2, y_2) to use as (x, y) and follow same steps as above.
   - *From a graph:* If the line crosses the y-axis at an integer value, use this value for b.

3. Write equation as $y = mx + b$ using your values of m and b.
   - Use the format $y = \square x + \square$ to help set up your equation
   - Check your equation for simplifying and formatting.

Example 4

Given the table of a linear function:

a   Find the slope.

x	-4	-2	0	2	4
y	-2	1	4	7	10

Create a strategy

The slope of a table is written as a simplified ratio of the change in y to the change in x.

Apply the idea

The x-values of {-4, -2, 0, 2, 4} have a change of +2.

The y-values of {-2, 1, 4, 7, 10} have a change of +3.

The slope of the table is $\frac{3}{2}$.

Purpose

Demonstrate to students that they can use the concept of slope as a ratio of the change in y to the change in x, to find the slope of a linear function represented in a table.

Expected mistakes

Students miaht ianore the x-values in the table onlv consider the difference between consecutive outputs listed in the table. Remind them that we need to determine the ratio of the change in y to the change in x, and the change in x is not 1 in this table.

b   Identify the y-intercept of the table and write the equation of the line represented by the table in slope-intercept form.

Create a strategy

The y-intercept is the ordered pair $(0, y)$ in a table.

The slope from part (a) will be used as m and the y-intercept of the table will be used as b in the slope-intercept form $y = mx + b$.

Apply the idea

The point $(0, 4)$ is in the table, so $b = 4$, and the slope from part (a) means $m = \frac{3}{2}$.

The equation of the line represented by the table is $y = \frac{3}{2}x + 4$.

Reflect and check

Graphing the points from the table would also let us find the equation of the line since we could use the graph to find the slope and y-intercept.

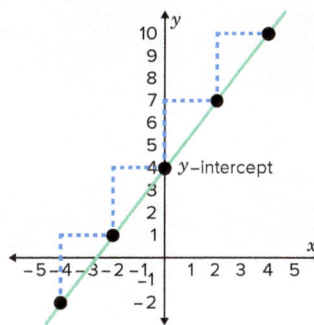

We see a y-intercept at $(0, 4)$ and a change of 3 squares up and 2 squares right, making the slope $\frac{3}{2}$. This would give the same values of m and b to write the equation $y = \frac{3}{2}x + 4$.

Purpose

Show students how to identify the y-intercept from a table and use this information, along with the slope, to write the equation of a line in slope-intercept form.

Example 5

The graph of $y = -\frac{1}{3}x$ is shown.

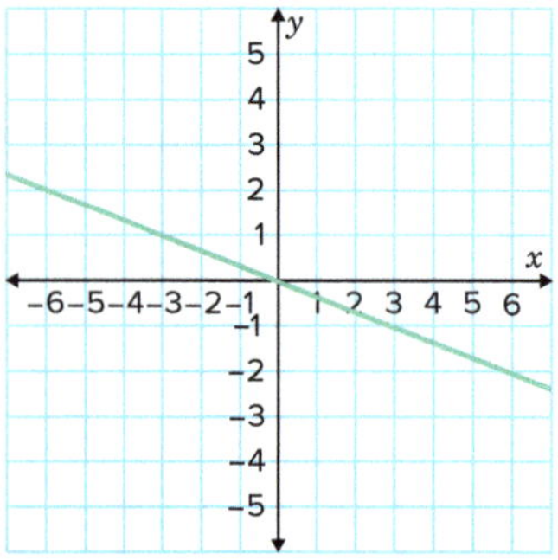

a The graph will be shifted 4 units down. Write the new equation of the line after the translation.

Create a strategy

If a line only shifts up or down, the y-intercept of the graph will change, but the slope will remain the same. In an equation, keep m the same, but find the new b.

Apply the idea

Moving 4 units down corresponds to -4 being added to b. The equation $y = -\frac{1}{3}x$ has a b of 0, making the full original slope-intercept form $y = -\frac{1}{3}x + 0$.

$$y = -\frac{1}{3}x + 0 - 4$$

$$y = -\frac{1}{3}x - 4$$

The new equation of the line will be $y = -\frac{1}{3}x - 4$.

Purpose

Show students how to determine the equation of a line after it has been vertically translated.

Students: Page 161–162

b Graph the equation from part (a).

Create a strategy

Use the m and b from the new slope-intercept form to graph the y-intercept and use slope to find additional points.

Apply the idea

The slope-intercept form is $y = -\frac{1}{3}x - 4$, so the line will cross the y-axis at $(0, -4)$ and move 1 unit down and 3 units right between points.

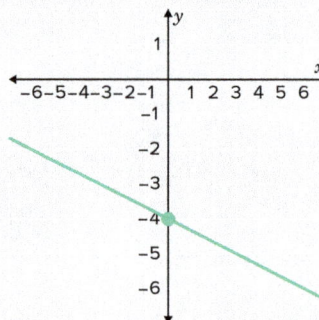

Reflect and check

Another way to graph the new equation would be to make a table of values from the original graph of $y = -\frac{1}{3}x$ and subtract 4 from each y-value. The table below shows pairs from $y = -\frac{1}{3}x$ and subtracting 4 from each y-value.

x	−6	−3	0	3	6
y	2 − 4	1 − 4	0 − 4	−1 − 4	−2 − 4

After simplifying the y-values, the new table matches the points seen in the graph.

x	−6	−3	0	3	6
y	−2	−3	−4	−5	−6

Purpose

Show students how to graph a linear equation by using the slope-intercept form and identifying the y-intercept and slope.

Reflecting with students

Ask students to share the strategies they used to graph the line. Some may have used a table of values, and some may have graphed using the slope and y-intercept. Spend time discussing how to use a translation to graph the line. For example, students can shift points on the original line 4 units down to graph the new line.

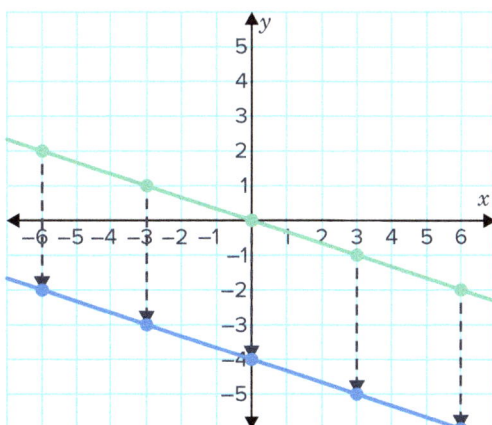

Students: Page 162

Students: Page 162

> 💡 **Idea summary**
>
> A linear equation is said to be in **slope-intercept form** when it is expressed as
>
> $$y = mx + b$$
>
> m is the slope.
>
> b is the y-intercept.
>
> The y-intercept will shift a line above the x-axis if b is positive and below the x-axis if b is negative.
>
> When given an equation in slope-intercept form, we can graph the line by plotting the y-intercept as the first point. Then we can use the slope to find additional points before drawing our line.

Practice

Students: Page 162–167

What do you remember?

1 State what the following variables stand for in the equation of a line $y = mx + b$:

   a m **b** b

2 Consider the following three linear equations and their corresponding graphs:

   $y = x + 4, y = 2x + 4, y = 4x + 4$

   a   What do all of the equations have in common?

   b   What do all of the graphs have in common?

   c   State the coordinates where the lines that have the form $y = mx + 4$ will pass through.

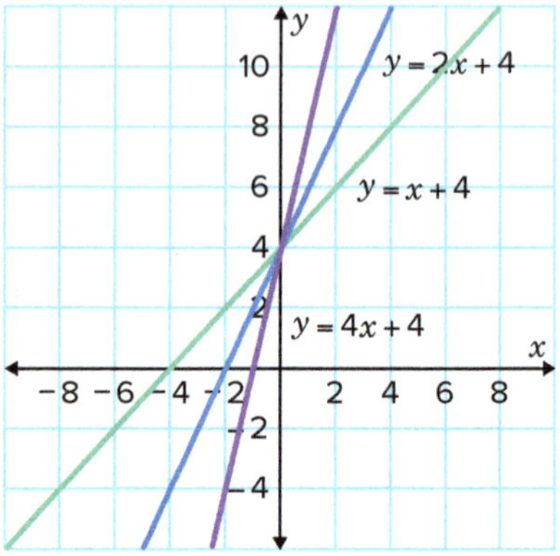

3 For each of the following equations:

   i   State the y-intercept.

   ii   State the slope.

   iii   Create a table of values.

   a $y = 4x - 6$ **b** $y = \frac{1}{2}x + 10$ **c** $y = 9 + 3x$ **d** $y = -12 - 3x$

   e $y = -\frac{x}{3} - 8$ **f** $y = -4x$ **g** $y = -8 - \frac{4x}{9}$ **h** $y = 5 + \frac{5x}{8}$

4 Given the slope and y-intercept, write the equation of each line in the form $y = mx + b$.

   a   Slope: 2, y-intercept: 3   **b**   Slope: $-\frac{3}{4}$, y-intercept: –2

   c   Slope: $\frac{5}{2}$, y-intercept: 0   **d**   Slope: –1, y-intercept: 5

5 For each of the following graphs:

   i   State the y-intercept.

   ii   State the slope.

   iii   Match the equation of the line in slope-intercept form to the graph.

   A $y = 2x - 4$ **B** $y = -\frac{2x}{5} - 4$ **C** $y = -\frac{x}{4} + 5$ **D** $y = \frac{5x}{4} + 5$

   a

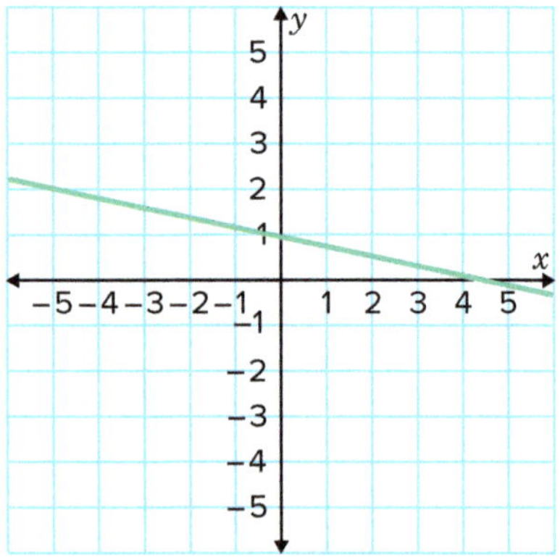

   b

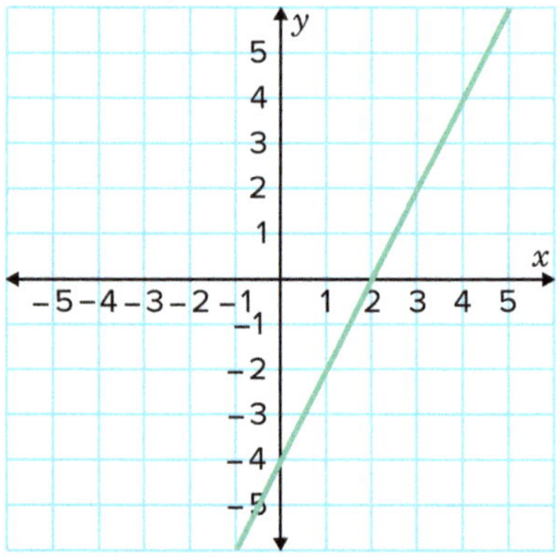

c

d

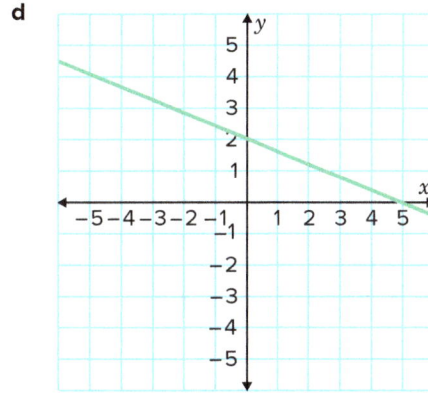

6 Graph each linear equation on a coordinate plane:

   a $y = 2x$ **b** $y = -\dfrac{3}{2}x$ **c** $y = \dfrac{1}{4}x$ **d** $y = -x$

Let's practice

7 Consider the equation of a line: $y = 4\,(4x - 3)$

   a Rewrite the equation in the form $y = mx + b$.   **b** State the y-intercept of the line.

8 State the value of the y-intercept for the following lines:

   a $y = \dfrac{2x}{3} - 5$ **b** $y + 4 = -5x$ **c** $y = 2$ **d** $5x = 4y$

   e $x = 1$ **f** $2x + 6y = 3$

9 Consider the following tables of values:

   i Find the coordinates of the y-intercept.

   ii Find the slope.

   iii Write the equation of the function represented by the table of values.

   a

x	−1	0	1	2
y	−1	0	1	2

   b

x	−4	0	4	8
y	−19	−3	13	29

   c

x	−10	−5	0	5
y	−44	−19	6	31

   d

x	1	2	3	4
y	−5	−10	−15	−20

10 Describe how changing the value of b in the equation $y = mx + b$ translates the line on a graph.

   a If $b = 0$, what is the y-intercept and how does the graph of the equation compare to the original equation?

   b How does the original graph change if $b = 5$?

   c How does the original graph change if $b = -3$?

   d Describe the translation of the original graph if $b = \dfrac{3}{2}$.

11 Consider the equation of a linear function $y = x$. For each of the following equations, describe the transformation that occurs when adding a constant (b) to the equation.

   a $y = x + 3$ **b** $y = x - 2$ **c** $y = x + 5$ **d** $y = x - 4$

12 Given the original equation $y = mx$, examine the graph provided for each part and identify the value of b that has been added to translate the line.

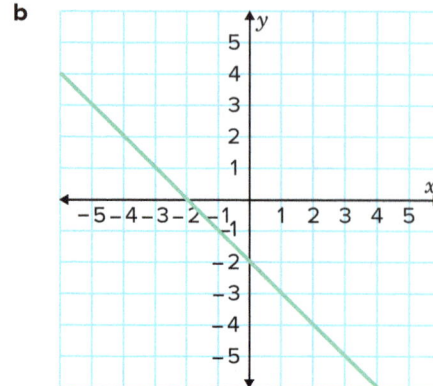

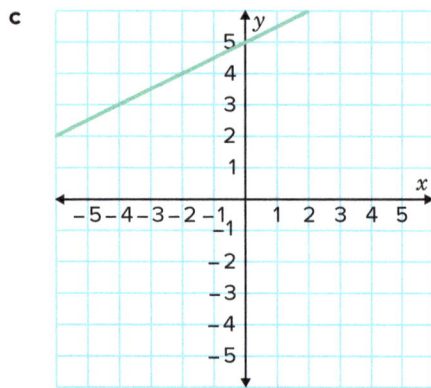

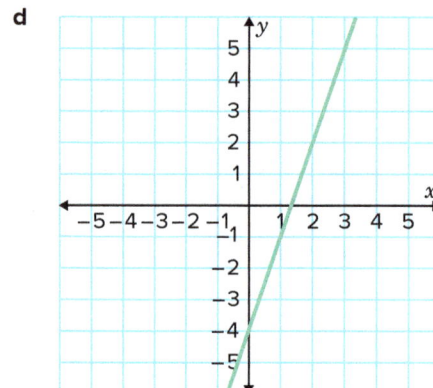

a

b

c

d

13 State whether the line represented by each the following equations crosses the y-axis at the origin, below the origin or above the origin.

a $y = -5x + 2$ **b** $y = 3x$ **c** $y = 3x - 20$ **d** $y = -8x$

14 Use the following table of values to:

   i Write the equation of the linear function

   ii Graph the function

a

x	-2	0	2	4
y	1	5	9	13

b

x	-1	1	3	5
y	6	4	2	0

c

x	1	2	3	4
y	-6	-3	0	3

d

x	-4	0	4	8
y	7	4	1	-2

15 Use the following equations to:

   i Identify the slope of the linear function.

   ii Identify the y-intercept of the linear function.

   iii Graph the function.

a $y = 2x + 3$ **b** $y = \dfrac{x}{3} - 2$ **c** $y = \dfrac{4x}{5} - 3$ **d** $y = -x + 5$

16 For each of the following graphs, write the equation of the line in slope-intercept form.

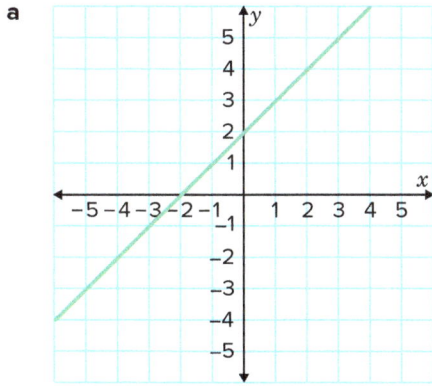

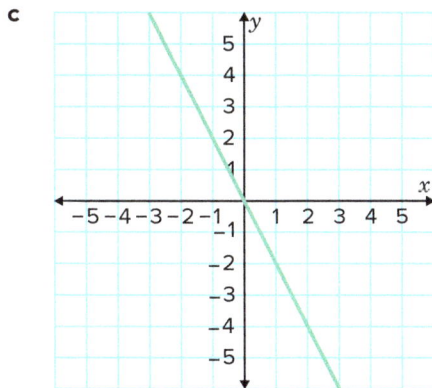

a

b

c

d

17 Each of the following shows an equation, table, or graph of a function. Represent each function in the two representations that have not been provided.

a Given the equation $y = 2x - 3$, create a table and graph of the function.

b Given the table of values, represent the function with an equation and graph.

c Given the graph of a function, create a table and an equation that represent this function.

x	1	3	5
y	2	4	6

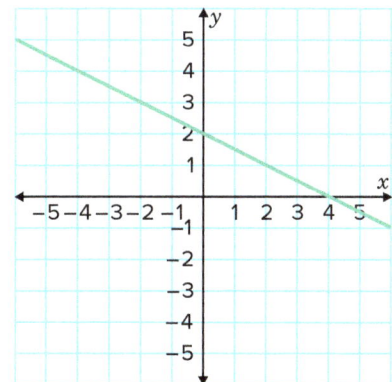

Let's extend our thinking

18 The table shows the water level of a well that is being emptied at a constant rate with a pump, where the time is measured in minutes and the water level in feet.

Time (x)	3	7	10
Water level (y)	56.7	48.3	42

a Write an equation in the form $y = mx + b$ to represent the relationship between the water level (y), and the minutes passed (x).

b Interpret the y-intercept of this linear function.

c Does the line cross the y-axis above the origin? Explain how you know without graphing the equation on the coordinate plane.

19 The total amount y, a carpenter charges his customer for every hour x, is represented by the graph:

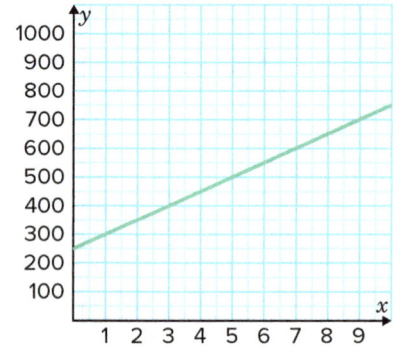

a Write an equation in slope-intercept form to represent the relationship between the total amount charged, y, by the carpenter and the number of hours worked, x.

b Interpret the y-intercept of this linear function.

20 Match the following functions with the correct context from the list provided.

a $y = 2x + 3$ **b** $y = \frac{1}{2}x - 5$ **c** $y = -3x + 4$ **d** $y = 4x - 2$

i The depth of water in a tank decreases linearly over time due to a leak, starting from a height of 4 meters.

ii The cost of entry to a concert is determined by a booking fee of 3 dollars plus twice the cost per ticket.

iii The temperature decreases by half a degree Celsius for every kilometer increase in altitude above sea level, starting from an initial temperature of $-5°C$ at sea level.

iv A car accelerates at a constant rate, covering more distance over the same time interval, starting from a point 2 meters from a reference point.

21 Consider the following graphs:

i State whether the line has a y-intercept.

ii Write an equation that could represent the graph.

a

b

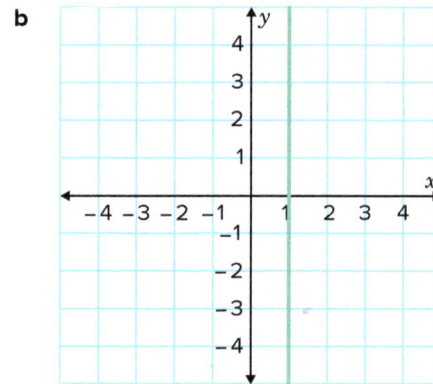

c

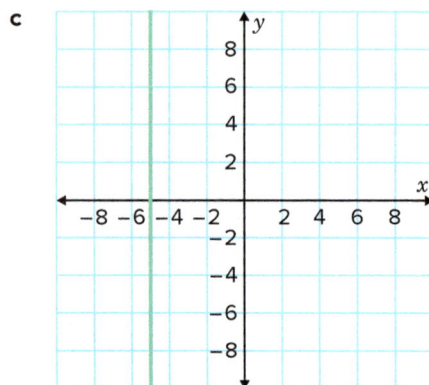

d

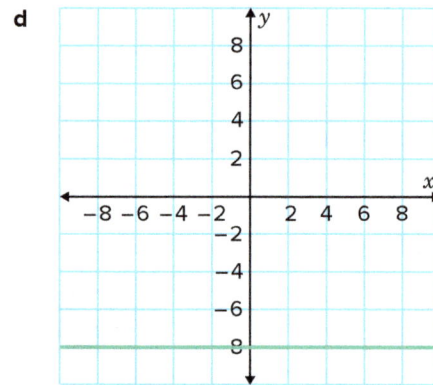

Answers

3.06 Slope-intercept form

What do you remember?

1 a Slope **b** y-intercept

2 a The constant term of 4 is the same.

  b All the graphs intersect the y-axis at the point (0,4).

  c The lines will pass through (0,4).

3 a i −6   **ii** 4

  iii
x	−2	−1	0	1	2
y	−14	−10	−6	−2	2

  b i 10   **ii** $\frac{1}{2}$

  iii
x	−4	−2	0	2	4
y	8	9	10	11	12

  c i 9   **ii** 3

  iii
x	−2	−1	0	1	2
y	3	6	9	12	15

  d i −12   **ii** −3

  iii
x	−2	−1	0	1	2
y	3	6	9	12	15

  e i −8   **ii** $-\frac{1}{3}$

  iii
x	−6	−3	0	3	6
y	−6	−7	−8	−9	−10

  f i 0   **ii** −4

  iii
x	−2	−1	0	1	2
y	8	4	0	−4	−8

  g i −8   **ii** $-\frac{4}{9}$

  iii
x	−18	−9	0	9	18
y	0	−4	−8	−12	−16

  h i 5   **ii** $\frac{5}{8}$

  iii
x	−16	−8	0	8	16
y	−5	0	5	10	15

4 a $y = 2x + 3$ **b** $y = \frac{3}{4}x - 2$

  c $y = \frac{5}{2}x$ **d** $y = -x + 5$

5 a i 5   **ii** $-\frac{1}{4}$ **iii** C

  b i −4   **ii** 2   **iii** A

  c i 5   **ii** 5   **iii** D

  d i 2   **ii** $-\frac{2}{5}$ **iii** B

6 a

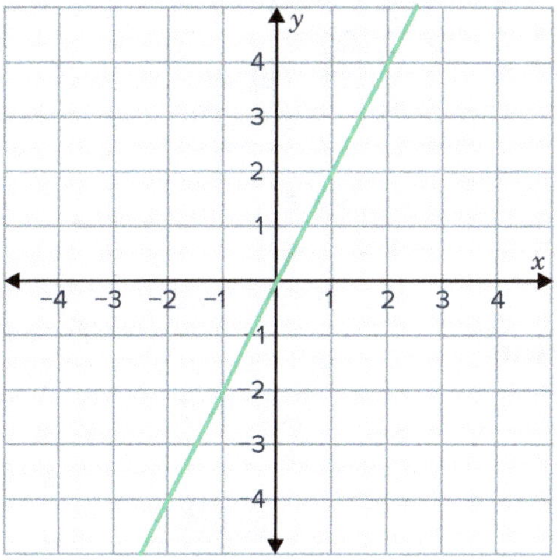

  b

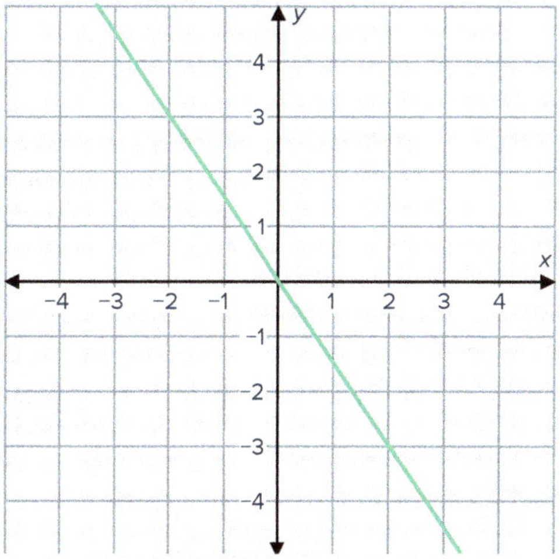

  c

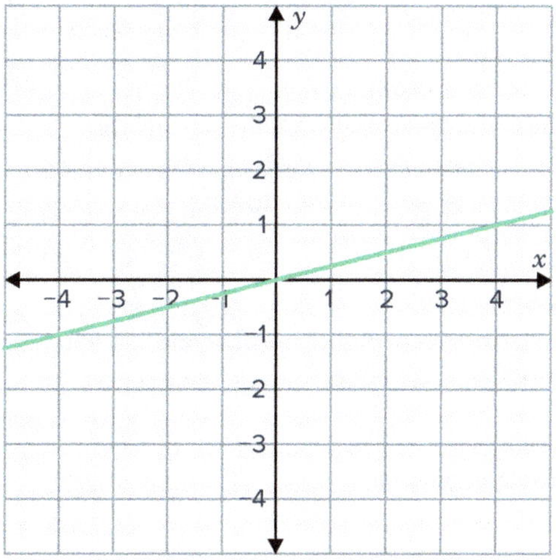

3.06 Slope-intercept form   293
mathspace.co

d

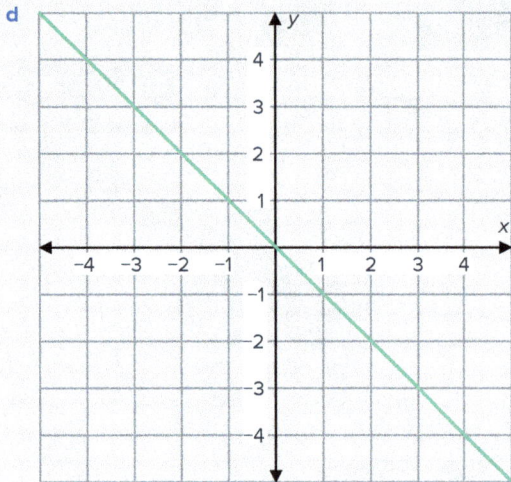

14 a i $y = 2x + 5$

ii

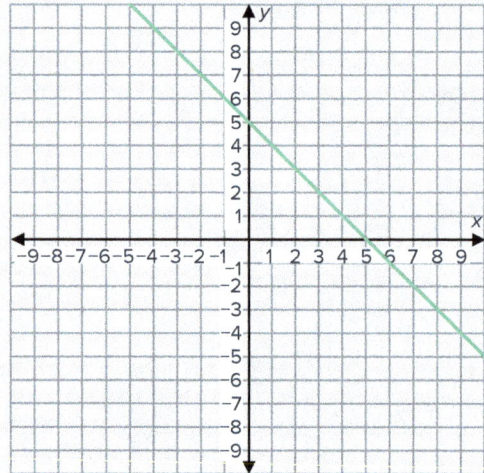

Let's practice

7 a $y = 16x - 12$ **b** -12

8 a $y = -5$ **b** $y = -4$ **c** $y = 2$

   d $y = 0$ **e** No y-intercept   **f** $y = \dfrac{1}{2}$

9 a i $(0, 0)$ **ii** 1 **iii** $y = x$

   b i $(0, -3)$ **ii** 4 **iii** $y = 4x - 3$

   c i $(0, 6)$ **ii** 5 **iii** $y = 5x + 6$

   d i $(0, 0)$ **ii** -5 **iii** $y = -5x$

10 a With $b = 0$, the y-intercept is at the origin $(0, 0)$, and the graph is not translated; it will pass through the origin.

   b With $b = 5$, the y-intercept is at $(0, 5)$. The entire graph shifts upwards by 5 units.

   c With $b = -3$, the y-intercept is at $(0, -3)$. The graph moves down by 3 units.

   d With $b = \dfrac{3}{2}$, the y-intercept is at $\left(0, \dfrac{3}{2}\right)$. The graph is translated $\dfrac{3}{2}$ units upwards.

11 a The graph is translated 3 units upwards.

   b The graph is translated 2 units downwards.

   c The graph is translated 5 units upwards.

   d The graph is translated 4 units downwards.

12 a The value of b is 3 because the line has been shifted upwards to pass through the point $(0, 3)$.

   b TThe value of b is -2 because the line has been shifted downwards to pass through the point $(0, -2)$.

   c The value of b is 5 because the line has been shifted upwards to pass through the point $(0, 5)$.

   d TThe value of b is -4 because the line has been shifted downwards to pass through the point $(0, -4)$.

13 a Above the origin   **b** At the origin

   c Below the origin   **d** At the origin

b i $y = -x + 5$

ii

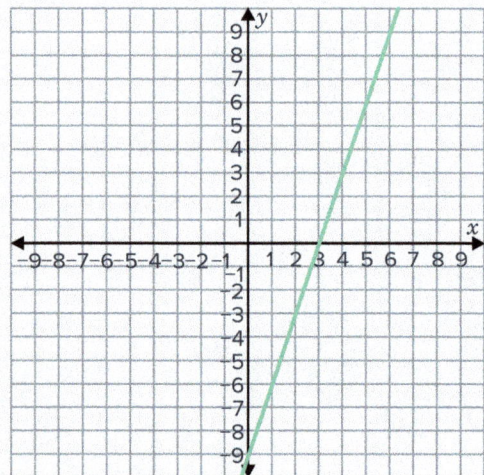

c i $y = 3x - 9$

ii

d i $y = -\frac{3x}{4} + 4$

ii

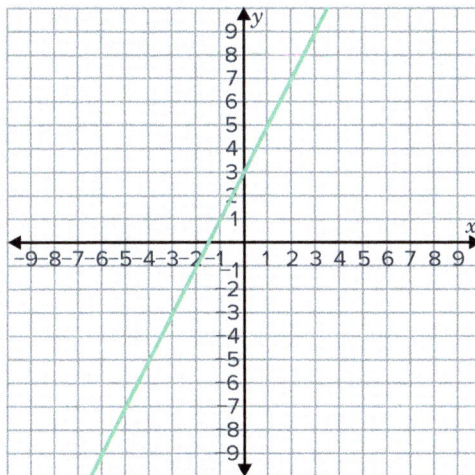

15 a i $m = 2$ **ii** $b = 3$

iii

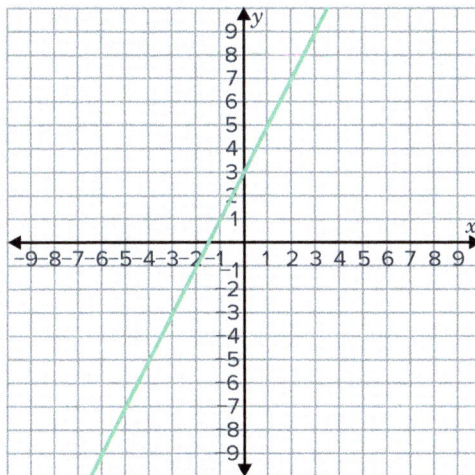

b i $m = -\frac{1}{3}$ **ii** $b = -2$

iii

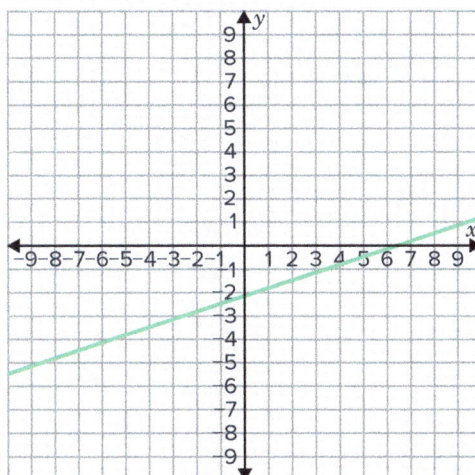

c i $m = \frac{4}{5}$ **ii** $b = -3$

iii

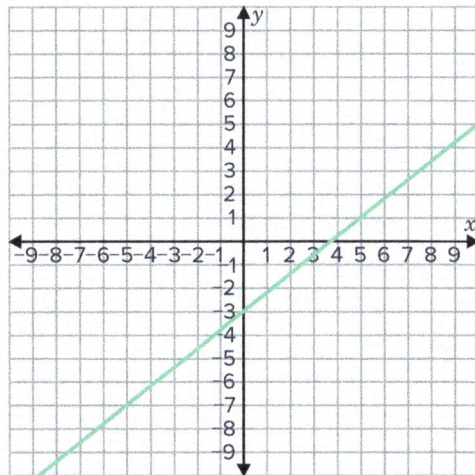

d i $m = -1$ **ii** $b = 5$

iii

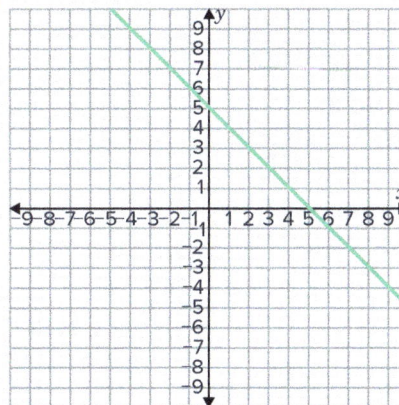

16 a $y = x + 2$ **b** $y = -\frac{3x}{5} + 6$

c $y = -2x$ **d** $y = -\frac{4x}{3} - 8$

17 a i

x	-1	0	1	2
y	-5	-3	-1	1

ii

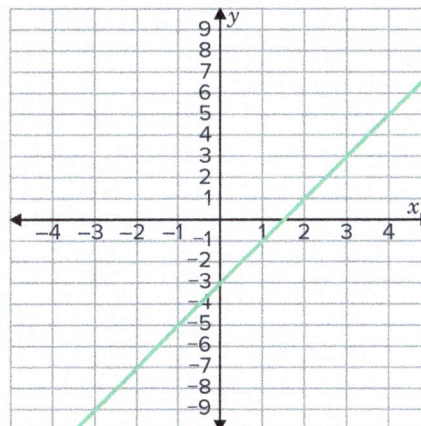

b **i** $y = x + 1$

ii

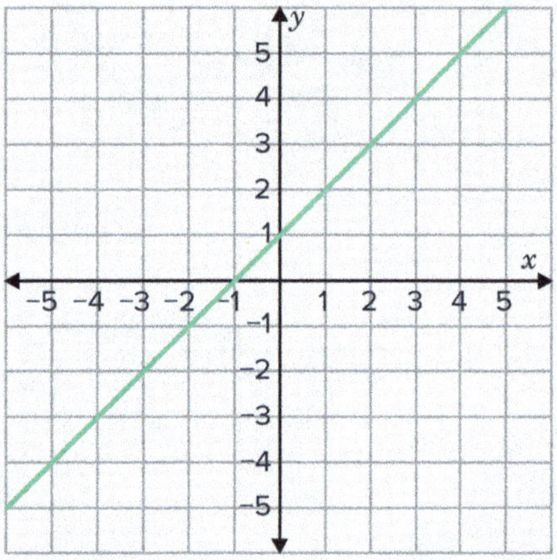

c **i**

x	−2	0	2	4
y	3	2	1	0

ii $y = -0.5x + 2$

Let's extend our thinking

18 **a** $y = -2.1x + 63$

b The water level was 63 feet before any water was pumped out.

c Yes, since the value of the y-intercept is positive then the line crosses the y-axis above the origin.

19 **a** $y = 50x + 250$

b The minimum amount charged by the carpenter is the initial fee of $50.

20 **a** ii. The cost of entry to a concert is determined by a booking fee of 3 dollars plus twice the cost per ticket.

b iii. The temperature decreases by half a degree Celsius for every kilometer increase in altitude above sea level, starting from an initial temperature of −5°C at sea level.

c i. The depth of water in a tank decreases linearly over time due to a leak, starting from a height of 4 meters.

d iv. A car accelerates at a constant rate, covering more distance over the same time interval, starting from a point 2 meters from a reference point.

21 **a** **i** Yes   **ii** $y = 4$

b **i** No   **ii** $x = 1$

c **i** No   **ii** $x = -5$

d **i** Yes   **ii** $y = -8$

3.07 Problem solving with linear functions

Subtopic overview

Lesson narrative

In this lesson, students will apply their understanding of linear functions to solve real-world problems. The lesson emphasizes the practical use of the slope-intercept form ($y = mx + b$) to model various scenarios. Students will learn that the slope (m) represents a constant rate of change, while the y-intercept (b) represents an initial value or starting point. The lesson includes several examples to illustrate these concepts. Students will engage in an exploration where they create their own realworld problems that can be modeled with linear functions. This hands-on activity involves identifying variables, determining rates of change, and writing equations in slope-intercept form. They will also graph these equations to interpret and solve the problems they created. By the end of the lesson, students will be adept at using linear functions to model and solve practical problems, understanding how the slope and y-intercept translate to real-world rates and starting values.

Learning objectives

Students: Page 168

After this lesson, you will be able to...

- graph a linear function given a table, equation, or contextual situation.
- create a table of values for a linear function given a graph, equation in the form $y = mx + b$, or context.
- describe the slope, y-intercept, independent and dependent variables of a linear function given a graph, table, equation, or contextual situation.
- write an equation of a linear function in the form $y = mx + b$, given a graph, table, or a situation in context.
- create a context for a linear function given a graph, table, or equation in the form $y = mx + b$.

Key vocabulary

- slope
- y-intercept

Essential understanding

Linear functions have a constant rate of change, this determines the real world situations that linear functions can model.

Standards

This subtopic addresses the following Virginia 2023 Mathematics Standards of Learning standards.

Mathematical Process Goals

MPG1 — Mathematical Problem Solving

Teachers can integrate this goal into their instruction by presenting students with real-life problems that can be solved using linear functions. For example, teachers can give students a scenario in which they need to calculate the cost of a product based on a given sales tax rate. The students will then use the slope and y-intercept concepts to determine the cost. Teachers can also encourage students to create their own real-world problems and solve them, thereby enhancing their problem-solving skills.

MPG4 — Mathematical Connections

Teachers can make mathematical connections by showing how linear functions apply to various real-world situations. For example, teachers can use situations like calculating the cost of a taxi ride or predicting the growth of a savings account to illustrate how linear functions are used in different contexts. By making these connections, students can see how the mathematical concepts they are learning have practical applications.

MPG2 — Mathematical Communication

Teachers can encourage students to articulate their mathematical reasoning both verbally and in writing. For example, teachers can begin by initiating a classroom discussion where students work in small groups to explain in their own words what the slope (m) and the y-intercept (b) represent in real-world contexts, using precise mathematical vocabulary and symbolic notation. Teachers can prompt the discussion with vocabulary like variable, rates of change, and initial value. Students may be asked to write a brief reflection summarizing their process and how the components of the slope-intercept form relate to their real-world context. By embedding these communication opportunities, students can clarify their thinking and enhance their ability to express mathematical ideas with precision.

Content standards

8.PFA.3 — The student will represent and solve problems, including those in context, by using linear functions and analyzing their key characteristics (the value of the y-intercept b) and the coordinates of the ordered pairs in graphs will be limited to integers).

8.PFA.3b — Describe key characteristics of linear functions including slope (m), y-intercept (b), and independent and dependent variables.

8.PFA.3c — Graph a linear function given a table, equation, or a situation in context.

8.PFA.3d — Create a table of values for a linear function given a graph, equation in the form of $y = mx + b$, or context.

8.PFA.3e — Write an equation of a linear function in the form $y = mx + b$, given a graph, table, or a situation in context.

8.PFA.3f — Create a context for a linear function given a graph, table, or equation in the form $y = mx + b$.

Prior connections

7.PFA.1 — The student will investigate and analyze proportional relationships between two quantities using verbal descriptions, tables, equations in $y = mx$ form, and graphs, including problems in context.

8.PFA.2 — The student will determine whether a given relation is a function and determine the domain and range of a function.

Future connections

A.F.1 — The student will investigate, analyze, and compare linear functions algebraically and graphically, and model linear relationships.

Lesson Preparation

Suggested review

Depending on your students' level of prior knowledge, consider revisiting the following lessons:

Grade 8 — 3.04 Independent and dependent variables
Grade 8 — 3.05 Characteristics of linear functions
Grade 8 — 3.06 Slope-intercept form

Tools

You may find these tools helpful:
- Graphing calculator

Student lesson & teacher guide

Problem solving linear functions

This lesson introduces how linear functions, represented as equations, tables, and graphs, are used to solve real-world problems. Students are reminded that the slope signifies a constant rate of change associated with the independent variable, while the y-intercept represents an initial or one-time value. Students examine key phrases like "each," "per," or "starting value" that can help them identify slope or y-intercept components in contextual situations.

Examples

Students: Page 169

Example 1

A diver starts at the surface of the water and begins to descend below the surface at a constant rate. The table shows the depth of the diver, in yards, over several minutes:

Time passed (min)	0	1	2	3	4
Depth (yds)	0	−1.4	−2.8	−4.2	−5.6

a Identify the independent and dependent variables.

Create a strategy

The value of the dependent variable changes based on the value of the independent variable.

Apply the idea

How far the diver has descended depends on the amount of time passed.

The independent variable is time passed in minutes, and the dependent variable is the depth in yards.

Reflect and check

The independent variable and dependent variable are often associated with consistent variables across different representations if specific variables are not listed.

The independent variable is an input, so it is associated with x. The dependent variable is an output, so it is associated with the variable y.

Purpose

Show students how to identify the independent and dependent variables in a given situation.

    b Determine the slope and y-intercept of the situation.

Create a strategy

The slope, or rate of change, is represented by the ratio $\dfrac{\text{change in } y}{\text{change in } x}$. The y-intercept is represented by the pair $(0, y)$ in a table.

Apply the idea

To find the slope, we see that the x-values change by 1, so we will need to find the change in y-values by subtracting one value by the previous, checking all pairs to make sure the rate is constant.

$-1.4 = -1.4 - 0$	Subtract 0 from -1.4
$-1.4 = -2.8 - (-1.4)$	Subtract -1.4 from -2.8
$-1.4 = -4.2 - (-2.8)$	Subtract -2.8 from -4.2
$-1.4 = -5.6 - (-4.2)$	Subtract -4.2 from -5.6

The rate of change is $\dfrac{-1.4}{1}$. The pair $(0, 0)$ is in the table, so this is the y-intercept.

The slope is -1.4, and the y-intercept is 0.

Purpose

Challenge students to determine the slope and y-intercept of the given situation.

    c Write an equation for the relationship between the number of minutes passed, x, and the depth, y, of the diver.

Create a strategy

We can use slope-intercept form, $y = mx + b$, where m is the change in depth per minute and b is the initial depth of the diver.

Apply the idea

$y = -1.4x + 0$	Substitute the value of m and b
$y = -1.4x$	Evaluate

Purpose

Show students how to translate the relationship between time and depth into a mathematical equation using the slope-intercept form.

Reflecting with students

Students may be unsure of how to write the equation when the y-intercept is 0 . Have a discussion about whether they should include the $+0$ in the equation, and help them make connections back to proportional equations.

    d Graph the relationship and determine the depth after 6 minutes.

Create a strategy

Graph using the original table or the slope-intercept form from part (c). Once graphed, find the point on the line with an x-value of 6.

Apply the idea

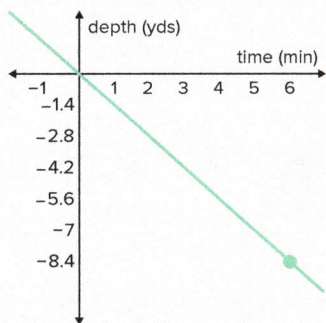

The point (6, –8.4) is on the graph. This means that after 6 minutes, the depth is –8.4 yards.

Reflect and check

If the table is used to graph, the graph could be used to determine the slope and y-intercept. The slope can be found using slope triangles, and the y-intercept is the point (0, y) where the function crosses the y-axis.

Purpose

Demonstrate to students how to graphically represent the relationship between time and depth, and how to use the graph to determine specific values.

Reflecting with students

Encourage students to use precise mathematical language and notation when working through this problem. For example, guide them to state, "After 6 minutes, the diver is at a depth of –8.4 yards," rather than simply saying "$y = -8.4$." When plotting the graph, remind students to label the axes clearly with "Time (minutes)" and "Depth (yards)."

Students: Page 170

Example 2

A carpenter charges a callout fee of $150 plus $45 per hour.

a   Write a linear equation to represent the total amount, y, charged by the carpenter as a function of the number of hours worked, x.

Create a strategy

The slope, m, represents a repeated change as x increases. The y-intercept, b, represents an initial value or constant value that is added once.

The slope-intercept form of a linear equation, $y = mx + b$, will use this rate of change, m, and constant, b, to write the equation.

Apply the idea

A callout fee is a one-time fee that is added to the charge. This value of $150 is b, the y-intercept. A charge of $45 per hour is repeated as the number of hours increases. This value is our slope, m.

$$y = mx + b \qquad \text{Slope-intercept form}$$
$$y = 45x + 150 \qquad \text{Substitute } m = 45 \text{ and } b = 150$$

The equation is $y = 45x + 150$.

Purpose

Show students how to create a linear equation to represent a real-world problem involving a rate and a constant.

Expected mistakes

Students may have trouble translating the scenario into an equation. Help them to identify the initial charge and the rate the carpenter charges, and help them relate these to the y-intercept and slope respectively.

Students: Page 170–171

b Use the function to complete the table shown. Use the table to find the total amount charged by the carpenter for 6 hours of work.

hours	0	2	4	6	8
total amount charged ($)					

Create a strategy

We can use the equation found from part (a), $y = 45x + 150$, and then substitute and evaluate the values of x.

Apply the idea

Each value of x from the table will be used as an input to the equation and evaluated.

$\$150 = \$45 \cdot 0 + \$150$	Substitute $x = 0$ and evaluate
$\$240 = \$45 \cdot 2 + \$150$	Substitute $x = 2$ and evaluate
$\$330 = \$45 \cdot 4 + \$150$	Substitute $x = 4$ and evaluate
$\$420 = \$45 \cdot 6 + \$150$	Substitute $x = 6$ and evaluate
$\$510 = \$45 \cdot 8 + \$150$	Substitute $x = 8$ and evaluate

Pairing each input with its output in the table:

hours	0	2	4	6	8
total amount charged ($)	150	240	330	420	510

For 6 hours of work, the total charged will be $420.

Reflect and check

We can also fill out the table by graphing the function in the coordinate plane using the equation $y = 45x + 150$ and finding the y-values paired with the x-values.

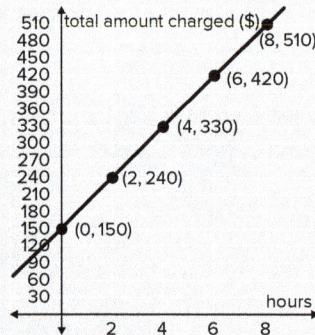

Purpose

Demonstrate to students how to use a linear function to complete a table and find a specific value.

For your advanced learners, consider presenting the problem without step-by-step guidance or scaffolding. Simply ask students to find the total amount the carpenter charges for 6 hours of work. Encourage them to interpret the problem, identify the variable components (the callout fee and the hourly rate), and determine the total on their own.

After they have found the solution using their chosen method, perhaps by writing and solving an equation, prompt them to verify their answer using a different strategy. For example, they can create a graph or a table to represent the linear relationship and confirm that their calculated total aligns with these representations.

This approach allows students to apply their understanding independently while also reinforcing their learning through multiple methods. By removing the scaffolding and encouraging verification through alternative strategies, you give advanced learners the opportunity to think critically, make connections, and deepen their conceptual understanding.

Students: Page 171–172

Example 3

Create a situation that could be modeled by the graph, describing the slope and y-intercept in context.

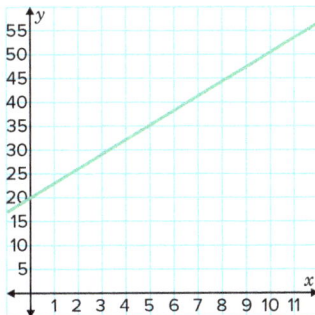

Create a strategy

The slope, m, describes the rate of change, and the y-intercept, b, gives an initial value. Whether the slope is positive or negative will tell you if the rate of change is increasing or decreasing.

Apply the idea

Using the points on the graph to find the slope and y-intercept, we will be determine the specific values needed and create a realistic situation.

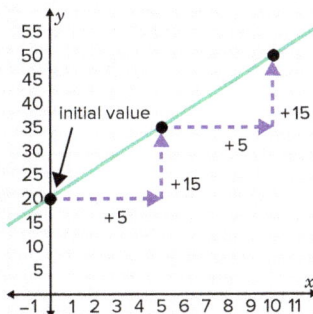

To find the slope of this graph, we must consider that the x-values have a scale of 1, while the y-values have a scale of 5.

The slope triangles each show a vertical change of 15, and a horizontal change of 5. To write this as slope, we will write

$$\frac{\text{vertical change}}{\text{horizontal change}} = \frac{+15}{+5} \qquad \text{Write as a ratio}$$

$$\frac{15 \div 5}{5 \div 5} = \frac{3}{1} \qquad \text{Reduce}$$

$$\frac{3}{1} = 3 \qquad \text{Divide}$$

The slope is 3, and the situation needs to describe increasing 3 vertical units for every 1 horizontal unit. The y-intercept of (0, 20) means the initial value of the situation is 20.

A possible situation could be a person will pay a one-time fee of $20 to enter an amusement park and pay $3 per ride. They want to determine how much money will be spent by going on different amounts of rides.

Reflect and check

Several other situations could be used to describe an increasing function.

Other common situations for increasing functions include relating money to time, distance traveled to time, or money spent to items purchased.

Purpose

Show students how to interpret a graph and create a realistic situation that models it.

Stronger and clearer each time
English language learner support   use with Example 3

Invite students to individually write their own real-world situation that could be modeled by the given graph. Encourage them to describe what the slope and y-intercept represent in their context, using mathematical terms like "slope," "rate of change," "initial value," and "y-intercept."

Then, have students share their situations with a partner. Encourage them to listen carefully, ask questions, and provide constructive feedback to help clarify and strengthen each other's explanations.

After the first round, have students meet with a new partner to share their revised situations and receive further feedback. Encourage them to refine their language, making their descriptions more precise and detailed, and to incorporate any new vocabulary or ideas they have learned.

Finally, ask students to revise their original written situations, incorporating the feedback and new ideas from their discussions. By iterating through this process, students will strengthen both their understanding of the mathematical concepts and their ability to express them in English.

Word bank to assist students in describing real-world scenarios   use with Example 3
Student with disabilities support

To support students who have difficulty connecting graphs to equations, begin by helping them identify the equation that models the given graph. Encourage them to select two points on the line to calculate the slope, and help them identify the y-intercept. This process will lead them to derive the equation $y = 3x + 20$, enhancing their understanding of how the graph translates into an algebraic expression.

Once the equation is established, to assist students who struggle with generating ideas for real-world contexts, provide a "variable bank" with a list of possible quantities that could represent the variables x and y. This bank might include items like:

Independent variable	Dependent variable
Hours worked	Number of items made
Number of items sold	Amount of money made
Number of items	Total cost
Hours traveled	Distance from home
Days	Total amount

Encourage students to choose variables from this bank to create their own scenarios. For example, they could select "days" for x and "total amount of blankets made" for y, where a knitting group starts with 20 blankets and knits 3 more each day. Providing this bank reduces the cognitive load and helps students focus on understanding the relationship between the variables and the constants in the equation.

Students: Page 172

> 💡 **Idea summary**
>
> Linear relationships (functions) as equations, tables, and graphs can be used to solve a variety of real-world problems.
>
> The slope, m, will represent a rate of change, and will be connected the the independent variable.
>
> The y-intercept, b, represents a value that occurs once and does not repeat over time.

Practice

Students: Page 173–178

What do you remember?

1   Consider the points in the table:

Time in minutes (x)	1	2	3	4	5
Temperature in °F (y)	8	13	18	23	28

    a   What are the independent and dependent variables?

    b   By how much does the temperature increase each minute?

    c   What would the temperature have been at time 0?

2   The equation $y = 25x + 85$ represents the total manufacturing cost, y, as a function of the number of plastic chairs manufactured, x.

    a   Find the cost of manufacturing each chair based on the slope of the function.

    b y-intercept of the function.

3 Match the equation of the line in slope-intercept form to the graph.

A $y = \dfrac{3}{2}x - 2$ **B** $y = -\dfrac{3}{4}x + 1$ **C** $y = \dfrac{x}{3} + 3$ **D** $y = -2x + 4$

a

b

c

d

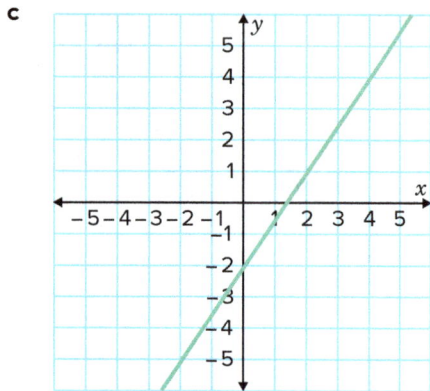

Let's practice

4 There are 20L of water in a rainwater tank. It rains for a whole day and during this time, the tank fills up at a rate of 8L per hour.

   a Complete the table of values:

Number of hours passed (x)	0	1	2	3	4	4.5	10
Amount of water in the tank (y)							

   b Write a linear equation linking the number of hours passed, x, and the amount of water in the tank, y.

5 The cost of a taxi ride is given by $C = 5.5t + 3$, where t is the duration of the trip in minutes.

   a Calculate the cost of an 11-minute trip.

   b For every extra minute the trip takes, how much more will the trip cost?

6 A racing car starts the race with 250L of fuel. From there, it uses fuel at a rate of 5L per minute.

   a Complete the table of values:

Number of minutes passed (x)	0	5	10	15	20	50
Amount of fuel left in the tank (y)						

   b Write a linear equation linking the number of minutes passed, x, and the amount of fuel left in the tank, y.

7 After Mae starts running, her heartbeat increases at a constant rate.

    a Complete the table:

Number of minutes passed (x)	0	2	4	6	8	10	12
Heart rate (y)	49	55	61	67	73	79	

    b What is the unit change in y for the above table?

    c Write a linear equation that describes the relationship between the number of minutes passed, x, and Mae's heartbeat, y.

8 It starts raining and an empty rainwater tank fills up at a constant rate of 2 gal per hour. By midnight, there are 20 gal of water in the rainwater tank. As it rains, the tank continues to fill up at this rate.

    a Complete the table of values:

Number of hours passed since midnight (x)	0	1	2	3	4	4.5	10
Amount of water in the tank (y)							

    b Sketch the graph depicting the situation on a coordinate plane.

    c Write a linear equation linking the number of hours passed since midnight, x, and the amount of water in the tank, y.

    d Determine the y-intercept of the line.

    e At what time prior to midnight was the tank empty?

9 A company decides to launch a new product, and the initial cost for production setup is $500. Each unit of the product is produced at a cost of $20. Graph a function that represents the given context.

10 Create a table of values for x values ranging from -2 to 2 for the following equations:

    a $y = 3x - 2$ **b** $y = -\dfrac{1}{2}x + 3$

11 Graph the following equations:

    a $y = 2x - 3$ **b** $y = -\dfrac{3}{4}x + 4$ **c** $y = \dfrac{5}{3}x - 3$ **d** $y = -x + 1$

12 The graph shows the relationship between water temperatures and surface air temperatures:

    a What are the independent and dependent variables?

    b Complete the table of values:

Water temperature (°C)	-3	-2	-1	0	1	2	3
Surface air temperature (°C)							

    c Write a linear equation representing the relationship between the water temperature, x, and the surface air temperature, y.

    d Find the surface air temperature when the water temperature is 14°C.

    e Find the water temperature when the surface air temperature is 23°C.

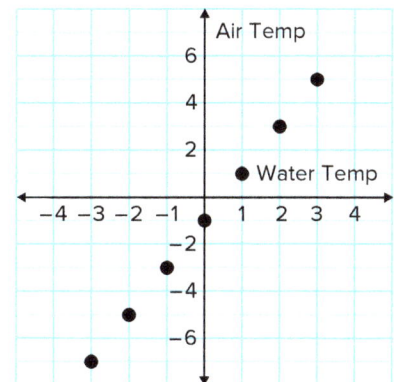

13 Create a table of values that corresponds to the graph.

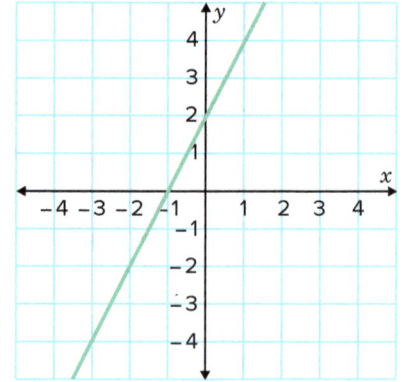

14 In a study, scientists found that the more someone sleeps, the quicker their reaction time. The table displays the findings:

Number of hours of sleep (x)	0	1	2	3	4	5
Reaction time in seconds (y)	6	5.8	5.6	5.4	5.2	5

    a   How much does the reaction time decrease for each extra hour of sleep?

    b   Write a linear equation relating the number of hours of sleep, x, and the reaction time, y.

    c   Calculate the reaction time for someone who has slept 4.5 hours.

    d   Calculate the number of hours someone sleeps if they have a reaction time of 5.5 seconds.

15 The graph shows the temperature of a room after the heater has been turned on:

    a   Determine the slope of the line.

    b   Determine the y-intercept.

    c   Find the temperature of the room after the heater has been turned on for 40 minutes.

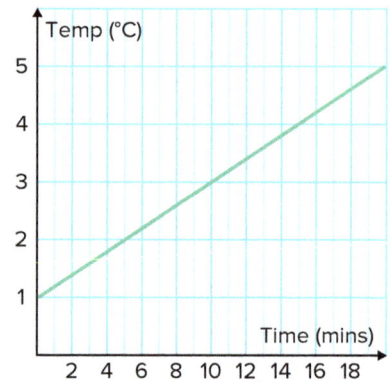

16 A cellphone salesman earned $600 in a particular week during which he sold 26 phones and $540 in another week during which he sold 20 phones.

    a   What are the independent and dependent variables?

    b   Write a linear equation to represent the weekly earnings of the salesman, y, as a function of the number of phones sold, x.

    c   Determine the slope of this function.

    d   Find the y-intercept.

    e   Find how much the salesman will earn in a week during which he sells 36 phones.

17 A particular restaurant has a fixed weekly cost of $1300 and receives an average of $16 from each customer.

    a   Write an equation to represent the net profit, y, of the restaurant for the week as a function of the number of customers, x.

    b   Find the slope of the function.

    c   Find the y-intercept.

    d   Find the restaurant's net profit if it has 310 customers for the week.

18 Mohamad is taking his new car out for a drive. He had only driven 50 mi in it before and is now driving it down the highway at 75 mi/h.

    a Write a linear equation to represent the total distance, y, that Mohamad had driven in his car as a function of the number of hours, x.

    b Determine the slope of the function.

    c Find the y-intercept.

    d Find the total distance Mohamad will have driven in his car if his current drive begins at 5:10 p.m. and finishes at 7:25 p.m.

19 Luigi is running a 100 mi ultramarathon at an average speed of 8 mi/h.

    a Write a linear equation to represent the distance Luigi has left to run, y, as a function of the number of hours since the start, x.

    b Determine the slope of the function.

    c Find the y-intercept.

    d Find the distance Luigi will have left to run after 4.5 hours.

20 A ball is rolled down a slope. The table shows the velocity, V, of the ball after a given number of seconds, t:

Time in seconds (t)	0	1	2	3	4	5
Velocity in m/s (V)	12	13.3	14.6	15.9	17.2	18.5

    a Sketch a graph that displays the ball's velocity against time.

    b Calculate the slope of the line.

    c Determine the y-intercept of the line.

    d Write a linear equation for the line, expressing V in terms of t.

    e Now, determine the velocity of the ball after 19 seconds.

21 The number of fish in a river is approximated over a five year period. The results are shown in the table:

Time in years (t)	0	1	2	3	4	5
Number of fish (F)	4800	4600	4400	4200	4000	3800

    a Sketch a graph corresponding to the given information.

    b Calculate the slope of the line.

    c Determine the value of F when the line crosses y-axis.

    d Determine an equation for the line using the given values.

    e Now, determine the number of fish remaining in the river after 13 years.

    f Find the number of years, t, until 2000 fish remain in the river.

22 The graph shows the amount of water remaining in a bucket that was initially full before a hole was made on its side:

    a Determine the slope of the line.

    b Find the y-intercept.

    c Explain the meaning of the slope in this context.

    d What does the y-intercept represent in this context?

    e Find the amount of water remaining in the bucket after 54 minutes.

23 Given the table create a real-world context that could be modeled by the function represented by the table. Describe the scenario in a few sentences.

Week	Money ($)
1	20
2	40
3	60
4	80

24 Given the equation $y = -15x + 300$, create a real-world context that could be modeled by this linear function. Describe the scenario in a few sentences.

25 Given the graph, create a real-world context that could be modeled by the linear function. Describe the scenario in a few sentences.

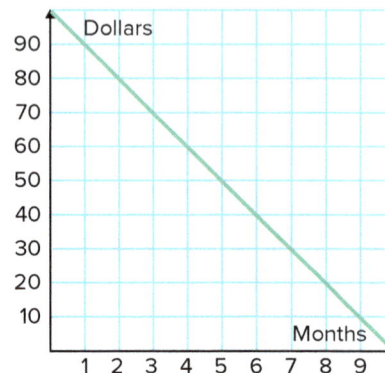

Let's extend our thinking

26 A baseball is thrown vertically upward by a baseball player when he is standing on the ground, and the velocity of the baseball, V (in yards per second), after T seconds is given by $V = 120 - 32T$.

    a What are the independent and dependent variables?

    b Complete the table of values:

Time	0	1	2	3	4
Vertical velocity					

    c Determine the slope of the linear function.

    d Explain the negative value of V when $T = 4$.

27 Gas costs a certain amount per liter. The table shows the cost of various amounts of gas in dollars:

Number of liters (x)	0	10	20	30	40
Cost of gas (y)	0	15	30	45	60

Calculate, if there is any, the change someone will receive if they use a $100 bill to pay for 47L of gas.

28 A racing car starts the race with 150L of fuel. From there, it uses fuel at a rate of 5L per minute.

How many minutes will it take for the car to run out of fuel?

29 Given the graph:

   i Predict the future value of the function at $x = 15$.

   ii Discuss how the value and slope of the function might represent changes over time in a real-world context. Consider what might happen if the trend continues beyond the shown graph.

Trend Over Time

Answers

What do you remember?

1 **a** Independent variable: Time in minutes

Dependent variable: Temperature in °F

b 5°F   **c** 3°F

2 **a** $25   **b** $85

3 **a** **B**   **b** **D**   **c** **A**   **d** **C**

Let's practice

4 **a**

Number of hours passed (x)	0	1	2	3	4	4.5	10
Amount of water in tank (y)	20	28	36	44	52	56	100

b $y = 8x + 20$

5 **a** $63.50   **b** $5.50

6 **a**

Number of minutes passed (x)	0	5	10	15	20	50
Amount of fuel left in tank (y)	250	225	200	175	150	0

b $y = 250 - 5x$

7 **a**

Number of minutes passed (x)	0	2	4	6	8	10	12
Heart rate (y)	49	55	61	67	73	79	85

b 3   **c** $y = 3x + 49$

8 **a**

Number of hours passed since midnight (x)	0	1	2	3	4	4.5	10
Amount of water in tank (y)	20	22	24	26	28	29	40

b

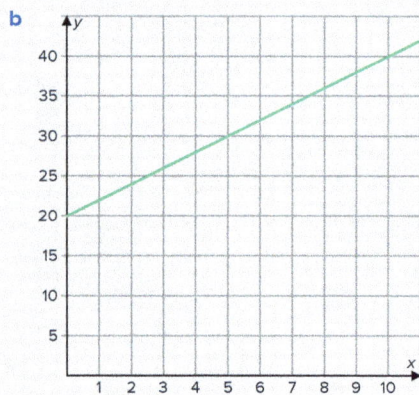

c $y = 2x + 20$ **d** 20   **e** 2 pm

9

10 **a**

x	y
−2	−8
−1	−5
0	−2
1	1
2	4

b

x	y
−2	4
−1	3.5
0	3
1	2.5
2	2

11 **a**

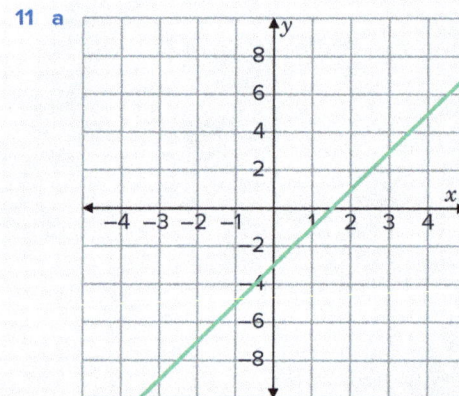

b

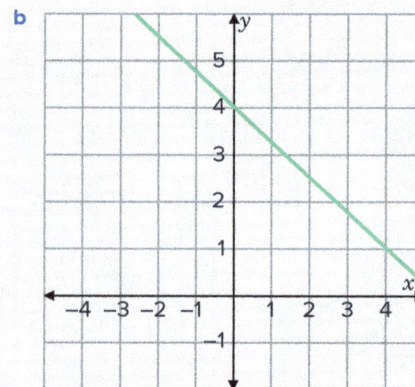

c

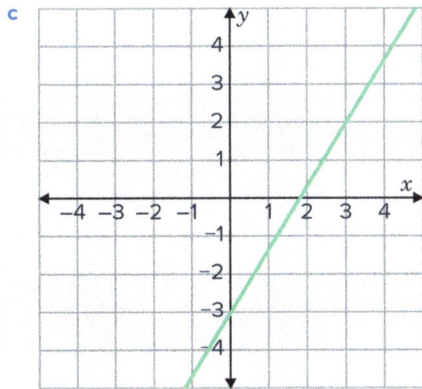

d

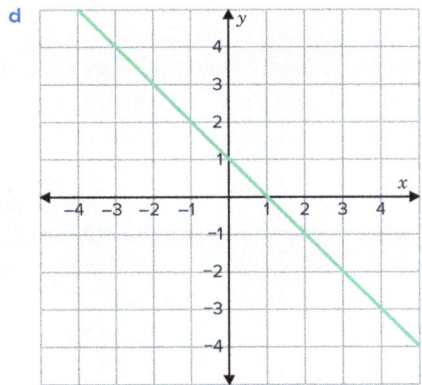

12 a Independent variable: Air temperature

Dependent variable: Water temperature

b

Water temperature (°C)	−3	−2	−1	0	1	2	3
Surface air temperature (°C)	−7	−5	−3	−1	1	3	5

c $y = 2x + 1$ **d** 27°C   **e** 12°C

13

x	y
−2	−2
−1	0
0	2
1	4
2	6

14 a 0.2 seconds   **b** $y = -0.2x + 6$

c 5.1 seconds   **d** $x = 2.5$ hours

15 a $\dfrac{1}{5}$ **b** 1   **c** 9°C

16 a Independent variable: Phones Sold

Dependent variable: Weekly earnings

b $y = 10x + 340$ **c** 10   **d** 340

e $700

17 a $y = 16x - 1300$ **b** 16

c −1 300   **d** $3 660

18 a $y = 75x - 50$ **b** 75

c 50   **d** 218.75 mi

19 a $y = 100 - 8x$ **b** −8

c 100   **d** 64 mi

20 a

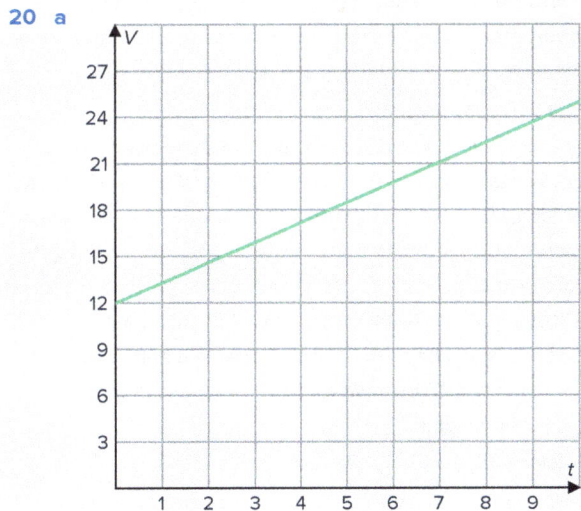

b $\dfrac{13}{10}$ **c** 12   **d** $V = 12 + \dfrac{13}{10}t$

e 36.7m/s

21 a

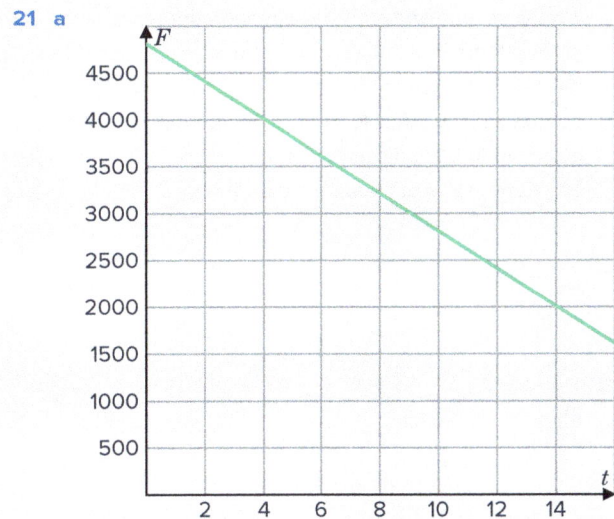

b −200   **c** $F = 4\,800$ **d** $F = -200t + 4\,800$

e 2 200 fish   **f** $t = 14$

22 a $\dfrac{1}{2}$ **b** 30

c The amount of water that is flowing out of the hole every minute.

d The capacity of the bucket.

e 3L

23 Answers may vary.

This table could represent the total amount of money saved by a student who saves $20 each week from their allowance. The "Week" column shows the number of weeks that have passed, and the "Money ($) column shows how much money the student has saved in total. For example, after 1 week, the student has saved $20, and by the 4th week, their savings have increased to $80, following a linear pattern of saving an additional $20 each week.

24 Answers may vary.

This equation could represent the remaining balance of a prepaid gift card that is used to make consistent purchases of $15 each week. The variable x represents the number of weeks, and y represents the remaining balance on the gift card in dollars. For example, the initial balance on the card is $300, and after 1 week of spending $15, the remaining balance would be $285. After 2 weeks, the balance would decrease to $270, showing a linear decrease in the card balance over time.

25 Answers may vary

This graph could represent the decreasing balance of a budget allocated for a monthly subscription service, where the service costs $10 each month. The x-axis represents the number of months that have passed, while the y-axis represents the remaining budget in dollars. Initially, there is a $100 budget, and for each month that passes, $10 is spent on the subscription, reducing the budget linearly. After 10 months, the budget would be depleted to $0, indicating that the service can no longer be afforded without adding more funds.

Let's extend our thinking

26 a Independent variable: Time
Dependent variable: Vertical Velocity

b

Time	0	1	2	3	4
Vertical Velocity	120	88	56	24	−8

c −32

d The baseball is moving downwards instead of upwards at T = 4.

27 $29.50

28 30 minutes

29 Answers may vary.

i At $x = 15$, the future value of the function can be predicted by the equation of the line, which is $y = 10x$. Substituting 15 for x, we get $y = 10 \cdot 15 = 150$. This suggests that, if the trend continues, the value after 15 months will be 150.

ii The slope of the function, which is positive, indicates a steady increase in the "Value" over "Time (Months)". In a real-world context, this could represent a scenario such as consistent monthly savings where the individual saves an additional $10 each month. If this trend continues beyond the shown graph, it could imply a successful savings strategy, leading to increased financial security over time.

Topic 3 Assessment: Relations & Functions

SOL **1** Graph each of the following relations represented by the tables of ordered pairs:

a

x	−2	5	7
y	6	−1	−3

b

x	1	2	3	4	5
y	5	1	−3	−7	−11

2 A function is defined as $y = -4x + 2$.

a Complete the table for this function.

x	−4	−3	−2	−1	0	1	2	3	4
y									

b Plot the points on a coordinate plane.

3 Carlos graphed the equation $y = 2x$ on a coordinate plane and then needed to graph the function $y = 2x - 3$ as well. Sam told him that was easy to graph because he could just shift his original line to the left 3 units. Is Sam correct? Justify your reasoning.

4 A function is defined by the equation:

$$y = 3x + 2$$

a Plot the points on a coordinate plane.

b Write a real-world context that this function could descibe.

5 Consider the following ordered pairs:

$$\{(-9,-5),(-5,-10),(-5,-4),(-3,7),(-2,-4),(-1,1)\}$$

a Plot the ordered pairs on a coordinate plane.

b Which ordered pair would need to be removed from the set so that the remaining ordered pairs represent a function?

6 Determine whether each of the following graphs is:
- a function,
- a relation,
- a relation and a function, or
- neither

a

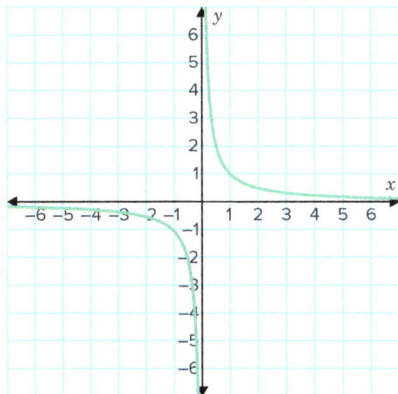

b

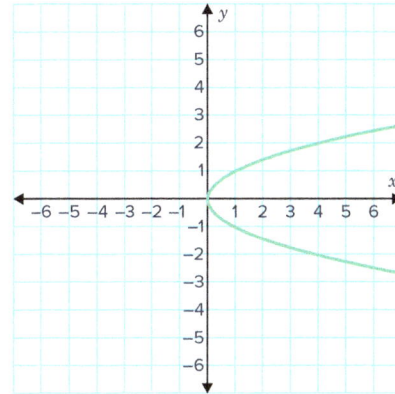

c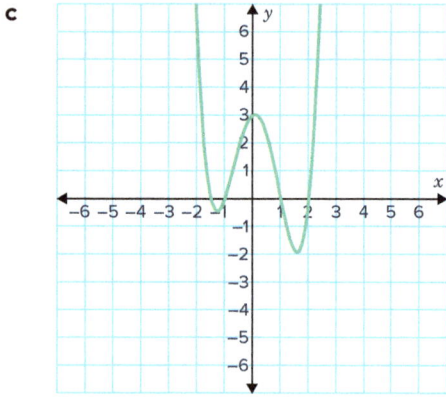

d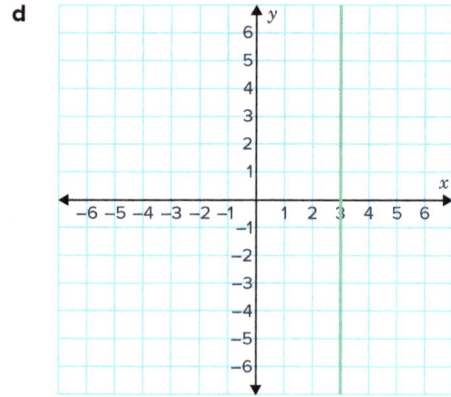

7 For each of the following relations:

i Find the domain.

ii Find the range.

iii Determine whether or not the relation is a function.

a

x	1	6	3	8	2
y	3	2	7	1	2

b

x	7	7	8	5	3
y	1	9	3	2	6

c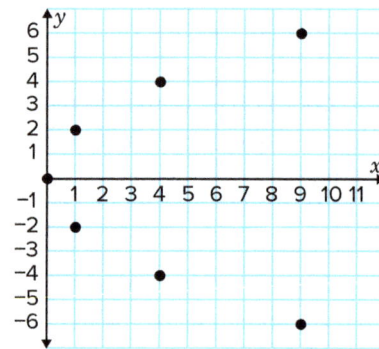

d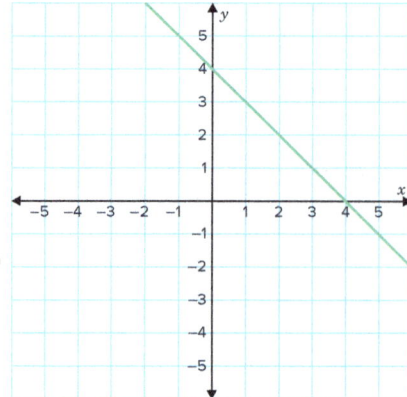

8 For each of the following pairs of linear functions, identify which function has the greater y-intercept:

a

- Function 1: The line with a slope of 4 that crosses the y-axis at (0, 6).
- Function 2: $y = x + 4$

b

- Function 1:

x	2	4	6
y	2	-2	-6

- Function 2:

c

- Function 1:

x	2	4	6
y	19	35	51

- Function 2: $y = 4x + 6$

9 For each of the following pairs of linear functions, identify which function has the greater slope:

a

- Function 1:

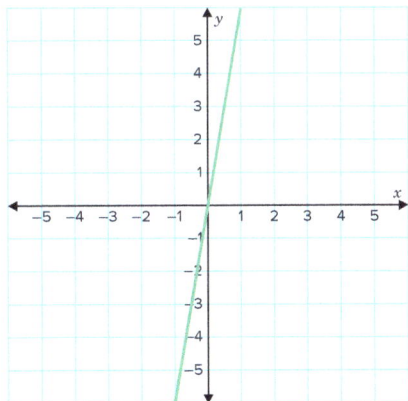

- Function 2:

x	0	1	2
y	0	4	8

b

- Function 3:

x	2	4	6
y	1	6	11

- Function 4: $y = 8x + 1$

10 A racing car starts the race with 150 L of fuel. From there, it uses fuel at a rate of 5 L per minute.

a Complete the following table of values:

Number of minutes passed (x)	0	5	10	15	20
Amount of fuel left in the tank (y)	150	☐	100	☐	☐

b Write a linear function linking the number of minutes passed, x, and the amount of fuel left in the tank, y.

c Explain the meaning of the slope in this context.

11 A carpenter charges a callout fee of $225 plus $45 per hour.

a Fill in the blank with the appropriate term:

The independent variable is represented by ☐, and the dependent variable is represented by ☐.

b Write a linear function to represent the total amount charged, y, by the carpenter as a function of the number of hours worked, x.

c State the y-intercept of the linear function.

12 Write an equation for each of the functions graphed below:

a

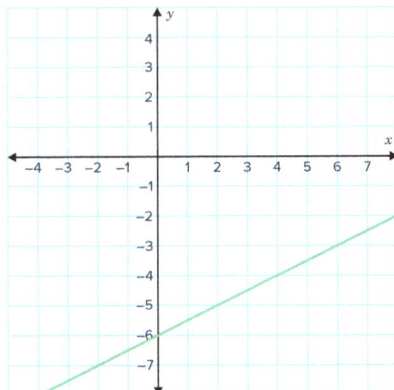

b

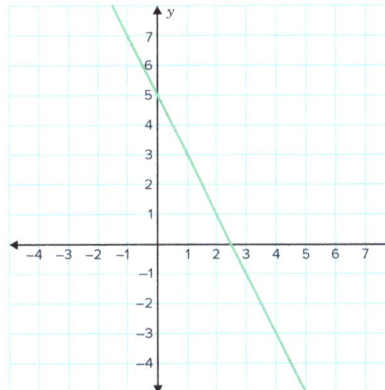

13 Mae's heartbeat was recorded below:

Number of minutes passed (x)	0	2	4	6	8	10	12
Heart rate (y)	49	55	61	67	73	79	85

Which of the following situations could explain her heart rate?

A Mae took an afternoon nap and her heart rate dropped by 3 every minute.

B Mae started sprinting and her heart rate increased by 6 every minute.

C Mae started playing tag and her heart rate increased by 3 every minute.

SOL 14 Identify the set that represents a function with a domain of {1, 4} and a range of {7, 9}?

A $\{(1,9),(4,7),(7,1)\}$

B $\{(4,9),(1,7)\}$

C $\{(1,7),(4,9),(9,4)\}$

D $\{(7,1),(9,4)\}$

SOL 15 A group of friends are organizing a charity run to support a local animal shelter. The total donation they collect depends on the number of kilometers they run. What is the independent variable in this scenario?

A The total amount of donations collected

B The number of friends participating

C The number of kilometers run

D The operational cost of the animal shelter

Performance task

16 You are entering a cup stacking competition. You want to order cups ahead of time to practice, and you would like the cups to be the same size as the competition. The competition doesn't provide a lot fo information, but you found this image:

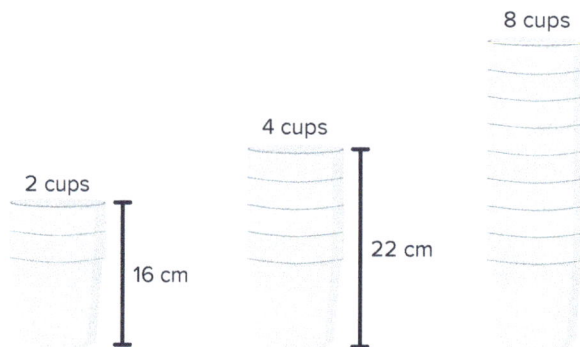

Your teacher wants you to help her get organized for when the cups arrive next week. Using only the information shown in the picture, she asks you to figure out some other specific measurements.

a How tall, in cm, is the stack of 8 cups?

b How tall, in cm, is 1 cup? Explain how you determined the height of 1 cup.

c Write an equation expressing the relationship between the height of the stack and the number of cups in the stack. Let h represent the height of the stack, in cm, and n the number of cups in the stack.

d The website selling these cups shows a stack that is 110 cm tall. Your friend Callum says, "That must be a misprint because a stack of that height is not possible." Do you agree or disagree with Callum? Explain your reasoning.

e The most plastic cups stacked in 30 seconds is 50 and was achieved by Silvio Sabba in Italy, on September 4, 2017. What is the minimum height for your town to break this record?

Answers

Topic 3 Assessment: Relations & Functions

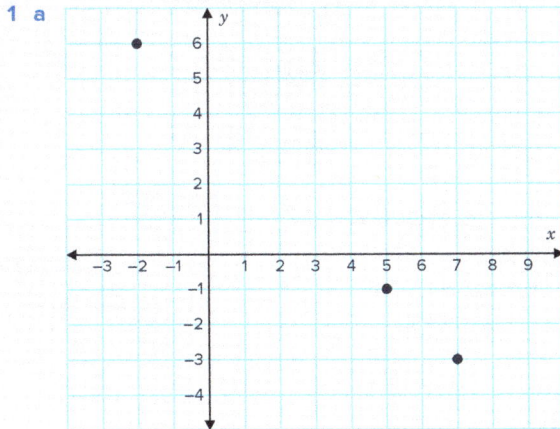

1 **a**

b

8.PFA.3c

2 **a**

x	−4	−3	−2	−1	0	1	2	3	4
y	18	14	10	6	4	0	−4	−8	−12

b

3 Sam is incorrect. While there is a simple way to graph this transformation, the −3 in the equation will shift the y-intercept down from 0 to −3 shifting the whole line down 3 units instead of shifting it to the left.

8.PFA.3a

4 **a**

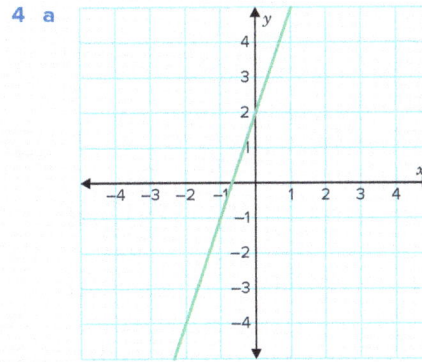

b Answers vary. Let x be time in weeks and y be height in inches. A plant starts at 2 inches tall and grows 3 more inches every week.

8.PFA.3c, 8.PFA.3f

5 **a**

b $(−5,−10)$ or $(−5,−4)$

8.PFA.2a

6 **a** Both relation and function

b Relation

c Both relation and function

d Relation

7 **a** **i** Domain $= 1, 6, 3, 8, 2$

ii Range $= 3, 2, 7, 1$

iii Function

b **i** Domain $= 7, 8, 5, 3$

ii Range $= 1, 9, 3, 2, 6$

iii Not a function

c **i** Domain $= −2, −1, 0, 1, 3, 4$

ii Range $= −4, −2, 0, 2, 6, 8$

iii Function

d **i** Domain $= 0, 1, 4, 9$

ii Range $= 0, 2, −2, 4, −4, 6, −6$

iii Function

8.PFA.2a, 8.PFA.2b

8 **a** Function 1

   b Function 1

   c Function 2

   8.PFA.3b

9 **a** Function 1

   b Function 4

   8.PFA.3b

10 **a**

Number of minutes passed (x)	0	5	10	15	20
Amount of fuel left in the tank (y)	150	125	100	75	50

   b $y = 150 - 5x$

   c The amount of fuel in the car is decreasing at a constant rate of 5 L per minute.

   8.PFA.3b, 8.PFA.3d, 8.PFA.3e

11 **a** The independent variable is represented by "x - the number of hours worked", and the dependent variable is represented by "y - the total money made".

   b $y = 45x + 225$

   c 225

   8.PFA.3b, 8.PFA.3e

12 **a** $y = \dfrac{1}{2}x - 6$

   b $y = -2x + 5$

   8.PFA.3e

13 **B**

   8.PFA.3f

14 **B**

   8.PFA.2b

15 **C**

   8.PFA.3b

Performance task

16 **a** 34 cm

   b One cup is 13 cm tall. The lip of each cup is 3 cm and the base is 10 cm.

   c $h = 10 + 3n$

   d Using our equation with $h = 100$, we have:
   $$h = 10 + 3n$$
   $$110 = 10 + 3n$$
   $$100 = 3n$$
   $$33.\overline{33} = n$$

   Callum is correct because this would require a partial cup to achieve this height.

   e Using our equation with $n = 51$, we have:
   $$h = 10 + 3(51)$$
   $$h = 10 + 153$$
   $$h = 163$$

   A height equal to or greater than 163 cm would set a new record.

   8.PFA.3d, 8.PFA.3e, 8.PFA.4c, 8.PFA.4e, MP1, MP2, MP4

4 Data Analysis

Big ideas

- Collecting and analyzing data can inform predictions and decisions, as long as the data is based on a valid sample.
- Different representations of data highlight different characteristics of the data.
- Many patterns can be found in sets of data. These patterns can be useful in making inferences but any inferences made from a set of data should not be taken as fact.
- Many sets of bivariate data can be modeled using familiar functions.

Chapter outline

Boxplots help detect outliers! An outlier can be a surprising piece of data, like finding a penguin in a grassy, rocky area —totally unexpected but real!

4. Data Analysis

Topic Overview

Foundational knowledge

Evaluating standards proficiency

The skills book contains questions matched to individual standards. It can be used to measure proficiency for each.

Students should be proficient in these standards.

6.PS.2 — The student will represent the mean as a balance point and determine the effect on statistical measures when a data point is added, removed, or changed.

7.PS.2 — The student will apply the data cycle (formulate questions; collect or acquire data; organize and represent data; and analyze data and communicate results) with a focus on histograms.

5.PS.2 — The student will solve contextual problems using measures of center and the range.

4.PS.1 — The student will apply the data cycle (formulate questions; collect or acquire data; organize and represent data; and analyze data and communicate results) with a focus on line graphs.

6.MG.3 — The student will describe the characteristics of the coordinate plane and graph ordered pairs.

8.PFA.3 — The student will represent and solve problems, including those in context, by using linear functions and analyzing their key characteristics (the value of the y-intercept b) and the coordinates of the ordered pairs in graphs will be limited to integers.

6.PS.1 — The student will apply the data cycle (formulate questions; collect or acquire data; organize and represent data; and analyze data and communicate results) with a focus on circle graphs.

Big ideas and essential understanding

Collecting and analyzing data can inform predictions and decisions, as long as the data is based on a valid sample.
4.01 — Representative samples are crucial if a data set will be used to make predictions and decisions.

Different representations of data highlight different characteristics of the data.
4.02 — A boxplot is a convenient and informative way to represent univariate data. It highlights the center and spread of the data.

4.04 — The way a set of data is represented affects the conclusions and generalizations that can be made.

4.06 — The way a set of data is represented affects the conclusions and generalizations that can be made.

Many patterns can be found in sets of data. These patterns can be useful in making inferences but any inferences made from a set of data should not be taken as fact.
4.03 — An outlier is an abnormal distance from the rest of the values in a data set. An outlier can sometimes give insight into the variable being explored but it can sometimes cause incorrect conclusions to be drawn from the data.

Many sets of bivariate data can be modeled using familiar functions.
4.05 — Bivariate data can often be approximated using a linear function.

Standards

8.PS.2 — The student will apply the data cycle (formulate questions; collect or acquire data; organize and represent data; and analyze data and communicate results) with a focus on boxplots.

8.PS.2a — Formulate questions that require the collection or acquisition of data with a focus on boxplots.
4.01 Data collection and sampling
4.02 Boxplots
4.03 Outliers
4.04 Compare data sets

8.PS.2b — Determine the data needed to answer a formulated question and collect the data (or acquire existing data) using various methods (e.g., observations, measurement, surveys, experiments).
4.01 Data collection and sampling
4.02 Boxplots
4.03 Outliers
4.04 Compare data sets

8.PS.2c — Determine how statistical bias might affect whether the data collected from the sample is representative of the larger population.
4.01 Data collection and sampling
4.02 Boxplots
4.04 Compare data sets

8.PS.2d — Organize and represent a numeric data set of no more than 20 items, using boxplots, with and without the use of technology.
4.02 Boxplots
4.03 Outliers
4.04 Compare data sets

8.PS.2e — Identify and describe the lower extreme (minimum), upper extreme (maximum), median, upper quartile, lower quartile, range, and interquartile range given a data set, represented by a boxplot.
4.02 Boxplots
4.03 Outliers
4.04 Compare data sets

8.PS.2f — Describe how the presence of an extreme data point (outlier) affects the shape and spread of the data distribution of a boxplot.
4.03 Outliers

8.PS.2g — Analyze data represented in a boxplot by making observations and drawing conclusions.

4.02 Boxplots
4.03 Outliers
4.04 Compare data sets

8.PS.2h — Compare and analyze two data sets represented in boxplots.
4.04 Compare data sets

8.PS.2i — Given a contextual situation, justify which graphical representation (e.g., pictographs, bar graphs, line graphs, line plots/dot plots, stem-and-leaf plots, circle graphs, histograms, and boxplots) best represents the data.
4.04 Compare data sets

8.PS.2j — Identify components of graphical displays that can be misleading.
4.06 Misleading data displays

8.PS.3 — The student will apply the data cycle (formulate questions; collect or acquire data; organize and represent data; and analyze data and communicate results) with a focus on scatterplots.

8.PS.3a — Formulate questions that require the collection or acquisition of data with a focus on scatterplots.
4.05 Scatterplots and lines of best fit

8.PS.3b — Determine the data needed to answer a formulated question and collect the data (or acquire existing data) of no more than 20 items using various methods (e.g., observations, measurement, surveys, experiments).
4.05 Scatterplots and lines of best fit

8.PS.3c — Organize and represent numeric bivariate data using scatterplots with and without the use of technology.
4.05 Scatterplots and lines of best fit

8.PS.3d — Make observations about a set of data points in a scatterplot as having a positive linear relationship, a negative linear relationship, or no relationship.
4.05 Scatterplots and lines of best fit

8.PS.3e — Analyze and justify the relationship of the quantitative bivariate data represented in scatterplots.
4.05 Scatterplots and lines of best fit

8.PS.3f — Sketch the line of best fit for data represented in a scatterplot.
4.05 Scatterplots and lines of best fit

Future connections

A.ST.1 — The student will apply the data cycle (formulate questions; collect or acquire data; organize and represent data; and analyze data and communicate results) with a focus on representing bivariate data in scatterplots and determining the curve of best fit using linear and quadratic functions.

A2.ST.1 — The student will apply the data cycle (formulate questions; collect or acquire data; organize and represent data; and analyze data and communicate results) with a focus on univariate quantitative data represented by a smooth curve, including a normal curve.

4.01 Data collection and sampling

Subtopic overview

Lesson narrative

In this lesson, students will learn about data collection and sampling methods used in statistical investigations. The lesson begins by explaining the difference between univariate and bivariate data. Univariate data involves a single characteristic and can be displayed using various graphs such as histograms or bar graphs. Bivariate data involves two variables and requires a table for comparison, such as time versus distance or age versus height. The lesson continues with an exploration of formulating statistical questions that can be answered by collecting data. Students will learn the criteria for a good statistical question: it should anticipate variability, be answerable through data collection, and result in data that can be visually displayed. Examples and exercises help students practice identifying and creating such questions. Students will also learn about different methods of data collection, including observation, measurement, surveys, experiments, and using existing data. The importance of collecting unbiased data is emphasized, and various types of bias such as sampling bias, undercoverage bias, exclusion bias, and self-selection bias are discussed. Students will explore strategies to avoid these biases to ensure their data is representative of the population. By the end of the lesson, students should understand how to formulate statistical questions, choose appropriate data collection methods, and recognize and avoid biases in sampling. They will be able to apply these skills to gather accurate and representative data for statistical analysis.

Learning objectives

Students: Page 182

> 🎓 **After this lesson, you will be able to...**
> - write statistical questions and use them to collect univariate data.
> - collect univariate data using observations, measurements, surveys, and experiments.
> - explain how bias affects how representative a sample is.

Key vocabulary

- bias
- measurement
- sample
- univariate data
- bivariate data
- observation
- secondary data
- census
- population
- statistical question
- experiment
- representative sample
- survey

Essential understanding

Representative samples are crucial if a data set will be used to make predictions and decisions.

Standards

This subtopic addresses the following Virginia 2023 Mathematics Standards of Learning standards.

Mathematical Process Goals

MPG2 — Mathematical Communication

Teachers can integrate this goal by encouraging students to communicate their thinking throughout the data cycle process. They can ask students to explain their reasoning when formulating questions, deciding on data collection methods, and analyzing the collected data. Teachers can also provide opportunities for students to present their findings and solutions, using appropriate mathematical terminology and notation.

MPG3 — Mathematical Reasoning

Teachers can foster mathematical reasoning by guiding students to use logical reasoning when formulating questions and determining the data needed to answer those questions. They can also encourage students to think critically about potential sources of bias in data collection and how these biases can affect the accuracy of their conclusions. Teachers can also reinforce the use of reasoning by discussing potential challenges and limitations in data collection and sampling.

MPG4 — Mathematical Connections

Teachers can help students make mathematical connections by linking their knowledge of data collection and analysis with real-world situations. For example, teachers can discuss how statistical bias impacts real-world data collection and analysis, such as in scientific research or public opinion polls. They can also connect the concepts of sample size and random sampling to students' everyday experiences, such as social media algorithms or music streaming playlists.

Content standards

8.PS.2 — The student will apply the data cycle (formulate questions; collect or acquire data; organize and represent data; and analyze data and communicate results) with a focus on boxplots.

8.PS.2a — Formulate questions that require the collection or acquisition of data with a focus on boxplots.

8.PS.2b — Determine the data needed to answer a formulated question and collect the data (or acquire existing data) using various methods (e.g., observations, measurement, surveys, experiments).

8.PS.2c — Determine how statistical bias might affect whether the data collected from the sample is representative of the larger population.

Prior connections

6.PS.2 — The student will represent the mean as a balance point and determine the effect on statistical measures when a data point is added, removed, or changed.

7.PS.2 — The student will apply the data cycle (formulate questions; collect or acquire data; organize and represent data; and analyze data and communicate results) with a focus on histograms.

Future connections

A.ST.1 — The student will apply the data cycle (formulate questions; collect or acquire data; organize and represent data; and analyze data and communicate results) with a focus on representing bivariate data in scatterplots and determining the curve of best fit using linear and quadratic functions.

A2.ST.1 — The student will apply the data cycle (formulate questions; collect or acquire data; organize and represent data; and analyze data and communicate results) with a focus on univariate quantitative data represented by a smooth curve, including a normal curve.

Lesson Preparation

Suggested review

Depending on your students' level of prior knowledge, consider revisiting the following lessons:

Grade 6 — 9.01 Formulate questions and collect data
Grade 7 — 5.04 Formulate questions and collect data

Lesson supports

The following supports may be useful for this lesson. More specific supports may appear throughout the lesson:

Compare and contrast the STEAM and data cycles
Targeted instructional strategies

Have students compare the data cycle to the STEAM cycle. Prompts could include:

- What similarities are there to the STEAM and data cycles?
- What types of contexts would you use the STEAM cycle versus the data cycle?
- What are the difference between the STEAM and data cycles?

Some similarities are:

- Similarities Iterative Process: Both cycles involve repeating steps to refine solutions or conclusions.
- Problem-Solving Focus: Both are frameworks for addressing problems and finding solutions.
- Data Use: Each cycle emphasizes the use of data in some capacity—STEAM in the "Create and Test" and "Improve" steps, and the Data Cycle throughout. Collaboration and Communication: Both encourage collaboration, sharing of ideas, and communication of results.
- Collaboration and Communication: Both encourage collaboration, sharing of ideas, and communication of results.

Some differences are:

STEAM cycle	Data cycle
Focuses on multidisciplinary approaches, incorporating science, technology, engineering, arts, and mathematics	Focuses on data-related tasks like collection, analysis, and interpretation
May require more creativity in the «Imagine» step	More structured in the approach
Results involve prototypes or tangible solutions to a problem.	Results involve data-driven conclusions or explanations
Encourages innovation through design, testing, and improvement of physical or conceptual solutions	Prioritizes understanding trends, patterns, and conclusions from data sets

Students may benefit from seeing contexts where the different approaches are beneficial. For example:

- Designing an energy-efficient school building would be applicable to both cycles.
  - STEAM: Brainstorming innovative design solutions, testing models, and improving them. testing models, and improving them.
  - Data Cycle: Analyzing energy consumption patterns, collecting data on sustainable materials, and evaluating potential environmental impacts.

- Building a robot for a robotics competition would only be applicable to the STEAM cycle.
  - This requires creative problem-solving, engineering, and testing prototypes, which are core to the STEAM cycle. The Data Cycle's focus on data analysis is less central here.
- Investigating the relationship between screen time and sleep duration among teenagers would only be applicable to the Data cycle.
  - This involves formulating questions, collecting survey data, creating graphs, and drawing conclusions—steps that fit neatly within the Data Cycle without needing the creative prototyping of the STEAM cycle.

The data cycle is cyclical
Address student misconceptions

Students may mistakenly believe that the data cycle is a linear process. It's crucial to correct this misconception by explaining that the data cycle is a circular process, with each step feeding into the next. If data is interpreted and new questions arise, the cycle begins again with new planning.

Student lesson & teacher guide

Formulate questions for univariate data

This introduces the statistical investigation process, guiding students through structuring data cycles to address real-world problems like statisticians. We focus on distinguishing between univariate and bivariate data, equipping students to craft precise questions for effective data collection. These skills are crucial for analyzing and presenting statistical data accurately.

Real world examples
Targeted instructional strategies

Encourage students to think of real-world situations or problems that could be investigated using statistical methods. This will help them understand the practical applications of the concepts they are learning. For example, they could consider how a company might use data to improve its products or services, or how a sports team might use data to improve its performance.

Examples
Students: Page 183–184

Example 1

Select the question(s) which could be answered by collecting univariate data. Select all that apply.

A What is the median house price in Virginia?

B Do I need to file taxes every year as a part-time employee?

C What is the relationship between reaction time and hours of sleep the previous night?

D What is the distribution of salaries of professional rugby players in North America?

Create a strategy

For univariate data each member of the sample or population would have one characteristic recorded.

Apply the idea

Let's go through each of the options:

Option A: The characteristic that would be collected for each house in the sample would be its selling price or value. This is a single characteristic and would be different for different houses. This could be answered by collecting univariate data.

Option B: There is no characteristic being collected here. This is a fact with a single answers, so would not have univariate data collected to answer it.

Option C: The characteristics being collected from each member of the sample would be reaction time and amount of sleep. There are two characteristics being compared, so this is bivariate data, not univariate.

Option D: The characteristic that would be collected for each North American professional rugby player in the sample would be their salary. This is a single characteristic and would be different for different players. This could be answered by collecting univariate data.

The correct answers are A and D.

Reflect and check

Although option A could be answered with univariate data, it is not a statistical question as there would be a single correct answer at a given point in time.

Purpose

Check students' understanding of univariate data, and their ability to identify questions that can be answered by collecting this type of data.

Expected mistakes

Students may select option C if they did not fully read the question and missed that it specified "univariate" or they may not understand the term. Ask students who selected option C to explain what univariate means to identify is there is a gap in knowledge or attention to detail.

Students: Page 184

Example 2

Determine whether or not each question is a statistical question. Explain why or why not.

a How many books did my teacher read last year?

Create a strategy

A statistical question is one that does not have a single possible answer, requires data collection, and be summarized with a data display.

Apply the idea

This question is not a statistical question because it has only one possible answer and there is not enough data to create a display.

Reflect and check

This could be reworded to "For students in my class, how did the number of books they read last year vary?"

Purpose

Show students how to distinguish between statistical and non-statistical questions.

b How many steps do most students in our school walk each day?

Create a strategy

This question has a clear population of "students in our school". We now need to consider if the data collected will have more than one possible answer and can be displayed.

Apply the idea

This is a statistical question because:

- We would need to collect data to answer this question.
- A variety of answers are possible- likely ranging from 2000 to 20 000 depending on level of activity.
- A histogram could be used to display the results.

Reflect and check

The data would be univariate data.

Purpose

Check students' understanding of statistical questions by having them determine whether a given question can be answered with collected data, if it has variability, and if it can be displayed in a graph.

Students: Page 185

c What is the range of money spent on snacks per week by those who pack a lunch?

Create a strategy

We need to look at if the data collected will have more than one possible answer and can be analyzed.

Apply the idea

This is a statistical question because:

- We would need to collect data to answer this question.
- A variety of answers are possible.
- A histogram could be used to display the results.

Reflect and check

The population would need to be made more specific before collecting data.

Purpose

Show students that they can analyze the range of a data set and understand the concept of a statistical question by considering different possible answers and data collection methods.

Facilitate a classroom discussion to help students distinguish between statistical and non-statistical questions using targeted sentence stems like:

- A statistical question requires data because...
- This question is a statistical question since it involves...
- This question is not a statistical question because...
- We can answer this question without collecting data because...
- A statistical question differs from a non-statistical question in that...

Encourage students to use these stems to explain their reasoning about each question in the problem. For example, when considering "What is the median house price in Virginia?", prompt students to discuss how this question involves collecting data on various house prices, highlighting the concept of variability.

By using these discussion supports, you help students articulate their thoughts in English while grasping the mathematical idea of statistical questions. This strategy reinforces key vocabulary such as "statistical question," "data collection," "variability," and "univariate data," supporting both their language development and mathematical understanding.

This math language routine could be paired with "compare and connect" for further support.

Students: Page 185

💡 Idea summary

Univariate data is data where only one attribute or characteristic is collected.

Before we can collect data, we need a clear **statistical question** that:

- Can be answered by collecting data
- Has some amount of variation - more than one possible answer
- Results in data that can be shown in a visual display - the type of display will depend on the type of data and the question

Collecting data without bias

Student will explore essential strategies for unbiased data collection in statistical investigations, covering methods like observation, measurement, surveys, experiments, and existing data. We emphasize the importance of selecting representative samples to prevent skewed results and discuss various biases and their mitigation. These concepts equip students with the skills needed to effectively gather and analyze data, ensuring accurate statistical conclusions.

Example 3

Determine whether each situation demonstrates a sample survey, an experiment, or an observational study.

   a A grocery store wants to know if their customers would use self-checkouts if they were added or if they prefer using the standard checkout lanes that are staffed.

Create a strategy

To determine which type of design is best for this situation, we need to determine how the data can be collected.

Apply the idea

Because the grocery store wants to know their customers opinions, they will need to ask their customers about their preference on checkout style. A sample survey is the best design for gathering this information.

Reflect and check

If the grocery store installed self-checkouts and wanted to know which types of checkouts were used more frequently, an observational study would be a suitable design.

Purpose

Show students how to identify the most suitable type of statistical study (sample survey, experiment, or observational study) to collect data for different types of research questions or situations.

   b A group of students wants to know how different levels of fertilizer affect plant growth.

Create a strategy

A sample survey would not make sense in this situation. An observational study determines correlation, but it cannot determine if the fertilizer was the cause of specific growth differences. An experiment can determine cause and effect relationships.

Apply the idea

Since the students want to know if different levels were the cause of plant growth levels, the students should design an experiment.

Reflect and check

Although an observational study could have identified a correlation between fertilizer and plant growth, it cannot distinguish between fertilizer levels and plant heights. Other factors (like the sun or water levels) may have also caused a difference in plant levels.

If they use an experiment, the students can control the other factors (like sun and water levels) to make sure all plants receive the same amount. This will help them determine if the fertilizer was truly the cause of the plant growth.

Purpose

Show students how to determine the best method to use (survey, observational study, or experiment) based on the research question and the type of conclusion they want to draw.

c   Endangered, wild wolves were reintroduced to Yellowstone National Park. The conservationists want to know if, and by how much, the population of wolves is growing.

Create a strategy

A sample survey would not make sense in this context. An observational study does not interfere with the lifestyle of the wolves. An experiment would use certain factors to attempt to change the population of the wolves with the purpose of determining if these factors had an effect on the population.

Apply the idea

Because the conservationists do not want to control the population or try to make it grow by using a certain tactic, an observational study is the best type of design for collecting data.

The conservationists would still need to find a way to track the population, like tagging the wolves or placing trackers on them, but this tactic does not try to increase the population. It is a method of observing how the population grows naturally.

Purpose

Show students how to identify an appropriate study design based on the context and objective of a research question.

Reflecting with students

Have students consider what observer bias would look like in this scenario. For example, the conservationists may only count wolves that appear to be healthy and not those that they expect not to survive the season, leading to an undercount.

Example 4

A city council wants to determine whether a new skateboard park or a new ice-skating rink should be built as the new community building project. The new project will be located in the city park.

a   Identify the target population.

Apply the idea

The target population will be the members of the community or residents of the city. These will be the people using, building, and paying for the upkeep of the park.

Reflect and check

The city may hope that residents of other cities will be drawn to the park because of the new project, but the target population will be the people most likely to use the skateboard park or ice skating rink.

Purpose

Teach students to identify the target population in a real-world scenario, helping them understand the importance of a well-defined target population in planning and decision-making processes.

b What design methodology would be best to find out how the community feels about the two proposed community building projects?

A Observation

B Measurement

C Survey

D Acquire secondary data

Create a strategy

We are looking for people's opinions on a project. We should consider if we could get public opinions from each method.

Apply the idea

Let's look at each option:

Option A: Can we watch the public and determine what they want? No, it would be difficult to just watch people and see if they would be interested in ice skating or not.

Option B: Can we measure the public and determine what they want? No, their interests in skateboarding or ice skating is not able to be measured using measurement tools.

Option C: Can we ask the public and determine what they want? Yes, by asking them a specific question about their preferences for skateboarding or ice skating we can get the data to answer this question.

Option D: Would there be existing data to answer this question? No, it is highly unlikely that this question has already been asked to a sample which represents the current population.

The answer is C: Survey.

Purpose

Demonstrate to students how to evaluate and select the most appropriate methodology for data collection based on the nature of the information needed.

c Explain why using the local ice hockey team as the sample would not give representative data.

Create a strategy

For the sample to be representative of the population, different types of community members of varying ages with various hobbies should have an opportunity to express their opinion.

Apply the idea

An ice hockey team will most likely prefer a new ice skating rink, but we do not know the preferences of the entire community.

The team members may also be around the same age, and this age does not represent all ages within the community.

Finally, there may not be many team members, but the population of the city might be relatively large in comparison.

Purpose

Show students the importance of having a representative sample when collecting data for research or decision making.

Reflecting with students

Have students consider what type of bias it would be if they did only survey ice hockey players (sampling bias - exclusion bias).

Encourage students to extend the problem by adding layers of complexity to the sampling methods and population considerations. After identifying the target population and discussing why the local ice hockey team isn't representative, prompt students to explore other groups that might also produce biased results, such as surveying only skateboarders or park visitors at a specific time of day.

Invite them to design their own sampling strategies - like sampling across different age groups or random sampling from the entire city population - to determine the community's preference accurately. Ask them to predict and analyze how each sampling method could influence the outcome between building a skateboard park or an ice-skating rink. This multilayered approach helps advanced learners critically examine the impact of sampling techniques on data representativeness and decision-making. If possible, illustrate these concepts with a diagram showing different sampling methods and how they relate to the entire population, highlighting potential biases visually.

Students: Page 189

Example 5

For each survey question and sample, determine whether the results are likely to be biased or not. Explain your answer.

a   To answer the question "How much time do students at my school spend practicing a musical instrument per week?", Yvonne surveys the people in her jazz quartet.

Create a strategy

First we can look at whether or not the sample is representative of the population. If it is a good sample, then we should consider if there are any ways the data that is collected could be biased by the data collector.

Apply the idea

The results would likely be biased.

A quartet only has four people, so that sample would be too small to represent the population. Also, this sample would miss out of people who play non-jazz instruments or do not play an instrument. It is also a convenience sample, so would not be random.

Purpose

Show students how to assess the potential for bias in survey results based on the chosen sample and methodology.

b To answer the question, "What range of speeds do people drive on I95 throughout the day?", Lachlan uses a radar gun to observe and measure the speeds of 100 cars in the right lane between the hours of 8 AM and 9 AM.

Create a strategy

We need to consider whether the cars whose speed was recorded represent all of the cars that travel on 195.

Apply the idea

The results would likely be biased.

One main cause of bias would be that between 8 AM and 9 AM, it is rush hour traffic, so the speeds at that point in time would be very different than midday or at night.

There is also experimenter bias as he only measures cars in the right lane which are usually slower than cars in the left lane.

Reflect and check

There could also be more experimenter bias if he chose cars that were easier to follow like red cars. If his radar gun was not working well, he could also have measurement bias.

Purpose

Make students aware of the potential biases that can occur in data collection and how these biases might affect the interpretation of the data.

Students: Page 189

c To answer the question "How much rain does Middleburg, VA get per month?" Tricia uses historical weather data from a reliable source for the past 20 years.

Create a strategy

Tricia is using secondary data, not primary data, so bias would come from using an unreliable source.

Apply the idea

The data would likely be unbiased.

She uses a reliable source which would likely have used proper measurement tools.

Reflect and check

When using secondary sources, we can compare the results from two different sources check how accurate the results are.

Purpose

Show students that the reliability of data sources is crucial when using secondary data for analysis.

Building and connecting concept maps to understand sampling bias

use with Example 5

Student with disabilities support

Encourage students to create visual concept maps that connect the idea of surveying to the concept of bias and the data collection methods. Some students may benefit from being given a completed mind map. Others may benefit from being given the structure with a few terms filled in and a word bank to fill in.

Under each branch, prompt them to add short notes or examples. By visually mapping these connections, students can better understand why surveying a small, specific group may lead to biased results.

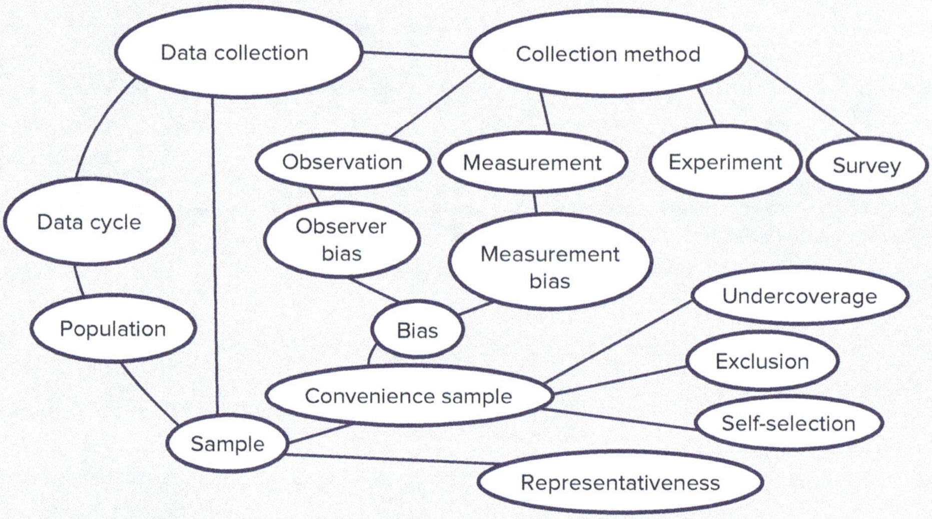

Students: Page 190

Idea summary

After we formulate a clear statistical question, we use the data cycle to collect, show, and explain information. To get data, we can use methods like:

- Watching **(Observation)**
- **Measuring**
- Asking questions **(Survey)**
- Doing **experiments**
- Acquiring existing **secondary data**

If the sample is representative of the population, the data may be used to understand the population. There are a number of potential sources of bias including:

- A sample that does not resemble the population.
- A sample that is too small to be representative.
- A sample that is not randomly selected, such as a convenience sample.

Practice

Students: Page 190–193

What do you remember?

1 Identify the four components of the data cycle. Explain each briefly. Correctly order the following steps as they should occur in the data cycle.

- Esther uses a histogram to make the data more readable.
- Esther determines that most computers last between 5 and 8 years.
- Esther is curious about computers.
- Esther asks every 5th person buying a new computer how long their previous computer lasted at five randomly selected computer stores in her area.
- Esther is now curious about different styles and types of computers and asks "Is there a relationship between the type or brand of computer and how long they last?"
- Esther writes the statistical question "How long does a computer last?"

2 Is each statement true or false?

a A random sample is usually representative of the population.

b The population size is always smaller than the sample size.

c The sample size should be large enough that the sample is representative of the population.

d Sampling processes can be biased or unbiased.

e Every sample will look the same.

3 Match type of bias to scenario

a Observer bias

b Sampling bias

c Exclusion bias

d Self-selection bias

e Measurement bias

f Undercoverage bias

    i After not being selected for a sample, Quang asks about the sampling process and determined that it was very unlikely that she would have been selected.

    ii A sample is selected to represent all Americans, but only includes one child.

    iii A questionnaire is sent out the in the mail and those who are interested can mail it back, but they must pay for the postage.

    iv A baker is doing an experiment where she uses different temperatures to make cookies and gets feedback on them. At the end of the study, she realizes that the thermometer was not properly calibrated.

    v A study on dog behavior in general, did not include those who have ever bitten a person before.

    vi In a study on stomach pain, a data collector will ask the patient if they are exaggerating when they say 10/10 unless the patient has appendicitis or a kidney stone.

4 Is each an example of univariate data or bivariate data?

a How heavy are poodle dogs?

b How many zucchinis can a single plant make in a season? Does it vary based on soil type?

c Does the iron level in the soil impact the growth rate of dandelions?

d Does the number of hours of natural sunlight a person gets impact their test scores?

e How does enrollment compare between all of the different math courses that are offered?

5 Is each statement true or false for a statistical questions?

a The question can be answered using the opinion of one person.

b A secondary source could be used to acquire data to help answer the question.

c When collecting data to answer the question, most respondents should have the same answer.

d The question can be answered by analyzing univariate or bivariate data.

6 Is each question a statistical questions?

  a What is a typical salary for people in California?

  b What superpower would you want to have?

  c How far do teachers have to walk to get from one class to the next?

  d How old was the oldest dog ever recorded?

  e How many eggs to chickens lay in a day?

  f Which American TV show is the most popular in Australia?

7 Determine which data collection method would best be used to collect data for each of these scenarios.

  a Asking people on the street about how long they have lived in the area and recording their answers.

  b Using a scale to weigh a puppy and track their growth over several months.

  c Planting the same type of seed in different types of soil, and observing the differences

  d Watching people on an quiet section of road to see how many cross at the light and how many jaywalk.

Let's practice

8 Write a statistical question that could be used to explore each scenario.

  a Leah notices that the cost for a loaf of bread is very different at the bakery, compared to the farmer's market, compared to the supermarket.

  b You got really into a novel and stayed up too late reading and are now curious if other people read a lot.

9 Lively formulated the statistical question "How much time do 8th graders spend being active or exercising each day?"

  Design a simple study that could be used to collect data to answer her question.

10 You are part of a school council and you are tasked with finding out how many corn cobs, hot dogs, and buns to order for an upcoming barbeque.

  a Explain how you could collect data to help answer this question.

  b What type of data would you be collecting?

  c How would you ensure that your data collection methods are unbiased and accurate?

11 A clothing company wants to understand their customer's shopping habits better to improve their sales.

  a What kind of data should they collect?

  b How could this data help improve their business strategy?

12 A radio station conducts a poll asking its listeners to call in to say if they are for or against restrictions on people re-selling their tickets for concerts at a higher price.

  Determine whether these are reasons why this is not an appropriate way to conduct a poll:

  a A large variety of people are likely to call.

  b A person can call more than once, so they could be counted more than once.

  c People with stronger views are more likely to call than those who don't have a strong view.

13 A large corporation is considering updating their work-from-home policy. They want to know if working from home will affect productivity. The management is concerned that some employees won't have reliable internet with fast enough speed.

  a Identify the target population.

  b What design methodology would be best to find out the home internet reliability of the target population?

  c Explain why using an online survey that closes after 5 minutes might lead to biased results.

14 Imogen gets to school many different ways. Some days she walks, some she bikes, and some she takes public transit. She is curious how long it takes students to get to school using different modes of transportation. Describe how she could do a survey and ensure the sample is representative of the entire student body.

15 Jamal wants to collect data on the heights of his classmates.

    a He is using two yard sticks to measure by stacking them on top of one another. What type of bias could this lead to and how could he avoid it?

    b To remove the bias from his previous approach, he borrowed a 60 inch measuring tape from his friend's sewing kit. There were some students who were taller than 60 inches, so he excluded them from the study because he didn't want to be inaccurate with his measurements. What type of bias could this lead to and how could he avoid it?

    c Jamal was able to accurately measure a random sample of 20 classmates, to the nearest inch and recorded the data in the table.

Height (inches)	62	55	55	59	54	50	61	61	54	54
	56	63	57	61	63	55	54	59	62	58

Organize the data using this grouped frequency table:

Height interval	Frequency
50–52	
53–55	
56–58	
59–61	
62–64	

    d Describe any findings you can see from the table.

Let's extend our thinking

16 Shintaro is booking a venue for graduation due to construction at the school. He needs an estimate of how many people will be there and has one day to complete his data collection and analysis. He quickly considers two options for his study.

Option A: Ask the senior students who have homeroom in the main hallway, this would give him time to survey about 100 students.

Option B: Assign each senior student a random number and randomly select students then find them in their classes, this would give him time to survey about 20 students.

    a What type of sample is used for Option A?

    b What type of sample is used for Option B?

    c Compare and contrast the benefits and downsides of the two options. Which option would you choose and why?

17 Jolene anonymously surveyed all 36 students in her class about which age range they were in when they took their first steps. Her results are shown in the table:

Age range	Frequency
9–11 months	11
12–14 months	14
15–17 months	6
18–20 months	5

a She made the conclusion that "most Americans can walk by the time they are 1 year old". Is this a valid conclusion based on the study? Explain.

b Describe her sampling technique.

c How could she have improved the study design?

18 Danielle just flew on an airplane for the first time. She started looking up prices to fly to the Jamaica and noticed that prices varied significantly depending on the website and route.

a Formulate a statistical question that Danielle could use to explore flight prices to Jamaica.

b Could she use observation, measurement, survey, experiment, or acquire secondary sources? Explain.

c Explain how Danielle could collect data that could be used to answer her question from part (a).

d Suppose the given data shows the cost of 20 different flights. Analyze the data to draw a conclusion for Danielle.

Cost in USD	121	170	191	191	193	207	214	221	221	226
	248	265	273	278	283	286	291	340	1542	2600

e Formulate another statistical question that could be used to explore prices to Jamaica even further.

19 Piper gets an allowance and is curious if other students at her school do as well. She saves some, donates some, and spends some of her allowance.

Go through the whole data cycle at least once using a context that involves allowance, pocket money, or savings.

Answers

4.01 Data collection and sampling

What do you remember?

1 The four components of the data cycle:

1. Formulate questions: This initial phase involves choosing the topic of interest and writing a statistical question that will be answered using the data cycle.

2. Collect data: Once the question is formulated, the next step is gathering the necessary data to answer them. Data collection can use a variety of methods such as surveys, experiments, observations, or retrieving information from existing secondary sources.

3. Organize and represent data: After data collection, the raw data need to be organized into a usable format. This step involves representing or summarizing it with tables, charts, graphs, or statistical summaries.

4. Analyze and communicate results: This final stage involves analyzing the data representations or summaries and drawing conclusions. The analysis should address the initial question. The population and those involved with the population can make informed decisions based on the data.

The correct order the following steps:

1. Esther is curious about computers.

2. Esther writes the statistical question "How long does a computer last?"

3. Esther asks every 5th person buying a new computer how long their previous computer lasted at five randomly selected computer stores in her area.

4. Esther uses a histogram to make the data more readable.

5. Esther determines that most computers last between 5 and 8 years.

6. Esther is now curious about different styles and types of computers and asks "Is there a relationship between the type or brand of computer and how long they last?"

2 a True   b False   c True
  d True   e False

3 a vi   b i   c v   d iii   e iv   f ii

4 a Univariate data
  b Bivariate data
  c Bivariate data
  d Bivariate data
  e Univariate data

5 a False   b True   c No   d True

6 a Yes   b No   c Yes
  d No   e Yes   f Yes

7 a Survey   b Measurement
  c Experiment   d Observation

Let's practice

8 a How much does a loaf of bread cost across the US?
  b How much time do people spend reading books each day?

9 1. Determine the population:
   • Target Population: All 8th-grade students in a specific school, district, or community.

   2. Develop a sampling strategy:
   • Sampling Method: Use random sampling to ensure diverse representation from various schools, demographics, and backgrounds. For instance, Lively could make the population 8th graders at her school and then randomly select 10 participants from each homeroom.

   • Sample Size: Depending on the resources available and the variability of exercise habits, aim for a sample size that is statistically significant yet feasible.

   3. Collect the data:
   • Distribution method: She could distribute the survey electronically through school email or giving a link to an online form to make data collection easy and efficient. Alternatively, she could provide paper surveys to all the homerooms with names on them.

   • Time frame: She could collect data over a period of one week to ensure that you capture variations in daily exercise habits.

10 a Conduct a survey that asks attendees about their participation and food consumption preferences, including the number of hot dogs, corn cobs, and buns they plan to eat, and how many people from their family they are expecting.

   b Discrete univariate numerical data.

   c To ensure that your data collection methods are unbiased and accurate, you could use random sampling by selecting a representative sample of students from each grade level, ensuring that the survey is anonymous, and encourages honest responses about whether or not they will come to the barbeque and how much they will eat.

11 a The clothing company should collect data on customer preferences, purchasing habits, frequency of visits to their store or website, and average spending per visit. They could also collect demographic data such as age, gender, and location.

   b This data could be used to tailor marketing campaigns, optimize inventory and product selection, and create targeted promotions to improve sales.

12 a No   b Yes   c Yes

13 a The target population for this study would be all employees of the corporation who are eligible to work from home. This includes all current remote workers and any others who might be considered for remote work under the updated policy.

b A survey or measurement would be the best. A survey could provide a link to an internet speed test and ask them to complete it at home and honestly answer. Or they could have a bot that pulls the IP address and internet speed when they open an email at home, but this may require additional consent.

c Using an online survey that automatically closes after 5 minutes could lead to biased results for several reasons:

- Undercoverage Bias: Employees with slower internet speeds may take longer to load the survey and navigate through the questions, possibly preventing them from completing it within the allotted 5 minutes. This would disproportionately exclude those with the very issues the survey aims to investigate.

- Non-response Bias: Quick closing times can lead to a higher non-response rate among employees who might be willing to participate but need more time due to their work schedule, slower reading speeds.

- Response Quality: Under time pressure, employees might rush through the survey, leading to less careful responses and potentially less accurate data.

14 Imogen can follow these steps:

1. Write the questions for the survey: Create questions about the mode of transportation, commute time, and any relevant factors like distance or traffic conditions.

2. Select a sample: Divide the student body into relevant groups based on factors like grade level or neighborhood, and ensure each group is proportionally represented in the sample.

3. Distribute the survey: Using an online survey that could be complete at school would be a way to ensure she reaches most students. She could provide options like large print copies or with a translator for students who may need alternatives.

15 a Measurement bias

To avoid measurement bias, use a single measuring tool like a 25-foot tape measure.

b Sampling bias, in particular, exclusion bias

To avoid exclusion bias, use an appropriate measuring tool like a 25-foot tape measure.

c

Height interval	Frequency
50–52	1
53–55	7
56–58	3
59–61	5
62–64	4

d Here are some key findings:

The most common height range is the height interval 53 – 55 inches.

The majority of the students are between 53 and 61 inches tall.

The median would be in the 56 – 58 inches interval.

There are 4 taller students in the 62 – 64 inch interval who are taller than the majority.

Let's extend our thinking

16 a Convenience sample

b Random sample

c Benefits of Option A:

- Speed and Ease: Surveying students in a single location simplifies the process and speeds up the data collection process, which is important with the one-day timeframe.

- Larger Sample Size: With about 100 students surveyed, this option potentially provides more data points, which might represent the population.

Downsides of Option A:

- Bias: It may not represent the entire student body as it only includes a specific group of students (those with homerooms in the main hallway), leading to biased results.

Benefits of Option B:

- Reduced Bias: This approach minimizes bias by giving all students an equal chance to be included in the sample, leading to a more representative subset of the population.

- Increased Accuracy: The data collected are likely to be more reliable in reflecting the true characteristics of the entire student body, making the findings more generalizable.

Downsides of Option B:

- Time and Effort: Randomly selecting and locating each student in various classes is more time-consuming and logistically challenging.

- Smaller Sample Size: With only about 20 students surveyed, the smaller sample size could mean less data to analyze, which might not provide as robust an insight as a larger sample could.

Either option could be used, with a justification. For example: Option B would be the preferable choice even though it has a smaller sample and is more complicated. The main advantage of Option B is that the sample is more likely to be more representative as there could be a particular group of seniors who have homeroom in that hallway due to course choices.

17 a This conclusion may not be valid because the survey only includes 36 students from a single class. This sample is not representative of the entire American population, which includes people from diverse backgrounds that can influence developmental milestones like walking. Concluding about "most Americans" based on such a small and non-representative sample is overgeneralization.

b Convenience sampling

c To get more reliable and accurate data, Jolene could have used a secondary data source like medical journals or pediatric studies. Another option would be to simplify her question and conclusion to only be about students in her region or school.

18 a Answers will vary: One possible question is "How much does it typically cost to fly to Jamaica?"

b Danielle could utilize observation by manually checking flight prices on various websites or acquire data from secondary sources like data from travel agencies or flight comparison websites. Experiments and surveys are less relevant or applicable to her situation.

c Danielle can collect flight price data by:
- Using flight comparison websites to gather and compare prices from different airlines and booking platforms.
- Recording price changes over time to analyze trends based on different factors such as booking time or seasonal changes.
- Collecting data on special promotions or discounts that may influence ticket prices.

d Answers will vary. Some possible conclusions include:
- Most flight prices are under $340, suggesting more affordable options are common.
- Two significant outliers ($1542 and $2600) indicate that prices can spike under certain conditions like last-minute bookings or more luxurious travel classes.
- The majority of options are more budget-friendly, but prices can vary greatly, recommending price comparison for the best deals.

e Answers will vary. A possible question is: "What factors influence the price variability of flights to Jamaica?"

19 Answers will vary. Here is a sample answer working through the data cycle:

1. Question Formulation
   - What is the typical allowance students at my school receive each week?

2. Collect or Acquire Data

   We can select a random sample using class lists for the whole school or giving three surveys to each teacher and having them randomly select using a random number generator and their class list. It would be very

important that the surveys be anonymous as students may not feel comfortable giving information about how much money they are given.

Since it is a relatively simple study, it may be possible to survey the whole population.

Suppose this data was collected from a simple random sample:

Allowance	$0	$0	$0	$0	$3	$7	$7	$8	$9	$10
	$11	$11	$12	$13	$13	$13	$14	$14	$16	$16
	$17	$18	$18	$18	$19	$20	$20	$21	$22	$22

3. Organize and Represent Data

   To organize and visualize the allowance data, we can use a frequency table and then represent it with a histogram or just use the table:

Allowance Range	Frequency
$0 – $4	5
$5 – $9	4
$10 – $14	9
$15 – $19	7
$20 – $24	5

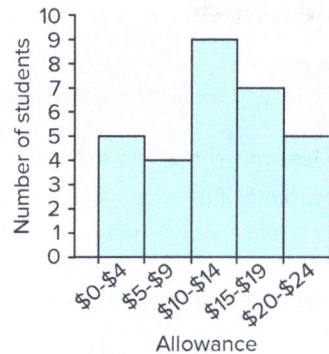

4. Analyze and Communicate Results

   Measures of center including zeros:
   - Mean: $12.4
   - Median: $13
   - Mode: $0
   - Modal class: $10–$14

   Measures of center not including zeros:
   - Mean: $14.3
   - Median: $14
   - Mode: $13 and $18
   - Modal class: $10 – $14

   Conclusion:
   - Piper can conclude that students at her school typically receive an allowance of around $14, if they receive an allowance, but that many students, 20% to be specific, do not receive an allowance. The range for those who receive an allowance is between $3 and $22, so there is quite a lot of variation overall.

5. This may lead us to formulate other questions like "What factors are related to the amount of allowance someone receives?" or "How do people use their pocket money?"

4.02 Boxplots

Subtopic overview

Lesson narrative

In this lesson, students will learn how to create and interpret boxplots, also known as box-and-whisker plots, to display the distribution of a data set. The lesson begins by reviewing key statistical concepts such as range, median, and quartiles. Students will learn that quartiles divide a data set into four equal parts and that each quartile represents 25% of the data. Students will engage in an exploration of the five-number summary, which includes the minimum, lower quartile (Q1), median (Q2), upper quartile (Q3), and maximum values. This summary is used to construct a boxplot. The lesson will guide students through the process of creating a boxplot.. The lesson includes examples and practice problems where students will find these critical values and use them to create boxplots. They will also learn to calculate the interquartile range (IQR), which measures the spread of the middle 50% of the data and is less affected by outliers compared to the range. By the end of the lesson, students will be able to interpret boxplots to understand the distribution and variability of data. They will be able to identify which quartile has the most spread and determine how much of the data falls within specified ranges. This skill will help them analyze data more effectively and make informed decisions based on statistical representations.

Learning objectives

Students: Page 194

After this lesson, you will be able to...
- create boxplots of data sets with and without technology.
- identify the minimum, maximum, median, upper quartile, lower quartile, range, and interquartile range from a boxplot.
- analyze data represented in a boxplot.

Key vocabulary

- boxplot
- lower extreme
- median
- third quartile

- first quartile
- lower quartile
- quartile
- upper extreme

- five-number summary
- measure of center
- range second quartile
- upper quartile

- interquartile range
- measure of spread

Essential understanding

A boxplot is a convenient and informative way to represent univariate data. It highlights the center and spread of the data.

Standards

This subtopic addresses the following Virginia 2023 Mathematics Standards of Learning standards.

Mathematical Process Goals

MPG1 — Mathematical Problem Solving

Teachers can utilize the concept of boxplots to enhance students' problem-solving skills by presenting them with real-world data sets and challenging them to represent and analyze the data using boxplots. For example, teachers can provide data on the heights of students in a school and ask students to create boxplots to determine the range, quartiles, and median. They can also encourage students to solve problems that require them to compare data sets represented by different boxplots.

MPG4 — Mathematical Connections

Teachers can establish mathematical connections by linking the concept of boxplots to students' prior knowledge of mean, median, mode, and range. They can also illustrate how boxplots are used in various fields such as statistics, data analysis, and research, thereby connecting mathematics to other disciplines and real-world contexts. For example, they could discuss how boxplots are used in scientific research to visualize the distribution of data.

MPG5 — Mathematical Representations

Teachers can integrate this goal into their instruction by teaching students different methods of representing data, including boxplots. They can guide students to understand that the five critical points in a boxplot represent a summary of the data set and that the rectangle and whiskers visually represent the distribution of the data. Teachers can also demonstrate how to create a boxplot both manually and using technology, emphasizing that the process of representation is as important as the final product.

Content standards

8.PS. 2 — The student will apply the data cycle (formulate questions; collect or acquire data; organize and represent data; and analyze data and communicate results) with a focus on boxplots.

8.PS.2a — Formulate questions that require the collection or acquisition of data with a focus on boxplots.

8.PS.2b — Determine the data needed to answer a formulated question and collect the data (or acquire existing data) using various methods (e.g., observations, measurement, surveys, experiments).

8.PS.2c — Determine how statistical bias might affect whether the data collected from the sample is representative of the larger population.

8.PS.2d — Organize and represent a numeric data set of no more than 20 items, using boxplots, with and without the use of technology.

8.PS.2e — Identify and describe the lower extreme (minimum), upper extreme (maximum), median, upper quartile, lower quartile, range, and interquartile range given a data set, represented by a boxplot.

8.PS.2g — Analyze data represented in a boxplot by making observations and drawing conclusions

Prior connections

5.PS. 2 — The student will solve contextual problems using measures of center and the range.

6.PS.2 — The student will represent the mean as a balance point and determine the effect on statistical measures when a data point is added, removed, or changed.

7.PS.2 — The student will apply the data cycle (formulate questions; collect or acquire data; organize and represent data; and analyze data and communicate results) with a focus on histograms.

Future connections

A2.ST. 1 — The student will apply the data cycle (formulate questions; collect or acquire data; organize and represent data; and analyze data and communicate results) with a focus on univariate quantitative data represented by a smooth curve, including a normal curve.

A.ST. 1 — The student will apply the data cycle (formulate questions; collect or acquire data; organize and represent data; and analyze data and communicate results) with a focus on representing bivariate data in scatterplots and determining the curve of best fit using linear and quadratic functions.

Lesson Preparation

Suggested review

Depending on your students' level of prior knowledge, consider revisiting the following lessons:

Grade 6 — 9.04 Review: measures of center and spread

Tools

You may find these tools helpful:

- Scientific calculator
- Statistics calculator
- Number line
- Ruler
- Grid paper

Student lesson & teacher guide

Boxplots

This lesson focuses on finding the five number summary and connecting these to boxplots representing a data set. Students are to interpret and create boxplots to make inferences and predictions about sets of data.

Exploration

This exploration is designed to help you understand the distribution of data through quartiles and the range, using an interactive applet.

Students: Page 194

Interactive exploration
Explore online to answer the questions

⊕ **mathspace.co**

Use the interactive exploration in 4.02 to answer these questions.

1. What percentage of the data lies below Q_1? Hence, what does Q_1 represent?
2. What percentage of the data lies below the median? Hence, what does the median represent?
3. What percentage of the data lies below Q_3? Hence, what does Q_3 represent?
4. What percentage of the data lies between the minimum value and the maximum value?
5. Which section of the data is the most spread out?
   - From the minimum to Q_1
   - From Q_1 to the median
   - From the median to Q_3
   - From Q_3 to the maximum

Suggested student grouping: Small groups

This exploration is designed to help you understand the distribution of data through quartiles and the range, using an interactive applet.

Ideal student responses

These ideal responses may differ from other correct student responses. Less formal responses can be connected with the more precise mathematical language presented here.

1. What percentage of the data lies below Q_1? Hence, what does Q_1 represent?

Typically, Q_1 (the first quartile) represents the 25th percentile of the data. This means that 25% of the data lies below Q_1.

2 What percentage of the data lies below the median? Hence, what does the median represent?

The median represents the 50th percentile, so 50% of the data lies below the median. The median splits the data set into two equal halves.

3 What percentage of the data lies below Q_3? Hence, what does Q_3 represent?

Q_3 (the third quartile) represents the 75th percentile, meaning that 75% of the data lies below Q_3.

4 What percentage of the data lies between the minimum value and the maximum value?

100% of the data lies between the minimum and maximum values as these are the extremes of the data set.

5 Which section of the data is the most spread out?

The most spread out section depends on the data set. Without being shown Q_1, the median, and Q_3 we cannot easily determine the spread of different section. However, if we are given these statistics, then each of the four sections has about 25% of the data, so the amount of data in each section is consistent. However, the spread depends on the specific data set. For example, the initial data set is most spread out between Q_1 and the median.

Purposeful questions

- How does dividing the data into various percentages help you understand the distribution of the dataset?
- Why is it important to know what percentage of data lies below each quartile, and how does this help in interpreting the data?

Possible misunderstandings

- Students might incorrectly assume that the data is always evenly spread across quartiles, not realizing that some sections of the data can be more concentrated or spread out, which affects the interpretation of the data's distribution.

Concrete-Representational-Abstract (CRA) Approach
Targeted instructional strategies

Concrete Stage: Engage students with physical manipulatives to explore boxplots. Use small objects like counters, beads, or blocks to represent individual data points from a data set. Have students arrange these objects in order along a large number line drawn on poster paper or taped on the floor. Guide them to count the total number of items (ideally a multiple of four), and use ribbons or string to group that data into four equal groups. Use pieces of paper or blocks to put a marker at the point where two groups meet. This hands-on activity helps students see how the data is distributed and understand the concept of dividing data into quartiles.

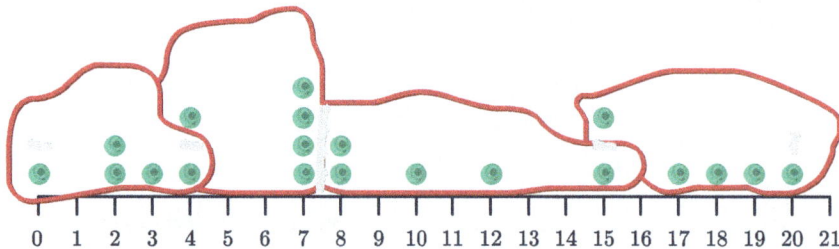

Representational: Transition to drawing boxplots on paper based on the physical models they created. Have students sketch a number line and above it, plot the minimum (lowest value), lower quartile (point where the first two groups meet), median (middle value), upper quartile (point where the last two groups meet), and maximum (largest value). Guide them in drawing the box from Q_1 to Q_3, with the median as a line in the box, and then adding whiskers to the minimum and maximum values. Encourage them to label each part of the boxplot clearly. Use graph paper to help them attend to precision and keep their drawings accurate and to scale. This visual representation connects the physical arrangement of data to a standard way of depicting data distribution.

Representational stage

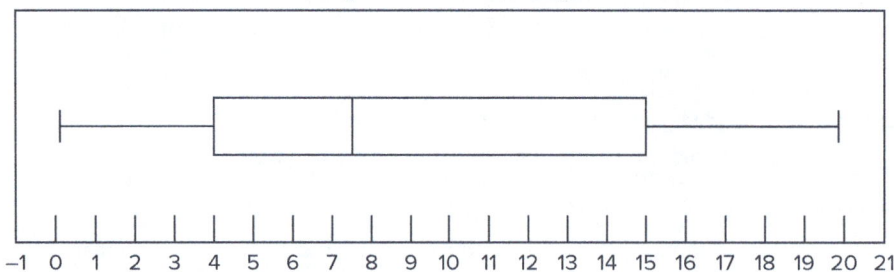

Abstract: Move to calculating and interpreting boxplots using abstract mathematical concepts. Teach students how to find the median and quartiles using numerical methods. Have them record key values for the boxplot without relying on physical objects or drawings. Introduce them to using technology, like graphing calculators or statistical software, to create boxplots from data sets. Engage them in interpreting boxplots by analyzing the range, interquartile range (IQR), and identifying any potential outliers. This stage develops their ability to work with statistical concepts abstractly and prepares them for more advanced analysis.

Connecting the Stages: Help students make connections between the concrete manipulatives, the drawings, and the abstract calculations. Discuss how the physical placement of objects on the number line relates to the positions on their drawn boxplots. Ask them to explain how the numbers in their five-number summary correspond to both the manipulatives and the drawings. Encourage them to reflect on how each representation shows the same data in different ways. By making these connections, students deepen their understanding and can choose the most helpful representation when analyzing data.

Collect and display
English language learner support

Encourage students to share their understanding of key statistical terms used in creating boxplots. As students discuss terms like "median," "quartiles," and "interquartile range" listen for their explanations and descriptions. Write down their words and phrases on chart paper or a visible display in the classroom. Organize the terms with definitions, diagrams, or examples provided by the students.

Connect informal language to precise mathematical vocabulary, and encourage students to refer back to the display during the lesson. This visual reference will help students internalize the vocabulary and concepts, supporting both their mathematical understanding and English language development."

Examples

Students: Page 196

Example 1

For the following boxplot:

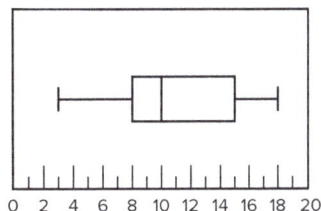

a Find the lower extreme.

Create a strategy

The lower extreme is the smallest value in the data set, also called the minimum. It is represented by the end of the left whisker.

Apply the idea

The lower extreme is 3.

Purpose

Students demonstrate that they can use a boxplot to find the lower extreme (lowest value).

b Find the upper extreme.

Create a strategy

The upper extreme is the largest value in the data set, also called the maximum. It is represented by the end of the right whisker.

Apply the idea

The upper extreme is 18.

Purpose

Students demonstrate that they can use a boxplot to find the highest value.

Students: Page 197

c Find the range.

Create a strategy

The range is the difference between the largest data value and the smallest data value.

Apply the idea

Range = 18 − 3   Find the difference of the extreme values

= 15   Evaluate the subtraction

Purpose

Students demonstrate that they can use a boxplot to find the range.

Students: Page 197

d Find the median.

Create a strategy

The median is marked by the line in the middle of the box.

Apply the idea

The median is 10.

Reflect and check

The median represents the middle value of the data.

Purpose

Students demonstrate that they can use a boxplot to find the median.

Expected mistakes

Students may confuse one of the quartiles for the median, or look at the center of the number line itself.

Reflecting with students

The interquartile range acts as the middle 50% of the data, while the median divides this 50% into a further split of the data. The median line being in different spots of the box can tell us about the spread of these quarters of data. What do we notice about the median of this set of data?

e   Find the interquartile range (IQR).

Create a strategy

The interquartile range (IQR) is the difference between the upper quartile and the lower quartile.0

Apply the idea

The upper quartile (Q_3) is marked by the right side of the box. In this boxplot, $Q_3 = 15$
The lower quartile (Q_1) is marked by the left side of the box. In this boxplot, $Q_1 = 8$.

$$IQR = \text{Upper quartile } - \text{ Lower quartile} \qquad \text{Formula for IQR}$$
$$= 15 - 8 \qquad \text{Substitute known values}$$
$$= 7 \qquad \text{Evaluate the subtraction}$$

Purpose

Students demonstrate that they can use a boxplot to find the interquartile range

Reflecting with students

The interquartile range tells you where the middle 50% of the data lies. What would a larger interquartile range indicate? What would a smaller interquartile range indicate?

Annotate boxplots to aid with visual processing   use with Example 1
Student with disabilities support

Encourage students to use a ruler to support with reading boxplots. Explicitly teach students how to line up a ruler with the key points on a boxplot and the number line axis to read off the five key values.

Similarly, have students annotate the IQR and Range on the boxplots. For example:

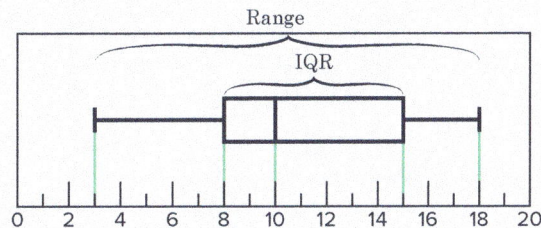

Example 2

You have been asked to represent this data in a boxplot:

20, 36, 52, 56, 24, 16, 40, 4, 28

a   Complete the table for the given data.

Minimum	
Lower quartile	
Median	
Upper quartile	
Maximum	
Interquartile range	

Create a strategy

To find the minimum, median, and maximum values, order the numbers from smallest to largest and find the first, middle, and last value. Then, find the quartiles.

Apply the idea

To find the minimum, median, and maximum values, first put them in order:

$$4, 16, 20, 24, 28, 36, 40, 52, 56$$
$$\text{Minimum} = 4$$
$$\text{Maximum} = 56$$

The middle value is: 28

$$\text{Median} = 28$$

To find the lower quartile, find the middle value of the lower half of the values: 4, 16, 20, 24

The middle values are: 16, 20

$$\text{Lower quartile} = \frac{16 + 20}{2} \qquad \text{Find the average of the middle values}$$

$$= \frac{36}{2} \qquad \text{Evaluate the addition}$$

$$= 18 \qquad \text{Evaluate the division}$$

To find the upper quartile, find the middle value of the upper half of the values: 36, 40, 52, 56
The middle values are: 40, 52

$$\text{Upper quartile} = \frac{40 + 52}{2} \qquad \text{Find the average of the middle values}$$

$$= \frac{92}{2} \qquad \text{Evaluate the addition}$$

$$= 46 \qquad \text{Evaluate the division}$$

The interquartile range is the difference between the upper and lower quartiles:

$$IQR = 46 - 18 = 28$$

Minimum	4
Lower quartile	18
Median	28
Upper quartile	46
Maximum	56
Interquartile range	28

Purpose

Challenge students to create a table of measures of spread and center of a set of data.

Students: Page 198–200

b   Construct a boxplot for the data.

Create a strategy

Use the the answer from part (a) to construct a boxplot.

Apply the idea

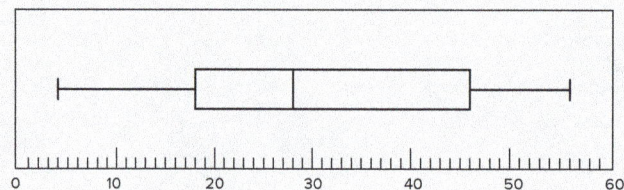

Reflect and check

We can use technology to check that we calculated the five critical points and constructed the boxplot correctly.

1 Enter the data in a single column.

	A	B	C	D	E	F	G	H	I
1	20								
2	36								
3	52								
4	56								
5	24								
6	16								
7	40								
8	4								
9	28								
10									
11									
12									

2 Select all of the cells containing data and choose "One Variable Analysis".

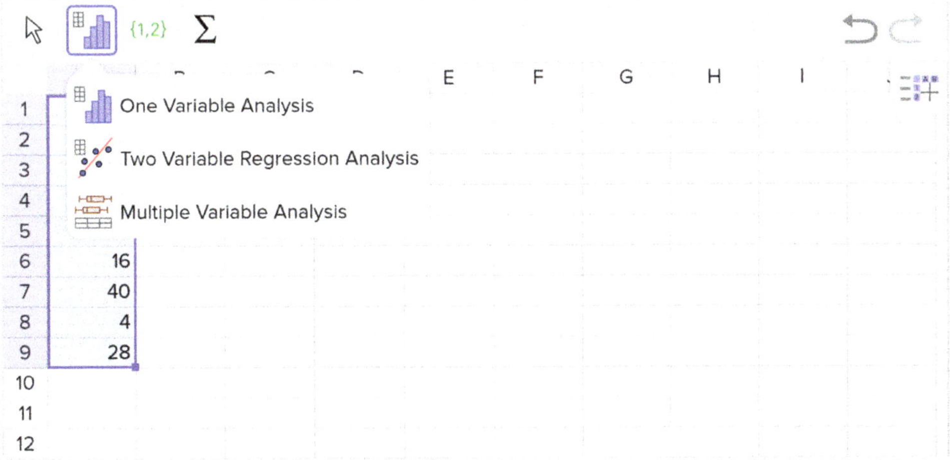

					E	F	G	H	I
1	One Variable Analysis								
2									
3	Two Variable Regression Analysis								
4									
5	Multiple Variable Analysis								
6	16								
7	40								
8	4								
9	28								
10									
11									
12									

3 In the dropdown menu, change the histogram to a boxplot.

4 Select "Show Statistics", the button with, Σx, to reveal a list of statistical values, including the five-number summary.

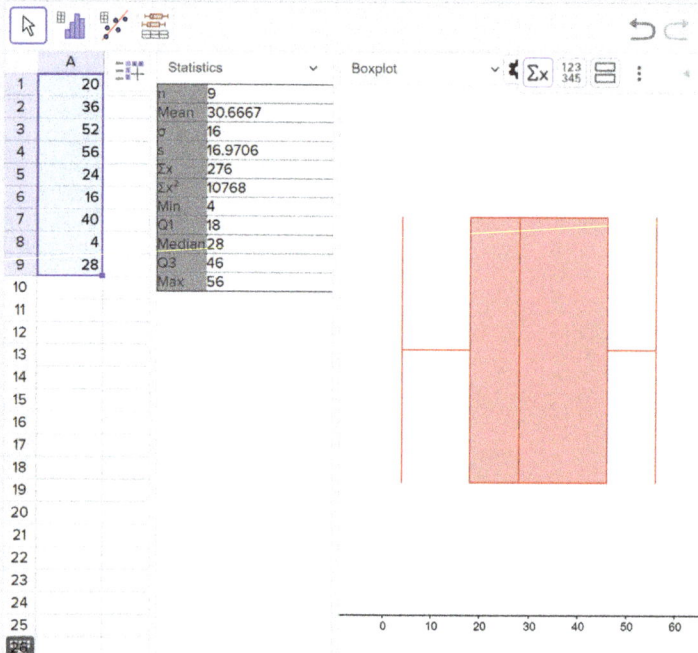

This confirms that the five critical points in our boxplot are correct.

Purpose
Students demonstrate that they can construct a boxplot from a summary of values.

Expected mistakes
Students may see the intervals on the bottom as being the only values included instead of a range of values, or use the data values instead of the frequency.

🎓 Thinking computationally
Targeted instructional strategies

use with Example 2

Encourage students to use decomposition to help understand how to represent the data in a boxplot. Incorporate algorithmic thinking by having them write or use a set of steps that connects the parts of the process to draw boxplots. By focusing on these individual parts, students will grasp how they connect to form the complete graphical representation.

1. Organize the data set in ascending order

2. Identify the minimum and maximum

3. Identify the median by counting the total number of elements in the set, n, then finding the middle value by counting inwards from the two ends or finding the $\left(\frac{n+1}{2}\right)^{th}$ value. If there are an even number of values, then find the average of the middle two values.

4. Identify the lower quartile (Q_1) by finding the median of the lower half of the data values. If there are an even number of values, then find the average of the middle two values.

5. Identify the upper quartile (Q_3) by finding the median of the upper half of the data values. If there are an even number of values, then find the average of the middle two values.

6. Draw a numbered line that goes at least from the minimum to the maximum.

7. Draw a line or point above the number line at the five key values, then join the middle three with the box and the outer two with whiskers.

8. Add titles to the graph and axis.

Students: Page 201

Example 3

The box-and-whisker plot represents the thickness of the glass on various dining tables.

Glass width (mm)

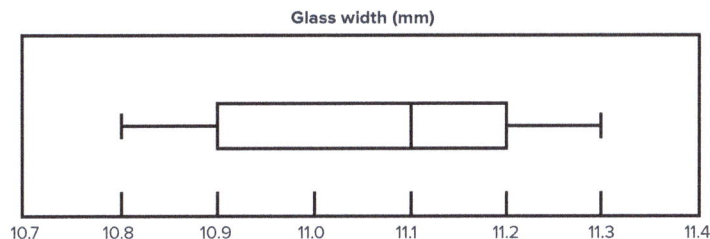

10.7   10.8   10.9   11.0   11.1   11.2   11.3   11.4

a Which formulated question could be answered by analyzing the given boxplot?

A How many dining room tables have a thickness of 11.1 mm?

B How does the size of a dining room table affect the thickness of the glass?

C What proportion of glass tables have a thickness of 11 mm?

D What range of thickness is most common for the glass of dining room tables?

Create a strategy

Boxplots represent univariate data, which means they are only used to represent one variable (in this case, glass thickness). They do not show individual data points, but they do show how the data values in the set vary (how spread out they are).

Apply the idea

Option A - this question requires us to know the number of individual glass tables have a thickness of 11.1 mm. Since boxplots do not show individual data points, this question cannot be answered by the boxplot.

Option B - The second question considers two different variables: the size of the table and the thickness of the glass. This boxplot only shows data on the thickness of the glass, so it cannot be used to answer this question.

Option C - Similar to the first question, the third question requires us to look at individual glass tables with a thickness of 11 mm, and compare that to the total number of data values in the set. This question cannot be answered by the boxplot.

Option D - The last question is related to a measure of spread, so the boxplot can be used to answer this question. In particular, we could use the interquartile range shown in the boxplot to answer the question.

Question D can be answered by analyzing the boxplot.

Reflect and check

According to the boxplot, half of the data lies between 10.9 mm and 11.2 mm. We can say that it is most common for the glass of a dining room table to have a thickness between 10.9 mm and 11.2 mm.

Purpose

Help students understand how to interpret and apply the information given in a box-and-whisker plot to answer relevant questions about a given dataset, specifically focusing on the concept of spread and how it relates to the data.

Students: Page 201–203

b   What percentage of values lie between:
- 10.9 and 11.2
- 10.8 and 10.9
- 11.1 and 11.3
- 10.9 and 11.3
- 10.8 and 11.2

Create a strategy

First, determine whether each of the values represents a critical point on the boxplot. Then, use the fact that one quartile represents 25% of the data set to find the percentages.

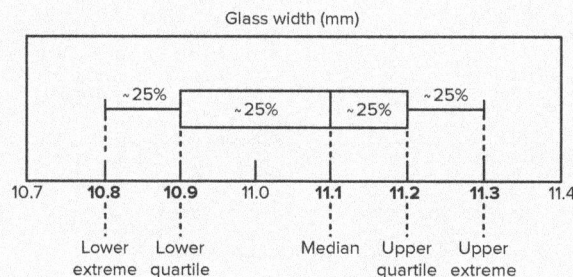

Glass width (mm)

Apply the idea

10.9 and 11.2 represent the lower and upper quartiles.

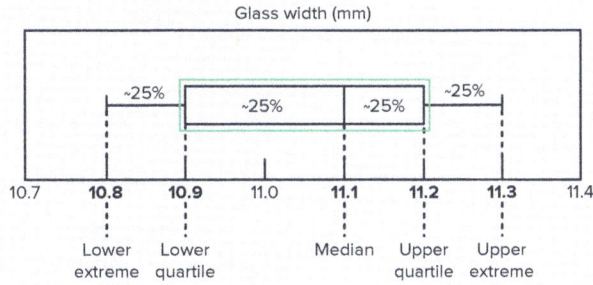

Glass width (mm)

50% of values lie between 10.9 and 11.2.

10.8 and 10.9 are the lower extreme and the lower quartile, respectively.

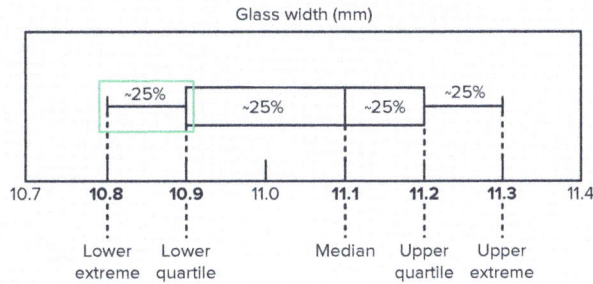

Glass width (mm)

25% of the values lie between 10.8 and 10.9.

11.1 and 11.3 are the median and the upper extreme, respectively.

Glass width (mm)

50% of values lie between 11.1 and 11.3.

10.9 is the lower quartile and 11.3 is the upper extreme.

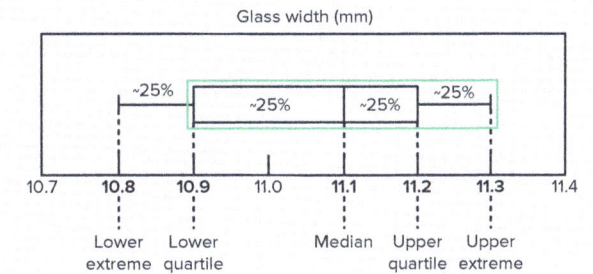

Glass width (mm)

75% of values lie between 10.9 and 11.3.

10.8 and 11.2 are the lower extreme and the upper quartile, respectively.

Glass width (mm)

75% of values lie between 10.8 and 11.2.

Purpose

Show students how to interpret a boxplot and how to calculate the percentage of data that falls within a given range, reinforcing their understanding of quartiles and percentage calculations.

Students: Page 203

c   In which quartile (or quartiles) is the data the most spread out?

Create a strategy

Which quartile takes up the longest space on the graph?

Apply the idea

The second quartile is the most spread out.

Reflect and check

Although the second quartile is the most spread out, it does not have more data values than the other quartiles. Each of the four quartiles have approximately the same number of data values, but the data values differ more in the second quartile.

Purpose

Show students how to interpret quartiles on a graph, and reinforce the concept that the spread of data within a quartile is independent of the number of data points it contains.

🎓 Advanced learners: Deepening boxplot interpretation through inquiry-based exploration
Targeted instructional strategies

use with Example 3

Encourage students to delve deeper into the concept of boxplots by having them investigate how different data sets affect the plot's shape and spread. After they answer the initial questions, prompt them to remove or add additional data values, perhaps by researching real-world measurements or generating hypothetical data that meets certain criteria.

Have students construct boxplots from the new data sets and compare them to the original plot, analyzing how changes in minimums, maximums, medians, and quartiles impact the representation. Encourage the use to appropriate tools including statistics calculators to facilitate this exploration.

Additionally, invite students to consider the effect of outliers on the box plot and discuss what this reveals about data variability and distribution. By exploring these extensions, students naturally deepen their understanding of statistical concepts and make meaningful connections to real-world applications.

Example 4

Lucille wants to track the amount of time she spends on her phone or tablet outside of school. Her goal is to only spend one to two hours on her devices each day. The statistical question she writes for her study is "How does the amount of time I spend on my phone or tablet vary each day?"

a Determine the data Lucille must collect to answer her statistical question.

Create a strategy

Consider whether Lucille needs to collect univariate or bivariate data, and determine the variable(s) of interest.

Apply the idea

Lucille must collect univariate data, and the variable of interest is the amount of time she spends on her phone or tablet every day.

Reflect and check

The way the statistical question is worded will have a great effect on the type of data that needs to be collected. For example, if Lucille's statistical question was, "When do I spend the most amount of time on my devices throughout the week?", she would need to collect data on the amount of time she spends on her devices and the time of day that she is on her devices.

Purpose

Show students the importance of understanding a statistical question thoroughly before gathering data. This will help students identify whether the data needed is univariate or bivariate and to determine the variable(s) of interest.

Students: Page 204

b Which method of data collection would lead to the least amount of statistical bias for Lucille's study?
  A Collecting data on the amount of time she spent on her phone and tablet in the past three weeks
  B Collecting data on the amount of time she spends on her phone and tablet over the next three weeks

Create a strategy

Determine how statistical bias might be introduced in the data collection process by considering whether the data will be representative of Lucille's typical amount of screen time. Also consider whether Lucille's overall goal of the study will impact the collection of the data.

Apply the idea

Because Lucille's goal is to reduce the amount of time she spends on her devices each day, she may begin trying to reduce her screen time from the moment she begins the study. This means the amount of time she spends on her devices in the next three weeks may not be representative of the amount of time she spends on her devices normally.

Collecting data on the amount of time she spent on her phone and tablet in the past three weeks would lead to less bias. The correct answer is option A.

Purpose

Show students that the timing and context of data collection can significantly impact the results of a study and potentially introduce bias. Make them aware that they should consider these factors when designing a study to ensure the data collected is representative and reliable.

c   According to Lucille's phone and tablet settings, the total amount of time she spent on her devices (in hours) each day over the past three weeks is shown:

$$\{3.2, 7.5, 6.1, 8.0, 1.8, 2.5, 4.8, 5.0, 3.2, 2.0, 0.5, 1.2, 2.8, 4.5, 3.6, 5.5, 7.1, 6.2, 4.2, 2.3\}$$

Represent the data in a boxplot.

Create a strategy

We can use technology to represent the given data in a boxplot. To do so, we will follow these steps:

1   Enter the data in a single column.

2   Select all of the cells containing data and choose "One Variable Analysis."

3   In the dropdown menu, change the histogram to a boxplot.

4   Select "Show Statistics" to reveal the five critical points of the boxplot.

Apply the idea

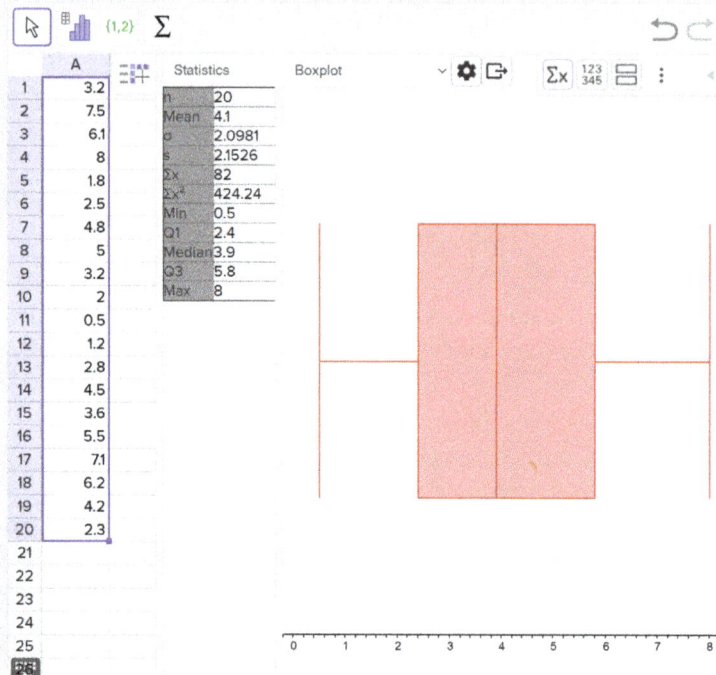

We can use the five-number summary to copy the boxplot.

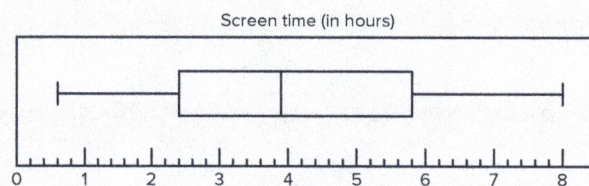

Screen time (in hours)

Purpose

Show students how to use technology to create a boxplot and interpret the five-number summary to understand data distribution.

d   How does the amount of time Lucille spends on her phone or tablet vary each day?

Create a strategy

Consider the spread of the data by looking at the range and interquartile range. The statistics tell us how much the data varies by.

Apply the idea

The lower extreme of the data is 0.5 hours, and the upper extreme is 8 hours. Overall, the data varies by

$$8 - 0.5 = 7.5 \text{ hours}$$

This shows that there is a big difference in the amount of time Lucille spends on her phone or tablet each day. To get a better idea of how the data varies between the lower and upper extrema, we can look at the interquartile range. The lower quartile is 2.4 hours, and the upper quartile is 5.8 hours. The interquartile range is

$$5.8 - 2.4 = 3.4 \text{ hours}$$

This shows that the middle 50% of the data varies by 3.4 hours. Although the difference here is smaller, it is still relatively large when considering Lucille's overall goal.

Because Lucille's goal is to spend only one to two hours on her devices each day, the data shows that her screen time varies a lot each day and it varies more than she would like.

Reflect and check

Going one step further, we can find what percentage of days Lucille meets her goal of spending one to two hours on her device daily. The lower quartile is 2.4 hours, and 25% of the data lies below this value. This means that Lucille achieves her goal less than 25% of the time.

Purpose

Show students how to use measures of spread, such as range and interquartile range, to interpret real-world data and draw conclusions.

Students: Page 206

💡 Idea summary

A list of the minimum, lower quartile, median, upper quartile, and maximum values is often called the **five-number summary.**

- The **lower extreme** (minimum) is the smallest value in the data set.
- The **lower quartile** (Q_1 or the **first quartile**) is the middle score in the bottom half of data.
- The **median** (Q_2 or the **second quartile**) is the middle value of a data set.
- The **upper quartile** (Q_3 or the **third quartile**) is the middle score in the top half of the data set.
- The **upper extreme** (maximum) is the largest value in the data set.

One quartile represents 25% of the data set.

These features are shown in a boxplot:

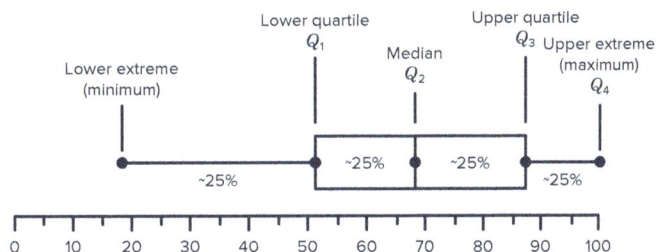

Creating a boxplot:

1. Put the data in ascending order (from smallest to largest).
2. Find the median (middle value) of the data.
2. To divide the data into quarters, find the middle value between the minimum value and the median, as well as between the median and the maximum value.

    To calculate the interquartile range:

$$IQR = Q_3 - Q_1$$

IQR is the interquartile range
Q_1 is the first quartile
Q_3 is the third quartile

Practice

Students: Page 207–212

What do you remember?

1 Find the median for each data set of scores:

    a 31, 32, 34, 42, 49, 50, 51   **b** 21, 23, 26, 31, 34, 38, 42, 47, 48

    c 2, 10, 28, 35, 50   **d** 8, 9, 11, 12, 13, 15

    e 3, 4, 6, 7, 7, 8, 12, 14   **f** 7, 8, 9, 12, 18, 21, 31, 33

2 For each data set, the median is shown. Find the upper and lower quartiles.

    a 0, 1, 4, ⑥ 8, 9, 11   **b** 22, 24, 25, 29, 32, 34

            Median   Median: 27

    c 5, 11, 19, 20, ㉒ 24, 26, 27, 27   **d** 14, 15, 16, 16, 17, 18, 19, 20, 24, 25, 30, 36

               Median   Median: 18.5

3 Match each term with its definition.

    i The largest data value   **a** Minimum

    ii 50% of the data lies on either side of this value   **b** Lower quartile, Q_1

    iii About 25% of the data is above this value   **c** Median

    iv The smallest data value   **d** Upper quartile, Q_3

    v About 25% of the data is below this value   **e** Maximum

4 50% of the data lies between which two values? Select all that apply.

    A 1 and 6   **B** 4 and 6   **C** 4 and 14.5

    D 6 and 14.5   **E** 14.5 and 16

$Q_1 = 4$ $Q_3 = 14.5$

1, 2, 3, 5, 5, 7, 11, 14, 15, 16

Minimum   Median = 6   Maximum

5 Fill in the missing numbers for each boxplot.

    a

Lower extreme	20
Lower quartile	28
Median	36
Upper quartile	42
Upper extreme	52

20   28   ☐   42   ☐

    b

Lower extreme	16
Lower quartile	33
Median	47
Upper quartile	61
Upper extreme	71

16   ☐   47   ☐   71

    c

Lower extreme	26
Lower quartile	33
Median	41
Upper quartile	44
Upper extreme	45

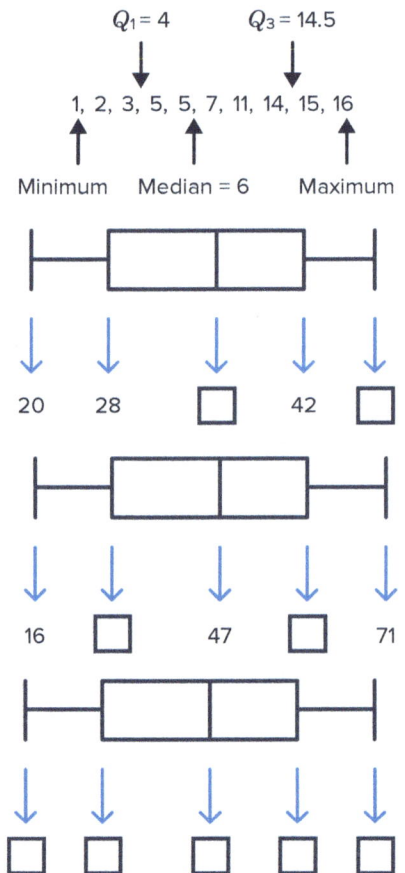

☐   ☐   ☐   ☐   ☐

6 For each five-number summary, find the:

i Range

ii Interquartile range

a

Minimum	26
Lower quartile	33
Median	41
Upper quartile	44
Maximum	45

b

Minimum	16
Lower quartile	27
Median	35
Upper quartile	42
Maximum	50

7 Determine whether the question would lead to data that could be displayed in a boxplot.

a How do the rings in a tree change over time?

b How many grams of sugar do most people consume daily?

c How does the height of middle school students vary?

d How many players are on a rugby team?

Let's practice

8 For each boxplot, find the:

i Lower extreme

ii Lower quartile

iii Median

iv Upper quartile

v Upper extreme

vi Interquartile Range

a

b

c

d

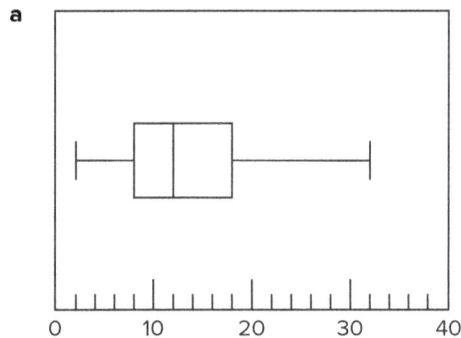

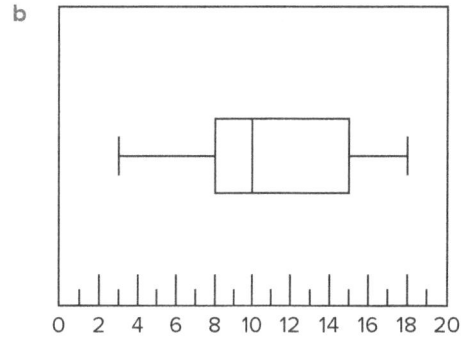

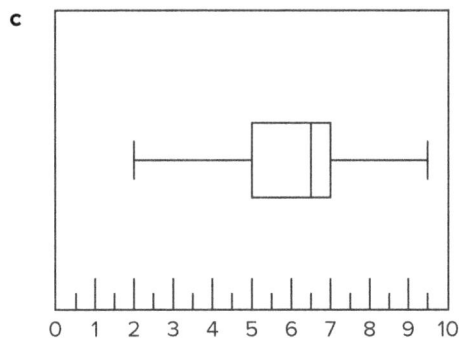

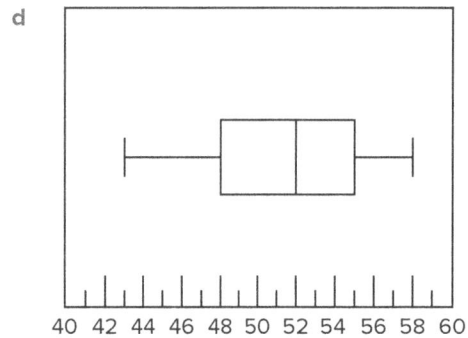

9 For each 5 number summary, create a boxplot to represent the data:

a

Minimum	5
Lower quartile	25
Median	35
Upper quartile	60
Maximum	75

b

Lower extreme	10
Lower quartile	16
Median	26
Upper quartile	38
Upper extreme	50

c

Minimum	8
Lower quartile	10
Median	14
Upper quartile	16
Maximum	19

d

Lower extreme	31
Lower quartile	34
Median	38
Upper quartile	41
Upper extreme	47

10 For each set of data:

i Complete the table for the given data.

ii Create a boxplot.

Minimum	
Lower quartile	
Median	
Upper quartile	
Maximum	
Interquartile range	

a 15, 8, 13, 15, 7, 15, 2, 7, 16

b 20, 36, 52, 56, 24, 16, 40, 4, 28

c 15, 17, 10, 18, 19, 11, 12, 13

d 71, 73, 79, 85, 89, 80, 79, 76

11 Consider the boxplot shown:

a State the percentage of values that lie between each of the following:

i 7 and 15   **ii** 1 and 7

iii 19 and 9   **iv** 7 and 19

v 1 and 15

b In which quartile is the data the least spread out?

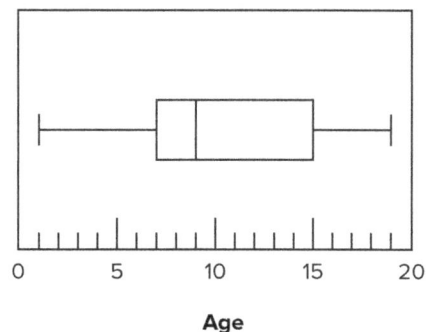

12 The boxplot represents data collected on the age that people had their first tablet device.

Lower extreme	
Lower quartile	
Median	
Upper quartile	
Upper extreme	
Interquartile range	

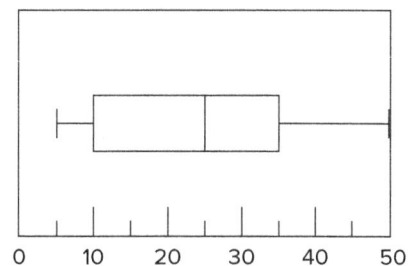

Age

a Complete the table using the information from the boxplot.

b Which question could be answered using this data?

A What is the average age that people get their first tablet device?

B How many people are 20 years old when they get their first tablet device?

C How does the age at which people get their first tablet device vary?

D How many tablets does a family own?

c How was the data most likely collected?

A Observation   **B** Measurement   **C** Survey   **D** Experiment

d In which quartile is the data the least spread out?

13 A cyclist recorded her heart rate immediately after finishing each event in which she competed. The results are recorded in the boxplot.

Heart rate

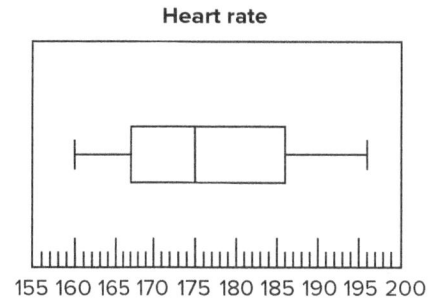

155 160 165 170 175 180 185 190 195 200

    a What method of data collection did the cyclist use?

        A Observation   **B** Measurement

        C Survey   **D** Experiment

    b Formulate a question that could be answered using this data.

    c Find and interpret the range.

    d Find and interpret the interquartile range.

14 A geography teacher wants to answer this question: "How does the number of states visited by the students in my class vary?"

    a What type of data does the teacher need to collect?

    b How should the teacher collect the data?

    c The teacher collected the data and organized it into the table shown. Create a boxplot to represent the data.

    d Use the boxplot to answer the teacher's question.

Minimum	3
Lower quartile	5
Median	8
Upper quartile	15
Maximum	24

15 Isabelle measures the length of the worms in her worm farm in centimeters. She has already sorted her data in the table shown.

    a Create a boxplot that represents Isabelle's data.

    b What percent of the worms are 6 cm or longer?

    c What percent of the worms are between 3 cm and 5 cm?

    d In which quartile is the data the most spread out?

Lower extreme	3
Lower quartile	5
Median	6
Upper quartile	7
Upper extreme	10

16 You asked some of your classmates, "How many times have you been assigned homework this week?" Their responses are shown:

$$7, 10, 7, 7, 3, 3, 10, 4, 1, 7, 4$$

    a Complete the given table:

    b Construct a boxplot for the data.

    c What is the range of the number of homework assignments received this week?

    d Could this data be used to analyze the number of homework assignments received by all students in your school? Explain your answer.

Minimum	
Lower quartile	
Median	
Upper quartile	
Maximum	
Interquartile range	

17 A group of friends are analyzing the ratings of various TV shows. The shows can be given a rating between 0 and 5, and half ratings are permitted. The data they collected is shown.

$$3, 1, 0.5, 2, 3, 2, 3.5, 2.5, 3, 4.5,$$

$$3, 4, 2, 3.5, 1, 0.5, 4.5, 2.5, 3, 4$$

    a Complete the given table.

    b Construct a boxplot for the data.

    c Which question can be answered by the boxplot?

        A What is the average customer satisfaction rating?

        B How many customer satisfaction ratings are above 3.5?

        C What is the range of the middle 50% of the customer satisfaction ratings?

        D How many ratings are below 2.0?

Minimum	
Lower quartile	
Median	
Upper quartile	
Maximum	
Interquartile range	

18 The scores on an end-of-year exam for a class of students are shown:

$$52, 56, 59, 64, 66, 77, 78, 80, 80, 80, 81, 84, 86, 90, 95, 96$$

    a   Construct a boxplot for the data.

    b   Calculate the interquartile range.

    c   What percentage of scores lie in the range 85 to 96?

    c   Which values do the lowest 75% of scores lie between?

    d   If the score 91 was added to the data set, what percentage of the scores are below Q_3? Why isn't this 75%?

19 The boxplot shows data collected to answer the question, "At what age do most people get their driver's license?"

    a   State the oldest age that someone got their license.

    b   State the youngest age that someone got their license.

    c   What percentage of people were 18 to 22 years old?

    d   The middle 50% of responders were within how many years of one another?

    e   In which quartile are the ages least spread out?

    f   The bottom 50% of responders were within how many years of one another?

20 In training, a fighter pilot measures the number of seconds he blacks out over a number of flights. He constructs the following box plot for his data:

As long as the pilot is not unconscious for more than 7 seconds, he will be safe to fly.

The pilot concludes that he is safe to fly all the time. Is his conclusion correct? Explain your answer.

Let's extend our thinking

21 Construct a boxplot for the data set:

- Range = 14
- Lowest score = 2
- IQR = 8
- 25% of scores are below 4
- 50% of scores are above 8

22 Justin recently started his job as a waiter at a restaurant. The restaurant tracks the amount of time each table's tab is open. They want Justin to try to close his tables' tabs within 48 minutes. He uses the data cycle to track his progress.

    a   Formulate a question which could be investigated using a boxplot.

    b   Describe a method which could be used to collect data.

    c   During Justin's first shift, the duration of his tables' open tabs is shown:

$$32, 35, 39, 40, 40, 43, 45, 46, 46, 46,$$

$$48, 49, 50, 50, 52, 54, 55, 57, 58, 60$$

    Organize his data using a boxplot.

    d   Use the boxplot to draw conclusions to answer your formulated question.

23 To gain a place in the main race of a car rally, teams must compete in a qualifying round. The median time in the qualifying round determines the cut-off time to make it through to the main race. Here are some results from the qualifying round:

- 75% of teams finished in 159 minutes or less.
- 25% of teams finished in 132 minutes or less.
- 25% of teams finished between with a time between 132 and 142 minutes.

a Find the cut-off time required in the qualifying round to make it through to the main race.

b Find the interquartile range in the qualifying round.

c In the qualifying round, the ground was wet, while in the main race, the ground was dry. To make the times more comparable, the finishing time of each team from the qualifying round is reduced by 5 minutes.

Find the new median time from the qualifying round.

24 Pranav enjoys reading, but he notices that a lot of his classmates do not enjoy reading as much as he does. He wants to investigate the reading habits of students in his middle school.

Go through the whole data cycle at least once to investigate the reading habits of students, where the data can be organized into a boxplot.

Answers

4.02 Boxplots

What do you remember?

1 **a** 42   **b** 34   **c** 28   **d** 11.5
    e 7   **f** 15

2 **a** $Q_1 = 1$, $Q_3 = 9$ **b** $Q_1 = 24$, $Q_3 = 32$
    c $Q_1 = 15$, $Q_3 = 26.5$ **d** $Q_1 = 16$, $Q_3 = 24.5$

3 **a** iv: Minimum - Lowest value

    b v: Lower quartile, Q_1 - At most 25% of the data is below
      this value

    c ii: Median - 50% of the data lies on either side of this
      value

    d iii: Upper quartile, Q_3 - At most 25% of the data is
      above this value

    e i: Maximum - Highest value

4 **A, C**

5 **a**

     20   28   36   42   52

    b

     16   33   47   61   71

    c

     26   33   41   44   45

6 **a** **i** 19   **ii** 11
   b **i** 34   **ii** 15

7 **a** No   **b** Yes   **c** Yes   **d** No

Let's practice

8 **a** **i** 2   **ii** 8   **iii** 12
     iv 18   **v** 32   **vi** 10
   b **i** 3   **ii** 8   **iii** 10
     iv 15   **v** 18   **vi** 7
   c **i** 2   **ii** 5   **iii** 6.5
     iv 7   **v** 9.5   **vi** 2
   d **i** 43   **ii** 48   **iii** 52
     iv 55   **v** 58   **vi** 7

9 **a**

Data

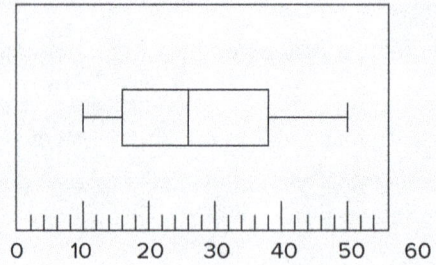

   b

Data

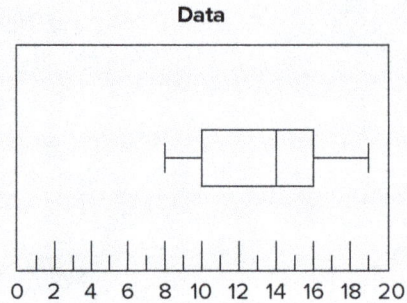

   c

Data

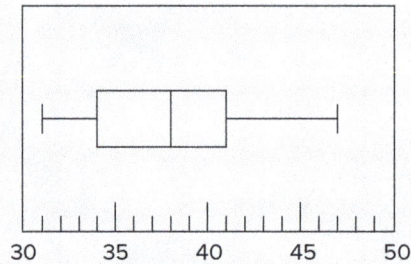

   d

Data

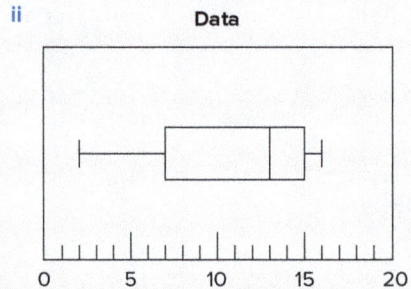

10 **a** **i**

Minimum	2
Lower quartile	7
Median	13
Upper quartile	15
Maximum	16
Interquartile range	8

    ii

Data

b i

Minimum	4
Lower quartile	18
Median	28
Upper quartile	46
Maximum	56
Interquartile range	28

ii

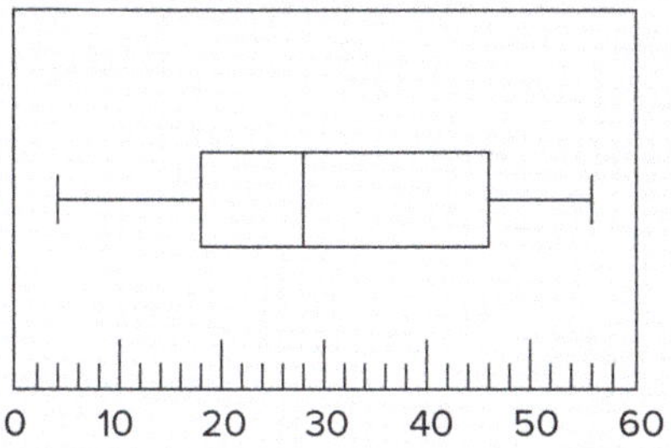

c i

Minimum	10
Lower quartile	11.5
Median	14
Upper quartile	17.5
Maximum	10
Interquartile range	6

ii

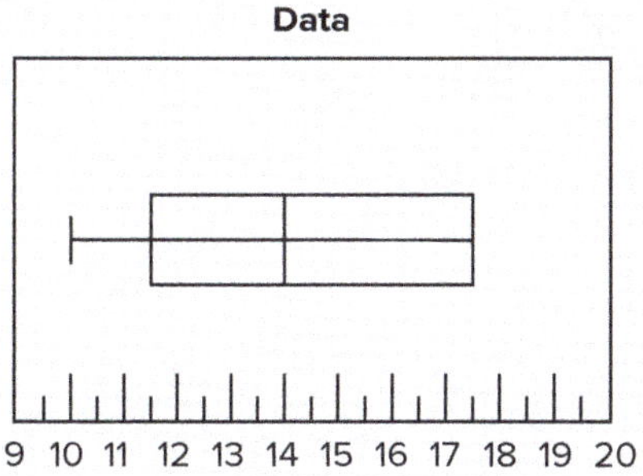

Data

d i

Minimum	71
Lower quartile	74.5
Median	79
Upper quartile	82.5
Maximum	89
Interquartile range	8

ii

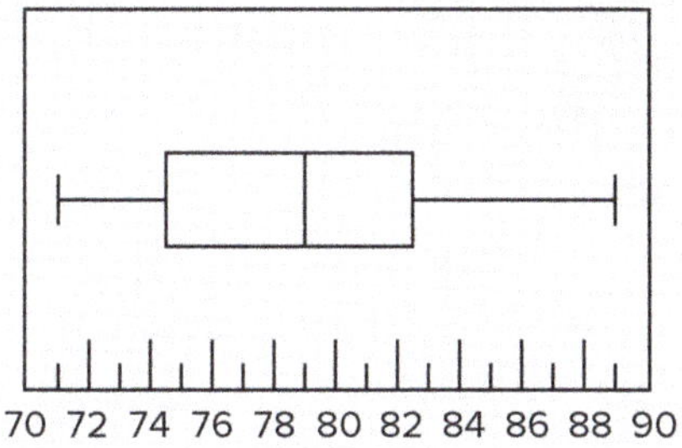

11 a i 50% **ii** 25% **iii** 50% **iv** 75%
   v 75%
 b 2nd

12 a

Lower extreme	5
Lower quartile	10
Median	25
Upper quartile	35
Upper extreme	50
Interquartile range	25

 b C   **c** C
 d The first or lower quartile

13 a B
 b Many possible answers. For example, "What is the cyclist's median heart rate after finish an event?"
 c The cyclist's heart rates varied by 36 beats per minutes (bpm).
 d The middle half of the cyclist's heart rates varied by 19 beats per minute.

14 a The teacher needs to collect univariate data on the number of states students have visited.
 b The teacher should survey the students in the class.
 c

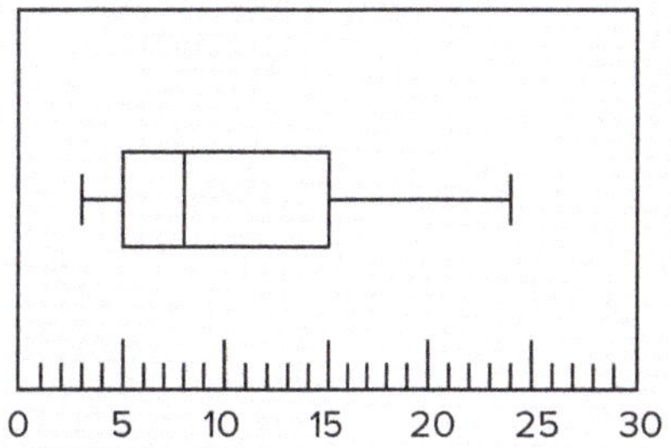

 d The number of states the students in the classed have visited varies by 24 − 3 = 21 states. However, the number of state visited by the middle 50% of students varies by only 15 − 5 = 10 states.

15 a

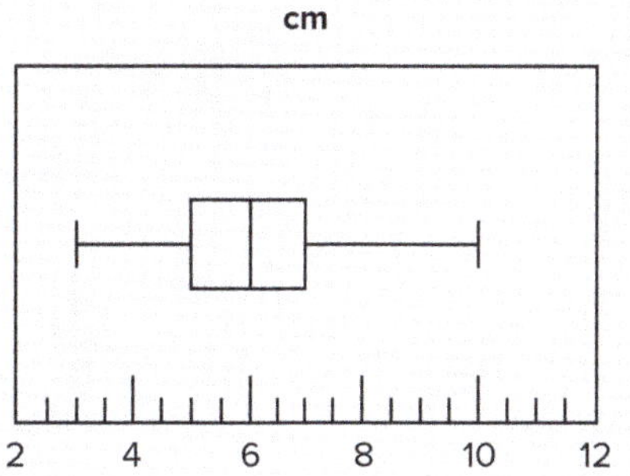

cm

 b 50%
 c 25%
 d In the fourth quartile, between the upper quartile and the upper extreme.

16 **a**

Minimum	1
Lower quartile	3
Median	7
Upper quartile	7
Maximum	10
Interquartile range	4

b

Times

c 9

d No, the students in my class are likely all in the same grade level, so their responses are not representative of the whole school.

17 **a**

Minimum	0.5
Lower quartile	2
Median	3
Upper quartile	3.5
Maximum	4.5
Interquartile range	1

b

Hours

c C

18 **a**

b 20

c 25%

d 52 and 85

e There are now 17 scores, so there are 13 values below Q_3 and $\frac{13}{17} = 76.5\%$. At least 75% of the data lies below Q_3.

19 **a** 31 years   **b** 17 years   **c** 25%   **d** 7

   e 1st   **f** 5

20 No. The upper extreme is 9 which means he has blacked out for longer than 7 seconds.

Let's extend our thinking

21

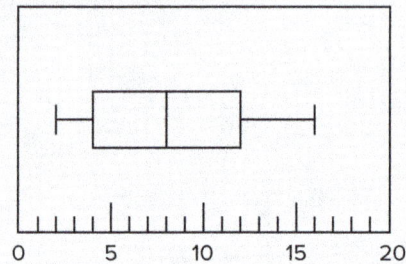

22 **a** There are many possible questions. For example, "What percentage of tables close their tabs within 48 minutes?"

   b As the restaurant already tracks the time for him, Justin does not need to collect the data himself. Instead, he can acquire the existing data from the restaurant.

   c

Duration of open tabs

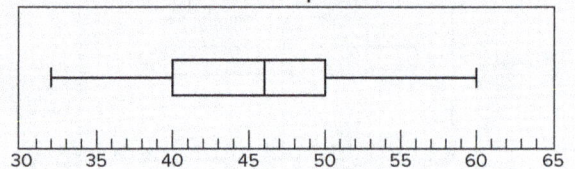

   d Answer will depend on the question from part (a). To answer the question, "What percentage of tables close their tabs within 48 minutes?" using the boxplot, we would first note the 48 falls halfway between the median and the upper quartile. Because approximately 25% of the data lies between the median and the upper quartile, we can say approximately $25 \div 2 = 12.5\%$ of the data lies between the median and 48 minutes. So, approximately $50 + 12.5 = 62.5\%$ of the tables close their tabs within 48 minutes.

23 **a** 142 minutes   **b** 27   **c** 137 minutes

24 1. **Formulate questions**

Possible questions:
- What is the difference in reading times of middle school students?
- What is the median reading time of middle school students?
- What is the difference in the reading times of the middle 50% of students?

2. **Collect data using the questions above.**

We can collect data by doing a survey of fellow students from various grade levels. Here is some sample data.

12, 17, 21, 23, 26, 28, 31, 33, 34, 36,
38, 40, 41, 42, 44, 46, 48, 52, 56, 60

3. **Create data displays using a boxplot.**

This boxplot summarizes the sample data from above.

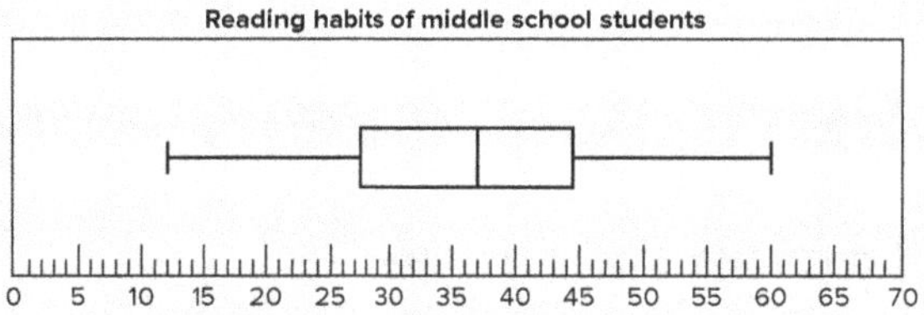

Reading habits of middle school students

4. **Analyze and explain the results.**

From the boxplot, we can see that the reading times of middle school students range from 12 minutes to 60 minutes. The median reading duration is 37 minutes. The reading times of the middle 50% of students differs by 17 minutes.

4.03 Outliers

Subtopic overview

Lesson narrative

In this lesson, students will learn about outliers and their impact on data sets and statistical measures. The lesson begins by discussing the shape and spread of data distributions, including symmetrical, uniform, negatively skewed (left skewed), and positively skewed (right skewed) distributions. Students will understand that outliers are data points significantly different from other observations and can skew or change the shape of data distributions. Students will learn to identify outliers and understand their effects on summary statistics such as range, median, mean, and mode. The lesson emphasizes that outliers can distort the representation of data, particularly in small data sets. For example, removing a low outlier decreases the range, might increase the median, and increases the mean, while removing a high outlier decreases the range, might decrease the median, and decreases the mean. The lesson includes examples and practice problems where students will calculate the range and interquartile range (IQR) of data sets and identify how these measures change when outliers are removed. They will also learn to construct boxplots, marking outliers as distinct points, and compare the shapes of boxplots before and after removing outliers. By the end of the lesson, students will be proficient in identifying outliers, understanding their impact on statistical measures, and using boxplots to visualize data distributions with and without outliers. They will be able to describe how outliers can affect data interpretation and make informed decisions about whether to include or exclude outliers in their analyses.

Learning objective

Students: Page 213

> 🎓 **After this lesson, you will be able to...**
> • explain how an outlier affects the shape and spread of a boxplot.

Key vocabulary

• negative skew   • outlier   • positive skew   • symmetrical   • uniform

Essential understanding

An outlier is an abnormal distance from the rest of the values in a data set. An outlier can sometimes give insight into the variable being explored but it can sometimes cause incorrect conclusions to be drawn from the data.

Standards

This subtopic addresses the following Virginia 2023 Mathematics Standards of Learning standards.

Mathematical Process Goals

MPG3 — Mathematical Reasoning

Teachers can foster Mathematical Reasoning by engaging students in discussions on why outliers occur and how they impact the measures of center in a dataset. Teachers can also encourage students to use logical reasoning to decide on the best strategy for handling outliers, whether it be removing them, adjusting the analysis, or investigating the cause of the outlier.

MPG4 — Mathematical Connections

Teachers can develop Mathematical Connections by relating the concept of outliers to students' prior knowledge of graphical representations and statistical measures, such as mean and median. Teachers can also connect the concept of outliers to other topics within mathematics, such as probability, and to real-world contexts where understanding outliers is important, like in data analysis for sports performance or climate changes.

MPG5 — Mathematical Representations

Teachers can facilitate Mathematical Representations by instructing students on how to visually represent outliers in box plots and how to adjust these representations to account for outliers. Teachers can also have students practice converting between different representations of the same data set, such as a table of values and a box plot, and discuss how each representation can help us understand the impact of outliers. The use of various representations will help students see the importance of mathematical representations in learning, doing, and communicating mathematics.

Content standards

8.PS.2 — The student will apply the data cycle (formulate questions; collect or acquire data; organize and represent data; and analyze data and communicate results) with a focus on boxplots.

8.PS.2a — Formulate questions that require the collection or acquisition of data with a focus on boxplots.

8.PS.2b — Determine the data needed to answer a formulated question and collect the data (or acquire existing data) using various methods (e.g., observations, measurement, surveys, experiments).

8.PS.2d — Organize and represent a numeric data set of no more than 20 items, using boxplots, with and without the use of technology.

8.PS.2e — Identify and describe the lower extreme (minimum), upper extreme (maximum), median, upper quartile, lower quartile, range, and interquartile range given a data set, represented by a boxplot.

8.PS.2f — Describe how the presence of an extreme data point (outlier) affects the shape and spread of the data distribution of a boxplot.

8.PS.2g — Analyze data represented in a boxplot by making observations and drawing conclusions.

Prior connections

5.PS.2 — The student will solve contextual problems using measures of center and the range.

6.PS.2 — The student will represent the mean as a balance point and determine the effect on statistical measures when a data point is added, removed, or changed.

7.PS.2 — The student will apply the data cycle (formulate questions; collect or acquire data; organize and represent data; and analyze data and communicate results) with a focus on histograms.

Future connections

A2.ST.1 — The student will apply the data cycle (formulate questions; collect or acquire data; organize and represent data; and analyze data and communicate results) with a focus on univariate quantitative data represented by a smooth curve, including a normal curve.

A.ST.1 — The student will apply the data cycle (formulate questions; collect or acquire data; organize and represent data; and analyze data and communicate results) with a focus on representing bivariate data in scatterplots and determining the curve of best fit using linear and quadratic functions.

Lesson Preparation

Suggested review

Students are introduced to the common shapes of data sets: symmetrical, uniform, negative (left) skew, and positive (right) skew.

Grade 6 — 9.07 Outliers

Student lesson & teacher guide

Shape and spread

This lesson focuses on understanding and interpreting boxplots, a type of graphical representation used to display the distribution of data.

Collect and display
English language learner support

Listen as students describe in their own words the shapes of different data distributions: symmetrical, uniform, positive skew, negative skew. Encourage students to share their thoughts with a partner or small group. As they discuss, listen for the specific language and descriptions they use. Collect these words, phrases, and any sketches they provide on the board or chart paper.

For example, a student might describe a symmetrical distribution as "the data is evenly spread out on both sides," a uniform distribution as "all the data values have about the same frequency," a positive skew as "the graph has a long tail on the right side," a negative skew as "the data tails off to the left". In the next section you can add outlier which they might describe as "a data point that's far away from the others."

Organize their contributions under headings such as "Symmetrical," "Uniform," "Positive Skew," "Negative Skew," and "Outliers." Include simple drawings next to each term to illustrate their meanings. Throughout the lesson, refer back to this display to reinforce vocabulary, clarify misunderstandings, and connect students' informal language to formal mathematical terminology. This approach helps students build their mathematical vocabulary and deepen their understanding of how different distribution shapes and outliers affect statistical measures and data interpretation.

Confusing left and right skew
Address student misconceptions

Students often confuse left skewed (negatively skewed) and right skewed (positively skewed) distributions when interpreting data sets and boxplots. They may incorrectly assume that "left (negative) skewed" means the data is piled up on the left side, or they might misinterpret which side the "tail" is on. This misunderstanding can lead to errors when describing the shape of data distributions.

To address this misconception, encourage students to focus on the direction of the *tail* in the distribution rather than where the data is concentrated. Explain that in a *left skewed* distribution, the tail extends to the *left* due to a few lower value outliers pulling the distribution in that direction. Conversely, in a *right skewed* distribution, the tail extends to the *right* because of higher value outliers.

Use visual aids such as histograms and boxplots to illustrate this concept. For example, display a histogram where the bars gradually decrease towards the left for a left skewed distribution. Show boxplots with a longer whisker on the left side to indicate left skewness. Encourage students to practice by matching different distributions with their corresponding histograms and boxplots, and have them identify the direction of skewness based on the tails. Including annotated images that highlight the tails and key features can greatly enhance their understanding.

Examples

Students: Page 214

Example 1

The stem-and-leaf plot displays the scores of students in a class on an exam.

	Leaf
6	7 7 9
7	0 0 2 3 4 5 5
8	0 1 3 3 5

Key: 6|1 = 61

a Construct the five-number summary.

Create a strategy

- To find the lower extreme, locate the smallest score.
- To find the upper extreme, locate the highest score.
- Calculate the median by finding the middle score if the number of scores is odd, or by averaging the two middle scores if even.
- Determine the lower quartile by identifying the middle score of the values below the median.
- Identify the upper quartile as the middle score of the values above the median.

Apply the idea

By listing the data, we can easily identify the quartiles:

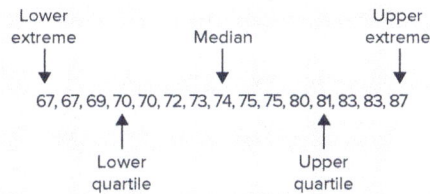

Since the lower quartile is between the same two values, the lower quartile is 70. For the upper quartile, we need to average the two values on either side

Here's the five-number summary:

Minimum	67
Lower quartile	70
Median	74
Upper quartile	81
Maximum	87

Purpose

Demonstrate to students how to construct a five-number summary from a stem-and-leaf plot, aiding in understanding of descriptive statistics and data distribution.

Reflecting with students

Discuss with students that the quartiles cut the data into fourths. The lower quartile may be called the first quartile and the upper quartile may be called the third quartile. The median may be thought of as the second or middle quartile.

b Construct a boxplot for the data.

Create a strategy

Use the five-number summary in part (a) to construct the boxplot.

Apply the idea

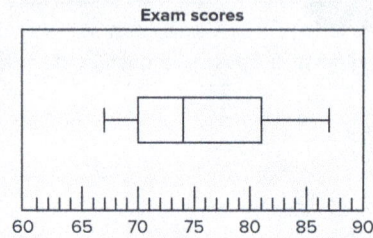
Exam scores

Purpose

Show students how to use a five-number summary to construct a boxplot, which provides a visual representation of data distribution.

c Describe the shape of the boxplot.

Create a strategy

Observe the distribution of data.

- Symmetrical boxplots are symmetrical about the median.
- Negatively skewed boxplots have the majority of data points with higher values.
- Positively skewed boxplots have the majority of the data points with lower values.
- Uniform boxplots have all quartiles the same width.

Apply the idea

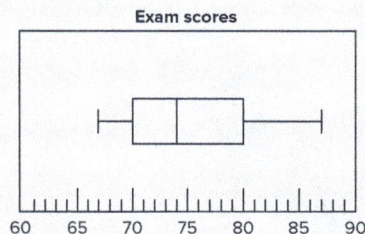
Exam scores

In this boxplot, most of the data have smaller values because the box and whisker above the median are more spread out. Because the data above the median is more spread out, the boxplot is positively skewed.

Reflect and check

The direction of the skew will always be the side that is more spread out. In this example, we said the data is positively skewed (or skewed right) because the right side is more spread out.

The reason for this is because the data points on the right side will make the distribution "crooked". If those data points were closer to the rest of the data, it would be more symmetric.

Purpose

Show students that boxplots can be used to understand the distribution of data and identify skewness in a dataset.

Students: Page 216

> **Idea summary**
>
> We can describe the shape based on the distribution of the data set.
> - Symmetrical boxplots are symmetrical about the median.
> - Uniform boxplots have all quartiles the same width.
> - Positively skewed boxplots have the majority of data points with higher values.
> - Negatively skewed boxplots have the majority of the data points with lower values.

Student lesson & teacher guide

Outliers

Students are introduced to the concept of outliers before engaging in an exploration to discover more.

Exploration

Students: Page 216

Interactive exploration
Explore online to answer the questions
mathspace.co

Use the interactive exploration in 4.03 to answer these questions.

1. Move the point to the position of a really low outlier, then move the point closer to the data. Complete the sentences in the table that follows:

Removing a really low outlier
The range will ⬚.
The median might ⬚.
The mean will ⬚.
The mode will ⬚.

2. Move the point to the position of a really high outlier, then move the point closer to the data. Complete the sentences in the table that follows:

Removing a really high outlier
The range will ⬚.
The median might ⬚.
The mean will ⬚.
The mode will ⬚.

Suggested student grouping: Small groups

This interactive exploration engages students in understanding the impact of outliers on data statistics such as the range, median, mean, and mode. Students are tasked with manipulating a point labeled P on a Geogebra applet to simulate the effect of introducing low and high outliers to a dataset.

Ideal student responses

These ideal responses may differ from other correct student responses. Less formal responses can be connected with the more precise mathematical language presented here.

1. **Move the point to the position of a really low outlier, then move the point closer to the data. Complete the sentences in the table that follows:**

   The range will decrease.
   The median might change slightly or not at all.
   The mean will increase.
   The mode will not change.

2. **Move the point to the position of a really high outlier, then move the point closer to the data. Complete the sentences in the table that follows:**

   The range will decrease.
   The median might change slightly or not at all.
   The mean will decrease.
   The mode will not change.

Purposeful questions

- How does moving point P to the position of a low outlier affect the overall spread (range) of the data, and why does this change occur?
- What impact does the placement of point P as a high outlier have on the mean and median of the data set?
- When point P is moved closer to the cluster of data points, how do the range, mean, median, and mode change? What does this reveal about the stability of these measures when outliers are minimized or removed?

Possible misunderstandings

- Students might incorrectly assume that the median will change significantly when an outlier is introduced or removed, not recognizing the median's resistance to outliers compared to the mean.
- Students may mistakenly believe that the mode will always change when an outlier is moved closer to the data set, misunderstanding that the mode is unaffected by outliers unless the outlier becomes the most frequent value.

Examples

Students: Page 217

Example 2

The number of fatal accidents from 2000 to 2014 for different airlines are listed in the set and displayed in the boxplot:

$$\{0, 0, 0, 0, 0, 0, 1, 1, 1, 1, 2, 2, 2, 2, 2, 2, 4, 4, 4, 5, 5, 5, 5, 5, 6, 7, 10, 11, 11, 12, 15, 24\}$$

Number of fatal accidents

a Identify and interpret the range of the data set.

Create a strategy

The range of the data set is the distance between the minimum value and the maximum value.

Apply the idea

The maximum of the data set is at 24 and the minimum is at 0. So, $24 - 0 = 24$ is the range.

From 2000-2014 the number of fatal accidents with different airlines varied by 24 accidents.

Reflect and check

All statements about the range that describe variance (how the data varies) should be sentences phrased in terms of the variable of interest and include the correct units of measurement.

Purpose

Show students how to identify and interpret the range of a data set in the context of a real-world situation, enhancing their understanding of how the range can represent variance in data.

Students: Page 218

b Identify and interpret the IQR of the data set.

Create a strategy

The IQR of the data set is the distance between the upper and lower quartiles.

Apply the idea

The upper quartile is at 5.5, and the lower quartile is at 1 so:

$$
\begin{aligned}
IQR &= Q_3 - Q_1 && \text{Formula for IQR} \\
&= 5.5 - 1 && \text{Substitute } Q_3 = 5.5 \text{ and } Q_1 = 1 \\
&= 4.5 && \text{Evaluate the subtraction}
\end{aligned}
$$

From 2000-2014 the number of fatal accidents for the middle half of all airlines varied by 4.5 accidents.

Reflect and check

Since each quartile of a boxplot represents about 25% of the data set, the IQR represents the middle 50% of the data set.

Purpose

Show students how to calculate and interpret the Interquartile Range (IQR) as a measure of statistical dispersion, providing insights into the spread of the middle half of a dataset.

Students: Page 218

c   Explain what will happen to the range and IQR if the outlier at 24 is removed.

Create a strategy

From parts (a) and (b) we know that the range and IQR are 24 and 4.5, respectively. We need to recalculate, or estimate, the new range and IQR without the point at 24.

Apply the idea

Without the point at 24, the maximum of the data set will change to 15 and the minimum is still 0. So, $15 - 0 = 15$ is the new range. This is a reduction in the range by 9 accidents.

With the point at 24 removed, the lower quartile remains at 1 and the upper quartile lowers slightly to 5. So, $5 - 1 = 4$ is the new interquartile range. This is a reduction by 0.5 accidents.

Reflect and check

Similar to mean, the range is greatly affected when extreme data points are added or removed. The median may change a little, but the interquartile range is least affected.

Purpose

Show students how the range and interquartile range of a data set are affected by the removal of an outlier.

Reflecting with students

Start a discussion with students about which measures would change the most significantly and which would chaneg the least. Outliers do not affect all measures equally.

Students: Page 218–219

Example 3

Yartezi works at a coffee shop and tracks the number of customers that come in each day. The data she collected is shown:

$$90, 85, 88, 86, 95, 101, 98, 84, 35, 82, 87, 90, 92, 97$$

a   Formulate a question that could be answered using a boxplot.

Create a strategy

Boxplots help us see how clustered or spread out the data is, so the question could focus on the spread of the data. Boxplots also help us see how the data breaks down into quarters (or quartiles).

Apply the idea

One possible question about this data that could be answered using a boxplot is, "What is the typical number of customers that come into the coffee shop each day?"

Reflect and check

Other questions might be, "How does the number of customers who come into the coffee shop in a day vary?" or "How often can I expect an unusually high number of customers to visit the coffee shop?"

Purpose

Show students how to use boxplots as a tool to analyse and interpret real-world data by formulating a meaningful question that a boxplot can answer.

Reflecting with students

Other questions might be, "How does the number of customers who come into the coffee shop in a day vary?" or "How often can I expect an unusually high number of customers to visit the coffee shop?"

Students: Page 219

b   Describe the data collection method that Yartezi used.

Create a strategy

Consider whether Yartezi collected the data herself or whether she acquired data that was already collected by someone else.

If Yartezi collected the data herself, decide whether she used an observation, measurement, a survey, or an experiment to collect the data.

Apply the idea

Since Yartezi tracks the number of customers herself, she collected the data rather than acquired it from elsewhere. Yartezi used an observation to collect the data by counting the number of customers as they came into the shop.

Reflect and check

Yartezi did not measuring anything, and she did not need to ask the customers anything. She did not use an experiment because she was not interested in the factors that caused customers to come into the shop.

Purpose

Show students how to identify different data collection methods and understand the process of data collection in a real-world context.

Students: Page 219–220

c   Construct a boxplot using the data points Yaretzi collected.

Create a strategy

To create a boxplot, we need to identify the five critical values:

- Lower extreme
- Upper quartile (Q_3)
- Lower quartile (Q_1)
- Upper extreme
- Median

Apply the idea

First, we need to put the data values in order:

$$35, 82, 84, 85, 86, 87, 88, 90, 90, 92, 95, 97, 98, 101$$

Now, we can find the lower extreme, upper extreme, and the median (the middle value). After that, we can find the middle value of the lower extreme and median (this will be the lower quartile) and the middle value of the median and the upper extreme (this will be the upper quartile).

Lower extreme → 35, 82, 84, 85, 86, 87, 88, 90, 90, 92, 95, 97, 98, 101 ← Upper extreme
Median, Lower quartile, Upper quartile

Number of customers

30 40 50 60 70 80 90 100 110

Purpose

Show students how to construct a boxplot from a set of data points in order to visualize the distribution and identify the key statistical properties including lower extreme, lower quartile, median, upper quartile and upper extreme.

Expected mistakes

Students might incorrectly calculate the quartiles when the dataset has an even number of data points. They may be unsure whether to include the median in both halves of the data or exclude it, leading to errors in determining the lower quartile, Q_1, and upper quartile Q_3. This misunderstanding can result in an inaccurate boxplot that does not accurately represent the data distribution.

To address this misconception, guide students through the proper method of calculating quartiles in an even-sized dataset. Explain that after finding the median, the dataset should be divided into two halves—lower and upper—with each half containing the same number of data points. Emphasize that the median is not included in either half when there is an even number of data points. Encourage students to list the data points in order and physically separate them into the lower and upper halves to visualize the division.

Students: Page 220

d Answer the formulated question from part (a) using the boxplot and explain whether the answer is reasonable.

Create a strategy

The question from part (a) is, "What is the typical number of customers that come into the coffee shop each day?" To answer this question, we could consider a measure of center, such as the median.

Apply the idea

The median is 89, so we could say that 89 customers typically come into the coffee shop each day.

This answer does seem reasonable because it is near the middle of the data set, so it is a reasonable estimate of all the data.

Reflect and check

The mean of this data set is

$$\frac{35 + 82 + 84 + 85 + 86 + 87 + 88 + 90 + 90 + 92 + 95 + 97 + 98 + 101}{14} \approx 86.4$$

This value would be an unreasonable estimate of the center of the data because 86 is toward the bottom 25% of the data.

Purpose

Show students how to interpret a boxplot and apply their understanding of measures of center, specifically the median, to answer questions about a dataset.

Reflecting with students

Encourage students to calculate the mean of the dataset in addition to the median they found. They will find that the mean is approximately 86.4, which is lower than the median of 89. Discuss with students why the mean is lower and how the extremely low (outlier) value of 35 influences it. Ask them how the mean and median would change if 35 were removed from the dataset. This will help students understand the impact of outliers on measures of center and why the median can sometimes be a more representative measure for skewed data or data with outliers.

Students: Page 220–221

e   Construct a second boxplot after the outlier has been removed, but mark the outlier as a point. Compare the shape of the new boxplot with the first boxplot.

Create a strategy

First, we need to identify the outlier. Since the left whisker is very long, the outlier will be the minimum value. Then, we will need to recalculate the five-number summary after removing the outlier.

Apply the idea

The outlier of the data set is 35 because it is much lower than the other data values.

We can now find the new five-number summary without the outlier.

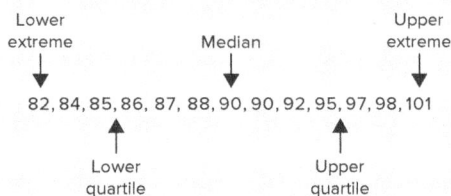

```
         Lower                    Median              Upper
         extreme                                      extreme
            ↓                        ↓                   ↓
     82, 84, 85, 86, 87, 88, 90, 90, 92, 95, 97, 98, 101
                     ↑                        ↑
                   Lower                    Upper
                   quartile                 quartile
```

The lower quartile is $Q_1 = \dfrac{85 + 86}{2} = 85.5$, and the lower quartile is $Q_3 = \dfrac{95 + 97}{2} = 96$.

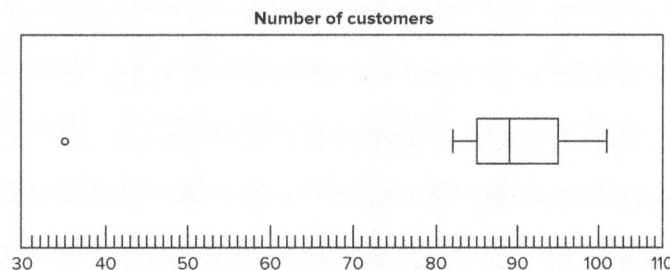

Number of customers

With the outlier, the original boxplot was very negatively skewed with a very large spread in the first quartile. The boxplot without the outlier has a slight positive skew and a much smaller spread.

Reflect and check

Although the outlier may skew the data, we should still represent it on the graph to represent the full picture of the data. Removing the outlier when calculating measures of center or spread give us a better understanding of the *majority* of the data, but we should not disregard the outlier completely.

Purpose

Demonstrate to students how removing an outlier from a dataset affects the shape and values of a boxplot.

f   Answer the formulated question from part (a) using the boxplot that represents the data set after the outlier was removed.

Create a strategy

Previously, we looked at the median to determine the typical number of customers that come into the coffee shop each day. We now want to reconsider that answer, as it is likely that the median has changed since the outlier was removed.

Apply the idea

Before the outlier was removed, the median was 89 customers. After removing the outlier, the median is 90 customers.

Now, we can estimate that 90 customers typically come into the coffee shop each day.

Reflect and check

Without the outlier, this new median is a better measure of the center of the data. However, we should not completely disregard the outlier. The outlier could potentially prompt us to ask a new question about the data, such as "What factors cause a drop in the number of customers?"

A number of factors could have caused the outlier in this context, such as a heavy snowstorm, a power outage, or a city event happening in a different part of the town. To determine the factors that cause outliers, we would need to run an experiment.

Purpose

Show students how the removal of an outlier can impact the median of a dataset and the importance of considering outliers in data analysis.

Use concept maps to support working through the data cycle   use with Example 3
Student with disabilities support

Encourage students to create visual concept maps that outline each step of the data cycle, helping them understand and connect the concepts involved. Start by placing "Data Cycle" in the center of the map. Have students branch out to the four key steps: Formulate Questions, Collect or Acquire Data, Organize and Represent Data, and Analyze and Communicate Results. Under each branch, students can add notes, examples, or illustrations related to the problem they are working on—like Yaretzi's coffee shop customer data.

This visual representation aids conceptual processing by breaking down the cycle into manageable parts and showing how each step leads to the next. Provide students with partially completed maps or a word bank of key terms to support them as they begin. For example, you might provide a concept map with the central idea and main branches drawn, and students fill in details like "Formulate Question: What is the typical number of customers?" or "Organize Data: Construct a boxplot of customer numbers." This approach helps students see the big picture and understand the interconnectedness of each step in the data cycle.

💡 **Idea summary**

An **outlier** is a data point that varies significantly from the body of the data. An outlier will be a value that is either significantly larger or smaller than other observations.

Removing outliers will have the following effects on the summary statistics:

A really low outlier	A really high outlier
The range will decrease	The range will decrease
The median might increase	The median might decrease
The mean will increase	The mean will decrease
The mode will not change	The mode will not change

The IQR is resistant to outliers because it describes the middle half of the data, not the extremes.

Practice

Students: Page 222–228

What do you remember?

1 Match each boxplot with the correct description.

   i Negative (left) skew   **ii** Positive (right) skew   **iii** Uniform   **iv** Symmetrical

a

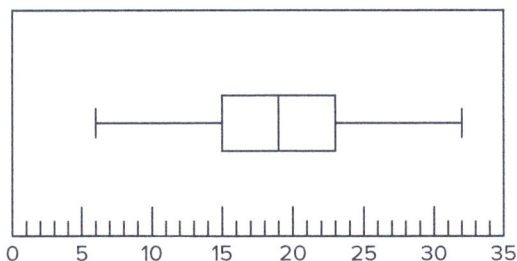

b

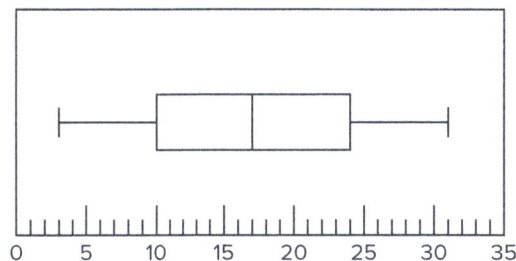

c

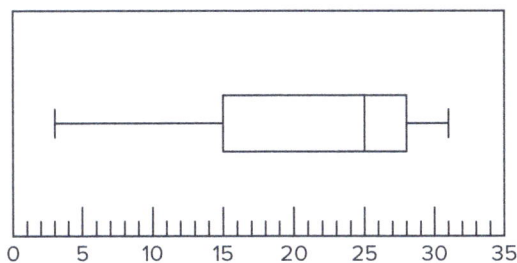

d

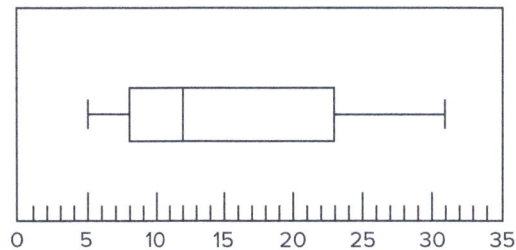

2 For each scenario, decide if the range would increase, decrease, or stay the same:

    a If 28 is added to the data set:

$$0, 1, 1, 2, 3, 3, 3, 4, 6$$

    b If 12 is removed from the data set:

$$12, 42, 55, 57, 59, 60, 62, 65, 70$$

    c If 5 and 27 are removed from the data set:

$$5, 12, 12, 13, 14, 14, 14, 15, 15, 27$$

    d If 3 is added to the data set:

$$21, 24, 25, 25, 26, 27, 27$$

3 For each scenario, an outlier was removed. Was the outlier smaller or larger than the values that remain?

    a The mean decreased after the outlier was removed.

    b The mean increased after the outlier was removed.

    c The median decreased after the outlier was removed.

    d The median increased after the outlier was removed.

4 Identify the outlier for each data set:

    a 104, 115, 215, 127, 109, 118, 121, 130, 116, 134, 131, 129, 125

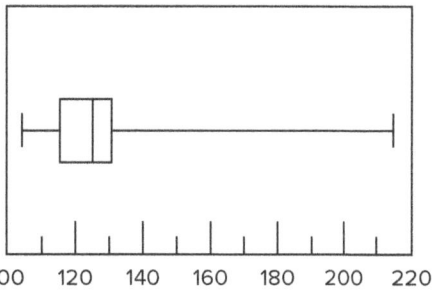

    b 4.0, 3.5, 5.1, 4.4, 0.3, 3.9, 3.5, 3.1, 4.9, 4.3, 4.7, 5.0, 3.7, 3.8

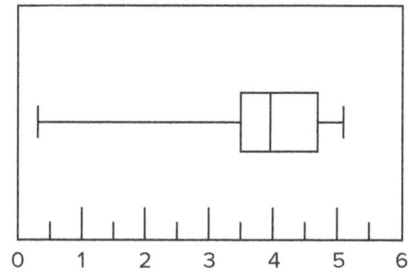

    c 61, 80, 108, 68, 70, 72, 63, 65, 75, 78

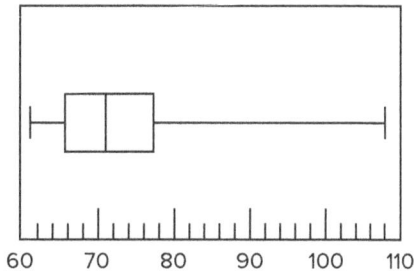

    d 22, 37, 39, 27, 29, 30, 60, 31, 32, 34, 28, 29, 35, 42, 25, 45

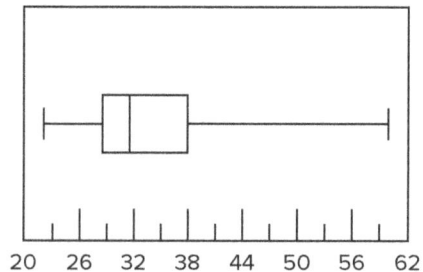

5 For the boxplot shown, find each of the following:

    a 5-number summary

    b Range

    c Interquartile range

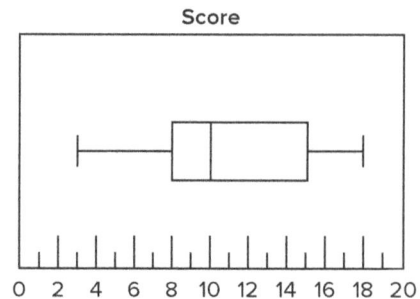

Score

Let's practice

6   Consider the data set represented by the boxplot:

   {1, 2, 2, 4, 4, 5, 6, 6, 8, 8, 8, 9, 9}

   a   Which option best describes the spread of the boxplot?

   A   The boxplot has a very large spread.

   B   The boxplot has a relatively small spread.

   C   The boxplot has a uniform spread across each quartile.

   b   Suppose one score of 8 is changed to a 19. Will these measures change or not?

   i   Median   **ii**   Interquartile range

   iii Range

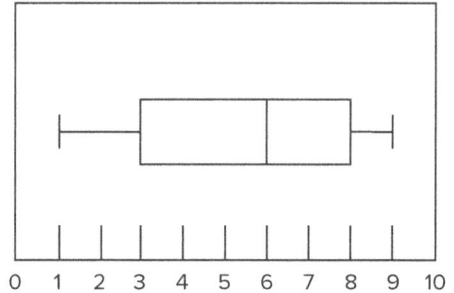

0   1   2   3   4   5   6   7   8   9   10

7   A boxplot was created for a data set. The value 23 was a typo and should have been 32.

   a   How does this change the range?

   b   With this change, in which quartile is the data most spread out?

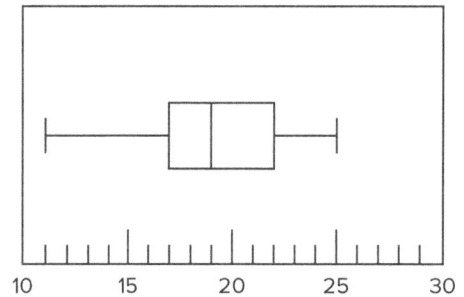

10   15   20   25   30

8   For each set of data:

   i   Find the five critical values of the boxplot.

   ii   Which data value is an outlier?

   iii Redraw the boxplot after the outlier is removed, but mark the outlier as a point.

   iv   Describe how the five critical points changed after removing the outlier.

   a   27, 50, 24, 37, 47, 41, 27, 126, 44, 27

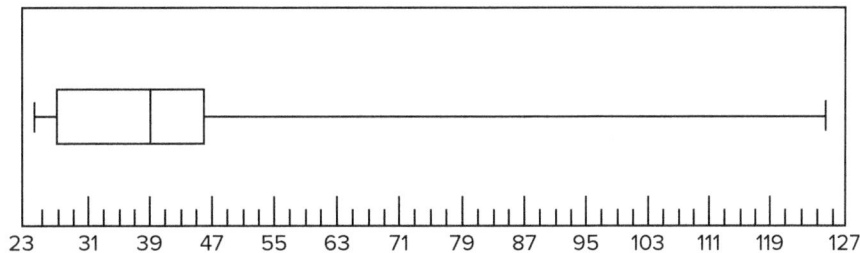

23   31   39   47   55   63   71   79   87   95   103   111   119   127

   b   4.7, 2.8, 1.6, 1.9, 0.9, 0.9, 2.2, 2.2, 1.2, 1.5, 0.9, 2.5

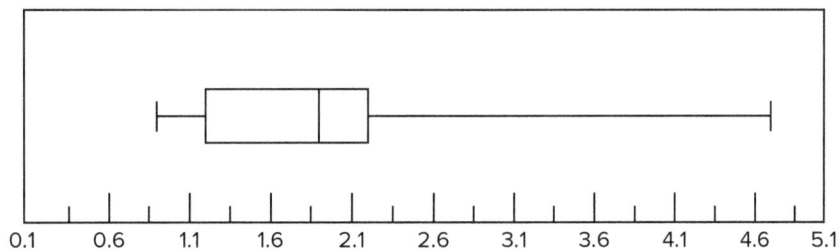

0.1   0.6   1.1   1.6   2.1   2.6   3.1   3.6   4.1   4.6   5.1

c 470, 505, 470, 485, 470, 450, 530, 490, 520, 480, 150, 510, 485, 495, 515

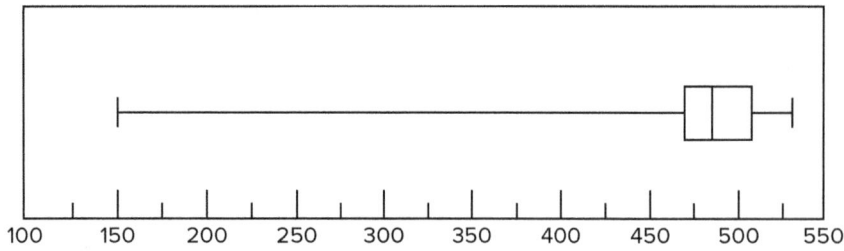

9 A manager at a particular company is interested in knowing how the commute times, in minutes, of her employees varies. She collects the data shown:

22, 30, 18, 19, 24, 18, 26, 24, 27, 29, 21, 23, 24, 28, 21, 25, 27, 68

a What method of data collection did the manager use?

b Find the median with and without the outlier.

c Find the IQR with and without the outlier.

d Construct a boxplot after the outlier has been removed, but mark the outlier as a point.

e Describe the shape of the boxplot.

f Describe the spread of the boxplot.

10 Suppose you are using the data cycle to answer the question, "How did the weather in my town vary for most days last month?"

a What type of data would you need to collect to answer this question?

b This boxplot represents the temperatures from a specific town in the month of May. Describe the shape of the boxplot.

c One day, the temperature dropped to 33°F. Redraw the boxplot with the outlier.

d Describe the shape of the boxplot with the outlier.

e Describe the spread of the boxplot with the outlier.

Temperature (Fahrenheit)

11 The number of hat tricks scored in a hockey season by several teams are shown in the table.

a Construct a boxplot to represent the data.

b Describe the shape of the boxplot.

c After reviewing the data, someone noticed a mistake. The 10 was supposed to be a 1. How do you think this will affect the shape of the boxplot?

d Redraw the boxplot and compare it to your hypothesis in part (c).

Hat tricks	Frequency
0	2
1	4
2	5
3	3
4	2
5	2
10	1

12 The selling prices of twelve houses in the same county are:

$360 000, $374 000, $467 000, $395 000, $413 000, $410 000,
$456 000, $422 000, $487 000, $129 000, $479 000, $408 000

Selling prices of homes (in thousands)

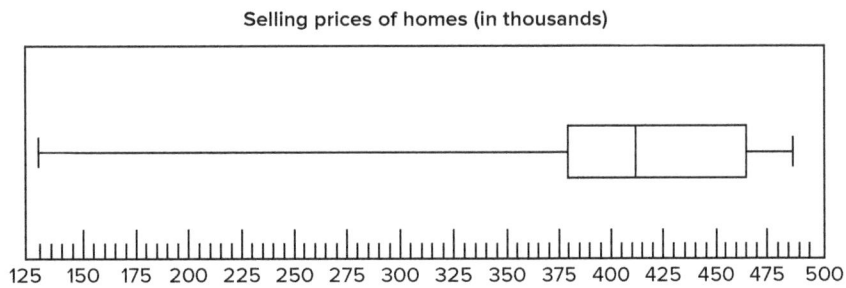

a Describe the shape of the boxplot.

b In which quartile is the data most spread out?

c Suppose the house that was sold for $129 000 was remodeled. It is now valued at $480 000. Describe how this would change the shape and spread of the boxplot.

13 The boxplot shows the age at which a group of people passed their driving license test:

a Formulate two questions that could be answered by the boxplot.

b Select the option that describes the shape of the boxplot.

A Symmetric

B Uniform

C Negatively skewed

D Positively skewed

Age

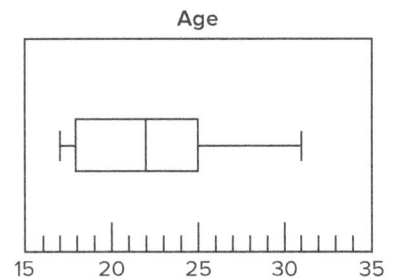

c Use center and spread to describe the typical age at which people in this group passed their driving test.

14 The average gas price of all 50 states in 2021 is shown in the boxplot:

Average gas price by state

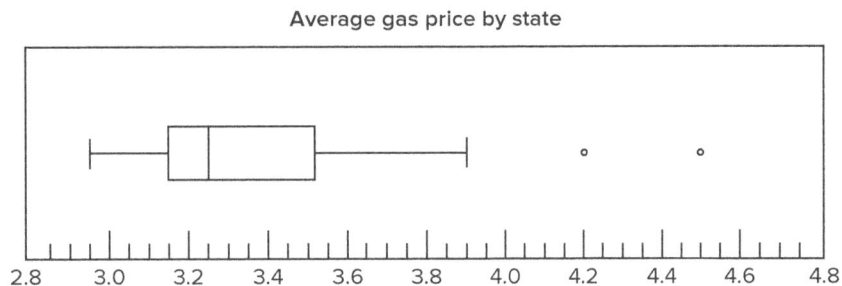

a Which quartile has the smallest spread?

b Describe the typical gas price for the United States.

c If the outliers of $4.20 and $4.50 were included when drawing the boxplot, how would the shape of the boxplot change?

15 Amarah volunteers at an animal shelter and collected data to answer the questions, "What is the typical weight of medium-sized dogs?" and "What is the difference in the weight of most medium-sized dogs?"

She created a boxplot to represent the data, then noticed an outlier and drew a second boxplot with the outlier as a point.

Amarah's boxplots are shown.

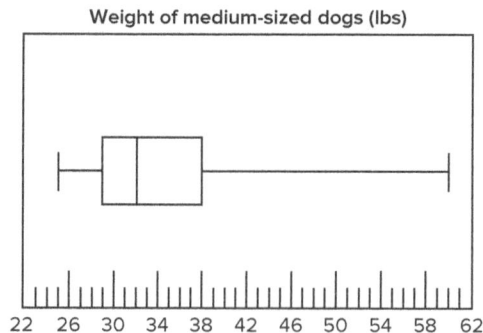

Weight of medium-sized dogs (lbs)

   a   What method of data collection did Amarah use?

      A Observation   **B** Measurement

      C Survey   **D** Experiment

   b   Find the median with and without the outlier.

   c   Find the range with and without the outlier.

   d   Find the interquartile range with and without the outlier.

   e   What is the typical weight of a medium-sized dog? Explain your answer.

   f   What is the difference in the weight of most medium-sized dogs? Explain your answer.

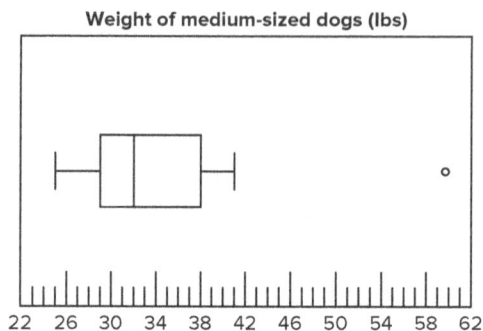

Weight of medium-sized dogs (lbs)

16 The donations a cheerleading team received for a fundraiser are shown:

$$\$80, \$100, \$70, \$50, \$80, \$75, \$50, \$100, \$60, \$75, \$80, \$65, \$85, \$90, \$85$$

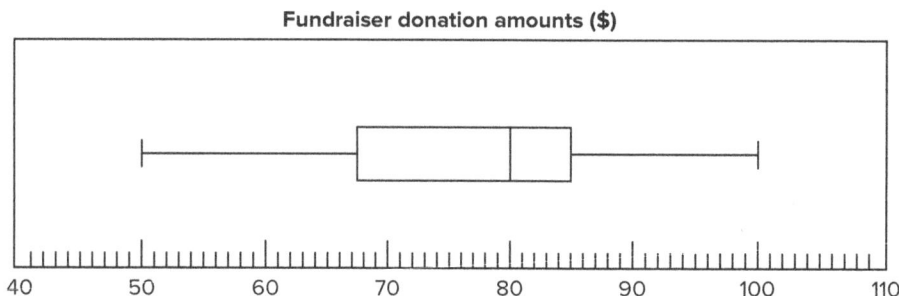

Fundraiser donation amounts ($)

   a   Formulate a question that could be answered with this data.

   b   Describe the data collection method used.

   c   Answer the question you formulated in part (a).

   d   If a last minute donation of $500 was added to the data set, would your conclusion from part (c) change? Explain your answer.

17 Eliot loves to read books and logs the number of pages he reads each week. Over several weeks, he recorded these page counts:

$$150, 160, 155, 152, 158, 155, 168, 157, 168, 156, 174, 170, 0, 155, 172, 166, 166$$

   a   Create a boxplot to represent Eliot's weekly page counts.

   b   Construct a second boxplot after the outlier has been removed, but mark the outlier as a point. Compare the shape of the new boxplot with the first boxplot.

   c   Using one of the boxplots, Eliot made the following conclusions:

   •   My weekly reading habits are relatively consistent.

   •   I typically read about 159 pages on a weekly basis.

   •   If I can't read up to 159 pages in a week, I still read within 9 pages of that amount.

Which boxplot did he use to make these conclusions?

Let's extend our thinking

18 A group in a study volunteer to take a test that assesses their reaction time. The participants clicked a button as soon as they heard a sound which was played at random intervals. The reaction time, in milliseconds, of each participant is shown below:

220, 280, 210, 220, 215, 180, 185, 190, 190, 195, 185, 190, 195, 195

a Identify any outliers.

b Which of the following is the most likely explanation for the outlier?

    A The participant was not focused.   **B** The sound took longer to play.

    C The participant missed the button.   **D** The timer malfunctioned.

c Explain how including the outlier could affect the study's conclusions.

d Explain how not including the outlier could affect the study's conclusions.

19 Akio wants to learn more about the number of first cousins people have. He wants to use the data cycle to investigate.

a Formulate a question about first cousins that could be answered using a boxplot.

b Describe a method which could be used to collect data.

c Collect up to 20 data values for your formulated question.

d Organize your data using a boxplot.

e Answer your formulated question using the boxplot and explain whether the answer is reasonable.

f The data shown represents the number of first cousins for a random sample of people.

6, 8, 33, 9, 28, 17, 0, 11, 32, 8, 20, 26, 13, 8, 12, 31, 13, 7, 17, 13

Assume that 0 cousins is an outlier. Construct a boxplot after the outlier has been removed, but mark the outlier as a point. Compare the shape and spread of this boxplot with your boxplot from part (d).

20 Go through the whole data cycle at least once to investigate the populations of various counties in Virginia, where the data can be organized into a boxplot. Randomly choose up to 20 counties to use in your investigation. If any counties appear to be outliers, explain how you accounted for them throughout the data cycle.

Answers

4.03 Outliers

What do you remember?

1 a iv   **b** iii   **c** i   **d** ii

2 a Increase   **b** Decrease   **c** Decrease   **d** Increase

3 a Larger   **b** Smaller   **c** Larger   **d** Smaller

4 a 215   **b** 0.3   **c** 108   **d** 60

5 a The minimum is 3, the lower quartile is 8, the median is 10, the upper quartile is 15 and the maximum is 18.

    b 15   **c** 7

Let's practice

6 a B

    b i No   **ii** Yes   **iii** Yes

7 a This changes the range from 15 to 22, an increase by 7.

    b Q_4

8 a i • Lower extreme: 24
- Lower quartile (Q_1): 27
- Median: 39
- Upper quartile (Q_3): 46
- Upper extreme: 126

    ii 126

    iii

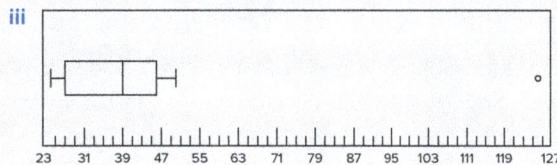

    iv The lower extreme and lower quartile stayed the same. The median, upper quartile, and upper extreme decreased.

    b i • Lower extreme: 0.9
- Lower quartile (Q_1): 1.2
- Median: 1.9
- Upper quartile (Q_3): 2.2
- Upper extreme: 4.7

    ii 4.7

    iii

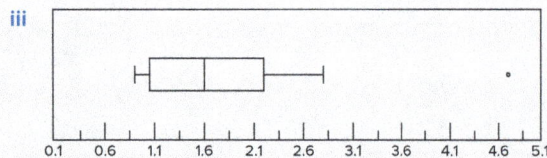

    iv The lower extreme and upper quartile stayed the same. The lower quartile and median slightly decreased. The upper extreme decreased.

c i • Lower extreme: 150
- Lower quartile (Q_1): 470
- Median: 485
- Upper quartile (Q_3): 507.5
- Upper extreme: 530

    ii 530

    iii

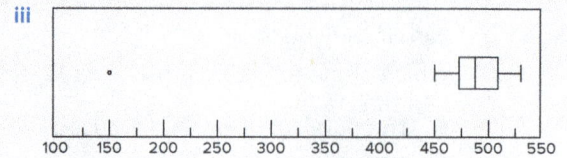

    iv The five critical points all increased.

9 a Survey

    b With the outlier: 24 minutes

    Without the outlier: 24 minutes

    c With the outlier: 6.25 minutes

    Without the outlier: 6 minutes

    d

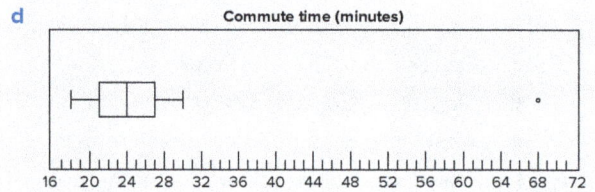

    e Uniform

    f The boxplot has a perfectly even spread across all four quartiles. The overall spread of the boxplot is relatively small, differing by only 12 minutes.

10 a One possible answer: Data on the temperature highs (or lows) over each day in the past month would need to be acquired from a reliable, local weather source.

    b Symmetrical

    c

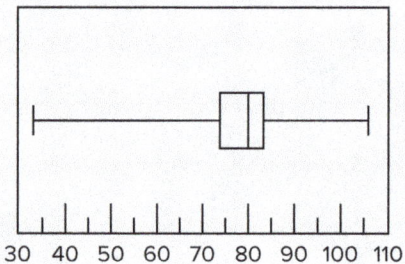

    d Negatively skew

    e The boxplot displays a moderately tight spread of the middle 50% of data. The rest of the data has a relatively large spread.

11 a

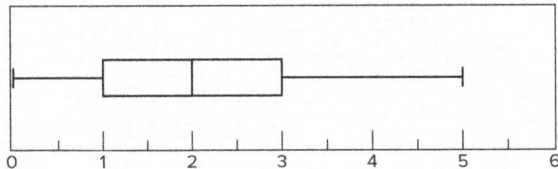

b Positive (right) skew

c Removing the outlier will reduce the spread and make the boxplot more symmetric.

d The new boxplot has a smaller spread and is more symmetric than the previous one, which confirms the hypothesis from part (c).

12 a Negative (left) skew

b First quartile

c The boxplot would become more symmetric and less spread out because the remaining data have values that are closer together.

13 a Many possible answers, such as "At what age do most people get their drivers license?" and "How does the age at which people get their drivers license vary?"

b D

c The typical age a person in this group was when they passed their driving test was 22 and the middle half of ages varied by 7 years as the middle half of people passed their driving test between 18 and 25 years old. The ages of all members of this group vary by 14 years with the youngest passing age at 17 and the oldest at 31.

14 a The second quartile

b The typical average gas price in the US can be estiamted by the median gas price, which is $3.25.

c If the outliers were included, there were be a small change in the location of the lower quartile, the median, and the upper quartile. These values would likely increase by a small amount. The biggest change will be in the fourth quartile, and the whisker would be much longer. This would make the boxplot positively skewed.

15 a B

b The median is 32 lbs with and without the outlier.

c With the outlier: 35 lbs

Without the outlier: 16 lbs

d The IQR is 9 lbs with and without the outlier.

e The typical weight of a medium-sized dog is 32 lbs. The typical weight can be represented by a measure of center, such as the median. The median was not affected by the outlier, so it is an accurate description of a typical weight for medium-sized dogs.

f The weight of most medium-sized dogs differs by about 9 lbs. Most medium-sized dogs' weights can be represented by the interquartile range, which is the middle 50% of the data, and the difference refers to how spread out the data is. The IQR was not affected by the outlier, so it is an accurate description of the difference of most medium-sized dogs.

16 a Many possible answers. One example is "What was the difference in the size of the donations?"

b Observation

c Answers will vary based on formulated questions. Using the example "What was the difference in the size of the donations?": The range of the data is $100 - 50 = 50$, so the difference in the size of the donations is $50.

d Yes, the range is largely affected by outliers. Now, the difference in the size of the donations is $500 - 50 = $450.

17 a

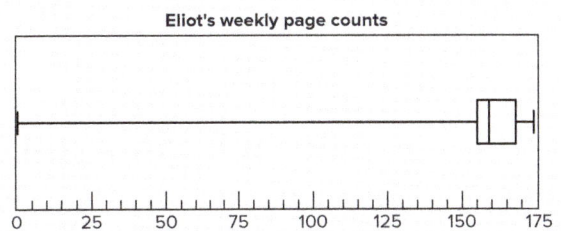

b

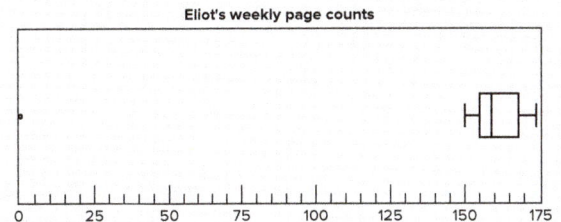

c The second boxplot

Let's extend our thinking

18 a 280

b A

c Including the outlier will result in a larger overall spread of the data and a higher mean value. It could cause researchers to conclude that in general reaction time varies more greatly or is overall slower than it actually is.

d Not including the outlier will result in a smaller spread of the data and a lower mean. It is possible that the data will be relatively symmetric. It could give a more accurate impression of the typical reaction time for the population, but we should also consider that the outlier is a valid reaction time and not completely leave it out of any decisions that are made using this data.

19 a Answers will vary, possible questions are "What is the typical number of first cousins people have?" or "What is the difference in the number of first cousins people have?"

b One method is to survey teachers and students from school.

c Answers will vary.

An example data set is

3, 2, 4, 15, 5, 2, 5, 4, 1, 6,
2, 7, 0, 2, 5, 4, 18, 1, 8, 2

d

Number of first cousins

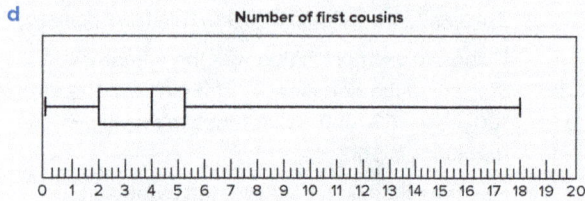

If 15 and 18 are considered outliers:

Number of first cousins

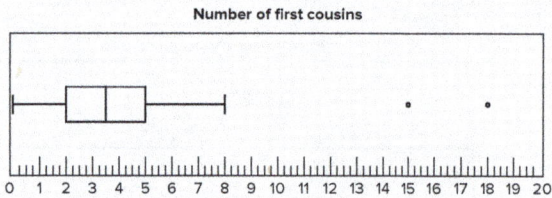

e Answers will vary.

Using the boxplot with the outliers marked as points and the example question, "What is the typical number of first cousins people have?": Most people have between 2 and 5 first cousins. This seems reasonable because it describes the center of the data after removing the outliers.

Using the boxplot that includes the outliers and the example "What is the difference in the number of first cousins people have?": The lowest number of first cousins is 0 and the highest is 18, so the data differs by 18 cousins. This is a relatively large spread, but it seems reasonable because the size of families varies a lot.

f

Number of first cousins

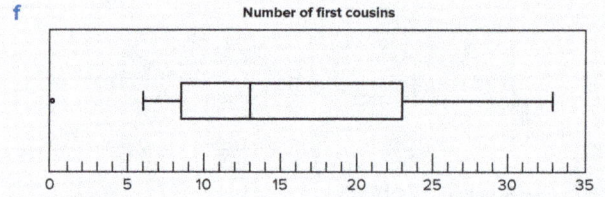

The original boxplot, where 15 and 18 cousins are marked as outliers, is approximately symmetric with a relatively small spread. This boxplot, where 0 cousins is an outlier, is positively skewed with a larger spread.

20

1. **Formulate questions**

   Possible questions:
   - What is the difference in populations of counties in Virginia?
   - What is the typical population of a county in Virginia?
   - What is the difference in the populations of most counties in Virigina?

2. **Collect data using the questions above.**

   We can collect data from https://www.census.gov/data.html. Here is some sample data from https://www.census.gov/library/stories/state-by-state/virginia-population-change-between-census-decade.html.

	Population in 2020
Fairfax county	1 150 309
Virginia Beach city	459 470
Loudon county	420 959
Henrico county	334 389
Arlington county	238 643
Richmond city	226 610
Spotsylvania county	140 032
Hanover county	109 979
Montgomery county	99 721
Roanoke county	96 929
Frederick county	91 419
James city county	78 254
York county	70 045
Franklin county	54 477
Charlottesville city	46 553
Manassas county	42 772
Tazewell county	40 429
Louisa county	37 596
Halifax county	34 022
Alleghany county	15 223

3. **Create data displays using a boxplot.**

This boxplot summarizes the sample data from above.

Population (in thousands)

However, Fairfax county appears to be an outlier. After removing the outlier, here is the new boxplot.

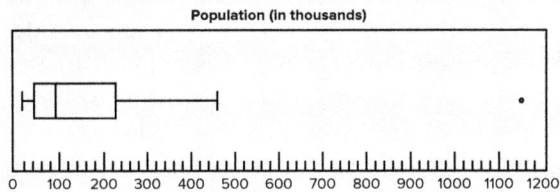

Population (in thousands)

4. **Analyze and explain the results.**

From the boxplot, we can see that the populations of counties in Virginia are positively skewed. The data from these 19 counties, not including Fairfax county, have a range of about 450 000 people. The median population is about 100 000 people. The population of the middle 50% of counties, not including Fairfax county, differs by about 180 000 people.

4.04 Compare data sets

Subtopic overview

Lesson narrative

In this lesson, students will learn how to compare data sets using various data displays. The lesson begins with an introduction to boxplots, which are used to visually compare the spread, median, quartiles, and range of different data sets. Students will create and interpret parallel boxplots, examining the differences between data sets and identifying which group performs better based on the median and consistency based on the interquartile range (IQR). Students will engage in an exploration activity where they will compare the heights of students in a class using a histogram, dot plot, and boxplot. They will analyze the strengths and weaknesses of each display and determine which aspects of the data are best represented by each type of graph. Through guided examples, students will practice creating boxplots from raw data, converting dot plots to boxplots, and selecting appropriate measures of center and spread to summarize and compare data sets. By the end of the lesson, students should be able to effectively use and interpret different data displays to compare data sets and understand the key features they reveal.

Learning objectives

Students: Page 229

After this lesson, you will be able to...
- analyze and compare two data sets represented in boxplots.
- justify whether a pictograph, bar graph, line graph, line plot/dot plot, stem-and-leaf plot, circle graph, histogram, or boxplot best represents a real-world data set.

Key vocabulary

- interquartile range
- measure of center
- measure of spread
- median
- parallel boxplots
- quartile
- range
- shape

Essential understanding

The way a set of data is represented affects the conclusions and generalizations that can be made.

Standards

This subtopic addresses the following Virginia 2023 Mathematics Standards of Learning standards.

Mathematical Process Goals

MPG2 — Mathematical Communication

Teachers can foster mathematical communication by encouraging students to share their observations and conclusions when comparing data sets using boxplots. They can also ask students to justify their choices of graphical representations, ensuring they use mathematical language and notation to express their ideas with precision.

MPG3 — Mathematical Reasoning

Teachers can stimulate mathematical reasoning by asking students to justify the steps they take when comparing data sets using boxplots. They can also encourage students to use logical reasoning to analyze arguments and determine whether conclusions drawn about the data sets are valid.

MPG5 — Mathematical Representations

Teachers can meet this goal by teaching students how to represent data sets using boxplots, as well as other graphical representations such as bar graphs, line graphs, histograms, etc. They can demonstrate that different representations can convey the same mathematical idea and discuss the advantages and limitations of each representation. Teachers can also emphasize the importance of accurately representing data to avoid misleading conclusions.

Content standards

8.PS.2 — The student will apply the data cycle (formulate questions; collect or acquire data; organize and represent data; and analyze data and communicate results) with a focus on boxplots.

8.PS.2a — Formulate questions that require the collection or acquisition of data with a focus on boxplots.

8.PS.2b — Determine the data needed to answer a formulated question and collect the data (or acquire existing data) using various methods (e.g., observations, measurement, surveys, experiments).

8.PS.2c — Determine how statistical bias might affect whether the data collected from the sample is representative of the larger population.

8.PS.2d — Organize and represent a numeric data set of no more than 20 items, using boxplots, with and without the use of technology.

8.PS.2e — Identify and describe the lower extreme (minimum), upper extreme (maximum), median, upper quartile, lower quartile, range, and interquartile range given a data set, represented by a boxplot.

8.PS.2g — Analyze data represented in a boxplot by making observations and drawing conclusions

8.PS.2h — Compare and analyze two data sets represented in boxplots.

8.PS.2i — Given a contextual situation, justify which graphical representation (e.g., pictographs, bar graphs, line graphs, line plots/dot plots, stem-and-leaf plots, circle graphs, histograms, and boxplots) best represents the data.

Prior connections

5.PS.2 — The student will solve contextual problems using measures of center and the range.

6.PS. 2 — The student will represent the mean as a balance point and determine the effect on statistical measures when a data point is added, removed, or changed.

7.PS.2 — The student will apply the data cycle (formulate questions; collect or acquire data; organize and represent data; and analyze data and communicate results) with a focus on histograms.

Future connections

A2.ST.1 — The student will apply the data cycle (formulate questions; collect or acquire data; organize and represent data; and analyze data and communicate results) with a focus on univariate quantitative data represented by a smooth curve, including a normal curve.

A.ST.1 — The student will apply the data cycle (formulate questions; collect or acquire data; organize and represent data; and analyze data and communicate results) with a focus on representing bivariate data in scatterplots and determining the curve of best fit using linear and quadratic functions.

Lesson Preparation

Suggested review

Depending on your students' level of prior knowledge, consider revisiting the following lessons:

Grade 6 — 9.02 Create and interpret circle graphs
Grade 7 — 5.05 Histograms
Grade 8 — 4.02 Boxplots

Tools

You may find these tools helpful:
- Boxplot template

Student lesson & teacher guide

Comparing data using boxplots

Students delve into the concept of using boxplots to measure and visually illustrate the spread of data, understanding the role of range, interquartile range, median, quartiles, and extreme values. They also learn about the usage of parallel boxplots for comparing two data sets and the importance of maintaining the same scale for accurate comparison.

Collect and display
English language learner support

The primary purpose of the strategy is to help students become familiar with and understand the context and meaning of specialised vocabulary or terminology.

Reserve a section of the classroom wall or a digital space to be a designated word wall. The terms to be placed here will come from the lessons on data comparisons, including terms like "median," "mode," "range," "boxplots," and any other relevant academic vocabulary.

During class discussions, presentations, or while students work, pay attention to the academic language used. Listen for key terms related to comparing data sets. Add these words to the word wall.

Encourage students to contribute to the word wall. They can add words or phrases they encounter in their readings, homework, or during discussions. Encourage them to also note any particularly useful sentence structures or idioms related to the discussion of data sets.

Each time a term is added to the wall, take a few minutes to discuss its meaning, its role in comparing data sets, and examples of its use. Encourage students to ask questions about the words on the wall.

Consistently refer back to the word wall during lessons. This will help students make connections between the academic language and the concepts they are learning.

Examples

Students: Page 229– 230

Example 1

These two boxplots show the data collected by the manufacturers on the lifespan of light bulbs, measured in thousands of hours.

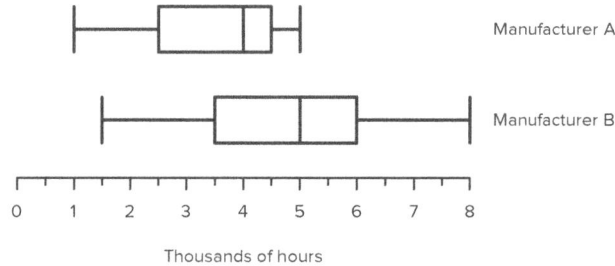

Manufacturer A

Manufacturer B

0 1 2 3 4 5 6 7 8

Thousands of hours

a   Complete this table using the two boxplots. Write each answer in terms of hours, remembering to multiply the values on the data display by 1000.

	Manufacturer A	Manufacturer B
Median		
Lower quartile		
Upper quartile		
Range		
Interquartile range		

Create a strategy

- To find the lower quartile, median, and upper quartile, find the corresponding values of the vertical lines of the box, respectively.
- For the range, use the formula: Range = Highest score – Lowest score
- For the interquartile range, use the formula: $IQR = Q_3 - Q_1$

Apply the idea

Since the scale is in thousands of hours, we can multiply the numbers on the scale by a thousand to find the number of hours.

	Manufacturer A	Manufacturer B
Median	$4 \cdot 1000 = 4000$	$5 \cdot 1000 = 5000$
Lower quartile	$2.5 \cdot 1000 = 2500$	$3.5 \cdot 1000 = 3500$
Upper quartile	$4.5 \cdot 1000 = 4500$	$6 \cdot 1000 = 6000$
Range	$(5 - 1) \cdot 1000 = 4000$	$(8 - 1.5) \cdot 1000 = 6500$
Interquartile range	$(4.5 - 2.5) \cdot 1000 = 2000$	$(6 - 3.5) \cdot 1000 = 2500$

Purpose

Show students how to interpret boxplots and fill out a table with the corresponding data.

b Which manufacturer produces light bulbs with the best lifespan?

Create a strategy

Choose the manufacturer with the greater median.

Apply the idea

The data set for Manufacturer B has a median of 5000 hours, while the median of the data set for Manufacturer A is 4000 hours, so Manufacturer B produces light bulb with the best lifespan.

Reflect and check

In fact, the best lightbulb produced by Manufacturer A has a lifespan of 5000 hours, which is the same as the median of Manufacturer B. This means that about half of the lightbulbs produced by Manufacturer B have a greater lifespan than all of the lightbulbs produced by Manufacturer A.

Purpose

Challenge students to interpret boxplot data to determine which manufacturer produces light bulbs with the best lifespan.

Reflecting with students

Encourage students to communicate their findings with precise language and proper use of units. For instance, when they mention the medians, prompt them to say "Manufacturer B has a median lifespan of 5000 *hours*," instead of just "5000." Emphasize the importance of units to give meaning to the numbers and avoid ambiguity. Guide them to use clear definitions and correct statistical terms, such as "median lifespan," "interquartile range," and "distribution."

Example 2

Sophie and Holly have been playing soccer for 20 years. These boxplots represent the total number of goals Sophie and Holly scored in each of their 20 seasons.

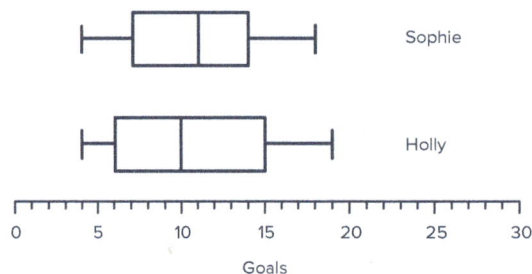

a Who had the highest scoring season?

Create a strategy

Compare the maximum value of both boxplots.

Apply the idea

By looking at the endpoints of the right whiskers, Sophie scored 18 goals, while Holly scored 19 goals. So, Holly had the highest scoring season.

Purpose

Demonstrate to students how to identify the highest season score from boxplots.

Expected mistakes

Students may struggle with which measurement to compare and may choose the median to compare instead fo the maximum. Discuss the meaning of each measure in context with students.

Students: Page 231

b How many more goals did Holly score in her best season compared to Sophie in her best season?

Create a strategy

Subtract Sophie's maximum from Holly's maximum.

Apply the idea

$$\begin{aligned} \text{Number of goals} &= 19 - 18 && \text{Subtract 18 from 19} \\ &= 1 && \text{Evaluate} \end{aligned}$$

Holly scored 1 more goal in her best season.

Purpose

Help students understand how to calculate the difference between maximum scores of two players.

Students: Page 231

c What is the difference between the median number of goals scored in a season by each player?

Create a strategy

Find the difference of the medians.

Apply the idea

Sophie's median is 11 and Holly's median is 10.

$$\begin{aligned} \text{Difference} &= 11 - 10 && \text{Subtract 10 from 11} \\ &= 1 && \text{Evaluate} \end{aligned}$$

Purpose

Teach students how to find the difference of the medians from boxplots.

Students: Page 231–232

d Which player was more consistent?

Create a strategy

We can look at the range and interquartile ranges and see whose is lower. The lower the spread, the more consistent.

We can use that: IQR = Upper quartile – Lower quartile

Apply the idea

$$\begin{aligned} \text{Sophie's IQR} &= 14 - 7 && \text{Substitute the quartiles} \\ &= 7 && \text{Evaluate} \end{aligned}$$

$$\begin{aligned} \text{Holly's IQR} &= 15 - 6 && \text{Substitute the quartiles} \\ &= 9 && \text{Evaluate} \end{aligned}$$

$$\begin{aligned} \text{Sophie's Range} &= 18 - 4 && \text{Substitute the upper and lower extremes} \\ &= 14 && \text{Evaluate} \end{aligned}$$

$$\begin{aligned} \text{Holly's Range} &= 19 - 4 && \text{Substitute the upper and lower extremes} \\ &= 15 && \text{Evaluate} \end{aligned}$$

Since Sophie has the lower IQR and range, her season goal totals are more consistent.

Purpose

Guide students in analyzing consistency in performance using the interquartile range and range from boxplots.

Reflecting with students

Encourage students to consider why the interquartile range (IQR) is a better measure of consistency than the median when analyzing Sophie and Holly's performances. Explain that the median shows the typical number of goals scored but does not indicate how much the scores vary around that median. The IQR, on the other hand, measures the spread of the middle 50% of the data, providing insight into the variability of their performances. Guide students to see that a smaller IQR means the data points are closer together, indicating greater consistency.

Students: Page 232–235

Example 3

The advertised fuel efficiency for 12 cars and 12 trucks was recorded in this table.

Cars	15	17	18	22	22	22	23	25	26	31	35	50
Trucks	12	13	13	14	15	15	15	16	16	17	19	27

The car data was represented with a boxplot.

Car fuel efficiency (Miles per gallon)

The truck data was represented with a dot plot.

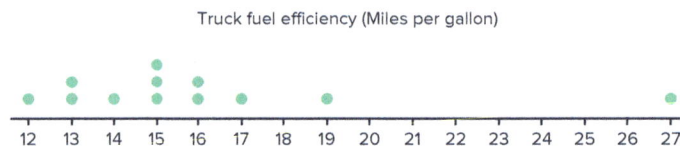

Truck fuel efficiency (Miles per gallon)

a Convert the dot plot to a boxplot and draw a parallel boxplot comparing cars to trucks.

Create a strategy

There are 12 data values on the dot plot, so we can start by identifying the upper and lower quartiles, extremes, and the median.

Apply the idea

It may be helpful to list the data from the dot plot to find the key values for the boxplot.

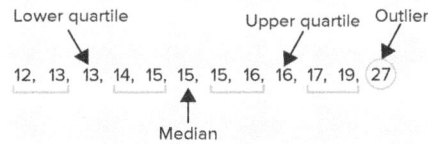

Lower quartile   Upper quartile   Outlier

12, 13, 13, 14, 15, 15, 15, 16, 16, 17, 19, (27)

Median

Lower extreme (min)	12
Lower quartile	13
Median	15
Upper quartile	16
Upper extreme (max)	19

The value of 27 mpg is very far from the rest of the data, so we can consider it to be an outlier (extreme value).

This means that the upper whisker would end at 19 and 27 would be shown as a dot.

Truck fuel efficiency (Miles per gallon)

When we create a parallel boxplot, we must use the same scale. This give us:

Fuel efficiency (miles per gallon)

Reflect and check

We can also use technology by

1 Inputting the data in two separate columns.

	A	B	C	D	E	F	G	H	I	
1	15	12								
2	17	13								
3	18	13								
4	22	14								
5	22	15								
6	22	15								
7	23	15								
8	25	16								
9	26	16								
10	31	17								
11	35	19								
12	50	27								
13										
14										

2 Highlighting the data and selecting "Multiple Variable Analysis" from the data drop down menu

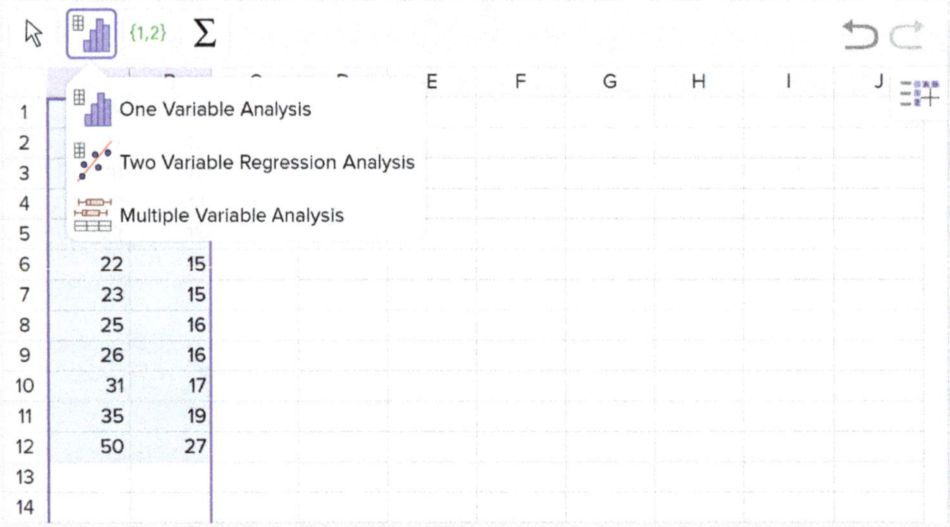

					E	F	G	H	I	J	
1	One Variable Analysis										
2											
3	Two Variable Regression Analysis										
4											
5	Multiple Variable Analysis										
6	22	15									
7	23	15									
8	25	16									
9	26	16									
10	31	17									
11	35	19									
12	50	27									
13											
14											

3 Adjusting the window size to get a good view.

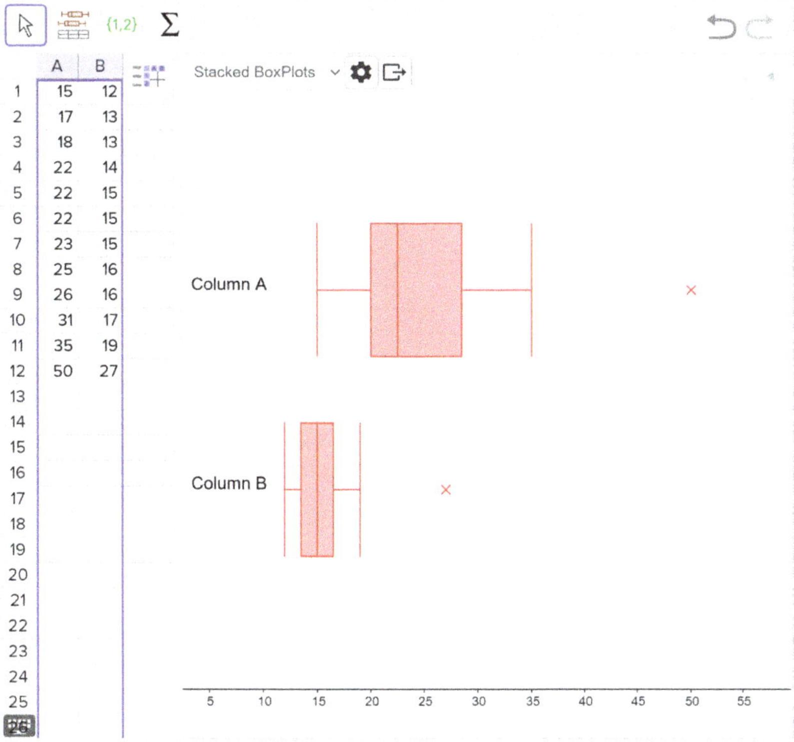

Purpose

Show students how to convert a dot plot to a boxplot and how to compare two boxplots with different scales.

Students: Page 235

b Select a measure of center from both data sets to compare the fuel efficiency of cars versus trucks.

Create a strategy

Since we know that outliers influence the mean of a data set, it's best to compare the median of each data set.

Apply the idea

The median miles per gallon of a car is 22.5 and the median miles per gallon of a truck is 15. That means that a car typically gets 7.5 miles more per gallon in fuel efficiency.

Reflect and check

Since both data sets have a higher-valued outlier, we can expect that the mean of each set is higher than the median and does not represent the typical value, as well as the median does.

Purpose

Show students how to select an appropriate measure of center to compare fuel efficiency between cars and trucks.

c Compare the spread of the data sets.

Create a strategy

Since the data sets have outliers, it is best to describe the spread of the data using the IQR.

Apply the idea

$$
\begin{aligned}
IQR \text{ for cars} &= Q_3 - Q_1 && \text{Formula for IQR} \\
&= 28.25 - 20 && \text{Substitute } Q_3 = 28.25 \text{ and } Q_1 = 10 \\
&= 8.25 && \text{Evaluate the subtraction}
\end{aligned}
$$

The IQR for the car fuel efficiency is approximately 8.25 miles per gallon, meaning the middle 50% of cars vary by 8.25 miles per gallon.

$$
\begin{aligned}
IQR \text{ for trucks} &= Q_3 - Q_1 && \text{Formula for IQR} \\
&= 16 - 13 && \text{Substitute } Q_3 = 16 \text{ and } Q_1 = 13 \\
&= 3 && \text{Evaluate the subtraction}
\end{aligned}
$$

The IQR for truck fuel efficiency is approximately 3 miles per gallon, meaning that the middle 50% of trucks vary by 3 miles per gallon.

While cars get better gas mileage overall, the spread is greater compared to trucks, so car mileage is less consistent.

Reflect and check

By comparing the range for cars and trucks, we could see that the spread for fuel efficiency for cars is signficantly larger, 35 for cars and 15 for trucks. But, once we examine the IQR, we can see that the difference in spread is not as drastic as it initially appears.

Purpose

Show students how to compare the spread of two data sets using the Interquartile Range (IQR) as a measure of spread, especially when data sets have outliers.

Including outliers within boxplot whiskers
use with Example 3
Address student misconceptions

Students may think that the whiskers of a boxplot should always extend to the minimum and maximum values in the dataset, including any outliers. They might not realize that outliers should be identified separately and plotted as individual points beyond the whiskers. This misconception can lead to boxplots that do not accurately represent the center and spread of the data.

To address this, consider introducing the 1.5IQR rule to identify outliers. Explain how to determine the thresholds for outliers by calculating 1.5 times the IQR above the upper quartile and below the lower quartile. Any values more than 1.5IQR above the upper quartile or below the lower quartile.

Emphasize that the whiskers should extend only to the highest and lowest values that are not outliers, and that correctly plotting outliers is important for accurately interpreting the data distribution.

Choosing the best data display

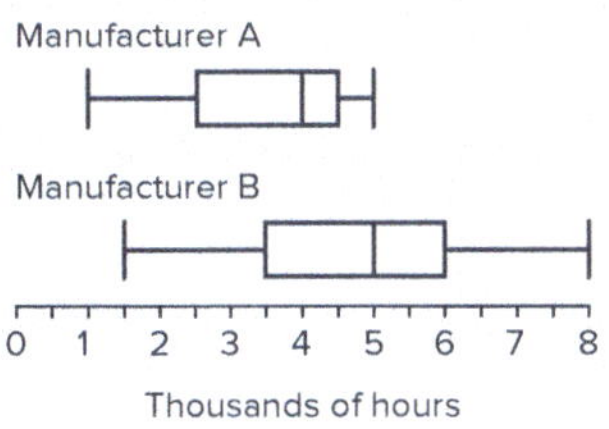

Students analyze different types of data displays such as histograms, boxplots, and dot plots, understanding their strengths and weaknesses in representing data. They explore measures of center and spread as tools to summarize data sets. In the exploration, students compare these different data displays using the same set of data.

🎓 Use a graphic organizer for possible data displays
Targeted instructional strategies

Have students organize the different types of data displays based on what types of information they best convey. Could organize into a graphic organizer for each display or do an overall summary for all displays with a flowchart based on data set size and/or data set range.

For example, for box plots:

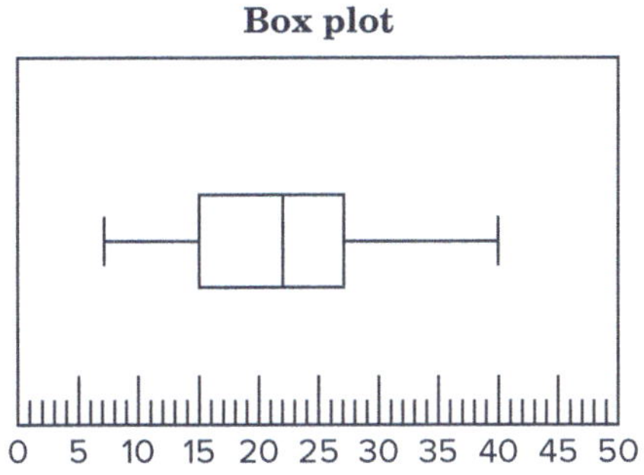

Box plot

0 5 10 15 20 25 30 35 40 45 50

Key features
- Used for numerical univariate data
- Shows the five-number summary and outliers
- Requires a scale
- The box shows the IQR (interquartile range) and the whiskers show the minimum and maximum (or the next value if they are an outlier)

Benefits:
- Easily identify the median and five-number summary including quartiles
- Can make very large data sets easier to interpret

Drawbacks:
- Loses the actual original data values
- Must be drawn precisely

Consider providing the following sentence frames as extra support for students to be able to use precise language in their conversations:

- "The best data display is a histogram because…"
- "We cannot use a box plot because…"
- "An advantage of a circle is graph is…"
- "A drawback of a line plot is…"

Review the different data displays

Student with disabilities support

Ensure that students have the prerequisite knowledge about the different types of data displays before beginning to analyze their appropriateness or interpreting them.

Displays for one variable data:

- Histogram

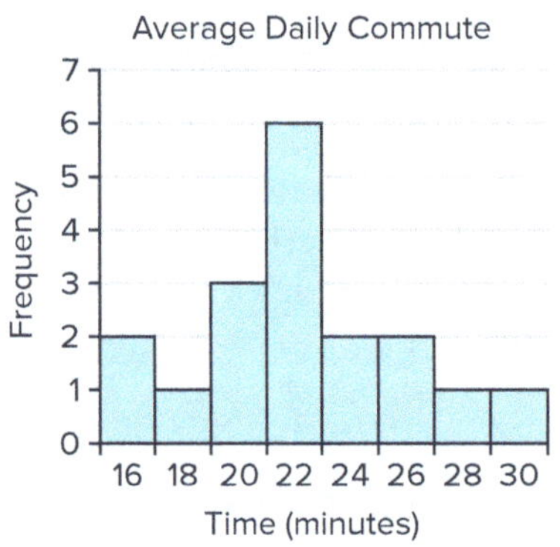

Average Daily Commute

- Stem-and-leaf plots

Stem	Leaf
3	0 1 3 7
4	2 9
5	1 8
6	0 7 9
7	8
8	
9	0

Key: 3 | 0 = 30

- Box plotsa

Double unders jumped

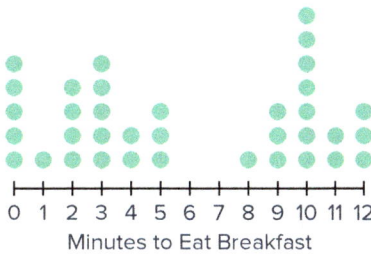

- Dot plots

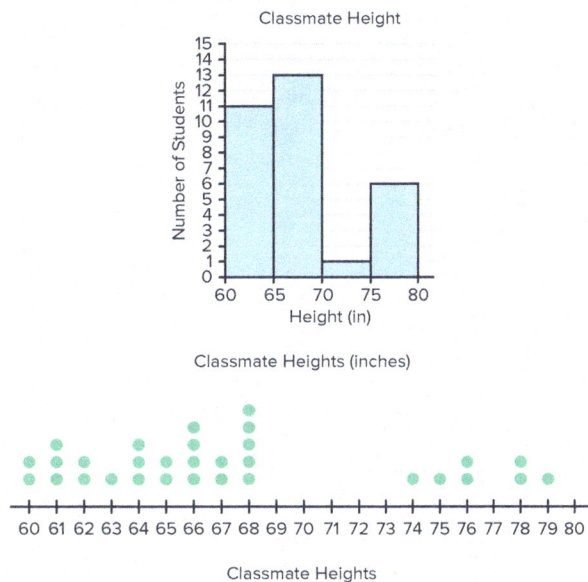

Minutes to Eat Breakfast

Exploration

Students: Page 238

Exploration

A high school has scolarship programs for gymnastics and basketball. The histogram, dot plot, and boxplot summarize the heights of 29 students in a class, in inches:

Classmate Height

Number of Students

Height (in)

Classmate Heights (inches)

60 61 62 63 64 65 66 67 68 69 70 71 72 73 74 75 76 77 78 79 80

Classmate Heights

Classmate Heights

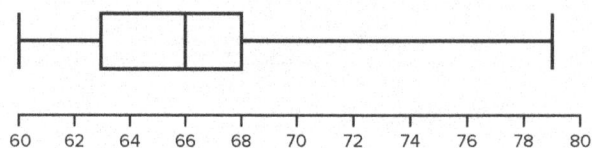

1. What do you notice about the different displays for the same data?
2. What can you see from the histogram and dot plot that you can't see from the boxplot?
3. What can you see from the boxplot and dot plot that you can't see from the histogram?

Suggested student grouping: Small groups

In this exploration, students are comparing and analyzing data on the heights of students in a class represented in three different ways: a histogram, a dot plot, and a boxplot. They will observe similarities and differences across these three representations and consider what information each type of display can provide.

Ideal student responses

These ideal responses may differ from other correct student responses. Less formal responses can be connected with the more precise mathematical language presented here.

1. **What do you notice about the different displays for the same data?**
   All three displays show the same data but in different ways. The histogram and dot plot show the number of students that fall into each height range, while the boxplot shows the median, quartiles, and outliers.

2. **What can you see from the histogram and dot plot that you can't see from the boxplot?**
   From the histogram and the dot plot, you can see the exact number of students at each height, which is not visible in the boxplot.

3. **What can you see from the boxplot and dot plot that you can't see from the histogram?**
   From the boxplot and dot plot, you can see the spread and outliers of the data, which is not as visible in the histogram.

Purposeful questions

- How do the three representations complement each other in understanding the data?
- What are some advantages and disadvantages of each type of display?

Possible misunderstandings

- Students might think that the height ranges with higher bars in the histogram correspond to individual data points rather than groups of data points. This could lead to confusion about the actual number of students within each range.

Example 4

Determine the best type of data display(s) for each statistical question:

a   How much variation is there in the number of zucchinis produced by a single plant?

Create a strategy

We should consider the size and possible range of values of the data set, as well as what key features the question is focusing on.

Apply the idea

A boxplot
Since we are focusing on variation, we want to quickly identify the spread of the data, this means a box plot is appropriate.

Reflect and check

If we used a histogram, then we would be able to estimate the range, but the interquartile range would not be easy to see and sometimes the range is very high due to extreme values.

Purpose

Challenge students to consider the most appropriate data display for a statistical question focusing on variation.

b   What types of vegetables are most popular to grow among urban gardeners?

Apply the idea

This is categorical data, so the options are pictograph, bar graph, line plot, or circle graph.

Since "most" is related to the mode or category with the largest frequency, we want a display where that is easy to see.

Depending on what the data set looks like, a bar graph or circle graph could be appropriate.

Reflect and check

If there were a very large number of categories, then the bar chart would be easier to read.

Purpose

Show students how to determine the most effective data display for a statistical question focusing on the mode of categorical data.

Advanced learners: Justify data display choices
Targeted instructional strategies

use with Example 4

Encourage students to not only identify the most appropriate data display but also to provide thorough justifications for their choices. Ask them to critically analyze different types of graphs and explain why one is more effective than others for a given data set.

For example, prompt students to compare how a box plot, histogram, and dot plot each display variation in zucchini production, and have them defend why a box plot best highlights the data's spread. Similarly, have students evaluate the merits of bar graphs versus circle graphs in representing the popularity of vegetables among urban gardeners.

By articulating their reasoning, students deepen their understanding of data representation and enhance their ability to make informed decisions about statistical displays.

Example 5

Shown are the quiz score percentages from Mr. Sanchez's first period math class:

$$\{20, 25, 26, 30, 30, 40, 43, 63, 65, 67, 70, 70, 75, 90, 93\}$$

a   Construct a boxplot of the quiz scores.

Create a strategy

Recall how to find the five-number summary using the data provided or use technology, as shown in the example:

1 Enter the data in a single column.

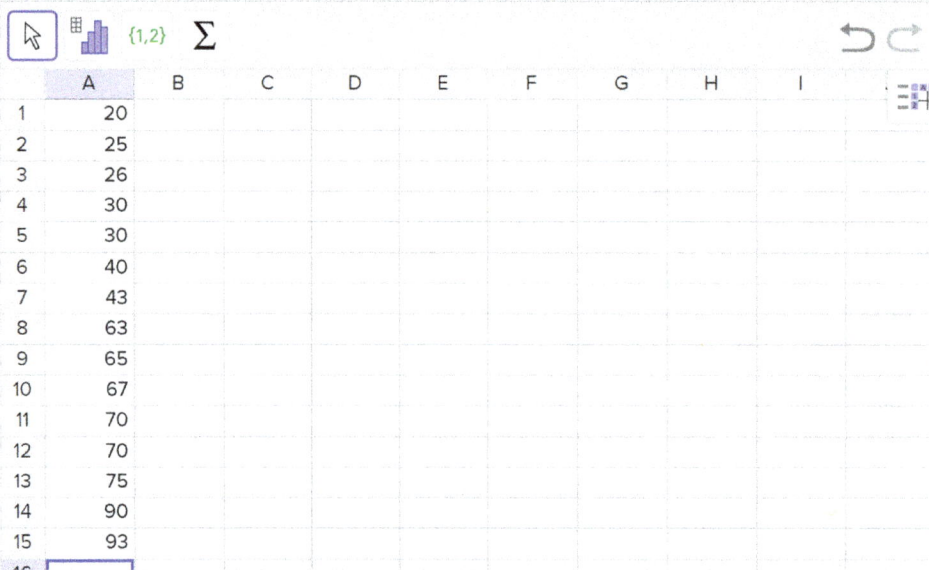

2 Select all of the cells containing data and choose "One Variable Analysis".

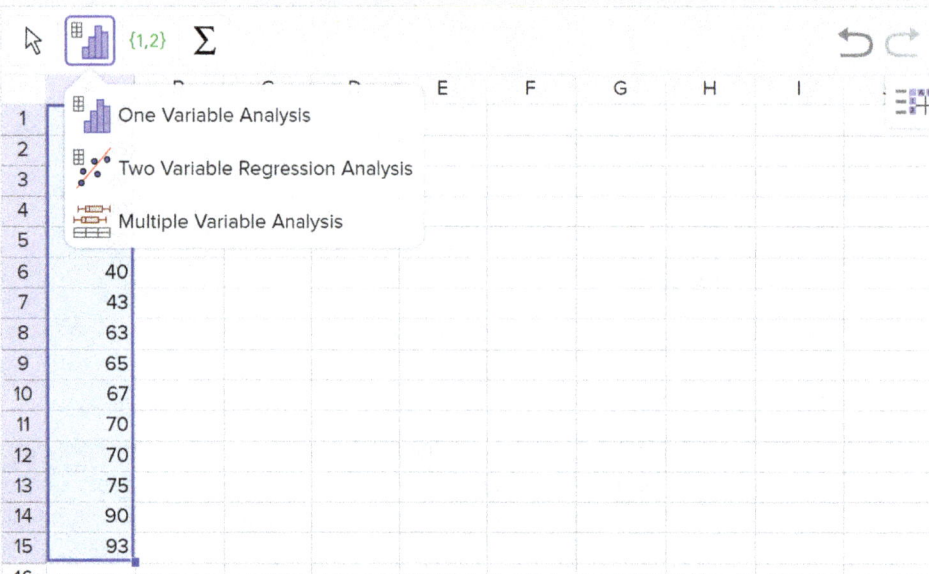

3 Select "Show Statistics" to reveal a list of statistical values, including the five-number summary.

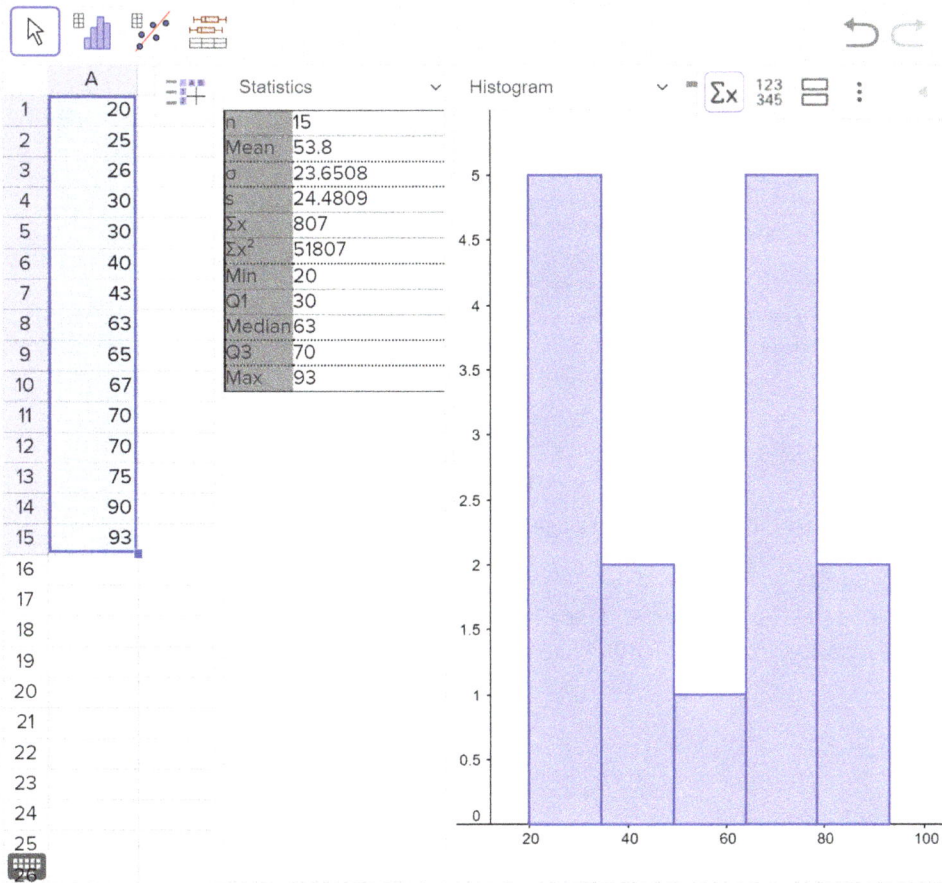

	A
1	20
2	25
3	26
4	30
5	30
6	40
7	43
8	63
9	65
10	67
11	70
12	70
13	75
14	90
15	93
16	
17	
18	
19	
20	
21	
22	
23	
24	
25	

Statistics

n	15
Mean	53.8
σ	23.6508
s	24.4809
Σx	807
Σx^2	51807
Min	20
Q1	30
Median	63
Q3	70
Max	93

Histogram

Use the minimum, Q_1 (first quartile), median, Q_3 (third quartile), and maximum from the statistics listed to create the boxplot.

Apply the idea

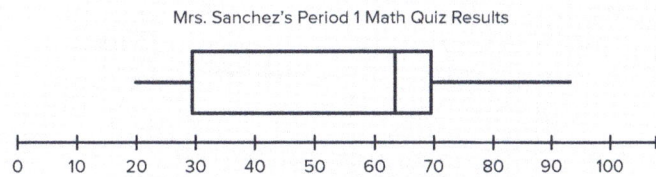

Mrs. Sanchez's Period 1 Math Quiz Results

Purpose

Show students how to construct a boxplot from a set of data and interpret the results.

> **b** What are the advantages and disadvantages of a boxplot?
>
> **Apply the idea**
>
> Boxplots visually summarize large sets of data, although they can be used for small sets too, like this one. In a boxplot, it is easy to see and estimate the shape, center (median), and spread of data. However, if we were not given the individual data points, we would not know how many students are in the set of data and their individual quiz scores.
>
> Even without that information, the boxplot can still provide important information about the quiz. We can see at a glance that most of the students did not do very well on the quiz. We can see that 75% of the students scored 70 or below, 50% of the students scored below about 63 and 25% got a score of less than 30.
>
> **Reflect and check**
>
> 25% of the quiz scores lie between the minimum and first quartile, the first quartile and the median, the median and the third quartile, and the third quartile and the maximum value. This is true even when the quarters of the boxplot are uneven in length.

Purpose

Challenge students to think critically about the usefulness and limitations of boxplots.

Reflecting with students

Have students consider the advantages and disadvantages of various data displays. In some situations there may not be one best data display, but in other situations a specific data display may better show the data of interest.

Students: Page 242

> **c** Explain whether a dot plot or a histogram could be a better display for the data.
>
> **Create a strategy**
>
> Use the size of the data set and its range to determine which would be better.
>
> **Apply the idea**
>
> While the data set is made up of only 15 students' scores, the scores are spread out from 25% to 93%, which would mean the dot plot is very long for the data points, so this display would not be suitable.
>
> A histogram could be graphed by organizing the data into 10% intervals. One advantage of the histogram is that we would be able to see the gaps in the 50s and 80s which are lost in the boxplot.
>
> **Reflect and check**
>
> A stem-and-leaf plot would be another good choice for this data set.

Purpose

Challenge students to compare and contrast different ways to display data.

Example 6

Match the boxplot shown to the correct histogram.

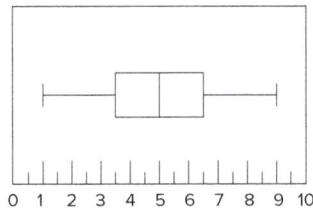

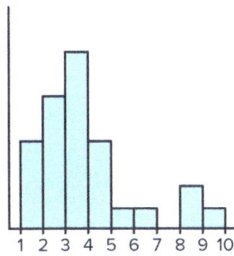

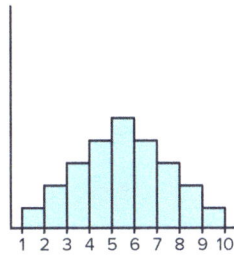

A   B   C   D

Create a strategy

Look for characteristics of the data set like the value it is centered around, the range of the data, and the shape - skewed or symmetrical.

Apply the idea

The given boxplot shows symmetrical data. Only option B shows symmetrical data. The correct answer is option B.

Reflect and check

If there were other histograms that were symmetrical, then we could match that the range of values should be 1 to 9, and the data set is centered around 5.

Purpose

Show students how to match a boxplot to a histogram by identifying key characteristics in distribution and range of data.

Students: Page 243

> 💡 **Idea summary**
>
> We should expect that the shape of data would be the same whether it is represented in a boxplot or histogram.
>
> The best display for a data set is one that reveals the information we want to share. Some displays hide key information like the individual data points, the total number of data points, or features like the shape, clusters, gaps, and spread.
>
> As a starting place, consider:
>
> - If it is categorical, look at bar graphs, dot plots, or circle graphs. However, for a younger or diverse audience, a pictograph might be appropriate.
> - For numerical data, if there is a small quantity and range of data, try a dot plot.
> - If the data has a large range or quantity of data, try a histogram or boxplot.
> - Choose a boxplot if you only need to see an overview of center, spread and shape.
> - Choose a histogram if, in addition to center, spread, and shape, you want to know the size of the data set and view any gaps or clusters among various intervals.
> - Choose a stem-and-leaf plot if you need to be able to see the actual data values.
> - A circle graph could be used for grouped numerical data if highlighting the proportions in each interval is important.

Practice

Students: Page 243–253

What do you remember?

1 Match each name to the correct data display.

   i Boxplot   **ii** Histogram   **iii** Circle graph   **iv** Line plot (dot plot)

a

Cookies sold

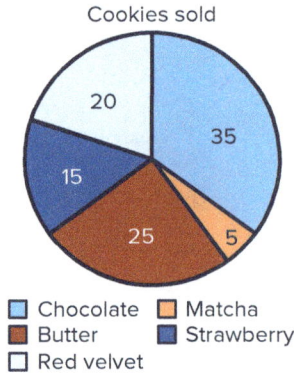

- Chocolate
- Matcha
- Butter
- Strawberry
- Red velvet

b

Average years of schooling completed

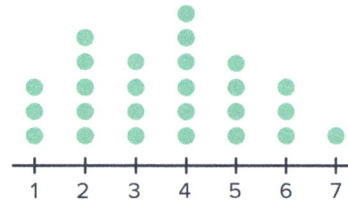

c

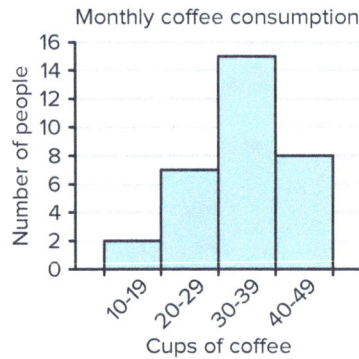

d

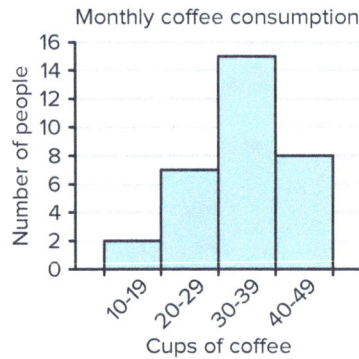

2 Is each statement true or false?

  a Dot plots and circle graphs could be used to display the same set of data.

  b A histogram shows individual data values.

  c A boxplot can't clearly show outliers like a stem-and-leaf plot can.

  d A boxplot and histogram of the same data set should have the same shape.

3 Which key features can be read off each display?

  a Histogram   **b** Boxplot   **c** Dot plot

4 The given histogram shows the time that people spent on a phone call.

  a Describe the shape of the data distribution.

  b Estimate the median of the data set.

  c Estimate the range of the data set.

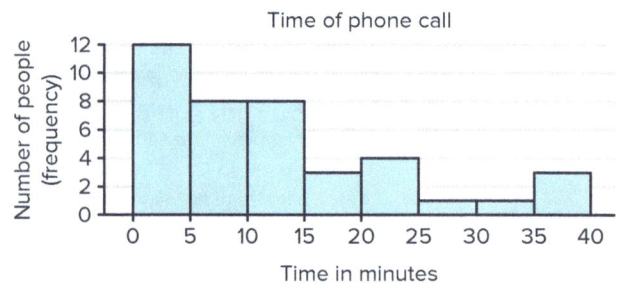

Time of phone call

5 20 people were asked how many hours of sleep they had gotten the previous night. The numbers shown are each person's response:

$$1, 6, 9, 8, 7, 9, 7, 10, 2, 3, 8, 7, 7, 3, 7, 3, 3, 7, 10, 9$$

Select the appropriate boxplot to represent the given set of data:

A

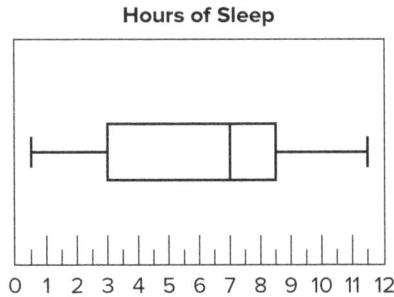

B

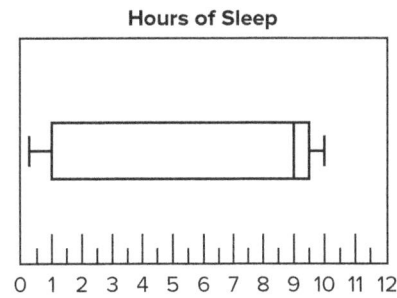

C

D

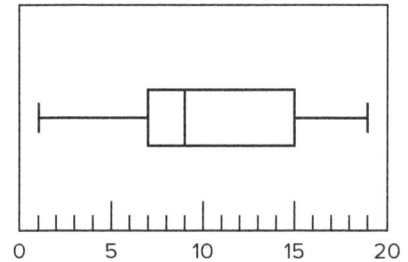

6 For the boxplot shown:

   a Approximately what percentage of scores lie between each of these values?

   i Upper and lower quartiles

   ii Lower extreme and lower quartile

   iii Median and upper quartile

   iv Lower quartile and upper extreme

   v Lower extreme and upper quartile

   b In which quartile(s) is the data the least spread out?

   c In which quartile(s) is the data the most spread out?

7 Match each of the following scenarios to a type of data display that would be appropriate:

   i Boxplot   **ii** Histogram   **iii** Circle graph   **iv** Dot plot

   a A market researcher asked 10 000 people what their preferred brand of soap is.

   b A baby headphone manufacturer wants to answer the question "What size of headphones would fit more than 75% of babies under the age of 1 year?"

   c An 8th grade student asks their class of 30 students what their favorite kind of juice they prefer?

   d An orthotics company is doing a study on foot size and measures the foot size of 200 people.

8 The histogram and the boxplot display the same data about ages of people on an airplane. Use either graph to answer each question.

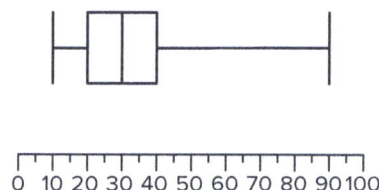

a What decade has the most people?

b What is the median age of people on the airplane?

c What is the range of ages on the airplane?

d How many people are in their 60s?

e What types of questions were more easily answered using the:

    i Histogram?   **ii** Boxplot?

Let's practice

9 A personal trainer has all her clients wear step counters. She asks the question "How varied are the steps that my clients take every day? Does it change week by week?"
Select the appropriate graph to represent the given set of data:

A Circle graph   **B** Boxplot   **C** Dot plot   **D** Histogram

E Stem-and-leaf plot

10 Over the whole year, the maximum temperature is recorded each day. The temperatures range from $45°F$ to $101°F$.

a Select all appropriate displays to represent the given set of data:

    A Circle graph   **B** Boxplot   **C** Dot plot   **D** Histogram

    E Stem-and-leaf plot

b What display would you use to represent this data? Explain your choice.

11 A group of people were surveyed on the number of websites they visit on a daily basis, and the results are displayed in the given line plot:

Number of Websites Visited Daily

a Construct a boxplot for this data.

b Which display would you use to find the median?

c Which display would you use to find the mode?

d Which display would you use to find the range?

e Which display would you use to find the interquartile range?

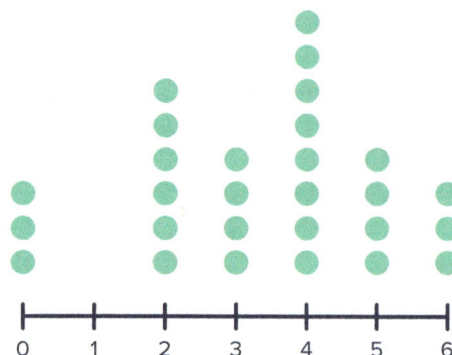

12 Shelby formulated the statistical question: "How much does a household or family typically spend on groceries per week?"

    a What display would you use to represent a data set that has 1000 data points that range from $126 to $213? Explain your choice.

    b What display would you use to represent a data set with only 20 data values in a range of $500. Explain your choice.

13 The given data set was collected to answer the statistical question "How long have the restaurants in my town been open? How long is typical?"

$$1, 1, 1, 1, 1, 2, 2, 2, 2, 2, 2, 3, 3, 3, 3, 3, 4, 4, 4, 4, 5, 5, 6, 10, 15, 16, 19, 22, 31, 114$$

What display would you use to represent this data? Explain your choice.

14 The parallel boxplots show the distances, in centimeters, jumped by two high jumpers.

    a Who had a higher median jump?

    b Who had the larger range of jumps?

    c Who made the highest jump?

    d Who made the smaller interquartile range?

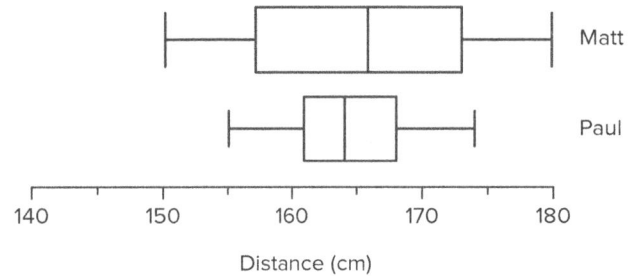

15 The following box plots shows the number of points scored by two basketball teams in each of their matches:

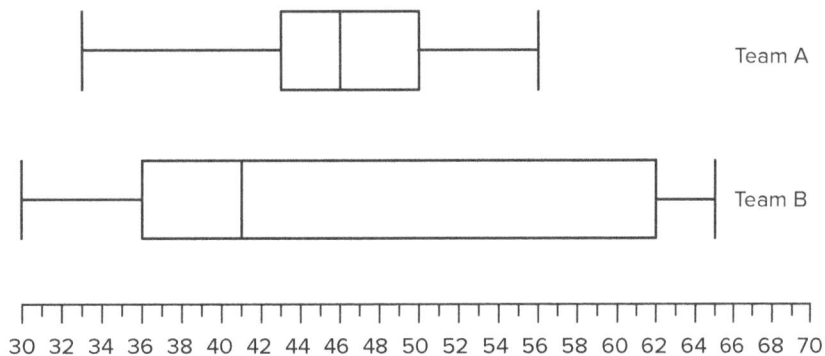

    a Compare the median score of Team A and Team B.

    b Compare the range of scores for Team A and Team B.

    c Compare the interquartile range for Team A and Team B.

    d If the two teams were to play against each other in the next game, determine which team you expect to win. Justify your response.

16 A class took an English test and a Mathematics test. Both tests had a maximum possible score of 20. The parallel box plots show the results of the tests:

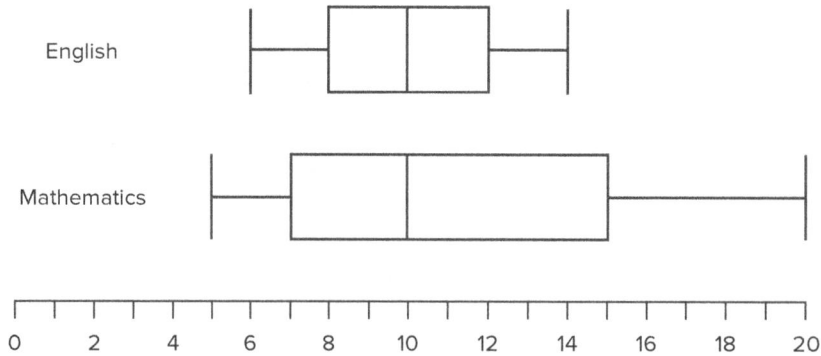

a Complete the table using the two boxplots.

	English	Mathematics
Median		
Lower quartile		
Upper quartile		
Range		
Interquartile range		

b In which test did the class tend to score better?

c Which class had symmetrical results?

d Which class had a smaller range, so was more consistent?

17 Two groups of people, athletes and non-athletes, had their resting heart rate measured. The results are displayed in the given pair of boxplots.

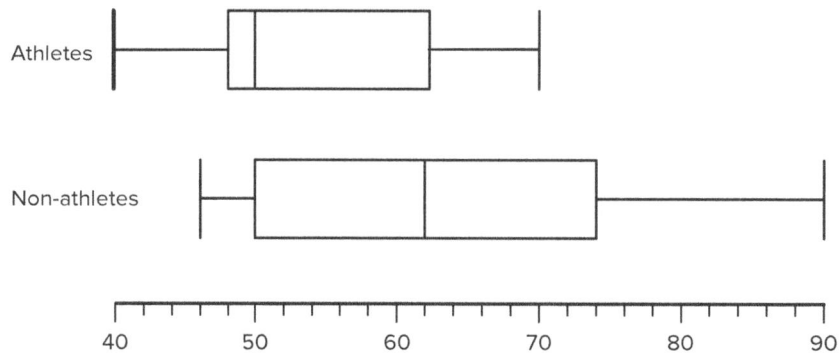

a What is the median heart rate of athletes?

b What is the median heart rate of the non-athletes?

c Using the median measure, which group has the lower heart rates?

d What is the interquartile range of the athletes' heart rates?

e What is the interquartile range of the non-athletes' heart rates?

f Using the interquartile range, which group has more consistent heart rate measures?

18   A mathematics test is given to two classes. The score out of 20 received by students in each class are represented in the boxplots.

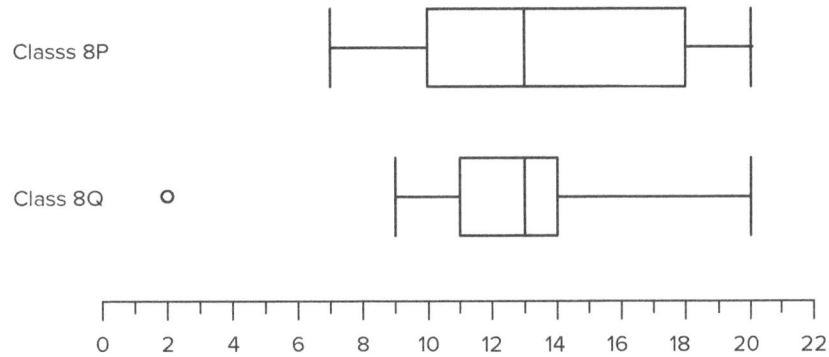

Classs 8P

Class 8Q   o

| | | | | | | | | | | | |
0   2   4   6   8   10 12 14 16 18 20 22

a   Which class tended to score better?
b   Which class(es) had an outlier?
c   Which class(es) had symmetrical shape?
d   Would the range or interquartile range be better for describing the spread of Class 8Q?
e   Which class had more consistent results?

19   The parallel boxplots show the prices, in dollars, of the items on the menu of an upmarket restaurant and the menu of a fast food restaurant.

Upmarket Restaurant

Fast Food

| | | | | | |
0   10 20 30 40 50 60

Price

a   Which restaurant has the higher median price for the items they sell?
b   What is the difference between the median prices of the items sold by each restaurant?
c   Which restaurant has a greater price range for the items on their menu?
d   What is the price difference between the most expensive items sold by each restaurant?
e   What amount of the cheapest item at the fast food restaurant could be bought for the same price as the most expensive item at the upmarket restaurant?

20   Two weightlifters both record their number of repetitions with a 70 kg bar over 30 days. The results are displayed in the boxplots:

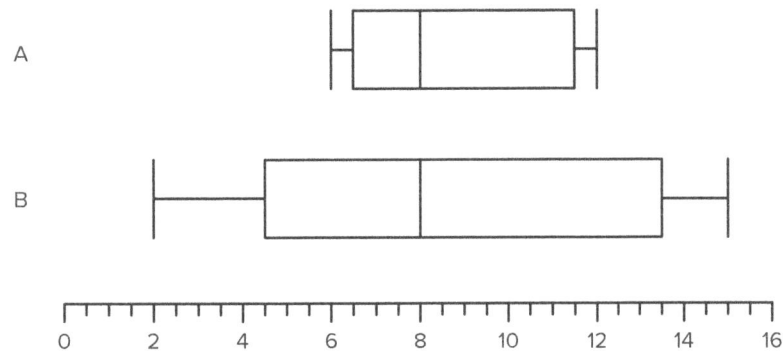

A

B

| | | | | | | | |
0   2   4   6   8   10 12 14 16

a   Which weightlifter has the more consistent results?
b   Which statistics determine which is more consistent?
c   Which statistic is the same for each weightlifter?
d   Which weightlifter can do the most repetitions of 70 kg?

21 Two bookstores recorded the selling price of all their books. The results are presented in the parallel boxplots.

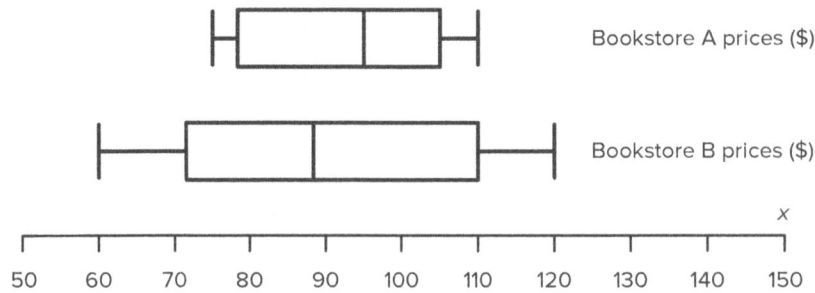

Bookstore A prices ($)

Bookstore B prices ($)

a Which bookstore had the more consistent prices? Explain your answer.

b Comparing the most expensive books in each store, how much more expensive is the one in store B?

c Is each statement true or false?

    i 25% of the books in Bookstore B are at least as expensive than the most expensive book in Bookstore A.

    ii 25% of the books in Bookstore B are cheaper than the cheapest book in Bookstore A.

22 Cooper and Marion are racing go-karts. The times (in seconds) for the 12 laps of their qualifying race are shown:
- Cooper: 58.9, 46.5, 52.6, 66.6, 58.4, 53.1, 45.0, 52.1, 52.4, 52.7, 44.8, 51.7
- Marion: 47.8, 54.6, 68.5, 68.0, 62.8, 57.2, 54.8, 63.4, 58.1, 64.3, 66.2, 47.1

a Complete this table with or without technology.

	Cooper	Marion
Lower extreme		
Lower quartile		
Median		
Upper quartile		
Upper extreme		

b Are there any outliers?

c Create a parallel boxplot of the two sets of times with the outlier displayed separately.

d Which racer will be in pole position for the final race, if it is given to the racer with the fastest qualifying lap time?

e Does spinning out on a lap, causing a high outlier, impact the selection for pole position? Explain your answer.

23 Ten participants had their pulse measured before, during, and 5 minutes after exercise with results shown in different boxplots.

Compare the pulse rates before, during, and after exercise for these participants.

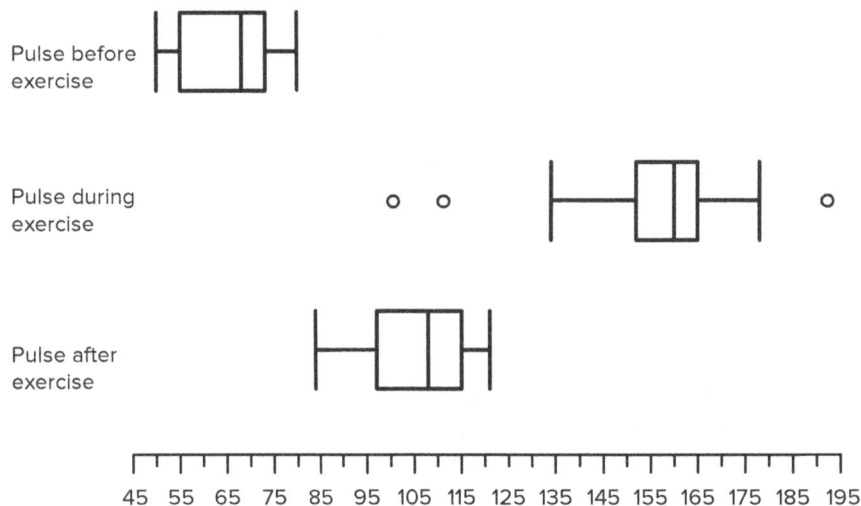

Pulse before exercise

Pulse during exercise

Pulse after exercise

Let's extend our thinking

24 Match the histograms to the corresponding boxplots.

Histogram A

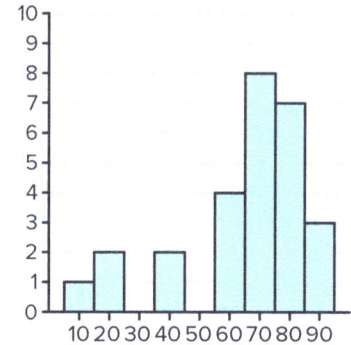

Histogram B

Histogram C

Boxplot 1

Boxplot 2

Boxplot 3

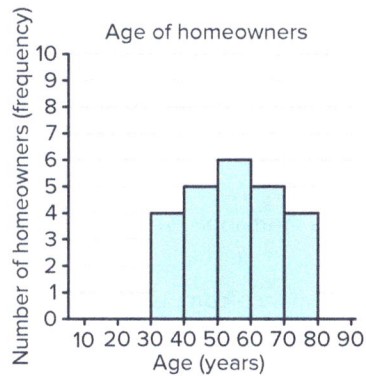

25 This histogram and boxplot represent the same data set that was collected to answer the question "How old are home owners?".

    a Describe the characteristics that are easier to identify from each display.

    b Draw at least one conclusion about the age of homeowners.

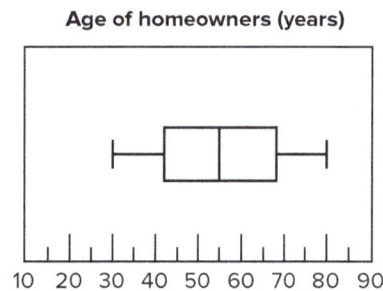

26 Identify and correct Gertie's error when she says "Class A donated higher amounts because the median donation is signficantly higher than Class B. The highest donation is also higher in Class A than for Class B."

Class A donations

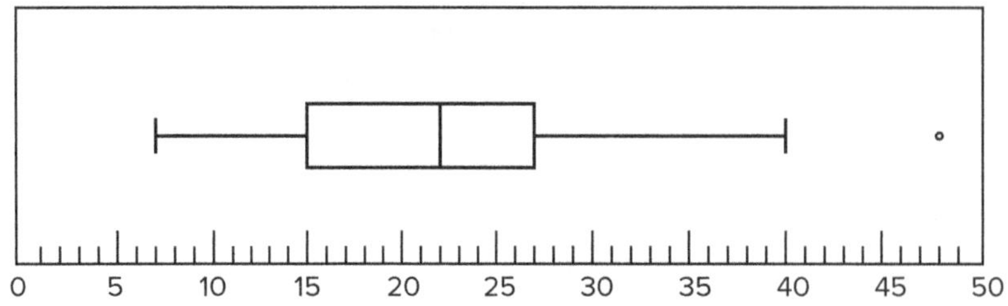

Class B donations

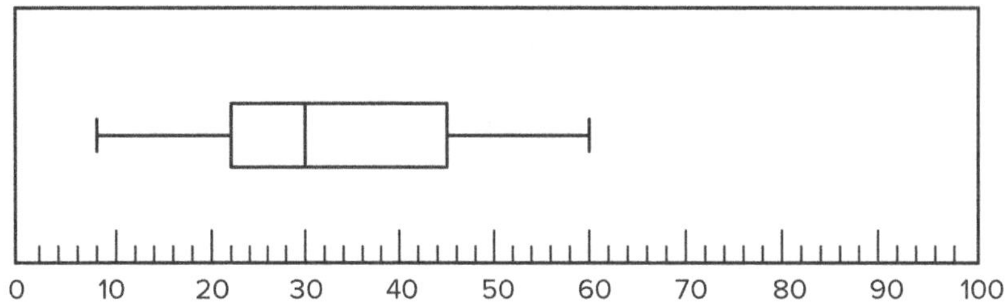

27 A dance studio is making an ad to put up on social media to boost enrollment. They want to include some graphics and data that shows they are the best dance studio.

One data set is the number of prizes they won at all competitions in the last 10 years and the other is the number of prizes they won at just the competitions that were hosted in their hometown in the last 10 years.

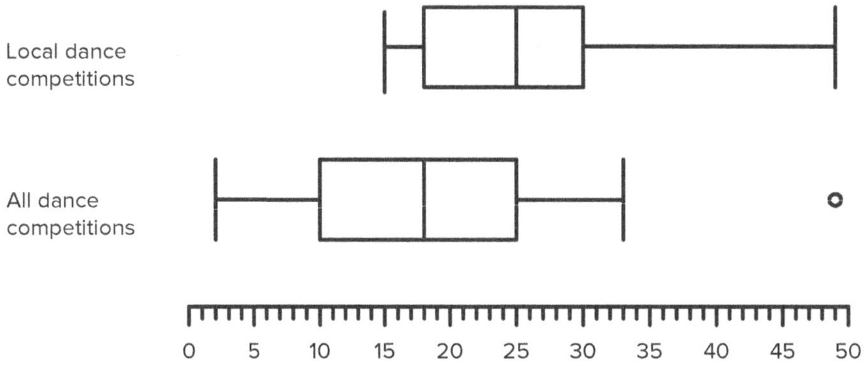

Local dance competitions

All dance competitions

a What type of bias is it if they only use the data that makes them look good?

b Compare and contrast the two data sets.

c Draw a conclusion that they could publish about the local dance competition results.

d Draw a conclusion that they could publish about the all of their dance competition results.

28 Larissa formulated the question "How many sinks are there in a typical residence?".

These two boxplots were taken from samples of different sizes from the same populations. Each person was asked "How many sinks are there where you live?"

Number of sinks in residence

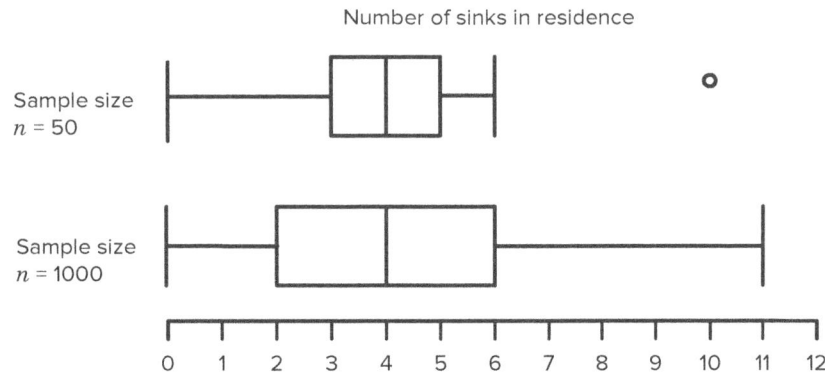

a Compare the two data sets.

b Which data set would lead to more valid conclusions?

c Xi said that "The two samples have a similar spread of values". Is that a valid statement? Explain.

d Lydia said that "All of the people in the smaller sample have fewer sinks than the top quarter of the people in the larger sample". Is that a valid statement? Explain.

29 There has been more flooding in Rosalind's neighborhood recently. She is curious about annual rainfall for her neighborhood compared to where she was born.

a Formulate a question about annual rainfall which could be investigated using a two or more boxplots.

b Describe a method which could be used to collect data.

c Collect data for your formulated question.

d Organize your data using an appropriate display.

e Draw at least one conclusion from the display.

f Create a different set of display from the same data set you collected. Draw an additional conclusion about rainfall from these displays.

Answers

4.04 Compare data sets

1 **a** iii: Circle graph   **b** i: Boxplot
   c ii: Histogram   **d** iv: Line plot (dot plot)

2 **a** True   **b** False   **c** False   **d** True

3 **a** • Frequencies for each group
      • Mode
      • Shape
      • Outliers
   b • Quartiles
      • Median
      • Shape
      • Outliers
      • Range and interquartile range
   c • Frequencies for each group
      • Mode
      • Shape
      • Outliers

4 **a** We can see that $0 \leq$ minutes < 5 has the highest number of people (frequency). Although the frequency fluctuates, the overall trend is a decrease as the number of minutes increases.
   b There are $12 + 8 + 8 + 3 + 4 + 1 + 1 + 3 = 40$ data values, so the median will be between the 20th and 21st values. The 20th value is in the $5 \leq$ minutes < 10 bin and the 21st value is in the $10 \leq$ minutes < 15 bin, so we can estimate that the median is 10 minutes.
   c As the histogram doesn't display the exact individual values, we can estimate the range by subtracting the lowest number of minutes from the highest number. So, the range is approximately 40.

5 **B**

6 **a** **i** 50%   **ii** 25%   **iii** 50%
     iv 75%   **v** 75%
   b Second
   c First and third

7 **a** iii: Circle graph   **b** i: Boxplot
   c iv: Dot plot   **d** ii: Histogram

8 **a** 30s
   b 30
   c 10 to 90, so range is 80 years.
   d None
   e **i** Questions about frequency or amounts
     ii Questions about spread or the median

9 **B**

10 **a** **B, D, E**
   b Answers vary depending on the display chosen.
   A histogram is an effective display for showing the distribution of daily temperatures. It shows the range, center, and shape of the data. By displaying the frequency of each interval of temperatures, a histogram provides a clear picture of the most common temperature ranges and helps in understanding patterns or trends.

11 **a**

Number of Websites Visited Daily

   b Boxplot   **c** Dot plot
   d Boxplot   **e** Boxplot

12 **a** Due to the large amount of data, a histogram would be a good choice. We can use the bins to divide the data and show the shape. Also, it is a visual representation of the frequency across various spending ranges, helping in the identification of patterns, trends, and center within the dataset.
   b For a small dataset of 20 with a range of $500, a stem-and-leaf plot would be a good choice. It would show how the data is spread between the hundreds and keep the original data.

13 Boxplot, because there is an outlier the best measure of center would be the median and that is easiest to identify from boxplots. The scale of the histogram would be too squished due to the outlier.

14 **a** Matt   **b** Matt   **c** Matt   **d** Paul

15 **a** The median score of Team A is 5 points higher than the median of Team B.
   b The range of Team B is 12 points wider than Team A.
   c The interquartile range of Team B is 19 points wider than the interquartile range of Team A.
   d Answers can support either team, but should use summary statistics to justify the response. For example:
     • We can expect Team A to win since the median number of points scored is 5 points higher than the median number of points scored by Team B, which tells us that they typically score higher, so would be more likely to win.
     • We can expect Team B to win since in about 25% of games they scored more points than the highest number of points scored by Team A.

16 a

	English	Mathematics
Median	10	10
Lower quartile	8	7
Upper quartile	12	15
Range	8	15
Interquartile range	4	8

b Mathematics

c English

d English

17 a 50 beats per minute   **b** 62 beats per minute

   c Athletes   **d** 14

   e 24   **f** Athletes

18 a Their medians are the same, so they tended to have similar scores.

   b Class 8Q

   c Neither

   d Interquartile range

   e Class 8Q, other than the outlier

19 a Upmarket restaurant   **b** $22

   c Upmarket restaurant   **d** $36

   e 26 items

20 a Weightlifter A

   b The range and interquartile range

   c The median

   d Weightlifter B

21 a Bookstore A, it has a lower range and interquartile range.

   b $10

   c **i** True   **ii** True

22 a

	Cooper	Marion
Lower extreme	44.8	47.1
Lower quartile	49.1	54.7
Median	52.5	60.45
Upper quartile	55.75	65.25
Upper extreme	66.6	68.5

b Yes, Cooper has an outlier of 66.6.

c

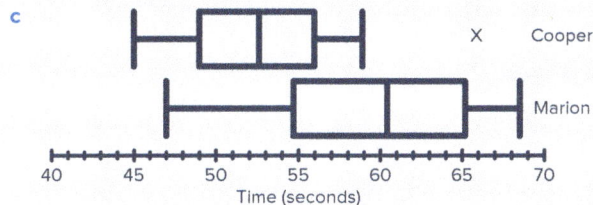

d Cooper

e No, a high outlier does not change what the minimum score is, hence would not alter the selection for pole position.

23 The typical pulse rates before exercise were about 68 bpm which is much lower than during (160 bpm) or after (108 bpm), with the highest heart rates happening during exercise.

Heart rates had similar ranges for before and after exercise, of 30 before and 37 after. While the range for during exercise was much larger at 92 bpm, the IQR is actually the lowest at 13, compared to 18 for the other two groups, so there are extreme values (outliers), but a consistent middle half for during exercise.

Let's extend our thinking

24

- Histogram A/ Boxplot 3
- Histogram B/ Boxplot 1
- Histogram C/ Boxplot 2

25 a It is easier to identify the shape and the mode of the dataset in a histogram. On the other hand, the range and median are easier to identify in a box plot.

   b One conclusion that we can draw from the given displays would be "The typical age of homeowners is 55 years, but they range from 30 to 80".

26 It looks like Gerlie assumed that the data scales of the two boxplots were the same because they were aligned, but the scales are totally different. The correct statement is "Class B donated higher amounts because the median donation ($30) is significantly higher than Class A ($22). The highest donation is also higher in Class B ($60) than for Class B ($48)".

Putting the two boxplots on the same scale, allows for easier comparison.

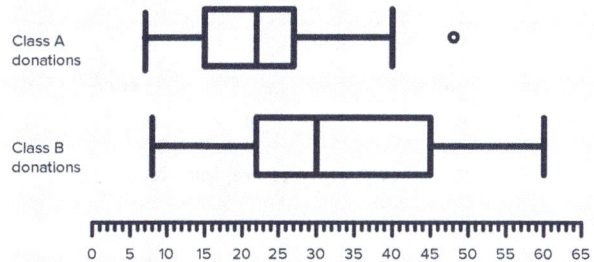

27 a Exclusion bias (cherry picking)

   b While both data sets share the highest value of 49, which is the maximum for the local dance competitions and an outlier for all dance competitions, the range of prizes won is higher in local dance competitions compared to all dance competitions. Additionally, the median number of prizes is higher in local dance competitions than in all dance competitions.

   c The dance studio typically wins 25 prizes at local dance competitions. One year they won 49 prizes.

   d The dance studio has received a prize at every competition it has attended. They have won up to 49 prizes, but usual win between 10 and 25 prizes.

28 **a** The data set with a larger sample has a wider range compared to the data set with a smaller sample size. However, they both have the same median value.

b The data set with a larger sample size of 1000 would likely lead to more valid conclusions, as its statistical values would tend to be closer to the population statistics. Unless there is a sample or selection bias.

c No, while the ranges are similar, 10 for the smaller data set versus 11 for the larger data set, the interquartile ranges tell a different story. An IQR of 4 for the larger data set versus 2 for the smaller data set means that the middle half of the larger data set is double as spread out as the smaller data set, so the larger data set has a larger spread.

d This statement is false, and would only be valid if we would excluded the outlier in the smaller data set. 75% of the larger data set is less than 6, so if this were true it would mean that all data points in the smaller data set were less than 6, but there is an extreme value of 10, so this is false. However, if the extreme value was excluded, then it would be true.

29 **a** A sample question would be, "Is the typical annual rainfall at my neighborhood higher than that at my hometown?" or "How does the annual rainfall at my neighborhood compare to that of my hometown?".

b Rosalind could acquire secondary data from sources like state and local government websites or reliable weather websites.

c The data tables show the monthly average precipitation (mm).

Neighborhood average precipitation (mm)

Jan	Feb	Mar	Apr	May	Jun
112.69	130.2	138.59	137.44	160.08	129.75
Jul	Aug	Sep	Oct	Nov	Dec
156.41	157.67	137.79	147.6	88.95	134.22

Hometown average precipitation (mm)

Jan	Feb	Mar	Apr	May	Jun
129.69	143.39	150.51	138.7	164.83	137.56
Jul	Aug	Sep	Oct	Nov	Dec
125.04	138.14	133.45	116.99	102.46	130.7

d

Neighborhood average precipitation

Hometown average precipitation

80 85 90 100 105 110 115 120 125 130 135 140 145 150 155 160 165 170

e One conclusion would be "Both my hometown and current neighborhood typically get around 135 mm of rain per month, but my hometown has more variation, while my current neighborhood had the driest month.".

f

Neighborhood monthly precipitation

Hometown monthly precipitation

The average precipitation for both my hometown and neighborhood is typically in the range of 130 – 139 mm.

4.05 Scatterplots and lines of best fit

Subtopic overview

Lesson narrative

In this lesson, students will explore scatterplots and lines of best fit. The lesson begins with an introduction to bivariate data, which involves collecting and organizing pairs of related data points to explore relationships between two variables. Students will learn how to formulate statistical questions that investigate these relationships, such as the impact of age on bone density or the relationship between education level and salary. Students will then create scatterplots to visually display bivariate data, identifying independent and dependent variables and examining patterns that suggest linear relationships. They will analyze scatterplots to determine whether relationships are positive, negative, or nonexistent. Students will participate in an exploration activity that involves examining data on the weight of someone's backpack versus their age, where students will evaluate different lines of fit to determine the best representation of the data. Through guided examples, students will practice plotting data points, drawing lines of best fit by balancing points above and below the line, and interpreting the trend lines to make predictions. By the end of the lesson, students should be proficient in using scatterplots to identify and describe relationships between two variables and drawing lines of best fit to model these relationships.

Learning objectives

Students: Page 254

After this lesson, you will be able to...
- write statistical questions and use them to collect bivariate data.
- collect bivariate data using observations, measurements, surveys, and experiments.
- create scatterplots of data sets with and without technology.
- analyze and interpret the relationship of bivariate data represented in a scatterplot using the shape and clustering to define the associations.
- sketch the line of best fit for data represented in a scatterplot.
- analyze and describe the association of the data based on the clustering of the data around the line.

Key vocabulary

- association
- bivariate data
- dependent variable
- independent variable
- line of best fit
- relationship
- scatterplot

Essential understanding

Bivariate data can often be approximated using a linear function.

Standards

This subtopic addresses the following Virginia 2023 Mathematics Standards of Learning standards.

Mathematical Process Goals

MPG1 — Mathematical Problem Solving

In teaching scatterplots, teachers can encourage mathematical problem-solving by providing students with real-world data sets and guiding them to apply their understanding of scatterplots to solve problems. For example, students could be given a data set representing the relationship between hours studied and test scores, and asked to create a scatterplot and interpret it to determine how study time affects test scores.

MPG2 — Mathematical Communication

Teachers can integrate this process goal by encouraging and facilitating conversations where students compare univariate and bivariate data. Students should use the bivariate data of scatterplots to discuss connections between the data and its characteristics. Students should then communicate and draw conclusions regarding the data and make predictions based on the scatterplots.

MPG3 — Mathematical Reasoning

Teachers can integrate this process goal by investigating questions of relationships of the bivariate data displayed in scatterplots. Students should consider the connections between the two characteristics represented in the scatterplot. Students should also use the labels and key features of the scatterplot to draw conclusions and make predictions using the data.

MPG4 — Mathematical Connections

Teachers can facilitate activities that link scatterplots and lines of best fit to other areas of mathematics and real-world contexts. Students can be guided to recognize how concepts from algebra, such as linear equations, are connected to statistical representations in scatterplots. Teachers can encourage discussions on how the line of best fit relates to the slope and intercept in linear functions, reinforcing their understanding of previous algebraic concepts. Additionally, students can explore how bivariate data analysis applies to other disciplines like science or social studies by examining real-world datasets relevant to those fields.

MPG5 — Mathematical Representations

Teachers can integrate the goal of mathematical representations by demonstrating how scatterplots serve as a visual representation of bivariate data. They can guide students to understand that scatterplots are an extension of what they learned about plotting points on a coordinate plane. Teachers can also show how to interpret these graphical representations in the context of real-world situations, such as analyzing the relationship between height and weight or temperature and ice cream sales. In this way, students can see representation as both a process and a product.

Content standards

8.PS.3 — The student will apply the data cycle (formulate questions; collect or acquire data; organize and represent data; and analyze data and communicate results) with a focus on scatterplots.

8.PS.3a — Formulate questions that require the collection or acquisition of data with a focus on scatterplots.

8.PS.3b — Determine the data needed to answer a formulated question and collect the data (or acquire existing data) of no more than 20 items using various methods (e.g., observations, measurement, surveys, experiments).

8.PS.3c — Organize and represent numeric bivariate data using scatterplots with and without the use of technology.

8.PS.3d — Make observations about a set of data points in a scatterplot as having a positive linear relationship, a negative linear relationship, or no relationship

8.PS.3e — Analyze and justify the relationship of the quantitative bivariate data represented in scatterplots.

8.PS.3f — Sketch the line of best fit for data represented in a scatterplot.

Prior connections

4.PS.1 — The student will apply the data cycle (formulate questions; collect or acquire data; organize and represent data; and analyze data and communicate results) with a focus on line graphs.

6.MG. 3 — The student will describe the characteristics of the coordinate plane and graph ordered pairs.

8.PS.2 — The student will apply the data cycle (formulate questions; collect or acquire data; organize and represent data; and analyze data and communicate results) with a focus on boxplots.

8.PFA.3 — The student will represent and solve problems, including those in context, by using linear functions and analyzing their key characteristics (the value of the y-intercept b) and the coordinates of the ordered pairs in graphs will be limited to integers).

Future connections

A.ST.1 — The student will apply the data cycle (formulate questions; collect or acquire data; organize and represent data; and analyze data and communicate results) with a focus on representing bivariate data in scatterplots and determining the curve of best fit using linear and quadratic functions.

Lesson Preparation

Suggested review

Depending on your students' level of prior knowledge, consider revisiting the following lessons:

Grade 8 — 4.01 Data collection and samplings

Student lesson & teacher guide

Collect and organize bivariate data

This section introduces the concept of bivariate data, where each study participant contributes two pieces of data, allowing for the exploration of relationships between two variables. Through examples and visual aids like scatterplots, students will learn how to formulate statistical questions about these relationships, collect appropriate data, and analyze it graphically to discern patterns.

Exploration

Students: Page 255

> **Interactive exploration**
> Explore online to answer the questions
>
> ⊕ **mathspace.co**

Use the interactive exploration in 4.05 to answer these questions.

1. What do you think a positive relationship means?
2. Describe what a negative relationship looks like.
3. For a particular type of relationship, if you change the number of data points, does this change the general shape of the scatterplot?

Suggested student grouping: Small groups

In this interactive exploration, students will engage with a dynamic tool to manipulate and observe changes in scatterplots based on bivariate data.

Ideal student responses

These ideal responses may differ from other correct student responses. Less formal responses can be connected with the more precise mathematical language presented here.

1. **What do you think a positive relationship means?**

   A positive relationship in bivariate data means that as one variable increases, the other variable also increases. In the context of a scatterplot, this is represented by a trend where the points slope upward from left to right.

2. **Describe what a negative relationship looks like.**

   A negative relationship is observed when one variable increases and the other decreases. On a scatterplot, this type of relationship is depicted by a downward sloping line from left to right, indicating that as the x-value (independent variable) increases, the y-value (dependent variable) decreases.

3. **For a particular type of relationship, if you change the number of data points, does this change the general shape of the scatterplot?**

Changing the number of data points in a scatterplot does not generally change the overall shape or the type of relationship depicted, whether it is positive, negative, or no relationship. It may make the relationship appear more clear or robust due to more data points either supporting or refuting the trend, but it does not alter the nature of the relationship itself (linear, non-linear, etc.). More data points can, however, lead to a better understanding of the variability and distribution within the dataset.

Purposeful questions

- How does increasing the number of data points affect your ability to identify the type of relationship in the scatterplot?
- Can you predict what will happen to the scatterplot if you increase the number of data points in a negative relationship?

Possible misunderstandings

- Students might mistakenly think that a more spread-out scatter of points (especially with fewer data points) indicates no relationship at all, rather than recognizing it as a weak or less obvious relationship.

Co-crafting statistical questions
English language learner support

Use the Co-Craft Questions routine to support students in formulating statistical questions about relationships between two variables. Begin by presenting a context involving bivariate data, such as age and bone density or education level and salary, without providing specific questions. Invite students to work in pairs to brainstorm possible statistical questions that investigate the relationship between the two variables.

Encourage them to use key vocabulary like "relationship," "impact," "association," "trend," "increase," and "decrease." As students share their questions, collect and display them on the board, highlighting the use of mathematical language. Then, guide the class in refining these questions for clarity and precision, ensuring they are suitable for data collection and analysis. This process helps students develop both their understanding of bivariate data and their ability to use mathematical language effectively.

Examples

Students: Page 256

Example 1

The scatterplot shows the relationship between sea temperature and the amount of healthy coral.

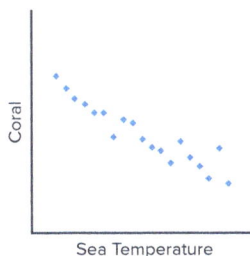

a   Which variable is the independent variable?

    A   Level of healthy coral   B   Sea temperature

Create a strategy

An independent variable is a variable that stands alone and is not changed by the other variables you are measuring. It is also typically on the x-axis.

Apply the idea

Sea temperature is the independent variable because it is on the x-axis and it would likely affect the coral, but the coral would not change the temperature. The correct answer is option B.

Purpose

Show students how to determine the independent variable from a scatterplot by recognizing it as the variable plotted on the x-axis and understanding its role in the relationship between two variables.

Students: Page 256

b   Which variable is the dependent variable?

    A   Level of healthy coral   B   Sea temperature

Create a strategy

The dependent variable is placed on the vertical axis and is affected by the independent variable.

Apply the idea

The level of healthy coral is determined by sea temperature and is on the vertical axis, making it the dependent variable. So, the correct answer is option A.

Purpose

Show students how to identify dependent and independent variables in a given context, which is crucial for understanding and interpreting data in various scientific and mathematical contexts.

c Describe the relationship between sea temperature the amount of healthy coral.

Create a strategy

We want to know if the relationship is negative, positive, or if there is no relationship.

Apply the idea

The scatterplot shows a negative (falling) relationship.

This means that as the sea temperature increases, the amount of coral decreases.

Purpose

Show students how to identify and describe the relationship between two variables on a scatterplot.

Guided scatterplot annotation
Student with disabilities support

use with Example 1

Provide students with copies of the scatterplot and guide them to annotate it directly. Begin by identifying and labeling the independent variable (sea temperature) on the x-axos and the dependent variable (amount of healthy coral) on the y-axos together. Encourage students to draw a trend line through the data points to visualize the negative correlation. Ask them to add notes or observations on the scatterplot, such as "Increasing temperature here" or "Coral decreases in this area."

This hands-on activity helps students, especially those with visual-spatial processing challenges, to engage more deeply with the graph. By annotating the diagram, students can better understand and remember how the variables are related, reinforcing their comprehension of independent and dependent variables in a scatterplot.

Example 2

The table shows the number of traffic accidents associated with a sample of drivers of different age groups.

Age	20	25	30	35	40	45	50	55	60	65
Accidents	41	44	39	34	30	25	22	18	19	17

a Construct a scatterplot to represent the above data.

Create a strategy

Draw the scatterplot by plotting each point from the table.

Apply the idea

Age is the independent variable, so should be put on the horizontal axis. So, Accidents should be put on the vertical axis.

So, the first row from the table corresponds to the point (20, 41) on the graph.

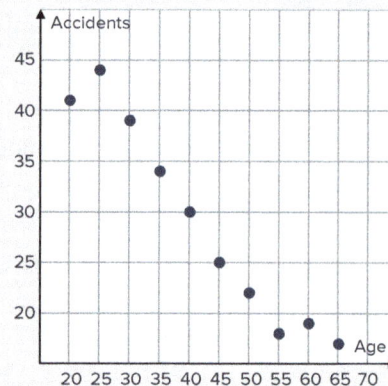

Purpose

Show students how to represent data from a table as a scatterplot, emphasizing the importance of correctly identifying and representing independent and dependent variables on the graph.

Expected mistakes

Students may mix up the independent and dependent variable, switching the placement of the axes. Have a conversation with students about how to determine which variable should go on each axis.

Students: Page 257

b Does the scatterplot show no relationship, a positive relationship, or a negative relationship?

Create a strategy

The points will look random if there is no relationship.

If the pattern of points slopes up from bottom left to top right, it indicates a positive linear relationship between the variables being studied.

If the pattern of points slopes down from top left to bottom right, it indicates a negative linear relationship.

Apply the idea

The points in the scatterplot are going down from left to right. This means this is a negative relationship.

Purpose

Show students how to identify the type of relationship presented in a scatterplot by analyzing the pattern of the points.

Students: Page 257

c As a person's age increases, does the number of accidents they are involved in increase, decrease, or neither?

Create a strategy

Check the trend of the data on the scatterplot.

Apply the idea

Based on the scatterplot, as one variable increases, the other one decreases. So as age increases, the number of accidents decreases.

Reflect and check

We cannot say that being younger causes you to have more accidents. However, a possible reason for this relationship could be that older drivers have had more experience or drive less often, so are less likely to get in an accident.

Purpose

Show students how to interpret the relationship between two variables using a scatterplot. This example emphasizes the importance of understanding and interpreting trends in data.

Example 3

Zenaida is curious about relationships involving follower to following ratio on social media for different levels of influencers.

a   Formulate a statistical question that could be used to explore this aspect of social media.

Create a strategy

A statistical question should require data to be collected, have some variety in the answers, and allow for a data display.

Apply the idea	Reflect and check
We could ask "For the students in my class that use social media, is there a relationship between the number of followers someone has and the number of people they are following?"	There is more than one possible statistical question for this context. Another possible statistical question is "For social media users, is there a relationship between the number of followers someone has and their follower to following ratio."

Purpose

Show students how to formulate a statistical question that requires data collection and analysis, using the context of social media as an example.

Reflecting with students

There is more than one possible statistical question for this context. Consider having students brainstorm as many statistical questions as they can. This will help cement the fact that there are multiple possible statistical questions for a given scenario.

b   Identify the independent and dependent variables.

Create a strategy

A social media user only has direct control over how many people they follow, not how many people follow them.

Apply the idea	Reflect and check
There are two attributes that need to be collected. Independent variable: Number of people they are following Dependent variable: Number of followers they have	They could ask each person to answer a survey with two questions. 1. How many people do you follow on this social media platform? 2. How many followers do you have on this social media platform?

Purpose

Show students how to identify independent and dependent variables in a real-world context, emphasizing the relationship between variables and the idea that the independent variable is what is manipulated or controlled, while the dependent variable is what is measured or observed.

    c   Collect data which could be used to help answer her question.

Create a strategy

We can ask each student if they have a particular social media platform. If they do, we can ask them to find the number of followers and following on their profile.

Apply the idea

For example, we could get this data set:

Following	316	514	326	95	551	407	221	492	234	736
Followers	344	178	132	211	290	452	323	251	257	494

Following	1056	518	162	621	684	286	156	433	33	318
Followers	424	218	305	473	433	280	76	177	19	316

Reflect and check

These numbers can change drastically, so the data would only be valid for a short amount of time, but the trend might still be relevant.

We could also sort the data into different sets based on social media platform to see if different trends arise for different platforms.

Purpose

Show students how to collect and organize data in a systematic way to answer a research question, highlighting the importance of considering the relevance and validity of the data.

Students: Page 259

    d   Based on the collected data, for students in your class, is there a relationship between the number of followers someone has and the number of people they are following?

Create a strategy

We should create a scatterplot to see if there is a positive (rising) pattern, negative (falling) pattern, or no pattern.

Apply the idea

This scatterplot shows that there is a positive (rising) relationship between the number of people you are following and the number of followers you have.

This means that as the number of people you are following increases, so does the number of followers you have.

Reflect and check

The scatterplot for your data may show a different relationship, or no relationship at all.

If the population was more broad and included influencers and celebrities, they may not follow this trend.

Purpose
Show students how to use a scatterplot to identify and interpret a positive relationship between two sets of data.

Students: Page 260

> 💡 **Idea summary**
>
> We can create scatterplots to help us identify patterns and relationships between two variables.
> A scatterplot can suggest different kinds of linear relationships between variables. Linear relationships can be postive (rising) or negative (falling). A scatterplot may also show no relationship if it is randomly scattered with no positive or negative pattern.

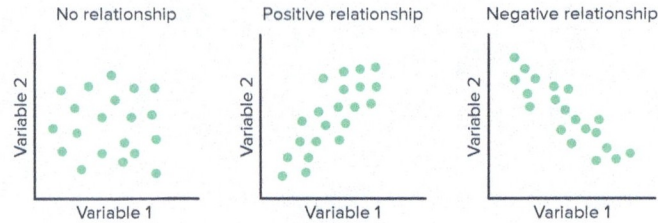

No relationship | Positive relationship | Negative relationship

- No relationship: The scatterplot suggests that there is no definite positive or negative pattern.
- Positive relationship: The scatterplot suggests that as variable 1 increases, the variable 2 also increases.
- Negative relationship: The scatterplot suggests that as variable 1 increases it has an opposite effect on variable 2, so as variable 1 increases, variable 2 decreases.

Line of best fit

This section explores the concept of the line of best fit, a crucial statistical tool used in analyzing scatterplots to discern the underlying trend of bivariate data. Students will learn how to visually determine and interpret lines of best fit, which may not pass through all data points but should capture the general trend of the dataset.

Exploration

Students: Page 260

Exploration

Data was collected on the weight of someone's backpack versus their age.

Each scatterplot shows the same data, but with different lines of fit.

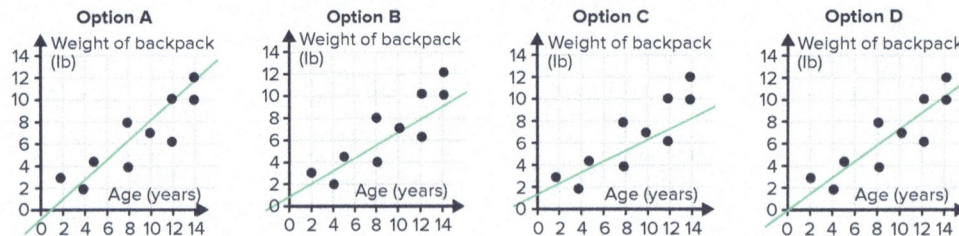

Option A | Option B | Option C | Option D

1. Wwhich line would you choose as a line of best fit? Explain.
2. Which one would you not choose? Explain.
3. What are the differences between the line you would choose and the line you would not choose?

Suggested student grouping: Small groups

This exploration involves analyzing four scatterplots, each depicting the relationship between the age of individuals and the weight of their backpacks. Each plot presents the same set of data but with different lines of best fit drawn through the points. The objective is to critically evaluate these lines to determine which one best represents the general trend of the data, and which one does not align well with the observed data distribution. Students are tasked with selecting the most appropriate line of best fit, explaining their choice, and discussing the shortcomings of the less suitable options. This exercise aims to enhance understanding of how lines of best fit are used to summarize trends in scatterplots and how they can be employed in making predictions and drawing conclusions from bivariate data.

Ideal student responses

These ideal responses may differ from other correct student responses. Less formal responses can be connected with the more precise mathematical language presented here.

1. **Which line would you choose as a line of best fit? Explain.**

    I would choose the line in Option A as the best fit. This line appears to be placed optimally among the data points, with a relatively even distribution of points above and below the line, and it follows the general upward trend of the data well. This suggests a positive relationship where the weight of the backpack tends to increase with age.

2. **Which one would you not choose? Explain.**

    I would not choose the line in Option D. This line does not capture the upward trend shown by the data points effectively. Most of the data points are above the line, suggesting it is not well-balanced. It also leaves too many points significantly distant from the line, particularly at the lower end of the age range, making it less reliable for predictions or interpretations about the relationship between age and backpack weight.

3. **What are the differences between the line you would choose and the line you would not choose?**

    The primary difference between the lines in Option A and Option D is in how they distribute the data points around themselves. The line in Option A has a distribution that suggests a balanced consideration of all points, maintaining proximity to a central trend, which indicates a more accurate reflection of the underlying relationship. In contrast, the line in Option D seems poorly positioned relative to the majority of data points and does not reflect the observed increase in backpack weight with age. This discrepancy makes it less effective for statistical analysis and predictions based on the data.

Purposeful questions
- What criteria did you use to determine which line of fit best represents the data?
- How does the placement of a line of fit affect the interpretation of the relationship between age and backpack weight?

Possible misunderstandings
- Students might think that any line that touches some data points can be considered a good fit, rather than understanding that the best line minimizes the distance between the line and all data points collectively.
- Students might incorrectly assume that a steeper or shallower slope automatically means a better fit without considering how well the line represents the overall trend of the data.

Apply decomposition to simplify complex tasks
Targeted instructional strategies

Guide students to decompose the complex process of analyzing bivariate data into smaller, manageable parts. Encourage them to break down the activity into specific steps, each focusing on a distinct component of the task. For example:

1. Formulate a statistical question about the situation
2. Choose an appropriate data collection method
3. Collect bivariate data through observations, measurements, surveys, or experiments
4. Plot the data points on a scatterplot
5. Draw the line of best fit, balancing points above and below the line
6. Interpret the line to make predictions
7. Analyze the clustering of data points around the line

Provide visual organizers or checklists to help students track each part of the process, reinforcing their understanding of how each component contributes to the overall analysis. This decomposed approach makes the complex task more approachable and helps students stay organized and focused.

Students: Page 261–262

Example 4

The following scatterplot shows the data for two variables, x and y.

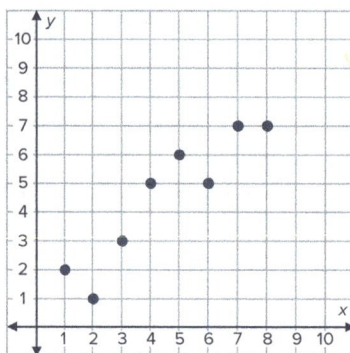

Draw a line of best fit for the data.

Create a strategy

Draw a line that follows the trend of the points and have the same number of points above and below the line.

Apply the idea

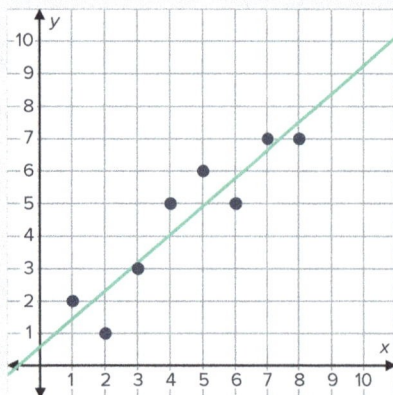

Here is an example of a line of best fit that follows the trend of the data and has 4 points above the line and 4 points below the line.

Purpose

Show students how to visualize the correlation between two variables by drawing a line of best fit on a scatterplot.

Decide and defend: Justifying the line of best fit
Targeted instructional strategies

use with Example 4

Use the 'Decide and Defend' strategy to deepen students' understanding of drawing a line of best fit. Begin by having students individually draw what they believe is the best-fitting line on the scatterplot provided. Next, have them exchange their scatterplots with a partner. Ask students to analyze their partner's line critically, deciding whether it accurately represents the data trend and annotating any points of agreement or concern.

Encourage each student to draft an argument defending their analysis, focusing on aspects like the balance of points above and below the line and the line's alignment with the overall data trend. Then, have pairs discuss their arguments, defending their perspectives and considering their partner's reasoning. This collaborative analysis promotes critical thinking and helps students articulate their understanding of how a line of best fit represents the relationship between variables.

Students: Page 262

Example 5

Amit is working on budgeting with his mom. He learned about the 30% guideline which says that housing costs should be at most 30% of your income.

He wants to explore the statistical question "What relationships exist between monthly housing costs and income?"

a   Explain how Amit could collect or aquire data to help answer his question, then collect or acquire some data.

Create a strategy

We can find a lot of aggregate (summary) data online, but finding raw data comparing the exact two variables of interest for a specific region can sometimes be difficult.

He could do a survey to answer this question where the population is his neighborhood. The survey should be anonymous as this could be very personal information

Apply the idea

He could choose a random house or apartment on each street and ask a resident to fill in an anonymous form with the questions "What is your monthly household income?" and "What are your monthly housing costs?"

His data could look something like this:

Monthly income	9831	7324	10332	6135	7717	5179	1627	5397	6313	4819
Housing cost	3000	1245	4200	1227	1852	1243	488	1619	1957	1494

Monthly income	2235	3706	1537	2200	4546	3552	1652	1958	8113	2950
Housing cost	693	1223	584	894	1864	1457	694	881	4462	1711

Reflect and check

This would be a fairly small sample for a whole neighborhood, but could indicate what the relationships might be.

Purpose

Show students that they can apply statistical methods to real-life situations, in this case, to understand the relationship between housing costs and income.

b Draw a line of best fit to summarize the data.

Create a strategy

To draw a line of best fit, we must first create a scatterplot. Then we can draw a line the goes through the data with about half of the points above and about half below the line.

We can start by opening a statistics calculator and typing the data into the spreadsheet.

	A	B	C	D	E	F	G	H	I
1	Income	Rent							
2	9831	3000							
3	7324	1245							
4	10332	4200							
5	6135	1227							
6	7717	1852							
7	5179	1243							
8	1627	488							
9	5397	1619							
10	6313	1957							
11	4819	1494							
12	2235	693							
13	3706	1223							
14	1537	584							
15	2200	894							
16	4546	1864							
17	3552	1457							
18	1652	694							
19	1958	881							
20	8113	4462							
21	2950	1711							
22									
23									
24									
25									

Next, we can highlight all the data and click Two Variable Regression Analysis. In other programs, we might need to insert a chart or zoom to see the data.

	A	B		E	F	G	H	I
1	In One Variable Analysis							
2	Two Variable Regression Analysis							
3								
4	Multiple Variable Analysis							
5								
6	7717	1852						
7	5179	1243						
8	1627	488						
9	5397	1619						
10	6313	1957						
11	4819	1494						
12	2235	693						
13	3706	1223						
14	1537	584						
15	2200	894						
16	4546	1864						
17	3552	1457						
18	1652	694						
19	1958	881						
20	8113	4462						
21	2950	1711						
22								
23								
24								
25								
26								

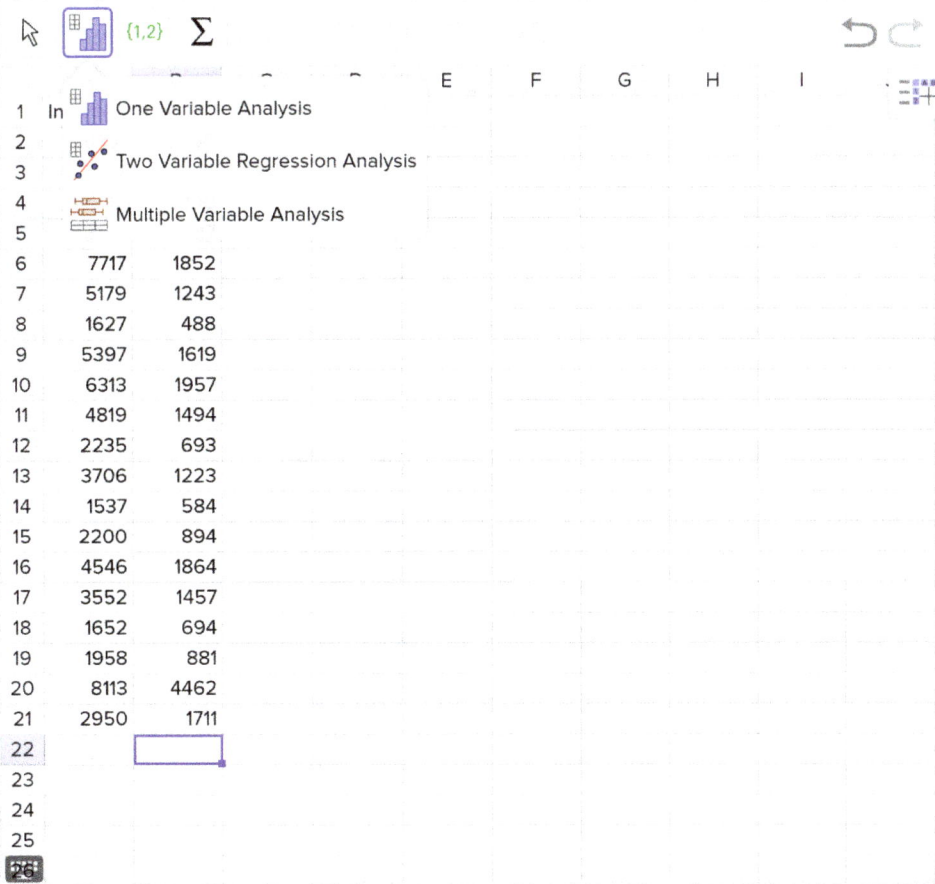

This will give us a scatterplot.

	A	B
1	Income	Rent
2	9831	3000
3	7324	1245
4	10332	4200
5	6135	1227
6	7717	1852
7	5179	1243
8	1627	488
9	5397	1619
10	6313	1957
11	4819	1494
12	2235	693
13	3706	1223
14	1537	584
15	2200	894
16	4546	1864
17	3552	1457
18	1652	694
19	1958	881
20	8113	4462
21	2950	1711
22		
23		
24		
25		

Scatterplot

Y: Column B

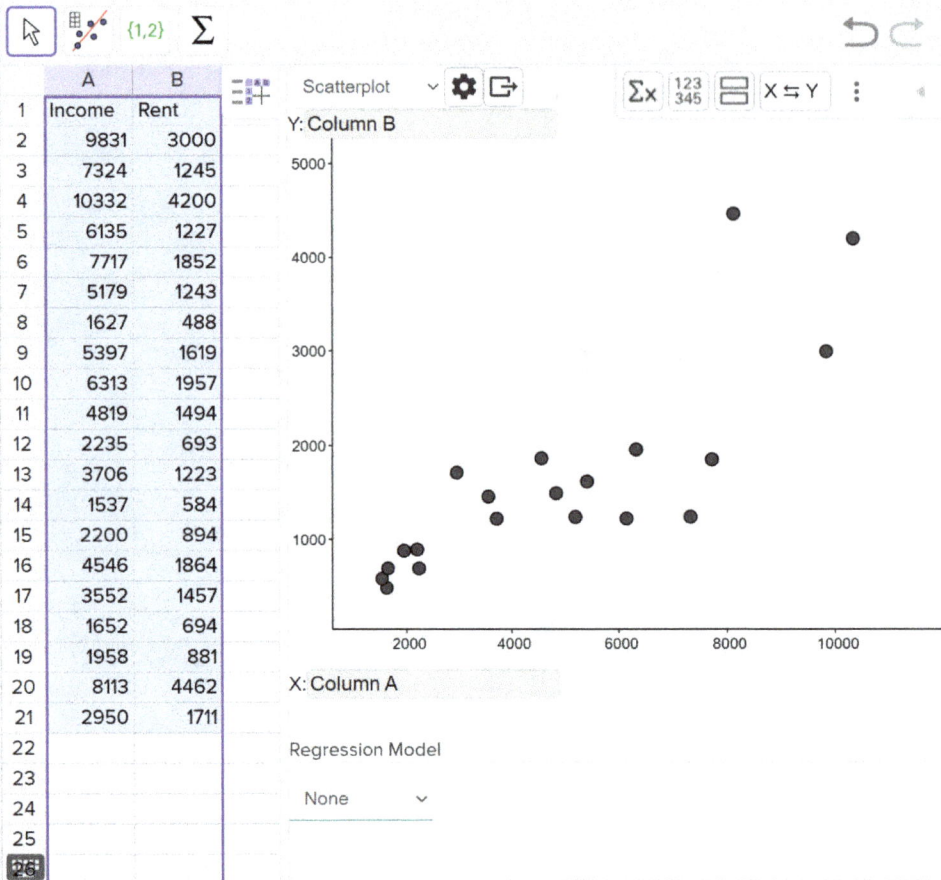

X: Column A

Regression Model

None

Apply the idea

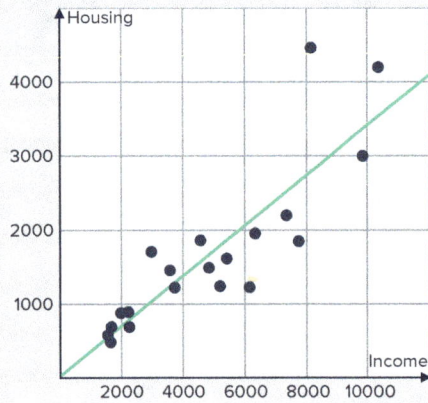

The independent variable is the income and the dependent variable is the housing cost.

There are about 10 points above and 10 points below the line.

Some of the points are very close to the line, while others are further.

Reflect and check

Most technology tools have the ability to plot a line of best fit for us. In GeoGebra, we use the drop down menu for Regression Model and click Linear. We can see that the line of best fit looks similar to the one done by eye.

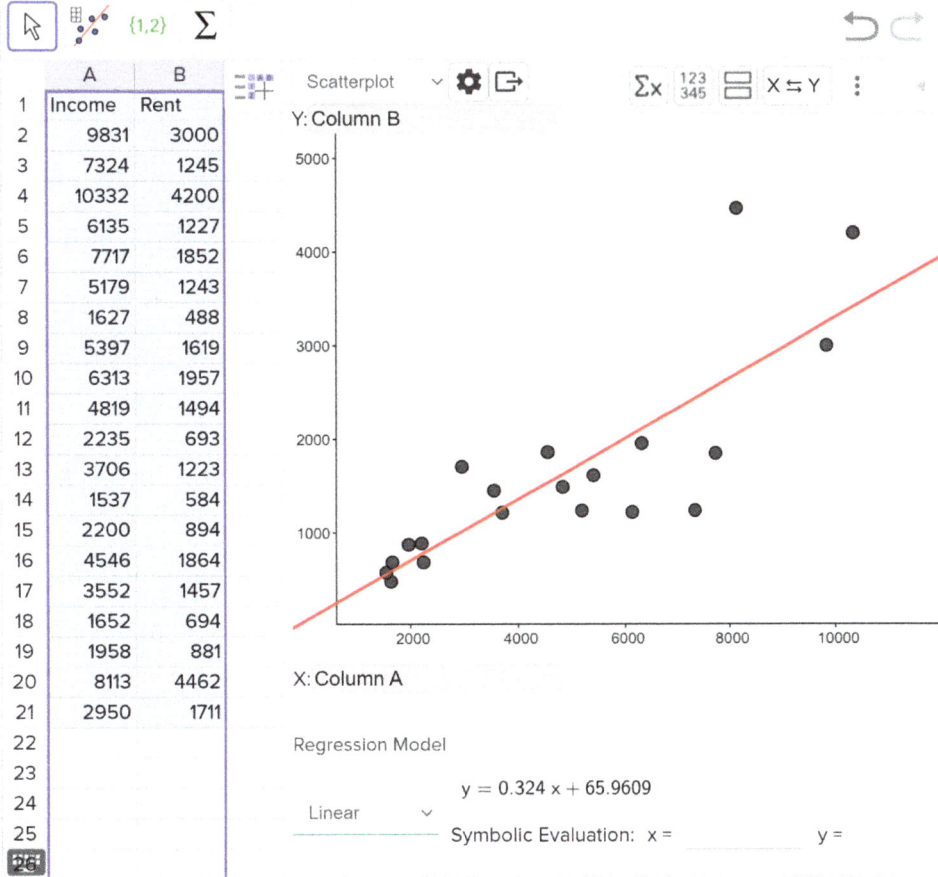

	A	B
1	Income	Rent
2	9831	3000
3	7324	1245
4	10332	4200
5	6135	1227
6	7717	1852
7	5179	1243
8	1627	488
9	5397	1619
10	6313	1957
11	4819	1494
12	2235	693
13	3706	1223
14	1537	584
15	2200	894
16	4546	1864
17	3552	1457
18	1652	694
19	1958	881
20	8113	4462
21	2950	1711
22		
23		
24		
25		

Scatterplot

Y: Column B

X: Column A

Regression Model

$$y = 0.324x + 65.9609$$

Linear

Symbolic Evaluation: x =   y =

Purpose

Show students how to interpret real-world data using scatterplots and lines of best fit, emphasizing the importance of visual data representation for understanding trends and patterns.

c Describe the relationship as positive, negative, or no relationship.

Create a strategy

A positive relationship will have the points rising from left to right.

A negative relationship will have the points falling from left to right.

If the points are randomly scattered, there is no relationship.

Apply the idea

As the income increases, the cost of housing also increases. This means that the points are rising, so there is a positive relationship.

Reflect and check

The slope of the line of best fit is related to the type of relationship. In this case, the slope of the line of best fit is positive, so the relationship is also positive.

Purpose

Help students understand how to describe the type of relationship between two variables by observing the trend in the data points. This helps students draw conclusions from data and understand the concept of correlation.

d Based on the clustering of the points around the line, what conclusions can you draw from the scatterplot.

Create a strategy

We can look at the direction of the line, how close the points are to the line, and patterns within the data set.

Apply the idea

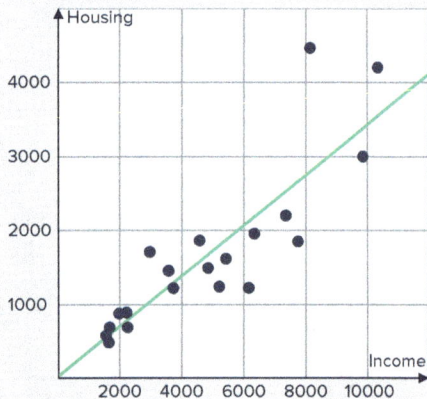

For those with monthly incomes around, $2000, the points are quite clustered around the line, so the relationship is stronger there.

However, some of the points are quite far from the line, so we cannot be completely confident any conclusions we make.

We could draw the conclusion that there is a moderately strong, positive relationship between income and housing costs. Also, we can say that for higher incomes, housing costs vary more than for lower incomes.

Reflect and check

This could lead to further questions about the proportion of income spent on housing, like "What is the distribution for percentage of income spent on housing costs?"

Purpose

Show students how to interpret scatterplots and draw conclusions about the relationship between two variables based on the trend and clustering of the points.

e How can we clearly communicate the results?

Create a strategy

We can first show the raw data, then the display, then show the conclusions.

Apply the idea

The raw data was collected using a random convenience sample in one specific neighborhood, so likely does not represent the wider US population. Surveys were done anonymously.

Monthly income	9831	7324	10332	6135	7717	5179	1627	5397	6313	4819
Housing cost	3000	1245	4200	1227	1852	1243	488	1619	1957	1494

Monthly income	2235	3706	1537	2200	4546	3552	1652	1958	8113	2950
Housing cost	693	1223	584	894	1864	1457	694	881	4462	1711

The data can be summarized with scatterplots. The first scatterplot shows the whole data set. The other two split it into incomes below and above $6000 per month.

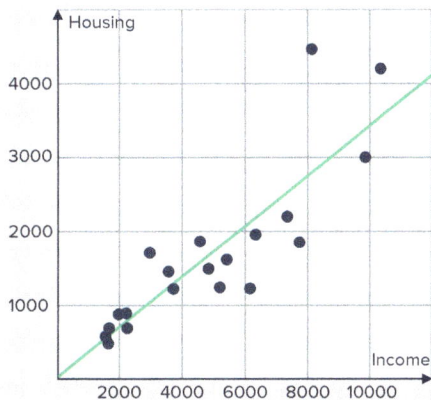

We can see that overall, there is a positive relationship between income and housing costs. For lower incomes, there is a stronger pattern that for higher incomes.

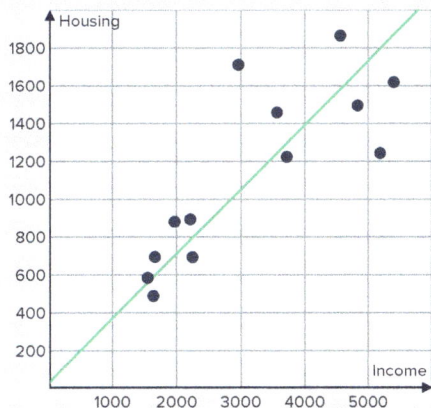

We can see that for the lower incomes, more points are above the line, so they are spending a higher proportion of their income on housing.

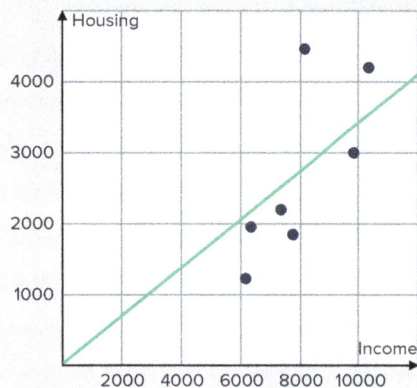

We can see that for the higher incomes, more points are below the line, so they are spending a lower proportion of their income on housing.

Overall, the higher your income, the more you spend on housing.

Reflect and check

This histogram shows the ratios of housing to income and could help to communicate the results. Different bin width can tell a slightly different story.

Purpose

Show students how to effectively communicate the results of a data analysis, including presenting raw data, displaying the data visually, and drawing conclusions.

Advanced learners: Design and investigate their own statistical questions

Targeted instructional strategies   use with Example 5

Encourage students to develop their own statistical questions based on topics that interest them. Building on Amit's exploration of the relationship between monthly housing costs and income, invite advanced learners to choose a real-world issue they're passionate about—such as the correlation between study habits and academic performance or the impact of screen time on sleep quality.

Guide them to design a data collection plan, considering variables, sampling methods, and ethical considerations like anonymity and consent. Support them as they gather and analyze their data, create visual representations like scatterplots, and draw conclusions from their findings.

This process not only deepens their understanding of statistical methods but also fosters independence and critical thinking. By investigating their own questions, students are more engaged and can appreciate the relevance of mathematics in analyzing and interpreting the world around them.

Idea summary

When drawing a line of best fit by eye, balance the number of points above the line with the number of points below the line, and place the line as close as possible to the points.

The closer the points fit to the line, the more confident we can be about the relationship between the two variables.

Practice

Students: Page 269–277

What do you remember?

1 Construct a scatterplot for this set of data:

x	1	3	4	5	6	7	8	9	12	15	17
y	3	7	9	11	12	15	16	16	21	22	25

2 For each pair of variables:

    i Identify the independent variable.   **ii** Identify the dependent variable.

    a Amount of sunlight and growth of plants

    b Time it takes to finish building a house and number of workers

    c Amount of rainfall and number of traffic accidents

    d Number of diseases and amount of pollution

3 Which question would lead to data that could be represented by a scatterplot?

    A How many hours does an average teenager sleep each night?

    B What is the difference in the number of hours teenagers sleep at night?

    C How does the number of hours of sleep impact the energy levels of a teenager?

    D When do most teenagers go to sleep each night?

4 You are a researcher studying the relationship between hours of sleep and academic performance among high school students.

    a What type of data would you need to collect to investigate this?

    b How would you go about collecting this data?

5 For each scatterplot, describe the linear relationship as positive or negative:

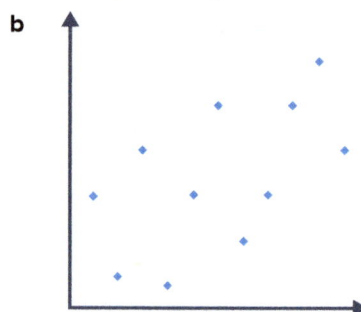

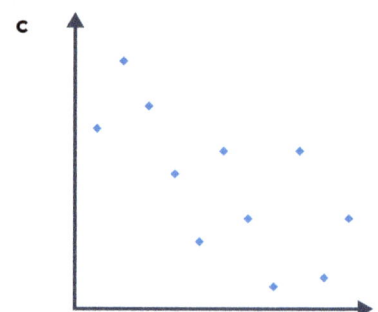

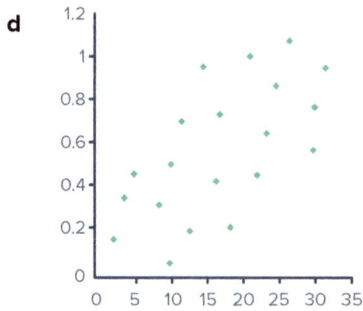

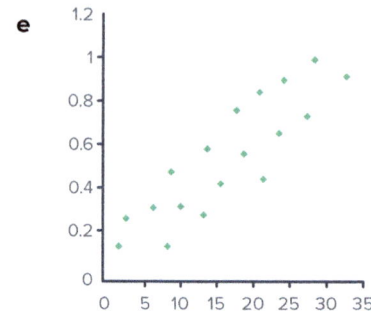

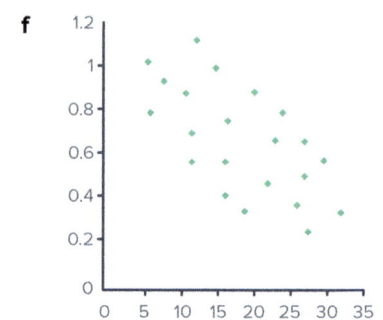

a

b

c

d

e

f

6 For each scatterplot:

i Identify the independent variable.

ii Identify the dependent variable.

iii Describe the relationship between the variables, if one exists.

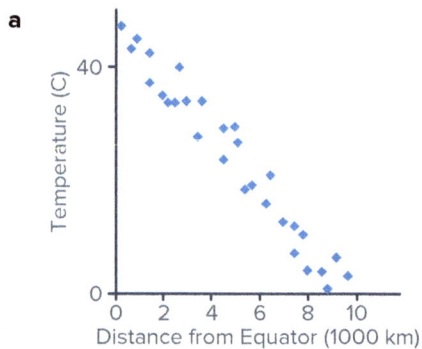

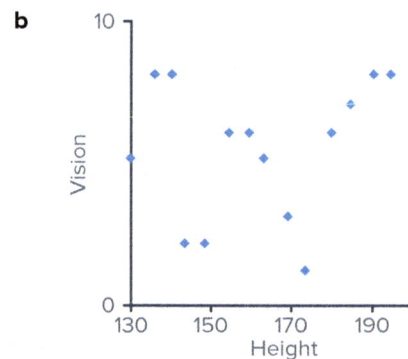

a

b

7 Does each of the following show a line of best fit?

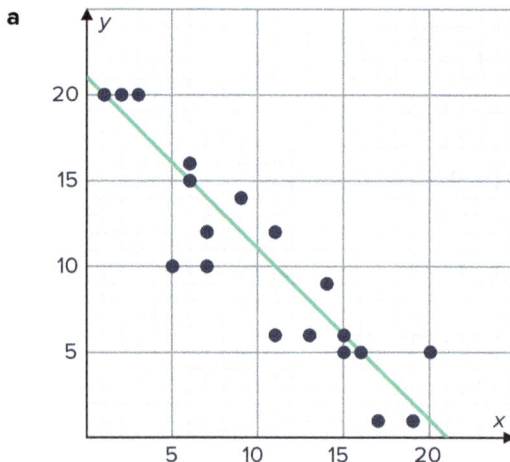

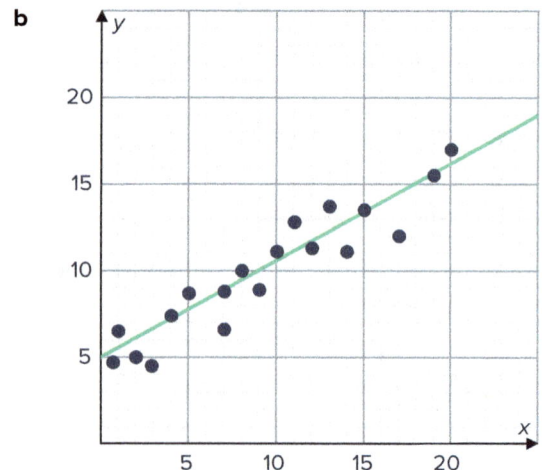

a

b

c

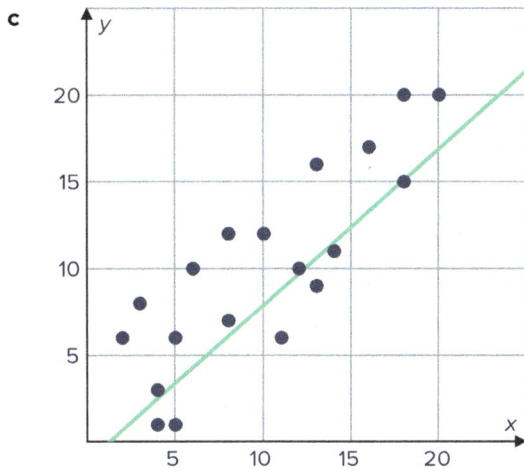

d

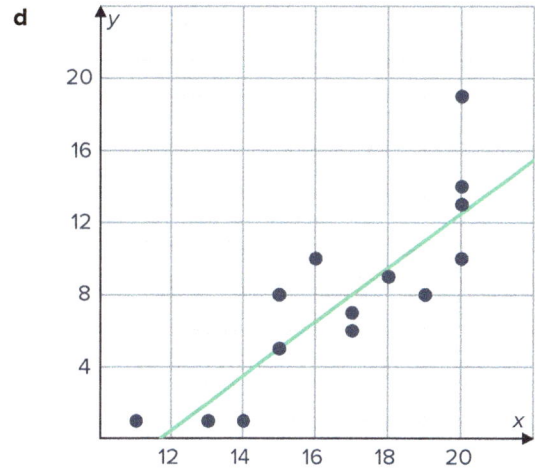

Let's practice

8 The table shows the average IQ and heights of a random group of people:

Height (cm)	140	145	150	155	160	165	170	175	180	185
IQ	103	95	98	111	85	89	108	145	110	93

a Identify the independent and dependent variables.

b Construct a scatterplot to represent this data.

c Which option describes the relationship between IQ and height?

    A Positive linear relationship   **B** Negative linear relationship   **C** No relationship

9 The table shows the time taken to finish a lap on a race track for various average speeds:

Speed	20	25	30	35	40	45	50	55	60	65
Time	85	87	75	82	69	73	60	57	45	49

a Formulate a question that could be answered using the given data.

b Identify the independent and dependent variables.

c Construct a scatterplot for the data.

d Is the relationship positive or negative?

10 Scientists wanted to answer the question, "Is there a relationship between the number of hours of sleep we receive and the effect it has on our motor and processing skills?"

Each participant was placed in a room under the same conditions and required to sleep for a certain amount of time. Then, they were all asked to undergo the same driving test in which their reaction time was measured.

The results are shown in the table:

a Identify the data collection method used.

    A Observation   **B** Measurement

    C Survey   **D** Experiment

b Construct a scatterplot for the observations in the table.

c Describe the relationship between the amount of sleep and reaction time.

Hours of sleep	Reaction time in seconds
9	3
6	3.3
4	3.5
10	3
3	3.7
7	3.2
2	3.85
5	3.55

11 Data was collected to answer the question, "Is there a relationship between students' math grade and the number of points they score in basketball?"

The Math grades of 12 students on the school basketball team and number of points made in the games were recorded in the table:

  a Construct a scatterplot for the students' math grade and the points they scored in the basketball season.

  b Describe the relationship between students' scores in Math and the number of points they scored in the games.

Student	Grade in Math	Basketball points scored
1	63	44
2	92	74
3	60	52
4	79	70
5	88	67
6	81	60
7	61	73
8	91	86
9	72	84
10	42	93
11	66	57
12	92	92

12 To determine whether the presence of sharks in a coastal region is influenced by cage diving, the number of nearby cage diving operations and the number of nearby shark sightings was recorded each month over several months. The results are shown in the table:

Cage diving operations	10	2	4	7	8	5	1	9	6	3
Shark sightings	4	3	5	4	1	4	5	7	10	8

  a Identify the data collection method used.

  b Construct a scatterplot using the data from the table.

  c Describe the relationship between the number of shark sightings and the number of cage diving operations.

  d According to the data, is it possible to determine whether cage diving operations encourage more sharks to come near the shoreline?

13 A study was conducted to compare running times in various outdoor temperatures. The table lists the time taken to sprint 400 meters by runners in different temperatures:

  a How many runners were tested in the study?

  b Is the relationship between the temperature and sprint times positive or negative?

  c If a runner wants a faster sprint time, what type of weather should they run in?

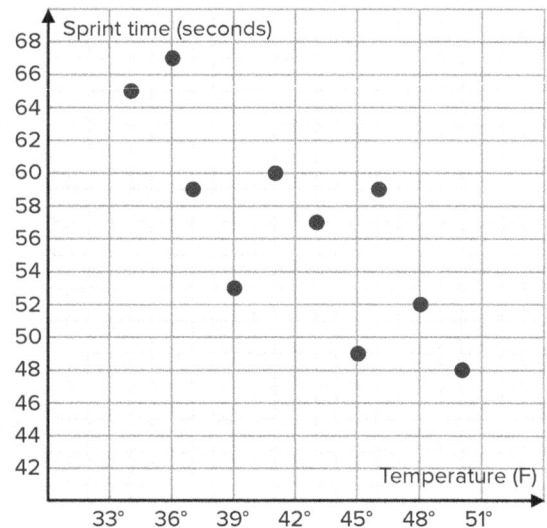

14 The scatterplot shows the grades of 15 students in English and French:

a Formulate a question that could be answered by the data.

b Describe the relationship between students' English and French grades.

c What conclusion can be drawn from this data?

 A If a student does well in English, it is likely they also do well in French.

 B A good grade in English will lead to a good grade in French.

 C If a student is failing English, they tend to be passing French.

 D No conclusions can be drawn because the points are too far from the line.

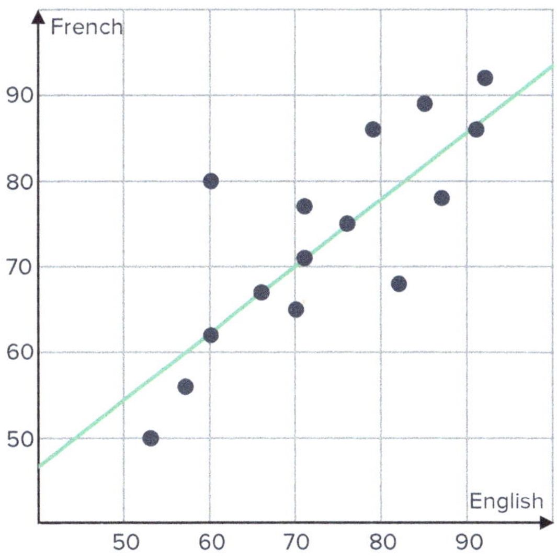

15 The scatterplot shows data for the number of ice blocks sold at a shop on days with different temperatures and the line of best fit.

a Explain why this is an example of a poor line of best fit.

b Draw the correct line of best fit.

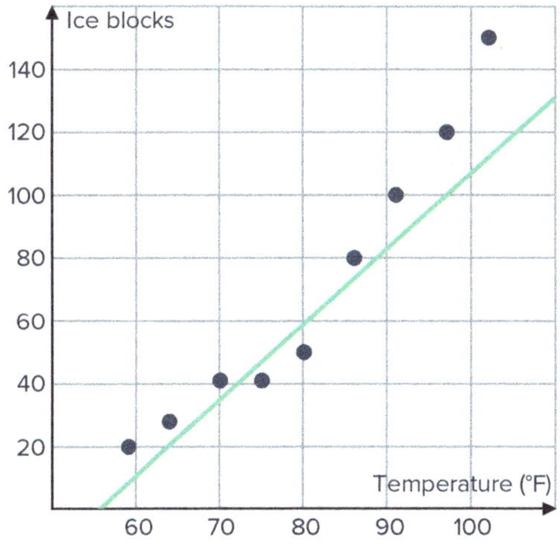

16 The scatterplot shows data for the number of people in a room and the room temperature collected by a researcher.

a Identify the independent and dependent variables.

b Sketch the line of best fit for this data.

c Is there a noticeable trend or pattern in the data? Explain your answer.

d Based on the clustering of the data, the relationship between the variables is:

 A Weak

 B Moderate

 C Strong

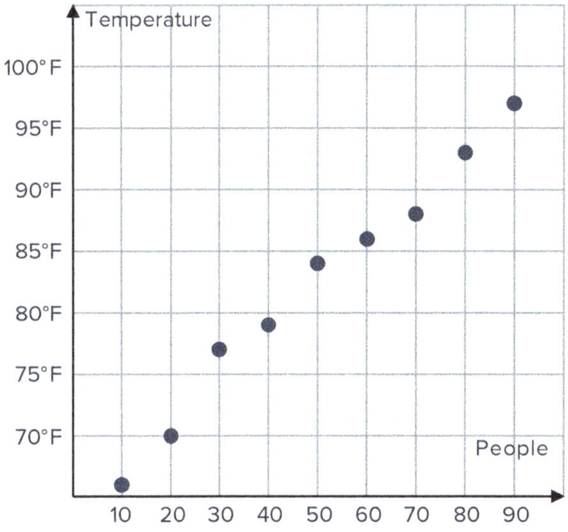

17 Rajah owns a book store and records data for the number of copies of a particular book sold at various prices. He organized the data into the scatterplot shown.

    a Which data collection method did Rajah use?

        A Observation   **B** Measurement

        C Survey   **D** Experiment

    b Is the relationship between the price of the book and the number of copies sold positive or negative? Justify your answer.

    c Sketch the line of best fit for this data.

    d How confident can we be in the scatterplot's ability to predict sales outcomes?

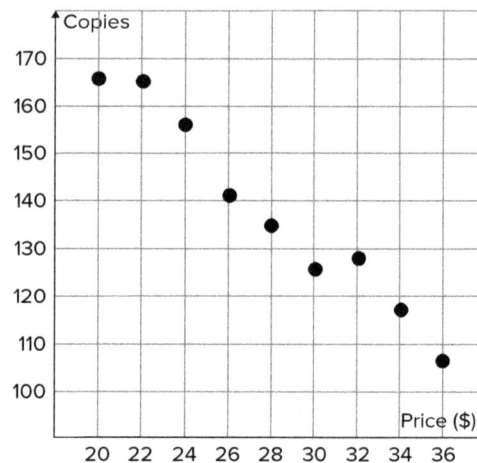

18 The scatterplot shows data for the number of balls hit and the number of runs scored by a baseball player:

    a Formulate a question that could be answered by the data.

    b Sketch the line of best fit for this data.

    c Is the relationship between the two variables positive or negative? Explain your answer.

    d Answer the question you formulated in part (a).

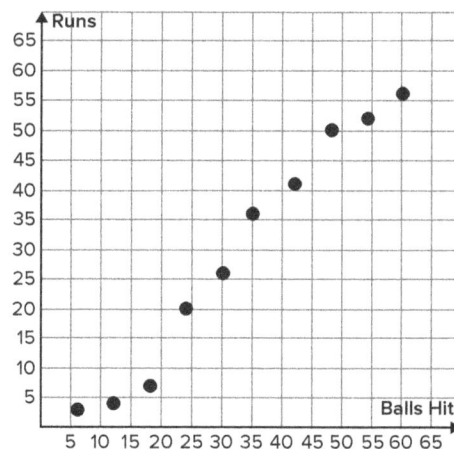

19 Jordano measures his heart rate at various times while running. The data is shown in the table:

Time (minutes)	0	1	2	3	4	5	6	7	8	9	10
Heart Rate (BPM)	58	57	55	61	57	69	67	64	68	73	71

    a Plot the data and sketch the line of best fit.

    b Which data point is closest to the line of best fit?

    c Which data point is furthest from the line of best fit?

    d Jordano concludes that running longer causes his heart rate to increase. Can he make this conclusion? Explain.

20 Scientists conducted a study where each person was asked to read a paragraph and then recount as much information as they can remember. Each participant was given a different paragraph, and the paragraphs varied in length.

    a Describe the type of data the scientists need to collect for this study.

    b Formulate a question that could be answered by the data.

c They found that the longer the paragraph, the less information each person could retain. Select the scatterplot that the scientists may have used to make this conclusion.

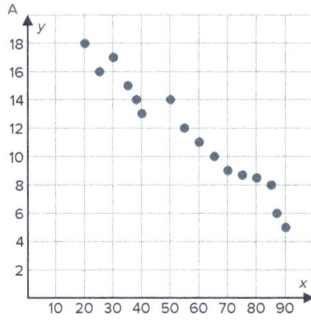

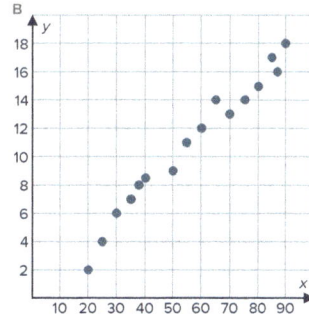

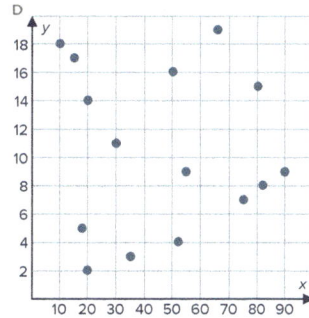

21 The land size of a sample of properties and their selling price were graphed on the scatterplot:

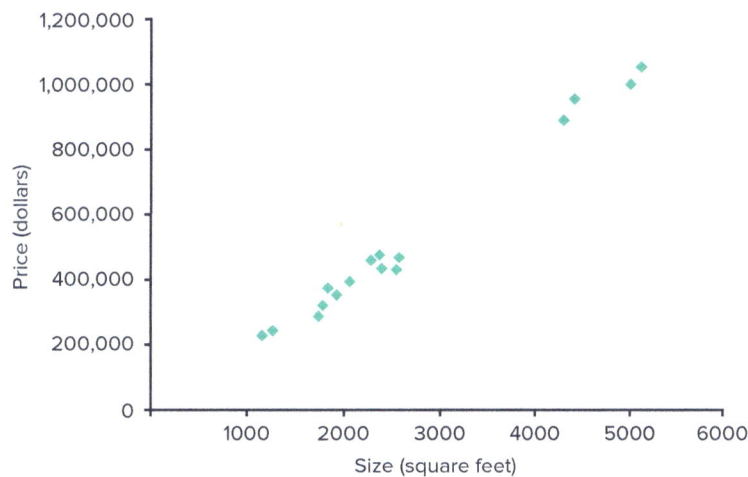

a Which size range contains the most properties?

   A 4500–5500 sq ft   **B** 1500–2500 sq ft   **C** 2500–3500 sq ft

b Another property that is not represented in the scatterplot sold for $704000 and measured 3600 square feet in size. Is this consistent with the relationship shown in the graph?

22 The market price of bananas varies throughout the year. Each month, a consumer group compared the average quantity of bananas supplied by each producer to the average market price (per unit).

a Describe the relationship between the supply quantity and the market price of bananas.

b Sketch the line of best fit.

c According to this data, when will a supplier of bananas receive a higher price per banana? Explain.

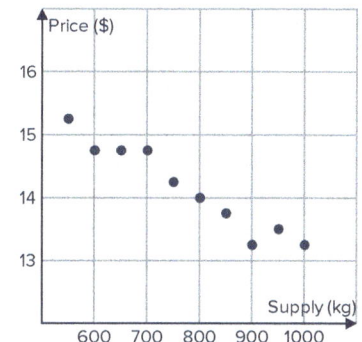

23 After a mathematics exam, Lachlan commented that he left out Question 10 because it did not fit in the same topic as the other questions. The teacher felt that Question 10 and Question 7 were assessing the same skills, so she compared the results of students across all her classes.

She plotted the percentage who had gotten Question 10 correct against the percentage who had gotten Question 7 correct as shown:

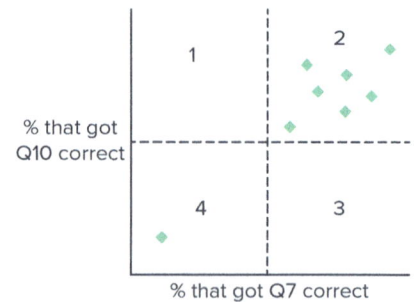

a According to the graph, who is correct: Lachlan or the teacher? Explain your reasoning.

b Let's assume that Lachlan performed exceptionally well on every other question. If a whole class of students had the same result as Lachlan, in which region would this class' results appear on the scatterplot? Explain.

24 A study found a strong positive relationship between the temperature and the number of beach drownings.

Does this mean that the temperature causes people to drown? Explain your answer.

25 Use the data cycle to investigate whether a relationship exists between the number of times someone uses the bathroom and the amount of water they drink.

a Formulate a question which could be investigated using a scatterplot.

b Describe a method which could be used to collect data.

c Collect data for your formulated question.

d Organize your data using a scatterplot. If a linear relationship exists, draw a line of best fit.

e Use the scatterplot to answer your formulated question.

26 The number of fish in a river is measured over a five year period:

Time in years (t)	0	1	2	3	4	5
Number of fish (F)	1903	1994	1995	1602	1695	1311

The data has been graphed along with a line of best fit:

a Predict the number of fish remaining in the river after 7 years.

b According to the line of best fit, after how many years will 900 fish be left in the river?

27 Suppose you want to determine if there is a relationship between the amount of daily screen time (time on a phone, laptop, tablet, or watching TV) and the hours of sleep you get at night.

Go through the whole data cycle at least once to investigate whether a relationship between the two exists.

Answers

4.05 Scatterplots and lines of best fit

What do you remember?

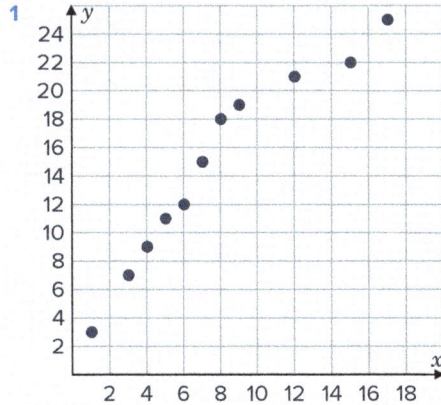

1

2 a i Amount of sunlight

    ii Growth of plants

  b i Number of workers

    ii Time to finish building a house

  c i Amount of rainfall

    ii Number of traffic accidents

  d i Amount of pollution

    ii Number of diseases

3 C

4 a To study the relationship between hours of sleep and academic performance, you would need to collect numerical data on the number of hours of sleep per night for each student and their academic performance (e.g., exam scores, GPA).

  b You could collect this data by using self-report questionnaires for hours of sleep and obtaining academic performance data from the school records.

5 a Negative   **b** Positive

  c Negative   **d** Positive

  e Positive   **f** Negative

6 a i Distance from the equator

    ii Temperature

    iii As the distance from the equator increases, the temperature decreases.

  b i Height

    ii Vision

    iii There is no relationship between height and vision.

7 a Yes   **b** Yes   **c** No   **d** Yes

Let's practice

8 a The independent variable is height, and the dependent variable is IQ.

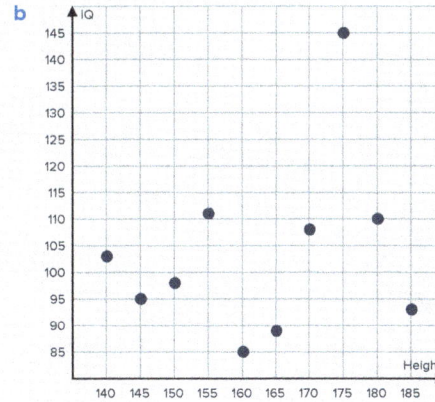

  b

  c C

9 a What is the relationship between average speed and time taken to finish a lap on a race track?

  b The independent variable is speed and the dependent variable is time.

  c

  d Negative

10 a D

  b

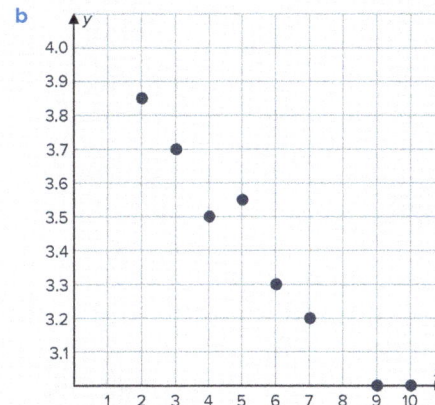

  c Sleeping longer improves reaction time.

11 a

Basketball points (scatterplot: Math grade vs Basketball points)

b As math score increases, students were more likely to score more points in the game.

12 a Observation

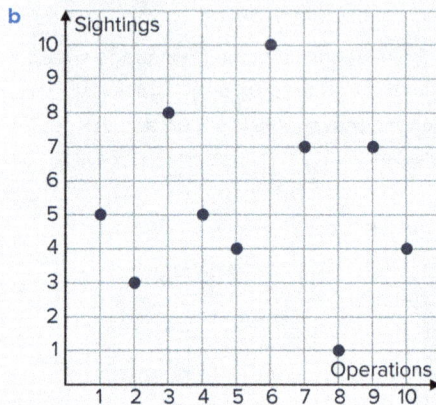

b

Sightings (scatterplot: Operations vs Sightings)

c No definite relationship

d No

13 a 10   **b** Negative   **c** Warmer weather

14 a What is the relationship between students' English and French grades?

b As their English grades increased, their French grades also tended to increase.

c A

15 a The drawn line of best fit in the given scatterplot has more points above it and does not accurately represent the general trend of the data.

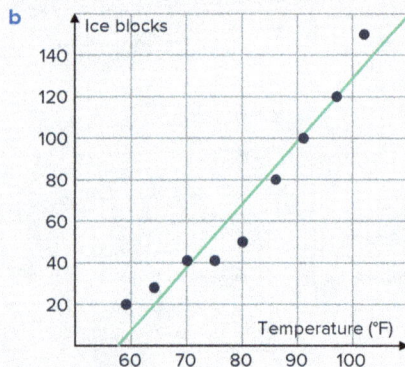

b

Ice blocks (scatterplot: Temperature (°F) vs Ice blocks)

16 a The independent variable is the number of people and the dependent variable is the room temperature.

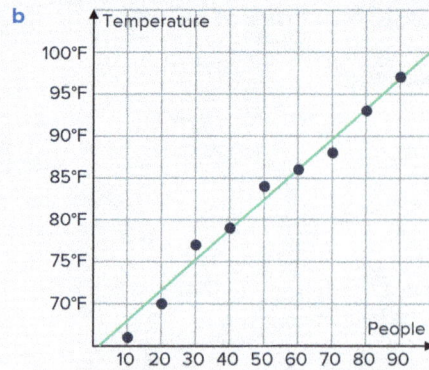

b

Temperature (scatterplot: People vs Temperature)

c Yes. As the number of people in a room increases, the room temperature also increases. The two variables show a positive relationship.

d C

17 a A

b Negative. The number of book copies sold decreases as the price increases. This shows a negative relationship between the two variables.

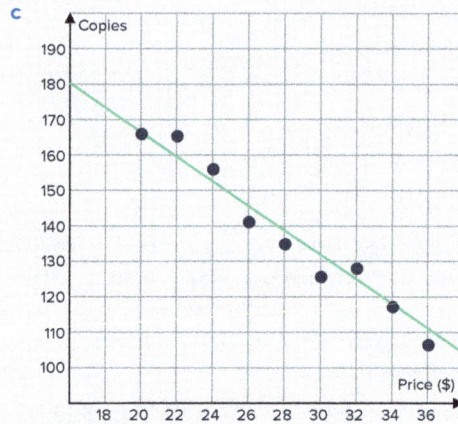

c

Copies (scatterplot: Price ($) vs Copies)

d The data points being close to the line of best fit suggest a strong relationship between the number of books sold and their price. This indicates that the scatterplot is likely to provide relatively accurate predictions. Thus, we can have a high level of confidence in its predictive capability for sales outcomes.

18 a Answers may vary. For example, "Does the number of balls hit impact the number of runs scored?"

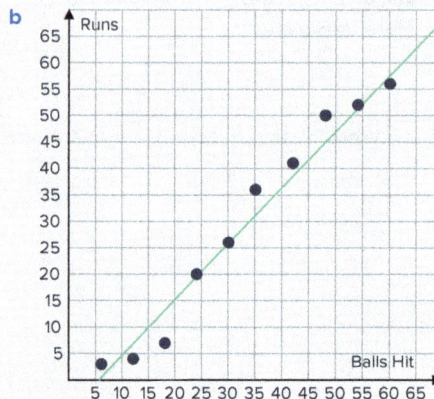

b

Runs (scatterplot: Balls Hit vs Runs)

c Positive: as the number of balls hit increases, the number of runs scored also increases. This implies a positive linear relationship.

d Yes, based on the relationship and clustering of the data, it appears that the number of balls hit has a positive impact on the number of runs scored.

19 a

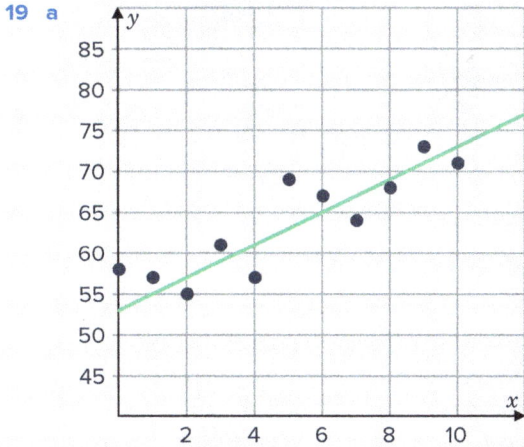

b (8, 67)

c (5, 69)

d No, Jordano cannot make this conclusion because he has not done a controlled experiment. His has identified that there is a relationship, but this does not mean that running longer has caused his heart rate to increase. Other factors could have contributed to his heart rate increasing, such as the amount of caffeine or water he drank or the temperature and humidity outside.

20 a The scientists need to collect numerical data on the length of each paragraph and the amount of information each person retained.

b Many possible answer, for example "Does the length of a paragraph impact the amount of information retained?"

c **A**

21 a **B**   b Yes

22 a As the supply of bananas increases the price decreases.

b

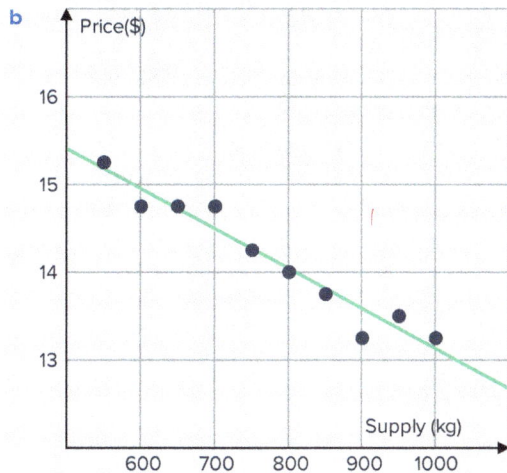

c When very few bananas are available to be sold. A negative linear relationship would mean that as the supply of bananas (independent variable) increases, the market price (dependent variable) decreases.

Let's extend our thinking

23 a The teacher. The data from the graph indicates that students perform similarly in Q7 and Q10.

b Region 3. Lachlan performed well in the rest of the exam, including Question 7, but didn't even attempt Question 10.

24 No. The high temperature may increase the number of people that go to the beach. This then increases the number of people that could drown.

25 a Many possible answers, for example, "How does the amount of water consumed affect the number of times someone uses the bathroom?"

b One possible method is to use a survey. Ask up to 20 randomly selected people to track their daily water consumption and the number of times they use the bathroom in a day. To respect the privacy of others, the data should be collected anonymously, such as using a online survey service to hide the participants' identities.

c Answers will vary.

An example sample of data is shown.

Sample	Water consumption (ounces)	Number of bathroom visits in a day
Person A	64	3
Person B	80	3
Person C	40	3
Person D	96	6
Person E	56	4
Person F	72	4
Person G	48	2
Person H	88	5
Person I	32	2
Person J	60	5
Person K	84	4
Person L	50	3
Person M	54	3
Person N	68	5
Person O	75	3
Person P	66	4

d

e We can observe that there is a positive linear relationship between water consumption and the number of restroom visits. So, as the amount of water consumed increases, the number of bathroom visits also increases.

26 a 1300 fish   **b** 11 years

27 1. Formulate questions

Possible questions:

- Does screen time decrease the amount of sleep we get at night?
- Do people who use screens more sleep less than those who don't?
- How does the amount of daily screen time affect the number of hours of sleep I get at night? We'll look at this question.

2. Collect data using the questions above.

We can collect data by measuring the daily amount of screen time and the number of hours slept at night. To get a good sample size, we can track this data for about 3 weeks.

The table shows an example sample of 20 days.

Screen time	Number of sleep hours
2.5	8.1
2	9
3	8.9
4	8.7
6	8.3
4	8.2
7	7
5.5	7.1
5	7.5
1.5	8.5

4	7.2
3.5	6
8	6.7
6	8.5
6.5	7.1
5.2	7.4
4.5	7.5
4	7
3.9	7.9
7	7.8

3. Create a data display using a scatterplot.

This scatterplot represents the sample data from the previous part. Since it shows an approximate linear relationship, a line of best fit has been drawn.

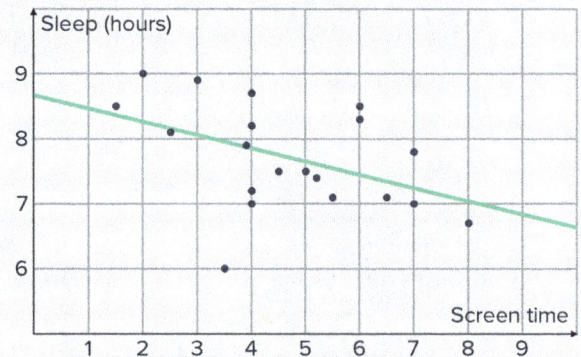

4. Analyze and explain the results.

From the scatterplot, we can observe that there is a weak, negative linear relationship between the screen time and number of hours of sleep. The relationship shows that as the screen time increases, the number of sleep hours tends to decrease.

The sample size is fairly small and only represents data for one person, but it gives an indication that the number of hours of sleep is affected by the amount of screen time. A larger sample collected from multiple people or a properly conducted experiment could have a different conclusion.

This could lead us to formulate a new question like "How does the amount of daily exercise affect the number of hours of sleep?" or "Does the amount of daily screen time affect the number of hours of sleep for adults?"

4.06 Misleading data displays

Subtopic overview

Lesson narrative

In this lesson, students will learn how to compare data sets using various data displays. The lesson begins with an introduction to boxplots, which are used to visually compare the spread, median, quartiles, and range of different data sets. Students will create and interpret parallel boxplots, examining the differences between data sets and identifying which group performs better based on the median and consistency based on the interquartile range (IQR). Students will engage in an exploration activity where they will compare the heights of students in a class using a histogram, dot plot, and boxplot. They will analyze the strengths and weaknesses of each display and determine which aspects of the data are best represented by each type of graph. Through guided examples, students will practice creating boxplots from raw data, converting dot plots to boxplots, and selecting appropriate measures of center and spread to summarize and compare data sets. By the end of the lesson, students should be able to effectively use and interpret different data displays to compare data sets and understand the key features they reveal.

Learning objectives

Students: Page 278

> **After this lesson, you will be able to...**
> * identify components of graphical displays that can be misleading.

Key vocabulary

* interval
* scale

Essential understanding

The way a set of data is represented affects the conclusions and generalizations that can be made.

Standards

This subtopic addresses the following Virginia 2023 Mathematics Standards of Learning standards.

Mathematical Process Goals

MPG3 — Mathematical Reasoning

Teachers can stimulate mathematical reasoning by asking students to identify the components of a data display that make it misleading. They can also encourage students to use logical reasoning to analyze arguments and determine whether conclusions drawn about the data sets are valid based on how the data was represented.

MPG5 — Mathematical Representations

Teachers can encourage students to critically examine graphs and charts used in media and promotions. Teachers can facilitate activities where students recreate misleading graphs with accurate scales, appropriate data displays, or complete data, highlighting the impact of proper representation. Additionally, students can take an appropriate display and attempt to make it misleading by using a representation that is not appropriate. Students can be guided to identify how manipulations of scale, intervals, and omission of data can lead to misleading interpretations. By analyzing real-world examples where graphical displays distort the true story of the data, students can learn to recognize and correct these misrepresentations.

Content standards

8.PS.2 — The student will apply the data cycle (formulate questions; collect or acquire data; organize and represent data; and analyze data and communicate results) with a focus on boxplots.

8.PS.2j — Identify components of graphical displays that can be misleading.

Prior connections

6.PS.1 — The student will apply the data cycle (formulate questions; collect or acquire data; organize and represent data; and analyze data and communicate results) with a focus on circle graphs.

7.PS.2 — The student will apply the data cycle (formulate questions; collect or acquire data; organize and represent data; and analyze data and communicate results) with a focus on histograms.

Future connections

A.ST.1 — The student will apply the data cycle (formulate questions; collect or acquire data; organize and represent data; and analyze data and communicate results) with a focus on representing bivariate data in scatterplots and determining the curve of best fit using linear and quadratic functions.

Lesson Preparation

Suggested review

Depending on your students' level of prior knowledge, consider revisiting the following lessons:

Grade 8 — 4.02 Boxplots
Grade 8 — 4.04 Compare data sets
Grade 8 — 4.05 Scatterplots and lines of best fit

Tools

You may find these tools helpful:
- Graph paper

Student lesson & teacher guide

Misleading graphs

Students: Page 278

Exploration

Graph 1 and Graph 2 show which stakeholders are in favor of a particular school policy.

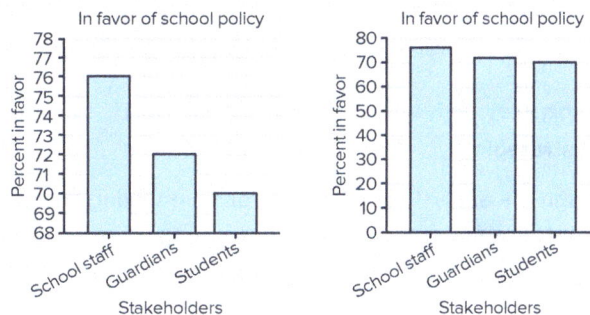

Graph 1

Graph 2

1. Are they showing the same data or different data? Are they conveying the same message? Explain.
2. What do you notice that is different about the graphs? Explain.
3. Which graph might the school use to represent the data in the city's local newspaper? Which graph might the students use to represent the data in the school newspaper? Explain.

Suggested student grouping: Small groups

Students will be examining two graphs which represent the same data - stakeholders in favor of a school policy - but are displayed differently. They will explore how the two graphs convey the same or different messages about the data, and discuss which graph might be used in different contexts.

Ideal student responses

These ideal responses may differ from other correct student responses. Less formal responses can be connected with the more precise mathematical language presented here.

1. **Are they showing the same data or different data? Are they conveying the same message? Explain.**

   They are showing the same data, but they are not conveying the same message. Graph 1 makes it look like there is a large difference between the stakeholders, while Graph 2 makes the differences look much smaller. This is because of the differences in the scale on the y-axis.

2. **What do you notice that is different about the graphs? Explain.**

   The main difference between the graphs is the scale on the y-axis. In Graph 1, the scale starts at 68 and goes to , while in Graph 2 the scale starts at 0 and goes to 80. This makes the differences between the stakeholders appear larger in Graph 1.

3. **Which graph might the school use to represent the data in the city's local newspaper? Which graph might the students use to represent the data in the school newspaper? Explain.**

The school might use Graph 2 in the city's local newspaper to show that there is broad, nearly equal support among all stakeholders. The students might use Graph 1 in the school newspaper to highlight the differences in support among the stakeholders.

Purposeful questions

- How can changing the scale on a graph affect the way the data is interpreted?
- Why might different stakeholders want to use different graphs to represent the same data?

Possible misunderstandings

- Students may think that the two graphs are showing different data because of the differences in the way the data is presented.

Teach critical thinking through the analysis of graphs
Targeted instructional strategies

Teaching critical thinking through the analysis of graphs involves enabling students to question, analyze, and interpret the information presented in the graph. It's about teaching them not to just accept the graph at face value but to critically examine the information being presented and consider if it is accurate or misleading. Here's a step-by-step process:

- Title: Begin with the title of the graph. The title should give a clear understanding of what information is being represented. Ask students to explain what they think the graph is about based on the title.

  Example: If the title is "Average Monthly Temperatures in City X", students should be able to state that the graph is about temperature trends in a specific city throughout the year.
- Labels: Next, look at the labels. These should help understand the variables that are being compared.

  Example: In the "Average Monthly Temperatures" graph, the x-axis may be labeled with months, and the y-axis might be labeled with temperatures in degrees.
- Scales: Discuss the importance of the scales on the axes. An accurate scale is critical to an honest representation of data. If the scale isn't uniform or is missing altogether, the graph might be misleading.

  Example: If the y-axis on the temperature graph jumps from 0 to 100 degrees, the temperature fluctuations might seem more dramatic than they actually are.
- Missing information: Finally, prompt students to look for any missing information. Are there any variables that aren't accounted for? Is there data missing for certain categories?

  Example: It could be misleading if the temperature graph only includes data for some months and not others.
- Accuracy: Encourage students to consider whether the graph is accurately representing what it claims to. Could the data be biased or skewed in any way? Are there any obvious distortions?

  Example: If the graph shows an unusual spike in temperature for a month, ask students to investigate whether it's an error or whether there was an actual heatwave during that period.

By following these steps, students will be encouraged to critically analyze graphical data, which is a vital skill in today's data-driven world.

Stronger and clearer each time
English language learner support

Introduce the concept of misrepresentations of data in graphs, discussing different methods through which data can be misleading.

Have students pair up. In each pair, one student explains a concept in their own words while the other listens.

Provide sentence stems to guide their explanations. For example, suggest they start their explanation with "This graph might be misleading because …".

Have partners switch roles and explain the same concept back. The listener should then provide their own explanation of the concept, ideally incorporating and building upon the first student's explanation.

After each round of explanations, bring the class together to discuss and clarify any misconceptions or confusing points. Share common mistakes or insightful observations from the pair's discussions.

Repeat this cycle of explanation and discussion with a new concept related to misleading data in graphs.

For instance, when discussing a graph that doesn't start its y-axis at zero, one student might say, "This graph might be misleading because it doesn't start at zero, which can make the differences between data points look larger than they actually are." After discussing and correcting if needed, the other student might offer their own interpretation or add more details.

Provide a checklist
Student with disabilities support

Guide students in creating a checklist or step-by-step guide to analyze graphs for potential misrepresentations. This should include steps like checking the axes labels and scales, considering the chosen format of the graph, and looking for any missing data or information. The checklist can include steps like:

- Check the title: Does it clearly state what the graph is about?
- Examine the axes labels: Are they clear and accurate?
- Look at the scales: Are they consistent and start from zero where necessary?
- Consider the chosen format of the graph: Is it the right format for the type of data being represented?
- Look for any missing data or information: Is all the relevant data included and represented correctly?

Encourage students to use this checklist whenever they encounter a graph. They should go through each step individually, checking off each step as they complete it. Initially, guide students through the checklist, explaining and demonstrating each step. As students become more comfortable with the process, allow them to use the checklist independently, but be available to offer support as needed.

Misconception about axis scales
Address student misconceptions

Students may think that the scale of the axes in a graph does not affect the interpretation of the data and that a graph is only misleading if data is omitted or false.

To address this, emphasise the impact of the scale on the appearance of the data and provide examples of graphs with different scales representing the same data. Show them how a non-uniform or inappropriate scale can exaggerate or downplay differences in the data, leading to misinterpretation.

Students explore how graphs can be used to misrepresent information intentionally or unintentionally. They are guided through different misleading features of graphs, such as misleading axes or scales and incorrect display creation, and engage with examples to solidify their understanding.

Examples

Students: Page 280–281

Example 1

Shawnte plays for the school soccer team, and she was one of the top scorers of the season. They played a total of 17 games, and Shawnte organized the number of goals she scored in the dot plot shown.

Goals scored

Select the option that describes why this graph is misleading.

A   This graph is misleading because there is no information on what the dots represent.

B   This graph is misleading because it does not show the data from all 17 games.

C   This graph is misleading because there is a very small interval for the number of goals scored.

D   The graph is misleading because a dot plot skews the results to make them appear better than reality.

Create a strategy

Check if each statement matches what the graph shows.

Apply the idea

The data represents the number of goals Shawnte scored throughout the season, which means each dot represents a game in the season. Option A is incorrect.

There are only 14 dots shown, but there were 17 games in the season. This means there are missing data values, so option B is correct.

It is not common in soccer to score a high number of goals, so a small range of goals is reasonable. Option C is incorrect.

Dot plots provide an ordered display of all values in a data set and shows the frequency of data on a number line. It is a reasonable choice for this data set, so option D is incorrect.

Purpose

Show students how to critically analyze data presented in graphical form.

Reflecting with students

Students may consider why a representation of the data would have ommitted 3 of the games played. Students may notice that since Shawnte represented his own data, the reasons could possibly include having games he did not want to include, games where no goals were scored, or a miscounting of goals scored. Students should wonder about the audience seeing the data display, as the misleading aspect of the graph may be impacted by the audience.

Example 2

A school newspaper article reads, "Students prefer pepperoni pizza over any other type of pizza topping."

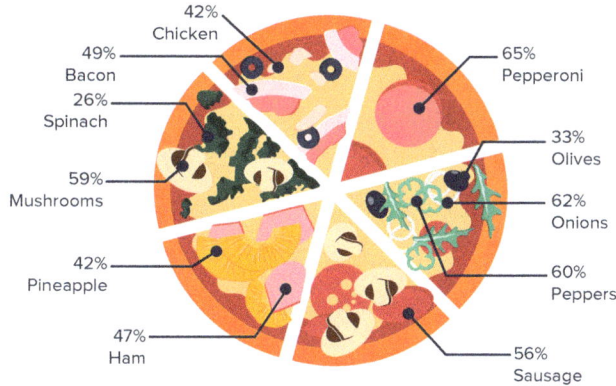

42%
Chicken

49%
Bacon

26%
Spinach

59%
Mushrooms

42%
Pineapple

47%
Ham

65%
Pepperoni

33%
Olives

62%
Onions

60%
Peppers

56%
Sausage

Explain why this graph does not best represent the data.

Create a strategy

Circle graphs are used to show a relationship of the parts to a whole. Consider what the article was trying to communicate and whether this graph clearly communicated those results.

Apply the idea

The article was comparing students' preferred type of pizza toppings. However, the percentages in this circle graph sum to a number larger than 100%, so it does not represent the proportion of students that prefer each type of topping.

Because the percentages do not represent parts of the whole population, it appears that students were able to choose more than one type of pizza topping, though not all type of toppings are represented.

Reflect and check

To compare the percentages of pizza topping preference, a bar graph would have been a better choice of display.

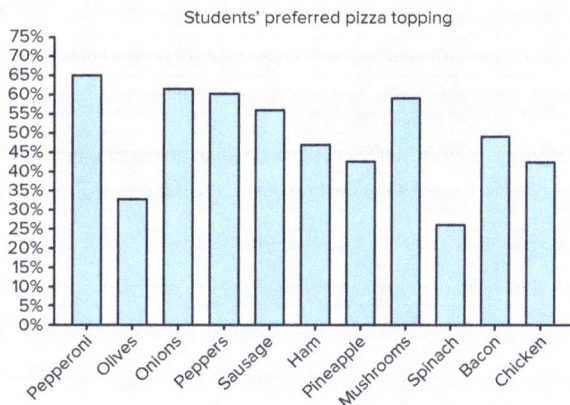

Students' preferred pizza topping

Purpose

Make students aware that data representation and interpretation are critical in conveying accurate information.

Expected mistakes

Students may assume that since a pizza is round and pizzas can have multiple toppings, that having the percentages not add to 100% could be possible. Encourage students to redraw graphs without additional graphics and ask them to double check that the values add to the correct amount.

To incorporate abstraction into this example, encourage students to focus solely on the numerical data rather than the decorative elements of the graph. Guide them to disregard the pizza image and toppings, which inaccurately suggest that the percentages represent parts of a whole, and extract the key information by listing the percentages associated with each topping.

By representing this data in a simplified form, such as a table or a bar graph, students can better analyze the information without distraction. This abstraction helps them recognize that the total percentage exceeds 100%, highlighting the misleading nature of the original graph and emphasizing the importance of appropriate data representation.

Students: Page 282–283

Example 3

Which feature of this graph is misleading?

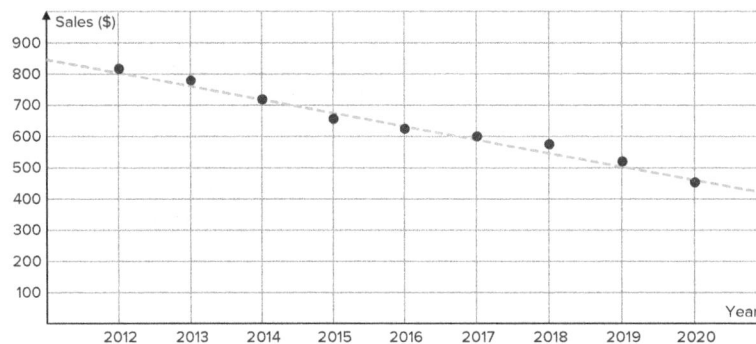

Create a strategy

Common ways for a graph to be misleading are:

- Manipulating the scale, such as not starting the scale at zero or using a scale that is not uniform
- Manipulating intervals that could exaggerate the distance between data points
- Omitting important information in titles and labels
- Omitting certain data points, such as outliers or values that do not align with the desired conclusion
- Choosing a graphical display that does not best represent the data

Apply the idea

When looking at the grid lines of the graph, notice that the vertical scale is not uniform with the horizontal scale. The distance between the years appears much larger then the distance between the sales amount, causing the decrease is sales to appear small.

Thus, the intervals on the horizontal axis exaggerate the distance between the data points, and the vertical scale is not uniform to the horizontal scale.

Reflect and check

Here's how the graph would look like with a uniform scale:

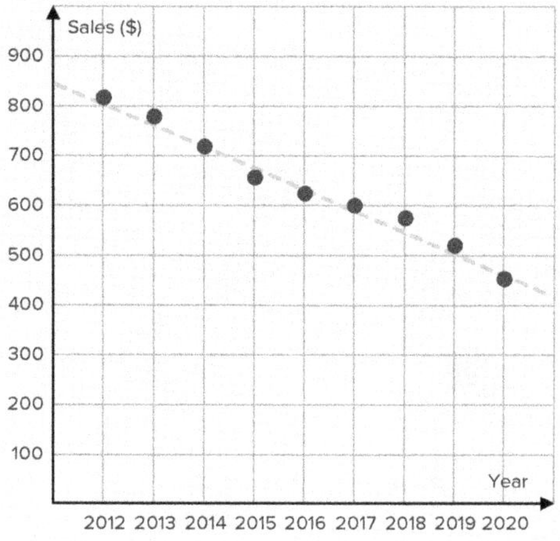

The decrease in sales appears to be faster in this graph.

Purpose

Show students how to identify misleading features in a graph.

Advanced learners: Design your own misleading graph   use with Example 2
Targeted instructional strategies

Encourage students to create their own misleading graphs using the same dataset provided. Invite them to manipulate elements such as the scales of the axes, intervals, or graphical representations to alter how the data trend appears. By designing graphs that exaggerate or minimize the decrease in sales, students will deepen their understanding of how graphical choices impact data interpretation.

This exercise empowers them to explore mathematical concepts like scaling and proportion while considering the ethical implications of data presentation. After students have created their graphs, have them present and explain the techniques they used to mislead the viewer. This will foster critical thinking and prompt discussions about accurate and ethical ways to represent data visually.

Example 4

In 2017, a state passed a "Seatbelt safety first" law in hopes that the law would reduce the number of deaths from vehicle accidents. The statistical question they want to answer is, "How does the number of deaths from vehicle accidents before 2017 compare to the number of deaths after the law was passed?"

a What type of data needs to be collected to answer their formulated question?

Create a strategy

Consider whether the data will be univariate or bivariate, and determine what variable(s) will be.

Apply the idea

The state department need to collect data on two different variables to answer their question, so they need to collect bivariate data. The first variable is the year, and the second variable is the number of deaths from vehicle accidents in each year.

Reflect and check

This data is generally collected by the state's department of transportation or the state's highway patrol. You can collect this data from either of these government websites.

Purpose

Show students that to answer a statistical question, proper identification of the type of data and the variables needed is crucial.

Students: Page 284

b The state collected the data and organized it into the graph shown. They concluded, "After passing the Seatbelt safety first law, the number of deaths related to vehicle accidents decreased."

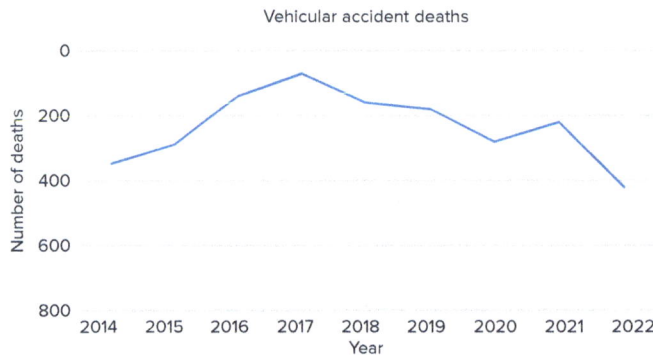

Vehicular accident deaths

Explain why this is misleading.

Create a strategy

When looking at the graph, it appears that the number of deaths does decrease after 2017. However, we should analyze the values on the vertical and horizontal scales carefully.

Apply the idea

The vertical axis on this graph does not increase from 0. Instead, the scale has been flipped, making an increase in deaths appear to be a decrease in deaths.

The correct graph would be flipped vertically, as shown.

Vehicular accident deaths

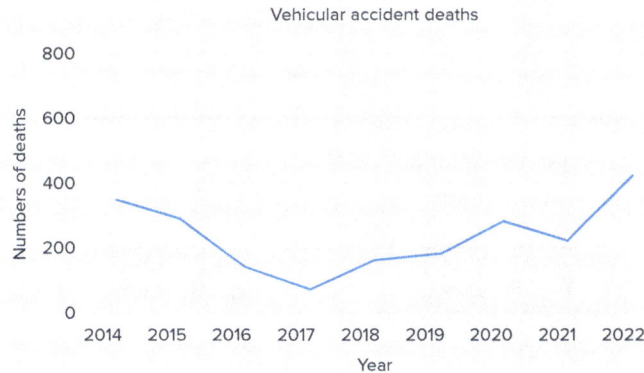

With this correct graph, we can see that the number of deaths related to vehicle accidents actually increased after the law was passed.

Purpose

Challenge students to critically analyze graphical data and understand that graph presentation can sometimes be misleading.

Reflecting with students

Encourage students to pay close attention to how data is presented, especially with labels and scales on graphs. In this example, guide them to notice that the vertical axis is inverted, which can lead to a misleading interpretation of the data. Emphasize the importance of accurately labeling axes and ensuring that scales increase in a logical manner, starting from zero if appropriate. Ask students to recreate the graph with the vertical axis correctly oriented, so that an increase in deaths is shown as an upward trend, accurately reflecting the data.

By stressing the need for precise labels and correct scales, you help students develop the habit of communicating mathematical information clearly and preventing misunderstandings. This attention to detail in labeling and scaling fosters their ability to present data precisely and interpret graphs accurately.

Students: Page 285

> 💡 **Idea summary**
>
> Some of the ways in which graphs can be used to mislead include:
> - Manipulating the **scale**, such as not starting the scale at zero or using a scale that is not uniform
> - Manipulating **intervals** that could exaggerate the distance between data points
> - Omitting important information in titles and labels
> - Omitting certain data points, such as outliers or values that do not align with the desired conclusion
> - Choosing a graphical display that does not best represent the data

Practice

Students: Page 285–292

What do you remember?

1 For each type of data display:

   i Is it used to represent univariate or bivariate data?

   ii Does it represent categorical or numerical data or both?

     a Pictograph   **b** Bar graph   **c** Line graph   **d** Stem-and-leaf plot

     e Circle graph   **f** Histogram   **g** Boxplot   **h** Scatterplot

2 300 people were asked if they liked breakfast, lunch, or dinner.

   a What should the percentages in a circle graph add up to?

   b What do the percentages in this circle graph add up to?

   c Is this an appropriate data display?

Favorite meal of the day

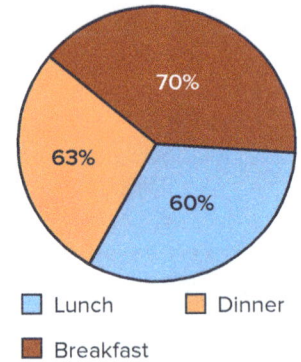

70%

63%

60%

■ Lunch   ■ Dinner

■ Breakfast

3 Select the statement that describes how this graph might be misleading.

   A The wrong graph type was used.

   B The graph doesn't start at zero.

   C The scale on the vertical axis is not uniform.

   D The intervals exaggerate the horizontal distance of the line graph.

Line graph

Let's practice

4 A student surveyed 75 other students and asked what type of pet they owned. The student drew this diagram and concluded that most students have dogs and fish.

a Why is this graph misleading?

b Suggest a way to adjust the graph so it is not misleading.

Pet	1 icon = 5 students
Dog	🐕🐕🐕🐕
Cat	🐈🐈🐈
Fish	🐠🐠🐠🐠🐠
Bird	🐦🐦
Hamster	🐹

5 What feature is misleading about the line graph?

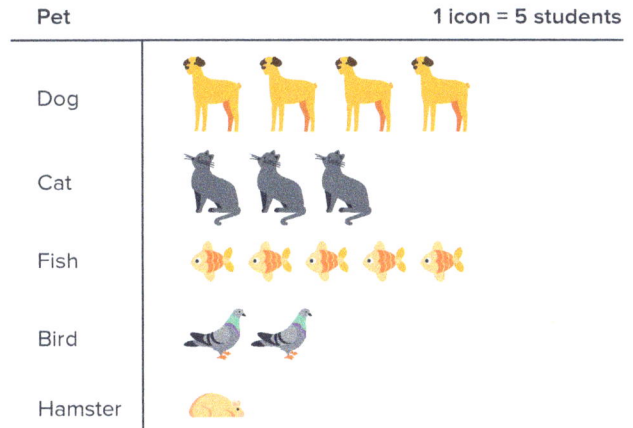

6 The figure shows the number of tourists from each country to the United States:

a What is the issue with the length of the line segments representing each country?

b What type of graph would have been better?

Top 20 tourist-generating countries to the United States of America

- Canada: 14 043 658
- Mexico: 4 295 124
- United Kingdom: 2 905 909
- Japan: 2 169 716
- Germany: 1 263 344
- France: 930 265
- Brazil: 613 347
- South Korea: 560 405
- Italy: 558 594
- Australia: 526 441
- China: 488 484
- Spain: 451 530
- India: 447 079
- Netherlands: 416 870
- Venezuela: 344 802
- Ireland: 296 771
- Colombia: 288 439
- Argentina: 271 737
- Switzerland: 262 595
- Israel: 237 818

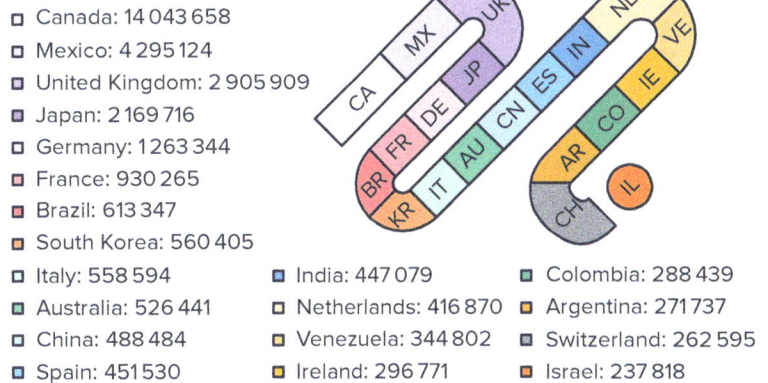

7   Consider the graph:

What two features of this graph might cause it to misrepresent the statistics?

Massive increase in house prices this year

8   The circle graph shows the most popular sports to watch on television.

Most popular sports to watch on TV

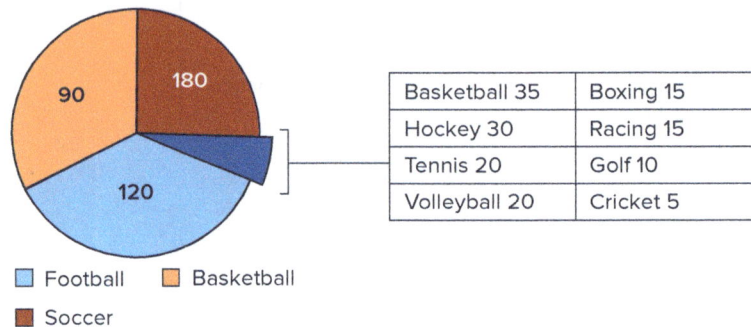

Basketball 35	Boxing 15
Hockey 30	Racing 15
Tennis 20	Golf 10
Volleyball 20	Cricket 5

☐ Football   ☐ Basketball
☐ Soccer

a   Which sport appears to be the most popular?

b   Looking closer at the numbers listed for each sport, which one is actually the most popular to watch on TV?

c   Has the circle graph been drawn accurately?

d   What type of graph would have been better for representing this data?

9   Kenau class collected data on the number of steps they took in a month. At the end of the month, each person organized their data into a histogram. The teacher planned to give out several awards, and Kenau was hoping to win the "High Stepper" award, given to the student with the most steps.

$$2000, 4000, 8000, 8000, 8000, 8500, 8500, 9000, 9500, 9500, 10000$$
$$10500, 11000, 11500, 11500, 12000, 12000, 12500, 13000, 13500$$
$$14000, 14000, 14000, 15000, 15000, 16000, 17000, 17000, 18000, 19000$$

Why is Kenau's graph misleading?

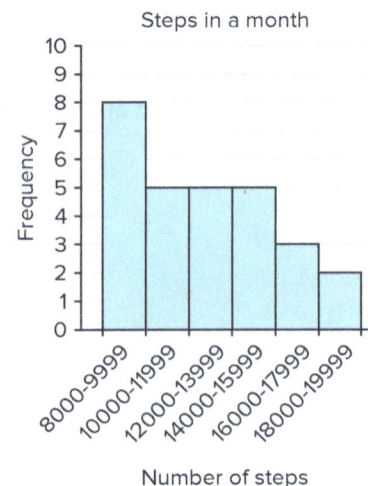

Steps in a month

10 In a study, a group of people were shown 30 names, and after one minute they were asked to list the names they could remember. The results are presented in the dot plot: After looking at the graph, Georgie concluded that most people rememebered a similar amount of names. Is she correct? Explain.

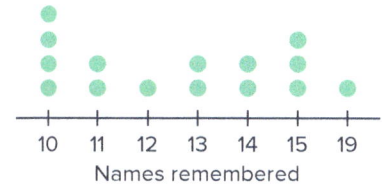

Names remembered

11 Consider the line graphs of the same set of sales figures:

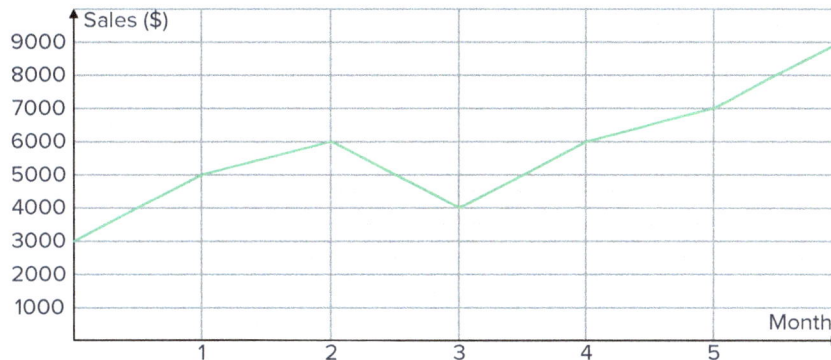

Describe how the difference in scale between the two graphs might be misleading.

12 Consider the given column graph:

a Why is this a poor example of a graph with broken axes?

b Describe a graphical display that would be better suited to this situation.

13 This histogram shows the number of accounts followed on a social media platform.

Can we conclude that most people follow over 1000 accounts on this social media platform? Explain your answer.

14 Compare the two graphs and explain which graph is misleading and why.

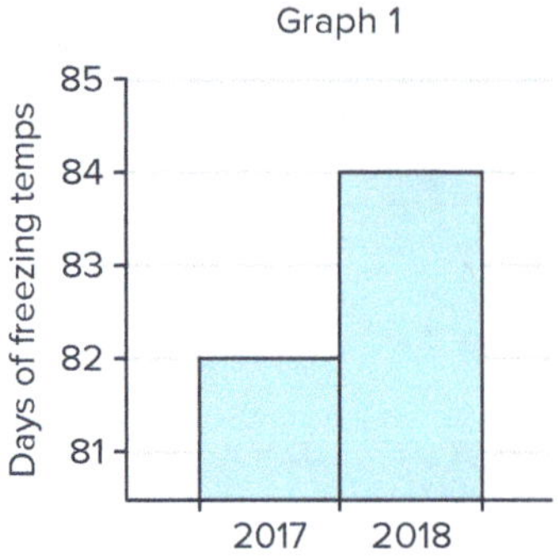

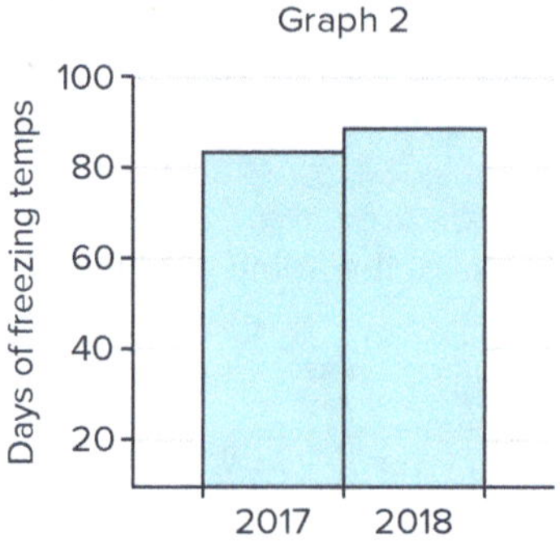

Graph 1

Graph 2

15 Data collected on the number of fights in a school district is displayed in the line graph.

a Select the option that describes why this graph is misleading.

A The y-axis scale does not increase from 0 because it is inverted.

B The x-axis scale exaggerates the distance between data points.

C Several outliers were ommitted from the graph.

D There is no information on what the data represents

School fights over time

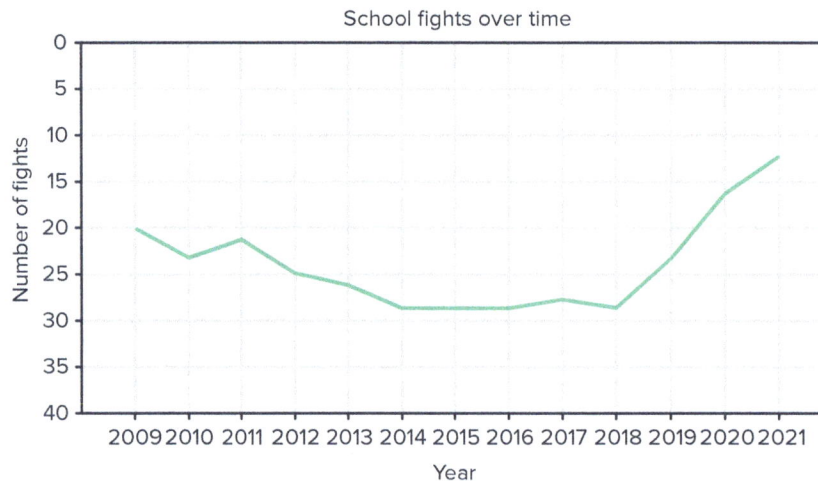

b Based on the graph, a newspaper reported "Fights in local schools have risen dramatically from 2009 to 2021". Is each statement true or false?

i The reporter has not included the data since 2014 accurately in their representation.

ii The reporter has represented the trend correctly.

iii The reporter has misinterpreted the graph because the number of fights has decreased from 2009 to 2021.

16 A new affordable housing initiative requires 10% of apartments in new buildings to cost $250000 or less. The apartments available for purchase in one particular building are priced at:

$250000, $250000, $250000, $600000, $605000, $620000
$630000, $640500, $645000, $650000, $660000, $662500
$668000, $675000, $680000, $695000, $700000, $710000
$715000, $725000, $740000, $744000, $750000, $755000
$759000, $760500, $765000, $770000, $772000, $775000

The marketer for the building says "Most apartments are under $630000" and uses this graph to support their statement.

Apartment prices (in thousands)

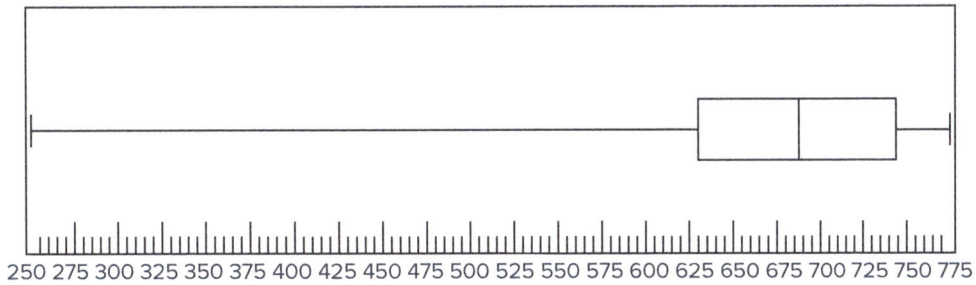

a Do the prices of the apartments meet the affordable housing initiative requirement?

b Is the marketer's statement accurate?

c What percentage of apartments are actually under $63000?

d The marketer decides to change his statement to "Units starting at $250000". Is this new statement accurate?

17 Consider the shown line graph:

The Main NAEP show dramatic gains in math.

From 1990 to 2000, fourth grade math scores increased fifteen points - equivalent to about 1.2 years of learning.

Source: Brookings report on American education

a The labels on the horizontal axis are not evenly spaced. Explain why this may be misleading.

b Is the scale on the vertical axis reasonable or misleading? Justify your reasoning.

Let's extend our thinking

18 The same information is displayed in two different ways.

a Which graph could be used to support the claim that the global temperature is increasing significantly?

b Which graph could be used to support the claim that the global temperature is not increasing significantly?

c At the United Nation's climate summit, it was stated that limiting the planet's temperature to a 1.5°C overall increase by 2100 could avoid catastrophic events. How does this claim help you understand what is considered a "significant increase" in temperature?

19 This graph was shown on an Argentinian news service with the comment that "Argentina's testing rate for Covid-19 is almost as high as that of the United States".

COVID-19 testing by country

Number of COVID-19 tests per million people

Source: Reddit May 2020

a Explain why the graph and comment are misleading.

b Redraw the graph correctly.

20 Jay is interested in understanding the relationship between how often students use a certain product and their satisfaction with it. He decides to use the data cycle to research this issue.

a Formulate a question which could be investigated using a scatterplot.

b Describe a method which could be used to collect data.

c Collect up to 20 data values for your formulated question.

d Organize your data using a scatterplot. If there appears to be a linear correlation, draw the line of best fit.

e Use the scatterplot to draw conclusions and to answer your formulated question from part (a).

f Suppose this graph is shown on the product's website. Beside the graph is the statement, "Customers are happiest when using our product on a daily basis." Determine whether this data display is misleading.

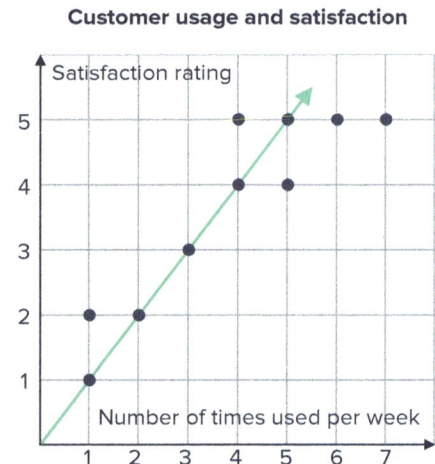

Customer usage and satisfaction

Answers

4.06 Misleading data displays

What do you remember?

1
  a i Univariate   ii Categorical

  b i Univariate   ii Categorical

  c i Bivariate   ii Numerical

  d i Univariate   ii Numerical

  e i Univariate

      ii Categorical and numerical

  f i Univariate   ii Numerical

  g i Univariate   ii Numerical

  h i Bivariate   ii Numerical

2
  a 100%   b 193%   c No

3
  C

Let's practice

4
  a The icons are not the same size, so some categories appear to have more icons than others.

  b Make all icons the same size and the same distance apart.

5
  The tick marks on the vertical axis are not evenly spaced.

6
  a They do not correspond to the tourist numbers they are supposed to represent. In particular, on the corners the sizes are exaggerated.

  b Bar graph

7
  The title is persuasive and the vertical axis scale doesn't start at 0. It also does not give an idea of the trend over time.

8
  a Football   b Soccer   c No   d Bar graph

9
  He left out the two low outliers.

10
  No, the dot plot should show a gap in the data between 15 and 19, showing that rememebering 19 names was better than the other candidates.

11
  By compressing or expanding the horizontal axis, it can appear as if the trend is gradual or quite abrupt.

12
  a The axis is broken too close to the top of the Mar 95 column, so it is not clear where the scale break is.

  b A line graph with clear indication of a broken axis below the lowest point in March.

13
  No, we cannot make that conclusion. Although following over 1000 accounts is the tallest column, the graph is missing a frequency scale on the vertical axis. If we knew the frequencies of the columns, we could find the total of all the frequencies and compare it to the frequency of the

tallest column to see if more than half of the data falls in the highest class. However, without a vertical scale, it is impossible to know where "most" of the data lies.

14
  Graph 1 is misleading because the vertical axis does not start at zero therefore the difference in the number of days is exaggerated.

15
  a **A**

  b i True   ii False   iii True

16
  a Yes   b No   c 25%   d Yes

17
  a It has made the increases in the 4-year intervals 1992-1996 and 1996-2000 appear faster than they really are (relative to the rate in the 2 -year interval 1990-1992).

  b Example answer:

    Although the scale on the y-axis does not begin from 0, it is reasonable because the difference in the data is small. If the scale began from 0, it would be difficult to choose a scale that would allow the reader to identify the exact values for each year.

Let's extend our thinking

18
  a The graph with the vertical axis that goes from 13 to 15.

  b The graph with the vertical axis that goes from 0 to 20.

  c Since 1920, the temperature has increased by around 0.8°C. The claim states that a further increase by 0.7°C in the next few decades could lead to an increase in catastrophic events. This shows that increases by less than 1°C are actually significant increases in temperature.

19
  a The y-axis scale is not shown and therefore the Argentinian column appears taller than it should be. There is actually a large difference in the rates of the USA and Argentina.

  b

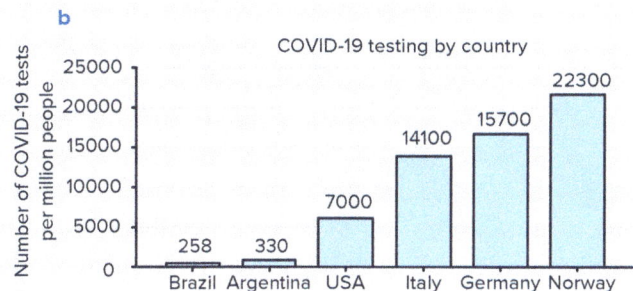

COVID-19 testing by country

Country	Number of COVID-19 tests per million people
Brazil	258
Argentina	330
USA	7000
Italy	14100
Germany	15700
Norway	22300

20
  a Many possible answers, for example, "Is there a relationship between usage frequency of a product and a user's satisfaction rating?"

  b One possible method is to create a survey asking students to record how many times they use the specified product each week and to rate their satisfaction with the product on a scale from 1 to 5.

c Answers will vary based on the method of data collection used.

An example data set based on a survey where participants recorded the number of times they used the product per week and rated their satisfaction on a scale of 1 to 5 is shown:

Usage Frequency (times per week)	Satisfaction Rating (1–5)
2	2
5	4
1	1
4	5
3	3
6	5
1	2
7	5
3	3
2	2
5	4
4	4
6	5
1	1
3	3
7	5
2	2
4	4
5	5
6	5

d Answers will vary. This scatterplot represents the example data from above.

Customer usage and satisfaction

e Answers will vary.

The scatterplot shows that there is a positive linear correlation between usage frequency and a user's satisfaction rating. As the number of times the product is used each week increases, the customer's satisfaction rating also increases.

f Yes, this graph is misleading because the line of best fit is drawn incorrectly. The line shows a steady, steep increase in satisfaction as the product is used more frequently. However, this is not the line of best fit because there are more data values below the line than above the line. The actual line of best fit still shows a positive relationship, but the incline is not as steep.

Topic 4 Assessment: Data Analysis

1 Select the question which could be answered by collecting data and representing it with a boxplot. Select all that apply.

   A What is the income range of the middle half of the population?

   B What is the mean age of buildings in my city?

   C What is the range of scores on the SOL test?

   D Why is the interquartile range of any data set less than or equal to the range?

2 Determine if each data set would allow you to answer these questions. If it would not be an appropriate data set, explain why not.

   Question 1:

   What is the distribution of annual spending for Americans?

   Question 2:

   What is the median amount of annual spending for Americans?

   a The results of a survey of 100 randomly selected people walking out of a mall in Richmond, VA.

   b Anonymized data from all major credit card companies with individual totals for December.

   c Anonymized income reports from the census.

   d The results of a survey of 100 000 randomly selected Americans.

3 A city mayor wants to estimate the number of residents who use public transportation daily. The city has 50 000 residents. For each of the given samples:

   i State if the sample is biased or unbiased.

   ii If biased, explain how the sample data might be different from a representative sample. If unbiased, explain how you know.

   a All residents who work as CEOs of their workplaces.

   b Fifteen residents at the public park at city center.

   c Fifty randomly selected residents from each of the 12 neighborhoods.

4 14 people were asked how many hours of sleep they had gotten the previous night. The numbers below are each person's response:

$$1, 6, 9, 8, 10, 9, 6, 12, 6, 6, 4, 7, 6, 10$$

   a Write a statistical question that could be answered using the given data.

   b Construct a boxplot for this data.

   c Answer the statistical question from part (a) using the boxplot and the original data set.

SOL **5** The boxplots shows the distances, in centimeters, jumped by two high jumpers, Paul and Matt:

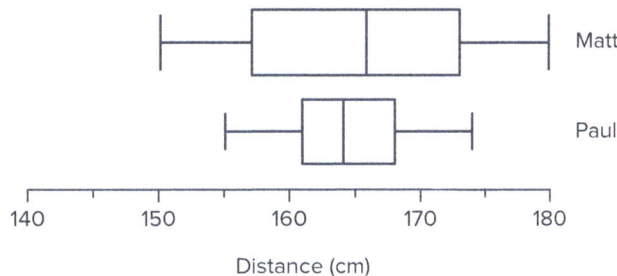

What conclusion(s) can we draw using the boxplots shown? Select all that apply.

   A Paul has a higher median jump than Matt.

   B Matt has a bigger range than Paul.

   C Paul has a lower mean than Matt.

   D 50% of Matt's jumps are higher than Paul's.

   E Matt's lower extreme is 5 cm smaller than Paul's.

6 A local interest group wants to poll citizens of the town to see how many minutes a day they spend commuting to and from work to see if more public transportion is needed. The group polled random citizens on a busy street and made a display of their results.

Which data display most likely represents the data collected from the poll? What are the advantages of this type of display?

a Set A

Score

b Set B

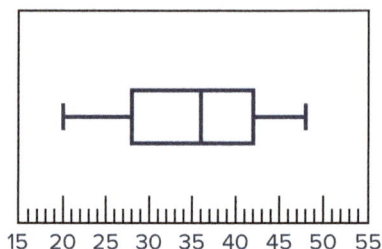

c Set C

6	5 8 8
7	0 2 5 5 6 8
8	0 1 2 3 3 5

Key 1|2 = 12

d Set D

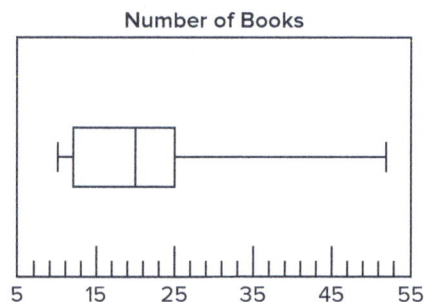

7 The boxplot shows the number of books a group of students read in a year:

a Find and interpret the range.

b What percentage of students read from 12 to 20 books?

c The middle 50% of students read within how many books of each other?

d In which quartile is the number of books read the most spread out?

e The bottom 50% of students read within how many books of each other?

Number of Books

8 The boxplots summarise the results from a botanical study. The treatment group received an experimental fertiliser to boost plant growth, and the control group received regular water. The boxplots show the number of weeks each group of plants took to reach maturity:

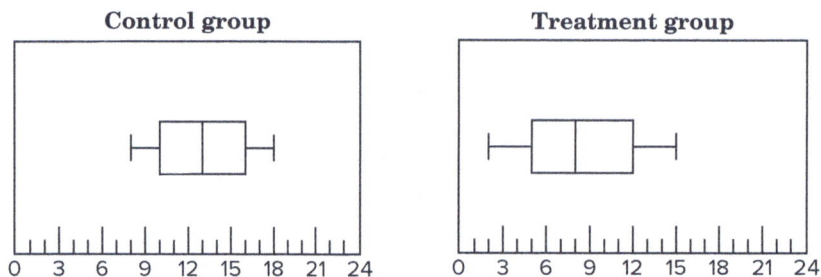

a Compare the medians of the control group and the treatment group.

b Compare the ranges of the control group and the treatment group.

c Compare the interquartile ranges of the control group and the treatment group.

d Does the fertiliser have a positive effect on plant growth speed? Explain your answer.

9 The average gas price of all 50 states in 2021 is shown in the boxplot:

Average Gas Price by State

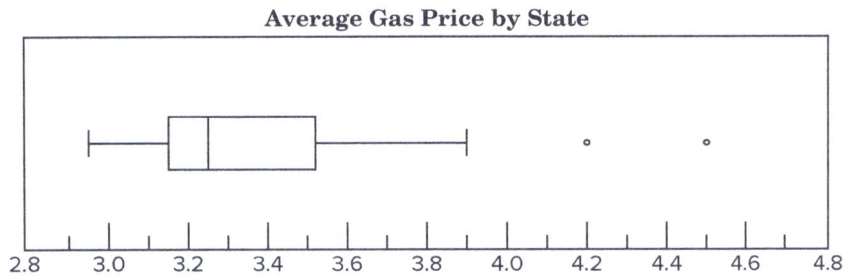

The two dots represent outliers that were not represented in the boxplot.

Describe which features of the boxplot would be most changed if these outliers were included in the boxplot, and which would not be likely to change.

10 What feature is misleading about the line graph?

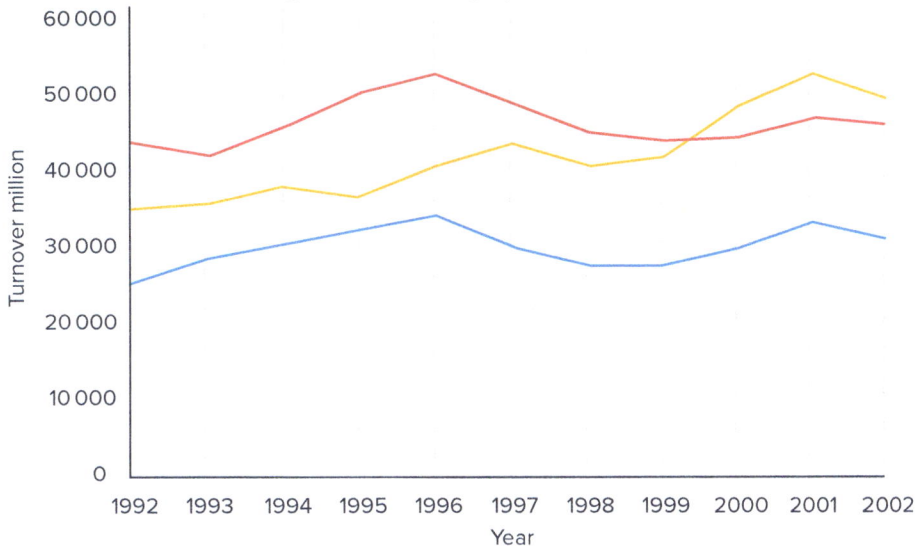

SOL **11** Which of these best describes the relationship of the data shown on the scatterplot?

A Constant relationship

B Negative relationship

C Positive relationship

D No relationship

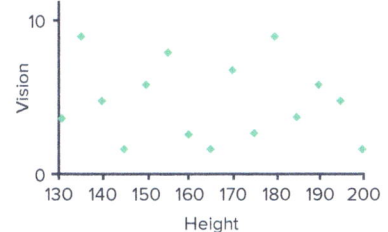

12 Use the scatterplots to answer the questions.

    a Which statistical question could be answered using the given scatterplot?

        A Does someone's shoe size affect the amount of time someone spends studying?

        B Is there a relationships between the number of times per day that people check their phones and their emails?

        C Is social media usage related to height?

    b Which line represents the line of best fit for the data shown?

13 The following scatterplot graphs data for the number of copies of a particular book sold at various prices.

    a Sketch the line of best fit for this data.

    b Use the line of best fit to find the number of books that will be sold when the price is:

        i $33

        ii $18

    c Is the relationship between the price of the book and the number of copies sold positive or negative?

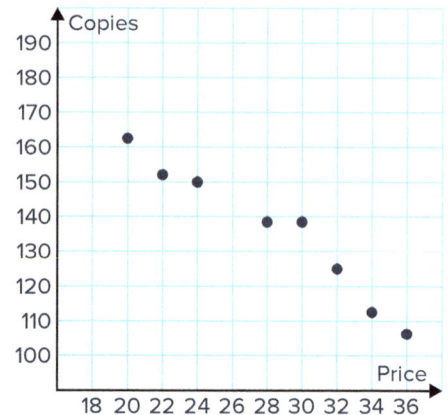

14 A student was performing an experiment to study the relationship between the current and voltage through a resistor. He noted his results in the following table:

Current	1	2	3	4	5	6	7	8	9	10
Voltage	1	6	3	10	13	10	16	15	21	17

    a Construct a scatterplot for the data in the table.

    b Describe the relationship between the data points.

15 Fidelia formulated the question "Is there a relationship between the time since a seed was planted and the plant's height?" Which table would best represent data collected to answer this question?

A

x	0	1	2	3	4	5	6
y	0	0.4	1.3	2.8	4.5	5.6	6.3

B

x	0	1	2	3	4	5	6
y	6.3	5.6	4.5	2.8	1.3	0.4	0

C

x	0	0.4	1.3	2.8	4.5	5.6	6.3
y	0	1	2	3	4	5	6

SOL **16** Fill in the blanks with the appropriate terms to complete the statement. Terms can be used more than once, or may not be used at all.

Possible terms:
- positive
- negative
- increases
- decreases

The graph shows a ☐ relationship since the value of y ☐ as x ☐.

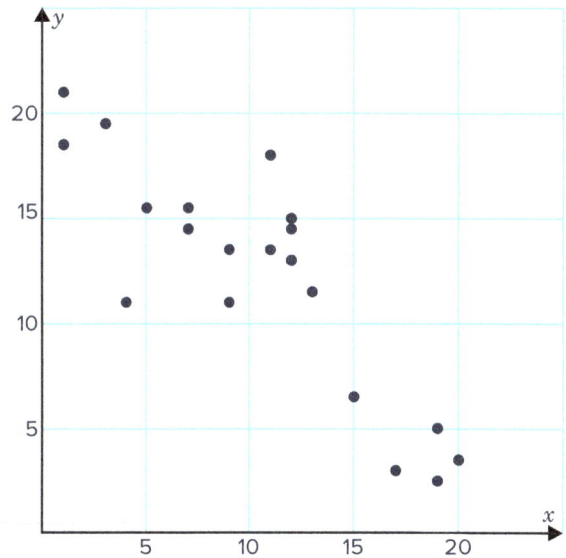

Performance task

17 A teacher plans to give the students in their 8th grade class a pre-test and a post-test on their statistics unit to see their teaching is effective.

 a Formulate a statistical question that could be answered using:

 i The bivariate data that would be collected.

 ii The data from the independent variable only.

 b Select a method of data collection that would be appropriate and how the teacher could collect the data.

 c Suppose this data was collected by the teacher. Represent the data for the pre-test using a boxplot.

Student	Pre-test	Post-text
Gemma	40	72
Hettie	37	77
Indu	45	93
Jaleesa	31	60
Keshawn	47	89
Lu	36	76
Morgan	39	64

Student	Pre-test	Post-text
Nadine	35	67
Oscar	43	89
Patrice	26	59
Quang	45	93
Ronaldo	27	72
Sadie	49	95
Troy	28	62

 d Analyze the boxplot and draw a conclusion to answer the statistical question from part (a)-i.

 e Complete the data cycle for the statistical question using it as bivariate data.

Answers

Topic 4 Assessment: Data Analysis

1 **A, C**

8.PS.2a

2 **a** Not appropriate, as the sample does not represent the diverse population of America and it is too small.

b Not appropriate, as spending in December is typically higher than the rest of the year and this would not account for cash or debit spending.

c Not appropriate, as income may be correlated with spending, but cannot be used to make conclusions about spending.

d Appropriate. This data would be more challenging to collect, but would allow us to answer both questions.

8.PS.2b

3 **a** **i** Biased

ii CEOs do not represent the general population. They would be more likely to have a higher income, so may be less likely to take public transit and more likely to use a car. The estimate from this sample would be too low.

b **i** Biased

ii Fifteen residents is not a large enough sample to represent the demographics of the larger population. Also, those at a public park at city center might live more centrally and have different transportation needs than those who live on the edges of town. The estimates from this sample would likely be inaccurate.

c **i** Unbiased

ii This sample would represent the different demographics across the city as all of the different neighborhoods are represented. The sample is also large enough to represent the population.

8.PS.2c

4 **a** Many possible answers. For example, "How much sleep do people typically get per night?"

b

Hours of Sleep

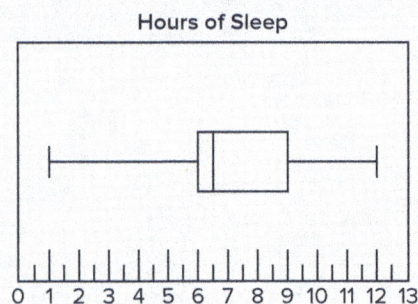

c This is a small data set, so this conclusion may not be widely applicable. The amount of sleep people get varies significantly with a range of 11 hours from 1 hour up to 12 hours. But a very small number slept less than 6 hours. Half of the people slept between 6 and 9 hours, with a median of 6.5 hours.

8.PS.1a, 8.PS.2d, 8.PS.1g

5 **B, D, E**

8.PS.2e, 8.PS.2h

6 The data is most likely represented by the boxplot in Set B. The other data displays have values that would either be too small for a commute, or are too large to consistently be a random group of people. The advantage to the boxplot is we can easily see how the data is concentrated and how spread out the values are.

8.PS.2g, 8.PS.2i

7 **a** The number of books read by students varied by 42 books.

b 25%

c 13

d 4th quartile

e 10

8.PS.2e, 8.PS.2g

8 **a** The median of the control group is 5 weeks higher than the median of the treatment group.

b The treatment group's range is 3 weeks wider than the control group's range.

c The treatment group's interquartile range is 1 week wider than the control group's range.

d Yes. The group that received the experimental fertiliser matured faster since it had a lower median, and most plants reached maturity in less than 12 weeks. In the control group, half of the plants took more than 13 weeks.

8.PS.2h

9 The upper extreme and 4th quartile will be most affected by the outliers. The median and upper quartile may change slightly, but it is unlikely that the lower quartile will change. The lower extreme will not change at all.

8.PS. 2 f

10 The tick marks on the vertical axis are not evenly spaced.

8.PS.2j

11 **D**, There is no relationship between someone's vision and their height.

8.PS.3d

12 **a** **B**

b Line 2

8.PS.3a, 8.PS.3f

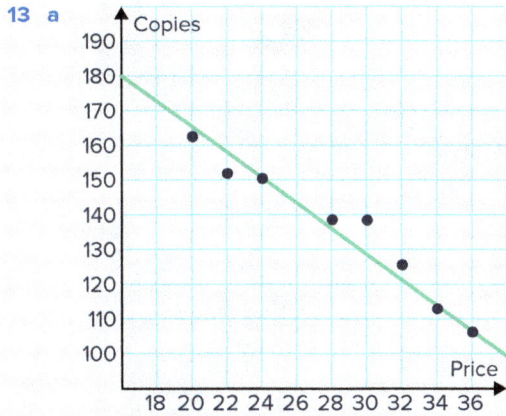

13 a

Copies / Price scatterplot

b i Approx. 117

  ii Approx. 173

c Negative

8.PS.3d, 8.PS.3e, 8.PS.3f

14 a

Voltage / Current scatterplot

8.PS.3c, 8.PS.3d

15 A

8.PS.3b

16 The graph shows a negative relationship since the value of y decreases as x increases.

8.PS.3d, 8.PS.3e

Performance task

17 a i Is there a relationship between the pre-test and post-test scores for 8th grade students on the statistics unit?

  ii How do 8th grade students do on the pre-test for a statistics unit? or How well do 8th grade students know the statistics content before doing the unit work?

b An experiment would be an appropriate data collection. The teacher can give the students the same test with different numbers or scenarios first before teaching the unit and then after teaching the unit. The data for each students should be recorded in a table.

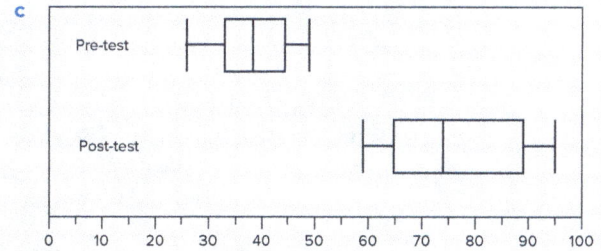

c

Pre-test / Post-test box plots

d The pre-test is evenly spread between 26 and 49 (range of 23), with a median around 38.

This means that students typically score around 40 on the pre-test and it is fairly consistent between 26 and 49 .

e We have already formulated the question and collected the data, so now we need to organize and represent the data, then draw conclusions.

We can create a scatterplot using technology or by hand.

From the scatterplot, we can see that there there is a positive relationship. It could be modeled with a linear function.

The relationship is fairly strong, but not a perfect model.

We can conclude that the pre-test is a predictor of the post-test and that their scores improve after the teaching, so it is effective.

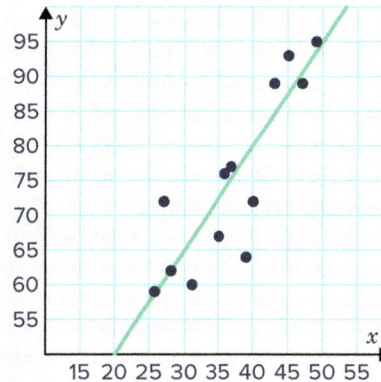

Scatterplot with trend line

8.PS.2a, 8.PS.2b, 8.PS.2d, 8.PS.2g, 8.PS.3a, 8.PS.3b, 8.PS.3c, 8.PS.3d, 8.PS.3e, 8.PS.3f

5 Probability

Big ideas

- Probability helps us analyze the chance that an event will occur and provides us with tools to make decisions about the future based on known information.

Chapter outline

When picking cards from a deck, if you pull out a face card first, the chance of drawing another face card changes with each draw!

5. Probability

Topic Overview

Foundational knowledge

Evaluating standards proficiency

The skills book contains questions matched to individual standards. It can be used to measure proficiency for each.

7.PS.1 — The student will use statistical investigation to determine the probability of an event and investigate and describe the difference between the experimental and theoretical probability.

Big ideas and essential understanding

Probability helps us analyze the chance that an event will occur and provides us with tools to make decisions about the future based on known information.

5.01 — Determining the independence or dependence of events is essential to drawing accurate conclusions.

5.02 — Determining the independence or dependence of events is essential to drawing accurate conclusions.

5.03 — Determining the independence or dependence of events is essential to drawing accurate conclusions.

5.03 — Knowing the outcome of one event can change the sample space from which the likelihood of a second event is calculated. If the sample space is influenced by the outcome of the first event, then the two events are dependent.

Standards

8.PS.1 — The student will use statistical investigation to determine the probability of independent and dependent events, including those in context.

8.PS.1a — Determine whether two events are independent or dependent and explain how replacement impacts the probability.
5.01 Introduction to independent and dependent events
5.02 Probability of independent events
5.03 Probability of dependent events

8.PS.1b — Compare and contrast the probability of independent and dependent events.
5.03 Probability of dependent events

8.PS.1c — Determine the probability of two independent events.
5.02 Probability of independent events

8.PS.1d — Determine the probability of two dependent events.
5.03 Probability of dependent events

Future connections

G.RLT.1 — The student will translate logic statements, identify conditional statements, and use and interpret Venn diagrams.

5.01 Introduction to independent and dependent events

Subtopic overview

Lesson narrative

In this lesson, students will learn about the concepts of independent and dependent events in probability. They will explore how the outcome of one event can affect the outcome of another, distinguishing between events that do not influence each other (independent) and those that do (dependent). The lesson includes various examples, practice problems, and visual aids such as tree diagrams to illustrate these concepts. By the end of the lesson, students should be able to identify and analyze independent and dependent events and understand the impact of replacement on probability.

Learning objectives

Students: Page 296

After this lesson, you will be able to...
- identify whether two events are independent or dependent.

Key vocabulary

- dependent events
- independent events
- sample space

Essential understanding

Determining the independence or dependence of events is essential to drawing accurate conclusions.

Standards

This subtopic addresses the following Virginia 2023 Mathematics Standards of Learning standards.

Mathematical process goals

MPG3 — Mathematical Reasoning

Teachers could enhance mathematical reasoning by engaging students in activities that require them to use inductive and deductive reasoning skills. For example, when discussing the concept of replacement and its impact on the probability of events, teachers can ask students to justify their reasoning on why when an item is replaced, the events become independent, and when an item is not replaced, the events become dependent.

MPG4 — Mathematical Connections

Teachers can provide opportunities for students to relate the concepts of independent and dependent events to other areas of mathematics and real-world contexts. Connections to statistics can be made by analyzing how dependent events affect data interpretation in experiments. Real-world examples from science, such as genetics with Punnett squares, can be used to illustrate dependent events, while examples from gaming or technology can be used to show independent events.

MPG5 — Mathematical Representations

Teachers can incorporate mathematical representations by encouraging students to utilize various methods to represent their understanding of independent and dependent events. This could include using diagrams, graphs, or symbolic notation. For instance, they could represent the scenario of choosing a marble from a bag without replacement in a tree diagram or a probability model, helping students to visualize and understand the concept better.

Content standards

8.PS.1 — The student will use statistical investigation to determine the probability of independent and dependent events, including those in context.

8.PS.1a — Determine whether two events are independent or dependent and explain how replacement impacts the probability.

Prior connections

7.PS.1 — The student will use statistical investigation to determine the probability of an event and investigate and describe the difference between the experimental and theoretical probability.

Future connections

G.RLT.1 — The student will translate logic statements, identify conditional statements, and use and interpret Venn diagrams.

Lesson Preparation

Suggested review

Depending on your students' level of prior knowledge, consider revisiting the following lessons:

Grade 7 — 5.03 Experimental probability

Student lesson & teacher guide

Independent and dependent events

The lesson formally defines the terms independent events and dependent events. Independent and dependent events can be determined by analyzing whether one event impacts the other. In an independent event, the two events do not impact the results of one another. In a dependent event, the first event impacts the results of the second.

Concrete-Representational-Abstract (CRA) Approach
Targeted instructional strategies

Concrete: Engage students with hands-on activities to explore independent and dependent events. Use two separate bags filled the same way with three different colored marbles or beads, for example, one blue, one red, and one yellow marble in each bag. For independent events, have students draw one marble from each bag at the same time and note the outcome of the color of the two marbles. Have them repeat this process 20 times. Next, for dependent events, ask students to draw a marble, and then, without replacing it, draw a second marble from the same bag. Have them note the outcome of the color of the two marbles. Have them repeat this process 20 times. Ask probing question like "Did you get two of the same color in either of the experiments? If so, which one? If not, why not?" or "For the different experiments, what came after/with a blue marble most often? Why do you think that is?"

To prepare for the representational stage, have students organize their data. For example, some students may benefit from being given these tables to record their results.

Outcome (Bag 1, Bag 2)	Number of occurences	Possible or impossible?
(B,B)		
(B,Y)		
(B,R)		
(R,R)		
(R,Y)		
(R,B)		
(Y,Y)		
(Y,B)		
(Y,R)		

Outcome (Draw 1, Draw 2)	Number of occurences	Possible or impossible?
(B,B)		
(B,Y)		
(B,R)		
(R,R)		
(R,Y)		
(R,B)		
(Y,Y)		
(Y,B)		
(Y,R)		

Representational: Have students dump out the marbles, placing them on top of the bags. Guide students to create tree diagrams or charts that represent the possible outcomes of the marble draws. For the independent events, have them draw branches showing all combinations of marbles from the two bags. Use colors in their diagrams to match the marbles, which helps make the connection clear. For dependent events, show how the branches change when a marble is not replaced, altering the probabilities. Provide examples on the board, illustrating how each choice leads to the next outcome. If possible, include images of tree diagrams with colored branches to match the marbles drawn.

Prepare students to move from visual diagrams to mathematical symbols. Explain that the diagrams represent probabilities that can be calculated using formulas. Highlight how each branch in the tree diagram corresponds to a specific probability expression.

Abstract: Introduce the formal probability notation and formulas for independent and dependent events. Explain that for independent events, the probability of both events occurring is the product of their individual probabilities:

$$P(A \text{ and } B) = P(A) \times P(B)$$

For dependent events, the probability changes based on the first event:

$$P(A \text{ and } B) = P(A) \times P(B \text{ after } A)$$

Work through examples using numbers, and encourage students to calculate probabilities using these formulas. Refer back to their diagrams and marble activities to make the abstract concepts more concrete.

Connecting All Stages: Throughout the lesson, consistently link the hands-on activities, visual representations, and mathematical formulas. Ask questions like, "How does the tree diagram represent the marble draws you did?" or "How does the formula relate to the choices shown in your diagram?" This helps students monitor their thinking and choose the best way to understand and solve probability problems. Encourage them to explain their reasoning at each stage, reinforcing the connections between doing, seeing, and symbolizing.

Examples

Students: Page 297

Example 1

Determine whether the selections in each experiment are independent or dependent.

a   A teacher has a "prize bag" filled with different prizes. The students form a line to draw a prize from the bag at random. Once a student has drawn a prize, they take it back to their desk.

Create a strategy

Determine if the chances for an event changes as a result of the previous event.

Apply the idea

Since each student's prize is kept and not returned to the bag, it will affect the chance of selecting a particular item from the prize bag. As a result, the events are dependent.

Purpose

Check students can distinguish between independent and dependent events in probability, using real-world scenarios.

Expected mistakes

Students might incorrectly identify the events as independent, thinking that the selection of a prize by one student does not affect the selection by the next student. They may not realize that once a prize is taken, it is no longer available for the following students, thus affecting the probabilities.

Reflecting with students

Have students consider how the teacher could change the scenario to be independent. For example, the teacher could have extra prizes and replace each one as it is chosen.

b A card is randomly selected from a normal deck of cards, and then returned to the deck. The deck is shuffled and another card is selected.

Create a strategy	Apply the idea
Determine if the chances for an event changes as a result of the previous event. In this case, will the card selected second be affected by the first card selected?	Each card drawn is returned back into the deck, so the chances of picking a certain card on the following draw are unaffected. The events are independent.

Purpose

Help students understand the concept of independent events in probability, and illustrate that if events do not affect each other, they are said to be independent.

Stronger and clearer each time
use with Example 1
English language learner support

Ask students to write an initial explanation of whether the selections in this experiment are independent or dependent, including their reasoning. Then, have students pair up to share their explanations with a partner. Encourage them to listen carefully and provide constructive feedback, focusing on the use of precise mathematical language such as "sample space," "independent events," "probability," and "outcomes." After the discussion, give students time to revise their original explanations, incorporating new vocabulary and clarifying their reasoning based on the feedback they received. This process allows students to refine their understanding and improve their ability to communicate mathematical ideas clearly and accurately.

Example 2

This spinner is spun and a six-sided die is rolled.

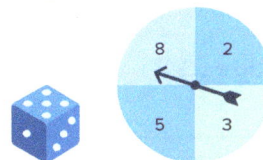

a Create an array that shows the sample space for the experiment.

Create a strategy

Place the outcomes of the die on the left side of the array and the outcomes for the spinner along the top of the array. Then, fill each cell with the outcomes of each.

Apply the idea

The array for the result of spinning the spinner and rolling a six-sided die is given by:

	8	2	3	5
1	1, 8	1, 2	1, 3	1, 5
2	2, 8	2, 2	2, 3	2, 5
3	3, 8	3, 2	3, 3	3, 5
4	4, 8	4, 2	4, 3	4, 5
5	5, 8	5, 2	5, 3	5, 5
6	6, 8	6, 2	6, 3	6, 5

Reflect and check

Recall that the number of outcomes can be determined using the fundamental (basic) counting principle. This principle tells us that the total number of outcomes in the sample space is the product of the outcomes for each event.

There are 6 possible outcomes of the die, and 4 possible outcomes of the spinner, so there are $6 \cdot 4 = 24$ total possible outcomes.

Purpose

Show students how to create a sample space for a compound event using an array, and to apply the fundamental counting principle to determine the total number of possible outcomes.

b Let A be the event that the die is rolled.

Let B be the event that the spinner is spun.

Are the events A and B independent or dependent?

Create a strategy

Consider whether the outcome from rolling the die will affect the outcome of spinning the spinner.

Apply the idea

The number rolled on the die will have no effect on which number is spun on the spinner.

Events A and B are independent.

Purpose

Help students understand the concept of independent and dependent events in probability. This example will show them how to determine whether two events are independent (i.e., the outcome of one event does not affect the outcome of the other).

Use color-coding to support creating sample space arrays
Student with disabilities support

use with Example 2

When guiding students to create the sample space array for this experiment, use color-coding to help them organize and differentiate the outcomes. Assign one color for the die outcomes, 1–6, and another color for the spinner outcomes 2, 3, 5, 8. Encourage students to set up a grid or table where the die results are listed along one axis in one color, and the spinner results along the other axis in a different color. As students fill in the grid with all possible outcome pairs, the consistent color scheme will help them visually parse the information and keep track of the combinations. This approach supports students who may struggle with visual-spatial processing by providing a clear and organized structure. To reinforce the concept of independence between events A and B, highlight how the outcomes from the die and the spinner combine without affecting each other.

Consider providing a template of the grid with the axes already color-coded, or offer colored pencils and markers for students to use. This visual aid can make the task more accessible and engaging for all students.

Idea summary

To determine whether two events are independent or dependent:

- If the events are affected by what has already happened, they are **dependent** upon each other.
- If a previous event makes no difference to what can happen in the future, they are **independent** of each other.

With and without replacement

This lesson focuses on understanding the concepts of independent and dependent events through the mechanism of selection with or without replacement.

Exploration

Students: Page 298

Exploration

The jar of marbles shown contains 7 red marbles, 3 blue marbles, and 10 green marbles.

1. A red marble is drawn from the jar, then placed back into the jar. How many marbles are in the jar?

2. A green marble is removed from the jar. How many marbles are left in the jar?

3. Suppose one marble is drawn and placed back into the jar, then a second marble is drawn. Let event A represent drawing a red marble first and event B represent drawing a green marble in the second draw. Are these events independent or dependent? Explain.

4. Suppose one marble is drawn, then a second marble is drawn. The first marble was not replaced. Are events A and B independent or dependent? Explain.

Suggested student grouping: Small groups

This exploration uses a jar of marbles to develop deeper understanding of independent and dependent events. By comparing different events, students are guided to understand the effects of replacing or not replacing marbles after they are drawn and how this affects the independence of the events.

Ideal student responses

These ideal responses may differ from other correct student responses. Less formal responses can be connected with the more precise mathematical language presented here.

1. **A red marble is drawn from the jar, then placed back into the jar. How many marbles are in the jar?**
   Since the red marble was replaced, the total number of marbles in the jar remains unchanged. There are still 20 marbles in the jar (7 red, 3 blue, 10 green).

2. **A green marble is removed from the jar. How many marbles are left in the jar?**
   After removing one green marble from the jar, there are now 19 marbles left in the jar (7 red, 3 blue, 9 green).

3. **Suppose one marble is drawn and placed back into the jar, then a second marble is drawn. Let event A represent drawing a red marble first and event B represent drawing a green marble in the second draw. Are these events independent or dependent? Explain.**
   Events A and B are independent because the replacement of the first marble ensures that the composition of the jar remains unchanged for the second draw. The probability of drawing a green marble (event B) is not affected by the outcome of the first draw (event A).

4. **Suppose one marble is drawn, then a second marble is drawn. The first marble was not replaced. Are events A and B independent or dependent? Explain.**

Events A and B are dependent in this case because the first marble drawn is not replaced. Therefore, the composition of the jar is altered after the first draw, affecting the probability of the second draw. The probability of drawing a green marble (event B) after drawing a red marble first (event A) and not replacing it will be based on a reduced total number of marbles, affecting the likelihood of the outcomes.

Purposeful questions

- When a red marble is drawn from the jar and then placed back into the jar, how does this action affect the total number of marbles? Why does the number of marbles remain the same in this scenario?
- How does the removal of a marble without replacement affect the total count and the probabilities of drawing another marble?

Possible misunderstandings

- Students might misunderstand the concept of replacement in probability problems, thinking that the total number of marbles in the jar changes even when a marble is placed back after being drawn. For example, they might incorrectly state that the number of marbles decreases in all scenarios, regardless of whether the marble was replaced.

Designing scenarios to deepen understanding of event independence
Targeted instructional strategies

Invite students to create their own examples of independent and dependent events to share with the class. Encourage them to think of real-world situations or experiments where one event affects another and where events are unrelated. For instance, they might consider how weather could impact outdoor activities (dependent events) versus rolling a die and flipping a coin (independent events).

By designing these scenarios, students actively engage with the definitions and nuances of event independence and dependence. This process helps them internalize the concepts as they explain their reasoning and analyze the relationships between events. Facilitating discussions around their examples will further deepen their understanding and allow them to explore a variety of contexts where these concepts apply. Have them consider events that involve some randomness, but also some reasoning like playing rock-paper-scissors

Examples
Students: Page 300

Example 3

The tree diagram shows all the ways a captain and a co-captain can be selected from Matt, Rebecca and Helen.

Are the events of selecting a captain and a co-captain independent or dependent?

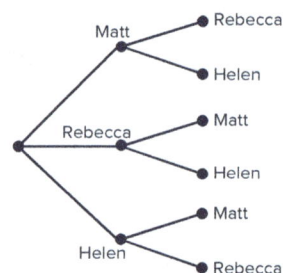

Create a strategy

Any of the three people can be chosen as captain. Once a captain has been selected, there are only two remaining people that can be selected as co-captain.

Apply the idea

The number of possible outcomes between the first selection and the second selection are different. The are 3 people that could be chosen as captain, but only 2 people can be chosen from for co-captain. This means the second selection is affected by the first selection.

Since the person that gets selected as co-captain will depend on who is selected as captain, the events are dependent.

Purpose

Show students how to use a tree diagram to determine whether events are independent or dependent.

Students: Page 300

Example 4

Four cards numbered 1 to 4 are placed face down on a table. Two cards are drawn with replacement.

a Construct a tree diagram of this situation.

Create a strategy

Since the two events occur with replacement, the number of options for both draws will be the same.

Apply the idea

For the first selection, we have 1, 2, 3, 4 as possible options. Since the first card is replaced, the options will stay the same for the second selection.

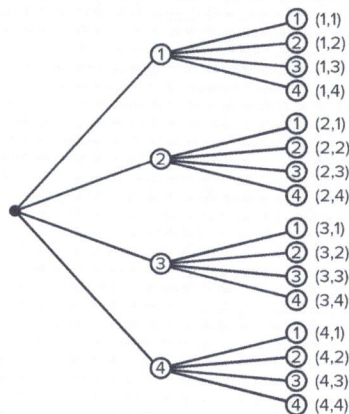

Reflect and check

We could also have used a chart or an array to determine the possible outcomes.

	1	2	3	4
1	(1, 1)	(1, 2)	(1, 3)	(1, 4)
2	(2, 1)	(2, 2)	(2, 3)	(2, 4)
3	(3, 1)	(3, 2)	(3, 3)	(3, 4)
4	(4, 1)	(4, 2)	(4, 3)	(4, 4)

Regardless of which method we use, the sample space is the same.

A good check is that the size of the sample space is the number of possibilities in the first step times the number of possibilities in the second step.

Sample space size = $4 \cdot 4 = 16$ which matches what we see in the tree diagram or array.

Purpose

Show students how to construct a tree diagram for a situation involving drawing cards with replacement, and how to determine the size of the sample space.

Reflecting with students

Have students consider how they could find the total number of outcomes without the tree diagram. Then ask them if it would be the same if the card was not repalced each time.

Students: Page 301

b Are the events of selecting two cards independent or dependent?

Create a strategy

We can use the fact that the cards are drawn with replacement to determine whether they are independent or dependent.

Apply the idea

From the tree diagram, we can see that the number of outcomes for both selections is the same. Because the first card is replaced, the outcome of the first draw will not affect the outcome of the second draw.

Therefore, the events are independent.

Purpose

Show students how to determine whether events are independent or dependent by analyzing the conditions of the situation.

🎓 **Advanced learners: Extend with coding**   use with Example 4
Targeted instructional strategies

To develop coding skills and help students understand the concept of listing all possible outcomes when two cards are drawn with replacement, guide them in writing a simple program using loops. Start by showing them how to create a list representing the cards and use nested loops to generate all possible pairs. This hands-on coding activity will reinforce their grasp of listing sample spaces with replacement and introduce fundamental programming concepts like loops and list manipulation.

For example, in Python we could use this code and compile with Google Colabs:

```
1   # List of card numbers
2   cards = [1, 2, 3, 4]
3
4   # Empty list to store possible outcomes
5   outcomes = []
6
7   # Loop for the first draw
8   for first_card in cards:
9     # Loop for the second draw
10    for second_card in cards:
11      # Create an ordered pair for the outcome of both draws
12      outcome = (first_card, second_card)
13      # Add the outcome to the list
14      outcomes.append(outcome)
15
16  # Print all possible outcomes
17  for outcome in outcomes:
18    print(outcome)
```

Students: Page 301

💡 **Idea summary**

When selecting items with replacement, the first item selected is replaced before another item is selected. The first selection will have no effect on the second selection, so the events are independent.

When selecting items without replacement, the first item selected is not replaced before the second item is selected. The first selection will have an effect on the second selection, so the events are dependent.

Practice

Students: Page 301–304

What do you remember?

1 Fill in the blank with "independent" or "dependent" to complete the definition:

Two events are ⬚ if the outcome of one event does not affect the outcome of the other event.

2 State whether the following describe independent events or dependent events.
  a It is possible for the events to happen in any order.
  b One event affects the outcome or the odds of the other event.
  c Sampling is done where each member of the population is replaced after it is picked.
  d Sampling is done where each member of a population may be chosen only once.

3 A player is rolling two dice and calculating their sum. They draw an array of all the possible dice rolls with the sum of two dice.

List the sample space for the sum of two dice.

	1	2	3	4	5	6
1	2	3	4	5	6	7
2	3	4	5	6	7	8
3	4	5	6	7	8	9
4	5	6	7	8	9	10
5	6	7	8	9	10	11
6	7	8	9	10	11	12

4 Xavier is choosing an outfit for the day and has 3 shirts (cyan, pink and white) and 4 ties (black, grey, red and yellow) to select from.
  a Complete the given array.
  b How many different outfits are possible?

	Cyan (C)	Pink (P)	White (W)
Black (B)	C, B	⬚, B	W, B
Grey (G)	C, G	P, G	W, ⬚
Red (R)	⬚, ⬚	P, R	W, R
Yellow (Y)	C, Y	⬚, ⬚	W, Y

5 Three cards labeled 2, 3, and 4 are placed face down on a table. Two of the cards are selected randomly to form a two-digit number. The outcomes are displayed in the tree diagram.
  a How many outcomes are possible for the selection of the first card?
  b How many outcomes are possible for the selection of the second card?
  c List the sample space of the two-digit numbers possible.

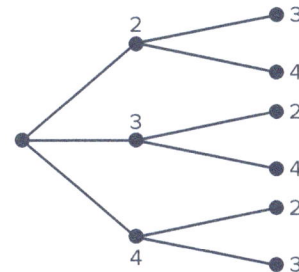

Let's practice

6 Ursula takes a bus to the station and then immediately gets on a train to work. Is the chance of her missing the train independent or dependent on her missing her bus?

7 A person runs a marathon, and then falls ill from exhaustion. Are the events independent or dependent?

8 These two spinners are spun and the result of each spin is recorded:
  a List the sample space.
  b Is the outcome of the second spinner independent of or dependent on the result from spinning the first spinner?

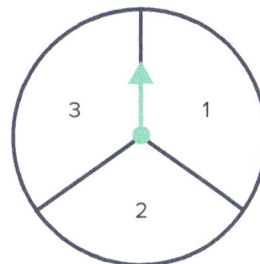

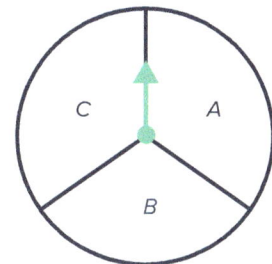

9   Aditya has three lollipops. One is blue, one is green, and one is yellow. They take one, eat it, and then eat one more.

   a   Are the events of eating the lollipops independent or dependent?

   b   List the sample space for the possible lollipops they eat.

10   State whether the following events or selections are independent or dependent.

   a   Two cards are randomly selected from a normal deck of cards without replacement.

   b   A card is randomly selected from a normal deck of cards, and then returned to the deck. The deck is shuffled and another card is selected.

   c   Parking a vehicle illegally, and get a parking ticket.

   d   Taking a cab home and finding your favorite book in bookshelf.

11   Select the scenario that describes two independent events.

   A   A card is randomly selected from a normal deck of cards, and then returned to the deck. The deck is shuffled and another card is selected.

   B   Each of four students is allowed to randomly pick an item from the teacher's prize bag (and keep it).

   C   Without looking, a child selects a chocolate from a box of milk and dark chocolates. They eat it, then select another.

   D   The selection of each ball in a game of bingo.

12   A project team consists of four members: Sarah, David, Olivia, and Micheal. The team needs to designate a leader and a coordinator. The tree diagram shows all the possible ways a leader and a coordinator can be chosen.

The two events represented in the diagram are:

   • Choosing a leader
   • Choosing a coordinator

Are the events independent or dependent? Explain your answer.

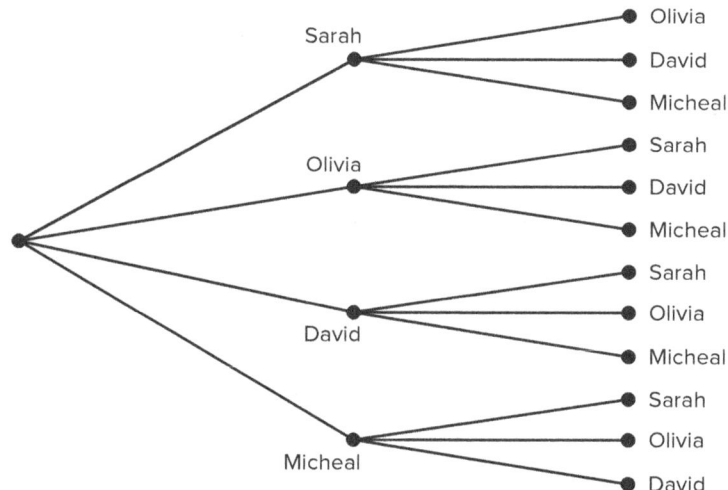

13   A container holds four ornaments that are painted different colors. One is red, one is blue, one is green and one is yellow. Two ornaments are randomnly selected, one after the other.

   a   Draw a tree diagram representing all possible outcomes when two ornaments are selected, and the first ornament is replaced before the second draw.

   b   When the ornament is replaced, are the selections independent or dependent?

   c   Draw a tree diagram representing all possible outcomes when two ornaments are selected, and the first ornament is not replaced before the second draw.

   d   When the ornament is not replaced, are the selections independent or dependent?

14   An experiment is done where two plants are randomly selected and their heights are measured. After the first plant is measured, it is replaced with the other plants, and a second plant is randomly selected again.

The same experiment is then done without replacement, so the same plant cannot be measured twice.

   a   When the experiment is done with replacement, will the number of outcomes for each selection be the same or different?

   b   Explain why selecting two plants are dependent events in the experiment conducted without replacement.

15 A bag has 2 blue marbles, 1 white marble, and 1 green marble. Select the tree diagram that shows the events of drawing two marbles as independent events.

A

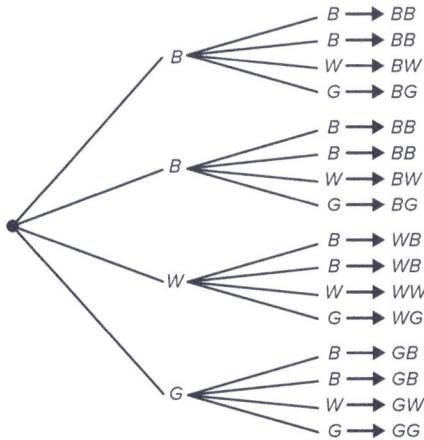

B

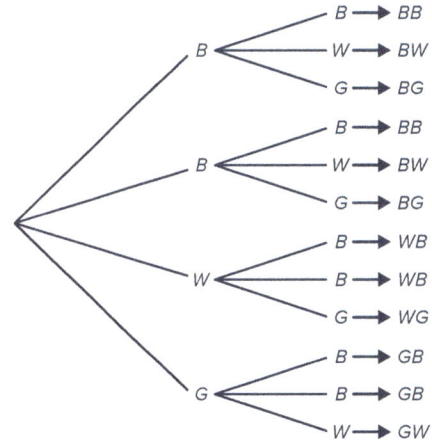

Let's extend our thinking

16 Each day Chris chooses one accessory, a hat or a tie. There is an even chance that Chris wears a tie or hat on any given day. The tree diagram represents the possibilities of Chris' choice on three consecutive days.

a Is Chris choosing an accessory independent or dependent of his choice on the other two days? Explain your answer.

b List the sample space for accessories Chris might choose all three days.

17 A die is rolled 691 times. The die lands on a six 14 times, and the probability that the next roll will land on a six is $\frac{1}{6}$.

State whether the outcome of the next roll is independent of or dependent on the outcomes of previous rolls.

18 In a lottery there are 37 balls.

• The probability of a particular ball being drawn first is $\frac{1}{37}$.

• A ball is discarded after it has been drawn. If ball number 22 is drawn on the first go, the probability of ball number 15 being drawn next is $\frac{1}{36}$.

Determine whether the draws are dependent or independent events.

19 A standard deck of cards is shuffled and placed face down. Dylan attempts to draw five cards randomly so as to get a royal flush (ten, jack, queen, king, ace all of the same suit) .

• He draws a queen of spades on the first go, and the probability that he draws a ten of spades on the second go is $\frac{1}{51}$.

• He draws a queen of spades on the first go and a ten of spades on the second go, the probability that he draws a jack of spades on the third go is $\frac{1}{50}$.

Is the probability of each draw dependent or independent of the previous draw? Explain your answer.

20 In a memory game, 16 pairs of identical cards are randomly placed face down. When someone has a go, they turn one card over and then turn a second card over to try to find an identical pair. If an identical pair are found, they are removed from the pack.

At the beginning of the game, Kathleen turns a card over. The probability that she will pick a card that will make an identical pair is $\frac{1}{31}$.

The first two cards Kathleen picks are identical. Is the probability of picking any more pairs of identical cards independent or dependent on the number of pairs picked before? Explain your answer.

Answers

5.01 Introduction to independent and dependent events

What do you remember?

1 Independent

2 **a** Independent events **b** Dependent events

   c Independent events **d** Dependent events

3 Sample space = {2, 3, 4, 5, 6, 7, 8, 9, 10, 11, 12}

4 **a**

	Cyan (*C*)	Pink (*P*)	White (*W*)
Black (*B*)	*C, B*	*P, B*	*W, B*
Grey (*G*)	*C, G*	*P, G*	*W, G*
Red (*R*)	*C, R*	*P, R*	*W, R*
Yellow (*Y*)	*C, Y*	*P, Y*	*W, Y*

   b 12

5 **a** 3

   b 2

   c {23, 24, 32, 34, 42, 43}

Let's practice

6 Dependent

7 Dependent

8 **a** 1A, 1B, 1C, 2A, 2B, 2C, 3A, 3B, 3C

   b Independent

9 **a** Dependent

   b {*BG, BY, GB, GY, YB, YG*}

10 **a** Dependent **b** Independent

   c Dependent **d** Independent

11 **A**

12 Dependent. As shown in the tree diagram, the person chosen to be the coordinator will dependend on who was chosen to be the leader. Once a leader is chosen, they cannot also be chosen to be a leader.

13 **a**

b Independent

c

d Dependent

14 **a** When the plant is replaced, the number of outcomes is the same for each selection.

   b The possible outcomes for the second plant is different depending on the first plant that was chosen to be measured. There will be one less plant to choose from in the sample space when compared to the first selection.

15 **A**

Let's extend our thinking

16 **a** Independent because there is an even chance that he selects a hat or a tie on all three days. The number of outcomes and the chance of his choice on each day does not change from one day to the next.

   b {*HHH, HHT, HTH, HTT, THH, THT, TTH, TTT*}

17 Independent

18 Dependent

19 Dependent, because the probability of drawing a particular card depends on the number of cards left in the deck. This means the number of cards previously drawn affects the probability of a particular card being drawn.

20 Dependent, because the probability of picking an identical pair depends on how many cards are left. Since the number of cards decreases as pairs are picked, the probability of picking a pair is affected by how many cards have been removed.

5.02 Probability of independent events

Subtopic overview

Lesson narrative

In this lesson, students will learn how to calculate the probability of independent events. They will explore the concept of events that do not influence each other, such as rolling a die and flipping a coin. The lesson includes various examples and practice problems, where students will multiply the probabilities of individual events to find the overall probability of combined independent events. By the end of the lesson, students should be able to identify independent events and accurately compute their combined probabilities.

Learning objectives

Students: Page 305

> **After this lesson, you will be able to...**
> - identify whether two events are independent or dependent.
> - calculate the probability of two independent events.

Key vocabulary

- independent events
- probability
- tree diagram

Essential understanding

Determining the independence or dependence of events is essential to drawing accurate conclusions.

Standards

This subtopic addresses the following Virginia 2023 Mathematics Standards of Learning standards.

Mathematical Process Goals

MPG1 — Mathematical Problem Solving

Teachers can enhance students' problem-solving skills by providing them with real-world scenarios involving independent events. For example, teachers can ask students to calculate the probability of getting a red card from a deck and rolling a 5 on a dice. This allows students to apply the formula for the probability of independent events and solve complex problems, thus honing their problem-solving skills.

MPG3 — Mathematical Reasoning

Teachers can help students develop mathematical reasoning by guiding them through the process of determining whether events are independent or dependent. Students should be encouraged to justify their responses using logical arguments. For example, if a student identifies two events as independent, they should be able to explain why the outcome of one event does not affect the outcome of the other.

MPG5 — Mathematical Representations

Teachers can use different tools and methods to represent the concepts of independent and dependent events. They can use physical props such as coins and dice, visual aids like charts and graphs, and symbolic notations like $P(A \text{ and } B) = P(A \cdot P(B))$. By displaying these concepts in various forms, teachers can help students understand that different representations can embody the same mathematical idea.

Content standards

8.PS.1 — The student will use statistical investigation to determine the probability of independent and dependent events, including those in context.

8.PS.1a — Determine whether two events are independent or dependent and explain how replacement impacts the probability.

8.PS.1c — Determine the probability of two independent events.

Prior connections

7.PS.1 — The student will use statistical investigation to determine the probability of an event and investigate and describe the difference between the experimental and theoretical probability.

Future connections

G.RLT.1 — The student will translate logic statements, identify conditional statements, and use and interpret Venn diagrams.

Lesson Preparation

Tools

You may find these tools helpful:
- Scientific calculator
- Dice

Student lesson & teacher guide

Probability of independent events

This lesson focuses on understanding the concept of independence between two events and how to calculate probabilities for single and combined events. We'll explore the definition of independent events–where the outcome of one event does not influence another–and illustrate this through practical examples such as rolling a die where each roll is independent of the previous ones.

Exploration

Students: Page 305–306

Exploration

Consider the tree diagrams shown.

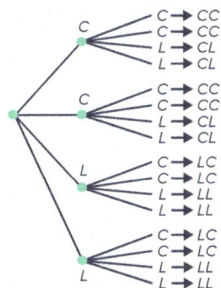

Without probabilities on
the branches

With probabilities on the
branches

1. Could these tree diagrams represent the same situation? Explain.
2. For the tree diagram without probabilities on the branches, how could you find the probability of CC?
3. Knowing the probability of CC from question 2, how could you find the probability of CC from the other tree diagram?
4. What are the benefits and drawbacks of each of the diagrams?

Suggested student grouping: Small groups

In this exploration, we delve into the analysis of tree diagrams, focusing on their use in probability calculations for different outcomes. Tree diagrams are instrumental in visualizing and calculating the probabilities of sequences of events. This particular exploration compares two tree diagrams—one with probabilities listed on the branches and one without.

Ideal student responses

These ideal responses may differ from other correct student responses. Less formal responses can be connected with the more precise mathematical language presented here.

1. **Could these tree diagrams represent the same situation? Explain.**

   Yes, these tree diagrams could represent the same situation. Both diagrams depict outcomes from two sequential choices, each with two possibilities: C and L. The diagram with probabilities explicitly shows the chance of each outcome at every branch, and the diagram without probabilities still implies the same sequence of events but writes out each possibility as a branch.

2. **For the tree diagram without probabilities on the branches, how could you find the probability of CC?**

   To find the probability of "CC" in the diagram without probabilities, we can assume each branch has an equal probability. If each branch is equally likely, the probability of "CC" would typically be $\frac{1}{4}$ $\left(\text{i.e } \frac{1}{2} \cdot \frac{1}{2}\right)$.

3. **Knowing the probability of CC from question 2, how could you find the probability of CC from the other tree diagram?**

   Using the other tree diagram, where probabilities are provided on the branches, you would multiply the probabilities along the path leading to "CC". If each branch for C and L has specified probabilities, you multiply these probabilities along the path of choosing C twice.

4. **What are the benefits and drawbacks of each of the diagrams?**

   The tree diagram with probabilities provides exact values for calculating the likelihood of each outcome, but potentially complicates the diagram with too much numerical data. The tree diagram without probabilities simplifies the diagram, making it useful for conceptual discussions or when probabilities are uniform. However, it lacks the necessary detail for precise calculations unless probabilities are known to be equal across branches.

Purposeful questions

- How can you determine if two tree diagrams represent the same situation, even if one includes probabilities on the branches and the other does not? What key aspects should you compare?
- How is the sample space in the first tree diagram related to the probabilities in the second tree diagram?

Possible misunderstandings

- Students may struggle to make sense of the branches of the tree diagram without probabilities written in. Have them compare the probabilities of each outcome in the first diagram to the probabilities in the second diagram.

Build and connect concept maps for probability events
Student with disabilities support

Encourage students to create visual concept maps to differentiate between independent and dependent events in probability. Begin with "Probability Events" at the center and branch out to "Independent Events" and "Dependent Events." Under each branch, have students add definitions, key characteristics, and examples like "with replacement" for independent events, and "without replacement" for dependent events.

This visual organization helps students understand how the events relate and when to multiply probabilities for combined independent events. Provide partially completed concept maps or a word bank of terms like "influence," "replacement," and "multiplication" to support students who may struggle with processing complex information. By connecting these concepts visually, students can better grasp and recall how to identify event types and calculate their probabilities accurately.

Students can continue to build on the concept map throughout the topic.

Examples

Students: Page 306–307

Example 1

A fair standard six-sided die is rolled 691 times.

If it lands on a six 108 times, what is the probability that the next roll will land on a six?

Create a strategy

To find the probability that the next roll will land on a six, we need to determine if the next roll is dependent on the previous rolls. This will help us determine whether we need to find the theoretical probability or the experimental probability.

Apply the idea

The previous results do not affect the outcome of the next roll, so we will find the theoretical probability of rolling a 6.

$$P(A) = \frac{\text{Number of favorable outcomes}}{\text{Total number of outcomes}} \qquad \text{Theoretical probability formula}$$

$$= \frac{1}{6} \qquad \text{Substitute the number of outcomes}$$

Purpose

Show students that the probability of a single event in a series of independent events is not affected by the outcomes of previous events.

Reflecting with students

Encourage students to communicate precisely by using clear definitions when solving this problem. For example, prompt them to define what a "fair standard six-sided die" means, emphasizing that each of the six outcomes has an equal probability.

Assuming past rolls affect future probabilities
Address student misconceptions

Students might think that because the die landed on a six 108 times out of 691 rolls, the probability of the next roll landing on a six is $\frac{108}{691}$. They may believe that the previous outcomes influence the next result, or that the experimental probability determines future probabilities. This reflects a misunderstanding of independent events and the nature of theoretical probability. The question states that this is a fair die, not a loaded die.

Emphasize to students that each roll of a die is an independent event, meaning the outcome of one roll does not affect the outcome of the next. Encourage students to compare the experimental probability with the theoretical probability by calculating both: the experimental probability is $\frac{108}{691}$, while the theoretical probability of rolling a six is always $\frac{1}{6}$. Encourage students to recall activities where they rolled dice and recorded outcomes to see how experimental probabilities varied, but got closer to the theoretical probability with more trials.

Students: Page 307

Example 2

Two events A and B are such that:

- $P(A) = 0.5$
- $P(B) = 0.7$
- $P(A \text{ and } B) = 0.3$

Determine whether events A and B are independent.

Create a strategy

Use the fact that if the events A and B are independent, then $P(A \text{ and } B) = P(A) \cdot P(B)$.

Apply the idea

$$
\begin{aligned}
P(A \text{ and } B) &= P(A) \cdot P(B) & \\
0.3 &= 0.5 \cdot 0.7 & \text{Substitute the values} \\
0.3 &\neq 0.35 & \text{Evaluate}
\end{aligned}
$$

This shows that events A and B are not independent events.

Purpose

Show students how to determine if two events are independent by applying the rule that if the events are independent, then the probability of both occurring is the product of their individual probabilities.

Reflecting with students

Have students consider what $P(B)$ would need to be to have the events be independent.

Collect and display
English language learner support

Introduce students to the key vocabulary in the problem by encouraging them to describe, in their own words, terms or expressions like:

- independent events
- dependent events
- probability

- intersection
- $P(A \text{ and } B)$
- $P(A) \times P(B)$

As students share their interpretations, listen and collect their phrases and explanations. Display these on a chart or board where everyone can see them, organizing their definitions alongside the formal mathematical terminology. Include visual aids such as Venn diagrams to represent events A and B, their intersection, and how independence relates to the multiplication of probabilities. This visual and linguistic reference will help students make connections between their own language and the mathematical concepts, supporting their understanding throughout the lesson. Refer back to this display as you work through the problem to reinforce the vocabulary and concepts.

Students: Page 307–308

Example 3

A standard deck of 52 cards has four queens. Ilham shuffles the deck then selects a card. He puts the card back in the deck, then selects a second card.

Let Q represent the event of drawing a queen and N represent the event of not drawing a queen. The tree diagram shows all the possible outcomes and probabilities:

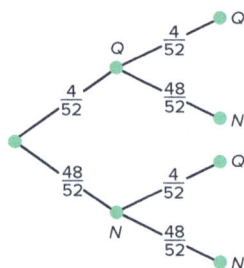

a   Find the probability that both cards are queens.

Create a strategy

Because the first card is replaced before drawing the second card, we know the events of drawing two cards are independent. The probability that both cards are queens can be represented by $P(Q \text{ and } Q) = P(Q) \cdot P(Q)$.

Apply the idea

To find $P(Q \text{ and } Q)$ using the tree diagram, we will multiply the probabilities on the branches with Q's. The tree shows that the probability of drawing a queen on each draw is $\dfrac{4}{52}$ because there are 52 cards in the deck and 4 of them are queens.

$$
\begin{aligned}
P(Q \text{ and } Q) &= P(Q) \cdot P(Q) && \text{Probability of drawing two queens} \\
&= \frac{4}{52} \cdot \frac{4}{52} && \text{Substitute known values} \\
&= \frac{1}{13} \cdot \frac{1}{13} && \text{Reduce the fractions} \\
&= \frac{1}{169} && \text{Multiply}
\end{aligned}
$$

Reflect and check

The probability of drawing two queens is about 0.6% which is very unlikely.

Purpose

Show students how to calculate the probability of independent events using a tree diagram and the rule of multiplication for independent events.

b Find the probability that neither card is a queen.

Create a strategy

The probability that neither card is a queen can be represented by $P(N \text{ and } N) = P(N) \cdot P(N)$. To find this probability using the tree diagram, we will multiply the probabilities on the branches with N's.

Apply the idea

The tree shows that the probability of not drawing a queen on each draw is $\dfrac{48}{52}$ because there are 52 cards in the deck and $52 - 4 = 48$ of them are not queens.

$$
\begin{aligned}
P\left(N \text{ and } N\right) &= P\left(N\right) \cdot P\left(N\right) && \text{Probability of drawing two queens} \\
&= \frac{48}{52} \cdot \frac{48}{52} && \text{Substitute known values} \\
&= \frac{12}{13} \cdot \frac{12}{13} && \text{Reduce the fractions} \\
&= \frac{144}{169} && \text{Multiply}
\end{aligned}
$$

Reflect and check

Reducing the fractions before multiplying helps us work with smaller values, which can also help us avoid mistakes. However, we could have multiplied first, then reduced the result. We should arrive at the same answer either way:

$$
\begin{aligned}
P\left(N \text{ and } N\right) &= P\left(N\right) \cdot P\left(N\right) && \text{Probability of drawing two queens} \\
&= \frac{48}{52} \cdot \frac{48}{52} && \text{Substitute known values} \\
&= \frac{2304}{2704} && \text{Multiply} \\
&= \frac{144}{169} && \text{Reduce by 16}
\end{aligned}
$$

Purpose

Demonstrate to students how to calculate the probability of compound events using the multiplication rule, and make them aware that reducing fractions before multiplying can help avoid mistakes and make calculations easier.

c Find the probability that Ilham selects only one queen.

Create a strategy

The order in which the queen is drawn has not been specified, so there are two ways that Ilham can select only one queen:

- Drawing the queen first, then not drawing a queen (Q and N)
- Not drawing a queen first, then drawing a queen (N and Q)

We will calculate the probability of each situation separately, then add the results.

Apply the idea

$$
\begin{aligned}
P\left(Q \text{ and } N\right) + P\left(N \text{ and } Q\right) &= P\left(Q\right) \cdot P\left(N\right) + P\left(N\right) \cdot P\left(Q\right) && \text{Probability of only one queen} \\
&= \left(\frac{4}{52} \cdot \frac{48}{52}\right) + \left(\frac{48}{52} \cdot \frac{4}{52}\right) && \text{Substitute known values} \\
&= \left(\frac{1}{13} \cdot \frac{12}{13}\right) + \left(\frac{12}{13} \cdot \frac{1}{13}\right) && \text{Simplify the fractions} \\
&= \frac{12}{169} + \frac{12}{169} && \text{Evaluate the multiplication} \\
&= \frac{24}{169} && \text{Evaluate the sum}
\end{aligned}
$$

Purpose

Check students can apply the rules of probability to a two-step event, and understand the concept of mutually exclusive events.

Reflecting with students

The probabilities we found in parts (a), (b), and (c) represent all the possible outcomes in the sample space. This means their probabilities will sum to 1.

$$\frac{1}{169} + \frac{24}{169} + \frac{144}{169} = 1$$

If students are ever unsure of their answer, encourage them to find the sum of the probabilities of all possible outcome in a sample space. If this total is not exactly equal to 1, they should review their work.

Advanced learners: Empowering students through designing their own probability problems

use with example 3

Targeted instructional strategies

Encourage your advanced learners to create their own probability scenarios based on the concept of drawing cards from a deck. After they solve the given problem, invite them to modify the conditions—such as changing the number of draws, altering the cards of interest (e.g., face cards instead of queens), or considering draws without replacement.

Writing their own problems allows them to explore how different variables affect the probabilities and to discover underlying patterns. For example, they might design a problem to find the probability of drawing at least one queen in three draws or compare the probabilities of drawing a queen when cards are not replaced. By designing their own problems, students delve deeper into probability concepts and enhance their critical thinking skills. This strategy empowers them to take ownership of their learning and fosters a more profound understanding of the subject matter.

Students: Page 309

Idea summary

Two events are independent if the outcome of the each event does not affect the outcome of the other event.

If two events are independent, the probability of both events occurring is:

$$P(A \text{ and } B) = P(A) \cdot P(B)$$

This formula can also be used to check if two events are independent.

Practice

Students: Page 309–313

What do you remember?

1   Determine whether or not these events will be independent. Justify your reasoning.

   a   Flipping a coin then rolling a die.

   b   Drawing a red card, then a black card without replacing the first from a standard deck of cards.

   c   Being taller than average and playing professional basketball.

   d   Rolling a die twice.

2   This spinner is spun and a six-sided die is rolled. List the sample space.

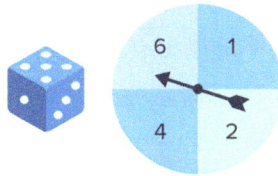

3   A bag has counters with different numbers on them. Two counters are drawn without replacement. The smaller number is recorded as the outcome.

   a   Describe an event that would have a probability of 0.

   b   Describe an event that would have a probability of 1.

   c   Describe an event that would have a probability of less than 0.5.

4   The probability that Donna will be elected school captain is 0.83. What is the probability that she won't be elected school captain?

5   On the island of Guam, the probability that it rains during rainy season is $P(R)$. This is shown in the tree diagram.

   a   What do the two sets of branches represent?

   b   Does the weather on the first day affect the chance of rain on the second day?

   c   Which two events have the same probability?

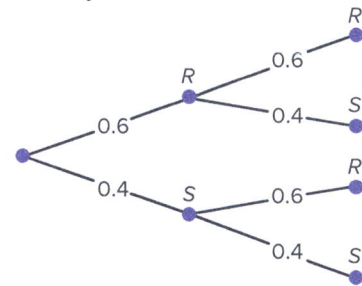

6   Given the probabilities, determine whether the events A and B are independent:

   a $P(A) = 0.8, P(B) = 0.4, P(A \text{ and } B) = 0.32$

   b $P(A) = 0.6, P(B) = 0.2, P(A \text{ and } B) = 0.14$

   c $P(A) = 0.15, P(B) = 0.6, P(A \text{ and } B) = 0.16$

   d $P(A) = 0.7, P(B) = 0.4, P(A \text{ and } B) = 0.28$

   e $P(A) = 0.4, P(B) = 0.1, P(A \text{ and } B) = 0.04$

Let's practice

7   A coin is tossed and a die is rolled.

   a   Find the probability of tossing heads.

   b   Find the probability of rolling a 5.

   c   Find the probability of tossing heads and rolling a 5.

   d   Are these events independent or dependent?

8 A bag contains 6 orange gems, 4 purple gems and 7 yellow gems.

    a Find the probability of selecting an orange gem.

    b Find the probability of selecting a purple gem.

    c You randomly select a gem from the bag, note the color, replace the gem, then randomly select another gem from the bag. What is the probability of getting an orange gem first and then a purple gem?

9 On a roulette table, a ball can land on one of 18 red or 18 black numbers.

    a If it lands on a red number on the first go, find the probability that it will land on a red number on the second go.

    b Are the successive events of twice landing on a red number dependent or independent?

10 A coin is flipped twice.

    a Find the sample space of this experiment.

    b Find the probability of flipping two tails.

    c Find the probability of flipping a tails and a head.

    d Find the probability of not flipping a tails and a head.

11 Two standard dice are rolled. One is red and one is white.

    a Are the results from the red die and the white die dependent or independent?

    b Calculate the probability that:

        i The same number is rolled.

        ii The sum of the two numbers exceeds 9.

        iii The red die is 4 and the white die 6.

        iv The red die is even and the white die is odd.

        v The sum of the two numbers is less than 2.

12 An ice cream shop offers one flavor of ice cream at a discounted price each day. There are 6 flavors that could be discounted; 3 of the flavors are sorbet and the other 3 are gelato. The three sorbet flavors are raspberry (R), lemon (L) and chocolate (C). The three gelato flavors are vanilla (V), mint (M) and chocolate (C). Each decision has equal probability.

Edge	a	b	c	d	e	f	g	h
Probability								

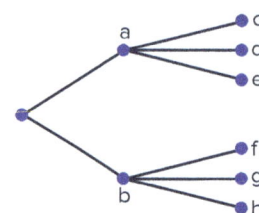

    a Complete the table with the probabilities matching the edges of the probability tree.

    b Find the probability that the ice cream type will be gelato and the flavor will be chocolate.

    c Find the probability that the lemon sorbet is selected.

13 A bucket contains 5 green buttons and 7 black buttons. Two buttons are selected one after another from the bucket. The first button is replaced before the second button is selected.

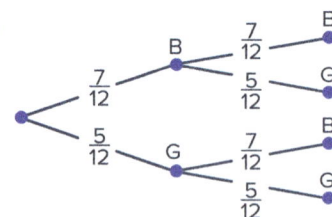

    a Find the probability of selecting a green button, then a black button.

    b Find the probability of selecting a black button, then a green button.

    c Are the probabilities from parts (a) and part (b) different or the same? Why?

14 In a jar, there are 3 red marbles and 7 green marbles. Assaf withdraws one marble from the bag, replaces the marble, then withdraws another marble. Let A be the event that Assaf draws a red marble first and B be the event Assaf draws a green marble second.

    a Determine whether A and B are independent events.

    b Find $P(A \text{ and } B)$.

    c Assaf repeats the experiment, but this time, he doesn't replace the first marble. Explain why the events are no longer independent.

15 On any given day, the probability that a student is absent is 0.05. If the attendance of each student is independent of the attendance of others, what is the probability that two specific students are both absent on the same day?

SOL **16** A deck of cards contains six equally shaped cards as shown. Joshua randomly selects one card from the deck, replaces the card, and then randomly selects a second card.

   a Does this situation describe independent or dependent events?
   b Find the probability of selecting a triangle first, then selecting a star.
   c Find the probability of selecting a circle for both cards.
   d What is the probability that neither card selected is a triangle?

17 Christa randomly selects two cards from a normal deck of cards.

Calculate the probability if the cards are selected with replacement.
   a The first card is a Queen of diamonds and the second card is a 10 of spades.
   b Both cards are red.
   c The first card is a Queen and the second card is a King.
   d The first card is a heart and the second card is not a spade.

Let's extend our thinking

18 Given events A and B are independent, find $P(B)$ for each set of probabilities:
   a $P(A \text{ and } B) = 0.1$ and $P(A) = 0.5$
   b $P(A \text{ and } B) = 0.3$ and $P(A) = 0.6$

19 A and B are two random events with these probabilities:
$$P(A) = 0.3 + x$$
$$P(B) = 0.2 + x$$
$$P(A \text{ and } B) = x$$

Find the value(s) of x if A and B are independent.

20 Han plays three tennis matches. In each match, he has $\frac{3}{5}$ chance of winning. When playing three matches in a row, the probability that he wins or not is presented in the following tree diagram:

Let A be the event where Han wins the third match and B be the event where Han loses the second match.

Explain whether A and B are independent events.

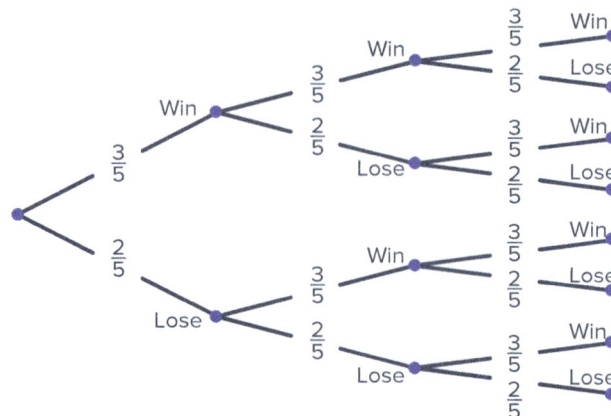

21 Three fair coins are tossed:

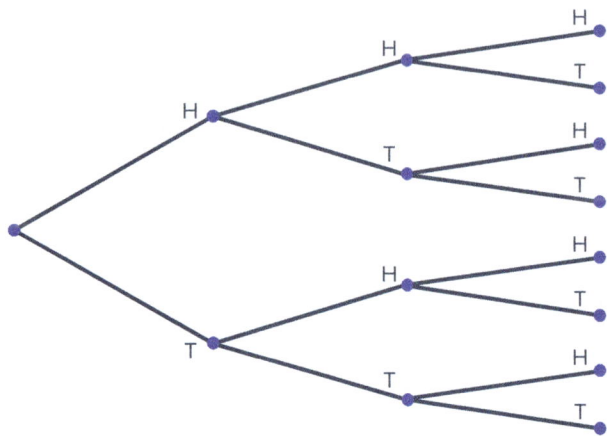

a Find the probability of obtaining at least 1 head.

b Find the probability of obtaining TTH in this sequence.

c Find the probability of obtaining THH in this sequence.

22 A basketball player is practicing his free throws. He makes the shot with a probability of $\frac{1}{5}$ and misses with the probability of $\frac{4}{5}$. The tree diagram shows all the possible outcomes and probabilities:

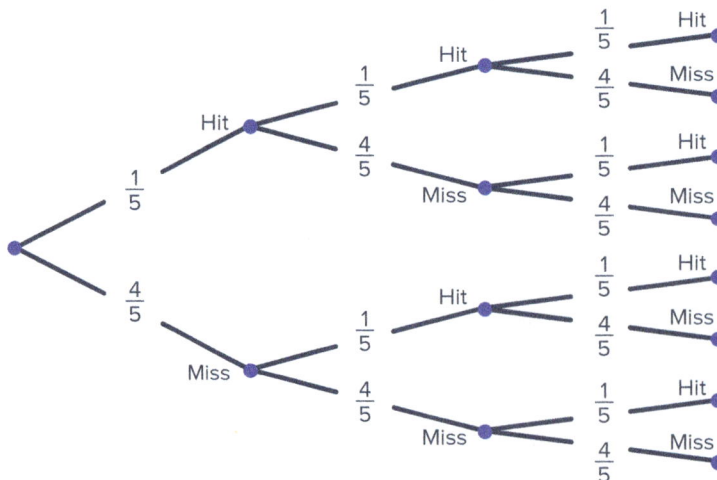

a Find the probability that he will make all three free throws.

b Find the probability that he will miss at least one free throw.

c Find the probability that he will make at least one free throw.

Answers

What do you remember?

1 **a** Independent, the outcome of the coin flip will have no effect on the outcome of the die roll.

b Not independent, the probability of the second draw is impacted by the outcome of the first draw.

c Not independent, the chances of playing professional basketball usually depend on the person's height.

d Independent, the outcome of the first roll does not affect the outcome of the second.

2 {(1, 1), (1, 2), (1, 4), (1, 6), (2, 1), (2, 2), (2, 4), (2, 6), (3, 1), (3, 2), (3, 4), (3, 6), (4, 1), (4, 2), (4, 4), (4, 6), (5, 1), (5, 2), (5, 4), (5, 6), (6, 1), (6, 2), (6, 4), (6, 6)}

3 **a** Answers may vary.

An event with a probability of 0 would be drawing two counters with the same number, for example, drawing both counters with the number 4, which is impossible as each number appears only once in the bag.

b Answers may vary.

An event with a probability of 1 would be drawing two counters with a sum greater than 8, as any pair of counters taken from the bag would have a sum larger than 8 (like $4 + 5 = 9$ and the sums only increase from there).

c Answers may vary.

An event with a probability of less than 0.5 would be drawing a smaller number like 4 as the outcome. The possible draws for this outcome are (4, 5), (4, 6), (4, 7) and (4, 8), which has a probability of $\frac{4}{20}$ or 0.2, considering all the 20 possible pairs of counters.

4 0.17

5 **a** Two days

b No

c Rain on the first day, sun on the second day (*RS*) and sun on the first day, rain on the second day *SR*.

6 **a** Yes   **b** No   **c** No   **d** Yes   **e** Yes

Let's practice

7 **a** $\frac{1}{2}$ **b** $\frac{1}{6}$ **c** $\frac{1}{12}$ **d** Independent

8 **a** $\frac{6}{17}$ **b** $\frac{4}{17}$ **c** $\frac{24}{289}$

9 **a** $\frac{1}{2}$ **b** Independent

10 **a** $S = \{HH, HT, TH, TT\}$ **b** $\frac{1}{4}$

c Note there has been no specification about the order of the events, so the events *HT* and *TH* are both favorable outcomes. The probability of flipping a tails and a head in any order is $\frac{1}{2}$.

d $\frac{1}{2}$

11 **a** Independent

b **i** $\frac{6}{36} = \frac{1}{6}$ **ii** $\frac{6}{36} = \frac{1}{6}$ **iii** $\frac{1}{36}$ **iv** $\frac{9}{36} = \frac{1}{4}$

12 **a**

Edge	a	b	c	d	e	f	g	h
Probability	$\frac{1}{2}$	$\frac{1}{2}$	$\frac{1}{3}$	$\frac{1}{3}$	$\frac{1}{3}$	$\frac{1}{3}$	$\frac{1}{3}$	$\frac{1}{3}$

b $\left(\frac{1}{2} \cdot \frac{1}{3}\right) = \frac{1}{6}$ **c** $\frac{3}{6} \cdot \frac{1}{3} = \frac{1}{6}$

13 **a** $\frac{35}{144}$ **b** $\frac{39}{144}$

c The same because the events are independent.

14 **a** The events are independent.

b $\frac{21}{100}$

c The events are no longer independent because the probability of drawing a green marble second will depend on what color marble was drawn first.

15 The probability that two specific students are both absent on the same day is 0.0025.

16 **a** Independent events   **b** $\frac{1}{9}$ **c** $\frac{1}{9}$ **d** $\frac{4}{9}$

17 **a** $\frac{1}{2704}$ **b** $\frac{1}{4}$ **c** $\frac{1}{169}$ **d** $\frac{3}{16}$

Let's extend our thinking

18 **a** 0.2   **b** 0.5

19 $x = \frac{3}{10}, x = \frac{1}{5}$

20 Yes, the events are independent events because for all three matches, the probability that Han wins or loses is always the same. It does not change based on what happened in the previous match. Whether Han wins the second match or loses the second match, the probability he wins the third match is still $\frac{3}{5}$.

21 **a** $\frac{7}{8}$ **b** $\frac{1}{8}$ **c** $\frac{1}{8}$

22 **a** $\frac{1}{125}$ **b** $\frac{124}{125}$ **c** $\frac{61}{125}$

5.03 Probability of dependent events

Subtopic overview

Lesson narrative

In this lesson, students will learn to determine the probability of dependent events. They will explore how the outcome of one event affects the probability of another and use conditional probability to solve problems. The lesson includes examples and practice problems where students calculate probabilities using formulas and tree diagrams. By the end of the lesson, students should understand how to distinguish between dependent and independent events and accurately calculate the probability of dependent events using various methods.

Learning objectives

Students: Page 314

> 🎓 **After this lesson, you will be able to...**
> - calculate the probability of two dependent events.
> - compare and contrast independent and dependent events.
> - explain how replacement impacts the probability of an outcome.

Key vocabulary

- dependent events
- sample space

Essential understanding

Determining the independence or dependence of events is essential to drawing accurate conclusions. Knowing the outcome of one event can change the sample space from which the likelihood of a second event is calculated. If the sample space is influenced by the outcome of the first event, then the two events are dependent.

Standards

This subtopic addresses the following Virginia 2023 Mathematics Standards of Learning standards.

Mathematical process goals

MPG1 — Mathematical Problem Solving

Teachers can guide students in applying the mathematical concepts of independent and dependent events in the process of solving probability problems. By providing real-world examples like choosing marbles from a bag without replacement or determining the probability of rain and a sports event being canceled, teachers can develop students' problem-solving skills.

MPG3 — Mathematical Reasoning

Teachers can help students to apply logical reasoning to justify their solutions when calculating the probability of two dependent events. By using the formula $P(A \text{ and } B) = P(A)$.

$P(B \text{ after } A)$, teachers can guide students in understanding the reasoning behind this method of calculation and the importance of distinguishing between independent and dependent events.

MPG5 — Mathematical Representations

Teachers can guide students in representing the probability of two dependent events using the formula P $P(A \text{ and } B) = P(A) \cdot P(B \text{ after } A)$. Students can be asked to represent these problems using symbolic notation and to interpret these representations in the context of the problem. For example, in the problem involving picking marbles from a bag, students can represent the problem and its solution using the formula and express it in a way that clearly communicates their understanding of the concept.

Content standards

8.PS.1 — The student will use statistical investigation to determine the probability of independent and dependent events, including those in context.

8.PS.1a — Determine whether two events are independent or dependent and explain how replacement impacts the probability.

8.PS.1b — Compare and contrast the probability of independent and dependent events.

8.PS.1d — Determine the probability of two dependent events.

Prior connections

7.PS.1 — The student will use statistical investigation to determine the probability of an event and investigate and describe the difference between the experimental and theoretical probability.

Future connections

G.RLT.1 — The student will translate logic statements, identify conditional statements, and use and interpret Venn diagrams.

Lesson Preparation

Student lesson & teacher guide

Probability of dependent events

This lesson explores the concept of dependent events in probability, emphasizing how the outcome of one event influences the outcome of another. Using practical examples such as drawing cards from a deck and selecting chores from a hat, the lesson delves into the difference between selections "with replacement" and "without replacement".

Exploration

Exploration

To decide which chores Jacques needs to do, he pulls out pieces of paper from a hat. The options are sweeping (S), mopping (M), or vacuuming (V).

Two chores on the same day   Two chores on different days

1. Which tree diagram shows the experiment with replacement?

2. What is the probability he will need to sweep and mop if they are done on the same day?

3. What is the probability he will need to sweep and mop if they are done on different days?

4. Is there a difference in the probabilities when selecting with or without replacement?

Suggested student grouping: Small groups

In this exploration, students examine the concept of dependent events through the practical scenario of Jacques selecting chores from a hat, focusing on the difference between choosing with and without replacement.

Ideal student responses

These ideal responses may differ from other correct student responses. Less formal responses can be connected with the more precise mathematical language presented here.

1. **Which tree diagram shows the experiment with replacement?**

   Tree diagram 2 shows the experiment with replacement. In this diagram, after selecting a chore (S, M, or V), the selected chore is still available for the second selection, which can be seen by the repetition of chores in the second set of branches.

2. **What is the probability he will need to sweep and mop if they are done on the same day?**

   In Tree diagram 1, which represents selection without replacement, the probability that Jacques will sweep (S) and mop (M) is two of the 6 total outcomes (S then M and M then S, depending on which one he does first). Therefore, the probability of sweeping and mopping in the same day is $\frac{2}{6} = \frac{1}{3}$.

3. **What is the probability he will need to sweep and mop if they are done on different days?**

In Tree diagram 2, with replacement, the probability of sweeping and mopping are 2 of the total 9 outcomes (S then M and M then S, depending on which one he does first). Thus, the probability of sweeping and mopping on different days is $\frac{2}{9}$.

4. **Is there a difference in the probabilities when selecting with or without replacement?**

Yes, there is a difference. When selecting without replacement, the probability is $\frac{1}{3}$. When selecting with replacement, this probability decreases to $\frac{2}{9}$.

Purposeful questions

- How can you determine which tree diagram represents the scenario with replacement and which represents the scenario without replacement? What key features or patterns should you look for in the diagrams?
- If the chores are done on the same day, is it likely that Jacques would do the same chore twice? Considering this, would this scenario be "with replacement" or "without replacement"?
- What is the difference between "A and B" versus "A then B"?
- How many events are in the sample space? How many of those events represent Jacques sweeping and mopping, in any order?

Possible misunderstandings

- Students might assume Jacques must sweep before mopping and not consider the event where Jacques mops before sweeping, leading to incorrect probability calculations.
- Students might incorrectly calculate the combined probabilities by adding instead of multiplying the probabilities of sequential events, leading to incorrect conclusions about the likelihood of certain outcomes.

Critique, correct, and clarify: understanding dependent probability
English language learner support

Provide students with the incorrect statement: "The probability of drawing two red cards in a row from a deck of cards without replacement is $\frac{1}{52} \cdot \frac{1}{52}$."

Ask students to work in pairs or small groups to critique this statement. Encourage them to discuss what is incorrect about the calculation and why. Prompt them with questions like, "What is the probability of drawing a red card from a full deck?", "Are the two events independent or dependent?" and "How does not replacing the card affect the probabilities?"

They should recognize that the probability of drawing a red card on the first draw is $\frac{26}{52}$, since there are 26 red cards in a standard deck of 52 cards. After one red card is drawn and not replaced, there are now 25 red cards left out of 51 total cards, so the probability of drawing a red card on the second draw is $\frac{25}{51}$.

Have students clarify the correct calculation by writing: "The probability of drawing two red cards in a row from a deck of cards without replacement is $\frac{26}{52} \cdot \frac{25}{51}$."

Encourage students to use precise mathematical vocabulary to explain how the outcome of the first event affects the second, reinforcing the concept of dependent events and how lack of replacement impacts the probability of subsequent outcomes.

Examples

Students: Page 315

Example 1

Esther was given four animal crackers as a snack. She has a donkey (*D*), an elephant (*E*), a goat (*G*), and a hippo (*H*). She eats one cracker and then another.

a Are the selections independent or dependent?

Create a strategy

Consider whether Esther replaced the first animal cracker before selecting the second animal cracker.

If Esther replaced the first animal cracker, the outcomes for the first and second choice will be the same. If Esther did not replace the first animal cracker, the outcomes for the first and second choice will be different.

Apply the idea

Since the first cracker is eaten, it is not available in the second round. This means the selections are dependent events because the outcome of the second selection will depend on which animal cracker was eaten first.

Purpose

Show students how to determine whether events are independent or dependent by considering whether the outcome of one event affects the outcome of another event.

Students: Page 316

b List the sample space for the possible two crackers she eats.

Create a strategy

We can make a list of all the outcomes or construct a tree diagram to visualize all the outcomes. Since the events are dependent, the number of options for the second cracker will reduce by 1.

Apply the idea

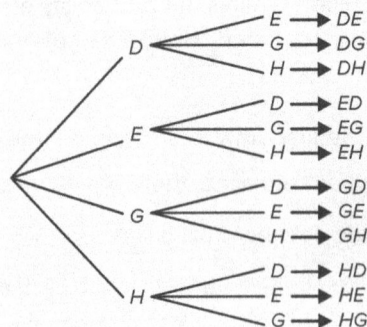

The sample space is *DE, DG, DH, ED, EG, EH, GD, GE, GH, HD, HE, HG*.

Reflect and check

For the first animal cracker, she has 4 choices. After eating one, she has 3 choices for the next one. By the fundamental counting priciple, there are 4 · 3 = 12 outcomes in the sample space.

Purpose

Show students how to determine the sample space for dependent events using both a list and a tree diagram.

Reflecting with students

Review the fundamental counting principle with students. Ask them how many outcomes there be would if there were 6 different animal choices or if Esther were choosing 3 cookies.

Students: Page 316

c Find the probability that Esther eats the elephant first, then the goat.

Create a strategy

We can use the tree diagram from part (a) to find the number of desired outcomes and the total number of outcomes in the sample space.

Apply the idea

From part (a), we found the total number of outcomes in the sample space is 12.

There is only one outcome where Esther eats the elephant first, then the goat (EG).

The probability is $\dfrac{1}{12}$.

Reflect and check

We can use the formula for dependent events to check our answer. There is only 1 elephant out of the 4 crackers, so the probability of eating the elephant first is

$$P(E) = \frac{1}{4}$$

Since Esther ate one, there are now only 3 crackers left, and 1 of them is the goat. The probability of eating the goat next is

$$P(G \text{ after } E) = \frac{1}{3}$$

The probability of eating the elephant then the goat is

$$P(E \text{ and } G) = P(E) \cdot P(G \text{ after } E) \qquad \text{Formula for dependent events}$$
$$= \frac{1}{4} \cdot \frac{1}{3} \qquad\qquad\qquad \text{Substitute the probabilities}$$
$$= \frac{1}{12} \qquad\qquad\qquad\quad \text{Multiply}$$

Purpose

Show students how to use a tree diagram to calculate the probability of compound events, and reinforce understanding of dependent events.

Provide students with a partially completed tree diagram showing Esther's options when eating her animal crackers. Start by drawing the first set of branches, without labels, from a starting point, each representing one of the four crackers she could eat first (donkey, elephant, goat, hippo).

Encourage students to then add notes or labels on these branches, including a heading above that set of branches with a title like "Choosing one of the four cookies and eating it". Then, guide them to complete the second set of branches for the second cracker she eats, remembering that one cracker has already been eaten, so they have one less option. As they work through the diagram, have them annotate it to highlight that the selections are dependent—each choice affects the next. Have them include a title above the second set of branches like "Choosing one of the three cookies that are left and eating it." This visual and interactive approach helps students understand the concept of dependent events and organize the sample space systematically.

To assist with part (c) of the problem, suggest that they highlight or color-code the specific path where Esther eats the elephant first, then the goat, making it easier to identify and calculate the required probability.

Students: Page 317

Example 2

A pile of playing cards has 4 diamonds and 3 hearts.

a   Find the probability of selecting two hearts if two cards are selected from the pile with replacement.

Create a strategy

To model the situation, a tree diagram can be drawn. The sample space is:

- Diamond, diamond
- Diamond, heart
- Heart, diamond
- Heart, heart

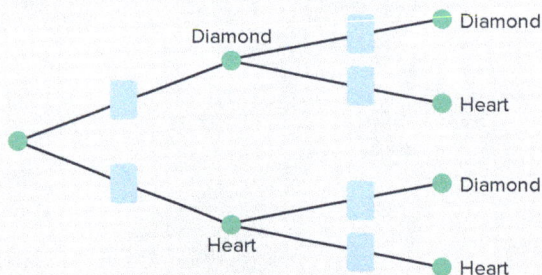

Then, we can use the diagram to find the probability of selecting a heart and another heart. If we let event A represent drawing a heart first and event B represent drawing a heart second, we can use the notation $P(A \text{ and } B)$.

Apply the idea

For the first card, the probability of drawing a diamond is $\frac{4}{7}$, and the probability of drawing a heart is $\frac{3}{7}$. The probabilities for the second card are independent of which card was drawn first, so they will be the same.

The probability tree shows this situation with the correct probability on each branch:

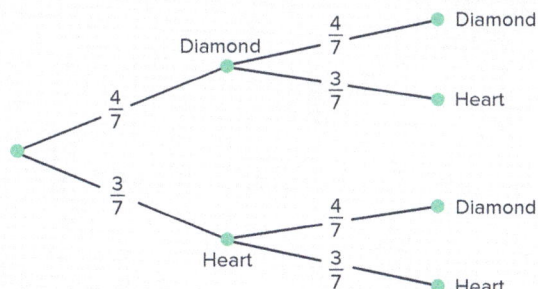

Because these events are independent, we must use the formula $P(A \text{ and } B) = P(A) \cdot P(B)$ to find the probability of drawing two hearts.

The probability of selecting a heart first is

$$P(A) = \frac{3}{7}$$

The probability of selecting a heart, given a heart was selected first, is

$$P(B) = \frac{3}{7}$$

The probability of selecting two hearts is

$$P(A \text{ and } B) = \frac{3}{7} \cdot \frac{3}{7}$$
$$= \frac{9}{49}$$

Purpose

Show students how to use a tree diagram to represent a multi-step probability event and calculate the probability of a specific outcome.

Students: Page 318

b   Find the probability of selecting two hearts if two cards are selected from the pile without replacement.

Create a strategy

We can use the tree diagram of the sample space from part (a) but change the probabilities in the second set of branches since the first card is not replaced.

Apply the idea

For the first card, the probability of drawing a diamond is $\frac{4}{7}$, and the probability of drawing a heart is $\frac{3}{7}$. The probabilities for the second card are dependent on which card was drawn first.

The total number of cards will reduce to 6 because the first card was not replaced. If a diamond was selected first, then the number of diamond cards is now 3 and hearts will still be 3. If a heart was selected first, then the number of diamond cards is still 4 and hearts will now be 2.

The probability tree shows this situation with the correct probability on each branch:

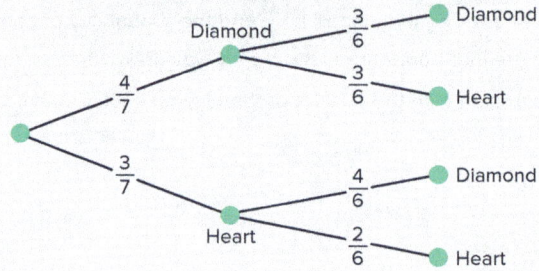

Because these events are dependent, we must use the formula $P(A \text{ and } B) = P(A) \cdot P(B \text{ after } A)$ to find the probability of drawing two hearts.

The probability of selecting a heart first is

$$P(A) = \frac{3}{7}$$

The probability of selecting a heart given a heart was selected first is

$$P(B \text{ after } A) = \frac{2}{6}$$

The probability of selecting two hearts is

$$P(A \text{ and } B) = \frac{3}{7} \cdot \frac{2}{6}$$
$$= \frac{6}{42}$$
$$= \frac{1}{7}$$

Purpose

Challenge students to apply their understanding of dependent events and probability trees to calculate the probability of a complex event.

Students: Page 318-319

c Is there a greater chance of selecting two hearts when the first card is replaced or when the first card is not replaced?

Create a strategy

To find which situation gives us a greater chance of selecting two hearts, we can compare the probabilities we found in part (a) and part (b).

Apply the idea

When the first card is replaced, the probability of selecting two hearts is $\frac{9}{49}$.

When the first card is not replaced, the probability of selecting two hearts is $\frac{1}{7}$.

To compare, let's rewrite $\frac{1}{7}$ as an equivalent fraction with a denominator of 49.

$$\frac{1}{7} \cdot \frac{7}{7} = \frac{7}{49}$$

Since $\frac{7}{49} < \frac{9}{49}$, there is a greater chance of selecting two hearts when the first card is replaced.

Purpose

Show students how to compare probabilities of dependent and independent events in a practical context, emphasising the importance of understanding the difference between events where an item is replaced or not replaced.

Expected mistakes

Students might forget that fractions should have the same denominator before making comparisons, which would cause them to struggle to compare the two different probabilities. Remind them that the size of the pieces represented by the fractions are different (i.e. the denominators are different), which is why it is difficult to compare them in their original formats.

Reflecting with students

After they find the specific probabilities for the given pile, prompt advanced learners to consider different scenarios by altering the number of hearts and diamonds or changing the total number of cards. Ask them to investigate how these changes affect the probabilities and to look for patterns in their results.

Encourage them to generalize their findings and develop a rule or set of conditions that determine when "without replacement" yields a higher probability than "with replacement" and vice versa. This inquiry-based approach allows students to actively engage with the concepts, make conjectures, test their ideas, and deepen their understanding of independent and dependent events in probability.

Students: Page 319

Example 3

Vanessa has 12 songs in a playlist. Four of the songs are her favorite. She selects shuffle and the songs start playing in random order. Shuffle ensures that each song is played once only until all songs in the playlist have been played. Find the probability that:

a   The first song is one of her favorites.

Create a strategy

The probability can be calculated using

$P \text{ (Event)} = \dfrac{\text{Number of favorable outcomes}}{\text{Total number of outcomes}}$.

Apply the idea

Vanessa has 4 favorite songs and there are 12 songs in the playlist.

The probability that the first song is one of her favorites is $\dfrac{4}{12} = \dfrac{1}{3}$.

Purpose

Show students how to calculate the probability of a single event happening from a total number of events.

Students: Page 319

b   Two of her favorite songs are the first to be played.

Create a strategy

The probability for the second song will be dependent on the first song that is played. We can calculate this probability using

$$P\big(A \text{ and } B\big) = P\big(A\big) \cdot P\big(B \text{ after } A\big)$$

where event A is a favorite song is played first and event B is a favorite song is played second.

Apply the idea

From part (a), we know the probability that the first song is one of Vanessa's favorites is $\frac{1}{3}$.

Because the songs cannot be repeated, this leaves 3 favorite songs out of 11 songs that have yet to play. So, the probability that the second song is also her favorite is $\frac{3}{11}$.

$$P(A \text{ and } B) = P(A) \cdot P(B \text{ after } A)$$ Probability of dependent events

$$= \frac{1}{3} \cdot \frac{3}{11}$$ Substitute known values

$$= \frac{1}{\cancel{3}} \cdot \frac{\cancel{3}}{11}$$ Divide out common factor

$$= \frac{1}{11}$$ Multiply

Purpose

Show students how to calculate the probability of dependent events, particularly in the context of a real-world scenario.

Use algorithmic thinking
Targeted instructional strategies

use with Example 3

Use algorithmic thinking by guiding students to develop a step-by-step method for calculating the probability of dependent events. Encourage them to outline each step, starting with determining the probability of the first event, then adjusting the probability of the second event based on the outcome of the first. Provide an exemplar set of steps:

1. Optional: Create the first branch of the tree diagram that shows all the possible options and their probabilities

2. Identify the probability of the first event occurring.

3. Adjust the sample space to reflect the outcome of the first event.

   - Optional: Create the second branch of the tree diagram that shows all the possible outcomes for the second event and their probabilities, based on the outcomes of the first event (branch).
   - If the events are independent, this will be the same as the first branch.
   - If the events are dependent, then this will be different from the first branch.

4. Calculate the probability of the second event given the first has occurred.

5. Multiply the probabilities of the first and second events to find the combined probability.

6. Check the reasonableness of the probability.

By following this algorithm, students will understand how each event impacts the next and accurately calculate the probability of dependent events.

Students: Page 320

> 💡 **Idea summary**
>
> Two events are dependent if the outcome of one event affects the outcome of the other event. The probability of two dependent events is given by:
>
> $$P(A \text{ and } B) = P(A) \cdot P(B \text{ after } A)$$
>
> Events "with replacement" occur when the item drawn is placed back into the group before each selection. Each selection is independent of the others.
>
> Events "without replacement" occur when the item remains outside of the group after selection. Each selection is dependent of the others. The probabilities of each selection will change depending on previous selections.

Practice

Students: Page 320–325

What do you remember?

SOL **1** Two bags contain marbles that are all the same size and shape.
- Bag A has 3 blue marbles and 5 red marbles
- Bag B has 4 blue marbles and 2 red marble

Which of these best describes dependent events?

A Randomly selecting one marble from Bag A, replacing the marble then randomly selecting another marble from Bag A

B Randomly selecting one marble from Bag B, not replacing the marble, then randomly selecting another marble from Bag B

C Randomly selecting one marble from Bag A, replacing the marble then randomly selecting one marble from Bag B

D Randomly selecting one marble from Bag B, not replacing the marble then randomly selecting one marble from Bag A

SOL **2** Mathias has 15 tiles in a bag that are the same size and shape.
- 6 green tiles
- 5 blue tiles
- 4 red tiles

Mathias will randomly select a green tile from the bag, not replace it, and then randomly select a red tile from the bag.

Does this scenario decribe two events that are independent or dependent?

3 A bag has 3 green marbles, 5 pink marbles, and 2 yellow marbles. Two marbles are drawn without replacement.

a How many marbles are left after the first draw?

b How many pink marbles are left if a green was drawn first?

c How many pink marbles are left if a pink was drawn first?

4 A bag contains 3 pink balls and 2 orange balls. You draw one ball, note its color, then draw another ball without putting the first ball back.

a Find the missing probability for drawing a second pink ball on the tree diagram.

b Which outcome can be found by multiplying $\frac{2}{5}$ by $\frac{3}{4}$?

c Which outcome(s) has a probability of $\frac{3}{10}$?

d Are the events of drawing the balls independent or dependent?

Let's practice

5   From a standard pack of cards, 1 card is randomly drawn and then put back into the pack. A second card is then drawn.

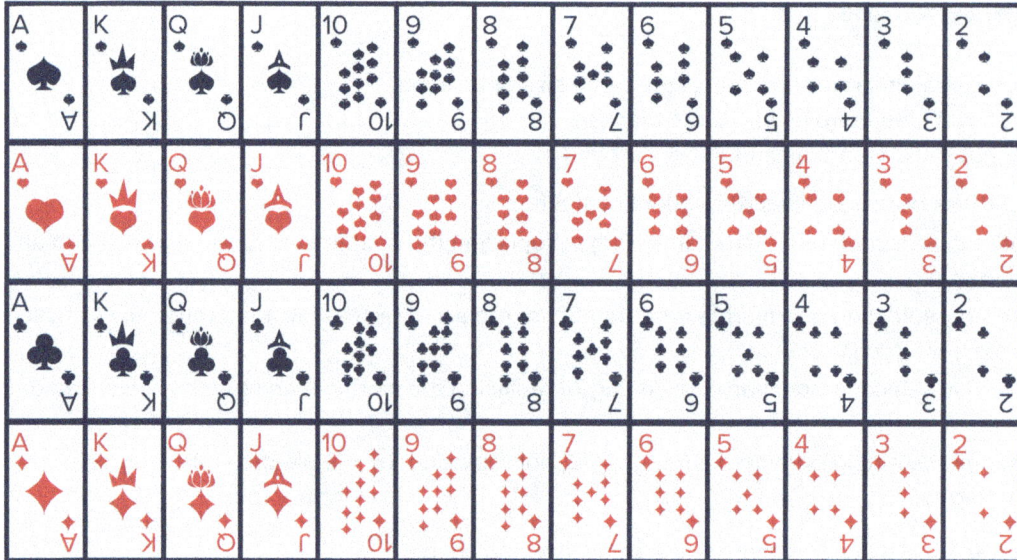

   a   Are the two draws independent or dependent?

   b   Find the probability that the first card is a spade, but the second card is not a spade.

   c   Find the probability that neither of the cards are spades.

6   Mae deals two cards from a normal deck of cards, without replacement.

   a   Are the events of drawing Card 1 and drawing Card 2 dependent or independent?

   b   Calculate the probability that she deals:

   i   Two 10s   **iii**   Two diamonds

   ii   Two red cards

7   A bag has five counters labeled 1 to 5. Two counters are drawn and the larger number is recorded as the outcome.

   a   Complete this array that lists the sample space if the experiment is done with replacement.

   b   Complete this array that lists the sample space if the experiment is done without replacement.

	1	2	3	4	5
1	1	2	3	4	5
2		2	3	4	5
3			3	4	5
4				4	5
5					5

	1	2	3	4	5
1		2	3	4	5
2			3	4	5
3				4	5
4					5
5					

   c   If the experiment is done without replacement, find the probability of the larger number being:

   i   5   **ii**   Odd

   iii   1   **iv**   Either even or prime

   v   Both even and prime

SOL **8** Elena has a collection of stickers that are the same size and shape.
- 5 purple stickers
- 8 yellow stickers
- 12 pink stickers

a What is the probability that Elena will randomly select a pink sticker, not replace it, then randomly select a yellow sticker?

b Select a point on the number line that represents this probability.

9 A hand contains a 10, a Jack, a Queen, a King and an Ace. Two cards are drawn from the hand at random, in succession and without replacement. Find the probability that:

a The Ace is drawn first, then the 10 is drawn.

b The King is not drawn.

c The Queen is the second card drawn.

10 In the "Black or Red" game, you guess whether the next card is black or red. You place your card back in the deck, and the deck of cards is re-shuffled every round. In this particular game, the previous 11 cards drawn were black.

a What is the probability that the next card is also black?

b Would this be the case if the cards drawn were put into a discard pile instead? Explain your answer.

11 A bag contains 5 green marbles, 7 red marbles and 4 blue marbles. If you want to find the probability of drawing two green marbles, one after the other, from the bag, explain how replacement affects the independence or dependence of the events.

SOL **12** A box of identical tokens contains 4 red tokens, 6 white tokens, and 5 black tokens. Lorena and Alejandro will each randomly select tokens from this box.
- Lorena will randomly select one token, replace it, then randomly select a second token.
- Alejandro will randomly select one token, not replace it, randomly select a second token.

Who has a greater probability of randomly selecting one black token and then one white token?

SOL **13** All the coins in a chest are in the same size and shape.

Jordan randomly selects a coin from the chest, does not replace it, and then randomly selects a second coin.

Taylor randomly selects a coin from the chest, replaces it, and then randomly selects a second coin.

Who has the least probability of selecting a silver coin and then a gold coin? Explain your thinking.

Type	Number
Copper	8
Bronze	20
Silver	12
Gold	16

SOL **14** Miranda has 20 colored markers in a box that are all the same size and shape.
- The probability of selecting a violet marker is 25%.
- The probability of selecting an orange marker is 15%.
- The probability of selecting a black marker is 20%.
- The probability of selecting a silver marker is 40%.

Carlos and Luis would like to determine the probability of selecting a violet marker, giving it away, and then selecting a silver marker.

Carlos determined the probability of the two events to be $\dfrac{1}{10}$.

Luis determined the probability of the two events to be $\dfrac{2}{19}$.

Who calculated the probability incorrectly? Explain what mistake the student made.

15 A bag contains 3 green marbles and 4 purple marbles. Two marbles are drawn at random from the bag.

   a If the first marble is replaced before the second marble is drawn complete the table with the probabilities matching the edges of the probability tree.

   b If the first marble is not replaced before the second marble is drawn complete the table with the probabilities matching the edges of the probability tree:

   c Determine the impact that replacing the marble before drawing a second has on the probability of these events:

   i Drawing two green marbles

   ii Drawing two purple marbles

   iii Drawing marbles of different colors

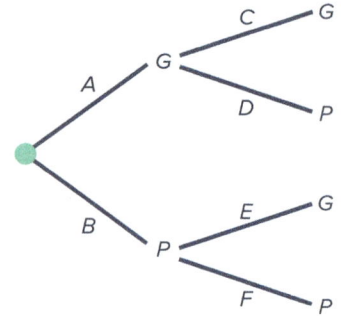

Edge	A	B	C	D	E	F
Probability	$\frac{3}{7}$					

16 James owns four green jackets and three blue jackets. He selects one of the jackets at random for himself and then another jacket at random for his friend:

   a Find the probability that James selects a blue jacket for himself.

   b Find the probability that both jackets James selects are green.

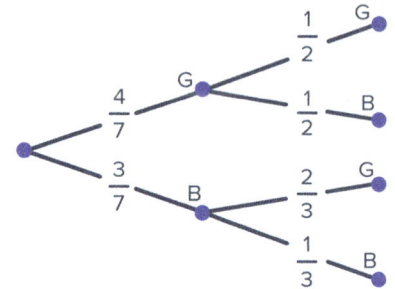

17 Han owns four green ties and three blue ties. He selects one of the ties at random for himself and then another tie at random for his friend.

   a Write the probabilities for the outcomes on the edges of the probability tree diagram.

   b Calculate the probability that:

   i Han selects a blue tie for himself.

   ii Han selects two green ties.

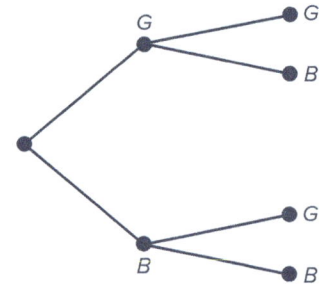

18 In tennis if the first serve is a fault (out or in the net), the player takes a second serve. A player serves with the following probabilities:

   • First serve in: 0.55

   • Second serve in: 0.81

The following probability tree shows the probability of the first two serves either being in or a fault:

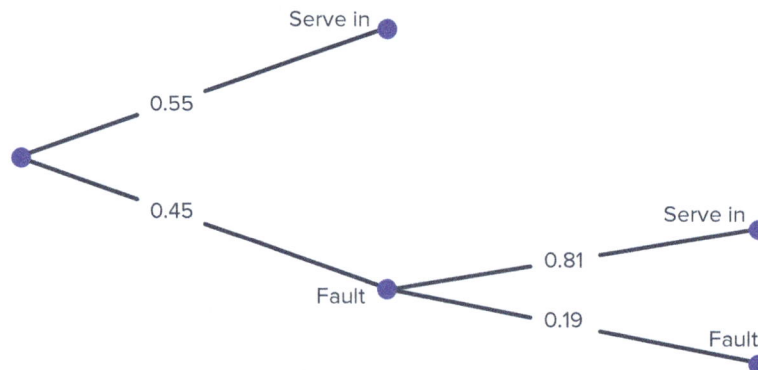

Find the probability that the player makes a double fault (both serves are a fault).

Let's extend our thinking

19 Using the scenario below, write a question for a classmate to solve that involves determining the probability of two dependent events.

Demi has a jar of 15 cookies. The jar contains 6 oatmeal cookies, 4 sugar cookies, and 5 peanut butter cookies.

20 A sailor has four meal packets to choose from on her last day but only wants to eat three meals. Two of the packets are oatmeal (*O*) and two of them are toasted muesli (*T*). She will eat three of them throughout the day, chosen at random. The probability tree shows the options for the day.

    a Fill in the probabilities matching the edges of the probability tree:

A	B	C	D	E	F

    b Find the probability that she has the same type of meal for breakfast and lunch.

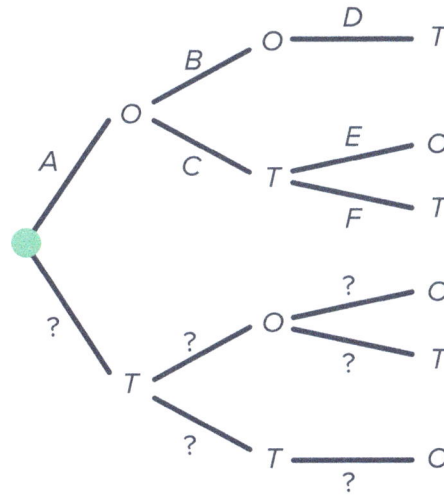

21 Valentina randomly selects three cards from a normal deck of cards.

Calculate the probability if the cards are selected:

    i With replacement   **ii** Without replacement

      a The cards are five of clubs, King of clubs, and Jack of spades, in that order.

      b The cards are all red.

      c The first card is the 2 of clubs, the second card is a spade and the third card is red.

      d The cards are all spades.

      e None of the cards is an 8.

22 Three marbles are randomly drawn without replacement from a bag containing 5 red, 4 yellow, 5 white, 4 black and 5 green marbles. Find the probability of drawing:

    a Three white marbles

    b Three black marbles

    c Zero green marbles

    d Zero yellow marbles

23 A number game uses a basket with 5 balls, all labeled with numbers from 1 to 5. Three balls are drawn at random, without replacement.

    a Find the probability that the ball labeled 4 is picked.

    b Find the probability that the ball labeled 4 is picked and the ball labeled 1 is also picked.

Answers

5.03 Probability of dependent events

What do you remember?

1 B

2 This scenario describes two events that are dependent. When Mathias randomly selects a green tile from the bag and does not replace it before selecting a red tile, the outcome of the first selection affects the probability of the second selection.

3
 a 9
 b 5
 c 4

4
 a $\frac{1}{2}$
 b Drawing an orange ball, then a pink ball
 c • Drawing a pink ball, then an orange ball
  • Drawing a pink ball, then a pink ball
  • Drawing an orange ball, then a pink ball
 d Dependent

Let's practice

5
 a Independent
 b $\frac{39}{52} \cdot \frac{39}{52} = \frac{9}{16}$
 c $\frac{13}{52} \cdot \frac{39}{52} = \frac{3}{16}$

6
 a Dependent
 b i $\frac{4}{52} \cdot \frac{3}{51} = \frac{1}{221}$
  ii $\frac{26}{52} \cdot \frac{25}{51} = \frac{25}{102}$
  iii $\frac{13}{52} \cdot \frac{12}{51} = \frac{1}{17}$

7
 a

	1	2	3	4	5
1	1	2	3	4	5
2	2	2	3	4	5
3	3	3	3	4	5
4	4	4	4	4	5
5	5	5	5	5	5

 b

	1	2	3	4	5
1		2	3	4	5
2	2		3	4	5
3	3	3		4	5
4	4	4	4		5
5	5	5	5	5	

 c
 i $\frac{2}{5}$ iii 0
 ii $\frac{3}{5}$ iv 1
       v $\frac{1}{10}$

8
 a $\frac{4}{25}$ or 0.16
 b

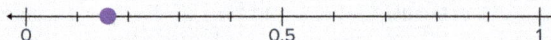

9
 a $\frac{1}{5} \cdot \frac{1}{4}\Big) = \frac{1}{20}$
 b $\frac{4}{5} \cdot \frac{3}{4} = \frac{3}{5}$
 c $\frac{4}{5} \cdot \frac{1}{4} = \frac{1}{5}$

10
 a $\frac{1}{2}$
 b No, the probabilities would change because there would be fewer black cards, and fewer cards in total, for the next draw.

11 With Replacement: If a marble is drawn and then replaced before drawing the second marble, the two events are independent. This means the outcome of the first draw does not affect the outcome of the second draw. The probability of drawing a green marble the second time remains the same as it was for the first draw because the total number of favorable outcomes (drawing a green marble) and the total number outcomes (number of marbles in the bag) are unchanged.

Without Replacement: If a marble is drawn and not replaced before drawing the second marble, the two events are dependent. The outcome of the first draw affects the probability of the second draw because the total number of marbles in the bag decreases by one. The number of green marbles remaining in the bag is also dependent on which color was drawn first. This changes the probabilities of drawing a green marble second marble compared to probability of drawing a green marble first.

12 Alejandro has a greater probability of randomly selecting one black token and then one white token without replacement compared to Lorena who replaces the token between selections. This difference arises because not replacing the first token slightly alters the total number of tokens and the composition of colors for the second selection, affecting the probability in favor of the desired sequence of selections for Alejandro.

13 Taylor has the least probability of selecting a silver coin and then a gold coin, as the probability does not increase with the replacement in his scenario.

14 Carlos made a mistake because he calculated the probabilities as if the events were independent, simply multiplying the chance of picking a violet marker by the chance of picking a silver marker. However, he didn't take into account the effect of not replacing the violet marker, which changes the total count and the odds for the second pick.

15 a

Edge	A	B	C	D	E	F
Probability	$\frac{3}{7}$	$\frac{4}{7}$	$\frac{3}{7}$	$\frac{4}{7}$	$\frac{3}{7}$	$\frac{4}{7}$

b

Edge	A	B	C	D	E	F
Probability	$\frac{3}{7}$	$\frac{4}{7}$	$\frac{1}{3}$	$\frac{2}{3}$	$\frac{1}{2}$	$\frac{1}{2}$

  c i Replacing the marble results in a higher probability by $\frac{2}{49}$

    ii Replacing the marble results in a higher probability by $\frac{2}{49}$

    iii Replacing the marble results in a lower probability by $\frac{4}{49}$

16 a $\frac{3}{7}$

  b $\frac{4}{7} \cdot \frac{1}{2} = \frac{2}{7}$

17 a

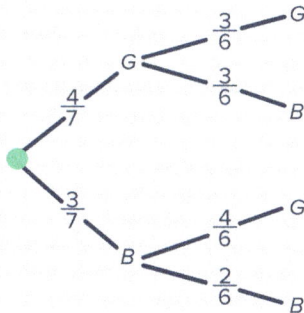

  b i $\frac{3}{7}$

    ii $\frac{4}{7} \cdot \frac{3}{6} = \frac{2}{7}$

18 0.0855

Let's extend our thinking

19 Answers may vary. An example problem can be:
Demi will take one cookie from the jar without looking, eat it, and then take a second cookie without looking. What is the probability that Demi first picks an oatmeal cookie and then picks a sugar cookie?

20 a

A	B	C	D	E	F
$\frac{2}{4}$	$\frac{1}{3}$	$\frac{2}{3}$	1	$\frac{1}{2}$	$\frac{1}{2}$

  b $\frac{1}{3}$

21 a i $\frac{1}{52} \cdot \frac{1}{52} \cdot \frac{1}{52} = \frac{1}{140608}$

    ii $\frac{1}{52} \cdot \frac{1}{51} \cdot \frac{1}{50} = \frac{1}{132600}$

  b i $\frac{26}{52} \cdot \frac{26}{52} \cdot \frac{26}{52} = \frac{1}{8}$

    ii $\frac{26}{52} \cdot \frac{25}{51} \cdot \frac{24}{50} = \frac{2}{17}$

  c i $\frac{1}{52} \cdot \frac{13}{52} \cdot \frac{26}{52} = \frac{1}{416}$

    ii $\frac{1}{52} \cdot \frac{13}{51} \cdot \frac{26}{50} = \frac{13}{5100}$

  d i $\frac{13}{52} \cdot \frac{13}{52} \cdot \frac{13}{52} = \frac{1}{64}$

    ii $\frac{13}{52} \cdot \frac{12}{51} \cdot \frac{11}{50} = \frac{11}{850}$

  e i $\frac{48}{52} \cdot \frac{48}{52} \cdot \frac{48}{52} = \frac{1728}{2197}$

    ii $\frac{48}{52} \cdot \frac{47}{51} \cdot \frac{46}{50} = \frac{4324}{5525}$

22 a $\frac{5}{23} \cdot \frac{4}{22} \cdot \frac{3}{21} = \frac{10}{1771}$

  c $\frac{18}{23} \cdot \frac{17}{22} \cdot \frac{16}{21} = \frac{816}{1771}$

  b $\frac{4}{23} \cdot \frac{3}{22} \cdot \frac{2}{21} = \frac{4}{1771}$

  d $\frac{19}{23} \cdot \frac{18}{22} \cdot \frac{17}{21} = \frac{969}{1771}$

23 a $\left(\frac{1}{5} \cdot \frac{4}{4} \cdot \frac{3}{3}\right) + \left(\frac{4}{5} \cdot \frac{3}{4} \cdot \frac{1}{3}\right) + \left(\frac{4}{5} \cdot \frac{1}{4} \cdot \frac{3}{3}\right) = \frac{3}{5}$

  b $2 \cdot \frac{3}{20} = \frac{3}{10}$

Topic 5 Assessment: Probability

1 State whether the following events or selections are independent or dependent.

   a A teacher has a "prize bag" filled with different prizes. The students form a line to draw a prize from the bag at random. Once a student has drawn a prize, they take it back to their desk.

   b Owning a cat and growing your own vegetable garden.

   c Buying lottery tickets and winning the lottery.

   d Driving a car and getting stuck in a traffic.

   e Drinking medicine and running out of milk.

2 A die is rolled 256 times. The die lands on a six 20 times, and the probability that the next roll will land on a six is $\frac{1}{6}$.

State whether the outcome of the next roll is independent of or dependent on the outcomes of previous rolls.

3 In a lottery there are 12 balls, each labeled with a different number.

   • The probability of a particular ball being drawn first is $\frac{1}{12}$.

   • A ball can be discarded after it has been drawn. If the ball is discarded, the probability of any remaining particular number being drawn next is $\frac{1}{11}$.

The lottery consists of 5 balls being chosen. Describe how this could be performed as an independent event. Then describe how this could be performed as a dependent event.

SOL **4** Tatiana has a bag of coins with different numbers printed on them.
   • The probability of selecting an even number is 40%.
   • The probability of selecting a 5 is 20%.

What is the probability that Tatiana draws a number that is not even, places it back in the bag, then draws a 5?

5 A fair die is rolled and then a coin is tossed. The tree diagram below shows the possible outcomes these events.

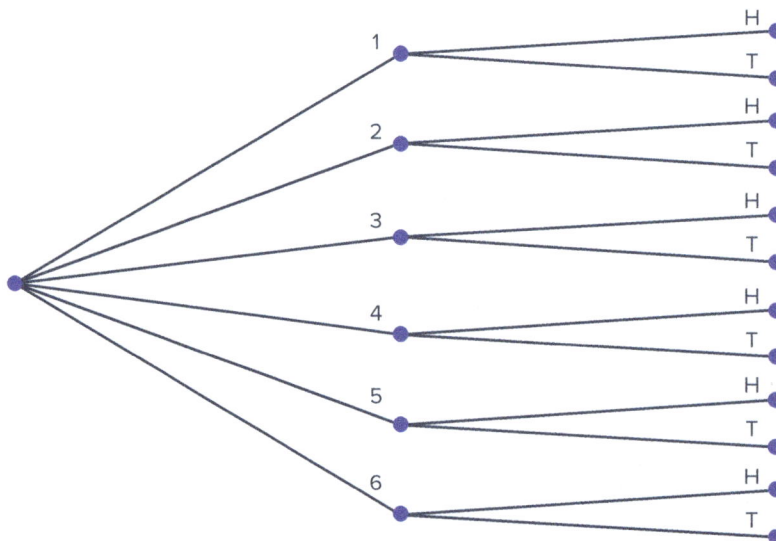

   a Find the probability of rolling a 3 and a heads as a whole percent.

   b Find the probability of getting an odd number and a tail as a decimal.

6 Two standard die are rolled. One is black and one is white. Find the probability as reduced fraction that:

    a   The black die is odd and the white die is even.

    b   The sum of the two numbers rolled is less than 2.

7 The 12 members of the city council, including the President and the Treasurer, have a row of 12 seats randomly designated for them at the party campaign launch.

Find the probability that the President is seated at either end of the row and the Treasurer is seated next to the President.

8 Vanessa has 10 songs in a playlist. 4 of the songs are her favorite. She selects shuffle and the songs start playing in random order. Shuffle ensures that each song is played once only until all songs in the playlist have been played. Find the probability as a reduced fraction that:

    a   The first song is one of her favorites.

    b   Two of her favorite songs are the first to be played.

SOL **9** The jar pictured contains:
- 5 red marbles
- 7 blue marbles
- 8 black marbles

Plot a point on the number line provided that represents the probability of selecting a red marble, placing it back in the jar, then selecting a marble that is NOT black.

10 Ms. Martin's math class is running a booth at the school carnival that gives out a prize for guessing the correct order of the colors of two balls drawn from a bag. The smaller the odds, the bigger the prize.

Color	Number of balls
Blue	7
Green	4
Yellow	6
Red	3

The class is having trouble deciding the best way to select the balls from the bag to make it more difficult to get the "big" prize.

Some students argue that replacement makes the prize harder to win, while others argued that not replacing the ball makes it harder to win.

Which students are correct? Justify your answer using the example of two red balls, writing any probabilities as percentages rounded to the nearest tenth.

Answers

Topic 5 Assessment: Probability

1 a Dependent   c Dependent   e Independent

   b Independent   d Dependent

   8.PS.1a

2 Independent

   8.PS.1a

3 The lottery could be performed as an independent event by replacing the ball after each draw. The lottery could be performed as a dependent event by discarding the ball after each draw.

   8.PS.1a

4 12%

   8.PS.1c

5 a 8%   b 0.25

   8.PS.1c

6 a $\dfrac{1}{4}$ b 0

   8.PS.1c

7 $\dfrac{1}{66}$

   8.PS.1d

8 a $\dfrac{2}{5}$ b $\dfrac{2}{15}$

   8.PS.1d

9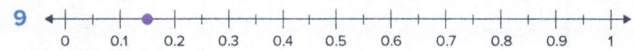

   8.PS.1c

10 Without replacement gives the smaller probability, thus making it hard to win the prize. Two red balls without replacement has a probability of 15%· 10.5%, or 1.6%. Two red balls with replacement has a probability of 15%· 15%, or 2.3%.

   8.PS.1b

6 2D Geometry

Big ideas

- Geometric figures are bound by properties that can be verified.
- Right triangles have special properties that allow their side lengths to be determined using the Pythagorean theorem or ratios.
- Physical objects can be modeled with 2D and 3D geometric figures whose properties can be applied to solve real-world problems.
- Geometric transformations are functional relationships that can be described in terms of angles, circles, parallel lines, perpendicular lines, and line segments.

Chapter outline

Architects and engineers rely on the Pythagorean Theorem to ensure buildings are safe and structures are sound.

6. 2D Geometry

Topic overview

Foundational knowledge

Evaluating standards proficiency

The skills book contains questions matched to individual standards. It can be used to measure proficiency for each.

5.MG.3 — The student will classify and measure angles and triangles, and solve problems, including those in context.

5.MG.2 — The student will use multiple representations to solve problems, including those in context, involving perimeter, area, and volume.

6.MG.2 — The student will reason mathematically to solve problems, including those in context, that involve the area and perimeter of triangles, and parallelograms.

6.MG.3 — The student will describe the characteristics of the coordinate plane and graph ordered pairs.

7.MG.4 — The student will apply dilations of polygons in the coordinate plane.

Big ideas and essential understanding

Geometric figures are bound by properties that can be verified.
6.01 — Certain angles have special relationships that allow us to find their measure in relation to another angle.

Right triangles have special properties that allow their side lengths to be determined using the Pythagorean theorem or ratios.
6.02 — The sides of any right triangle share the same relationship defined by the Pythagorean theorem.

6.03 — The sides of any right triangle share the same relationship defined by the Pythagorean theorem.

Physical objects can be modeled with 2D and 3D geometric figures whose properties can be applied to solve real-world problems.
6.04 — The properties of polygons can be applied to find their area and perimeter.

Geometric transformations are functional relationships that can be described in terms of angles, circles, parallel lines, perpendicular lines, and line segments.
6.05 — A translation preserves distance and angle measure because it moves all of the points on a figure the same distance and direction.

6.06 — A reflection preserves distance and angle measure, but reverses orientation. The preimage and the image of a figure are symmetric about a line of reflection.

6.07 — A translation preserves distance and angle measure because it moves all of the points on a figure the same distance and direction.

6.07 — A reflection preserves distance and angle measure, but reverses orientation. The preimage and the image of a figure are symmetric about a line of reflection.

Standards

8.MG.1 — The student will use the relationships among pairs of angles that are vertical angles, adjacent angles, supplementary angles, and complementary angles to determine the measure of unknown angles.

8.MG.1a — Identify and describe the relationship between pairs of angles that are vertical, adjacent, supplementary, and complementary.
6.01 Angle relationships

8.MG.1b — Use the relationships among supplementary, complementary, vertical, and adjacent angles to solve problems, including those in context, involving the measure of unknown angles.
6.01 Angle relationships

8.MG.3 — The student will apply translations and reflections to polygons in the coordinate plane.

8.MG.3a — Given a preimage in the coordinate plane, identify the coordinates of the image of a polygon that has been translated vertically, horizontally, or a combination of both.
6.05 Translations in the coordinate plane

8.MG.3b — Given a preimage in the coordinate plane, identify the coordinates of the image of a polygon that has been reflected over the x- or y-axis.
6.06 Reflections in the coordinate plane

8.MG.3c — Given a preimage in the coordinate plane, identify the coordinates of the image of a polygon that has been translated and reflected over the x- or y-axis, or reflected over the x- or y-axis and then translated.
6.07 Translations and reflections in the coordinate plane

8.MG.3d — Sketch the image of a polygon that has been translated vertically, horizontally, or a combination of both.
6.05 Translations in the coordinate plane
6.07 Translations and reflections in the coordinate plane

8.MG.3e — Sketch the image of a polygon that has been reflected over the x- or y-axis.
6.06 Reflections in the coordinate plane
6.07 Translations and reflections in the coordinate plane

8.MG.3f — Sketch the image of a polygon that has been translated and reflected over the x- or y-axis, or reflected over the x- or y-axis and then translated.
6.07 Translations and reflections in the coordinate plane

8.MG.3g — Identify and describe transformations in context (e.g., tiling, fabric, wallpaper designs, art).

6.05 Translations in the coordinate plane
6.06 Reflections in the coordinate plane
6.07 Translations and reflections in the coordinate plane

8.MG.4 — The student will apply the Pythagorean Theorem to solve problems involving right triangles, including those in context.

8.MG.4a — Verify the Pythagorean Theorem using diagrams, concrete materials, and measurement.
6.02 Pythagorean theorem

8.MG.4b — Determine whether a triangle is a right triangle given the measures of its three sides.
6.02 Pythagorean theorem

8.MG.4c — Identify the parts of a right triangle (the hypotenuse and the legs) given figures in various orientations.
6.02 Pythagorean theorem

8.MG.4d — Determine the measure of a side of a right triangle, given the measures of the other two sides.
6.02 Pythagorean theorem

8.MG.4e — Apply the Pythagorean Theorem, and its converse, to solve problems involving right triangles in context.
6.02 Pythagorean theorem
6.03 Applications of the Pythagorean theorem

8.MG.5 — The student will solve area and perimeter problems involving composite plane figures, including those in context.

8.MG.5a — Subdivide a plane figure into triangles, rectangles, squares, trapezoids, parallelograms, circles, and semicircles. Determine the area of subdivisions and combine to determine the area of the composite plane figure.
6.04 Perimeter and area of composite shapes

8.MG.5b — Subdivide a plane figure into triangles, rectangles, squares, trapezoids, parallelograms, and semicircles. Use the attributes of the subdivisions to determine the perimeter of the composite plane figure.
6.04 Perimeter and area of composite shapes

8.MG.5c — Apply perimeter, circumference, and area formulas to solve contextual problems involving composite plane figures.
6.04 Perimeter and area of composite shapes

Future connections

G.RLT.2 — The student will analyze, prove, and justify the relationships of parallel lines cut by a transversal.

G.TR.1 — The student will determine the relationships between the measures of angles and lengths of sides in triangles, including problems in context.

G.TR.3 — The student will, given information in the form of a figure or statement, prove and justify two triangles are similar using direct and indirect proofs, and solve problems, including those in context, involving measured attributes of similar triangles.

G.TR.4 — The student will model and solve problems, including those in context, involving trigonometry in right triangles and applications of the Pythagorean Theorem.

G.DF.1 — The student will create models and solve problems, including those in context, involving surface area and volume of rectangular and triangular prisms, cylinders, cones, pyramids, and spheres.

G.DF.2 — The student will determine the effect of changing one or more dimensions of a three-dimensional geometric figure and describe the relationship between the original and changed figure.

G.RLT.3 — The student will solve problems, including contextual problems, involving symmetry and transformation.

6.01 Angle relationships

Subtopic overview

Lesson narrative

In this lesson, students will explore various types of angle relationships, including adjacent, vertical, complementary, and supplementary angles. The lesson begins with definitions and visual examples to help students understand that adjacent angles share a common ray and vertex, while vertical angles are congruent and formed by intersecting lines. Complementary angles sum to 90 degrees, and supplementary angles sum to 180 degrees. Students will practice naming angles using different methods and apply their understanding to solve problems involving these angle relationships. Through examples, they will calculate unknown angle measures by recognizing relationships such as supplementary angles forming a straight angle and complementary angles forming a right angle. By the end of the lesson, students should be able to identify and solve for unknown angles in various configurations, reinforcing their understanding of these fundamental geometric concepts.

Learning objectives

Students: Page 328

After this lesson, you will be able to...
- identify vertical, adjacent, supplementary, and complementary angles.
- describe the relationship between pairs of angles that are vertical, adjacent, supplementary, and complementary.
- use the relationships among supplementary, complementary, vertical, and adjacent angles to find missing angle measurements in mathematical and real-world problems.

Key vocabulary

- adjacent angles
- complementary
- supplementary
- vertical angles

Essential understanding

Certain angles have special relationships that allow us to find their measure in relation to another angle.

Standards

This subtopic addresses the following Virginia 2023 Mathematics Standards of Learning standards.

Mathematical process goals

MPG1 — Mathematical Problem Solving

Teachers can integrate this goal into their instruction by guiding students through step-by-step problem-solving strategies. For instance, when solving problems involving angle relationships, teachers can demonstrate how to use the properties of angles and their relationships to calculate the measure of unknown angles. Teachers can also encourage students to apply these strategies to solve real-world problems, such as calculating the angle between the hands of a clock at a particular time.

MPG2 — Mathematical Communication

Teachers can integrate this goal into their instruction by encouraging students to use correct mathematical terminology when communicating their understanding of angle relationships. For example, after learning about vertical, complementary, supplementary, and adjacent angles, students should be prompted to use these terms accurately when explaining their problemsolving process. Teachers can also foster mathematical discussions where students can exchange ideas, ask clarifying questions, and justify their reasoning using these terms. This not only deepens their understanding of the concepts but also builds their mathematical vocabulary.

MPG5 — Mathematical Representations

Teachers can integrate this goal into their instruction by teaching students how to represent angle relationships using a variety of methods, such as diagrams, symbolic notation, or real-world examples. For instance, after explaining the concept of complementary angles, teachers can show students how to represent these angles in a diagram and how to denote them symbolically. Teachers can also guide students in interpreting these representations, and in understanding how they convey the same mathematical idea.

Content standards

8.MG.1 — The student will use the relationships among pairs of angles that are vertical angles, adjacent angles, supplementary angles, and complementary angles to determine the measure of unknown angles.

8.MG.1a — Identify and describe the relationship between pairs of angles that are vertical, adjacent, supplementary, and complementary.

8.MG.1b — Use the relationships among supplementary, complementary, vertical, and adjacent angles to solve problems, including those in context, involving the measure of unknown angles.

Prior connections

5.MG.3 — The student will classify and measure angles and triangles, and solve problems, including those in context.

Future connections

G.RLT.2 — The student will analyze, prove, and justify the relationships of parallel lines cut by a transversal.

G.TR.1 — The student will determine the relationships between the measures of angles and lengths of sides in triangles, including problems in context.

Lesson Preparation

Suggested review

Depending on your students' level of prior knowledge, consider revisiting the following lessons:

    Grade 8 — 2.04 Solve multistep equations

Tools

You may find these tools helpful:
- Scientific calculator
- Protractor
- Tracing paper
- Dynamic geometry software

Student lesson & teacher guide

Angle relationships

Students are shown examples of non-overlapping angle pairs and are given the definitions of adjacent, vertical, complementary, and supplementary angles. They are then taught the naming convention of angles using points on the rays in addition to the vertex. Lastly, students learn that angle relationships can be used to solve for missing angle values.

🎓 Connecting angle relationships to classifying angles
Targeted instructional strategies

When classifying and measuring angles in prior years, students were exposed to different types of angles such as acute, right, obtuse, and straight. Review the definitions of a right angle and straight angle, showing examples of each in different orientations.

An angle with the notation shown, regardless of orientation, measures 90°.

A straight angle measures 180°, regardless of orientation.

Show students a more complex with different angle relationships, such as the one shown.

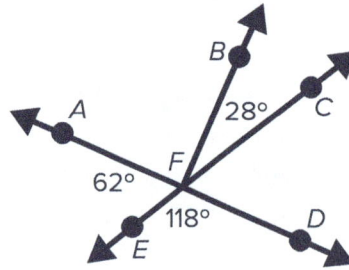

Demonstrate that the figure can be thought of as many different pairs of angles, highlighting different pairs of angles, like ∠AFE and ∠EFD.

Demonstrate the following connections among the angle relationships:
- ∠EFD also forms a straight angle with ∠CFD.
- ∠CFD is also congruent to ∠AFE.
- ∠AFE and ∠BFC would form a right angle since their measures add to 90°.
- ∠AFB is a right angle since it forms a straight angle with ∠AFE and ∠BFC.

Students may observe other relationships between pairs or groups of angles, and refer back to these relationships throughout the lesson.

Critique, correct, and clarify
English language learner support

Present a diagram with multiple angle relationships that contains intentional errors. For example, the diagram might show two angles labeled as complementary, but they actually add up to 180° and should be labeled as supplementary.

Students will individually analyze the diagram to identify the errors, then discuss their findings in pairs. They might note, for instance, that "these angles are supplementary, not complementary, because they add up to 180°.

Following the pair discussions, the class will come together to share corrections and update the diagram collectively. To conclude, students write reflections on how this activity helped them understand angle relationships more deeply. This process supports the development of critical thinking and peer collaboration.

Possible examples with errors to provide to students include:

1. Original Error: Two angles labeled as complementary that sum to 180°.
   Correction: Relabel these angles as supplementary, with the explanation that complementary angles sum to 90°, while supplementary angles sum to 180°.

2. Original Error: Two non-adjacent angles labeled as adjacent.
   Correction: Correct the labels to reflect that adjacent angles share a common vertex and side.

3. Original Error: Angles labeled as vertical that do not share the same vertex.
   Correction: Correctly identify vertical angles, explaining that they must be opposite each other when two lines intersect.

Provide a list of vocabulary terms and their definitions, and students can draw an example for each to keep as a reference during the lesson. An example is shown:

Relationship	Definition	Example
Complementary	Two angles whose measures add up to 90°	
Supplementary	Two angles whose measures add up to180°	
Vertical	Angles opposite each other when two lines cross	
Adjacent	Angles that share a common side and vertex	

Confusion with the terms and notation
Address student misconceptions

Students may confuse the definitions of terms, such as complementary and supplementary. The notation also may be missed, such as angle markings noting congruency.

Remind students that angle markings are used to indicate which angles are congruent. Encourage students to create a table listing angle relationships, their defining characteristics, and possible examples.

Examples

Students: Page 329

Example 1

Which of these diagrams shows a pair of adjacent angles?

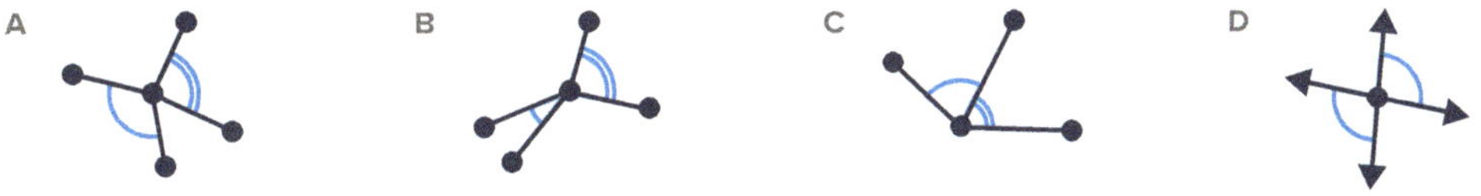

A   B   C   D

Create a strategy

We should look for two non-overlapping angles that share a common ray and a common vertex.

Apply the idea

The option that shows the two angles touching along a common segment is option C.

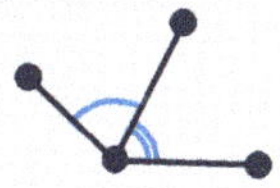

Purpose

Students demonstrate that they can identify adjacent angles given a diagram.

Reflecting with students

Encourage students to analyze why the other diagrams do not represent adjacent angles. Ask them to identify the key characteristics of adjacent angles: two angles that share a common vertex and a common side without overlapping. Have students discuss how some diagrams may show angles that only share a vertex or overlap, and why these do not meet the precise definition.

To deepen understanding, you might have students draw their own examples of adjacent and non-adjacent angles, highlighting the shared ray and vertex in their adjacent angle examples. Encourage them to think of real-world examples where adjacent angles occur, such as the hands of a clock (if it has a seconds hand) or the angles formed in a tiled floor pattern. This reflection will reinforce their comprehension and enable them to apply the concept to various contexts.

Students: Page 330

Example 2

Name an angle that is supplementary with $\angle 3$ in the figure shown:

Use the angle symbol $\angle$ in your answer.

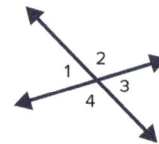

Create a strategy

Supplementary angles are two angles forming a straight angle.

Apply the idea

Both $\angle 2$ and $\angle 4$ are supplementary to $\angle 3$.

Reflect and check

Notice the supplementary angles are adjacent to $\angle 3$.

If we wanted to find the angles congruent to $\angle 3$, it would be $\angle 1$ since they are vertical angles.

Purpose

Demonstrate to students how to identify supplementary angles in a pair of intersecting lines.

Students: Page 330

Example 3

The angles in the diagram are complementary. What is the value of x?

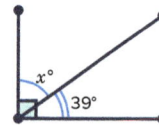

Create a strategy

Complementary angles are two angles forming a right angle equivalent to 90°.

Apply the idea

$x + 39 = 90$ Equate the sum of the angles to 90

$x = 51$ Subtract 39 from both sides

Purpose

When given a complementary pair of angles, students demonstrate that they can solve for a missing angle when given one.

Encourage students to deepen their exploration of complementary angles by adding layers of complexity to the original problem. After they find that $x - 51°$, prompt them to consider what would happen if the given angle measure changes: "If the 39° angle were increased by 10°, how would that affect the value of x?"

Ask students to generalize a formula for finding one complementary angle given the other $x = 90° - y$ and explore its applications with different values. Introduce additional scenarios where more than two angles sum to 90°, and challenge students to solve for unknown variables in these contexts.

Students: Page 330

Example 4

If the measure of angle $\angle BAD$ is $5x - 4°$, find the measure of $\angle CAD$.

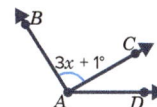

Create a strategy

Since $\angle BAC$ and $\angle CAD$ are adjacent:

$$m\angle BAD = m\angle BAC + m\angle CAD$$

Apply the idea

$$m\angle BAC + m\angle CAD = m\angle BAD$$

$$3x - 1° + m\angle CAD = 5x - 4°$$

Substitute the value of each angle

$$m\angle CAD = 5x - 4° - (3x - 1°)$$

Isolate the unknown angle

$$m\angle CAD = 2x - 3°$$

Subtract the expressions

Purpose

Demonstrate to students how to calculate an unknown angle using angle relationships and algebraic expressions.

Expected mistakes

Students may assume that $5x - 4°$ is the measure of $\angle CAD$ and add the two expressions together to equal a value that students estimate based on the appearance of the angle. Ask students to use their solved value of x to determine if the angle measurements would make sense in the context of the problem as a way to check their thinking.

Students may also assume that the letters in $m\angle CAD$ represent variables rather than a naming convention for the angle. Show students another way to represent $m\angle CAD$ in the problem is

$$3x - 1° + \square = 5x - 4°$$

Students: Page 331

Idea summary

Adjacent angles are two non-overlapping angles that share a common ray and a common vertex.

Vertical angles are a pair of nonadjacent angles formed by two intersecting lines. Vertical angles are congruent and share a common vertex.

Complementary angles are angles whose sum is 90°.

Supplementary angles are angles whose sum is 180°.

Practice

Students: Page 331–336

What do you remember?

1 An angle can be formed by:

    A Two intersecting lines   **B** Two rays with a common endpoint

    C Two segments that share an endpoint   **D** All of the above

2 Describe the angle measures of the following angles:

    a Acute angle   **b** Right angle   **c** Obtuse angle   **d** Straight angle

3 Select all of the correct ways to name the given angle.

    A $\angle PQR$ **B** $\angle RQP$

    C $\angle Q$ **D** All of the above

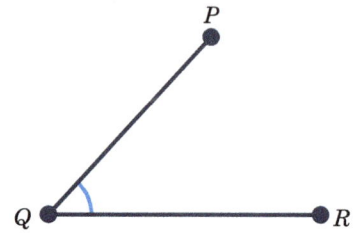

4 State whether the following diagrams indicate a pair of adjacent angles:

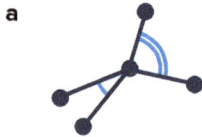

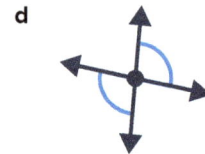

    a   **b**   **c**   **d**

5 State whether the following diagrams indicate a pair of vertical angles:

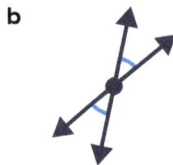

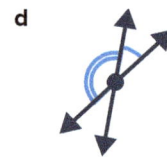

    a   **b**   **c**   **d**

6 State whether the following diagrams show a pair of complementary angles:

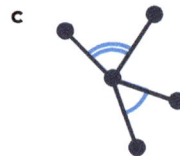

    a   **b**   **c**   **d**

7 State whether the following diagrams show a pair of supplementary angles:

    a   **b**   **c**   **d**

8 Complete each statement using a word or value from the table:

Acute	Obtuse	Right	Straight
60	90	180	360

    a Supplementary angles form a ☐ angle because they add to ☐ degrees.

    b Complementary angles form a ☐ angle because they add to ☐ degrees.

    c If two adjacent angles form an obtuse angle, their sum must be greater than ☐ degrees

Let's practice

9 Identify the missing angle for the angle pair.

    a ☐ forms a straight angle with $\angle QOT$.   **b** ☐ and $\angle BEH$ are vertical angles.

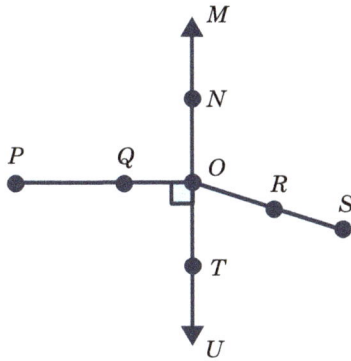 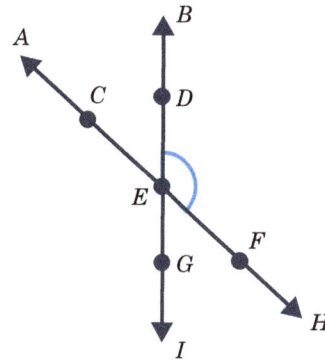

10 Among the marked angles in the diagram, identify:

    a A pair of adjacent angles.

    b A pair of vertical angles.

    c A pair of complementary angles.

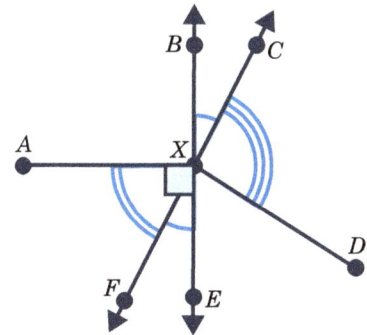

11 In the given diagram, identify:

    a An angle that is vertical to $\angle AXB$.

    b An angle that is supplementary to $\angle CXD$.

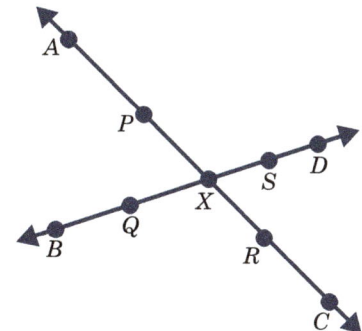

12 Which two angles are complementary?

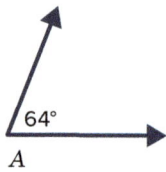

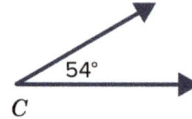

64°   *A*   36°   *B*   54°   *C*   144°   *D*

A   ∠*A* and ∠*B*   **B**   ∠*B* and ∠*C*   **C**   ∠*A* and ∠*C*   **D**   ∠*D* and ∠*A*

13 State whether the following angles are complementary, supplementary, or neither.

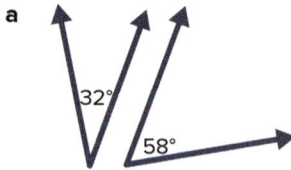

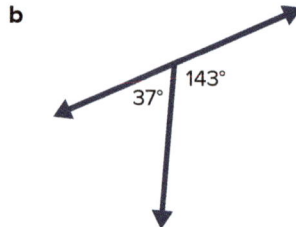

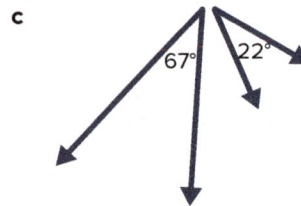

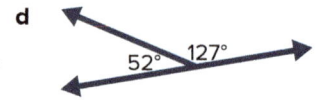

a   b   c   d

32°   58°   37°   143°   67°   22°   52°   127°

14 State whether each of the following pairs of angles are:

- Complemetary angles
- Supplementary angles
- Neither

a

25°   65°

b

160°   19°

c

34°   55°

d

40°   140°

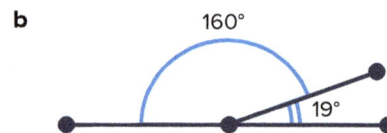

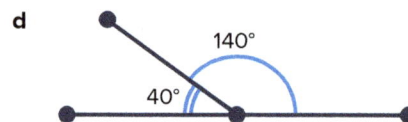

15 Determine whether each of the following diagrams shows the given angle pair. Explain.

a   Complementary angles

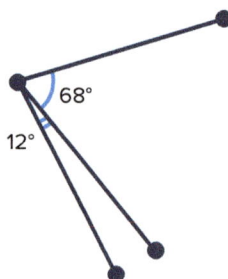

68°   12°

b   Vertical angles

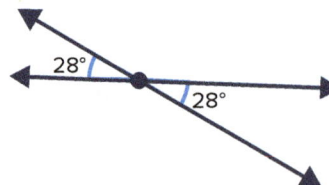

28°   28°

16 Find the value of x in each of the following diagrams:

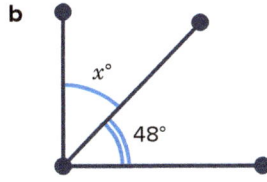

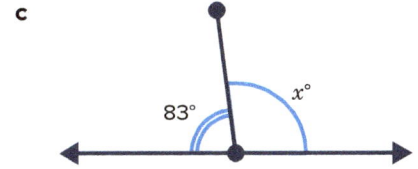

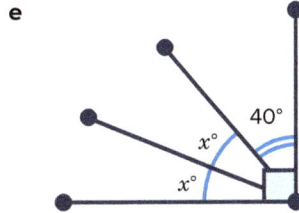

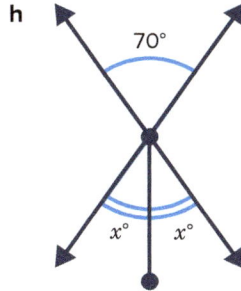

a

$x°$ 39°

b

$x°$ 48°

c

83° $x°$

d

94° $x°$

e

$x°$ 40° $x°$

f

10° $x°$ $x°$ $x°$ $x°$

g

70° $x°$

h

70° $x°$ $x°$

17 Find the value of the variable in each of the following diagrams:

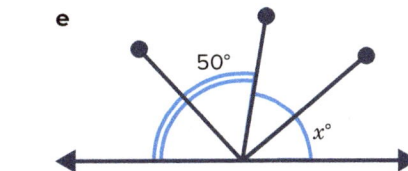

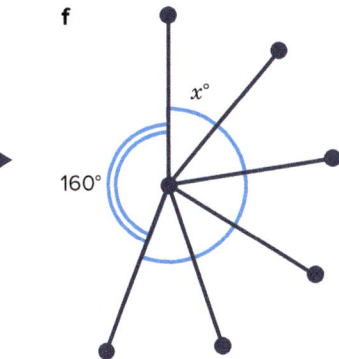

a

$x°$ $x°$ $x°$

b

$y°$ $y°$ $y°$ $y°$ $y°$

c

$z°$ $z°$ $z°$ $z°$ $z°$ $z°$ $z°$ $z°$ $z°$

d

50° $x°$ 55°

e

50° $x°$

f

$x°$ 160°

18 In the diagram shown, the given angle and the angle the ladder makes with the ground are supplementary.

A ladder needs to make a 75° angle with the ground for it to be safe to stand on. Is the ladder shown safe to stand on? Explain.

115°

19 On the clock shown, the angle measures formed by the hour hand and the minute hand are complementary.

If the angle formed by the minute hand and the second hand is 60°, calculate the angle measure between the hour hand and the second hand.

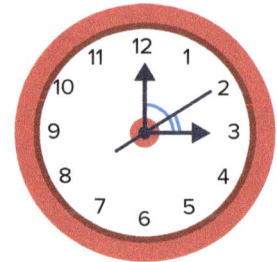

20 Find the measure of the angle formed by the open scissor blades.

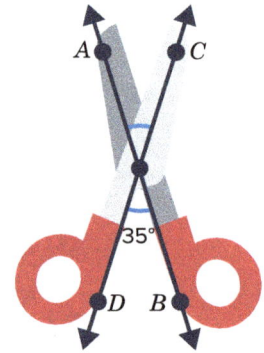

A   *C*

35°

D   *B*

Let's extend our thinking

21 Two angles (in degrees) are supplementary, and one angle is 3 times larger than the other.

Let x be the measure of the smaller angle and y be the measure of the larger angle.

a Solve for x.   **b** Solve for y.

22 The measure of an angle's complement is five less than half the measure of the angle's supplement. What is the measure of the angle?

23 For each diagram:

  i Write an equation that models the relationship shown in the diagram.

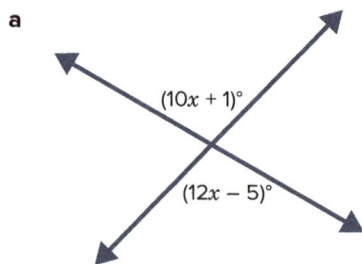

  ii Solve for x.

a

$(10x + 1)°$

$(12x - 5)°$

b

$(4x + 7)°$

$(2x + 5)°$

24 In the given picture, point E represents the point where a car would disapper from sight.

  a The measure of $\angle AED$ is 3.5 times greater than the measure of $\angle CED$. Calculate the measure of $\angle CED$.

  b $\angle AEF$ and $\angle DEB$ are complementary angles. If $\angle AEF$ is 10° smaller than $\angle DEB$, find the measure of each angle.

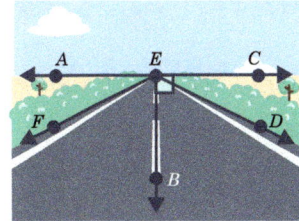

25 A city planner is drafting an intersection design for a new urban development. The design will involve the construction of three roads intersecting at one point.

The measures of the angle between any two roads must be at least 50° in order for vehicles to turn safely.

Does the design shown meet the requirement?

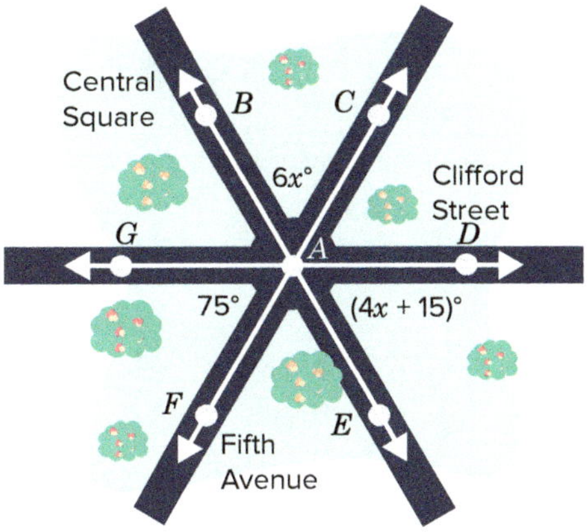

26 Determine if the statement is always, sometimes, or never true. Create diagrams to justify your answers.

  a Complementary angles are also adjacent.

  b Vertical angles are supplementary.

27 Determine if the information given is enough to justify the conclusion. Draw a diagram to support your solution.

Given: $\angle ABC \cong \angle DBF$

Conclusion: $\angle ABC$ and $\angle DBF$ are vertical angles

Answers

6.01 Angle relationships

What do you remember?

1 D

2 a An angle with a measure larger than 0° but smaller than 90°.

   b An angle with a measure exactly 90°.

   c An angle with a measure larger than 90° but smaller than 180°.

   d A straight angle is an angle with a measure exactly 180°.

3 D

4 a No   b Yes   c No   d No

5 a No   b Yes   c No   d No

6 a Yes   b No   c No   c No

7 a Yes   b No   c No   d No

8 a Supplementary angles form a straight angle because they add to 180°.

   b Complementary angles form a right angle because they add to 90°.

   c If two adjacent angles form an obtuse angle, their sum must be greater than 90°.

Let's practice

9 a $\angle QON$ b $\angle AEI$

10 a $\angle BXC$ and $\angle XD$ or $\angle AXF$ and $\angle EXF$

    b $\angle BXC$ and $\angle EXF$

    c $\angle AXF$ and $\angle$EXF or $\angle AXF$ and $\angle BXC$

11 a $\angle CXD$ b $\angle AXD$ or $\angle BXC$

12 a C

13 a Complementary   b Supplementary

    c Neither   d Neither

14 a Complementary angles   b Neither

    c Neither   d Supplementary angles

15 a These angles are not complementary angles because 68° + 12° = 80° and complementary angles must add up to 90°.

    b These are vertical angles because they are opposite angles formed by intersecting lines and have equal measures.

16 a $x = 51$ b $x = 42$ c $x = 97$ d $x = 86$

    e $x = 25$ f $x = 20$ g $x = 70$ h $x = 35$

17 a $x = 60$ b $y = 72$ c $z = 10$ d $x = 75$

    e $x = 40$ f $x = 40$

18 The ladder shown makes a 65° angle with the ground, which is not the required 75° for it to be safe to stand on. Therefore, the ladder is not safe to stand on as depicted.

19 30°

20 35°

Let's extend our thinking

21 a $x = 45$ b $y = 135$

22 10°

23 a i $10x + 1 = 12x - 5$ ii $x = 3$

    b i $4x + 7 + 2x + 5 = 180$ ii $x = 28$

24 a $x = 40°$

    b $\angle DEB = 50°$ and $\angle AEF = 40°$

25 Since the sum of angles at a point is 360°, we can set up the following equation:

$6x + 75° + (4x + 15)° + 6x + 75° + (4x + 15)° = 360°$

Solving this equation, we find $x = 9$.

We need to substitute the value of x to find the measure of each angle:

For $6x = 54°$

For $4x + 15 = 51°$

We have the six angles: 75°, 54°, 51°, 75°, 54° and 51°. Each of these angles is greater than 50°, so the design meets the requirement that the measures of the angle between any two roads must be at least 50° for vehicles to turn safely.

26 a Sometimes

    b Sometimes. If the vertical angles are 90° then they will also be supplementary.

27 This is not enough information to make the conclusion. The angles share a common vertex and are congruent, but vertical angles are only formed from intersecting lines. Here is a counterexample:

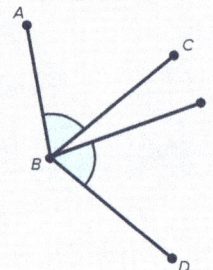

6.02 Pythagorean theorem

Subtopic overview

Lesson narrative

In this lesson, students will explore the Pythagorean Theorem, which states that in a right triangle, the square of the hypotenuse is equal to the sum of the squares of the other two sides. This theorem is presented as $(a^2 + b^2 = c^2)$, where (c) is the hypotenuse and (a) and (b) are the legs of the triangle. Students will learn to apply this theorem to find the length of any side of a right triangle when the lengths of the other two sides are known. The lesson also covers the converse of the Pythagorean Theorem, which helps determine if a given set of three side lengths form a right triangle. Additionally, students will explore Pythagorean triples, sets of three positive integers that satisfy the theorem, and how to generate new triples from known ones using scaling. Through various examples, students will practice identifying right triangles, calculating missing side lengths, and verifying Pythagorean triples. By the end of the lesson, students should be proficient in using the Pythagorean Theorem to solve geometric problems involving right triangles.

Learning objectives

Students: Page 337

🎓 **After this lesson, you will be able to...**
- use diagrams, concrete materials, and measurement to show the Pythagorean Theorem is true for any right triangle.
- use the converse of the Pythagoren theorem to determine if three given side lengths form a right triangle in mathematical and real-world problems.
- identify the the hypotenuse and the legs of a right triangle given in various orientations.
- use the Pythagorean Theorem to find unknown side lengths of right triangles in mathematical and real-world problems.

Key vocabulary

- converse of the Pythagorean theorem
- hypotenuse
- legs (of a right triangle)

- Pythagorean theorem
- Pythagorean triple
- right triangle

Essential understanding

The sides of any right triangle share the same relationship defined by the Pythagorean theorem.

Standards

This subtopic addresses the following Virginia 2023 Mathematics Standards of Learning standards.

Mathematical process goals

MPG1 — Mathematical Problem Solving

Teachers can integrate this goal by prompting students to apply the Pythagorean Theorem to solve complex problem situations. For instance, teachers can introduce real-world scenarios, such as calculating the shortest distance between two points on a map, where students would need to apply the Pythagorean Theorem to find the solution. This way, students are not only applying mathematical concepts but also relating them to real-world situations.

MPG3 — Mathematical Reasoning

Teachers can present examples and non-examples of right triangles, prompting students to use the Pythagorean Theorem or its converse to explain why the triangles are or are not right. This can help students understand the reasoning behind the theorem and its converse, and further develop their mathematical reasoning skills.

MPG4 — Mathematical Connections

Teachers can show students how the Pythagorean Theorem is connected to their prior knowledge about right triangles and properties of their sides and angles. Teachers can also create mathematical connections by linking the concept of the Pythagorean Theorem to realworld topics and situations. For example, discussing how this theorem is used in various careers such as engineering, architecture, and construction, thus establishing a connection between mathematics and the real world.

Content standards

8.MG.4 — The student will apply the Pythagorean Theorem to solve problems involving right triangles, including those in context.

8.MG.4a — Verify the Pythagorean Theorem using diagrams, concrete materials, and measurement.

8.MG.4b — Determine whether a triangle is a right triangle given the measures of its three sides.

8.MG.4c — Identify the parts of a right triangle (the hypotenuse and the legs) given figures in various orientations.

8.MG.4d — Determine the measure of a side of a right triangle, given the measures of the other two sides.

8.MG.4e — Apply the Pythagorean Theorem, and its converse, to solve problems involving right triangles in context.

Prior connections

5.MG.2 — The student will use multiple representations to solve problems, including those in context, involving perimeter, area, and volume.

Future connections

G.TR.3 — The student will, given information in the form of a figure or statement, prove and justify two triangles are similar using direct and indirect proofs, and solve problems, including those in context, involving measured attributes of similar triangles.

G.TR.4 — The student will model and solve problems, including those in context, involving trigonometry in right triangles and applications of the Pythagorean Theorem.

Lesson Preparation

Suggested review

Depending on your students' level of prior knowledge, consider revisiting the following lessons:

Grade 8 — 1.02 Estimate square roots

Tools

You may find these tools helpful:
- Scientific calculator
- Frayer model graphic organizer
- Dynamic geometry software

Student lesson & teacher guide

Pythagorean theorem

Exploration

Students: Page 337

> **▶ Interactive exploration**
> **Explore online to answer the questions**
>
> 🌐 **mathspace.co**

Use the interactive exploration in 6.02 to answer these questions:

1. What do you notice about the sum of the area of the squares of the two shorter sides and the area of the square of the hypotenuse?
2. If you subtracted the areas of the shorter sides' squares from the area of the largest side's square, what do you get?
3. Can you find another set of sides, where all three sides have integer values?
4. Try to write an equation that will always be true using the variables a, b, and c.

Suggested student grouping: In pairs

In this exploration, students will adjust the lengths of the legs of a right triangle using sliders. They will observe the relationship between the areas of squares drawn on each of the sides of the triangle. This activity is designed to help students discover the Pythagorean theorem.

Ideal student responses

These ideal responses may differ from other correct student responses. Less formal responses can be connected with the more precise mathematical language presented here.

1. **What do you notice about the sum of the area of the squares of the two shorter sides and the area of the square of the hypotenuse?**

   The sum of the areas of the squares of the two shorter sides is equal to the area of the square of the hypotenuse. This is true no matter how the leg lengths of the triangle are adjusted.

2. **If you subtracted the areas of the shorter sides' squares from the area of the largest side's square, what do you get?**

   If the area of the squares of the two shorter sides is subtracted from the area of the square of the hypotenuse, the result is zero. This means that the area of the square of the hypotenuse is equal to the sum of the areas of the squares of the two shorter sides.

3. **Can you find another set of sides, where all three sides have integer values?**

   Yes, besides $a = 6$ and $b = 8$, the only other possible combination is when $a = 8$ and $b = 6$. For all other values, c is not an integer.

4. **Try to write an equation that will always be true using the variables *a*, *b*, and, *c*.**

   The equation that always holds true for any right triangle is $a^2 + b^2 = c^2$, where a and b are the lengths of the two shorter sides and c is the length of the hypotenuse.

Purposeful questions
- What are the areas of each of the squares in terms of $a + b$, and c?
- Using your answer to the first question, can you write an equation that will always be true for the side lengths of a right triangle?
- Can you think of a real-world scenario where you might need to use this relationship?

Possible misunderstandings
- Students might think that this relationship holds true for all triangles, not just right triangles. It is important to reinforce that the Pythagorean theorem only applies to right triangles.

Concrete-Representational-Abstract (CRA) approach
Targeted instructional strategies

Concrete: Begin by engaging students with physical manipulatives to explore the Pythagorean Theorem. Provide students with cutouts of right triangles and square tiles or blocks. Have them place the triangle on a flat surface and build squares off each side using the tiles. Encourage them to count the number of tiles in each square, observing that the largest square (on the hypotenuse) has an area equal to the sum of the areas of the other two squares. Use tools like geoboards with rubber bands to create right triangles and construct squares on each side.

Representational: Explain how the physical models they've created can be represented through drawings. Guide them to sketch the triangles and the squares they built, helping them transition from concrete materials to visual representations. Encourage them to label the lengths of the sides and calculate the area of each square by counting the squares on the grid. Use shading or different colors to highlight each square.

Abstract: Discuss how the areas they've calculated correspond to mathematical expressions. Introduce the symbolic form of the Pythagorean Theorem: $a^2 + b^2 = c^2$. Teach students to identify the legs (a and b) and the hypotenuse (c) in right triangles, even when they are oriented differently. Emphasize how the abstract equation represents the relationship they've seen with the manipulatives and drawings.

Create a graphic organizer for Pythagorean triples
Student with disabilities support

Have students create a Frayer model graphic organizer, like the one shown.

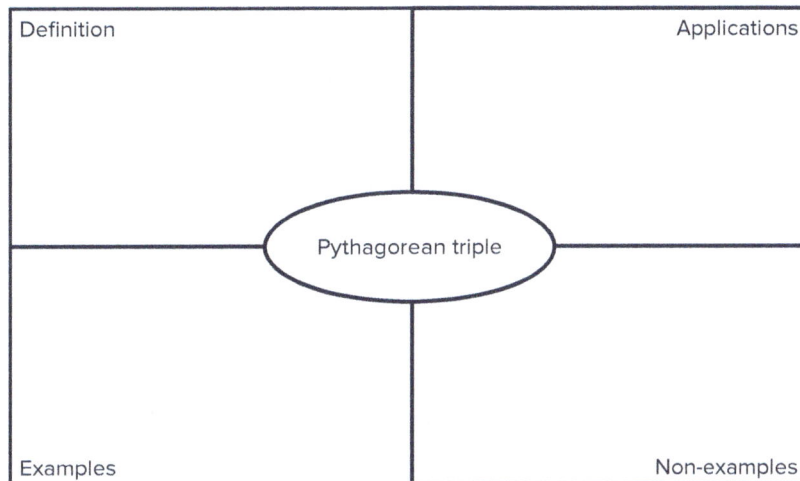

Definition	Applications

(center ellipse: Pythagorean triple)

Examples	Non-examples

In the top left box, have students define a Pythagorean triple. In the top right box, have students describe how Pythagorean triples are used in math. In the bottom left box, write a few examples of Pythagorean triples. In the bottom right box, have students create an example of side lengths that would not be a pythagorean triple and show why.

Avoid substituting incorrectly into the Pythagorean theorem
Address student misconceptions

Students may substitute incorrectly into the Pythagorean theorem. In particular, they may always assume that the given sides are a and b, and the missing side is c. Challenge this misconception by having students label the hypotenuse as c and the legs with a and b before substituting or by reinforcing that the term that is on its own is the hypotenuse. Students can check their answers by confirming that the side labeled as the hypotenuse is indeed the longest side.

Students will learn that the Pythagorean theorem can be used to solve for missing side lengths in a right triangle. They will also be introduced to a special subset of side lengths called Pythagorean triples.

Examples

Students: Page 338–339

Example 1

Determine if the triangle is a right triangle. If so, label the hypotenuse and legs.

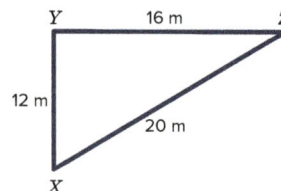

Create a strategy

Using the longest side as the hypotenuse, substitute the values into the Pythagorean Theorem to see if it results in a true statement.

Apply the idea

From the given triangle, we will use $a = 12$, $b = 16$, and $c = 20$.

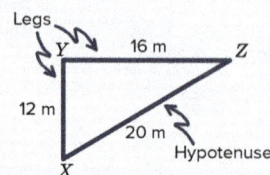

$c^2 = a^2 + b^2$	Use the Pythagorean theorem
$20^2 = 12^2 + 16^2$	Substitute the values
$400 = 144 + 256$	Evaluate the exponents
$400 = 400$	Evaluate the addition

Yes, this is a right triangle.

Purpose

Show students how to apply the Pythagorean Theorem to verify if a triangle is a right triangle.

Critique, correct, and clarify
English language learner support

use with Example 1

Provide students with a sample solution to the problem that contains intentional errors or misconceptions. For example, present a solution where the student incorrectly applies the Pythagorean Theorem, such as calculating $c^2 = a^2 - b^2$ instead of $c^2 = a^2 + b^2$. Ask students to work individually or in pairs to critique the solution by identifying and correcting the mistakes. Encourage them to clarify the correct method and explain their reasoning using precise mathematical vocabulary like "hypotenuse," "legs," and "right triangle."

This activity allows students to engage deeply with the content, fosters critical thinking, and supports the development of mathematical language as they articulate their critiques and corrections. By analyzing and improving upon the flawed solution, students strengthen their understanding of how to properly determine if a triangle is a right triangle using the Pythagorean Theorem.

Example 2

Find the length of the hypotenuse, c in this triangle.

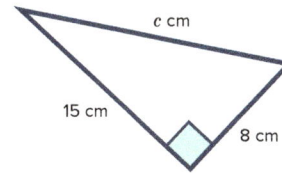

Create a strategy

Use the Pythagorean theorem: $c^2 = a^2 + b^2$.

Apply the idea

Based on the given diagram, we are given $a = 8$ and $b = 15$.

$$c^2 = a^2 + b^2 \quad \text{Use the Pythagorean theorem}$$
$$= 8^2 + 15^2 \quad \text{Substitute the values}$$
$$= 64 + 225 \quad \text{Evaluate the squares}$$
$$= 289 \quad \text{Evaluate}$$
$$\sqrt{c^2} = \sqrt{289} \quad \text{Apply the square root to both sides}$$
$$c = 17 \text{ cm} \quad \text{Evaluate the square root}$$

Reflect and check

Because all the lengths for the sides of this triangle are whole numbers and they satisfy Pythagorean theorem, the numbers (8, 15, 17) form a Pythagorean triple.

Purpose

Students will demonstrate that they can apply the Pythagorean theorem to solve for missing side lengths in a right triangle.

Expected mistakes

Students may forget to find the square root of the sum of a^2 and b^2 as the final step. Point out that the side length is c, but they have found c^2.

Reflecting with students

This problem can be extended for advanced students, or any students who are ready, by introducing students to Pythagorean triples. Show students that because the side lengths are whole numbers that satisfy the Pythagorean theorem, they are considered a Pythagorean triple.

Students may assume that the side lengths for a right triangle will always be Pythagorean triples. Show students exmaples of right triangles with at least one side length that is not a whole number.

Example 3

Calculate the value of a in the triangle shown.

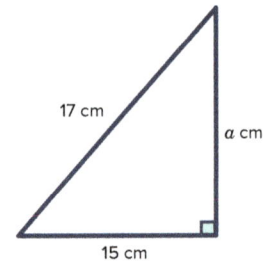

17 cm
a cm
15 cm

Create a strategy

Use the Pythagorean theorem: $c^2 = a^2 + b^2$

Apply the idea

We are given $b = 15$ and $c = 17$.

$$c^2 = a^2 + b^2 \quad \text{Use the Pythagorean theorem}$$
$$17^2 = a^2 + 15^2 \quad \text{Substitute the values}$$
$$17^2 - 15^2 = a^2 \quad \text{Subtract } 15^2 \text{ from both sides of the equation}$$
$$289 - 225 = a^2 \quad \text{Evaluate the squares}$$
$$64 = a^2 \quad \text{Evaluate the subtraction}$$
$$\sqrt{64} = \sqrt{a^2} \quad \text{Apply the square root to both sides}$$
$$8 = a \quad \text{Evaluate the square root}$$

Purpose

Students will demonstrate that they can apply the Pythagorean theorem to solve for missing side lengths in a right triangle. In this example, they will focus on finding a shorter side.

Expected mistakes

Students may substitute the values into the formula incorrectly. For example, they may write $17^2 + 15^2 = c^2$. Write the Pythagorean theorem, labeling "a" and "b" as the shorter sides and "c" as the hypotenuse. Remind students to notate which side is which before substituting the values into the equation.

Reflecting with students

Encourage students to check their answer by substituting all side lengths into the Pythagorean theorem and seeing if it produces a true statement.

Example 4

A triangle has side lengths of 6, 7, and 11. Do these three side lengths make a Pythagorean triple?

Create a strategy

Using the longest side as the hypotenuse and substitute the values into the Pythagorean Theorem to see if we get a true statement.

Apply the idea

We will use $a = 6$, $b = 7$, and $c = 11$.

$$c^2 = a^2 + b^2 \quad \text{Use the Pythagorean theorem}$$
$$11^2 = 6^2 + 7^2 \quad \text{Substitute the values}$$
$$121 = 36 + 49 \quad \text{Evaluate the exponents}$$
$$121 \neq 85 \quad \text{Evaluate the addition}$$

No, this is not a right triangle so the numbers cannot be a Pythagorean triple.

Purpose

Show students how to verify if a set of numbers forms a Pythagorean triple using the Pythagorean Theorem.

🎓 **Write code to check Pythagorean triples**
Targeted instructional strategies

use with Example 4

As an extension, have students write a simple program that takes three numbers as input and then outputs whether the set is a Pythagorean triple or not. A visual interface like Scratch is nice for beginner coders and for more advanced coders different languages could be used. For example, this Python code could be compiled in Google Colab.

```
1   def is_pythagorean_triple(a, b, c):
2     # Sort the numbers to ensure c is the largest (hypotenuse)
3     a, b, c = sorted([a, b, c])
4     # Check if the Pythagorean theorem holds
5     return a**2 + b**2 == c**2
6   print("Enter three numbers to check if they form a Pythagorean
      triple:")
7   a = int(input("Enter the first number: "))
8   b = int(input("Enter the second number: "))
9   c = int(input("Enter the third number: "))
10  if is_pythagorean_triple(a, b, c):
11    print(f"{a}, {b}, and {c} form a Pythagorean triple.")
12  else:
13    print(f"{a}, {b}, and {c} do not form a Pythagorean triple.")
```

Encourage students to clearly comment each code block as shown. An alternative activity could be to remove the comments from the above code and ask students to describe how the program works.

Example 5

Luke knows the two largest numbers in a Pythagorean Triple, which are 37 and 35. What number, a, does Luke need to complete the triple?

Create a strategy

Use the Pythagorean theorem: $c^2 = a^2 + b^2$.

Apply the idea

We are given $b = 35$ and $c = 37$.

$$c^2 = a^2 + b^2 \qquad \text{Use the Pythagorean theorem}$$
$$37^2 = a^2 + 35^2 \qquad \text{Substitute the values}$$
$$37^2 - 35^2 = a^2 \qquad \text{Subtract } 35^2 \text{ from both sides of the equation}$$
$$1369 - 1225 = a^2 \qquad \text{Evaluate the squares}$$
$$144 = a^2 \qquad \text{Evaluate the subtraction}$$
$$\sqrt{144} = a^2 \qquad \text{Apply the square root to both sides}$$
$$12 = a \qquad \text{Evaluate the square root}$$

So, the number that will complete the triple is 12.

Purpose

Show students how to use the Pythagorean theorem to find the missing number in a Pythagorean Triple.

Example 6

A sports association wants to redesign the trophy they award to the player of the season. The front view of one particular design is shown in the diagram:

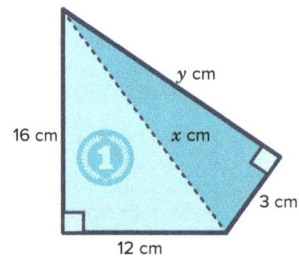

a   Find the value of x.

Create a strategy

We can use the Pythagorean Theorem: $a^2 + b^2 = c^2$.

Apply the idea

We can apply Pythagorean theorem to the bottom triangle that has side lengths of 16 cm and 12 cm, since the only unknown side is the hypotenuse.

$$12^2 + 16^2 = x^2 \qquad \text{Substitute } a = 12, b = 16, \text{ and } c = x$$
$$144 + 256 = x^2 \qquad \text{Evaluate the squares}$$
$$400 = x^2 \qquad \text{Evaluate the addition}$$
$$\sqrt{400} = x \qquad \text{Take the square root of both sides}$$
$$20 \text{ cm} = x \qquad \text{Evaluate the square root}$$

Purpose

To ensure students can apply the Pythagorean theorem in context.

b Find the value of y, rounded to two decimal places.

Create a strategy	Apply the idea	
We can use the Pythagorean Theorem: $a^2 + b^2 = c^2$.	$y^2 + 3^2 = 20^2$	Substitute $a = y$, $b = 3$, and $c = 20$
	$y^2 + 9 = 400$	Evaluate the squares
	$y^2 + 9 - 9 = 400 - 9$	Subtract 9 from both sides
	$y^2 = 391$	Evaluate the subtraction
	$y = \sqrt{391}$	Take the square root of both sides
	$y = 19.77$ cm	Evaluate and round to two decimal places

Purpose

To ensure students can apply the Pythagorean theorem in context.

Drawing clear diagrams
Targeted instructional strategies   use with Example 6

Encourage students to draw and label the right triangles separately from the overall design. Have them clearly mark all known side lengths and indicate the right angles. This helps students visualize the triangles in isolation, making it easier to identify which sides correspond to a, b, and c in the Pythagorean theorem.

Encourage them to use color coding to denote the parts that are shared between the diagrams.

Idea summary

The Pythagorean theorem states that in a right triangle the square of the **hypotenuse** (the side opposite the right angle) is equal to the sum of the squares of the other two sides:

$$a^2 + b^2 = c^2$$

c is the length of the hypotenuse

a is one of the shorter side lengths

b is the other shorter side length

A **Pythagorean triple** is an ordered triple (a, b, c) of three positive integers that represent the side lengths of a right triangle.

Practice

Students: Page 342–347

What do you remember?

1 Suppose c represents the length of the hypotenuse of a right triangle, and a, b are the two legs. Write the equation that represents the Pythagorean theorem.

2 Use the Pythagorean theorem to determine whether this is a right triangle.

    a Let a and b represent the two shorter side lengths. First, find the value of $a^2 + b^2$.

    b Let c represent the length of the longest side. Find the value of c^2.

    c Is the triangle a right triangle?

3 Consider the right triangle with sides (8, 15, 17).

    a What is the length of the hypotenuse?

    b What lengths are the two sides that are next to the right angle?

    c Do these side lengths represent Pythagorean triple?

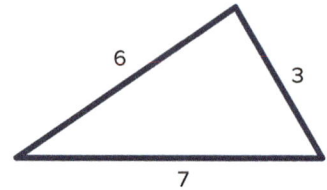

4 Identify the hypotenuse of each triangle:

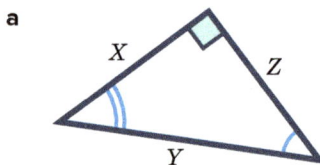

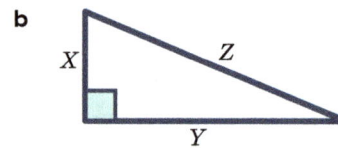

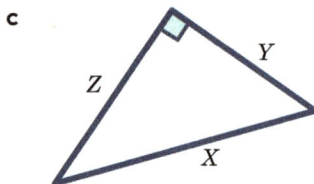

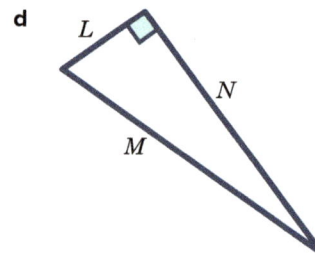

    a

    b

    c

    d

5 Write the equation for each triangle that represents the Pythagorean theorem.

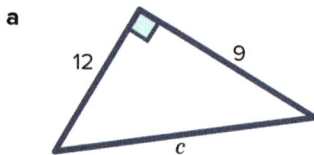

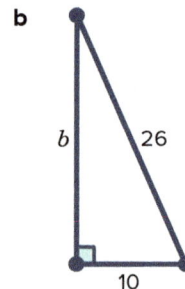

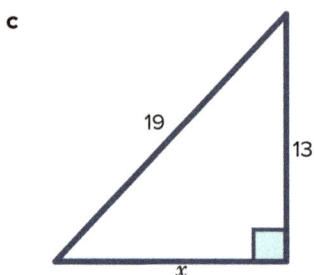

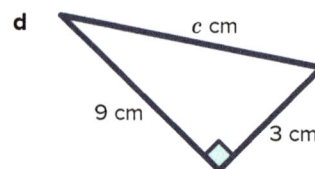

    a

    b

    c

    d

Let's practice

6 Consider the following figure:

    a Find the area of square with the side length of the shortest side.

    b Find the area of the square with the length of the next shortest side.

    c How do the areas of the two squares relate to the square with the side length of the longest side?

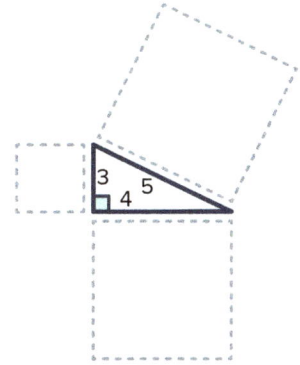

7 Calculate the value of c for the following triangles:

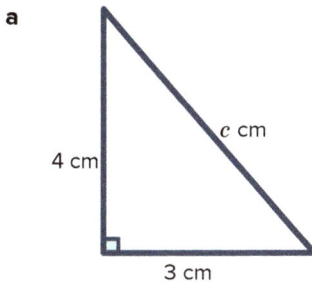

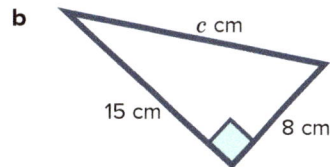

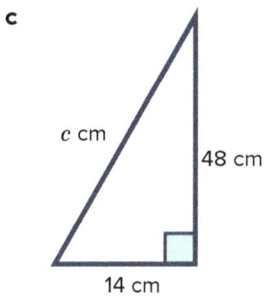

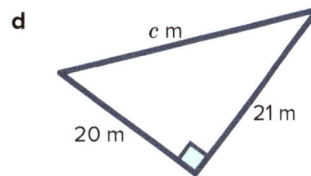

    a 4 cm, 3 cm, c cm

    b 15 cm, 8 cm, c cm

    c 48 cm, 14 cm, c cm

    d 20 m, 21 m, c m

8 Calculate the value of the variable for the following triangles:

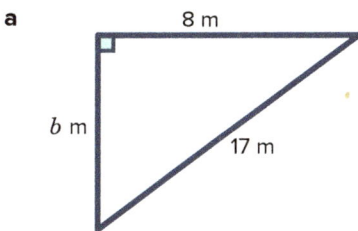

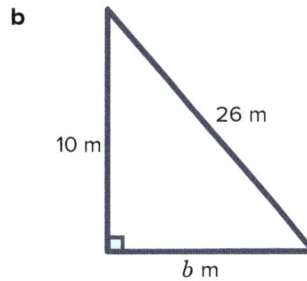

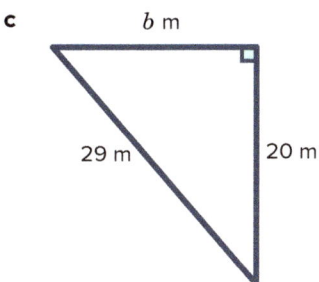

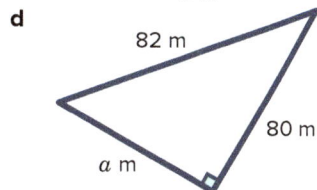

    a 8 m, 17 m, b m

    b 10 m, 26 m, b m

    c 29 m, 20 m, b m

    d 82 m, 80 m, a m

9 Calculate the value of the missing side:

a

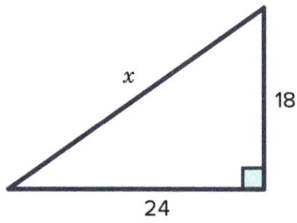

b

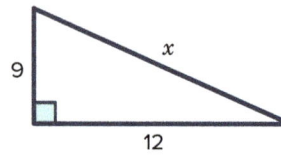

c

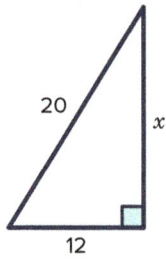

d

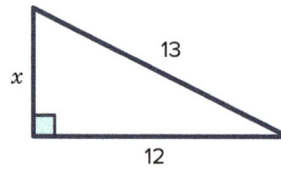

10 Can the following side lengths form a right triangle?

    a $(12, 5, 13)$ **b** $(8, 14, 18)$ **c** $(9, 16, 25)$ **d** $(60, 80, 100)$

    e $(9, 12, 15)$ **f** $\left(\dfrac{1}{3}, \dfrac{1}{4}, \dfrac{1}{5}\right)$ **g** $\left(5, \sqrt{12}, \sqrt{37}\right)$ **h** $\left(2, \sqrt{2}, \sqrt{2}\right)$

11 Use the converse of the Pythagorean theorem to determine whether the triangles are right triangles:

a

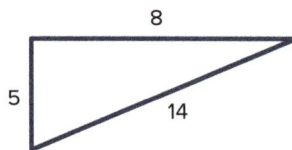

b

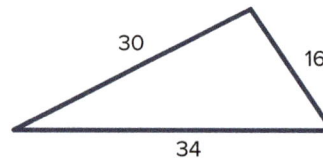

c

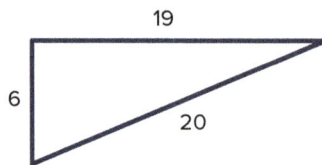

d

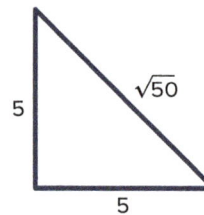

12 Which set of side lengths can represent the measures of the sides of a right triangle?

    A $(8, 10, 14)$ **B** $(3, 6, 8)$ **C** $(14, 22, 36)$ **D** $(12, 35, 37)$

13 Select the measures that could be the three side lengths of a right triangle.

| 7 in | 15 in | 24 in | 25 in |

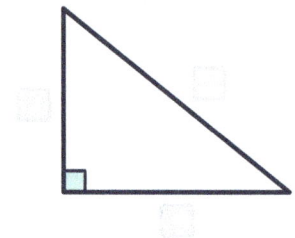

14 Three towns Farmville, Richmond and Petersburg are positioned as shown in the diagram:

Which two towns are furthest apart, assuming a direct route is taken? Explain your answer.

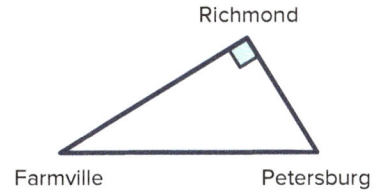

Richmond

Farmville   Petersburg

15 Consider the given triangle:

   a Find the length of the hypotenuse.

   b Do the sides of the triangle represent a Pythagorean triple?

   c Find the perimeter of the triangle.

8 m

15 m

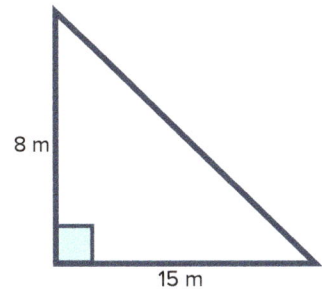

16 A 150 cm ladder is leaned against a wall. The base of the ladder is 80 cm from the wall.

   a Describe a way to find the height on the wall that the top of the ladder reaches without directly measuring the distance.

   b Find the height on the wall that the ladder reaches.

h

150 cm

80 cm

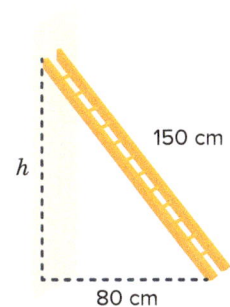

17 Javier is standing on a water tower. He is looking at a point on the ground, P. Javier is 15 meters from point P, and the distance from the base of the water tower to point P is 9 meters. What is Javier's height, h, from the ground?

   A 3.9 m   **B** 9.0 m

   C 12.0 m   **D** 15.0 m

h

15 m

9 m P

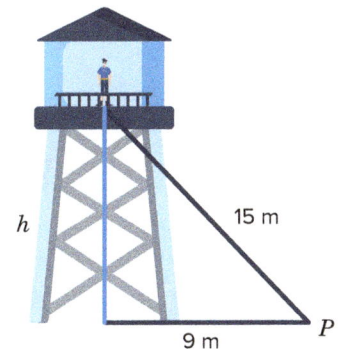

18 A group of engineering students have made a triangle out of some wooden strips. The side lengths are 20, 48, and 52 m.

   a Is the triangle they made a right triangle?

   b How many meters of wooden strips did they use to make the triangle?

   c Determine whether right triangles can be created using these lengths of wooden strips:

      i (30, 40, 50)   **ii** (26, 41, 53)   **iii** (14, 48, 50)   **iv** (24, 45, 51)

19 Consider the given triangle:

Find the following, rounding your answer to two decimal places:

   a   The value of x.

   b   The value of y.

   c   The length of the base of the triangle.

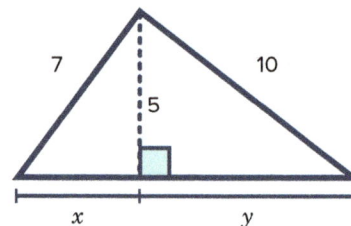

20 Given the given diagram, find the following, correct to two decimal places:

   a   The value of x.   **b**   The value of y.

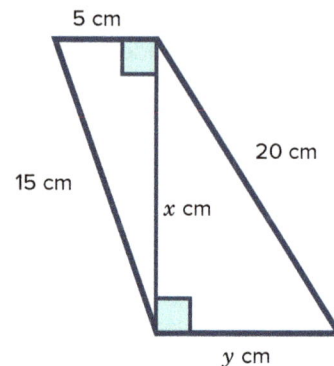

21 Consider the following trapezoid:

   a   Find the value of a.

   b   Find the value of b.

   c   Find x, correct to two decimal places.

   d   Find the perimeter of the trapezoid, correct to two decimal places.

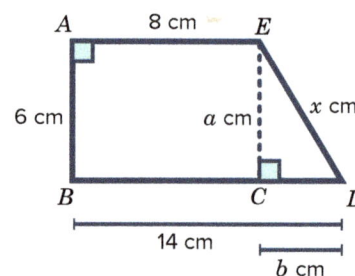

Let's extend our thinking

22 Find the value of k in the given figure. Round your answer to two decimal places.

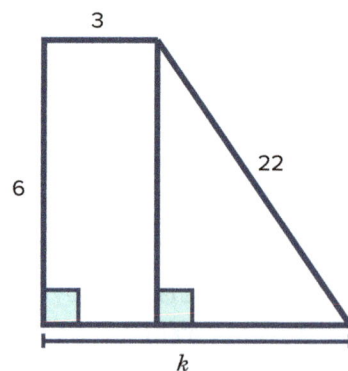

23 For each of the following figures, find the length of the unknown side x, correct to two decimal places:

a

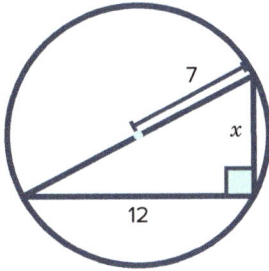

b

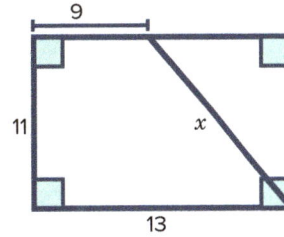

c

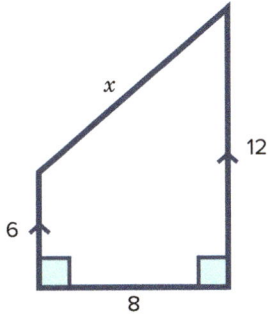

d

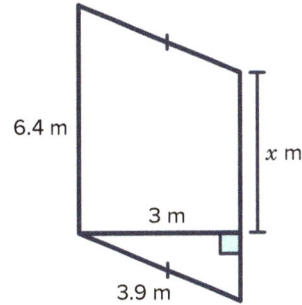

24 If $x, y,$ and z are three sides of a right triangle, will side lengths of $6x$, $6y$ and $6z$ make a right triangle as well? Explain your reasoning.

25 Consider the following triangle:
    a Show that the triangle is a right triangle.
    b State the size of the two acute angles.
    c Show that any triangle whose side lengths are k, k and $k\sqrt{2}$ are the sides of a right triangle.

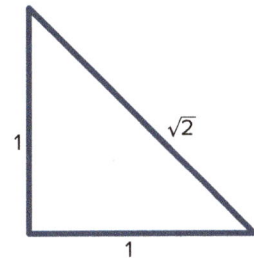

26 $VUTR$ is a rhombus with perimeter 112 in. The length of diagonal $\overline{RU}$ is 46 in.
    a Find the length of $\overline{VR}$.
    b Find the length of $\overline{RW}$.
    c Find the length of $\overline{VW}$, correct to two decimal places.
    d Now, find the length of the other diagonal $\overline{VT}$. Round your answer to two decimal places.

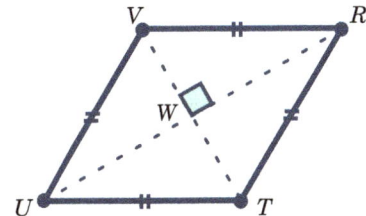

Answers

6.02 Pythagorean theorem

What do you remember?

1 $c^2 = a^2 + b^2$

2 a 45   **b** 49

   c No, it is not a right triangle since $45 \neq 49$.

3 a 17   **b** 8 and 15   **c** Yes

4 a Y **b** Z **c** X **d** M

5 a $c^2 = 12^2 + 9^2$ **b** $26^2 = b^2 + 10^2$

   c $13^2 + x^2 = 19^2$ **d** $3^2 + 9^2 = c^2$

Let's practice

6 a $A = 9$ **b** $A = 16$

   c The sum of the area of the smaller squares equals the area of the square with the longest side.

7 a $c = 5$ **b** $c = 17$ **c** $c = 50$ **d** $c = 29$

8 a $b = 15$ **b** $b = 24$ **c** $b = 21$ **d** $a = 18$

9 a $x = 30$ **b** $x = 15$ **c** $x = 16$ **d** $x = 5$

10 a Yes   **b** No   **c** No   **d** Yes

   e Yes   **f** No   **g** Yes   **h** Yes

11 a No   **b** Yes   **c** No   **d** Yes

12 D

13

14 a Farmville and Petersburg. The longest side of a right triangle is hypotenuse, which is the side that is opposite the right angle. The distance between Farmville and Petersburg is opposite the right angle, so these towns are the furthest apart.

15 a 17 m   **b** Yes   **c** 40 m

16 a The Pythagorean theorem can be used to find the height on the wall the top of the ladder reaches. This theorem applies because the ladder, the wall, and the ground form a right triangle, where:

- The ladder is the hypotenuse
- The distance from the base of the ladder to the wall is one leg
- The height on the wall that the ladder reaches is the other leg

   b $h = 126.89$ cm

17 C

18 a Yes   **b** 120 m

   c i Yes   **ii** No   **iii** Yes   **iv** Yes

19 a $x = 4.90$ **b** $y = 8.66$ **c** 13.56

20 a $x = 14.14$ **b** $y = 14.14$

21 a $a = 6$ **b** $b = 6$ **c** $x = 8.49$ **d** 36.49 cm

Let's extend our thinking

22 $k = 24.17$

23 a $x = 7.21$ **b** $x = 11.70$ **c** $x = 10.00$ **d** $x = 3.9$

24 Yes. Suppose z is the hypotenuse and x and y are the two shorter legs. By the Pythagorean theorem, $z^2 = x^2 + y^2$. We know that $6z$ is larger than $6x$ and $6y$ since we said z is hte hypotenuse. Observe that

$$(6x)^2 + (6y)^2 = 36x^2 + 36y^2 = 36(x^2 + y^2) = 36z^2 = (6z)^2$$

Thus, side lengths of $6x$, $6y$ and $6z$ make a right triangle.

25 a Using the inverse of Pythagorean Theorem for verification, the sum of the squares of the two legs should be equal to the square of the hypotenuse.

$$1^2 + 1^2 = \left(\sqrt{2}\right)^2 \Leftrightarrow 2 = 2$$

   b 45°

   c $k^2 + k^2 = \left(k\sqrt{2}\right)^2 \Leftrightarrow 2k^2 = 2k^2$

26 a 28 in   **b** 23 in

   c $x = 15.97$ in   **d** 31.94 in

6.03 Applications of the Pythagorean theorem

Subtopic overview

Lesson narrative

In this lesson, students will apply the Pythagorean Theorem to solve real-world problems. The lesson begins with a review of the theorem: $(a^2 + b^2 = c^2)$, where (a) and (b) are the legs of a right triangle, and (c) is the hypotenuse. Students will learn to identify right triangles in various contextual situations such as architecture, construction, sailing, and more. Through examples, students will practice solving for unknown side lengths using the theorem. For instance, they will calculate the direct distance a sailor travels when moving at an angle, determine the length of a roof in construction, and find the radius and diameter of a cone's base. Additional examples include calculating the diagonal of a handheld device's screen and the height a ladder reaches against a wall. By the end of the lesson, students should be proficient in recognizing right triangles in practical scenarios and using the Pythagorean Theorem to find missing lengths accurately.

Learning objectives

Students: Page 348

🎓 **After this lesson, you will be able to...**
  • solve mathematical and real-world problems involving right triangles using the Pythagorean theorem and its converse.

Key vocabulary

• Pythagorean theorem

Essential understanding

The sides of any right triangle share the same relationship defined by the Pythagorean theorem.

Standards

This subtopic addresses the following Virginia 2023 Mathematics Standards of Learning standards.

Mathematical process goals

MPG1 — Mathematical Problem Solving

Teachers can foster mathematical problem solving by having students apply the Pythagorean Theorem to solve real-world problems. This could involve calculating distances, determining the height of a building or finding the length of a ladder. Students can be encouraged to use a variety of problem solving strategies and to reflect on the effectiveness of their chosen strategy.

MPG4 — Mathematical Connections

Teachers can establish mathematical connections by linking the Pythagorean Theorem to prior knowledge of angle relationships and triangle properties. Teachers can also show how the theorem is connected to real-world situations such as engineering, architecture, surveying, and navigation. Furthermore, the relationship between the theorem and different careers can be highlighted to show its relevance and applicability.

MPG5 — Mathematical Representations

Teachers can foster understanding of mathematical representations by demonstrating how the Pythagorean Theorem can be visualized through geometric representations and symbolically through the formula $(a^2 + b^2 = c^2)$. Teachers can also encourage students to use different representations to solve problems and to understand how these different representations connect to the same mathematical idea.

Content standards

8.MG.4 — The student will apply the Pythagorean Theorem to solve problems involving right triangles, including those in context.

8.MG.4e — Apply the Pythagorean Theorem, and its converse, to solve problems involving right triangles in context.

Prior connections

5.MG.2 — The student will use multiple representations to solve problems, including those in context, involving perimeter, area, and volume.

Future connections

G.TR.3 — The student will, given information in the form of a figure or statement, prove and justify two triangles are similar using direct and indirect proofs, and solve problems, including those in context, involving measured attributes of similar triangles.

G.TR.4 — The student will model and solve problems, including those in context, involving trigonometry in right triangles and applications of the Pythagorean Theorem.

Lesson Preparation

Suggested review

Depending on your students' level of prior knowledge, consider revisiting the following lessons:

Grade 8 — 6.02 Pythagorean theorem

Tools

You may find these tools helpful:
- Scientific calculator
- Highlighters

Student lesson & teacher guide

Applications of the Pythagorean theorem

Students are shown the Pythagorean theorem and are given a list of steps to follow to solve real-life application problems using the formula. They are shown visuals using these steps with diagrams and explanations of solutions.

> 🎓 **Steps for applying the Pythagorean theorem to real-world problems**
> **Targeted instructional strategies**

When solving application problems with the Pythagorean theorem, problems may or may not include a diagram as a visual for determining which values are legs or the hypotenuse. As an example, give students the following problem:

Kylie is flying a kite with a 31 foot long string and is standing 10 feet away from her friend, Kingston. If Kingston is standing directly beneath the kite, find k, the height of the kite in feet rounded to the nearest hundredth.

Since the problem did not have a visual diagram with the values, show students a right triangle with blanks for the needed values, and show students how to use key words in the problem to fill in the blanks, such as "feet away," "beneath," and "find the height."

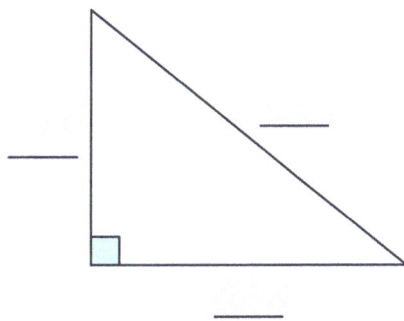

| Before values are filled in | Right triangle with values filled in |

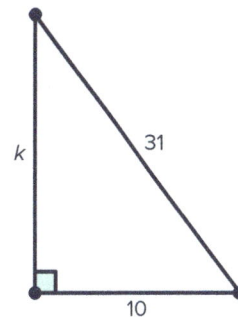

Tell students that while the triangle can face in different orientations, the values of a and b will be the lengths sides connected to the right angle, and c will be the length of the side opposite the right angle.

$$a^2 + b^2 = c^2 \qquad \text{Pythagorean theorem}$$
$$(10)^2 + (k)^2 = (31)^2 \qquad \text{Substitute values}$$
$$100 + k = 961 \qquad \text{Evaluate squares}$$
$$k^2 = \sqrt{861} \qquad \text{Subtract 100 from each side}$$
$$\sqrt{k} = \sqrt{861} \qquad \text{Take the square root of each side}$$
$$k^2 \approx 29.3428 \ldots \qquad \text{Evaluate}$$

The kite is approximately 29.34 feet above Kingston.

Providing this template method for filling in values will help students connect real-world context problems to the Pythagorean theorem.

Three reads
English language learner support

Choose real-life application problems for Pythagorean theorem, both with diagrams and without. For each problem:

1. *First Read*: Understanding the Context

   Begin by reading the problem aloud to the class, focusing on understanding the real-world scenario without considering the numbers or calculations. For instance, if the problem involves finding the height of a ladder against a wall, guide students to visualize the situation and describe it in their own words.

2. *Second Read*: Identifying Key Information

   Read the problem again, this time asking students to identify the important mathematical information. Have them underline or highlight the numbers, labels, and any given measurements. For example, they should recognize which side of the triangle represents the height, base, and hypotenuse. Discuss how these relate to the right triangle that will be used in the Pythagorean theorem.

3. *Third Read*: Formulating the Mathematical Approach

   In the third reading, guide students to apply the steps needed to solve the problem. Encourage them to articulate how they will apply the Pythagorean theorem to the identified sides of the triangle. Ask guiding questions like, "What do we need to find?" and "Which formula will help us solve this?" Finally, have students write out the equation they will use and predict the next steps in solving for the unknown side.

Color-code diagrams and values in the Pythagorean theorem
Student with disabilities support

As students work through problems with the Pythagorean theorem, provide different colors of highlighters or colored pencils to show which sides of the right triangle correspond to a, b, and c. This color-coding system should correspond to both any visual diagrams and the formula itself.

An example of colors to use is shown.

Substituting values for the incorrect variables in the Pythagorean theorem
Address student misconceptions

When solving a problem using the Pythagorean theorem, students may substitute the incorrect values of the variables. An example of incorrectly substituting values is shown:

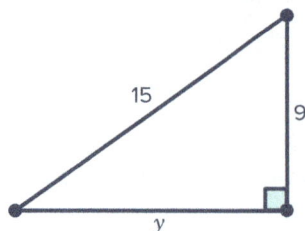

$$a^2 + b^2 = c^2$$
$$(9)^2 + (15)^2 = y^2$$
$$81 + 225 = y^2$$
$$\sqrt{306} = y^2$$
$$17.4939... \approx y$$

Students should label their triangles with the variables a, b, and c, or write out what each variable equals before substituting.

Examples

Example 1

Consider a cone with slant height 13 m and perpendicular height 5 m.

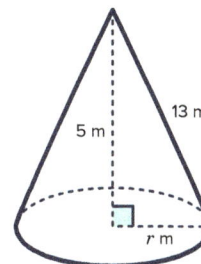

a Find the length of the radius, r, of the base of this cone.

Create a strategy

Use the Pythagorean theorem: $a^2 + b^2 = c^2$.

Apply the idea

We let $b = r$. Based on the given diagram, we are given $a = 5$ and $c = 13$.

$a^2 + b^2 = c^2$	Use the Pythagorean theorem
$5^2 + r^2 = 13^2$	Substitute $a = 5$, $b = r$, and $c = 13$
$25 + r^2 = 169$	Evaluate the exponents
$r^2 = 169 - 25$	Subtract 25 from both sides
$= 144$	Evaluate the subtraction
$\sqrt{r^2} = \sqrt{144}$	Take the square root of both sides
$r = 12$	Evaluate the square root

The radius is 12 m.

Reflect and check

Since this is a right triangle where all three sides are positive integers, this represents a Pythagorean triple.

Purpose

Check that students can use the Pythagorean theorem to solve for a missing dimension of a 3-D shape. In this example, they need to solve for the radius of a cone.

Expected mistakes

Students may substitute the values into the Pythagorean theorem equation incorrectly since the word "height" describes both given values. They may write $5^2 + 13^2 = c^2$ and solve for the hypotenuse, not realizing that they should be solving for a leg of the right triangle. Students should draw a picture of the triangle and label the sides with the appropriate values and their corresponding variable.

Reflecting with students

In the previous lesson, students were introduced to the concept of Pythagorean triples. After solving and ordering the sides from least to greatest, we have the Pythagorean triple (5, 12, 13). Show students other examples Pythagorean triples and how they could be used to help solve application problems of right triangles.

Encourage advanced learners to explore ways to generate infinitely many Pythagorean triples. They should discover that multiplying any Pythagorean triple by a whole number scale factor creates a new Pythagorean triple.

b Find the length of the diameter of the cone's base.

Create a strategy

Use the fact that the diameter is double the radius.

Apply the idea

diameter = 2 · 12   Multiply the radius by 2

= 24 m   Evaluate the multiplication

Purpose

Students will apply what they've learned about the Pythagorean theorem to solving problems in two and three dimensions. In this example, students will use the radius value they calculated from example a to find the diameter of the base.

Students: Page 349–350

Example 2

The screen on a handheld device has dimensions 8 cm by 4 cm, and a diagonal of length x cm.

Find the value of x, correct to two decimal places.

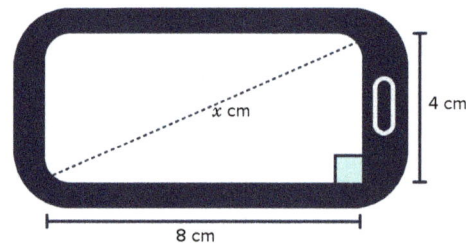

Create a strategy

Use the Pythagorean theorem: $a^2 + b^2 = c^2$.

Apply the idea

We let $c = x$. Based on the given diagram, we are given $a = 8$ and $b = 4$.

$a^2 + b^2 = c^2$ Use the Pythagorean theorem

$8^2 + 4^2 = x^2$ Substitute $a = 8$, $b = 4$, and $c = x$

$64 + 16 = x^2$ Evaluate the exponents

$x^2 = 80$ Reverse and evaluate the addition

$\sqrt{x^2} = \sqrt{80}$ Take the square root of both sides

$x = 8.94$ cm   Evaluate the square root and round

Purpose

Students will apply what they've learned about the Pythagorean theorem to solving problems in two and three dimensions. In this example, students will find the length of a diagonal within a rectangle.

Reflecting with students

Bring students' attention to precision while working with decimals. When checking to see if the values in the triangle represent a right triangle, a rounded decimal answer may not give a true statement. In the problem above, when we substitute the side lengths of the triangle into the Pythagorean theorem, we see

$$a^2 + b^2 = c^2$$
$$(8)^2 + (4)^2 = (8.94)^2$$
$$64 + 16 = 79.9236$$
$$80 \neq 79.9236$$

When students check their work with rounded decimal answers, they should notice that the two sides of the equation will be close in value, but may not be exactly equivalent.

Students: Page 350

Example 3

A ladder is leaning against a wall. The base of the ladder is 6 feet away from the wall, and the ladder is 7 feet long. How far up the wall does the ladder reach? Round your answer to two decimal places.

Create a strategy

Sketch a diagram to help you identify the hypotenuse and the legs. Use the Pythagorean Theorem to find missing side.

Apply the idea

We have that $a = 5$, $c = 7$, and we need to find b.

$a^2 + b^2 = c^2$	Use the Pythagorean theorem
$5^2 + b^2 = 7^2$	Substitute $a = 5$ and $c = 7$
$25 + b^2 = 49$	Evaluate the exponents
$b^2 = 49 - 25$	Subtract 25 from both sides
$= 24$	Evaluate the subtraction
$\sqrt{b^2} = \sqrt{24}$	Take the square root of both sides
$b \approx 4.9$	Evaluate the square root and round

The ladder will reach about 4.9 feet up on the wall.

Purpose

Show students how to apply the Pythagorean theorem to solve real world problems.

Students: Page 350

> 💡 **Idea summary**
>
> To apply the Pythagorean theorem to real-life situations:
>
> 1. Look for right triangles
> 2. Choose which side, hypotenuse or a shorter side, you are trying to find
> 3. Substitute the known side lengths in to the Pythagorean theorem
> 4. Solve for the unknown length

Practice

Students: Page 351–354

What do you remember?

1   Consider the following triangle:

   a   Name the hypotenuse.

   b   Write an equation for this triangle using the Pythagorean theorem.

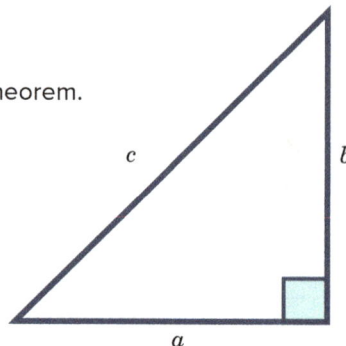

   c   Is $a + b > c$?

2   Find the value of the variable in the following triangles. Round your answer to two decimal places.

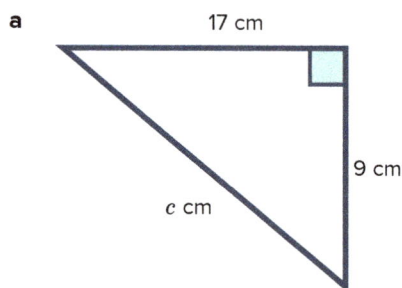

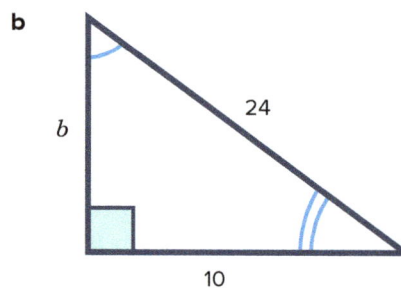

   a   17 cm, 9 cm, c cm

   b   24, b, 10

3   Are these statements true or false?

   a   The Pythagorean theorem can only be applied to a right triangle.

   b   If the side lengths of a triangle have decimals, the triangle cannot be a right triangle.

   c   To find the unknown length in a right triangle, we must know the length of the other two sides.

   d   If $\triangle ABC$ is a right triangle, where $\angle B$ is the right angle, the shortest distance between A and C can be represented by the hypotenuse.

4   For each situation:

   i   Can the problem be solved using the Pythagorean theorem?

   ii   If the Pythagorean theorem is applicable, what are the values of a, b, and c?

   a   Finding the diagonal length (x cm) of a television with dimensions of 55.9 cm and 48.2 cm.

   b   Finding the distance between a building and a fire hydrant (x m) if the building height is 72 m and the distance from the hydrant to the building is 26 m.

   c   Finding the length of the third side of a triangular fence (x m) given the lengths of opposite sides that are equal (36 m).

   d   Finding the height of the wall reached by an leaning ladder (x m) given the ladder length (3 m) and the angle between the ground and the ladder (75°).

Let's practice

5 Calculate the length of the diagonal of the handheld device:

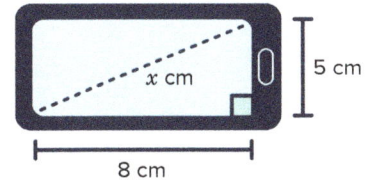

6 Iain's car has run out of gas. He walks 12 miles west and then 9 miles south looking for a gas station as shown in the diagram:

If he is now h mile directly from his starting point, find the value of h.

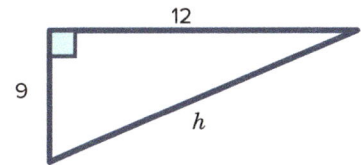

7 The city of Norfolk is 6 miles directly north of Chesapeake and 18 miles directly west of Virginia Beach. Find the shortest distance from Chesapeake to Virginia Beach.

8 The distance from the top of a flag pole to the end of its shadow is 9.85 m. If the shadow cast by the flag pole is 9 m long, find the height of the flag pole.

9 A ladder of height h cm is placed against a vertical wall. The bottom of the ladder is 80 cm from the base of the wall and the top of the ladder touches the wall at a height of 150 cm.

   a Dana draws the diagram shown. Explain Dana's mistake.

   b Find h.

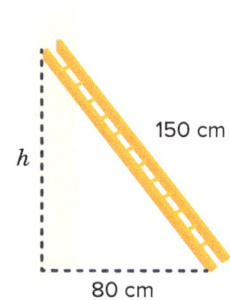

10 A landscaper wants to verify that the corner of a garden bed is a right angle. If they measure the sides of the triangular space and find them to be 5 feet, 12 feet, and 13 feet, can they conclude that one corner is a right angle?

11 Use the distances on the map to tell whether the triangle formed by the three cities is a right triangle.

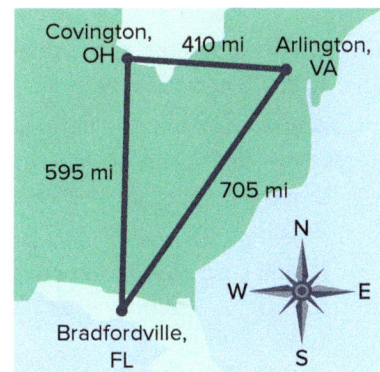

12 Consider the cone with slant height of 13 m and perpendicular height of 12 m:

    a Find the length of the radius, r.

    b Now, find the length of the diameter of the cone's base.

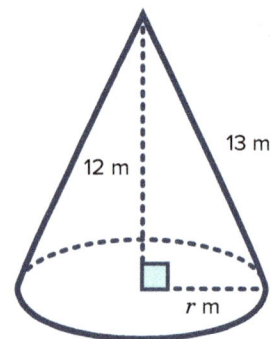

13 A farmer has a square field of side length 12 yards. He wants to build a fence along the diagonal of the field. How long will the fence be, to the nearest yard?

14 Archeologists have uncovered an ancient pillar which, after extensive digging, remains embedded in the ground. The lead researcher wants to record all of the dimensions of the pillar, including its height above the ground.

However, the team can only take certain measurements accurately without risking damage to the artifact. These measurements are shown in the diagram.

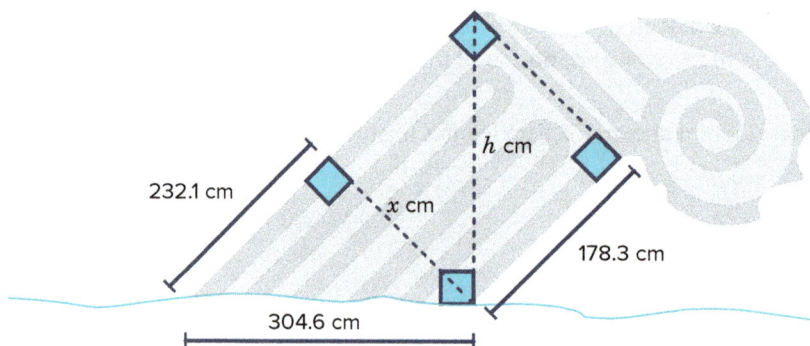

    a Find the value of x, correct to two decimal places.

    b Now, find h, the height of the pillar, correct to two decimal places.

15 On one particular night, the light from a lighthouse can be seen up to 19.3 kilometers away.

A boat traveling due north sails past the lighthouse. The boat is closest to the lighthouse when it is due east of it, as shown in the diagram. At that point, the boat is 16.1 kilometers from the lighthouse.

Find the distance in which the light will be visible from the boat, d. Round your answer to one decimal place.

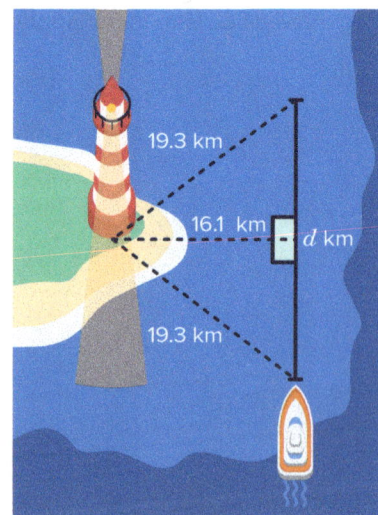

16 A pole is being held upright by two cables attached to the top of the pole and to the ground either side by pegs, as shown in the diagram. The length of the cables are 10 feet and 17 feet. The height of the pole perpendicular to the ground is 8 feet.

   a Skye notices that she can divide the diagram into two smaller right triangles, each with a height of 8. What are the base lengths of these right triangles?

   b How far apart (measured along the ground) are the pegs at the ends of the cables?

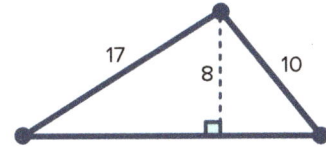

Let's extend our thinking

17 Fiona had asked her friend Doris if she can construct picture frame in the shape of a right triangle with side lengths of 16 ft, 10 ft, and 19 ft. Doris said it is possible. Describe Doris' error and correct her mistake.

1 $a^2 + b^2 = c^2$ Pythagorean theorem

2 $16^2 + 10^2 = 256 + 100$

3 $ = 356$

4 $c = \sqrt{356}$

5 $ \approx 19$

18 Two buildings, one of height 300 m and the other of height 450 m, stand a distance apart on level ground. If a cable joining the tops of the buildings is 200 m long, find to the nearest meter the distance between the buildings.

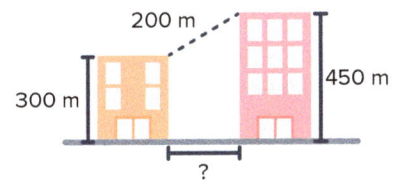

19 Two flagpoles of height 14 m and 19 m are 22 m apart. A length of string is connected to the tops of the two flagpoles and pulled taut.

Find the length of the string, rounded to one decimal place.

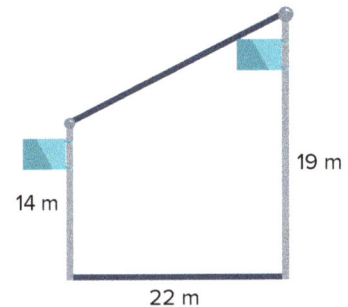

20 You are playing capture the flag. You are positioned 70 yards north and 30 yards east of your team's base. The opposing team's base is 110 yards north and 10 yards west of your base.

What is the shortest distance from you to the other base, rounded to two decimal places?

21 Jessica and Brielle are meeting each other for a mountain-top picnic. Brielle took a wrong turn and is walking on a trail that is parallel to the trail that Jessica is on. Brielle is at a height of 1870 yards and Jesmond is at a height of 2340 yards. The horizontal distance between them is 1080 yards.

   a What is the direct distance between the two friends? Round your answer to the nearest yard.

   b Going off trail, how long will it take for Brielle to walk directly to Jessica at a speed of 1.42 yd/s? Round your answer to the nearest minute.

Answers

6.03 Applications of the Pythagorean theorem

What do you remember?

1 a c b $c^2 = a^2 + b^2$ c Yes

2 a $c = 19.24$ b $b = 21.82$

3 a True   b False   c True   d True

4 a i Yes   ii $c = x, a = 48.2, b = 55.9$
   b i Yes   ii $c = x, a = 26, b = 72$
   c i No
      ii No direct application of the Pythagorean theorem is possible without additional information.
   d i No   ii $x = 2.90$

Let's practice

5 9.43 cm

6 $h = 15$

7 10 miles

8 4 m

9 a Dana labeled the ladder with 150 cm, but the height of the ladder is unknown. The ladder should be labeled as h and the wall should be labeled as 150 cm.

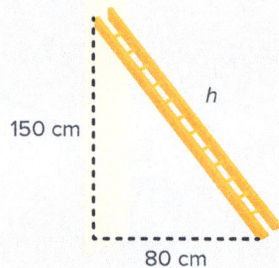

   b $h = 170$ cm

10 Yes, the corner of the garden bed is a right angle because the sides form a Pythagorean triple.

11 The triangle formed by Covington, OH, Arlington, VA, and Bradfordville, FL with the given distances is not a right triangle. This conclusion is based on the Pythagorean theorem, which states that in a right triangle, the square of the length of the hypotenuse (the side opposite the right angle) is equal to the sum of the squares of the lengths of the other two sides. Since $595^2 + 410^2 \neq 705^2$, the triangle does not meet this criterion.

12 a $r = 5$ b 10 m

13 17 yd

14 a $x = 197.26$ b $h = 265.90$ cm

15 $d = 21.3$

16 a 6 ft and 15 ft
    b 21 ft

Let's extend our thinking

17 Doris rounded her answer when solving for the length of the c in the Pythagorean theorem. We can see that
$$16^2 + 10^2 = 356$$
$$19^2 = 3616$$
$$356 \neq 361$$
so the three side lengths will not form a right triangle.

18 132 m

19 22.6 m

20 $c = 56.57$ yd

21 a 1178 yd   b 14 minutes

6.04 Perimeter and area of composite shapes

Subtopic overview

Lesson narrative

In this lesson on perimeter and area of composite shapes, students will learn to break down complex figures into simpler components to calculate their perimeter and area. A composite shape is any figure that can be subdivided into two or more simpler shapes like triangles, rectangles, and circles. To find the perimeter of a composite shape, students will use known side lengths and relationships, such as opposite sides of rectangles being equal. They will also learn strategies to rearrange the shapes to form simpler rectangles or other polygons to make calculations easier. For finding the area of composite shapes, the lesson covers both the addition and subtraction methods. The addition method involves calculating the area of each simpler shape and summing them up. The subtraction method is used when it's easier to find the area of a larger encompassing shape and subtract the areas that are not part of the composite figure. Students will engage with several examples to practice these concepts, such as breaking a shape into rectangles to find unknown side lengths and calculating the total area by summing the areas of smaller shapes within a composite figure. By the end of the lesson, students should be adept at both determining the perimeter and calculating the area of various composite shapes using these strategies.

Learning objectives

Students: Page 355

After this lesson, you will be able to...
- identify triangles, rectangles, squares, trapezoids, parallelograms, circles, and semicircles that make up a composite shape.
- use knowledge of the area of basic shapes to find the area of composite shapes.
- use knowledge of the perimeter of basic shapes to find the perimeter of composite shapes.
- apply perimeter, circumference, and area formulas to solve contextual problems involving composite plane figures.

Key vocabulary

- area
- composite figure
- perimeter

Essential understanding

The properties of polygons can be applied to find their area and perimeter.

Standards

This subtopic addresses the following Virginia 2023 Mathematics Standards of Learning standards.

Mathematical process goals

MPG1 — Mathematical Problem Solving

Teachers can encourage students to apply their mathematical problem-solving skills by posing various real-world problems involving composite shapes. For example, a teacher could present a problem involving the construction of a park that includes various shaped gardens and ask students to calculate the total area and perimeter. This encourages students to apply their knowledge of composite shapes and the formulas for area and perimeter to a complex, practical problem.

MPG2 — Mathematical Communication

Teachers can encourage mathematical communication by asking students to summarize their understanding of the main concepts and strategies used in the lesson in their own words. This could involve explaining the process of subdividing a composite shape, the strategies for calculating perimeter and area, or the application of these concepts in real-world situations. Teachers can also prompt students to reflect on their understanding and share their thought processes and problem-solving strategies with their peers. This allows them to articulate their mathematical thinking and deepen their understanding through discussion and dialogue.

MPG5 — Mathematical Representations

Teachers can help students meet this goal by demonstrating how to represent composite shapes and their subdivisions using diagrams. They can also show how to represent the process of calculating area and perimeter using mathematical notation. By guiding students on how to visually represent their solutions, teachers can help them make connections between different representations and understand that each representation conveys the same mathematical idea.

Content standards

8.MG.5 — The student will solve area and perimeter problems involving composite plane figures, including those in context.

8.MG.5a — Subdivide a plane figure into triangles, rectangles, squares, trapezoids, parallelograms, circles, and semicircles. Determine the area of subdivisions and combine to determine the area of the composite plane figure.

8.MG.5b — Subdivide a plane figure into triangles, rectangles, squares, trapezoids, parallelograms, and semicircles. Use the attributes of the subdivisions to determine the perimeter of the composite plane figure.

8.MG.5c — Apply perimeter, circumference, and area formulas to solve contextual problems involving composite plane figures.

Prior connections

6.MG.2 — The student will reason mathematically to solve problems, including those in context, that involve the area and perimeter of triangles, and parallelograms.

Future connections

G.DF.1 — The student will create models and solve problems, including those in context, involving surface area and volume of rectangular and triangular prisms, cylinders, cones, pyramids, and spheres.

G.DF.2 — The student will determine the effect of changing one or more dimensions of a three-dimensional geometric figure and describe the relationship between the original and changed figure.

Lesson Preparation

Suggested review

Depending on your students' level of prior knowledge, consider revisiting the following lessons:

Grade 6 — 7.03 Perimeter of triangles and parallelograms

Grade 6 — 7.04 Area of parallelograms

Grade 6 — 7.05 Area of triangles

Grade 6 — 8.02 Circumference and pi

Grade 6 — 8.03 Area of a circle

Tools

You may find these tools helpful:
- Scientific calculator
- Plastic or paper shapes

Lesson supports

The following supports may be useful for this lesson. More specific supports may appear throughout the lesson:

Color-coding figures for clarity
Student with disabilities support

As students break down composite figures, consider color-coding the individual shapes within the figure to make it easier for students to see how the composite figure is composed of different shapes, and to break down the problem into manageable parts.

An example is shown below:

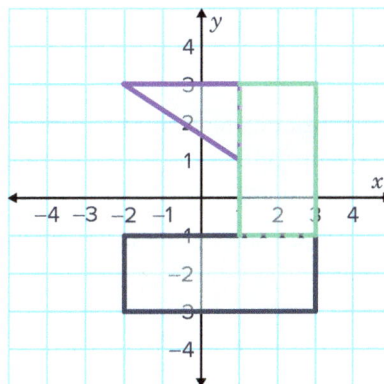

Determining unknown lengths
Address student misconceptions

Students may miss values needed for a perimeter or area if they are not pre-labeled with a numerical value. They may not realize that other values in the shapes can be used to solve for missing side lengths. Encourage students to look for relationships in side lengths, such as opposite sides in a rectangle or square being equivalent, or having the legs of a right triangle and needing the hypotenuse.

Students can label side lengths with equivalent or known values as a reminder to include all sides in the calculation of the perimeter. These values can then be used to determine the lengths of the unknown sides. An example of labeling a composite figure to find a missing length is shown.

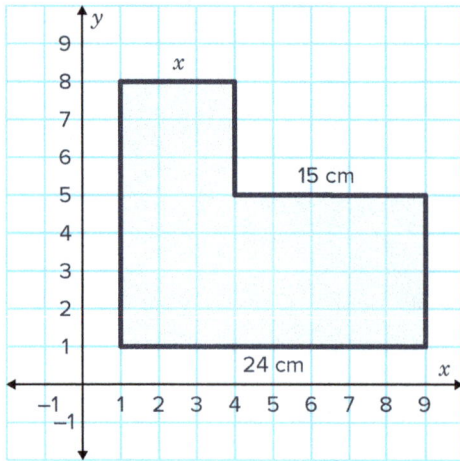

Original figure

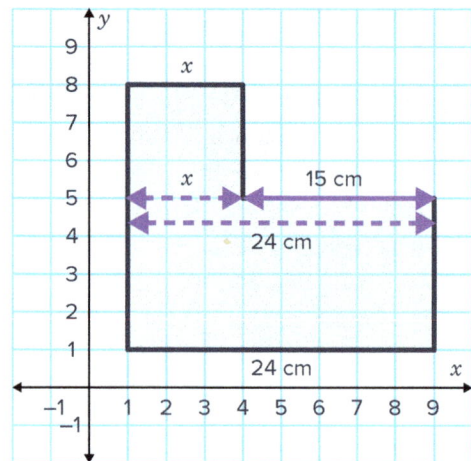

Labeled figure

Student lesson & teacher guide

Perimeter of composite shapes

Students are introduced to the definition of a composite figure and shown examples with different basic shapes. They are reminded of the definition of perimeter, and are shown how to divide a composite figure into its basic shapes. Students then learn how to find the perimeter of composite figures, including solving for missing side lengths and different strategies for solving.

Determining values from basic shapes needed for perimeter
Targeted instructional strategies

When finding perimeter, students need to consider which sides of the basic shapes in a composite figure will be needed since perimeter is the distance around the exterior. Show students two rectangles on a coordinate plane whose perimeter can be found using the unit squares.

Find the perimeter of the graphed shapes

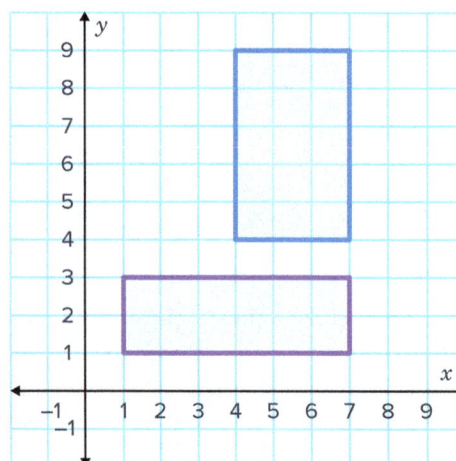

After reviewing the perimeter of the rectangles on the original graph, show students two graphs composed of the same rectangles, but rearranged on the graph to show different composite figures. The overlapping sides should be marked in a different color.

Ask students to answer the following questions:

- Will the two composite figures have the same perimeter?
- What is the perimeter of each composite figure?
- How can the perimeter of the original rectangles and the length of the common side be used to find the perimeter of the composite figure?

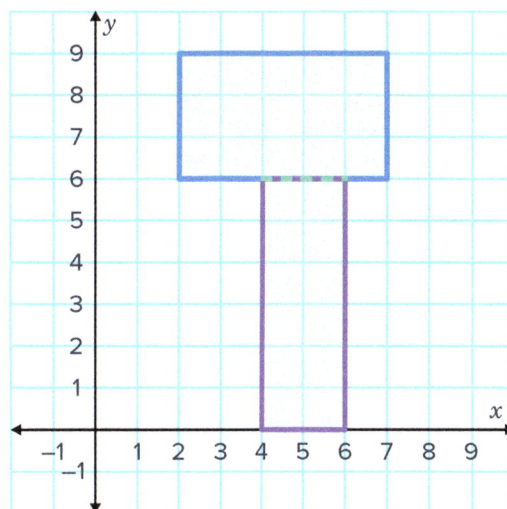

Composite figure 1

Composite Figure 2

Discuss that the arrangement of the basic shapes can change the perimeter of the composite figure and only exterior sides should be considered.

Information gap
English language learner support

Students should be divided into pairs or small groups. Assign each student different pieces of information about the same composite figure, such as side lengths or partial perimeter details.

Students must communicate effectively to share their information, with the goal of calculating the total perimeter together. Before they begin, model clear and precise communication by demonstrating how to ask specific questions and give detailed explanations. Emphasize the importance of using accurate mathematical language.

Then, allow students to engage in their collaborative problem-solving, encouraging them to check their understanding and ensure all information is correctly integrated. After the exercise, conduct a class discussion where pairs share their approaches and solutions, reinforcing the mathematical vocabulary and concepts learned.

Examples

Example 1

Consider the following figure.

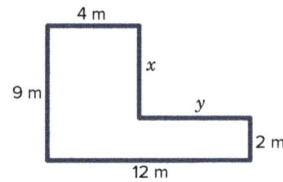

a Find the length x.

Create a strategy

The vertical side on the left has the same length as the two vertical sides on the right added together.

Apply the idea

The side length x will be equal to the difference between the other two vertical side lengths.

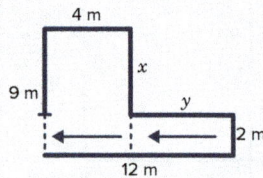

$x = 9 - 2$ m   Subtract the lengths of two vertical sides

   $= 7$ m   Evaluate

Purpose

Students demonstrate an understanding of congruent sides and dividing shapes into familiar shapes to find missing values.

Expected mistakes

Students may look at the side x is on and decide it is a little longer than the side with 4 m so $x = 5$ m. Help students see the division of the two shapes and how to use the 9 m and 2 m to find x.

---

b Find the length y.

Create a strategy

The horizontal side on the bottom has the same length as the two horizontal sides on the top added together.

Apply the idea

The side length y will be equal to the difference between the other two horizontal side lengths.

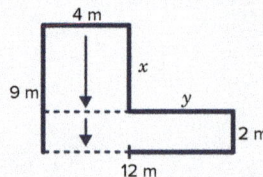

$y = 12 - 4$ m   Subtract the lengths of two horizontal sides

   $= 8$ m   Evaluate

Purpose

Students demonstrate an understanding of congruent sides and dividing shapes into familiar shapes to find missing values.

Students: Page 357

c   Calculate the perimeter of the figure.

Create a strategy

Add all the side lengths in the figure.

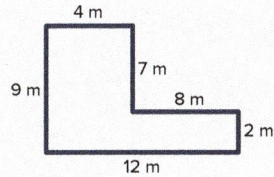

4 m
7 m
9 m
8 m
2 m
12 m

Apply the idea

Substitute $x = 7$ m and $y = 8$ m.

Perimeter $= 2 + 12 + 9 + 4 + 7 + 8$ m   Add all the side lengths

$= 42$ m   Evaluate

Purpose

Students will demonstrate an understanding of finding the missing sides on a congruent shape and then calculating the perimeter.

Students: Page 358

Example 2

Find the perimeter of the figure.

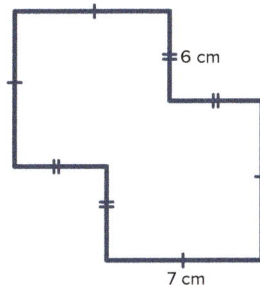

6 cm

7 cm

Create a strategy

Use the markings on all sides of the figure to identify each side length. Add all the side lengths in the figure.

Apply the idea

Four of the sides have length 6 cm and the other four have length 7 cm.

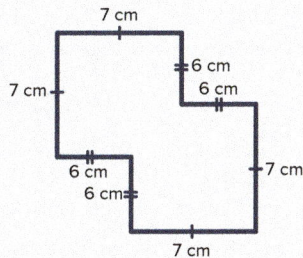

7 cm
6 cm
7 cm
6 cm
6 cm
7 cm
6 cm
7 cm

Perimeter $= 4 \cdot 6 + 4 \cdot 7$ cm   Add all the side lengths

$= 24 + 28$ cm   Evaluate the multiplication

$= 52$ cm   Evaluate the sum

Purpose

Using their knowledge of congruent segments, students will find the perimeter of a composite shape.

Reflecting with students

Discuss with students how they found the perimeter for the composite shape. Have students share the different way(s) to calculate the perimeter. Some examples are:

$$4 \cdot 6 + 4 \cdot 7$$
$$6 + 6 + 6 + 6 + 7 + 7 + 7 + 7$$

Students: Page 358–359

Example 3

Find the perimeter of the following figure. Use the π button on your calculator, rounding your final answer to one decimal place.

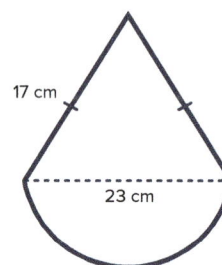

17 cm

23 cm

Create a strategy

Add the two straight sides of the triangle and the circumference of the semicircle.

Apply the idea

First, let's find the partial perimeter of the triangle. The tick marks show that both sides are 17 cm. Keep in mind we are not including the third, 23 cm side in this calculation because it is not part of the perimeter of the composite figure.

$$\text{Perimeter}_\triangle = 17 + 17 \qquad \text{Add the two given sides of triangle}$$
$$= 34 \text{ cm}$$

Now, let's find the circumference of the semicircle. The circumference of a circle is given by $C = 2\pi r$, but since we have a semicircle, half of a circle, we also need to divide by $C = 2\pi r$.

To find the radius for the formula we will find half of the diameter. The diameter is 23 cm, so the radius is 11.5 cm.

$$\text{Circumference} = \frac{2\pi r}{2} \qquad \text{Use circumference formula divided by 2}$$
$$= \frac{2\pi \cdot 11.5}{2} \qquad \text{Substitute } r = 11.5$$
$$= \pi \cdot 11.5 \qquad \text{Divide out the 2}$$
$$\approx 36.1 \text{ cm} \qquad \text{Evaluate to one decimal place}$$

Now, that we have found the partial perimeter of the triangle and the circumference of the semicircle, we can add the two perimeters together to get the perimeter of the entire figure.

$$\text{Perimeter} \approx 34 \text{ cm} + 36.1 \text{ cm} \qquad \text{Add the two perimeters together}$$
$$\approx 70.1 \text{ cm} \qquad \text{Evaluate the sum correct to one decimal place}$$

Reflect and check

Notice that our answer is approximate, indicated by $\approx$. This is because we used an approximate value of π. If we wanted an exact answer we could have written our answer in terms of π as $(34 + 11.5\pi)$ cm.

Purpose

Students will find the perimeter of a composite shape. The composite shape will involve students showing an understanding of what to do when only half of a shape is shown.

Expected mistakes

When finding the perimeter for the composite shape a student may divide the entire problem by 2 or multiply the entire problem by $\frac{1}{2}$ because of the semicircle.

$$\frac{2 \cdot 17 + \pi \cdot 23}{2} \text{ or } \frac{1}{2}(2 \cdot 17 + \pi \cdot 23)$$

If a student does divide the *entire* expression by 2 or multiply by $\frac{1}{2}$, instead of just the part that represents the area of the circle. Ask the student if only half of both shapes are showing. Discuss the where we can place the $\frac{1}{2}$ or divide by 2 and why.

Students: Page 359

Example 4

Here is an outline of a block of land owned by a farmer, who wants to put up fencing along the land to help keep all their cattle safe.

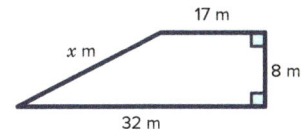

a Find the length of the side labeled x m.

Create a strategy

Visualize the block of land as combination of triangle and rectangle as shown.

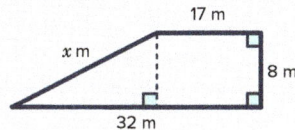

Use the Pythagorean theorem to find the length x.

Apply the idea

The two sides would be 8 m and $32 - 17 = 15$ m.

$a^2 + b^2 = a^2$	Write the formula of Pythagorean theorem
$8^2 + 15^2 = x^2$	Substitute x, 8 and 15 into the formula
$\sqrt{15^2 + 8^2} = x$	Take the square root of both sides
$\sqrt{225 + 64} = x$	Evaluate the square roots
$\sqrt{289} = x$	Evaluate the addition
$17 \text{ m} = x$	Evaluate the square root
$x = 17 \text{ m}$	Symmetric property

Purpose
Show students how to apply the Pythagorean theorem to find the unknown side in a right trapezoid.

Students: Page 360

b Find the perimeter of the block of land in meters.

Create a strategy

Add up all the lengths of the sides.

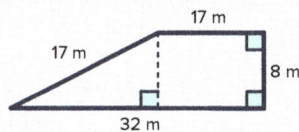

Apply the idea

Perimeter $= 17 + 17 + 8 + 32$ m	Add together the four straight side lengths
$= 74$ m	Evaluate

Purpose
Show students how to calculate the perimeter of a right trapezoid shape.

> ### 💡 Idea summary
>
> The perimeter of the composite shape can be found by determining the length of each side and then summing them up.
>
> We can divide composite figures into more familiar figures to help with finding unknown measures.
>
> Formulas that can help us: the perimeter of a rectangle, $P = 2 \cdot (l + w)$, and the circumference of a circle, $C = 2\pi r$.

Area of composite shapes

Exploration

> ### ▶ Interactive exploration
> **Explore online to answer the questions**
>
> 🌐 **mathspace.co**

Use the interactive exploration in 6.04 to answer these questions:

1. Identify the basic shapes that make up the composite shape shown in the applet.
2. How does the total area of the composite shape compare to the sum of the areas of the individual shapes?

Suggested student grouping: In pairs

In this exploration, students will be interacting with an applet that breaks down a composite shape into its basic shapes. They will identify these basic shapes and compare the total area of the composite shape to the sum of the areas of the individual shapes.

Ideal student responses

These ideal responses may differ from other correct student responses. Less formal responses can be connected with the more precise mathematical language presented here.

1. **Identify the basic shapes that make up the composite shape shown in the applet.**

   The composite shape is made up of rectangles and triangles.

2. **How does the total area of the composite shape compare to the sum of the areas of the individual shapes?**

   The total area of the composite shape is equal to the sum of the areas of the individual shapes that make it up.

Purposeful questions

- Why is it useful to break down a composite shape into its basic shapes when finding the area?
- What are some other ways we could divide this composite shape? Would it change the total area?

Possible misunderstandings

- Students may think that the way a composite shape is divided can change its total area. It's important to clarify that the total area remains the same regardless of how the shape is divided.

Breaking down area of composite shapes
Targeted instructional strategies

As a way for students to become familiar with finding area of composite shapes, show students a graph with familiar shapes where area can be counted with unit squares, such as the one shown:

Find the area of the graphed shapes

Find the area of the graphed shapes

After reviewing the area of the shapes on the original graph, show students two graphs composed of the same basic shapes, but rearranged on the graph to show different composite figures.

Ask students to answer the following questions:
- Determine the area of each of the composite figures
- Are the areas of the two composite figures different or the same?
- How do these composite figures compare to the shapes in the first graph?

Composite figure 1

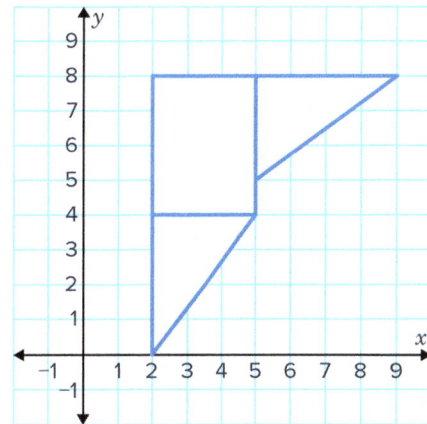

Composite Figure 2

Discuss how the arrangement of the basic shapes does not change the process of finding the area of the composite figure.

Students are shown how to break down composite figures into basic shapes in order to find their area. Students are shown different methods of solving for unknown lengths and finding the area of the composite figures, including examples where only part of a basic shape is used. Students are given area formulas for several basic shapes to help them in their calculations.

Examples

Students: Page 362

> ### Example 5
>
> Consider the composite shape.
>
> 7 cm
>
> 5 cm
>
> 5 cm
>
> 15 cm
>
> **a** Which basic shapes make up this composite shape?
>
> **A** A rectangle minus two triangles   **B** One rectangle and two trapezoids
>
> **C** Two parallelograms   **D** Two trapezoids
>
> **Create a strategy**
>
> We can construct a line on the composite shape to break it up into its components.
>
> **Apply the idea**
>
> As we can see there are two trapezoids in this construction.
>
> So, the correct option is D.

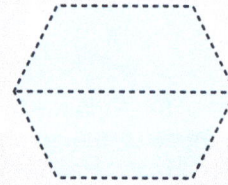

Purpose
Show students how to identify basic shapes within a composite shape.

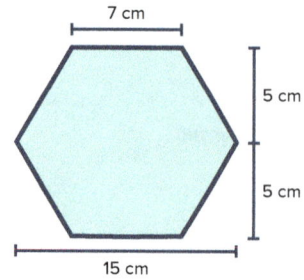

Reflecting with students
If the problem had not suggested the basic shapes within the composite shape, students could split the shape up in multiple possible configurations. Have advanced learners, or any students who are ready, use the same composite figure to find how many different ways the figure can be composed of:
- Triangles
- Rectangles
- Parallelograms
- Trapezoids

Students: Page 362

> **b** Find the area of the composite shape.
>
> **Create a strategy**
>
> We can find the area of the composite shape by adding the areas of the two trapezoids using formula
>
> $A = \frac{1}{2}(b_1 + b_2)h.$

Apply the idea

Since the two trapezoids are identical we can multiply the trapezoid formula by 2.

$$\text{Area} = \frac{1}{2}h(b_1 + b_2) \cdot 2 \qquad \text{Formula for the area of a trapezoid times 2}$$

$$= \frac{1}{2}(5)(7 + 15) \cdot 2 \qquad \text{Substitute the values}$$

$$= 110 \text{ cm}^2 \qquad \text{Evaluate}$$

Purpose

Show students how to calculate the area of a composite shape made up of two identical trapezoids.

Students: Page 363

Example 6

Find the total area of the figure shown.

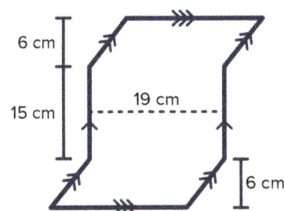

Create a strategy

Divide the shape into three shapes, two paralellograms and one rectangle.

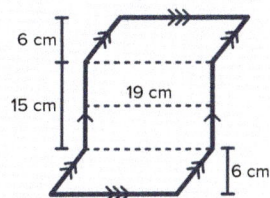

Apply the idea

Area of parallelograms:

$$A = b \cdot h \qquad \text{Use the area of parallelogram formula}$$

$$= 19 \cdot 6 \qquad \text{Substitute } b = 19 \text{ and } h = 6$$

$$= 114 \qquad \text{Evaluate}$$

$$\text{Both parallelograms} = 114 \cdot 2 \qquad \text{Multiply the area by 2}$$

$$= 228 \qquad \text{Evaluate}$$

Area of rectangle:

$$A = l \cdot w \qquad \text{Use the area of rectangle formula}$$

$$= 19 \cdot 15 \qquad \text{Substitute } l = 19 \text{ and } w = 15$$

$$= 285 \qquad \text{Evaluate}$$

Total Area:

$$A = 228 + 285 \qquad \text{Add the areas of the parallelograms and the rectangle}$$

$$= 513 \text{ cm}^2 \qquad \text{Evaluate}$$

Purpose

Show students how to find the area of a composite shape by dividing it into familiar shapes (parallelograms and rectangles) and finding their areas separately.

Expected mistakes

Students may not realize that the 19 cm value applies to the base of the parallelograms since the dashed line does not line up with the sides marked with three ticks. Students should be encouraged to annotate the composite figure with any lines needed to break down the figure into its basic shapes.

Students: Page 364

Example 7

Find the shaded area in the figure shown.

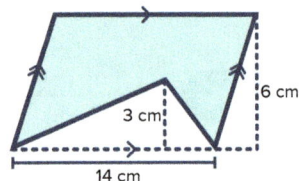

Create a strategy

Subtract the area of the triangle from the area of the parallelogram.

Apply the idea

Area of triangle:

$$A = \frac{1}{2} \cdot b \cdot h \qquad \text{Use the area of triangle formula}$$

$$= \frac{1}{2} \cdot 14 \cdot 3 \qquad \text{Substitute } b = 14 \text{ and } h = 3$$

$$= 21 \qquad \text{Evaluate}$$

Area of parallelogram:

$$A = b \cdot h \qquad \text{Use the area of parallelogram formula}$$

$$= 14 \cdot 6 \qquad \text{Substitute } b = 14 \text{ and } h = 6$$

$$= 84 \qquad \text{Evaluate}$$

Area of composite shape:

$$A = 84 - 21 \qquad \text{Subtract the area of triangle from the area of parallelogram}$$

$$= 63 \text{ cm}^2 \qquad \text{Evaluate}$$

Purpose

Show students how to find the area of a composite shape by subtracting the area of the triangle from the area of the parallelogram.

💡 Idea summary

The area of the composite shape can be found by finding the area of each of the smaller shapes, and then adding them to get the total area.

Sometimes it is easier to take a subtractive approach and find the area of a larger, familiar figure and subtract the area of shapes that are not part of the composite figure.

Some formulas we often use are:

- Area of Circle: $A = \pi \cdot r^2$
- Area of a Trapezoid: $A = \dfrac{1}{2}h(b_1 + b_2)$
- Area of a Parallelogram: $A = b \cdot h$
- Area of a Triangle: $A = \dfrac{1}{2} \cdot b \cdot h$
- Area of a Rectangle: $A = b \cdot h$
- Area of a Square: $A = s^2$

Practice

Students: Page 365–372

What do you remember?

1 Determine if the following scenarios require calculating area or perimeter.

    a Finding the amount of carpet needed for a room.

    b The amount of fence required to enclose a garden.

    c The amount of seed needed to plant a flower bed.

    d The amount of wood needed to lay a boarder around a bedroom.

2 For each of the following rectangles:

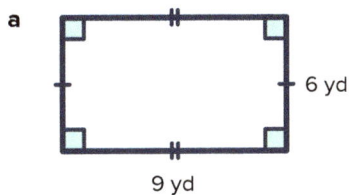

    i Find the perimeter.   **ii** Find the area.

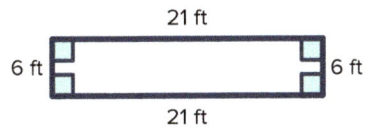

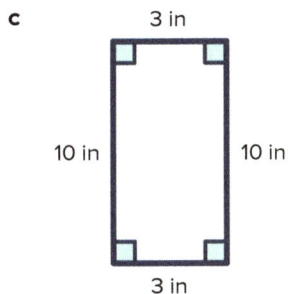

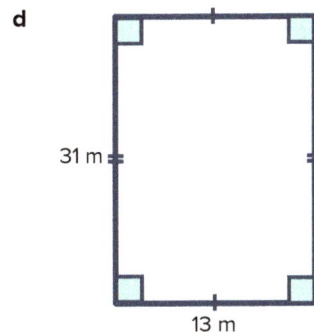

    a
6 yd
9 yd

    b
21 ft
6 ft   6 ft
21 ft

    c
3 in
10 in   10 in
3 in

    d
31 m
13 m

3 Find the circumference of the following circles. Use 3.14 for π and round your answers to two decimal places:

a

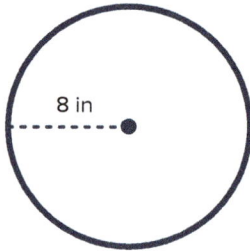

8 in

b

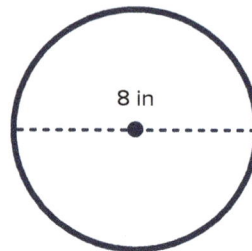

8 in

4 Find the area of the following circles. Use 3.14 for π and round your answers to two decimal places:

a

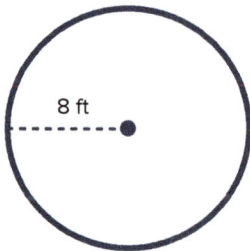

8 ft

b

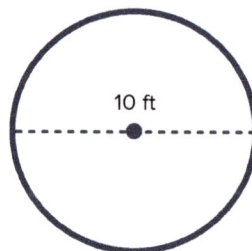

10 ft

5 For each of the following composite shapes:

  i State the basic shapes that make up the composite shape.

  ii Find the exact perimeter.

a

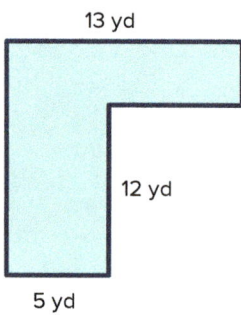

13 yd

12 yd

5 yd

b

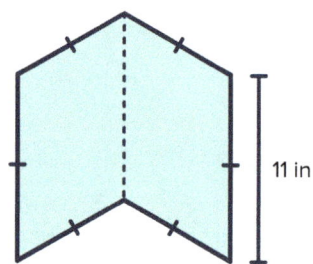

11 in

c

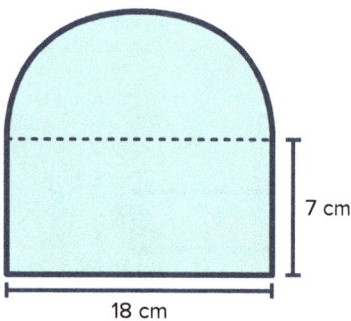

7 cm

18 cm

d

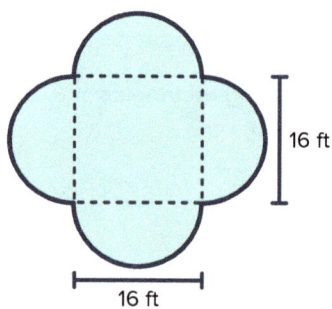

16 ft

16 ft

6 **a** Name the two geometric shapes that together form this composite shape.

  b Explain how to find the total area.

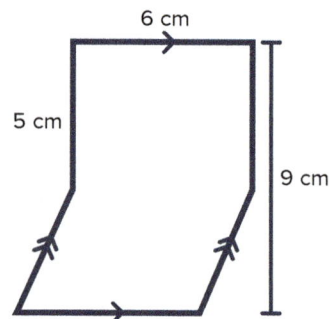

6 cm

5 cm

9 cm

7 How would you calculate the total area of this shape?

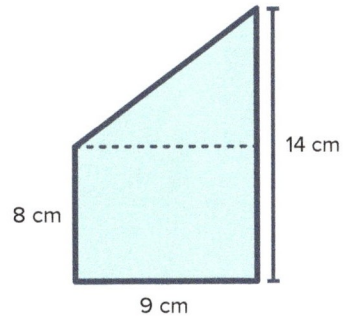

8 For each of the following figures:

   i Find the area of rectangle A.   **ii** Find the area of rectangle B.

   iii Find the total area of the composite shape.

a

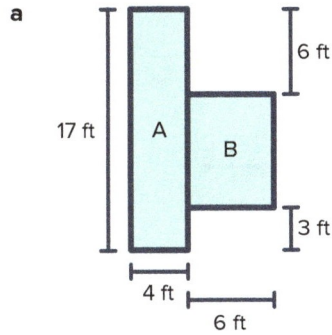

b

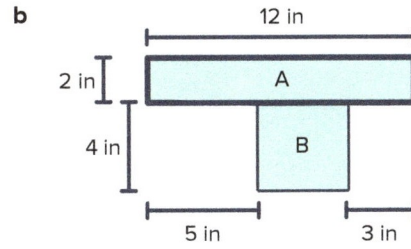

c

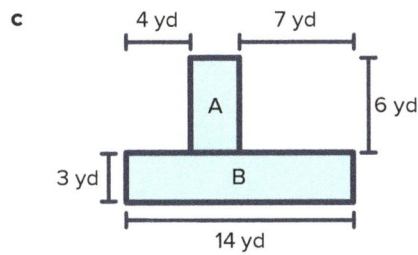

Let's practice

9 Find the perimeter of the following figures:

a

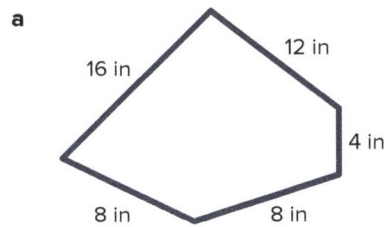

b

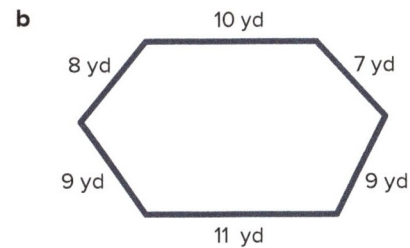

c

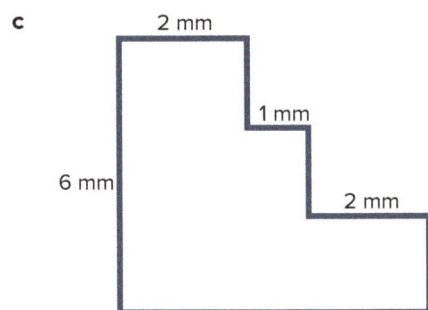

d

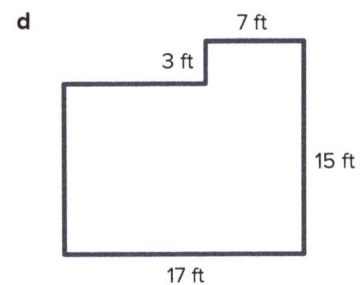

10 For each of the following figures:

    i Find the length x.   **ii** Find the length y.   **iii** Find the perimeter.

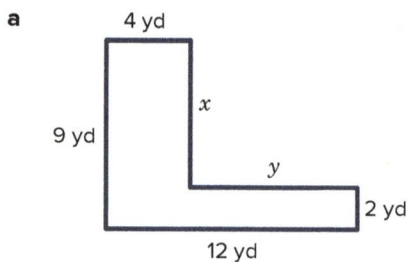

a

4 yd

x

9 yd

y

2 yd

12 yd

b

18 m

5 m

12 m

x

y

3 m

3 m

11 An outline of a block of land is pictured:

    a Find the value of x.

    b Find the perimeter of the block of land.

11 m

x m

6 m

19 m

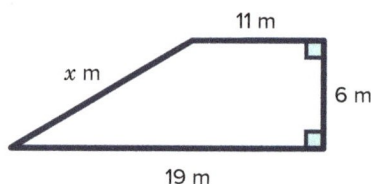

12 Describe and correct the error in finding the perimeter of the figure:

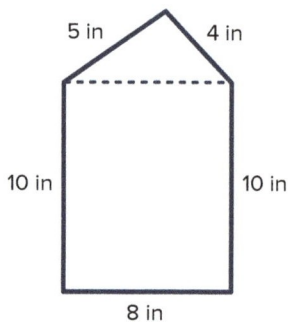

5 in   4 in

10 in   10 in

8 in

Perimeter $= 5 + 4 + 10 + 8 + 10 + 8$

$= 45$ in

13 Katrina has a garden bed that is shaped as shown in the following figure: Find the area of the garden bed.

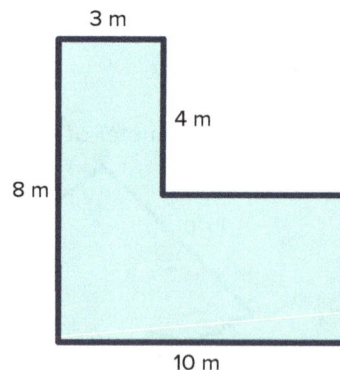

3 m

4 m

8 m

10 m

14 Find the area of the following composite shapes:

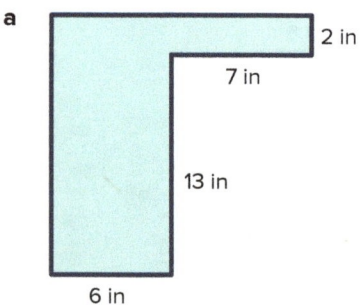

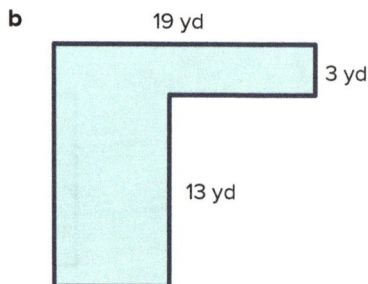

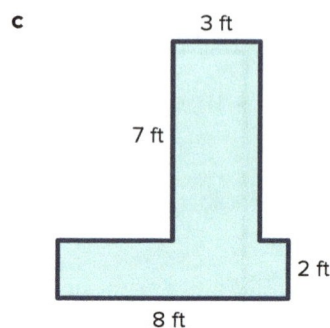

a

2 in

7 in

13 in

6 in

b

19 yd

3 yd

13 yd

c

3 ft

7 ft

2 ft

8 ft

15 Find the area of the following composite shapes:

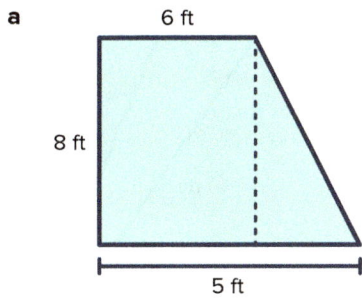

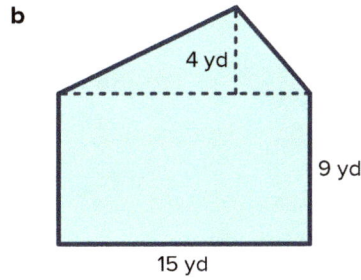

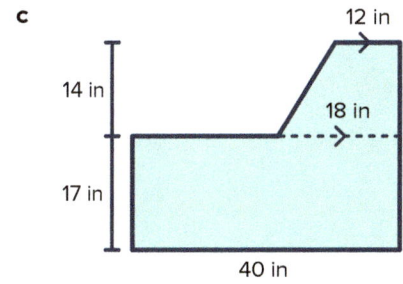

a

6 ft

8 ft

5 ft

b

4 yd

9 yd

15 yd

c

12 in

14 in

18 in

17 in

40 in

16 Find the perimeter of the following figures, correct to one decimal place:

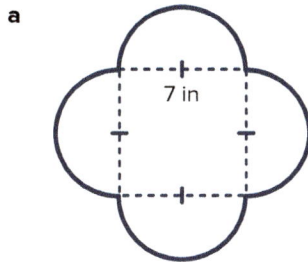

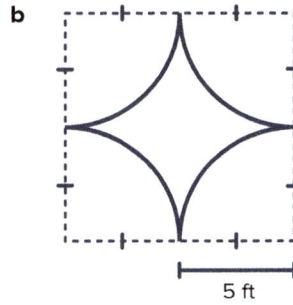

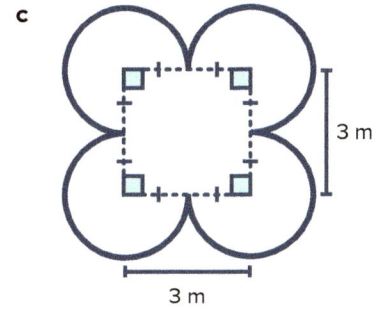

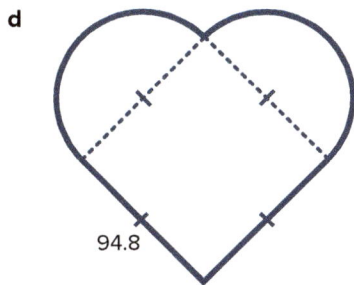

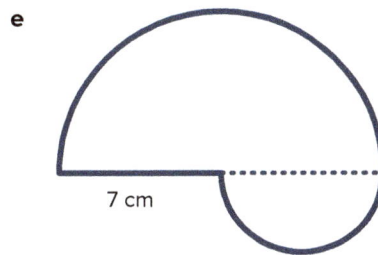

a

7 in

b

5 ft

c

3 m

3 m

d

94.8

e

7 cm

17 Find the perimeter of the shaded arc, correct to two decimal places:

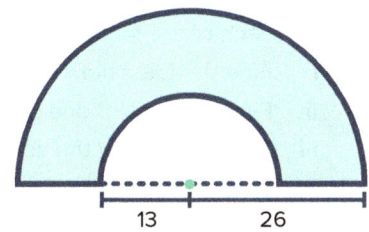

13   26

18 A rectangle has two semicircles of radius 3 m cut out at each end as shown:

Find the area of the shape, correct to two decimal places.

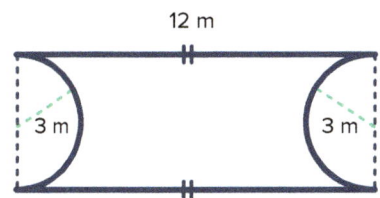

12 m

3 m   3 m

19 The quadrilateral in the diagram has a diagonal measuring 14 mi.

Calculate the total area of the quadrilateral.

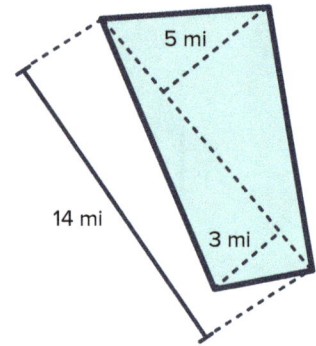

20 Consider the following figure:

    a Find the value of x.

    b Find the area of the shaded region.

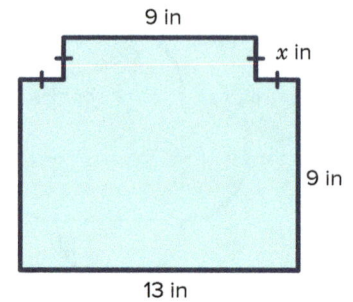

21 Lachlan draws the plot of land which contains his house and garden.

    a Find the total area of the plot of land.

    b Find the area covered by the house.

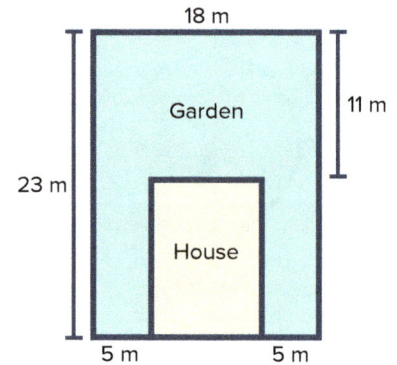

22 For each of the following regular polygons:

    i Find the least number of equal triangles that the polygon can be divided into.

    ii Find the area of one of the polygon's triangles.

    iii Find the area of the polygon.

    a

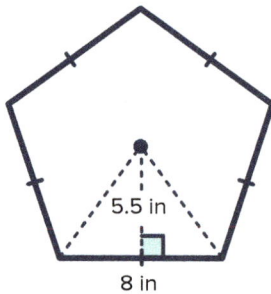

    b

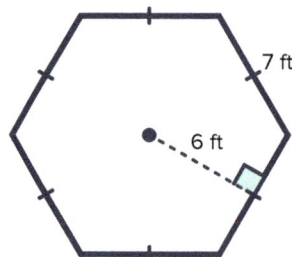

    c

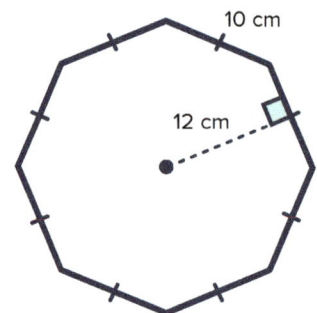

Let's extend our thinking

23 Calculate the perimeter in yards of the plot of land pictured here on this site plan. All measurements are given in meters.

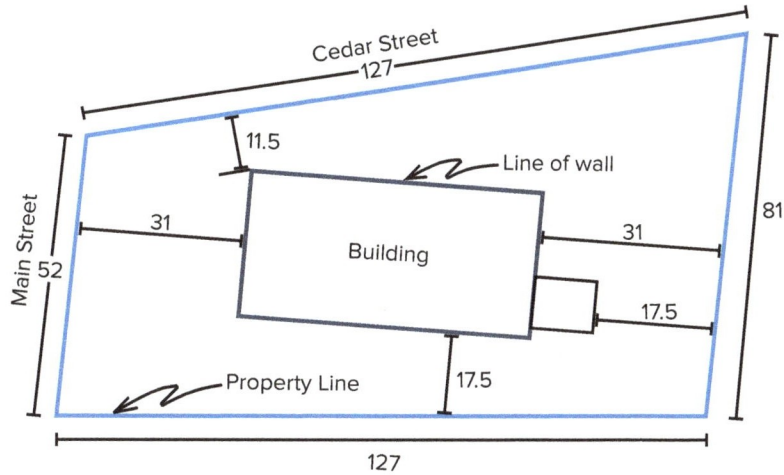

24 A farmer wants to build a fence around the entire perimeter of his land, as shown in the diagram. The fencing costs $37 per meter.

   a Find the value of x, to two decimal places.

   b Find the value of y, to two decimal places.

   c How many meters of fencing does the farmer require, if fencing is sold by the meter?

   d At $37 per meter of fencing, how much will it cost him to build the fence along the entire perimeter of the land?

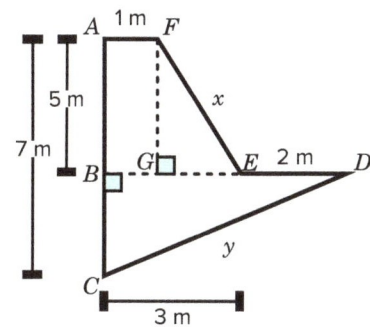

25 According to regulations, a standard 400 m running track consists of a straight section with length 84.39 m and two semicircular sections. The radius of the inner edge of Lane 1 is 36.80 m and the radius of the inner edge of Lane 2 is 38.02 m.

   a Calculate the inside perimeter of Lane 1 in meters. Round your answer to two decimal places.

   b Calculate the inside perimeter of Lane 2 in meters. Round your answer to two decimal places.

26 A piece of paper in the shape of a parallelogram is folded along its shortest diagonal, as shown. Find the total area covered by the folded paper.

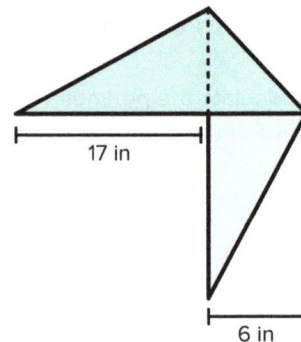

17 in

6 in

27 A farmer is going to fertilize his corn field, which has the following shape:

a Find the area of the corn field.

b If it takes 150 kg of fertilizer to fertilize 100 square meters, how many kilograms of fertilizer should the farmer purchase?

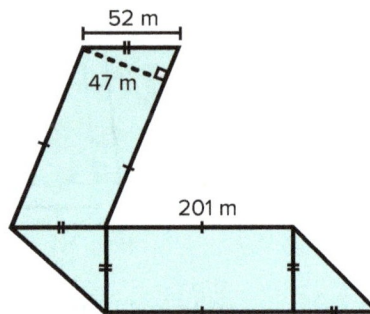

52 m

47 m

201 m

28 A city council is creating a new pathway in the local park. The pathway is planned to be a composite shape consisting of a rectangle and a trapezoid at one of the rectangle's longer sides. The total perimeter of the pathway can be no more than 100 meters.

Design a path that will meet the city council's criteria.

Answers

6.04 Perimeter and area of composite shapes

What do you remember?

1 a Area   b Perimeter   c Area   d Perimeter

2 a i 30 yd   ii 54 yd^2
   b i 54 ft   ii 126 ft^2
   c i 26 in   ii 30 in^2
   d i 88 m   ii 403 m^2

3 a 50.24 in   b 25.12 in

4 a 201.1 ft^2 b 78.5 ft^2

5 a i Two rectangles   ii 58 yd
   b i Two rhombuses   ii 66 in
   c i One semicircle and one rectangle
     ii $9\pi + 32$ cm
   d i One square and four semicircles
     ii 32π ft

6 a Rectangle and parallelogram
   b Example answer: Add the area of the rectangle and parallelogram together.

7 To calculate the total area of a composite shape composed of a rectangle and a triangle, you would find the area of the rectangle and the area of the triangle separately, using their respective formulas, and then add the two areas together.

8 a i 68 ft^2 ii 48 ft^2 iii 116 ft^2
   b i 24 in^2 ii 16 in^2 iii 40 in^2
   c i 18 yd^2 ii 42 yd^2 iii 60 yd^2

Let's practice

9 a 48 in   b 54 yd   c 22 mm   d 64 ft

10 a i $x = 7$ yd   ii $y = 8$yd   iii 42 yd
   b i $x = 4$ m   ii $y = 10$ m   iii 60 m

11 a 10 m   b 46 m

12 The calculation included an inner side length of 8 inches that isn't part of the outside edge we measure for the perimeter. When finding a shape's perimeter, we only add up the lengths around the outside. The correct perimeter is $P = 5 + 4 + 10 + 8 + 10$ which is equal to 37 in.

13 52 m^2

14 a 104 in^2 b 135 yd^2 c 37 ft^2

15 a 44 ft^2 b 165 yd^2 c 890 in^2

16 a 44.0 in   b 31.4 ft   c 28.3 m
   d 487.4 units   e 40.0 cm

17 148.52 units

18 43.73 m^2

19 56 mi^2

20 a $x = 2$ b 135 in^2

21 a 414 m^2 b 96 m^2

22 a i 5   ii 22 in   iii 110 in^2
   b i 6   ii 21 ft   iii 126 ft^2
   c i 8   ii 60 cm   iii 480 cm^2

Let's extend our thinking

23 387 yd

24 a $x = 5.39$ b $y = 5.39$
   c 21 m   d $777

25 a 400.00 m   b 407.67 m

26 120 in^2

27 a 22 603 m^2 b 33 904.5 kg

28 Possible answers:

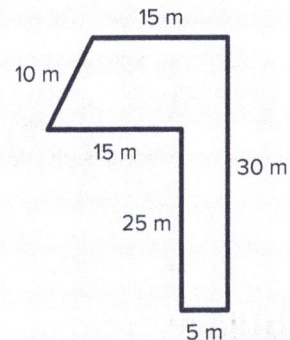

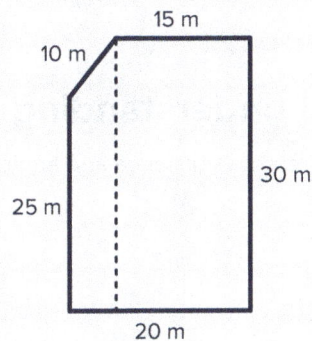

6.05 Translations in the coordinate plane

Subtopic overview

Lesson narrative

In this lesson on translations in the coordinate plane, students will engage in an exploration activity to deepen their understanding of geometric translations. The exploration focuses on identifying and describing translations by observing the movement of shapes on the coordinate plane. Students will begin by examining preimage and image pairs to understand how each point of a figure moves the same distance in the same direction. The exploration will guide students through the steps of recognizing corresponding points, determining whether the movement is horizontal, vertical, or both, and counting the number of units the figure has moved. The activity includes practical examples where students will translate various geometric figures, such as triangles and quadrilaterals, by specific units to the left, right, up, or down. They will plot the preimage and image on a coordinate plane and describe the translation in precise terms.

Learning objectives

Students: Page 373

> 🎓 **After this lesson, you will be able to...**
> - identify the coordinates of the image of a polygon that has been translated vertically, horizontally, or a combination of both when given a preimage.
> - sketch the image of a polygon that has been translated vertically, horizontally, or a combination of both.
> - identify and describe transformations in context.

Key vocabulary

- image
- preimage
- transformation
- translation

Essential understanding

A translation preserves distance and angle measure because it moves all of the points on a figure the same distance and direction.

Standards

This subtopic addresses the following Virginia 2023 Mathematics Standards of Learning standards.

Mathematical process goals

MPG1 — Mathematical Problem Solving

Teachers can incorporate this goal by presenting students with problem situations that require them to apply their understanding of the coordinate plane, polygons, and transformations. For example, they could provide a real-world problem where a figure needs to be translated in a certain way on the coordinate plane, and students would need to identify the new coordinates and sketch the solution.

MPG2 — Mathematical Communication

Teachers can integrate this process goal by encouraging students to communicate the transformations occurring as well as the impact of the transformations on the image's shape, size and position.

MPG3 — Mathematical Reasoning

Teachers can facilitate mathematical reasoning by asking students to justify the steps they used in identifying the coordinates of a translated image and sketching it on the coordinate plane. They could also present students with incorrect solutions and ask them to use logical reasoning to identify the mistakes.

MPG4 — Mathematical Connections

Teachers can establish mathematical connections by linking the concept of transformations and translations to other topics within mathematics and to real-world situations. For instance, they could relate the concept of transformations to the prior knowledge of students about polygons and the coordinate plane. Furthermore, they could use real-world examples like tiling, fabric design, wallpaper patterns, and art to illustrate the practical applications of translations.

Content standards

8.MG.3 — The student will apply translations and reflections to polygons in the coordinate plane.

8.MG.3a — Given a preimage in the coordinate plane, identify the coordinates of the image of a polygon that has been translated vertically, horizontally, or a combination of both.

8.MG.3d — Sketch the image of a polygon that has been translated vertically, horizontally, or a combination of both.

8.MG.3g — Identify and describe transformations in context (e.g., tiling, fabric, wallpaper designs, art).

Prior connections

6.MG.3 — The student will describe the characteristics of the coordinate plane and graph ordered pairs.

7.MG.4 — The student will apply dilations of polygons in the coordinate plane.

Future connections

G.RLT.3 — The student will solve problems, including contextual problems, involving symmetry and transformation.

Lesson Preparation

Suggested review

Depending on your students' level of prior knowledge, consider revisiting the following lessons:

Grade 6 — 7.06 Polygons in the coordinate plane
Grade 8 — 3.01 Review: Plot points and represent relations

Tools

You may find these tools helpful:
- Dynamic geometry software
- Graph paper
- Transparent sheets

Student lesson & teacher guide

Translations

Students are introduced to the vocabulary term transformation, and reminded of the definition of preimage and image. Students then engage in an exploration that translates a figure horizontally and vertically in the coordinate plane.

Use a Concrete-Representational-Abstract (CRA) approach
Targeted instructional strategies

Concrete: Engage students with physical manipulatives to explore translations on the coordinate plane. Provide each student with cut-out polygons, such as triangles and quadrilaterals, and a large grid mat representing the coordinate plane. Have students place their shapes on the grid and physically slide them to model translations. Encourage them to move the shapes horizontally, vertically, or both, ensuring that every point on the shape moves the same distance and in the same direction. As they manipulate the shapes, prompt them to observe how the figures maintain their size and shape, emphasizing that translations preserve distance and angle measures.

Representational: Guide students to transition from physical movement to drawing representations of translations. Have them draw the coordinate plane on graph paper and sketch the preimage of the polygon. Instruct them to use arrows to show the direction and distance of the translation from the preimage to the image. Encourage students to label corresponding points on the preimage and image, reinforcing that each point moves identically. Use diagrams to illustrate vertical, horizontal, and combined translations.

Abstract: Introduce the algebraic notation for translations on the coordinate plane. Teach students how to describe a translation using coordinate rules, such as $(x, y) \rightarrow (x + a, y + b)$, where a and b represent the horizontal and vertical shifts. Provide examples and have students practice finding the coordinates of the image after applying a translation to the preimage. Encourage them to solve problems by calculating new coordinates and writing equations that represent the translation.

Emphasize that this abstract representation captures the same movement they observed with the manipulatives and drawings, reinforcing that translations move all points the same distance and direction. Help students make connections between the concrete manipulatives, the drawings, and the algebraic expressions. Discuss how the physical movement of shapes corresponds to the arrows and shifts in their drawings, and how these relate to the coordinate rules in algebraic form. Encourage students to explain how each representation shows that translations preserve distance and angle measures by moving all points identically.

> ### ▶ Interactive exploration
> Explore online to answer the questions
>
> ### 🌐 mathspace.co

Use the interactive exploration in 6.05 to answer these questions:

1. How did the object change as you changed the horizontal slider on the applet?
2. How did the object change as you changed the vertical slider on the applet?
3. What remained the same between the object and image as you used either or both of the horizontal and vertical sliders?

Suggested student grouping: In pairs

In this exploration, students will interact with an applet that shows the translation of an object to its image. They will adjust the horizontal and vertical sliders to manipulate the object and observe how its image changes. Students will be developing their understanding of translation in geometry.

Ideal student responses

These ideal responses may differ from other correct student responses. Less formal responses can be connected with the more precise mathematical language presented here.

1. **How did the object change as you changed the horizontal slider on the applet?**

   The object moved side to side (left or right) as the horizontal slider changed. The distance it moved corresponded to the value on the slider.

2. **How did the object change as you changed the vertical slider on the applet?**

   The object moved up and down as changed the vertical slider is changed. The height it moved corresponded to the value on the slider.

3. **What remained the same between the object and image as you used either or both of the horizontal and vertical sliders?**

   The shape and size of the object remained the same, no matter how the horizontal and vertical sliders are moved. Only the position of the object changed.

Purposeful questions
- What would happen if you moved both the horizontal and vertical sliders at the same time?
- How did the sign of the number on the sliders affect the direction the object moved?

Possible misunderstandings
- Students may not connect that the object stayed the same orientation and size as it moved since the first two questions ask how the object changed. Students should be reminded to focus on direction, size, and position when answering the exploration questions.

Characteristics of translations using a figure in the coordinate plane
Targeted instructional strategies

After introducing the concept of rigid transformations, guide the students through a translation using the following steps:

1. Show students a rectangle in the coordinate plane such as the one below, and ask them to determine the area of the rectangle and the length of its sides using the unit squares.

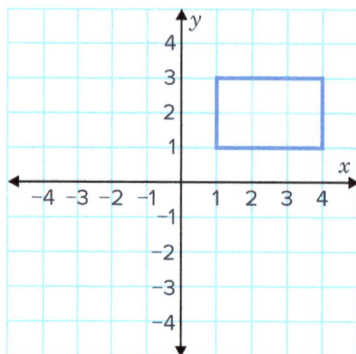

2. Show students moving the rectangle to a new location, such as 3 units to the left and 5 units down.

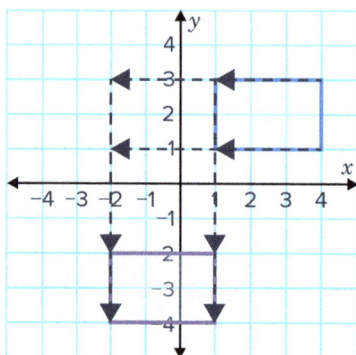

3. Ask students to find the area and side lengths of the rectangle after it was translated. Students should predict whether this will always be true for all translated figures before teaching the lesson.

Information gap
English language learner support

Provide students with the following:
- Figures graphed in the coordinate plane with vertices labeled with their coordinates.
- Cards with specific translations such as "Translate 3 units to the right and 2 units down."

Pair the students and assign roles: one student will be the Describer and the other the Drawer. The Describer reads the translation instructions without showing the shape card, while the Drawer plots the original shape and performs the translation based on these instructions.

The Describer begins by stating the coordinates of the shape's vertices, such as "(2, 3), (4, 3), and (3,5)," and then provides the translation instructions, like "Translate 3 units to the right and 2 units down." The Drawer plots the original shape on the graph paper and applies the translation accordingly.

After completing the task, both students compare the translated shape with the original on the Describer's card to check for accuracy, discussing any discrepancies and refining their descriptions and interpretations.

After one round, students switch roles and repeat the activity with new shapes and instructions.

Use technology to visualize translation
Student with disabilities support

Explicitly show students how to use technology to show the transformation in real-time, which can significantly aid students in understanding the concept. For example, explore transformations using $\triangle ABC$.

1. Before starting, it can be helpful to show the coordinate plane. To do this, right-click on the white space on the right side. Check Show Axes, and under the Show Grid, check Major Gridlines.

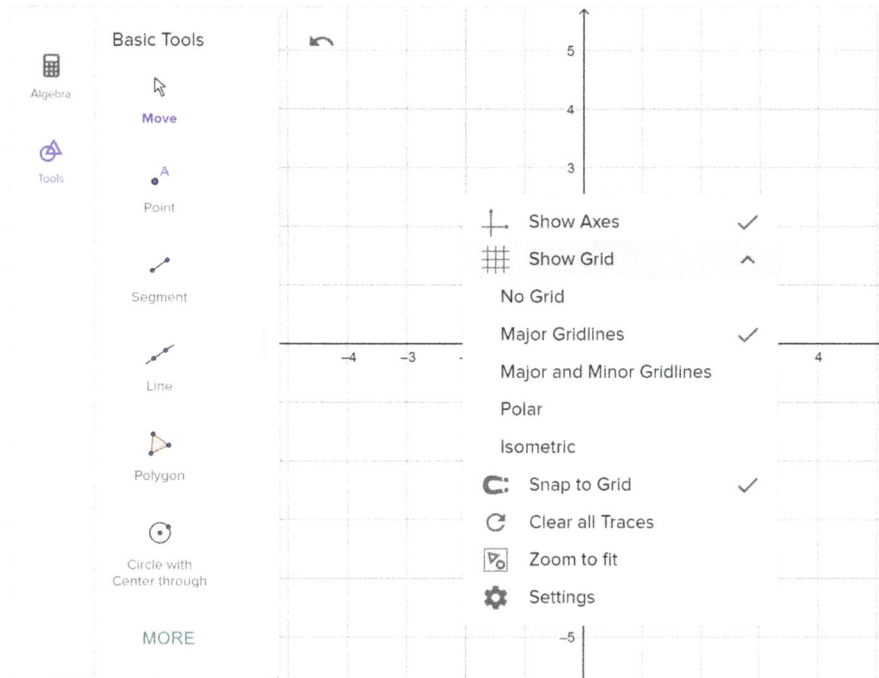

Basic Tools	
Algebra	
Tools	
	Move
	Point
	Segment
	Line
	Polygon
	Circle with Center through
	MORE

Show Axes ✓
Show Grid ^
No Grid
Major Gridlines ✓
Major and Minor Gridlines
Polar
Isometric
Snap to Grid ✓
Clear all Traces
Zoom to fit
Settings

2. Use the Polygon tool to draw triangle ABC.

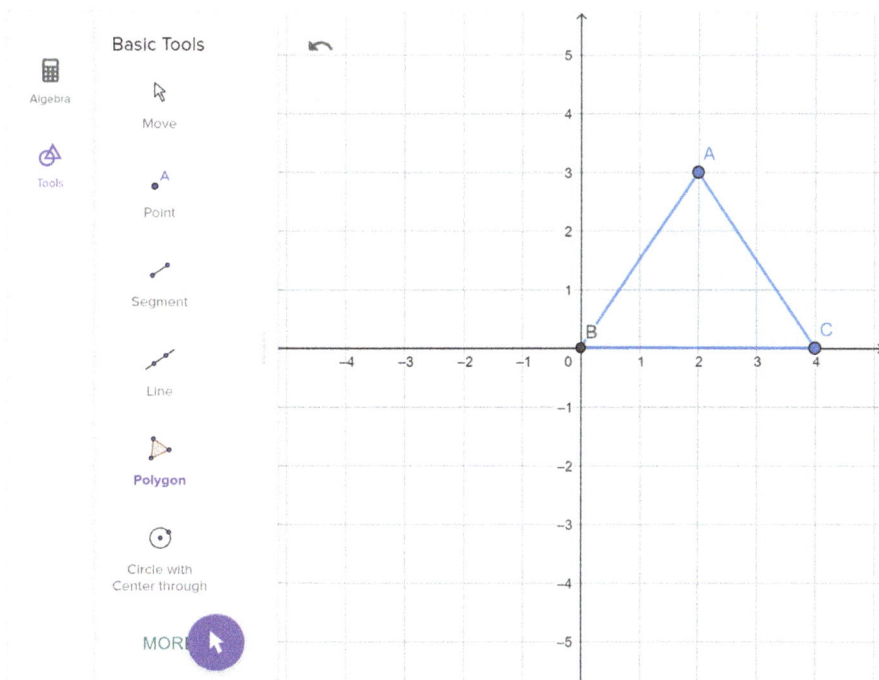

3. To show a translation of the triangle, use the Translate by Vector tool. However, before using the tool, create first a vector that will show the direction and the magnitude of the translation. Use the Vector tool to create a vector. The vector DE indicates that triangle ABC will be translated 2 units to the left.

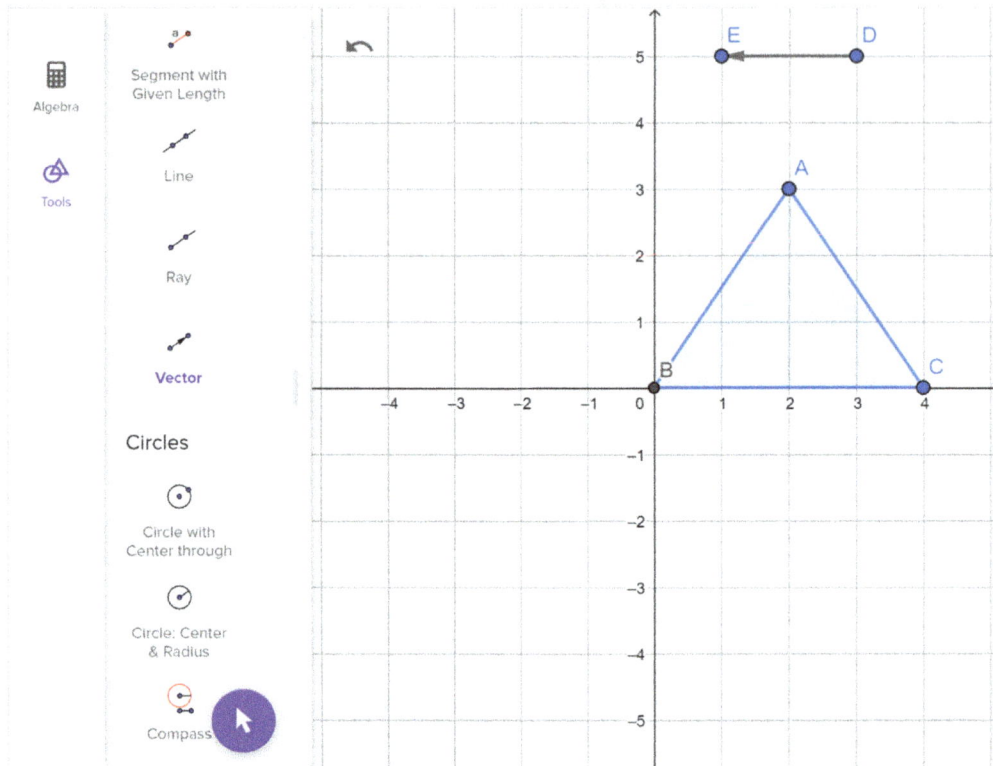

4. Click on the Translate by Vector tool, then select the triangle first, then the vector. A resulting triangle $A'B'C'$ is shown.

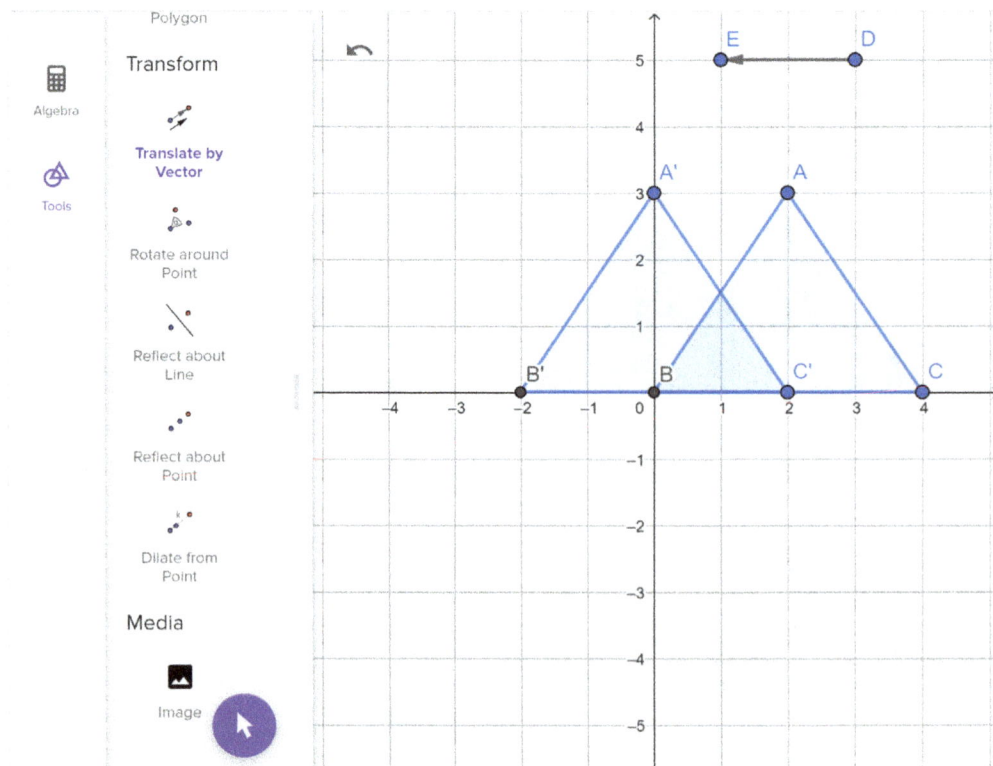

Counting lines instead of unit squares

Address student misconceptions

When moving the vertices of a figure in the coordinate plane, students may count the grid lines insetad of the unit squares, which leads to a shape moved an incorrect number of units.

Relating performing transformations by moving points using unit squares to the students' experience with number lines may help students move points correctly and consistently.

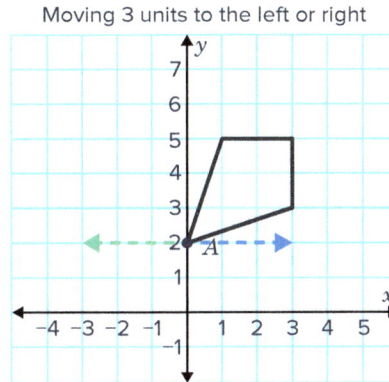

Moving 3 units to the left or right

Moving point A left or right 3 units is visually similar to using a number line to determine the absolute value of 3.

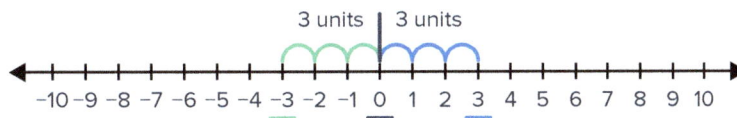

After the exploration, students are shown the notation for indicating image points after a transformation. The definition and an example of a translation is shown, and students are given the steps for translating a figure in the coordinate plane.

Examples

Students: Page 374–375

Example 1

What is the translation from triangle ABC to triangle $A'B'C'$?

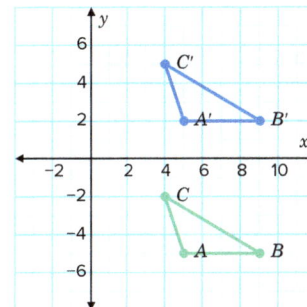

Create a strategy

Consider just one vertex from triangle ABC and the corresponding vertex from triangle $A'B'C'$. Count the units between each vertex and consider the direction of movement.

Apply the idea

Consider vertices C and C'. We count 7 units up from C to C'.

Triangle ABC is translated 7 units up to form triangle $A'B'C'$.

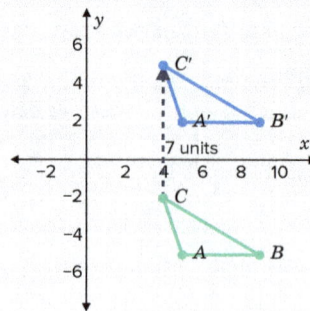

Reflect and check

We count 7 units up from A to A'.

We count 7 units up from B to B'.

We can count 7 units up from any point on triangle ABC to the triangle $A'B'C'$. For example, the point $(7, -5)$ on triangle ABC corresponds to point $(7, 2)$ on triangle $A'B'C'$.

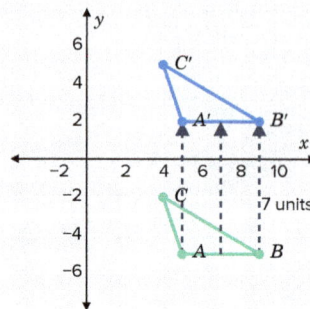

Purpose

Show students how to determine the translation of a triangle in a coordinate plane.

Expected mistakes

Students may assume that $A'B'C'$ is the preimage and ABC is the image, which reverses the direction of the described transformation. Remind students that points with zero (0) additional notation comes before points with one (1) additional mark.

Use abstraction to simplify the transformation
Targeted instructional strategies

use with Example 1

Encourage students to abstract the concept of translation by focusing on the movement of a single vertex to understand the shift of the entire figure. Show them that when vertex C moves up 7 units to become vertex C', this movement represents the translation applied to the whole triangle since we were given that the entire shape was translated.

By abstracting from the movement of one point, students can grasp that triangle ABC has been translated up 7 units without examining each coordinate individually or creating general rules. This approach simplifies the concept of translation and helps students see how a single movement affects the entire shape on the coordinate plane.

Example 2

What is the translation from quadrilateral A to quadrilateral B?

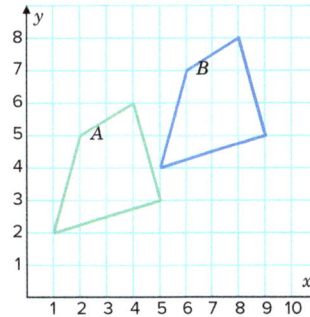

Create a strategy

Identify the directions of the movements of the corresponding vertices and count the number of steps.

Apply the idea

We can see that the vertices moved 4 units right and 2 units up.

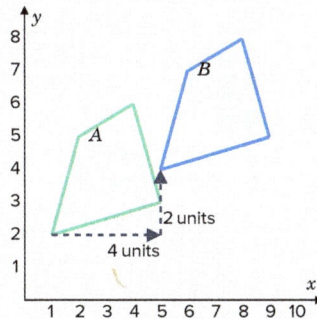

So, the translation is 4 units right and 2 units up.

Reflect and check

We count 4 units right and 2 units up from any point on A to any point on B.

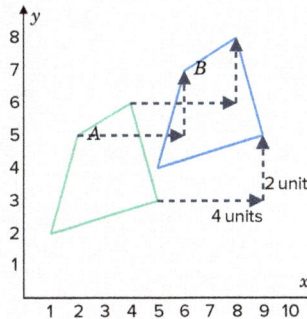

Purpose

Show students how to identify the direction and number of steps in a translation.

Reflecting with students

The translation of 4 units right and 2 units up can be thought of as $(+4, +2)$ to each of the vertices of the preimage. Writing out the vertices of quadrilateral A, we see:

$A = (1, 2), (5, 3), (4, 6), (2, 5)$

$\quad = (1 + 4, 2 + 2), (5 + 4, 3 + 2), (4 + 3, 6 + 2), (2 + 4, 5 + 2)$ Add $(+4, +2)$

$B = (5, 4), (9, 5), (7, 8), (6, 7)$ Simplify to get the coordinates of B

Students can use the coordinates of the preimage and the ordered pair representing their translation as a way to check if the image corresponds to their thinking.

Example 3

Quadrilateral $PQRS$ is to be translated 3 units to the left and 2 units down.

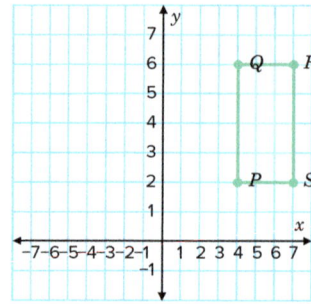

a   Determine the coordinates of vertex P' of the translated shape.

Create a strategy

Vertex P needs to be translated 3 units to the left and 2 units down.

Apply the idea

Count 3 units to the left and 2 units down from P to form P'.

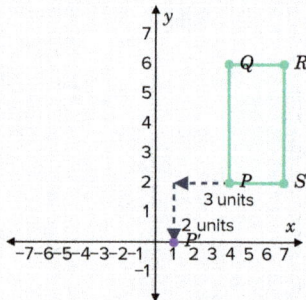

Reflect and check

We can check the ordered pair by subtracting 3 units from the x-coordinate and subtracting 2 units from the y-coordinate.

Vertex P is at (4, 2), so vertex P' is at $(4 - 3, 2 - 2) = (1, 0)$

Vertex P' is at (1, 0).

Purpose

Show students how to determine the coordinates of a translated point on a graph.

Students: Page 377

b   Sketch the original and translated quadrilateral $PQRS$ and $P'Q'R'S'$ on the same coordinate plane.

Create a strategy

Every point on quadrilateral $PQRS$ is translated 3 units to the left and 2 units down.

Apply the idea

Count 3 units to the left and 2 units down for each vertex on $PQRS$ to form $P'Q'R'S'$.

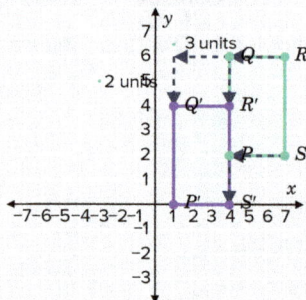

Reflect and check

We can check each ordered pairs by subtracting 3 units from each x-coordinate and subtracting 2 units from each y-coordinate.

Vertex Q is at (4, 6), so vertex Q' is at $(4 - 3, 6 - 2) = (1, 4)$

Vertex R is at (7, 6), so vertex R' is at $(7 - 3, 6 - 2) = (4, 4)$

Vertex S is at (7, 2), so vertex S' is at $(7 - 3, 2 - 2) = (4, 0)$

Purpose

Challenge students to apply their understanding of translations to sketch the original and translated shapes on the same plane.

🎓 **Advanced learners: Explore coordinate rules**   use with Example 3
Targeted instructional strategies

Encourage students to derive the algebraic rules that govern translations on the coordinate plane by examining how each coordinate changes. Instead of relying solely on counting units left or down, prompt students to express the translation as an operation on the coordinates: moving 3 units left and 2 units down transforms any point (x, y) to $(x - 3, y - 2)$. Ask students to justify why this rule applies to all points in the figure and how it reflects the movement on the plane.

Students: Page 377–378

Example 4

You're helping a friend navigate a theme park using a map on a coordinate plane. The starting point is the fountain, located at $A(1, 2)$. The goal is to reach three attractions in a specific order: the roller coaster, the Ferris wheel, and the food court.

a From the fountain, walk 5 units to the right and 3 units down to reach the roller coaster. Identify the coordinates of the roller coaster.

Create a strategy

First plot the fountain at (1, 2) on the coordinate plane. Then translate it right 5 units and down 3 units.

Apply the idea

Plot the point of the fountain on the graph.

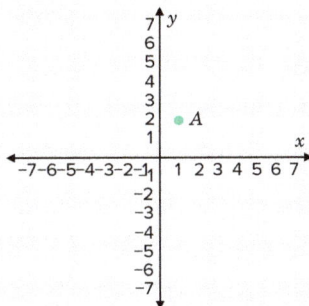

Then, translate point A to the right 5 units and down 3 units.

Name the point of the roller coaster point B.

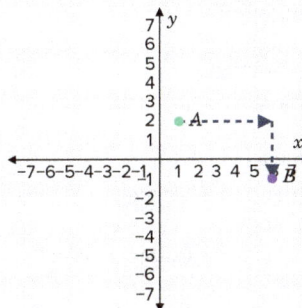

Point B, the location of the roller coaster, is at (6, −1).

Reflect and check

We can check the ordered pairs by adding 5 units to the x-coordinate and subtracting 3 units from the y-coordinate.

The fountain is at (1, 2), so the roller coaster is at $(1 + 5, 2 - 3) = (6, -1)$.

Purpose

Show students how to navigate on a coordinate plane and understand the concept of translation in a real world context.

b From the roller coaster, walk 4 units to the left and 6 units up to reach the Ferris wheel. Identify the coordinates of the Ferris wheel.

Create a strategy

Translate the ordered pair of the roller coaster left 4 units and up 6 units.

Apply the idea

Translate point B, the roller coaster, to the left 4 units and up 6 units.

Name the point of the Ferris wheel point C.

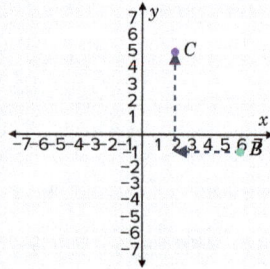

Reflect and check

We can check the ordered pairs by subtracting 4 units from the x-coordinate and adding 6 units to the y-coordinate.

The roller coaster is at (6, –1), so the roller coaster is at $(6 - 4, -1 + 6) = (2, 5)$.

Point C, the location of the Ferris wheel, is at (2, 5)

Purpose

Show students how to apply translation on a coordinate plane in a real-world context.

c From the Ferris wheel, walk 3 units to the right and 2 units down to reach the food court. Identify the coordinates of the food court.

Create a strategy

Translate the ordered pair of the Ferris wheel right 3 units and down 2 units.

Apply the idea

Translate point C, the Ferris wheel, to the right 3 units and down 2 units.

Name the point of the food court point D.

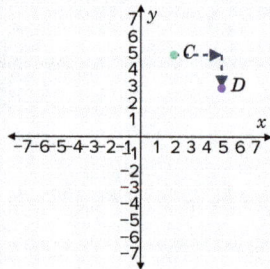

Reflect and check

We can check the ordered pairs by adding 3 units to the x-coordinate and subtracting 2 units from the y-coordinate.

The Ferris wheel is at (2, 5), so the roller coaster is at $(2 + 3, 5 - 2) = (5, 3)$.

Point D, the location of the food court, is at (5, 3)

Purpose

Help students understand how to translate points on a coordinate plane and determine new coordinates.

d   Describe the path taken from the fountain to the food court if the friend decides to avoid the roller coaster and the Ferris wheel.

Create a strategy

Plot the points of the fountain and the food court on a graph. Calculate the distance needed to travel to get from the fountain to the food court on the graph.

Apply the idea

Plot the points of the fountain, point A, and the food court, point D.

Count the horizontal and vertical distance from point A, the fountain, to point D, the food court.

To get from the fountain at $A(1, 2)$ to the food court at $D(5, 3)$ the friend needs to travel right 4 units and up 1 unit.

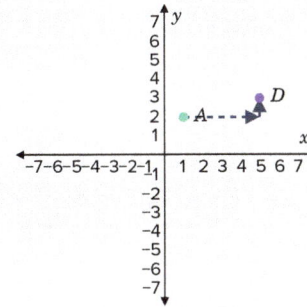

Reflect and check

Since we already knew the ordered pairs of the fountain and the food court, we could have found the difference between the points to calculate the translation.

The x-coordinate of the food court is 5 and the x-coordinate of the fountain is 1.

The difference between 5 and 1 is a positive 4. So, the friend needs to travel 4 units to the right.

The y-coordinate of the food court is 3 and the y-coordinate of the fountain is 2.

The difference between 3 and 2 is a positive 1. So, the friend needs to travel 1 unit up.

Therefore, the complete translation is 4 units to the right and 1 unit up.

Purpose

Demonstrate to students how to plot points on a coordinate plane and calculate the distance between them.

Step-by-step guide to translations
Student with disabilities support

use with Example 4

To support students in understanding translations in the coordinate plane, begin by introducing the concept with a labeled diagram of a coordinate plane showing a shape and its translated image. Explain that a translation slides a shape without changing its size, shape, or orientation.

Provide a step-by-step guide to translating a shape in the coordinate plane:

1. Identifying the vertices of the shape

2. Count the units it moves horizontally (left/right) and vertically (up/down)

   - Use sentence starters such as "The shape moves ⬚ units to the ⬚."

3. Repeat the first two steps for each point on the shape.

4. Draw lines connecting the points of the image and label the points using prime notation.

Students: Page 379

💡 **Idea summary**

To identify and describe a translation:

1. Look for corresponding points on the object

2. Identify if the object has moved horizontally or vertically, and then describe that movement as left/right or up/down

3. Count the number of places (or units) that the object has moved

---

Practice

Students: Page 380–386

What do you remember?

1 The point indicated on the coordinate plane is translated 4 units up. Determine the coordinates of its new position.

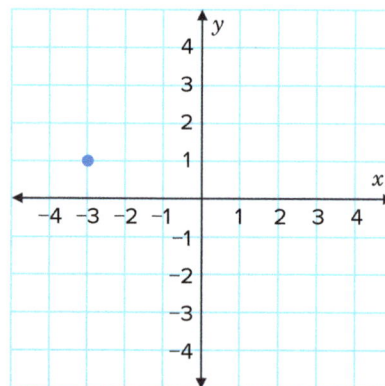

2 Determine the coordinates of each of the following translations of the point $A(3, -4)$:

a Point A is translated 5 units to the right. **b** Point A is translated 5 units to the left.

c Point A is translated 5 units upward. **d** Point A is translated 5 units downward.

---

3 Plot the following translations:

a The point (−8, 9) is translated 18 units to the right.

b The point (−6, 5) is translated 12 units to the right and 7 units down.

c The point (6, −2) is translated 15 units to the left and 6 units up.

d The point (8, 5) is translated 11 units to the left and 9 units down.

4 Determine whether each graph shows a dilation or a translation.

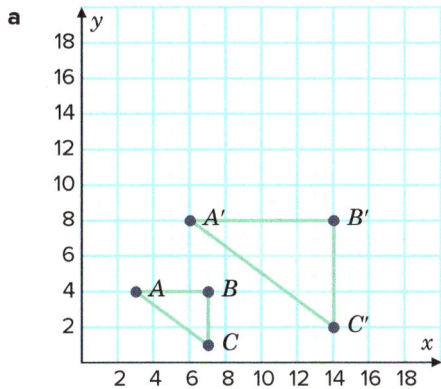

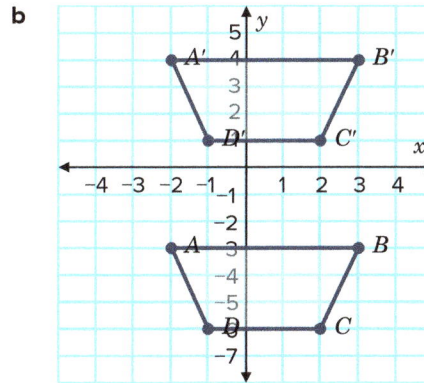

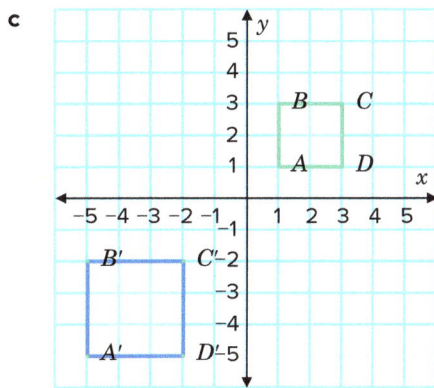

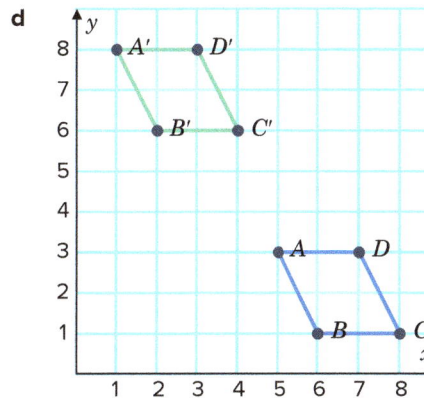

a

b

c

d

5 Identify whether the following statements are true or false:

a A translation will result in an image that is the same shape as its preimage.

b A translation will result in an image that has a different orientation than its preimage.

c A translation will result in an image that has congruent side lengths to its preimage.

d A translation on a preimage can result in an image that is a different size.

e A translation will move every point on a preimage the same distance and direction.

SOL **6** Which grid shows only a translation of the shaded polygon to create the unshaded polygon?

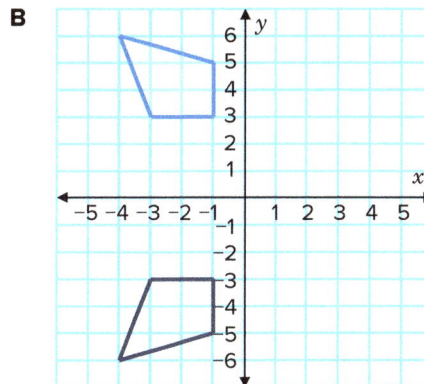

A

B

C

D

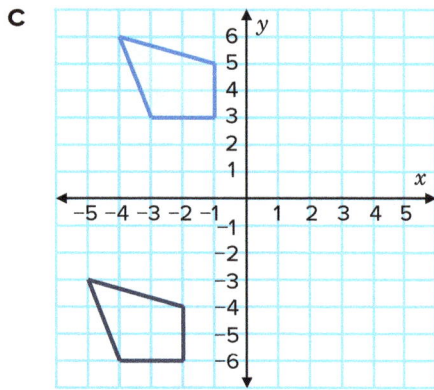

Let's practice

7 Describe the translation involved in each of the following:

   a Moving from square A to square B.

   b Moving from square A to square C.

   c Moving from square A to square D.

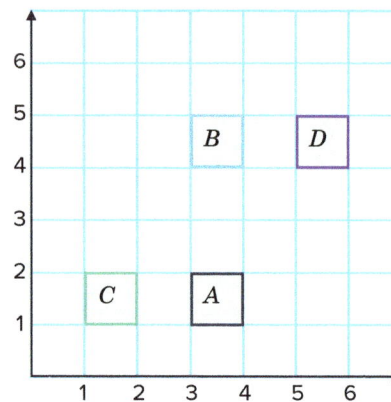

8 Determine the translation from polygon A to polygon B.

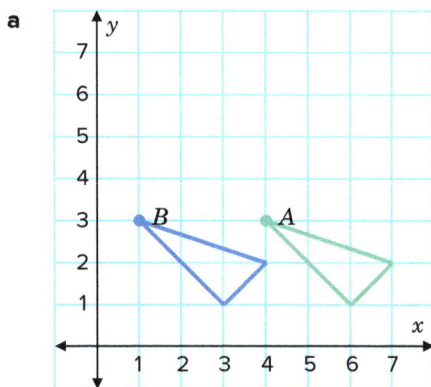

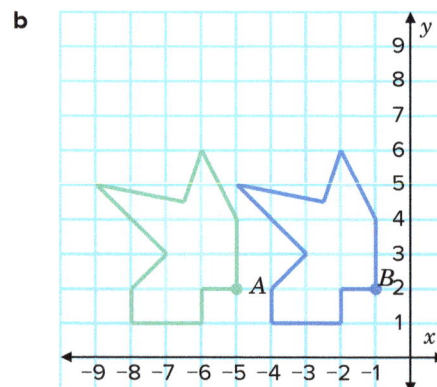

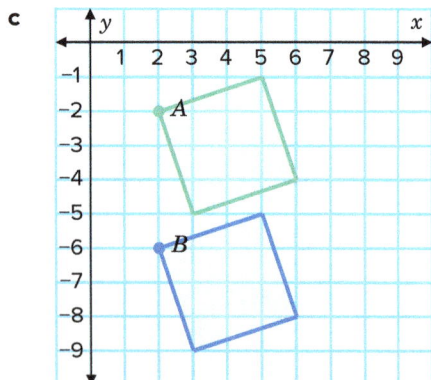

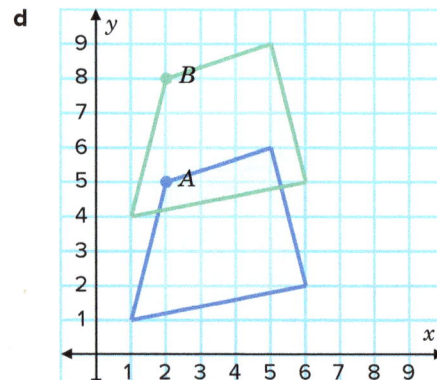

a

b

c

d

e

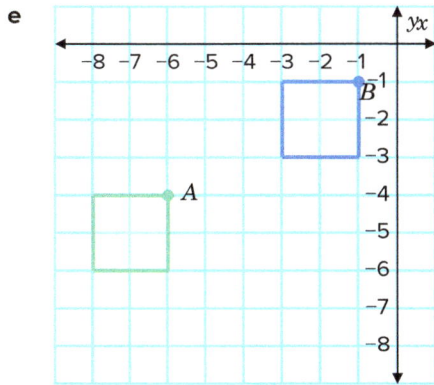

f

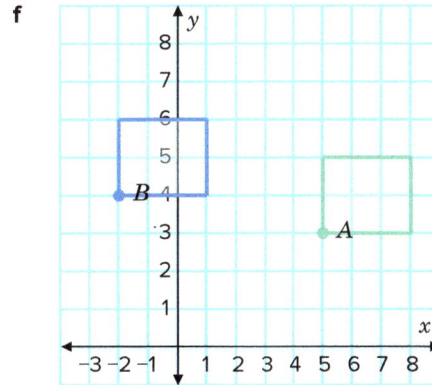

9 Quadrilateral $ABCD$ is to be translated 2 units up and 4 units to the left. Determine the coordinates of vertex A' of the translated shape.

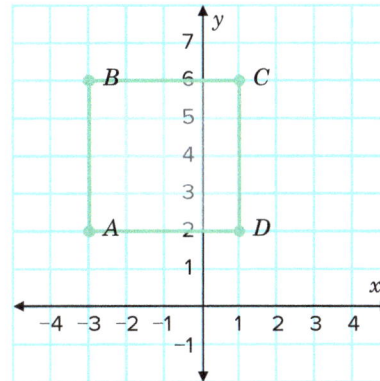

10 Triangle ABC is translated 4 units left and 2 units down. What could be the coordinates of B'?

A $(-1, 4)$ **B** $(-3, -1)$ **C** $(3, 2)$

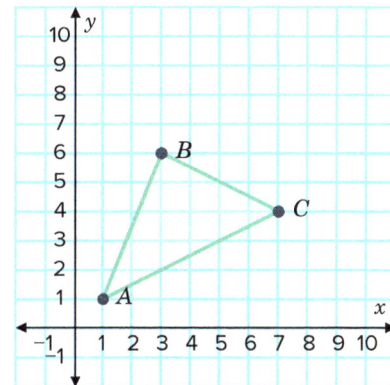

11 Translate the given flag 1 unit up and 2 units to the right on the coordinate plane:

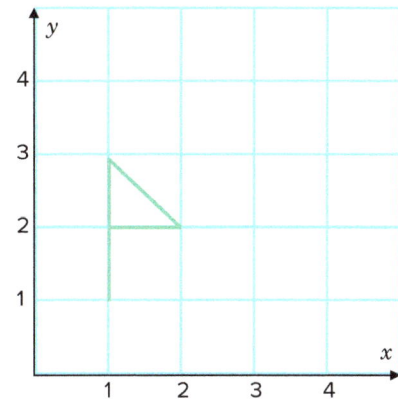

12 For each polygon, apply the given translation and graph the resulting image on the coordinate plane provided.

a Translate left 7 units

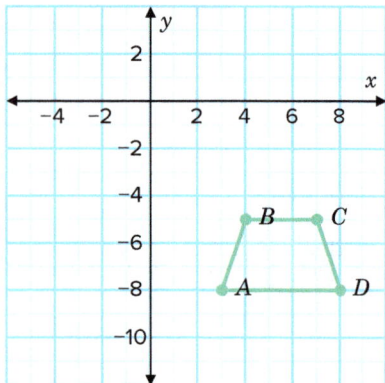

b Translate right 10 units

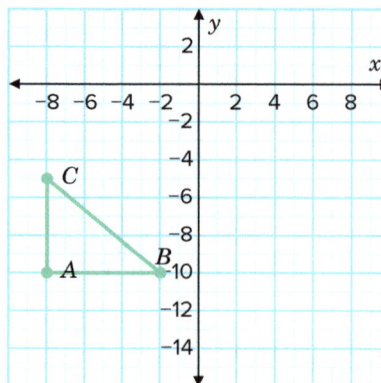

c Translate right 8 units

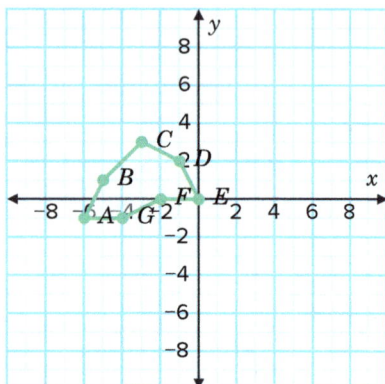

d Translate up 11 units

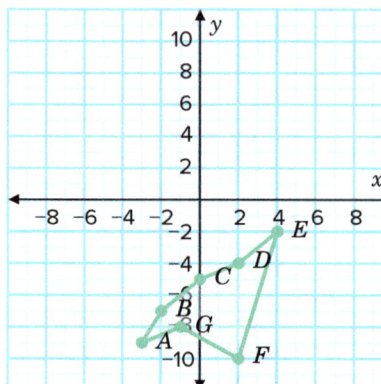

e Translate right 3 units and down 8 units

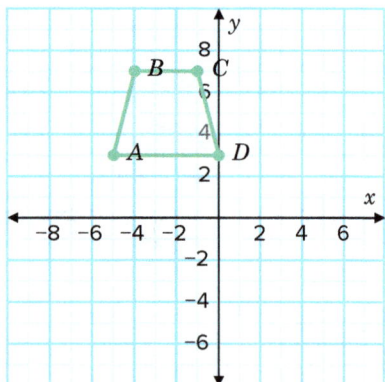

f Translate left 4 units and up 7 units

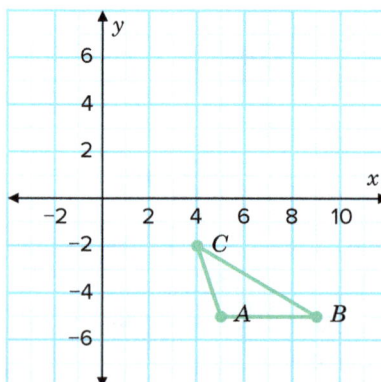

g Translate left 9 units and up 3 units

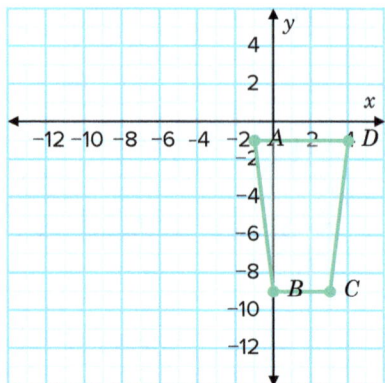

h Translate left 11 units and up 8 units

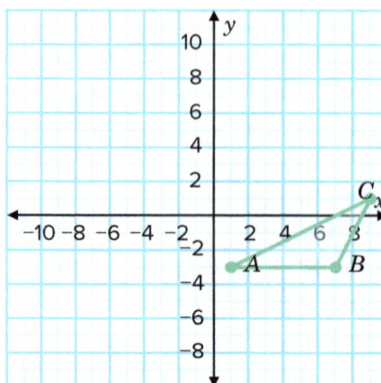

13 A triangle has vertices at (−2, 2), (−2, 5), and (−5, 2). If the triangle is moved 3 units to the right and 2 units up, what are the new coordinates of the vertices?

14 Draw the figure and its image after the translation.

  a Triangle LMN with vertices $L(4, 2)$, $M(8, 2)$, and $N(3, 5)$ is translated 1 unit left and 6 units up.

  b Square $ABCD$ with vertices $A(2, −2)$, $B(5, −2)$, $C(5, −5)$, and $D(2, −5)$ is translated 5 units right.

  c Trapezoid $EFGH$ with vertices $E(−6, 6)$, $F(−3, 6)$, $G(−2, 3)$, and $H(−7, 3)$ is translated 2 units right and 2 units up.

  d Rhombus $WXYZ$ with vertices $W(−2, −1)$, $X(−4, −1)$, $Y(−5, −7)$, and $Z(−2, −4)$ is translated 5 units left and 4 units down.

15 A polygon is translated twice. In the first translation, vertex D moves from (−4, −1) to (0, −2). The second translation is shown.

Are these the same translations? Why or why not?

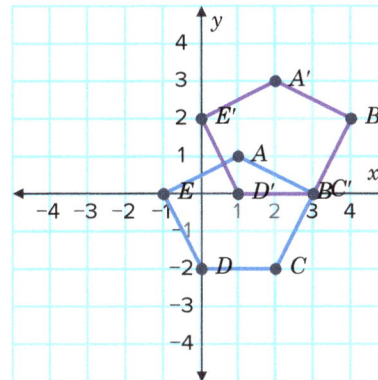

16 Is clicking and dragging an icon on a computer an example of a translation? Explain.

17 Imagine you have a treasure map, as shown in the diagram. The buried treasure chest is located at (−5, 3). You are currently standing at (2, −3).

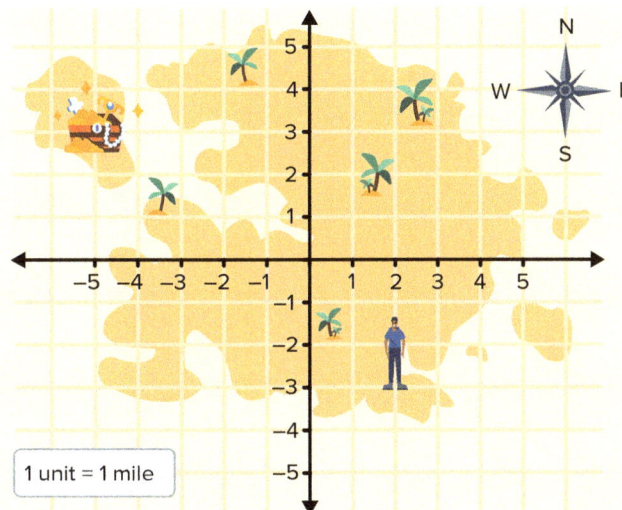

Describe a series of translations that will help you reach the buried treasure. Use the map key to describe the translations in context.

Let's extend our thinking

18 Point A' is located at (6, −2). Point A was translated 4 units left and 3 units up. What is the coordinate of point A?

19 Do translations always result in congruent figures? Explain.

20 In a game of chess, a knight can move around the board in an *L*-shape. That is, 1 unit in any direction and then 2 units in a perpendicular direction.

If a knight is located at square *E* now:

a How many possible places could it have been prior to its latest move?

b List all possible places it could have been prior to its latest move.

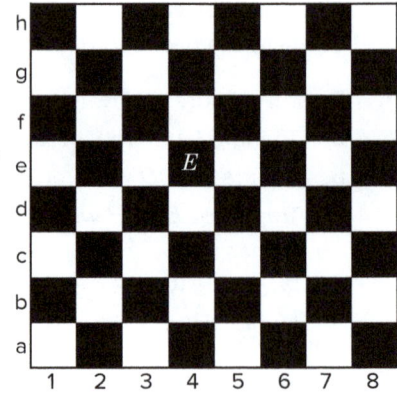

21 Riccardo has been creating a tower with building blocks.

Is it possible to use the same transformation mapping to place block *E* or block *N* at the highest point on the tower? If yes, describe the mapping. If not, explain how you know.

22 Follow these steps to create a picture:

   1. Translate square A down 2 units.
   2. Translate square B up 4 units.
   3. Translate hexagon C right 3 units.
   4. Translate hexagon D left 1 unit and down 1.5 units.
   5. Translate hexagon E left 1 unit and up 1.5 units.

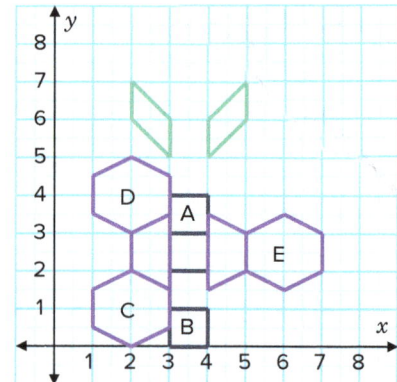

Answers

6.05 Translations in the coordinate plane

What do you remember?

1 (−3, 5)

2 a (8, −4) b (−2, −4) c (3, 1) d (3, −9)

3 a

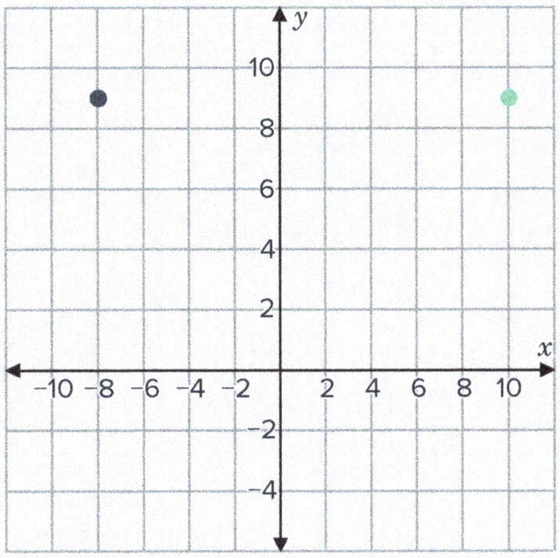

b

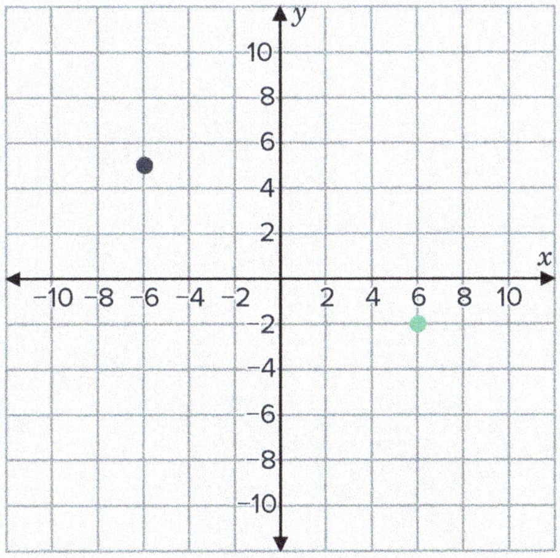

c

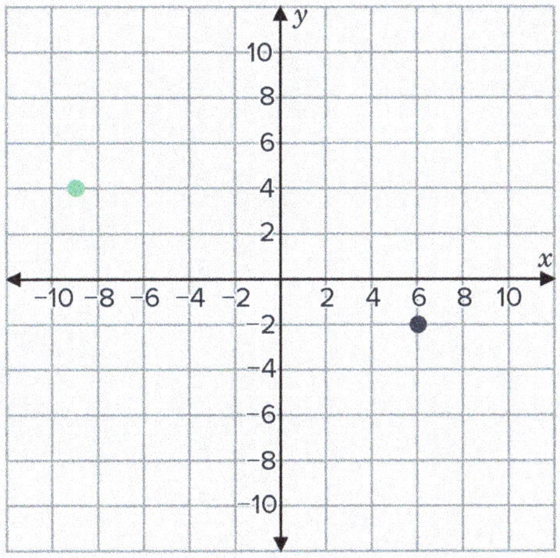

d

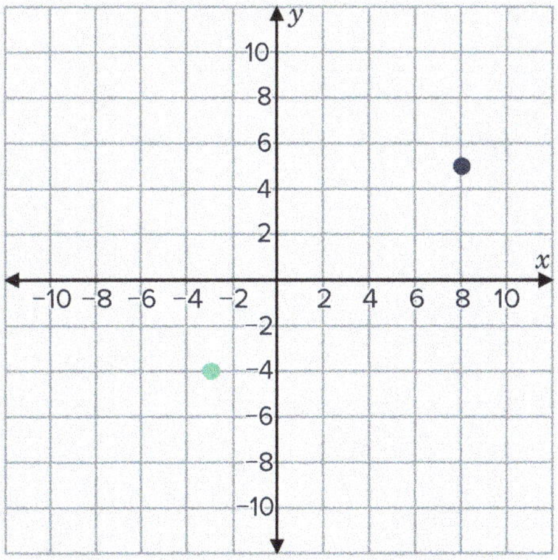

4 a Dilation b Translation
c Dilation d Translation

5 a True b False
c True d False
e True

6 C

Let's practice

7 a 3 units up
b 2 units to the left
c 3 units up and 2 units to the right

8 a 3 units left
b 4 units right
c 4 units down
d 3 units up
e 5 units right and 3 units up
f 7 units left and 1 unit up

9 (−7, 4)

10 A

11

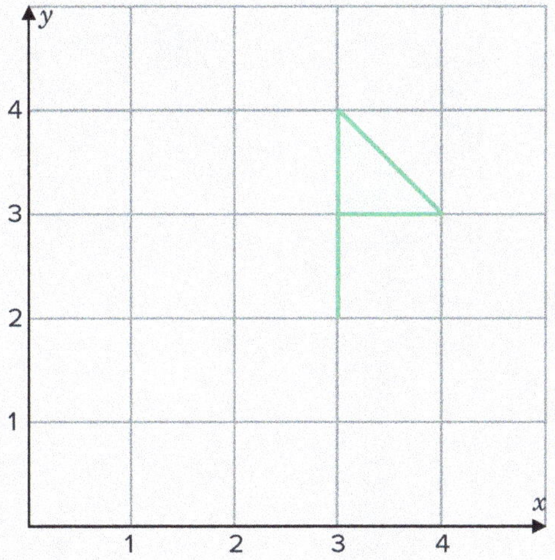

12 **a**

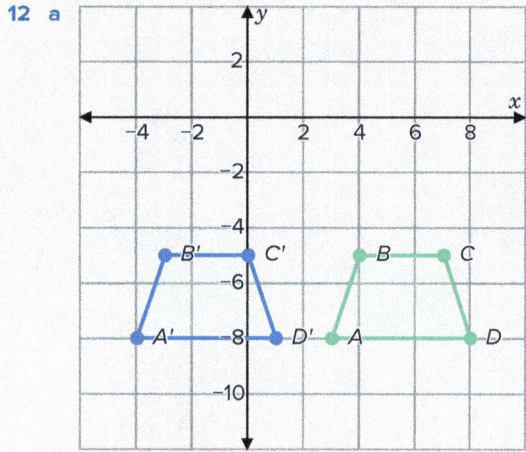

b

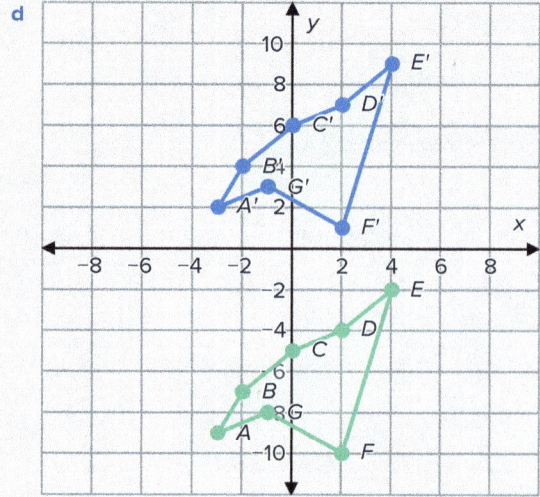

c

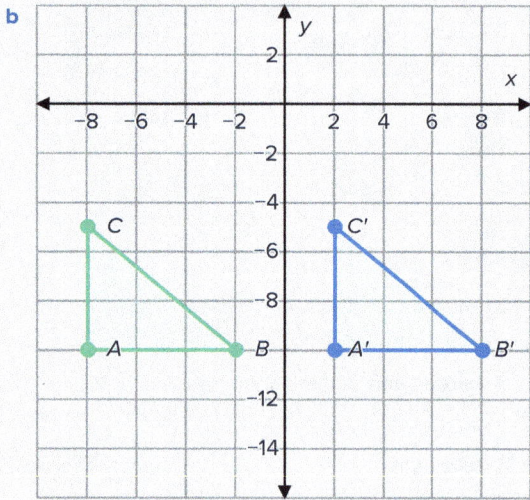

d

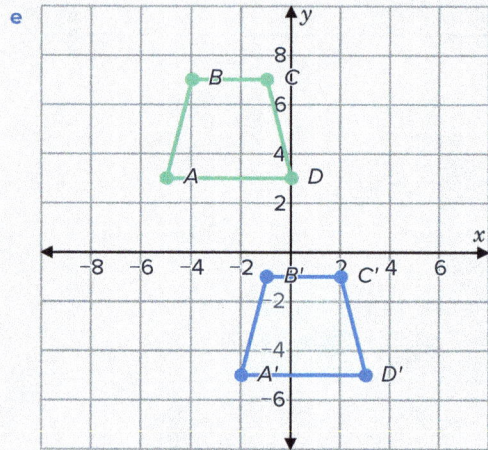

e

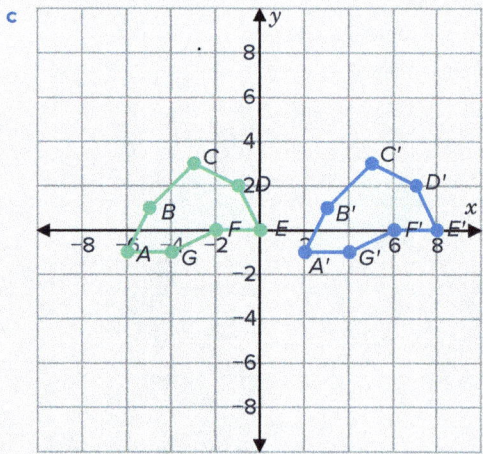

f

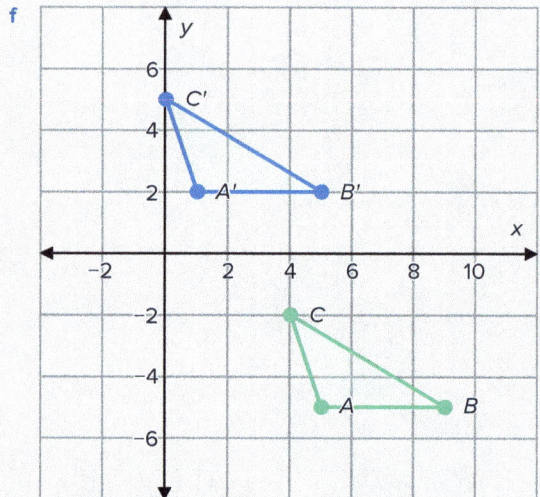

g

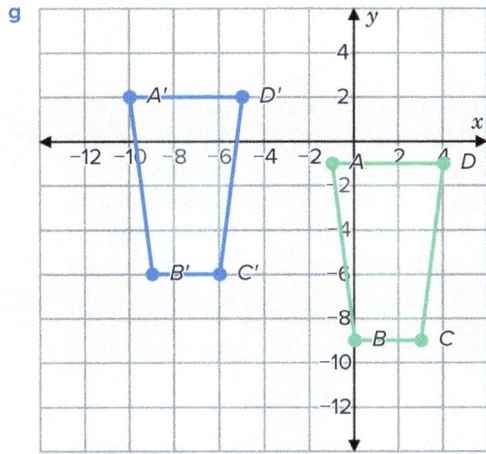

b

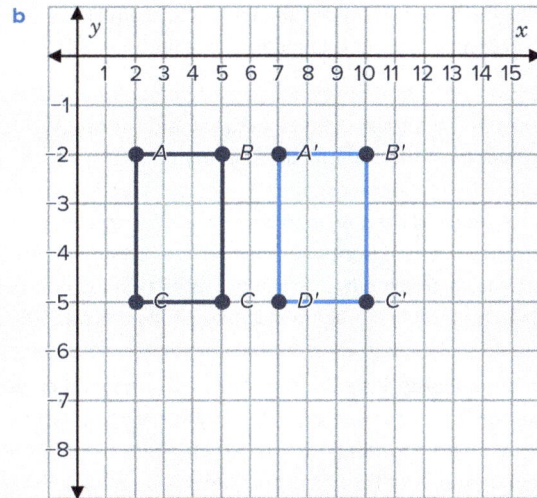

h

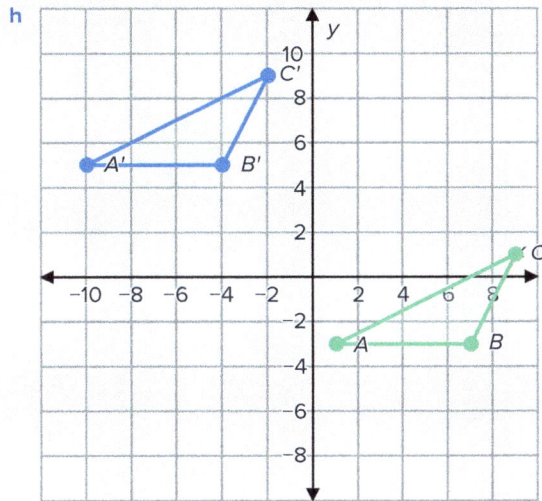

c

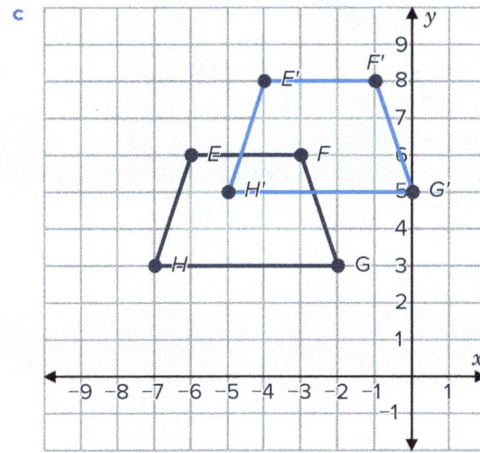

13 (1, 4), (1, 7), (−2, 4)

14 a

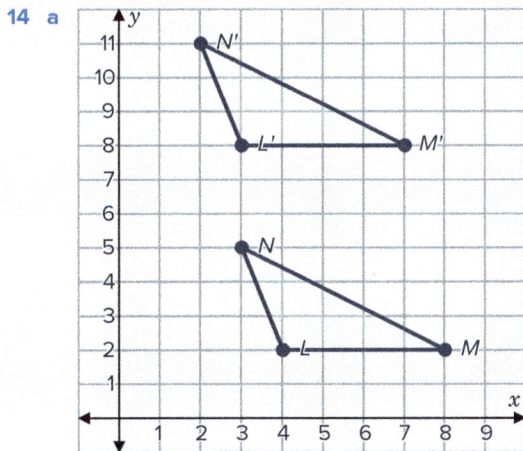

d

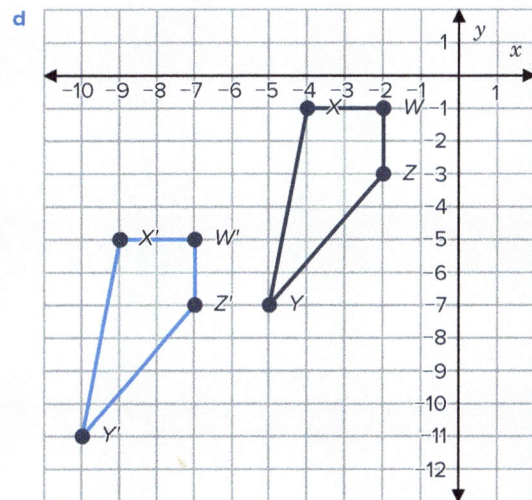

15 Yes, the two translations are the same. Both times, the polygon moved right 1 unit and up 2 units.

16 Clicking and dragging an icon on a computer is indeed an example of a translation in the geometric sense. In geometry, a translation moves every point of a figure or a space by the same distance in a given direction. When you click and drag an icon on a computer screen, you are moving the icon from one position to another without rotating it, flipping it, or changing its size. Every point of the icon moves the same distance and in the same direction, which is exactly what constitutes a translation.

In other words, if the icon started at one location on your screen (say, coordinate (x, y)) and you dragged it to a new location (to coordinate $(x + a, y + b)$), every point of the icon underwent the same horizontal change (a units) and vertical change (b units), which is consistent with the definition of a translation.

17 One possible series of translations:
1. Move left 7 units
2. Move up 6 units

In context, this means you would walk 7 miles west, then walk 6 miles north.

Let's extend our thinking

18 $A(10, -5)$

19 Yes, translations always result in congruent figures. A translation is a type of transformation that moves every point of a figure the same distance in the same direction. It does not change the size or shape of the figure. Because of this, the pre-image and the image are exactly the same shape and size; they are merely shifted to a different location on the coordinate plane.

20 **a** 8

b $c3, c5, d2, d6, f2, f6, g3, g5$

21 No. Block E and block N have the same vertical distance to the top of the tower, but different horizontal distances.

22

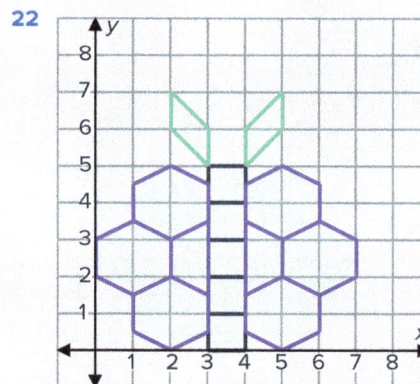

6.06 Reflections in the coordinate plane

Subtopic overview

Lesson narrative

In this lesson, students will explore reflections in the coordinate plane. They will learn that a reflection is a transformation that flips a preimage over a line, called the line of reflection, forming a mirror image where each point in the image is the same distance from the line of reflection as its corresponding point in the preimage. The lesson will focus on reflections over the x-axis and y-axis, highlighting how the coordinates of points change: reflecting over the x-axis changes the sign of the y-coordinate, and reflecting over the y-axis changes the sign of the x-coordinate. Students will practice identifying reflections and non-reflections, plotting points and their reflections over the axes, and reflecting geometric shapes such as trapezoids, hexagons, and triangles. The exploration activity involves students plotting and reflecting various shapes on a coordinate plane and checking their accuracy by ensuring the reflected points are equidistant from the line of reflection. This hands-on exploration reinforces the concept by having students apply the idea to multiple shapes and verify their work. By the end of the lesson, students should be able to accurately reflect shapes over the x-axis and y-axis and understand the properties of reflections in the coordinate plane.

Learning objectives

Students: Page 387

After this lesson, you will be able to...
- identify the coordinates of the image of a polygon that has been reflected over the x- or y-axis when given a preimage.
- sketch the image of a polygon that has been reflected over the x- or y-axis.
- identify and describe transformations in context.

Key vocabulary

- reflection

Essential understanding

A reflection preserves distance and angle measure, but reverses orientation. The preimage and the image of a figure are symmetric about a line of reflection.

Standards

This subtopic addresses the following Virginia 2023 Mathematics Standards of Learning standards.

Mathematical process goals

MPG1 — Mathematical Problem Solving

Teachers can integrate problem-solving by asking students to solve problems related to reflections in the coordinate plane. Students can be asked to identify the coordinates of the image of a polygon that has been reflected over the x-axis or the y-axis. Teachers can provide real-world problems, such as reflections of architectural structures or natural phenomena, and ask students to apply their understanding of reflections to solve these problems.

MPG2 — Mathematical communication

Teachers can integrate this process goal by encouraging students to communicate the transformations occurring as well as the impact of the transformations on the image's shape, size and position.

MPG4 — Mathematical Connections

Teachers can guide students to connect the concept of reflections to symmetry and congruence in geometry, deepening their understanding of how transformations preserve certain properties of figures. Teachers can encourage discussions on real-life applications of reflections, such as in mirror images, architectural designs, and art patterns, highlighting the relevance of mathematics in various fields. Connections to algebra can also be made by exploring how the coordinate changes during reflections relate to function transformations.

MPG5 — Mathematical Representations

Teachers can integrate mathematical representations by guiding students in representing the reflection process visually on the coordinate plane. They can show students how to sketch the preimage and the reflected image and how to identify their coordinates. Teachers can also encourage students to use symbolic representations, such as coordinate pairs, to represent the preimage and the image. They can also show students how to use these representations to understand and solve problems involving reflections.

Content standards

8.MG.3 — The student will apply translations and reflections to polygons in the coordinate plane.

8.MG.3b — Given a preimage in the coordinate plane, identify the coordinates of the image of a polygon that has been reflected over the x- or y-axis.

8.MG.3e — Sketch the image of a polygon that has been reflected over the x- or y-axis.

8.MG.3g — Identify and describe transformations in context (e.g., tiling, fabric, wallpaper designs, art).

Prior connections

6.MG.3 — The student will describe the characteristics of the coordinate plane and graph ordered pairs.

7.MG.4 — The student will apply dilations of polygons in the coordinate plane.

Future connections

G.RLT.3 — The student will solve problems, including contextual problems, involving symmetry and transformation.

Lesson Preparation

Suggested review

Depending on your students' level of prior knowledge, consider revisiting the following lessons:

Grade 6 — 7.06 Polygons in the coordinate plane

Grade 8 — 6.05 Translations in the coordinate plane

Tools

You may find these tools helpful:

- Dynamic geometry software
- Graph paper
- Transparency sheets
- Highlighter

Student lesson & teacher guide

Reflections

Students are given the definition of a reflection before engaging in an exploration where students examine how a line of reflection transforms a given polygon.

Exploration

Students: Page 387

> **Interactive exploration**
> Explore online to answer the questions
>
> ⊕ **mathspace.co**

Use the interactive exploration in 6.06 to answer these questions:

1. Start with a vertical line of reflection. What do you notice about the shape of preimage and image?
2. What do you notice about the size of the preimage and image?
3. What do you notice about the distance between the image and preimage to the line of reflection?
4. Change the line of reflection so it is horizontal? Are your observations still true?

Suggested student grouping: Small groups

In this exploration, students will manipulate a preimage triangle and a line of reflection on a Geogebra applet. They will observe and explore the properties of the image produced after the reflection. This activity helps students understand the concept of reflection in geometry and observe its properties such as maintaining the shape and size of the preimage and equal distance of the image and preimage from the line of reflection.

Ideal student responses

These ideal responses may differ from other correct student responses. Less formal responses can be connected with the more precise mathematical language presented here.

1. **What do you notice about the shape of preimage and image when the line of reflection is vertical?**

   The shape of the preimage and the image remain the same even after reflection. They are congruent.

2. **What do you notice about the size of the preimage and image?**

   The size of the preimage and the image are the same. The reflection does not change the size.

3. **What do you notice about the distance between the image and preimage to the line of reflection?**

   The distance between the image and the preimage to the line of reflection is the same. This is the property of reflection.

4. **Change the line of reflection so it is horizontal? Are your observations still true?**

   Yes, the observations still hold true. The shape and size of the image remain the same as the preimage, and the distance from the preimage and image to the line of reflection is still the same.

Purposeful questions

- Why does the shape and size of the preimage remain the same in the image after reflection?
- What would happen if we move the line of reflection closer or further from the preimage? How would it affect the position of the image?

Possible misunderstandings

- Students may reverse their definitions of horizontal and vertical when answering the questions, or misjudge the distance from the line of reflection if it is not a vertical or horizontal line. Display definitions for students, and encourage students to use the grid lines to help confirm distances.

Identifying and creating reflections
Targeted instructional strategies

Students will be reflecting polygons over the x-axis and y-axis, so a helpful exercise is to show students polygons that have already been reflected, and ask students to determine and generalize the rules to reflections.

Display a graph such as the one below and ask students to predict what key components of the graph were used to map A to A' in each case. Then ask students to answer each of the following questions, walking them through the visual on the graph after giving them time to respond.

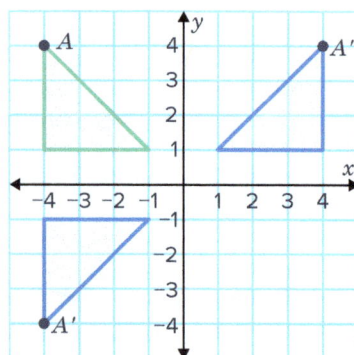

1. Count the number of units from A to A' for both directions.

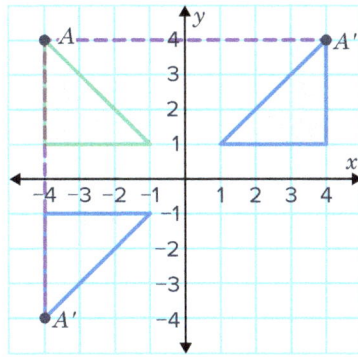

2. What similarities can we notice counting the distance between the points and the distance of the points to each axis?

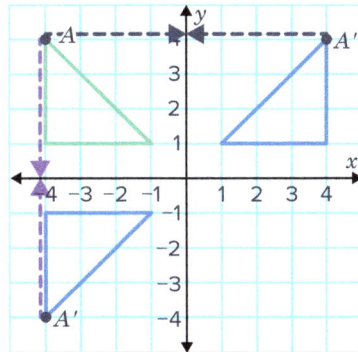

3. This type of transformation is called a *reflection*. How does the name connect to your observations from questions 1 and 2?

Collect and display
English language learner support

Begin by showing a triangle reflected over each axis and prompt students to observe changes in coordinates. For instance, a triangle at (2, 3), (4, 5), and (3, 1) becomes (2, −3), (4, −5), and (3, −1) when reflected over the x-axis, and (−2, 3), (−4, 5), and (−3, 1) over the y-axis.

Engage students in discussing how each type of reflection alters the coordinates, emphasizing that x-axis reflections change the sign of y-coordinates and y-axis reflections change the sign of x-coordinates.

Connect these reflections to real-world examples, such as their use in graphic design and architecture. Finally, have students articulate their understanding by writing a brief explanation, reinforcing precise mathematical language.

Use technology to visualize reflection
Student with disabilities support

Explicitly show students how to use technology to show the transformation in real-time, which can significantly aid students in understanding the concept. For example, explore transformations using $\triangle ABC$.

1. Before starting, it can be helpful to show the coordinate plane. To do this, right-click on the white space on the right side. Check Show Axes, and under the Show Grid, check Major Gridlines.

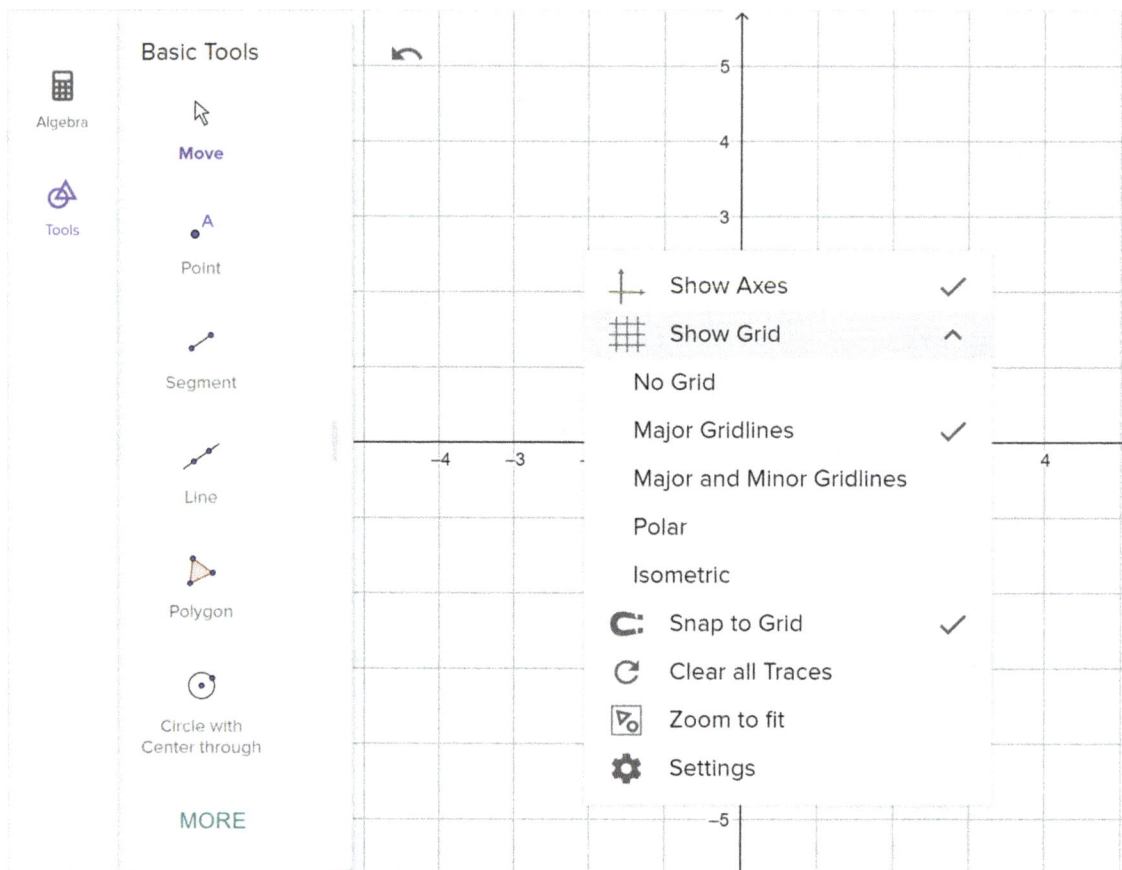

2. Use the Polygon tool to draw triangle ABC.

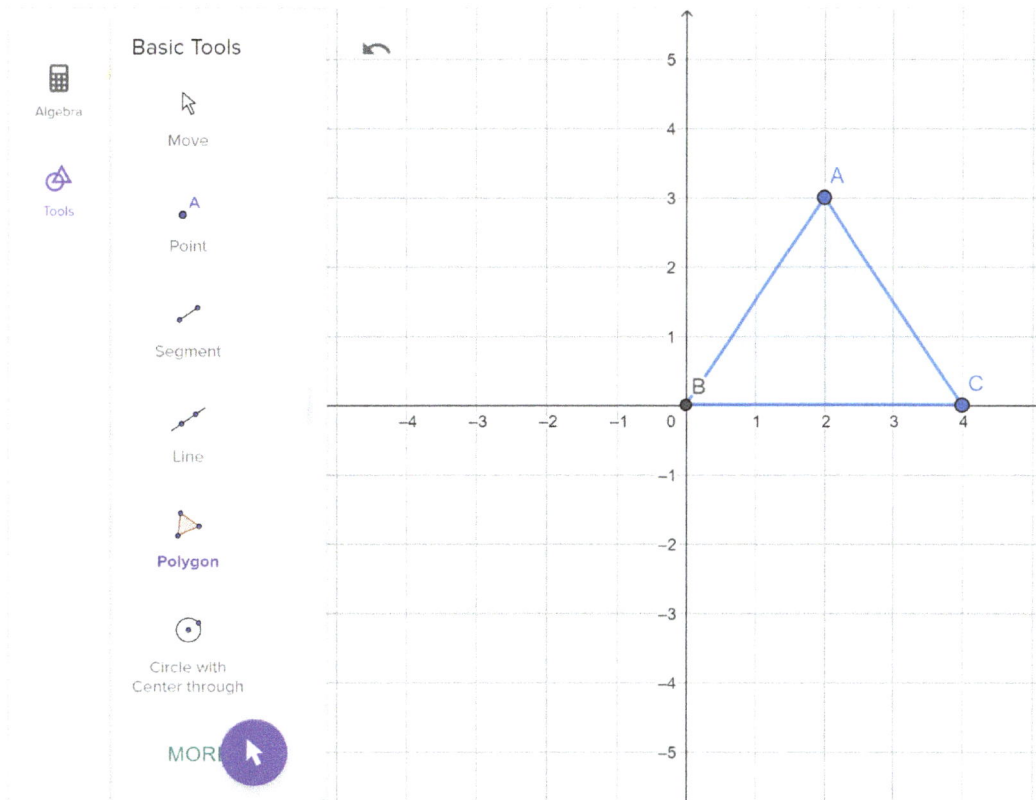

Basic Tools

Algebra

Tools

Move

Point

A

Segment

Line

Polygon

Circle with
Center through

MOR

3. To show a reflection of the triangle, use the Reflect about Line tool. However, before using the tool, create first a line that will serve as the line of reflection. Use the Line tool to create a line.

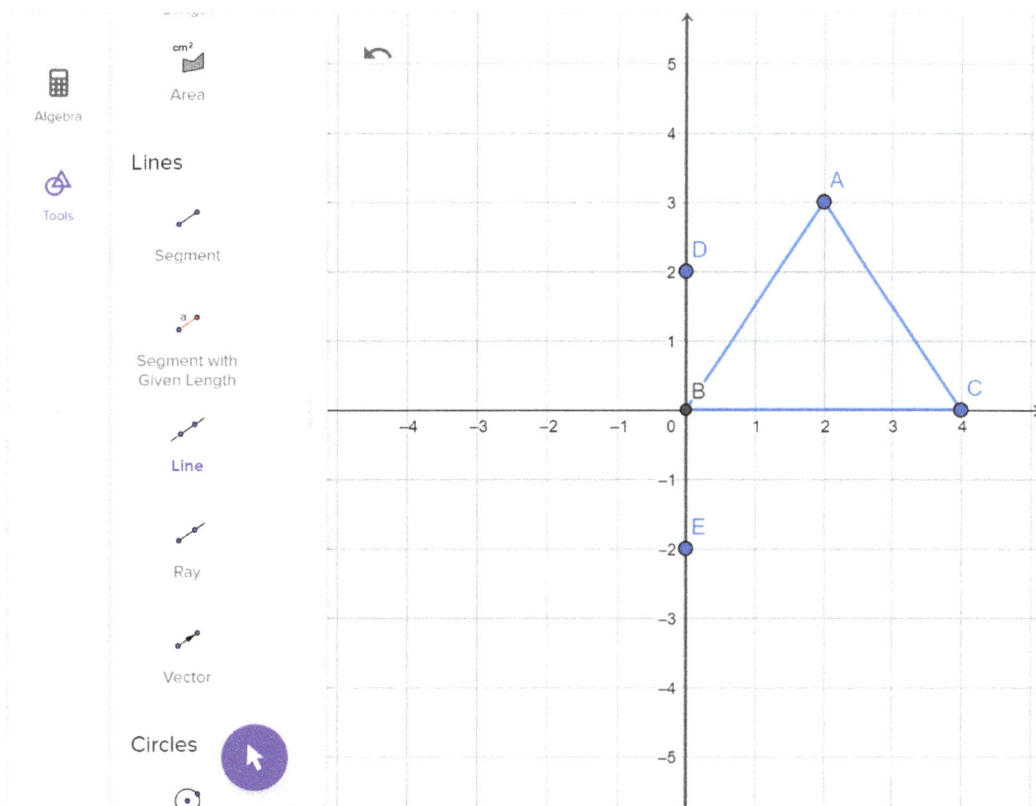

Area

Algebra

Lines

Tools

Segment

Segment with
Given Length

Line

Ray

Vector

Circles

4. Using the triangle formed in Step 2, reflect triangle ABC across line DE. Click on the Reflect about Line tool, select the triangle, and select the line. A resulting triangle A′B′C′ is shown.

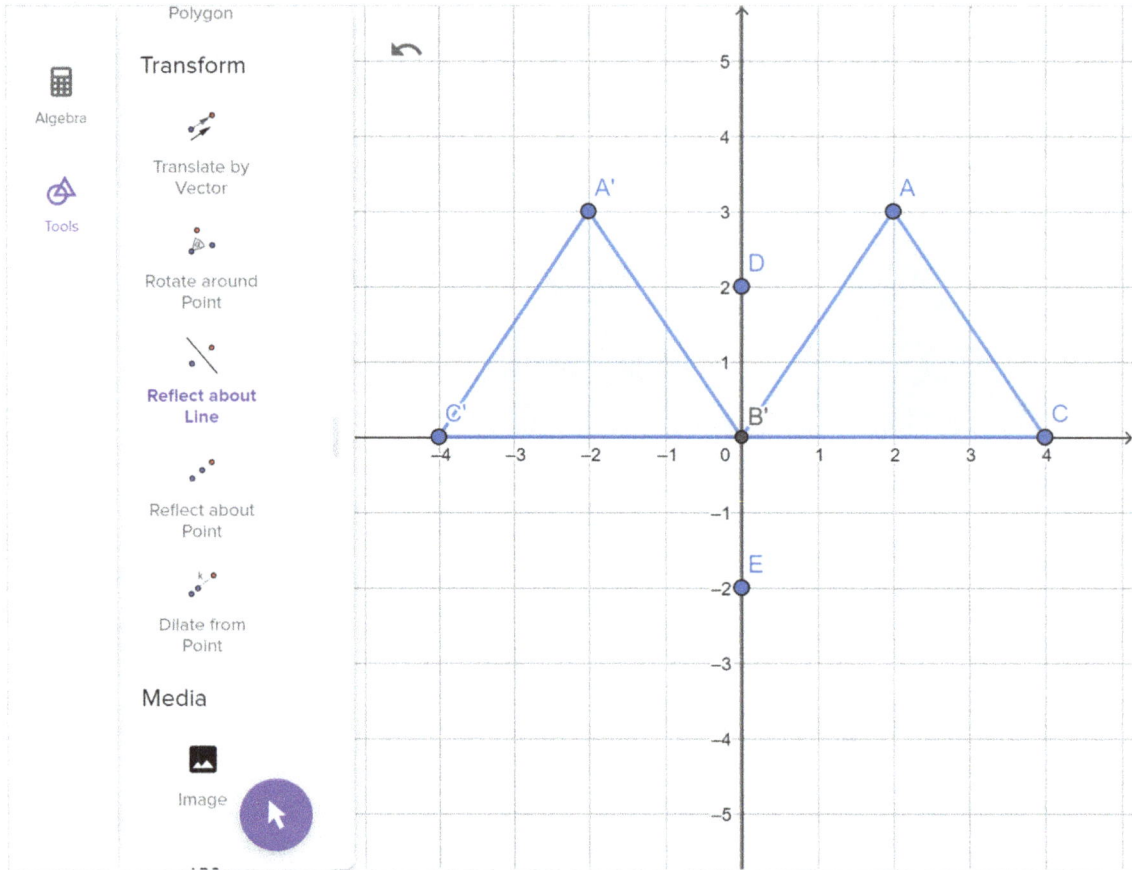

Reversing the x-axis and y-axis
Address student misconceptions

Students may reflect a polygon over the wrong axis, or assume that the coordinate that matches the axis is the coordinate that changes during the reflection. Students should notice that a reflection over each axis tells them which coordinate stays the same during a reflection.

Highlighting the axis helps students visualize the line of reflection. Student can also note that the axis that represents the line of reflection is the coordinate that does not change.

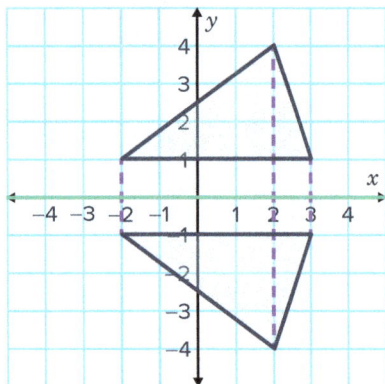

Reflection: x-axis

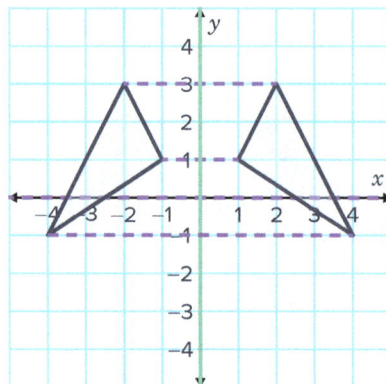

Reflection: y-axis

Students are given visual examples of reflections in the coordinate plane, and are shown how the coordinates of a reflected polygon change from the original preimage. They are then given rules for the coordinates of specific types of reflections.

Examples

Students: Page 388

Example 1

Which of the following does not show a reflection?

A  B 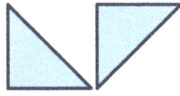 C  D

Create a strategy

Remember, a reflection is when the object flipped or reflected across a line.

Apply the idea

Which of object has not been flipped?

The correct answer is option B.

Purpose

Check that students can accurately identify a reflection as a transformation from a set of transformed figures.

Expected mistakes

Students may confuse a reflection with a rotation or turn. Remind students that in order for figures to be reflected, they must have a line of reflection that could be drawn somewhere between the figures.

Students: Page 388–390

Example 2

Trapezoid $ABCD$ is plotted on the coordinate plane shown.

Plot trapezoid $A'B'C'D'$, as a reflection of trapezoid $ABCD$ over the x-axis.

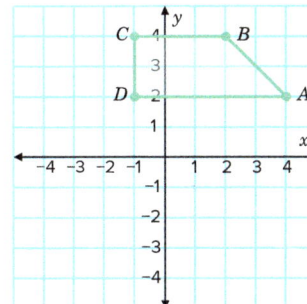

Create a strategy

A reflection over the x-axis creates an image on the other side of the x-axis where corresponding points on the pre-image and image are the same distance away from the x-axis.

Apply the idea

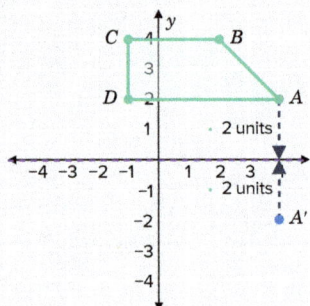

Point A is located 2 units above the x-axis, therefore A' will be 2 units below the x-axis.

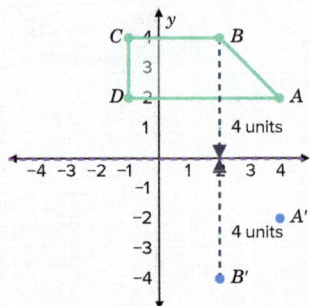

Point B is located 4 units above the x-axis, therefore B' will be 4 units below the x-axis.

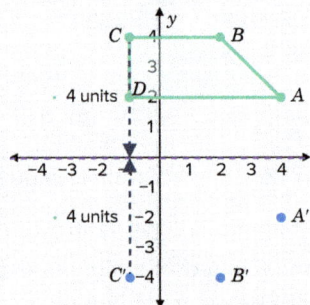

Point C is located 4 units above the x-axis, therefore C' will be 4 units below the x-axis.

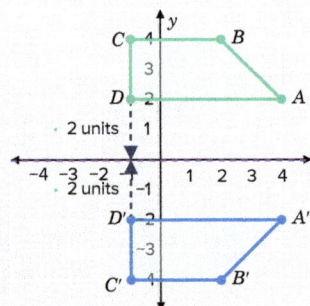

Point D is located 2 units above the x-axis, therefore D' will be 2 units below the x-axis.

Reflect and check

We can also look at how the signs of the x and y-coordinates change for a reflection over the x-axis.

When reflecting a point across the x-axis, the x-coordinate stays the same and the sign of the y-coordinate changes.

$A(4, 2)$'s reflection is $A'(4, -2)$.

$B(2, 4)$'s reflection is $B'(2, -4)$.

$C(-1, 4)$'s reflection is $C'(-1, -4)$.

$D(-1, 2)$'s reflection is $D'(-1, -2)$.

Purpose

To ensure students can reflect a polygon across an axis in a coordinate plane.

Advanced learners: Connecting reflections to coordinate rules
Targeted instructional strategies

Encourage students to explore the coordinate changes that occur when reflecting a figure over the x-axis. Have them create a table listing the coordinates of trapezoid $ABCD$ and its reflected image $A'B'C'D$. Prompt students to identify patterns between the original and reflected points. Encourage students to test their obeservations with other quadrilaterals.

Students might notice that the x-coordinates remain the same while the y-coordinates become their opposites. Ask them to explain why this occurs, guiding them to understand how reflecting over the x-axis affects the position of points on the coordinate plane. This connection between geometric transformations and algebraic rules deepens their comprehension of reflections.

Students: Page 390

Example 3

Plot the following.

a   Plot the polygon $PQRS$, where the vertices are $P(-3, -1)$, $Q(-2, 2)$, $R(0, 2)$, and $S(0, 0)$.

Create a strategy

Plot and connect the vertices.

Apply the idea

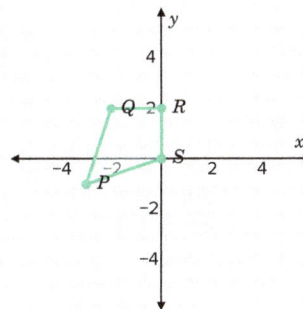

Purpose

This example allows students to practice plotting points on a graph and connecting vertices to form a polygon.

Students: Page 390–391

b   Plot the reflection of the polygon over the y-axis.

Create a strategy

A reflection over the y-axis creates an image on the other side of the y-axis where corresponding points on the pre-image and image are the same distance away from the y-axis.

Apply the idea

Point P is located 3 units to the left of the y-axis, so its reflection, P' will be 3 units to the right of the y-axis.

Point Q is located 2 units to the left of the y-axis, so its reflection, Q' will be 2 units to the right of the y-axis.

Point R on the y-axis, so it is 0 units from the y-axis. Its reflection, R' will be 0 units from the y-axis. This means it will also be on the y-axis.

Point S is located on the y-axis, so it is 0 units from the y-axis. Its reflection, S' will be 0 units from the y-axis. This means it will also be on the y-axis.

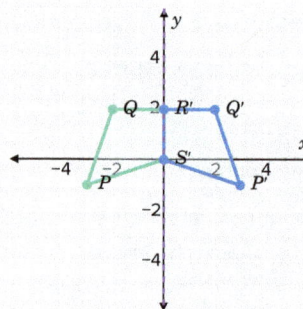

Reflect and check

We can also look at how the signs of the x-coordinates change for a reflection over the y-axis.

When reflecting a point across the y-axis, the y-coordinate stays the same and the sign of the x-coordinate changes.

$P(-3, -1)$'s reflection is $P'(3, -1)$.

$Q(-2, 2)$'s reflection is $Q'(2, 2)$.

$R(0, 2)$'s reflection is $R'(0, 2)$.

$S(0, 0)$'s reflection is $S'(0, 0)$.

Purpose

Show students how to plot the reflection of a polygon over the y-axis by demonstrating the concept of reflection and the rules for a reflection over the y-axis.

Reflecting with students

Encourage students to verify the accuracy of their reflected polygon by checking the congruency between the original and the reflected figures. Have them measure the lengths of corresponding sides and the measures of corresponding angles to confirm that they are equal. This process reinforces the understanding that reflections are rigid motions that preserve distance and angle measures.

Students: Page 391–393

Example 4

A fabric designer is creating a new pattern for a popular textile brand. The initial pattern is designed on a coordinate plane and consists of several geometric shapes arranged in a unique layout.

The designer plans to reflect the pattern across the y-axis to create a symmetrical design that will be printed on fabric.

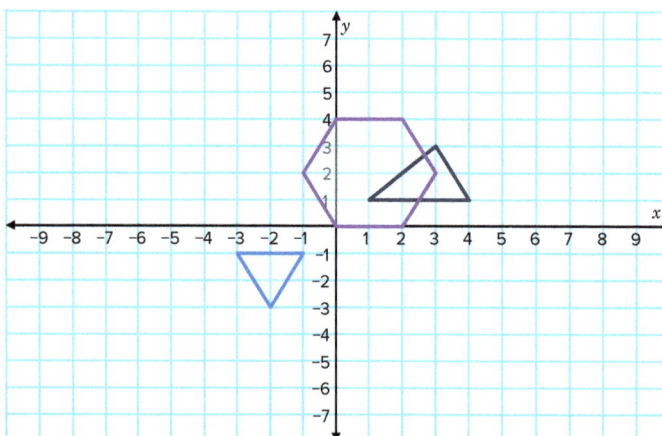

Draw the original pattern and its reflection across the y-axis on the same coordinate plane to visualize the complete design.

Create a strategy

When reflecting over the y-axis the points should be equidistant to the y-axis on either side.

The y-axis acts as the mirror line of the image and preimage.

Apply the idea

Start with the hexagon. Each point needs to be reflected across the y-axis and be the same distance from the y-axis as the original point.

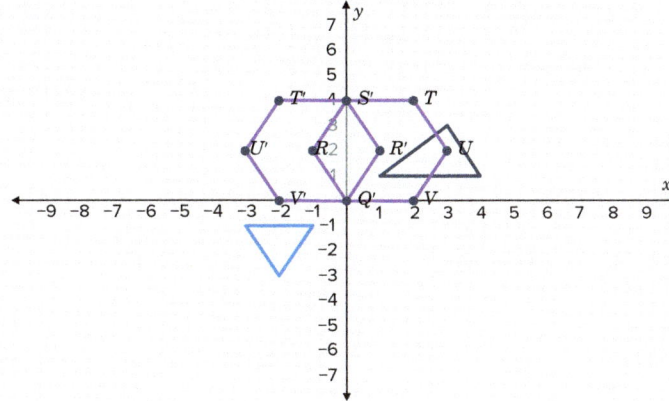

Next, reflect the equilateral triangle.

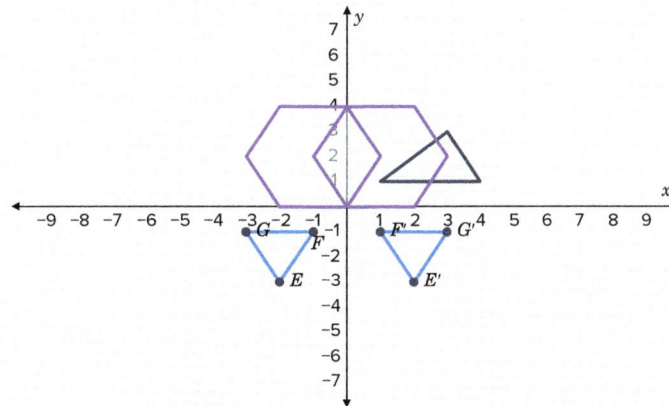

Then, reflect the right triangle.

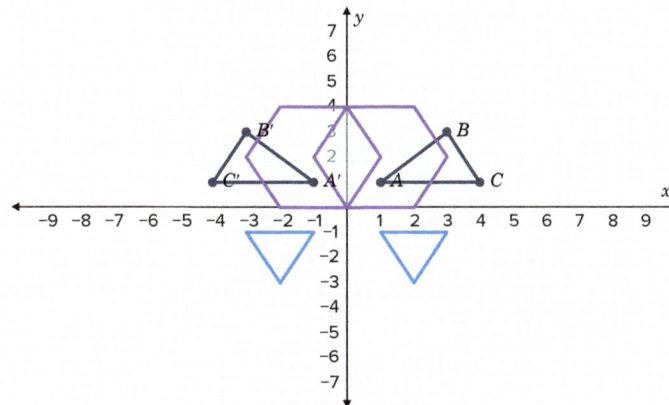

Finally, reflect the right triangle. Each point needs to be reflected to the other side of the y-axis and be equidistant to the y-axis as the original point.

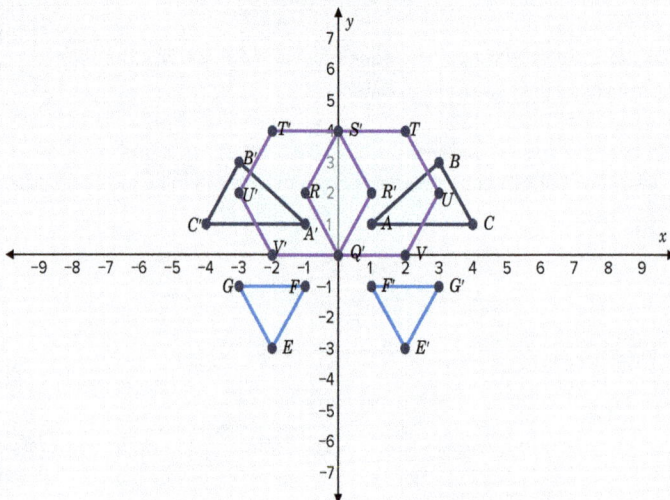

Reflect and check

We can check that each ordered pair has been reflected correctly by verifying that signs of the y-coordinates have changed while the signs of the y-coordinates have stayed the same.

For the hexagon:

(0, 0)'s reflection is '(0, 0).

(2, 0)'s reflection is (−2, 0).

(3, 2)'s reflection is (−3, 2).

(2, 4)'s reflection is (−2, 4).

(0, 4)'s reflection is (0, 4).

(−1, 2)'s reflection is (1, 2).

For the equilateral triangle:

(−1, −1)'s reflection is '(1, −1).

(−3, −1)'s reflection is (3, −1).

(−2, −3)'s reflection is (2, −3).

For the right triangle:

(1, 1)'s reflection is '(−1, 1).

(4, 1)'s reflection is (−4, 1).

(3, 3)'s reflection is (−3, 3).

Purpose

Show students how to reflect geometric shapes across an axis on a coordinate plane.

Physically transform graphs with transparency sheets   use with example 4
Student with disabilities support

To help students reflect polygons in the coordinate plane, students can use transparency sheets and grap paper to physically reflect given shapes.

Begin by drawing a simple shape, such as a triangle, on the grid paper and labeling its vertices with coordinates. Instruct students to place the transparent sheet over the grid paper. Have them use a dry-erase marker to trace the shape and then reflect it over the x-axis by flipping the sheet along the horizontal axis. Students should mark the new coordinates on the transparent sheet. Repeat the process for the y-axis.

For example, if reflecting triangle ABC with vertices at (2, 3), (4, 5), and (3, 1), students should trace and flip to see the reflection at (−2, 3), (−4, 5), and (−3, 1). This hands-on approach helps visualize how coordinates change during reflection. When the image may overlap the preimage, this strategy can be especially helpful for visualization.

> ### 💡 Idea summary
>
> A **reflection** is a transformation in which an image is formed by reflecting the preimage over a line called the line of reflection.
>
> Every point on the object or shape has a corresponding point on the image, and they will both be the same distance from the reflection line.
>
> For any figure reflected horizontally across the y-axis, the y-values of the coordinates will stay the same and the x-values will have opposite signs.
>
> For any figure reflected vertically across the x-axis, the x-values of the coordinates will stay the same and the y-values will have opposite signs.

Practice

What do you remember?

1 Choose the picture that shows a reflection.

   A 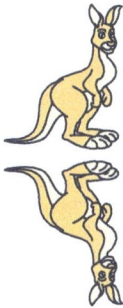   **B**    **C**

2 Choose the picture that shows a reflection.

   A  **B**    **C**    **D**

3 For each of the following:

- Plot point A on a coordinate plane.
- Plot point A', a reflection of point A across the x-axis.

   a $A(7, 3)$ **b** $A(-5, -6)$ **c** $A(2, -2)$ **d** $A(-4, 3)$

4 **a** Choose the transformation that is a reflection across the y-axis.

    A $P(-3, 4)$ is transformed to $P'(-4, 3)$ **B** $P(-3, 4)$ is transformed to $P'(-3, -4)$

    C $P(-3, 4)$ is transformed to $P'(3, 4)$ **D** $P(-3, 4)$ is transformed to $P'(4, -3)$

  b Choose the transformation that is a reflection across the x-axis.

    A $P(-3, 4)$ is transformed to $P'(-4, 3)$ **B** $P(-3, 4)$ is transformed to $P'(-3, -4)$

    C $P(-3, 4)$ is transformed to $P'(3, 4)$ **D** $P(-3, 4)$ is transformed to $P'(4, -3)$

5 Calculate the coordinates of the following points after being reflected over the y-axis.

  a $A(6, 9)$ **b** $A(9, -10)$ **c** $A(-1, -5)$ **d** $A(2, -10)$

6 A triangle is formed by the points $A(-2, -2)$, $B(1, -4)$ and $C(5, 2)$.

  a Plot the triangle on a coordinate plane.

  b Triangle ABC is reflected across the x-axis. The vertices of the reflected triangle are $A'(-2, 2)$, $B'(1, 4)$, and $C'(5, -2)$. Plot $\triangle A'B'C'$.

  c Compare the coordinates of the preimage and the image.

7 Determine whether each of the following statements are true or false:

  a When a figure is reflected across the x-axis, only the sign of the y-value changes.

  b When a figure is reflected across a vertical line, only the y-value changes.

  c When a figure is reflected, the preimage and image have the same size and shape.

  d When a figure is reflected, the preimage and image have the same orientation.

Let's practice

8 Describe each of the following reflections:

  a

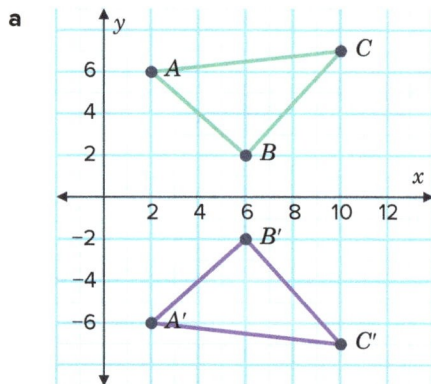

  b

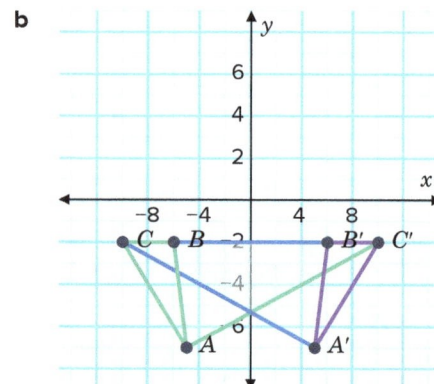

  c

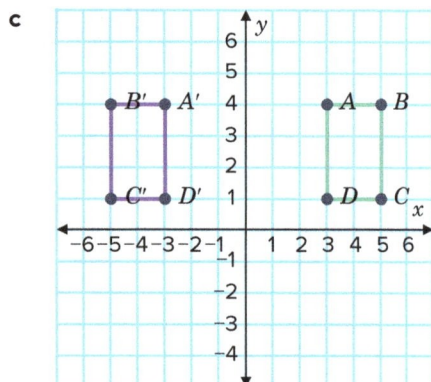

  d

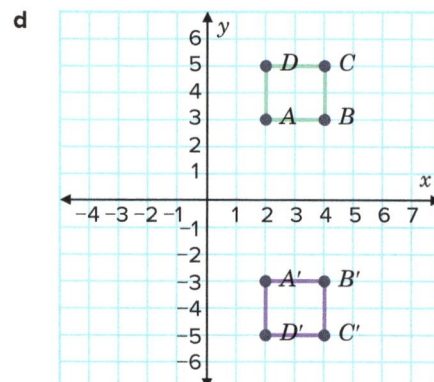

9   The quadrilateral shown is reflected across the x-axis.

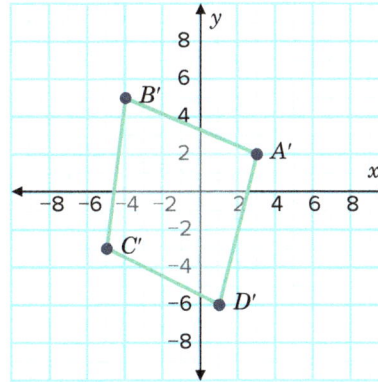

Select the graph of quadrilateral $A'B'C'D'$, which is the result of the transformation.

A

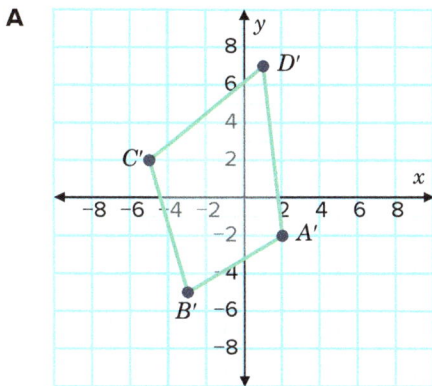

B

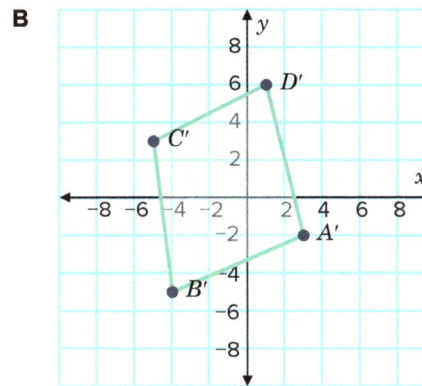

C

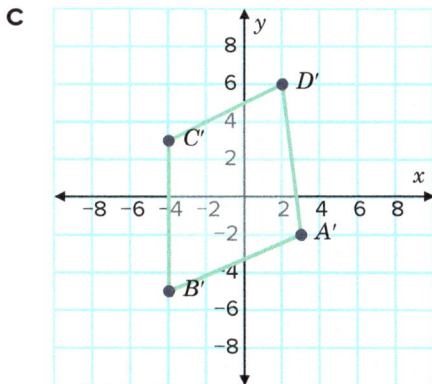

D

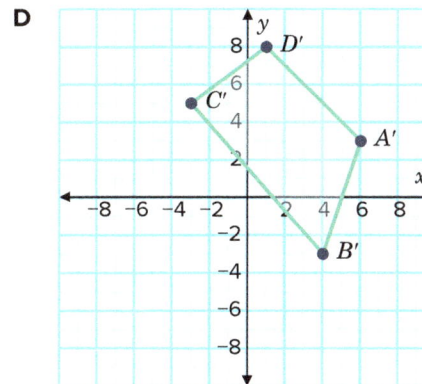

10   Describe both reflections that were applied to the preimage, in the correct order.

a

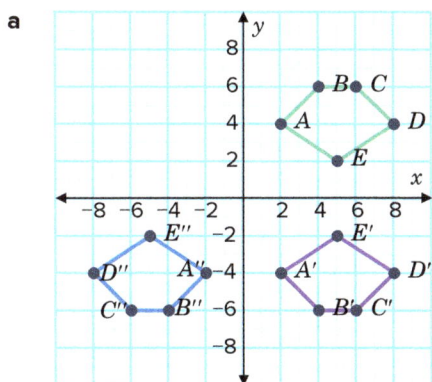

b

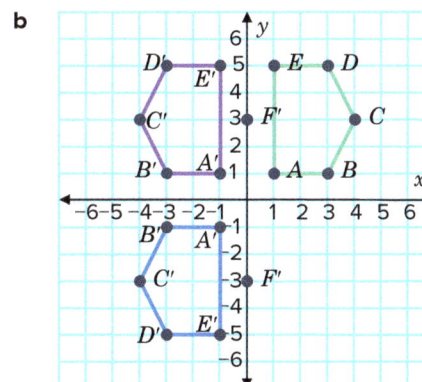

11 Which shape results when the purple triangle is reflected across the y-axis and then reflected across the x-axis?

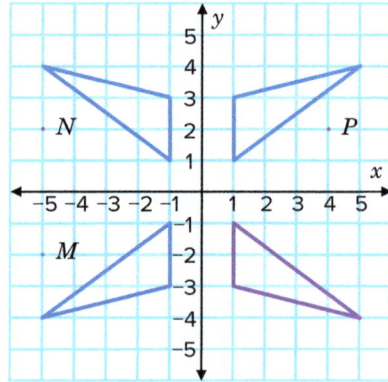

12 Consider the quadrilateral $ABCD$.

    a If quadrilateral $ABCD$ is reflected across the x-axis, find the coordinates of A'.

    b How many units would point A need to be translated to overlap point A'?

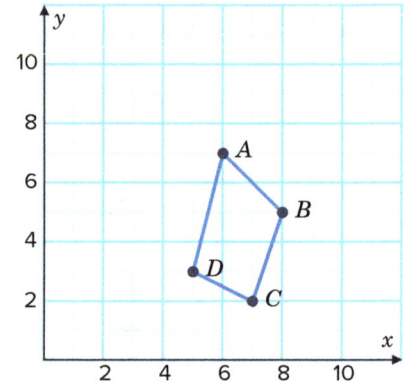

13 Graph the reflection of side AC across the y-axis.

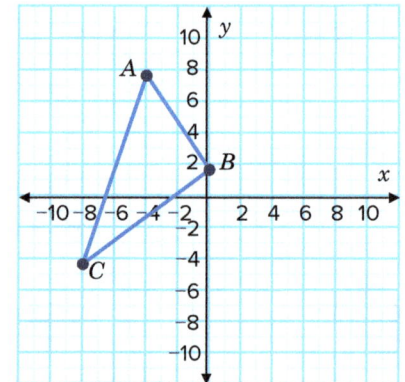

14 Reflect the following shapes over the y-axis:

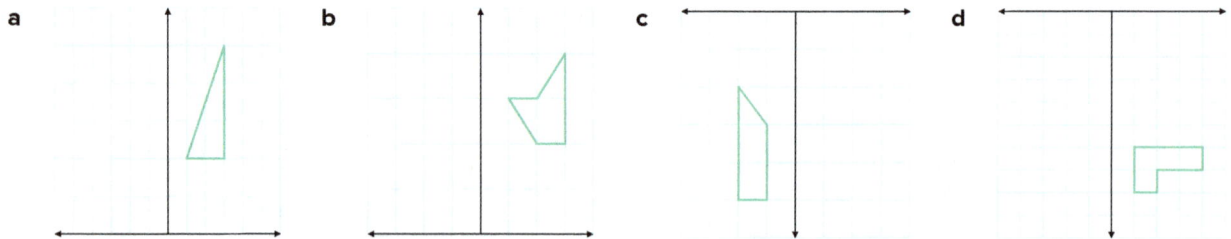

    a   **b**   **c**   **d**

15 Reflect the following shapes over the x-axis:

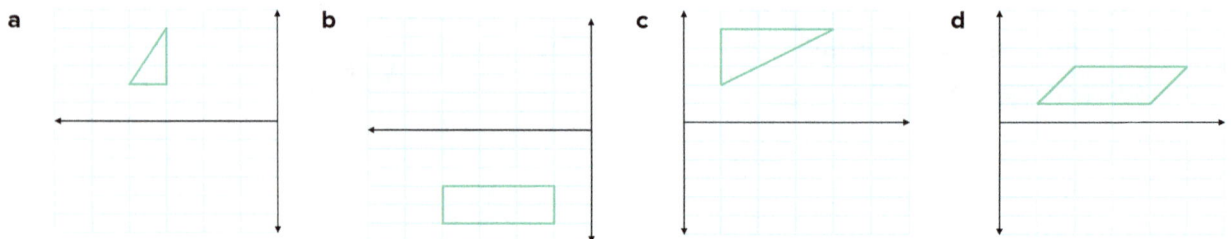

    a   **b**   **c**   **d**

16 A triangle has vertices at (2, 1), (2, 3), and (4, 1). What would be the coordinates of the vertices of the triangle after it is reflected over the x-axis?

17 A community in a bustling city has decided to redesign their urban garden to make it more symmetrical and visually appealing. The garden is originally laid out in the shape of a polygon on the coordinate plane, with vertices at points $A(0, 3)$, $B(3, 5)$, $C(3, 1)$, and $D(0, -1)$.

The redesign plan involves creating a mirror image of this garden across the y-axis to double its size and improve its visual appeal.

a Calculate the coordinates of the vertices of the polygon after it has been reflected across the y-axis. Label these new vertices A', B', C', and D'.

b Sketch the original and reflected polygons on the same coordinate plane.

18 An art teacher challenges their class to create a symmetrical mosaic design to be installed on a large wall at the school.

The students drafted their initial design on a coordinate plane, and they intend to reflect the design across the x-axis to create the full mosaic.

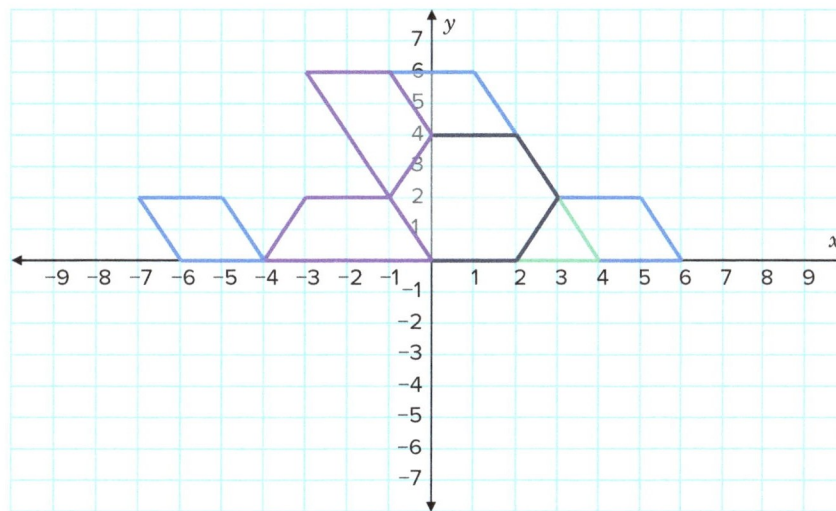

Draw the shapes and their reflections on the same coordinate plane to visualize the symmetrical mosaic design.

Let's extend our thinking

19 If a figure is reflected across a line and reflected across the same line again, will the reflection return to the original figure? Explain.

20 a What kind of transformation is applied to quadrilateral $ABCD$?

b Will the reflection result in congruent figures? Explain.

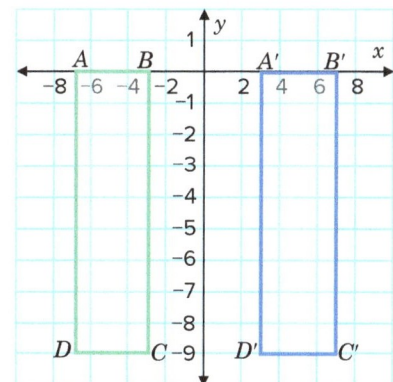

21 The following diagram shows reflections over line L:

a Find the value of the following:

    i a **ii** b **iii** c **iv** d **v** e

b Where will the reflection of point K lie?

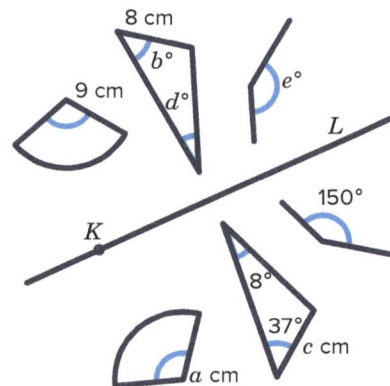

22 Sokayna drew figure A and Roly drew figure B.

a Can Sokayna's triangle be reflected to produce Roly's? If so, describe the reflection.

b Roly thinks his triangle has a greater perimeter than Sokanya's. Is this true? Explain why or why not.

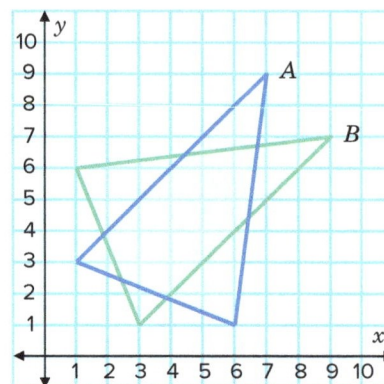

23 **a** Identify the capital letters in the English alphabet which can be reflected across a vertical or horizontal line and appear the same.

b Of the letters you found in part (a), identify which have:

    i A vertical line of symmetry

    ii A horizontal line of symmetry

    iii Both a vertical and a horizontal line of symmetry

Answers

6.06 Reflections in the coordinate plane

What do you remember?

1 A

2 C

3 a

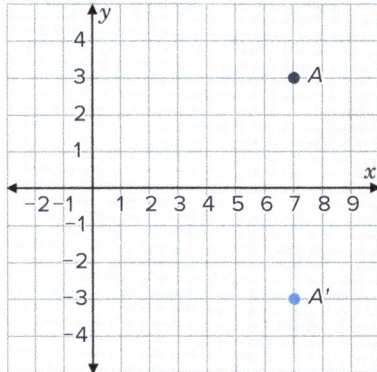

b

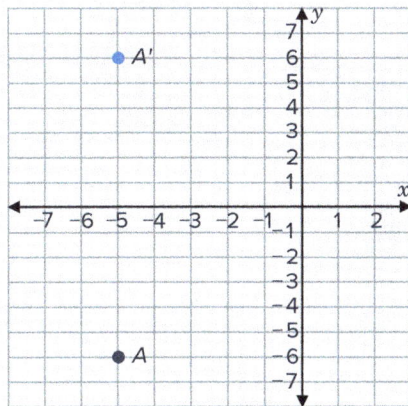

c

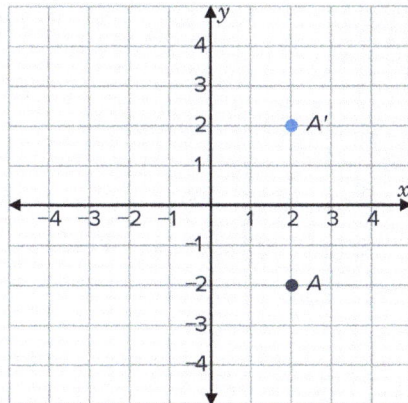

d

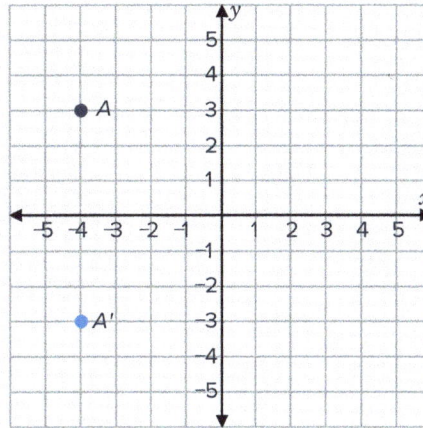

4 a C   **b** B

5 a

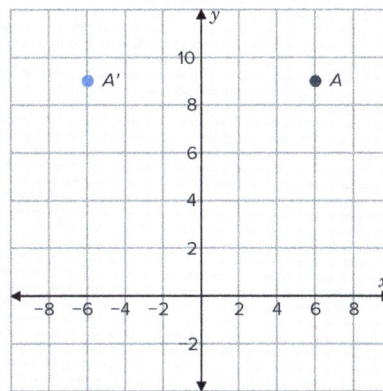

b

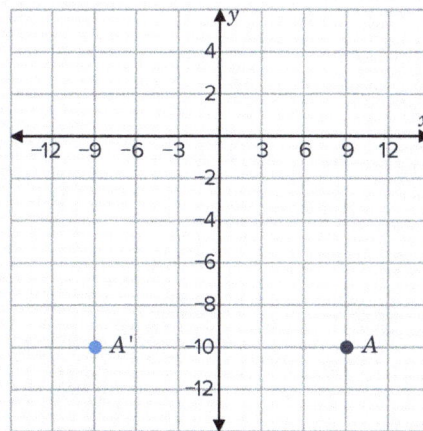

c

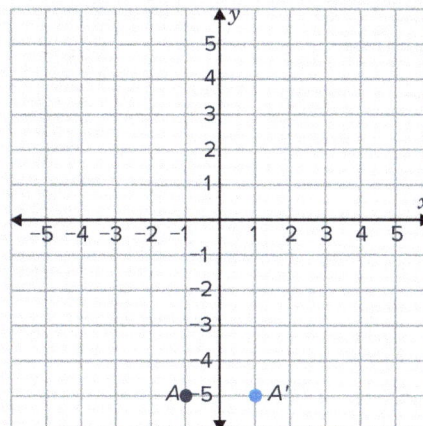

d

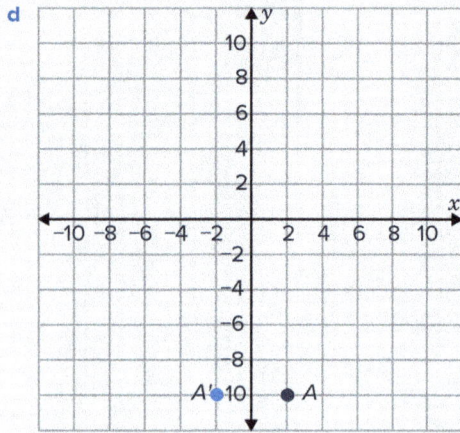

6 a

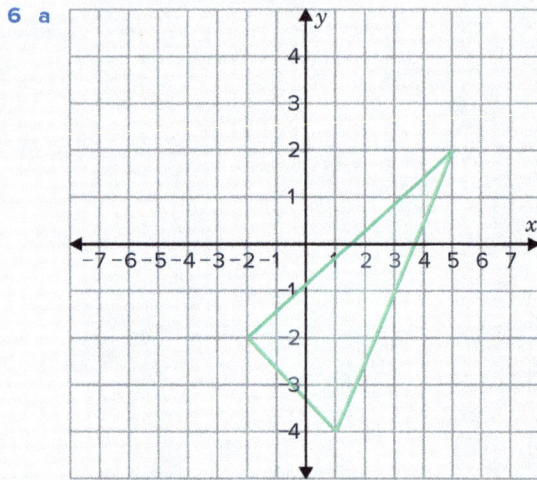

b

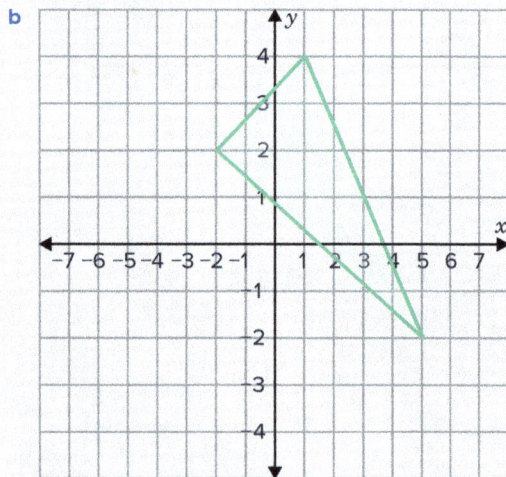

c After the reflection, the x-coordinates remain the same, while the y-coordinates are the negatives of the original y-coordinates.

7 a True   **b** False   **c** True   **d** False

8 a Reflection across the x-axis

b Reflection across the y-axis

c Reflection across the y-axis

d Reflection across the x-axis

9 B

10 a First is reflection across the x-axis, then reflection over the y-axis.

b First is reflection across the y-axis, then reflection over the x-axis.

11 N

12 a $B(6, -7)$ **b** 14 units down

13

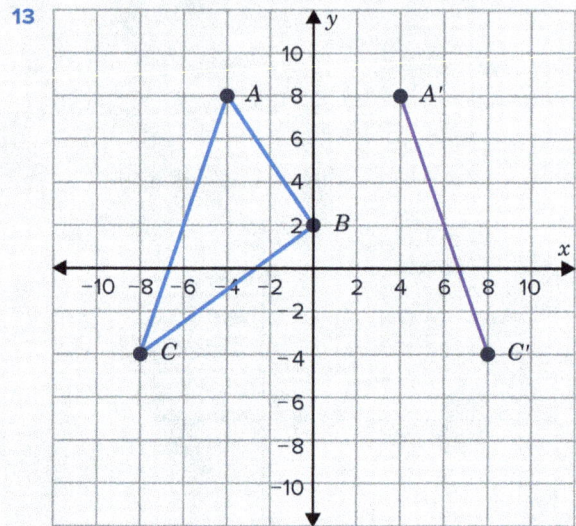

14 a

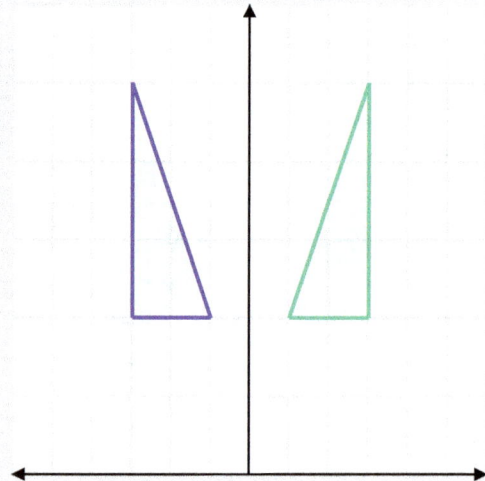

b

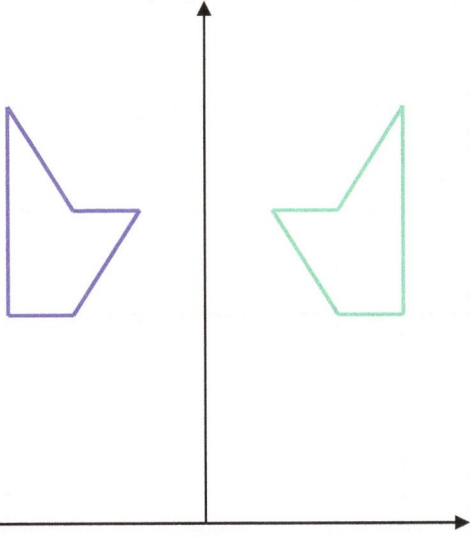

c

d

15 a

b

c

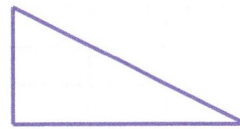

d

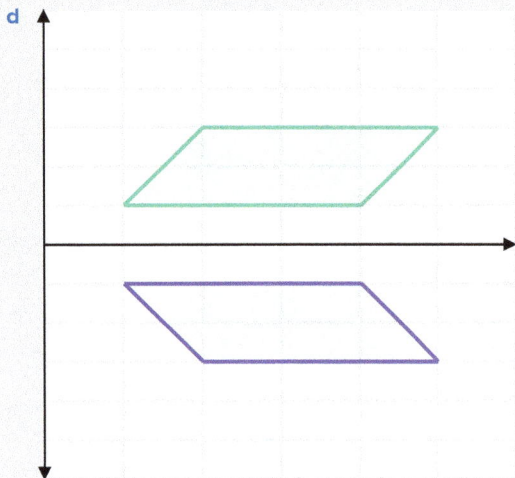

16 The coordinates of the vertices of the triangle after reflection over the x-axis are (2, –1), (2, –3), and (4, –1).

17 **a** The coordinates of the vertices of the polygon after reflection across the x-axis are $A'(0, 3)$, $B'(–3, 5)$, $C'(–3, 1)$, and $D'(0, –1)$.

b

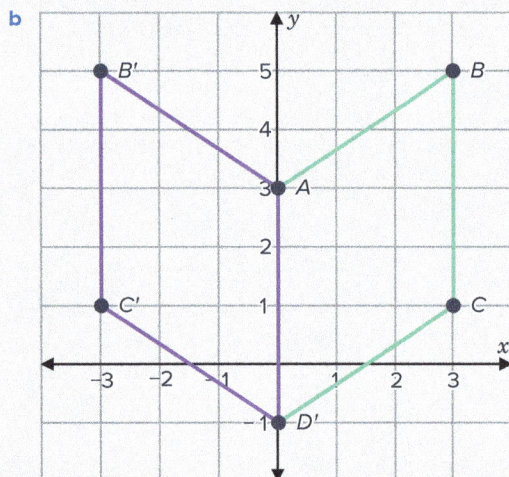

18

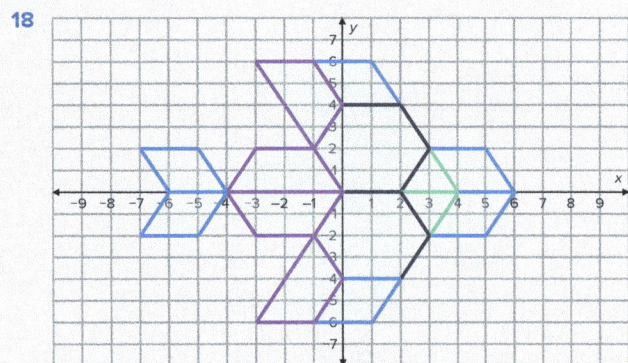

19 Yes, the reflection will return to the original figure. When a figure is reflected across a line, each point of the figure is mapped to a new point on the other side of the line, at the same distance but in the opposite direction. Reflecting the figure again across the same line reverses this process for each point, effectively undoing the first reflection and returning each point to its original position. Thus, reflecting a figure twice across the same line results in the figure being identical to its original state.

20 **a** Reflection

b Reflecting a figure across an axis, such as the y-axis here, produces a congruent figure to the original. This mirrored quadrilateral retains identical size and shape, with equal corresponding sides and angles, though its orientation is reversed.

21 **a i** $a = 9$ **ii** $b = 37$ **iii** $c = 8$
        iv $d = 8$ **v** $e = 150$

b On itself

22 **a** Yes. Sokayna›s triangle is Roly›s triangle reflected over the line $y = x$.

b No, both triangles have the same perimenter. This is because reflections do not change the size of a shape.

23 **a** A, B, C, D, E, H, I, K, M, O, T, U, V, W, X, Y There may be some variation depending on how letters are drawn. For example, if "B" is drawn with the top bump smaller than the bottom it would not have any reflectional symmetry.

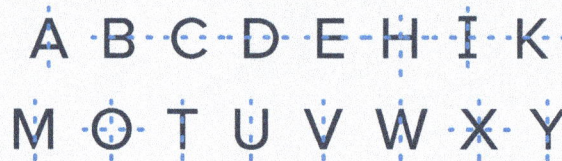

b i A, H, I, M, O, T, U, V, W, X, Y
    ii B, C, D, E, H, I, K, O, X
    iii H, I, O, X

6.07 Translations and reflections in the coordinate plane

Subtopic overview

Lesson narrative

In this lesson, students will explore translations and reflections in the coordinate plane. They will learn that a translation is a transformation that moves every point of a figure the same distance in the same direction, while a reflection is a transformation that flips a figure over a line, called the line of reflection. The lesson combines these transformations to show how different sequences of movements can create various outcomes. The exploration component involves using a polygon on a graph to practice combining translations and reflections. Students will reflect a polygon over the x-axis, then translate it up a specified number of units, and vice versa. They will compare the final positions to see how the order of transformations affects the outcome. This activity helps students understand that changing the sequence of transformations can lead to different results. Students will engage in problems where they apply specific transformations to shapes and observe the results. They will reflect and translate shapes such as triangles and quadrilaterals and analyze the impact of transformation order. By the end of the lesson, students should be able to perform and differentiate between translations and reflections and understand the significance of the order in which these transformations are applied.

Learning objectives

Students: Page 400

After this lesson, you will be able to...
- identify the coordinates of the image of a polygon that has been translated and reflected over the x- or y-axis in any order when given a preimage.
- sketch the image of a polygon that has been translated and reflected over the x- or y-axis in any order.
- identify and describe transformations in context.

Key vocabulary

- reflection

- translation

Essential understanding

A translation preserves distance and angle measure because it moves all of the points on a figure the same distance and direction. A reflection preserves distance and angle measure, but reverses orientation. The preimage and the image of a figure are symmetric about a line of reflection.

Standards

This subtopic addresses the following Virginia 2023 Mathematics Standards of Learning standards.

Mathematical process goals

MPG1 — Mathematical Problem Solving

Teachers can integrate this goal by challenging students to solve problems involving combined translations and reflections on the coordinate plane. For instance, they can present a scenario where a figure needs to be moved to a specific position on the coordinate plane using a combination of translations and reflections. Students can then apply their understanding of these transformations to solve the problem.

MPG2 — Mathematical communication

Teachers can integrate this process goal by encouraging students to communicate the transformations occurring as well as the impact of the transformations on the image's shape, size and position.

MPG3 — Mathematical Reasoning

Teachers can integrate this goal by guiding students to use logical reasoning in determining the sequence of transformations needed to achieve a certain result. For instance, they can pose a problem that requires students to first translate then reflect a figure, and another problem that involves reflecting then translating. Students can then analyze and reason the difference in results between the two sequences.

MPG4 — Mathematical Connections

Teachers can show students how the concept of translations and reflections connects to other areas of mathematics and real-world situations. For instance, they can discuss how these transformations are used in computer graphics, architecture, and design. They can also connect this lesson to previous lessons on polygons and the coordinate plane, and how the skills learned there apply here.

Content standards

8.MG.3 — The student will apply translations and reflections to polygons in the coordinate plane.

8.MG.3c — Given a preimage in the coordinate plane, identify the coordinates of the image of a polygon that has been translated and reflected over the x- or y-axis, or reflected over the x- or y-axis and then translated.

8.MG.3d — Sketch the image of a polygon that has been translated vertically, horizontally, or a combination of both.

8.MG.3e — Sketch the image of a polygon that has been reflected over the x- or y-axis.

8.MG.3f — Sketch the image of a polygon that has been translated and reflected over the x - or y-axis, or reflected over the x- or y-axis and then translated.

8.MG.3g — Identify and describe transformations in context (e.g., tiling, fabric, wallpaper designs, art).

Prior connections

6.MG.3 — The student will describe the characteristics of the coordinate plane and graph ordered pairs.

7.MG.4 — The student will apply dilations of polygons in the coordinate plane.

Future connections

G.RLT.3 — The student will solve problems, including contextual problems, involving symmetry and transformation.

Lesson Preparation

Suggested review

Depending on your students' level of prior knowledge, consider revisiting the following lessons:

Grade 8 — 6.05 Translations in the coordinate plane
Grade 8 — 6.06 Reflections in the coordinate plane

Tools

You may find these tools helpful:
- Graph paper
- Dynamic geometry software
- Tracing paper

Student lesson & teacher guide

Combinations of translations and reflections

Students are reminded of the definition of translations and reflections before engaging in an exploration that performs multiple transformations on a polygon in the coordinate plane.

Use the STEAM cycle with transformations in real-world designs
Targeted instructional strategies

Engage students by introducing the context of creating a decorative wall mural, highlighting how geometric transformations like translations and reflections are essential in art and design.

Ask: Present a problem like: "How can we design a visually appealing wall mural using geometric shapes and transformations?" Guide students to define the design challenge, consider various shapes and patterns, and discuss how transformations can be applied to create symmetry and interest in the mural.

Imagine: Encourage students to brainstorm different design ideas, sketching how shapes can be translated and reflected across the wall space. Have them predict the outcomes of combining transformations and how these will affect the overall aesthetics of the mural, fostering creativity and spatial reasoning.

Plan: Assist students in creating a detailed plan for their mural by selecting specific geometric shapes (like triangles, squares, or hexagons) and determining which transformations to apply. Help them establish criteria for their design (such as achieving symmetry or repeating patterns), acknowledge constraints (like the size of the wall or number of shapes), and decide how they will document the coordinate changes—possibly by using graph paper or design software.

Create and Test: Support students as they construct their mural designs on graph paper or with digital tools, applying translations and reflections to their chosen shapes. They should sketch the design after each transformation and record the coordinates of key points, observing how each movement affects the placement and orientation of the shapes within the mural.

Improve: Facilitate a session where students review their designs, identify areas for enhancement, and ask questions about the effectiveness of their transformation sequences. Encourage collaboration as they discuss how altering the order or type of transformations impacts the final design. Support them in refining their murals by making justified adjustments based on mathematical reasoning and discussing how these changes improve the visual appeal.

By guiding students through this STEAM cycle, you help them apply their understanding of translations and reflections to a real-world design challenge, enhancing their problem-solving skills and demonstrating the practical applications of geometric transformations in art and design.

Exploration

Students: Page 400

Exploration

Use the graph of polygon $ABCDE$ to explore combining translations and reflections.

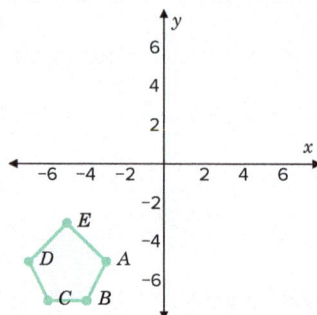

Apply each of the following pairs of transformations and identify the coordinates of the image:
- Reflect polygon $ABCDE$ over the x-axis. Then, translate the polygon up 3 units.
- Translate polygon $ABCDE$ up 3 units. Then, reflect the polygon over the x-axis.

1. Did changing the order of the transformations affect the image?
    - Reflect polygon $ABCDE$ over the y-axis. Then, translate the polygon up 6 units.
    - Translate polygon $ABCDE$ up 6 units. Then, reflect the polygon over the y-axis.

2. Did changing the order of the transformations affect the image?
    - Choose your own translation and reflection and test your hypothesis.

Suggested student grouping: Small groups

In this exploration, students will have an opportunity to discover how changing the order of transformations, such as reflection and translation, can affect the image of a polygon on a graph. Students will reflect and translate polygon ABCDE over both x and y axes, then observe how the coordinates of the image change based on the order of transformations applied.

Ideal student responses

These ideal responses may differ from other correct student responses. Less formal responses can be connected with the more precise mathematical language presented here.

1. **Did changing the order of the transformations affect the image?**

    Yes, changing the order of transformations did affect the image. When the polygon was reflected first and then translated, the final image was different than when the polygon was translated first and then reflected.

2. **Did changing the order of the transformations affect the image?**

    Yes, similar to the first case, changing the order of transformations altered the final image of the polygon. Reflecting first and then translating resulted in a different image than translating first and then reflecting.

Purposeful questions

- What does this tell you about the relationship between transformations and the order in which they are applied?
- Can you predict the final image of a polygon if given a series of transformations? What information would you need?

Possible misunderstandings

- Students might think that the order of transformations does not matter. It's important to clarify that while some transformations might not change the final image regardless of order, others certainly will. The order in which transformations are applied can significantly alter the final image.

Practicing translations and reflections
Targeted instructional strategies

Before combining translations and reflections together, practicing each transformation separately with the same starting polygon is a helpful exercise. Show students a shape in the coordinate plane labeled as the pre-image, such as the one shown:

Starting polygon

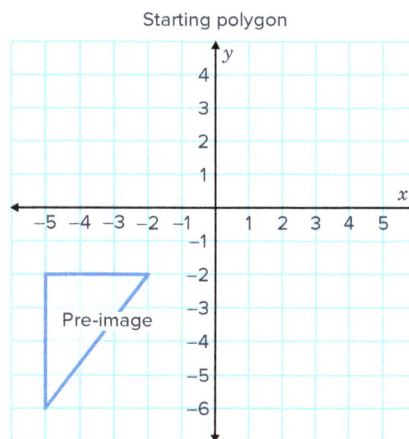

Show students a reflection of this polygon over the y-axis and translating the original polygon 6 units to the right and 4 units up.

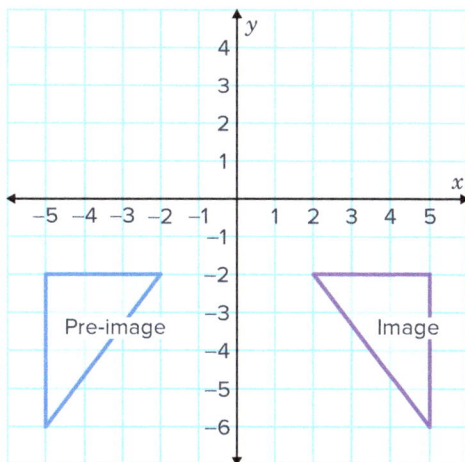

Reflect over the y-axis

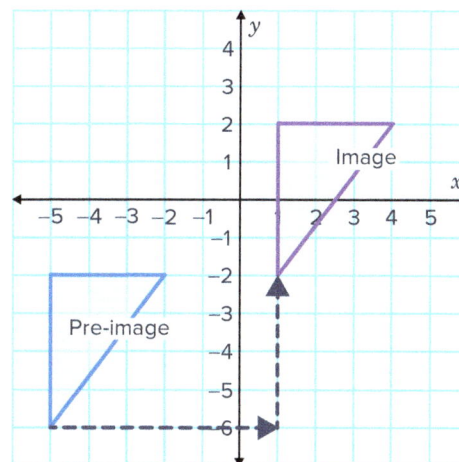

Translate 6 units right and 4 units up

Point out that in the reflection, the orientation of the polygon changes, while in the translation, the polygon stays in the same orientation but moves location.

These polygons transform in different ways in relation to where they started, so the order of their movement affects where they move to. Use this as an introduction prior to teaching the content of the lesson.

Compare and connect

English language learner support

After students complete the exercises on reflections and translations, prompt them to compare different approaches to solving the same problem.

For example, ask students to discuss how reflecting a shape over the x-axis before translating it differs from translating it first and then reflecting it. Encourage them to explore why these different sequences might result in the same or different final images.

Prompt questions could include:

- How does reflecting before translating compare to translating before reflecting? Can you explain why the final image changes (or doesn't change) depending on the order?
- What patterns do you notice when comparing different sequences of transformations?
- Can you connect this process to any other mathematical operations where the order matters?

These questions will help students develop meta-awareness of mathematical concepts and language by identifying similarities and differences in their problem-solving strategies.

Visual supports for translations and reflections

Student with disabilities support

Provide visual aids and manipulatives to support students with understanding translations and reflections. For example, use tracing paper or digital tools that allow students to physically or virtually move shapes to perform translations and reflections.

Encourage students to draw arrows showing the direction and distance of translations, and to draw the line of reflection for reflections. This can be particularly helpful for students with visual-spatial processing challenges.

An example is shown.

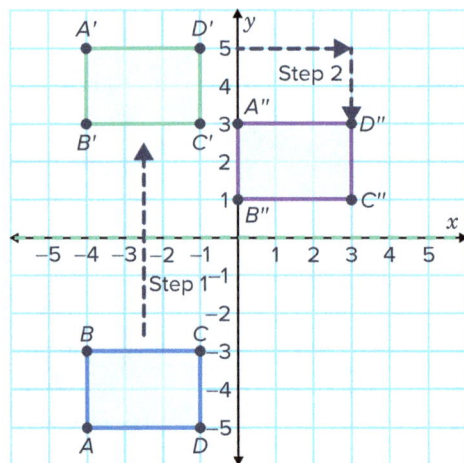

Reflect over the x-axis and translate $(4, -2)$

Order of Transformations
Address student misconceptions

A common misconception is that the order of transformations does not matter. Students may think that translating a shape first and then reflecting it will give the same result as reflecting it first and then translating it.

To correct this misconception, provide students with examples where they can clearly see that the order of transformations does matter. Encourage them to try performing the transformations in a different order to see if they get the same result.

An example is shown.

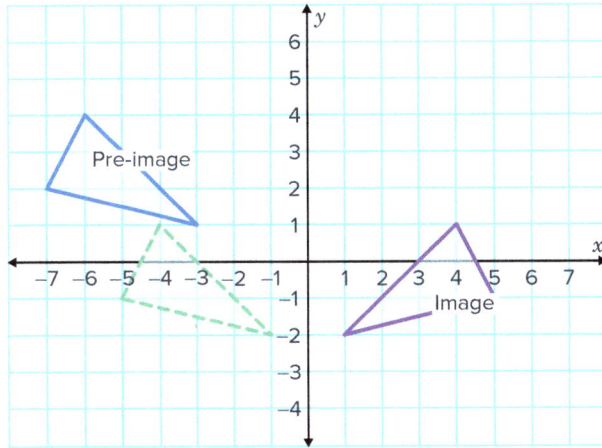

Translate (2, −3) and reflect over the y-axis

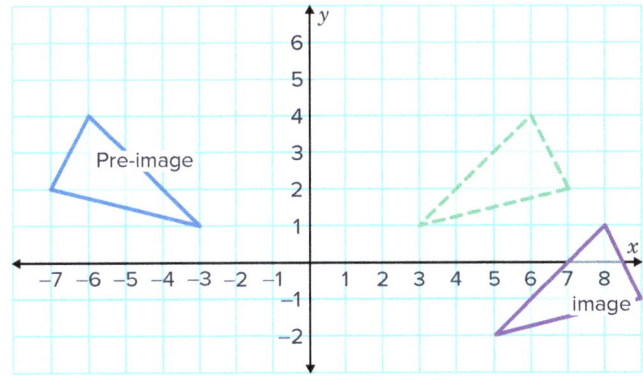

Reflect over the y-axis and translate (2, −3)

Students are then shown an example of the order of the transformations impacting the location of the image of a transformed polygon.

Examples

Example 1

Consider $\triangle ABC$ and $\triangle A''B''C''$.

Describe the combination of transformations to get from the preimage to the image.

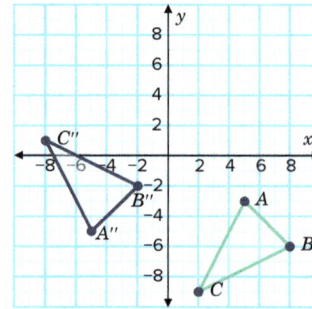

Create a strategy

We will use some combination of reflections and translations to align the two shapes exactly. There are multiple correct solutions.

Apply the idea

Find some combination of transformations to align the two shapes.

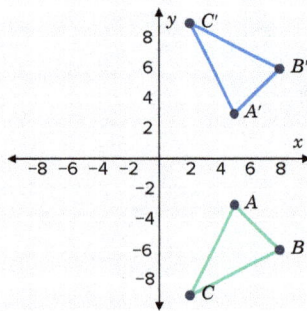

Reflect $\triangle ABC$ across the x-axis, to get $\triangle A'B'C'$.

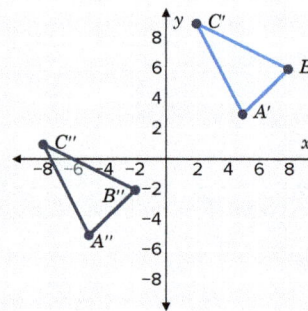

Then, translate $\triangle A'B'C'$, 10 units to the left and 8 units down, to get $\triangle A''B''C''$.

Reflect and check

Find a new combination of transformations to align the two shapes.

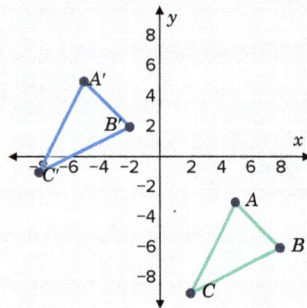

Translate $\triangle ABC$, 10 units to the left and 8 units up to get $\triangle A'B'C'$.

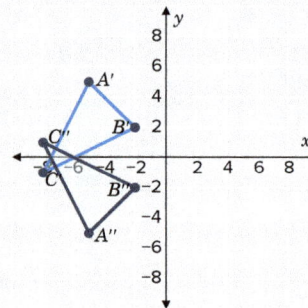

Then, reflect $\triangle A'B'C'$ over the x-axis to get $\triangle A''B''C''$.

Purpose

Show students how to apply a combination of transformations to align geometric shapes.

Example 2

Use △RST to answer the following questions.

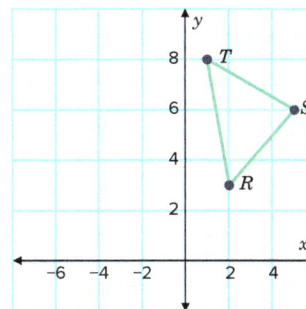

a   Translate △RST down 3 units and left 2 units. Then, reflect the triangle over the y-axis. Graph the image on the same coordinate plane as the preimage of △RST.

Create a strategy

A translation shifts the every point on the triangle the same distance in the same direction. Then, a reflection over the y-axis flips the triangle onto the other side of the y-axis.

Apply the idea

First, perform the translation on the graph down 3 units and left 2 units to create △$R'S'T'$.

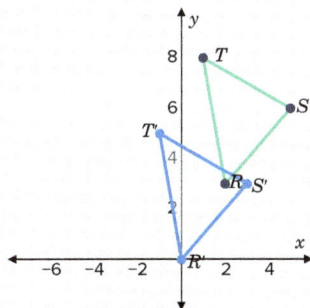

Then, reflect △$R'S'T'$ over the y-axis.

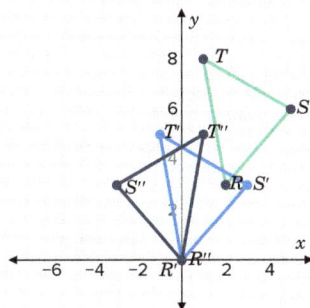

Reflect and check

We can check the ordered pairs algebraically.

The first transformation translated △RST down 3 units and left 2 units. We can subtract 3 from each y-coordinate and subtract 2 from each 2 from each x-coordinate to find the ordered pairs for △$R'S'T'$.

$R(2, 3) \rightarrow R'(2 - 2, 3 - 3) \rightarrow R'(0, 0)$

$S(5, 6) \rightarrow S'(5 - 2, 6 - 3) \rightarrow S'(3, 3)$

$T(1, 8) \rightarrow T'(1 - 2, 8 - 3) \rightarrow T'(-1, 5)$

The second transformation reflected △$R'S'T'$ over the y-axis. This reflection does not change any of the y-coordinates, but all of the x-coordinates will change signs to form the ordered pairs for △$R''S''T''$.

$R'(0, 0) \rightarrow R''(0, 0)$

$S'(3, 3) \rightarrow S''(-3, 3)$

$T'(-1, 5) \rightarrow T''(1, 5)$

Purpose

Show students how to perform sequential transformations on a geometric figure and understand the effect of each transformation.

Expected mistakes

Students may miss the words that indicate direction for each number since it is not stated in (x, y) order, and move 3 horizontal units and 2 vertical units instead of down and left. As students read the problem, they can label key words that indicate direction with arrows, for example "← 2" for "left 2 units."

Reflecting with students

Encourage advanced learners to represent transformations using algebraic coordinates.

A translation of 3 units down and 2 units left would be $(x - 2, y - 3)$ to each vertex R (2, 3), S (5, 6), and T (1, 8).
A reflection over the y-axis would take each resulting ordered pair (x, y) of $\triangle R'S'T'$ and rewrites it as $(-x, y)$.

Students: Page 404

b   If the order of those transformations are reversed, would it result in the same image?

Create a strategy

Transform $\triangle RST$ in reverse order to see if the same image is created.

Apply the idea

First, we will reflect $\triangle RST$ over the y-axis to create $\triangle R'S'T'$.

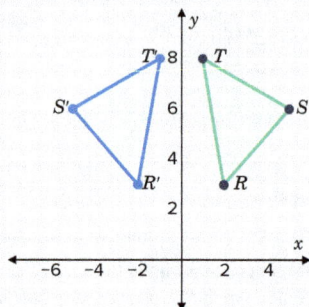

Then, we translate $\triangle R'S'T'$ down 3 units and left 2 units to create $\triangle R''S''T''$.

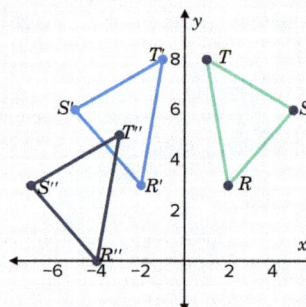

When the order of the transformations is reversed, it results in a different image.

Purpose

Challenge students to understand that the order of transformations can affect the resulting image.

Students: Page 404–405

Example 3

A local community is planning to revamp an old park by introducing a creative playground shaped as a polygon. The current design proposal includes a quadrilateral shaped play area with coordinates at $P(2, 3)$, $Q(5, 3)$, $R(5, 1)$, and $S(3, 1)$ on a grid system that represents a scaled down version of the park. To integrate this new structure seamlessly with the park's layout, two specific transformations are required:

1.   Mirror the proposed playground layout across the y-axis to better align with other park facilities.
2.   The mirrored layout should be moved 3 units downward and 4 units to the right to fit precisely in the designated area for playgrounds.

Identify the final coordinates of the vertices after applying these transformations.

Create a strategy

To reflect a point across the y-axis, we change the sign of the x-coordinate while keeping the y-coordinate the same.
To translate a point, add the translation values to the coordinates.

Apply the idea

Reflecting each vertex:

$P(2, 3) \rightarrow P'(-2, 3)$

$Q(5, 3) \rightarrow Q'(-5, 3)$

$R(5, 1) \rightarrow R'(-5, 1)$

$S(3, 1) \rightarrow S'(-3, 1)$

After reflecting across the y-axis, the vertices are: $P'(-2, 3)$, $Q'(-5, 3)$, $R'(-5, 1)$, $S'(-3, 1)$

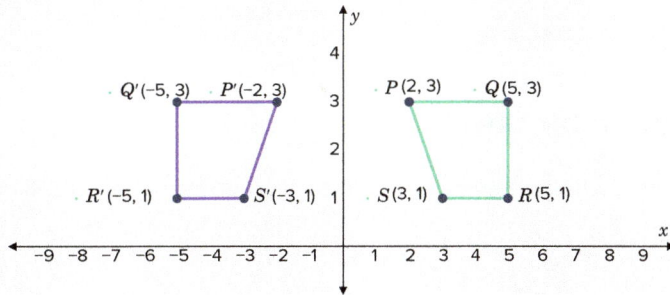

Translating each mirrored vertex:

$P'(-2, 3) \rightarrow P''(2, 0)$

$Q'(-5, 3) \rightarrow Q''(-1, 0)$

$R'(-5, 1) \rightarrow R''(-1, -2)$

$S'(-3, 1) \rightarrow S''(1, -2)$

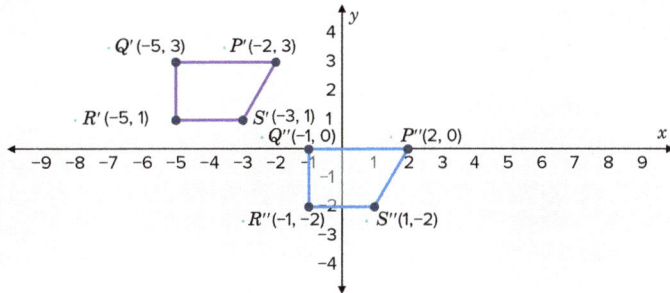

After translating, the final coordinates of the vertices are:

$P''(2, 0)$, $Q''(-1, 0)$, $R''(-1, -2)$, $S''(1, -2)$

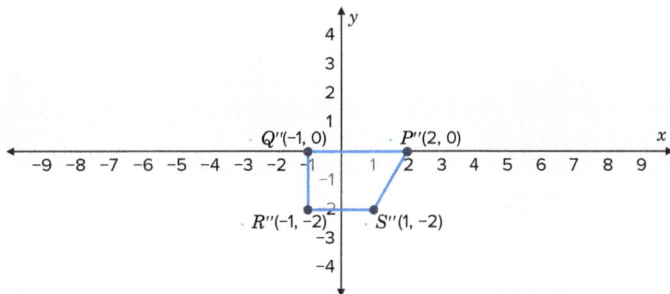

Purpose

Challenge students to interpret and apply transformations in context.

Students: Page 405

💡 Idea summary

When we translate and then reflect a shape over the x-or y-axis, it may not give the same result as when we reflect and then translate.

If we translate, and then reflect:

1. Translation: Move the shape to a new location.
2. Reflection: Flip the moved shape over the x- or y-axis.
3. Result: The shape's final position depends on where it was moved before being flipped.

If we reflect, and then translate:

1. Reflection: Flip the shape first over the x- or y-axis.
2. Translation: Move the flipped shape to a new location.
3. Result: The shape's final position depends on where it was flipped before being moved.

Practice

Students: Page 405–410

What do you remember?

1 Determine whether the pairs of triangles are a translation or reflection of one another:

a

b

2 Consider the graphed figures.

    a Describe how to transform one shape onto another using a translation.

    b Describe how to transform one shape onto another using a reflection.

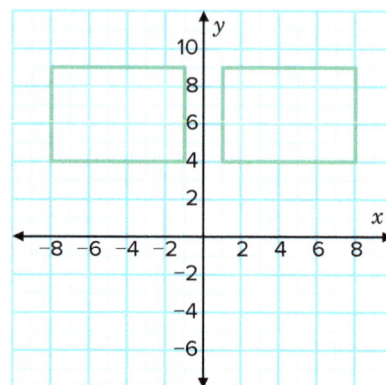

3 Describe the transformation from quadrilateral $ABCD$ to quadrilateral $A'B'C'D'$.

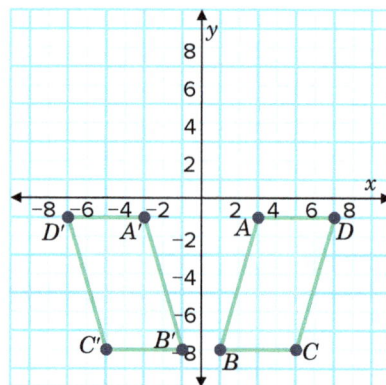

4 Consider the figures:

    a What type of transformation can take triangle A to triangle B?

    b Describe the transformation(s) from triangle A to triangle B.

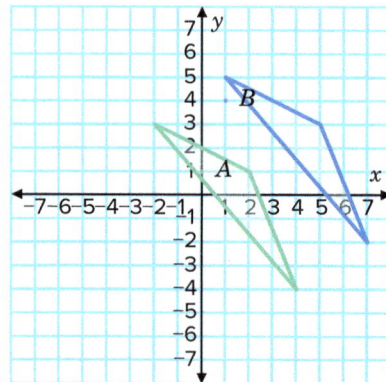

5 Point $A(3, -2)$ is reflected over the y-axis and then translated 4 units left and 1 unit down. What are the coordinates of A'?

Let's practice

6 Consider the figures:

a Identify the transformation from quadrilateral A to quadrilateral B.

b Identify the coordinates of quadrilateral C if quadrilateral C is a translation of quadrilateral B, 4 units to the right.

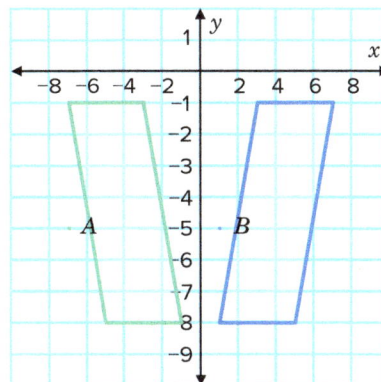

7 Triangle ABC is reflected across the x-axis and then translated 3 units left. What could be the coordinates of B'?

A (3, −6)
B (−2, −1)
C (0, −6)
D (4, −4)

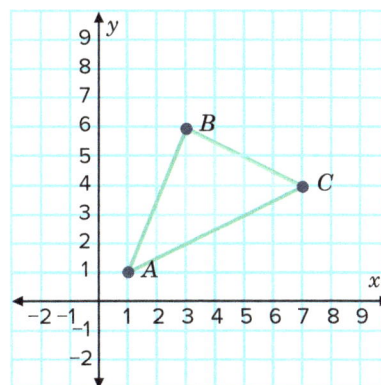

8 Consider quadrilateral $ABCD$.

If the quadrilateral is reflected over the x-axis and translated 2 units up, identify the coordinates of:

a A'
b B'
c C'
d D'

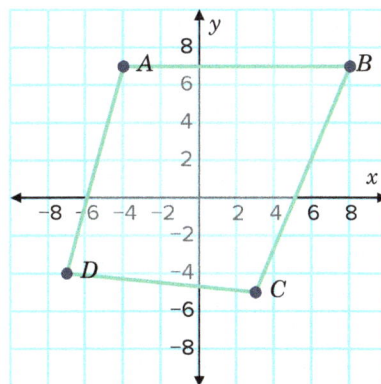

9 Triangle ABC is graphed with vertices at $A(−4, 4)$, $B(−1, 4)$ and $C(−4, 1)$. Identify the points of A', B' and C' after being reflected over the x-axis and translated 3 units down.

$$A'(\square, \square), B'(\square, \square) \text{ and } C'(\square, \square)$$

10 Quadrilateral A on the plane is to undergo two successive transformations:

a Plot quadrilateral A', which is the result when quadrilateral A is translated 4 units left.

b Plot quadrilateral A', which is the result from reflecting quadrilateral A' across the y-axis.

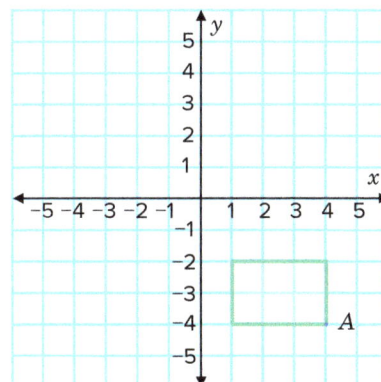

11   Consider the following graph:

   a   Reflect the figure over the y-axis and then translate 2 units right.

   b   Translate the figure 2 units right and then reflect over the y-axis.

   c   Is the order in which we perform the transformations important? Explain.

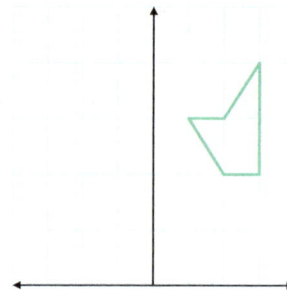

12   Plot the given shape that will result from reflecting it across the x-axis followed by a translation of 6 units right.

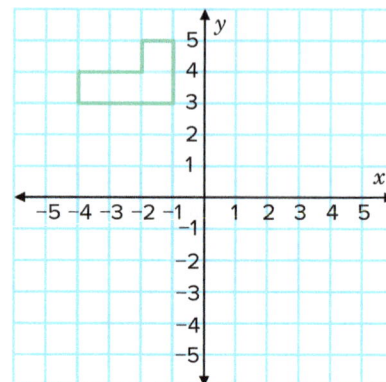

13   Draw the image of a square with vertices at (1, 1), (1, 3), (3, 3), and (3, 1) after it has been reflected over the y-axis and then translated 2 units to the right.

14   Reflect quadrilateral $ABCD$ across the y-axis, and translate 10 units left and 11 units up.

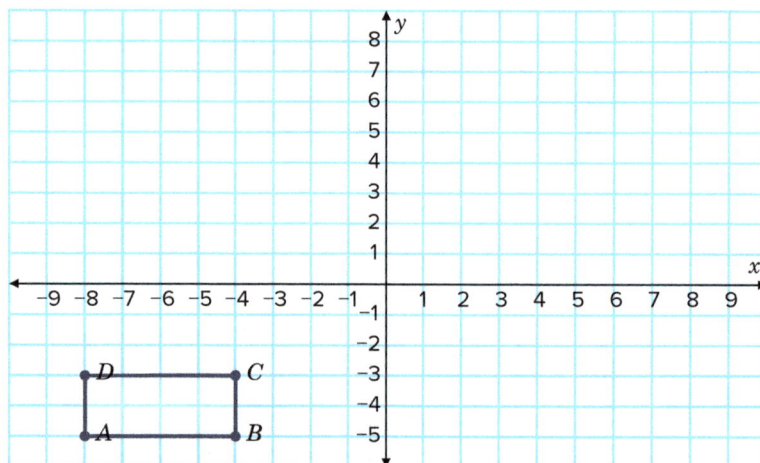

15   A city council has approved the construction of a new playground in a local park. The playground's initial design is based on a polygonal shape that needs to be transformed to fit the designated area in the park. The original playground layout includes a quadrilateral play area with vertices at $P(2, 3)$, $Q(5, 3)$, $R(5, 1)$, and $S(3, 1)$ on a coordinate plane. To finalize the design, the playground needs to undergo two transformations to better utilize the space and ensure safety:

   1.   The entire playground layout should be reflected across the y-axis to accommodate the existing park infrastructure.

   2.   After reflection, the playground should be translated 4 units down and 3 units to the right to align with the park's designated space for new constructions.

   Calculate the coordinates of the quadrilateral play area vertices after undergoing the combined transformations of reflection across the y-axis and the specified translation.

16 Chantelle wants to animate an object for her slideshow presentation.

    a She wants the woman in the image to be looking at the apple. How can she achieve this effect?

    b She wants to add the effect so that the apple falls to the ground. How can she achieve this?

Let's extend our thinking

17 Describe a sequence of transformations required to get from quadrilateral $ABCD$ to quadrilateral $A''B''C''D''$.

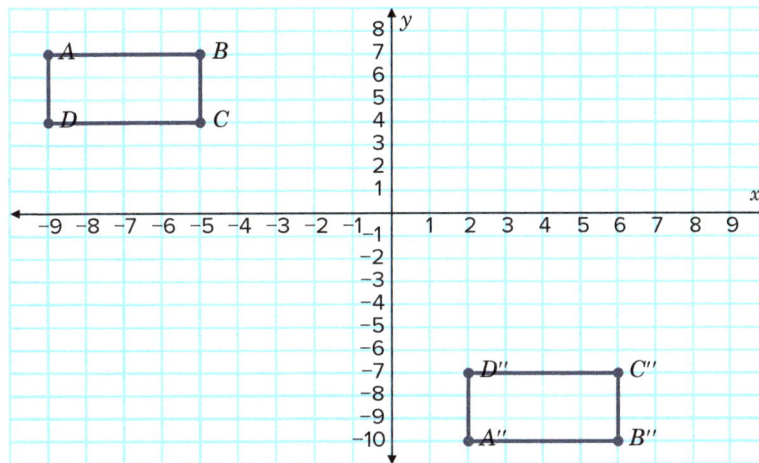

18 Describe a sequence of transformations required to get from triangle ABC to triangle $A''B''C''$. Use triangle $A'B'C'$ as a guide.

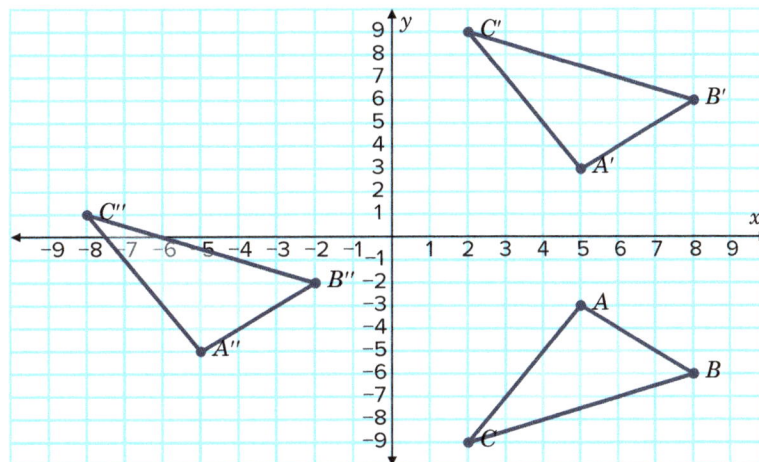

19 Describe a sequence of transformations required to get from triangle ABC to triangle $A''B''C''$.

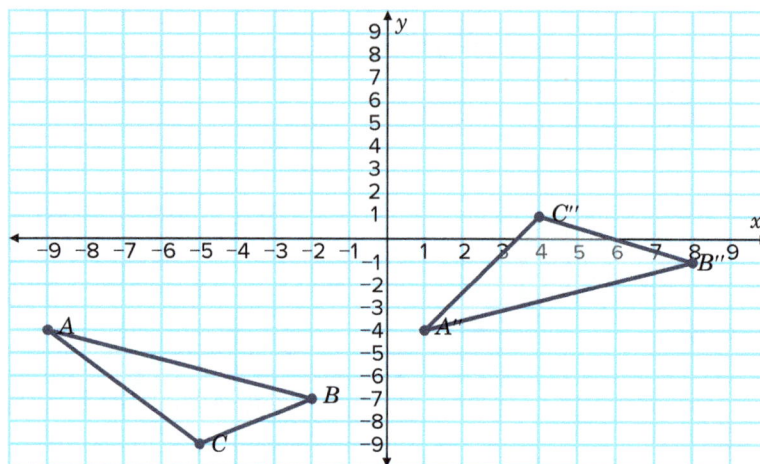

20 The picture shows the tiles in the school cafeteria. Arisha thinks that the tiles are being reflected while Lucia thinks that they are being translated.

Explain who you think is correct, taking into account the color of the tiles.

21 Explain why performing a combination of translations and reflections will always produce a congruent figure.

22 If the same transformation are applied to two shapes, but in different orders, will they end up with the same image? Why or why not?

Answers

6.07 Translations and reflections in the coordinate plane

What do you remember?

1 **a** Translation **b** Reflection

2 **a** To translate the left rectangle to the position of the right rectangle, we need to move it horizontally to the right. The left rectangle's bottom left vertex is at (−8, 4), and the right rectangle's bottom left vertex is at (1, 4). Thus, the horizontal distance to translate the left rectangle to the right is 1 − (−8) = 9 units to the right. Since the vertical positions of the two rectangles are the same, there is no need for vertical translation.

Therefore, the transformation using a translation can be described as moving the shape 9 units to the right.

b To transform the left rectangle onto the position of the right rectangle using a reflection, we would reflect it across a vertical line that is equidistant from both rectangles. The left rectangle's rightmost edge is at −1 on the x-axis, and the right rectangle's leftmost edge is at 1 on the x-axis. The line that is equidistant from these edges would be the y-axis, which is the line $x = 0$.

Therefore, the transformation using a reflection can be described as reflecting the left rectangle across the y-axis ($x = 0$). This reflection will mirror the left rectangle to the position of the right rectangle on the coordinate plane.

3 A reflection across the y-axis.

4 **a** Translation
   b A translation 3 units right and 2 units up.

5 $A'(−7, −3)$

Let's practice

6 **a** Reflection
   b (5, −8), (7, −1), (11, −1), (9, −8)

7 C

8 **a** $A'(−4, −5)$
   b $B'(8, −5)$
   c $C'(3, 7)$
   d $D'D(−7, 6)$

9 $A'(−4, −7)$, $B'(−1, −7)$ and $C'(−4, −4)$

10 **a**

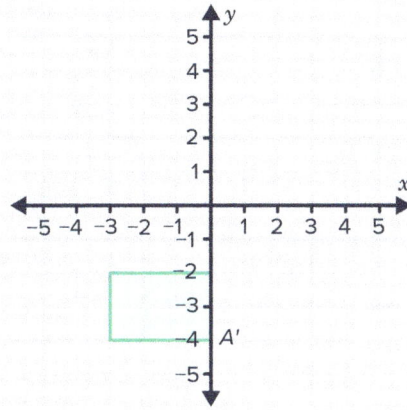

   b

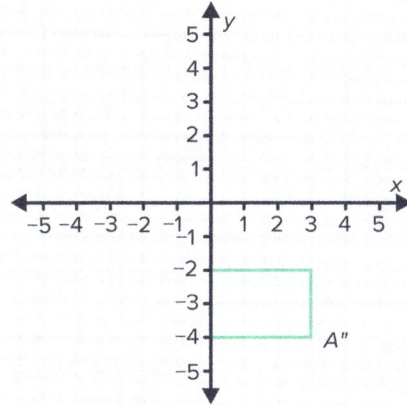

11 **a**

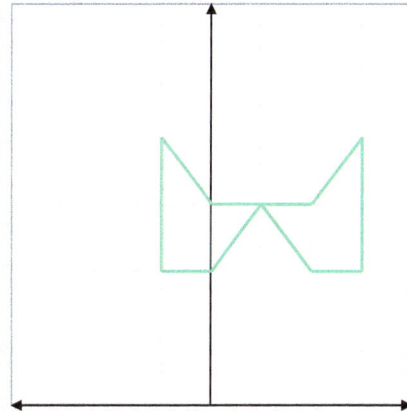

   b

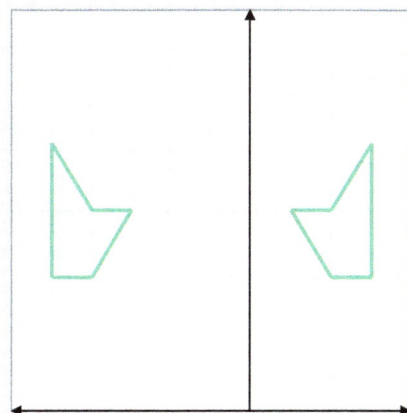

c The final positions of the points will be different for each sequence because the reflection changes the direction of the translation effect. If we reflect first, the translation moves the figure to the right of the y-axis. If we translate first and then reflect, the figure moves to the left of the y-axis, due to the change in sign of the x-coordinates after reflection. Thus, the order of transformations affects the final position of the figure.

12

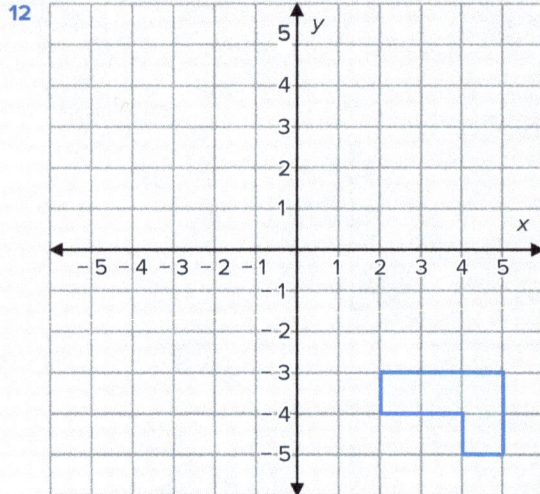

13

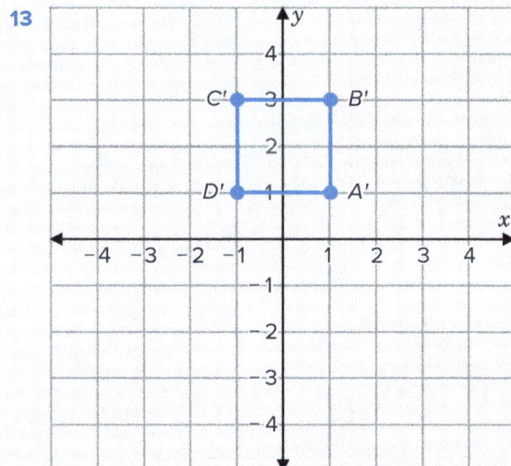

14

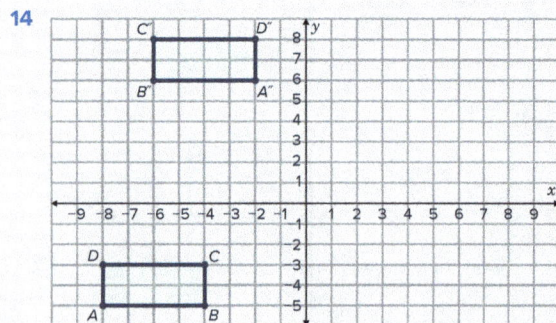

15 $P''(1, 1)$, $Q''(-2, -1)$, $R''(-2, -3)$, $S''(0, -3)$

16 a Example answer: She can reflect the apple to the left side of the woman by drawing an imaginary, vertical line down the center of the woman, then reflecting the apple across that line.

b Translate the apple downward

Let's extend our thinking

17 A translation 11 units right and 3 units up, and a reflection across the x-axis.

18 A reflection across the x-axis. and a translation 10 units left and 8 units down.

19 A translation 9 units right and 8 units up, and a reflection across the x-axis.

20 Answers may vary. Example answer:

Lucia is correct in thinking that the tiles are being translated. Each black and white tile is translated vertically and horizontally to create the checkered pattern.

In reflection, the image is flipped across a line (the line of reflection), creating a mirror image. If you take adjacent black tile and white tiles together and reflect them across a line, the same color tiles will be adjacent rather than alternating.

Therefore, based on the uniform, repeating pattern of the tiles, they are being translated, not reflected.

21 These three types of transformations are all rigid transformations, meaning that they preserve segment lengths and angle measures. So if we use one or more of these transformations on a figure to produce a new figure, the new figure will have identical angle measures and side lengths to the original shape.

22 The final images of two shapes subjected to the same transformations in different orders may or may not be the same, depending on the nature of the transformations. For commutative transformations like certain translations and rotations, the order does not impact the result. However, for non-commutative combinations, such as between reflections and translations or rotations and reflections, the order can significantly change the outcome.

Topic 6 Assessment: 2D Geometry

1 Write your answer in the appropriate space.

Two angles are complementary if their sum is ⬚. Two angles are sumentary if their sum is ⬚.

SOL **2** Given the diagram below, name

    1. an angle vertical to ∠2

    2. an angle adjacent to ∠3

    3. an angle supplementary to ∠4

    4. the value of ∠3 if the measure of ∠1 = 65°

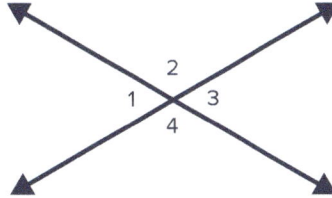

3 For each diagram shown, solve for the specified value.

    a Find the measures of ∠GJH and ∠IJL.

    b Solve for x.

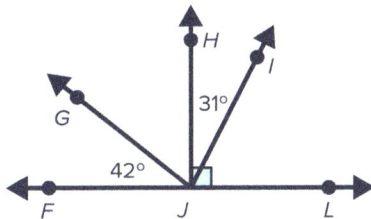

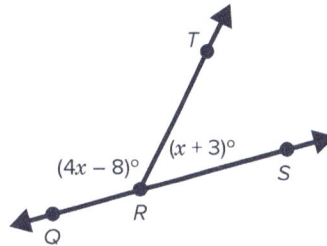

4 A triangle has side lengths of 10, 24, 26. Determine if these sides form a right triangle. Justify your answer.

5 Identify the legs and hypotenuse of the right triangle below.

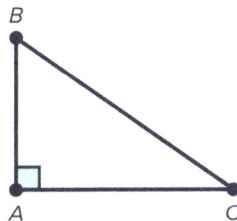

6 For each right triangle, solve for the specified value, rounding to the nearest tenth if necessary.

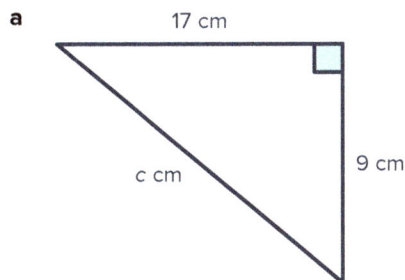

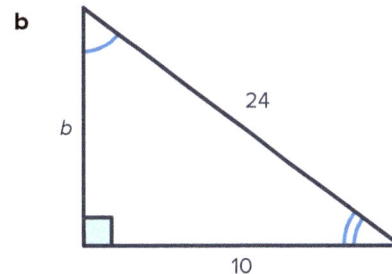

    a 17 cm, 9 cm, c cm

    b 24, 10, b

SOL **7** Determine the length of x.

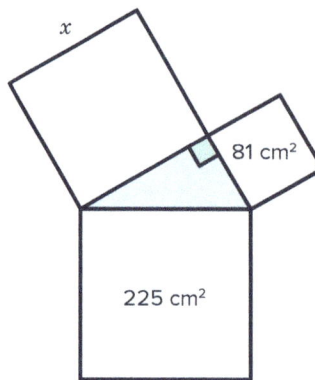

8 A sports association wants to redesign the trophy they award to the player of the season. The front view of one particular design is shown in the diagram.

    a Find the value of x.

    b Find the value of y. Round your answer to two decimal places.

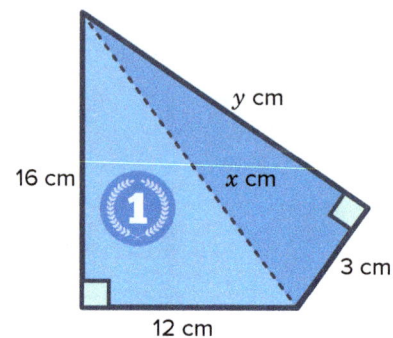

9 For each composite figure, find the following to the nearest hundredth, using 3.14 for π where necessary:

    i Perimeter   **ii** Area

    a

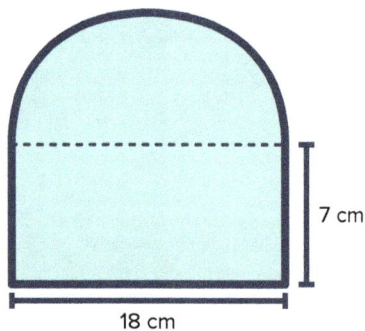

    b

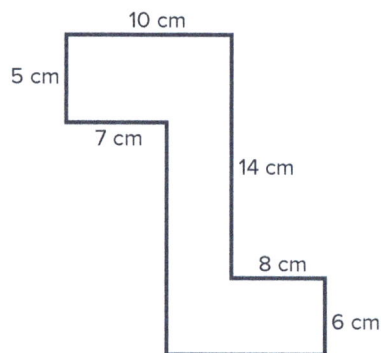

    c

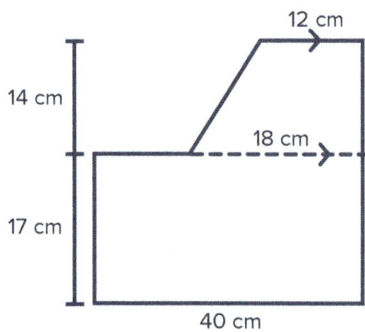

    d

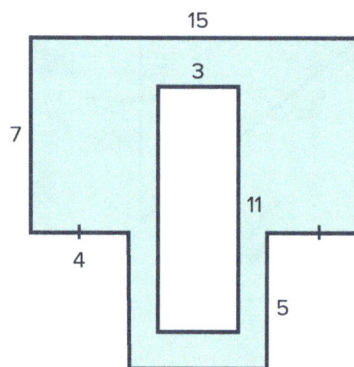

A city planner wants to add a walking path around a circular fountain in a new park. Her design is shown below.

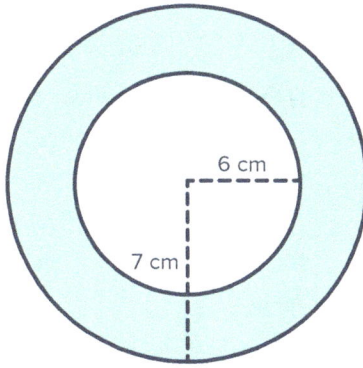

6 cm

7 cm

a The path will be covered in pavers for foot traffic. She wants to determine the total coverage of these pavers before submitting her design to the city council. Find the coverage of the pavers for her design, rounding to the nearest hundredth.

b In the city planner's design, 1 cm = 2.5 ft. Each paver will cover 4 square feet. What is the minimum number of pavers that the city will need for the path?

SOL 11 For each polygon, sketch the shape after the given transformation.

a A translation of two units down and four units to the right.

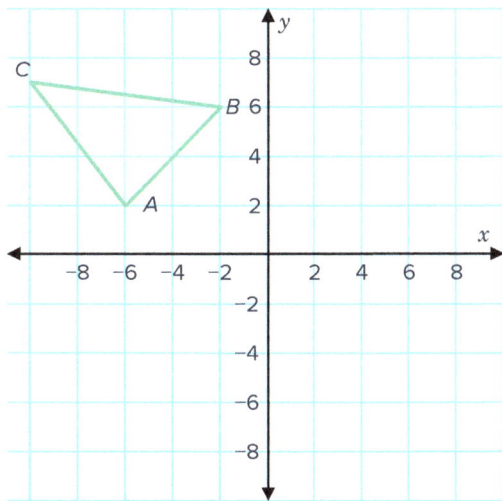

b A reflection over the x-axis.

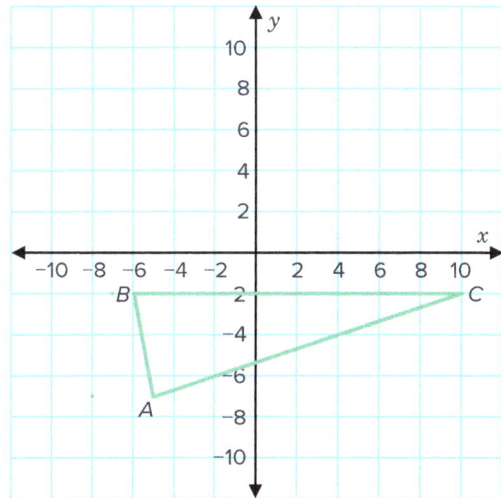

SOL 12 A chessboard currently has a knight on the space labeled E.

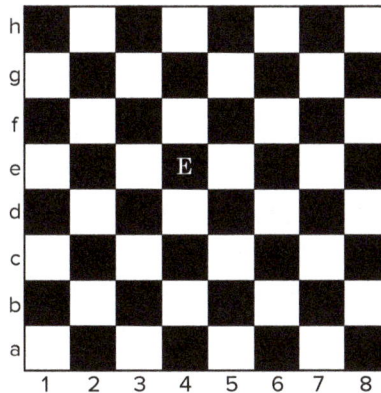

If a player wants to move the knight to space $f2$ on their next turn, what transformation would describe this move?

SOL **13** For each of the following polygons, name the coordinates of the image after each transformation.

    a A reflection over the y-axis for a rectangle with vertices M(–2,–3), A(–2,3), T(–6,3), and H(–6,–3).

    b A translation of 2 units right and 4 units up for a triangle with vertices C(2, 4), A(–1, 4), and T(–1, 0).

SOL **14** The polygon below will be translated (–8, 7) followed by a reflection over the y-axis.

    a Select all of the ordered pairs that represent the location of the new figure.

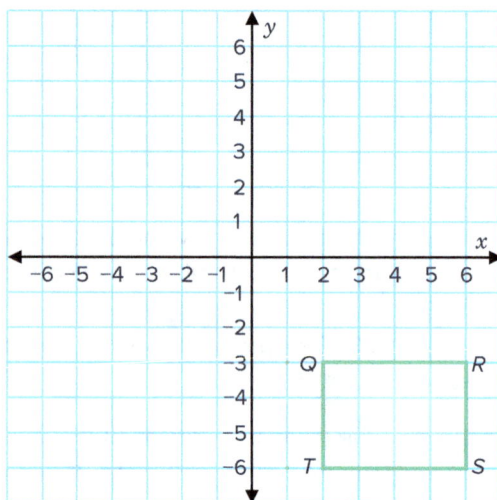

    A (6,4)   **B** (2, 4)   **C** (–2, 4)   **D** (2, 1)

    E (6, –6)   **F** (–6, 1)   **G** (6, 1)

    b Sketch the graph of the image after the polygon has been translated and reflected.

Performance task

15 The table shows several combinations of side lengths that form right triangle. Knowing the Pythagorean Theorem, we can verify that $a^2 + b^2 = c^2$ is true for each set.

a	b	c
3	4	5
5	12	13
7	24	25
9	40	41

Notice that each of these numbers is an integer. When three integers form the side lengths of a right triangle, we call them a Pythagorean Triple.

    a What is one thing you notice in common with all the values of a?

    b What relationship do you notice between b and c?

    c Can you write expressions for b and c using only the value of a?

    d Use your pattern to find two more Pythagorean Triples.

    e Do you think any Pythagorean Triples exist that do not follow your pattern? If so, find one.

Answers

Topic 6 Assessment: 2D Geometry

1 Two angles are complementary if their sum is 90°. Two angles are supplementary if their sum is 180°.

8.MG.1a

2 $\angle 4$

$\angle 2$ or $\angle 4$

$\angle 1$ or $\angle 3$

$\angle 3 = 65°$

8.MG.1a

3 a $\angle GJH = 48°$ and $\angle IJL = 59°$

b $x = 37$

8.MG.1b

4 Since the sides are in ascending order, this will make the values used in the Pythagorean theorem $a = 10$, $b = 24$, and $c = 26$. Applying this to the theorem gives $10^2 + 24^2 = 26^2$, which evaluates to $676 = 676$. Since the two sides of the equation are the same, it is a right triangle.

8.MG.4b

5 The legs of the triangle are $\overline{AB}$ and $\overline{AC}$. The hypotenuse is $\overline{BC}$.

8.MG.4c

6 a 19.2 cm   b 21.8

8.MG.4d

7 12 cm

8.MG.4a

8 a $x = 20$ b $y = 19.77$

8.MG.4e

9 a The perimeter is 60.26 cm. The area is 253.17 cm^2.

b The perimeter is 76 cm. The area is 125 cm^2.

c The perimeter is 137.23 cm. The area is 890 cm^2.

d The perimeter is 54 units. The area is 107 units^2.

8.MG.5a, 8.MG.5b

10 a 40.84 cm^2

b 26 pavers

8.MG.5a, 8.MG.5c

11 a

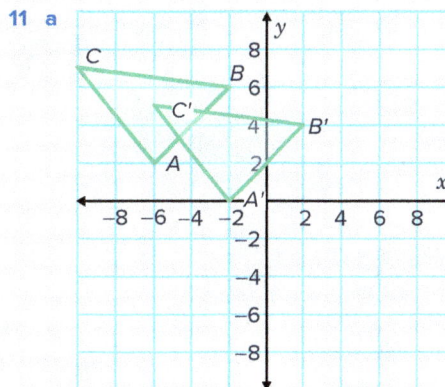

b

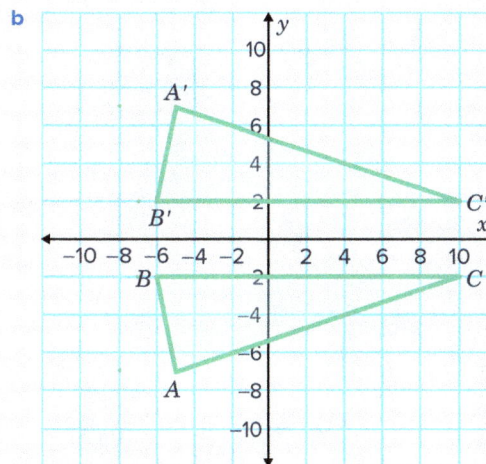

8.MG.3d, 8.MG.3e

12 A translation

8.MG.3g

13 M$\prime$(2, −3), A$\prime$(2, 3), T$\prime$(6, 3), H$\prime$(6, −3), C$\prime$(4, 8), A$\prime$(1, 8), T$\prime$(1, 4)

8.MG.3a, 8.MG.3b

14 a **A, B, D, G**

b

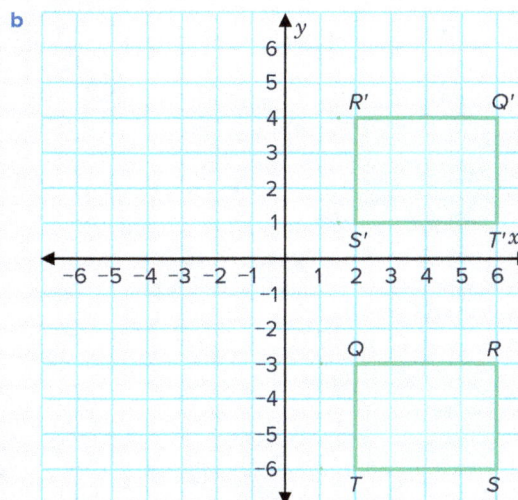

8.MG.3c, 8.MG.3f

7 3D Geometry

Big ideas

- Physical objects can be modeled with 2D and 3D geometric figures whose properties can be applied to solve real-world problems.

Chapter outline

Ever wonder why ice cream comes in a cone? Its shape keeps the ice cream from dripping everywhere and lets you enjoy a delicious, portable treat—just like a sweet, crunchy holder for your favorite flavor!

7. 3D Geometry

Topic overview

Foundational knowledge

🎧 Evaluating standards proficiency

The skills book contains questions matched to individual standards. It can be used to measure proficiency for each.

Students should be proficient in these standards.

7.MG.1 — The student will investigate and determine the volume formulas for right cylinders and the surface area formulas for rectangular prisms and right cylinders and apply the formulas in context.

5.MG.2 — The student will use multiple representations to solve problems, including those in context, involving perimeter, area, and volume.

6.MG.1 — The student will identify the characteristics of circles and solve problems, including those in context, involving circumference and area.

6.MG.2 — The student will reason mathematically to solve problems, including those in context, that involve the area and perimeter of triangles, and parallelograms.

Big ideas and essential understanding

Physical objects can be modeled with 2D and 3D geometric figures whose properties can be applied to solve real-world problems.
7.01 — A net is the 2D version of a solid shown by unfolding the edges of a solid that can be used to find the surface area of the solid.

7.02 — The volume of a square-based pyramid is related to the volume of the cube whose dimensions are the same as the square base of the pyramid.

7.03 — The volume of a cone is related to the volume of the cylinder with the same height and radius.

7.04 — Volume and surface area of 3D figures can be used to solve real-world problems.

Standards

8.MG.2 — The student will investigate and determine the surface area of square-based pyramids and the volume of cones and square-based pyramids.

8.MG.2a — Determine the surface area of square-based pyramids by using concrete objects, nets, diagrams, and formulas.
7.01 Surface area of square-based pyramids

8.MG.2b — Determine the volume of cones and square-based pyramids, using concrete objects, diagrams, and formulas.
7.02 Volume of square-based pyramids
7.03 Volume of cones

8.MG.2c — Examine and explain the relationship between the volume of cones and cylinders, and the volume of rectangular prisms and square based pyramids.

7.02 Volume of square-based pyramids
7.03 Volume of cones

8.MG.2d — Solve problems in context involving volume of cones and square-based pyramids and the surface area of square-based pyramids.

7.01 Surface area of square-based pyramids
7.02 Volume of square-based pyramids
7.03 Volume of cones
7.04 Problem solving with volume and surface area

Future connections

G.DF.1 — The student will create models and solve problems, including those in context, involving surface area and volume of rectangular and triangular prisms, cylinders, cones, pyramids, and spheres.

7.01 Surface area of square-based pyramids

Subtopic overview

Lesson narrative

In this lesson, students will learn how to calculate the surface area of square-based pyramids. A square-based pyramid consists of a square base and four triangular faces that meet at a common vertex, the apex. To find the surface area, students will add the area of the square base to the area of the four triangular faces. This involves using the formula: (Surface Area = Base Area + 4 × Triangle Area). Students will engage in exercises where they first visualize the pyramid using its net, which helps in understanding the components of the surface area. They will calculate the area of the square base using (s^2) and the area of each triangular face using $\left(\frac{1}{2} \times \text{base} \times \text{slant height}\right)$. Additionally, if only the perpendicular height is given, students will use the Pythagorean theorem to find the slant height. The exploration part includes solving problems where students create nets, calculate the areas of the base and triangular faces, and combine these to find the total surface area. By the end of the lesson, students should be able to accurately calculate the surface area of square-based pyramids and understand the relationships between the base, slant height, and the overall geometry of the pyramid.

Learning objectives

Students: Page 414

After this lesson, you will be able to...
- use concrete objects, nets, diagrams, and formulas to find the surface area of square-based pyraminds.
- solve real-world problems involving the surface area of square-based pyramids.

Key vocabulary

- net
- polyhedron
- pyramid
- square-based pyramid
- surface area

Essential understanding

A net is the 2D version of a solid shown by unfolding the edges of a solid that can be used to find the surface area of the solid.

Standards

This subtopic addresses the following Virginia 2023 Mathematics Standards of Learning standards.

Mathematical Process Goals

MPG1 — Mathematical Problem Solving

Teachers can integrate this goal into their instruction by providing students with problems that involve finding the surface area of square-based pyramids. They can challenge students to apply their knowledge of the formula for surface area and their understanding of the pyramid's structure. Teachers can also encourage students to develop problem-solving strategies by providing real-world contexts, such as determining the amount of material needed to construct a pyramid.

MPG5 — Mathematical Representations

Teachers can integrate this goal into their instruction by guiding students to represent the square-based pyramid and its surface area in different ways. This can include drawing the net of the pyramid, labeling its dimensions, and visually demonstrating how these relate to the surface area formula. Teachers can also encourage students to use different representations such as physical models, diagrams, and symbolic notation, and to connect these representations to real-life contexts.

MPG3 — Mathematical Reasoning

Teachers can integrate this process goal by guiding students to make connections to pyramids and prisms. Students should use reasoning to establish the relationship between the two shapes.

MPG4 — Mathematical Connections

Teachers can help students make mathematical connections by linking the concept of square-based pyramids and their surface areas to prior knowledge about the surface area of rectangular prisms and right cylinders. Teachers can also create connections to real-world situations such as architecture, engineering, and packaging design, showing how the mathematical concept applies in these fields.

Content standards

8.MG.2 — The student will investigate and determine the surface area of square-based pyramids and the volume of cones and square-based pyramids.

8.MG.2a — Determine the surface area of square-based pyramids by using concrete objects, nets, diagrams, and formulas.

8.MG.2d — Solve problems in context involving volume of cones and square-based pyramids and the surface area of square-based pyramids.

Prior connections

7.MG.1 — The student will investigate and determine the volume formulas for right cylinders and the surface area formulas for rectangular prisms and right cylinders and apply the formulas in context.

Future connections

G.DF.1 — The student will create models and solve problems, including those in context, involving surface area and volume of rectangular and triangular prisms, cylinders, cones, pyramids, and spheres.

Lesson Preparation

Suggested review

Depending on your students' level of prior knowledge, consider revisiting the following lessons:

Grade 6 — 7.03 Perimeter of triangles and parallelograms
Grade 6 — 7.04 Area of parallelograms
Grade 6 — 7.05 Area of triangles

Tools

You may find these tools helpful:
- Physical models of pyramids
- Printed nets
- Cardboard boxes that can be cut
- Scissors

Student lesson & teacher guide

Surface area of square-based pyramids

Students are introduced to square based pyramids before engaging in an exploration to investigate the surface area.

Use a Concrete-Representational-Abstract (CRA) approach
Targeted instructional strategies

Concrete: Begin by providing students with physical models of square-based pyramids. Use manipulatives like plastic shapes or have students build their own pyramids using cardboard or cardstock. Allow them to explore these models, touching and examining the square base and the four triangular faces that meet at the apex. Engage students in measuring the sides and slant heights with rulers to gather real measurements. You can include an activity where students cut out a net of a square-based pyramid and fold it into the 3D shape.

Representational: Transition to drawing by having students sketch the nets of square-based pyramids on graph paper. Guide them to accurately draw the square base with the four attached triangles, ensuring the sides match in length to form the proper shape when folded. Encourage students to label each side with the correct measurements they gathered from their models. Use diagrams on the board to show how the net folds into the 3D pyramid, helping them visualize the connection.

Abstract: Move on to teaching the formula for calculating the surface area of a square-based pyramid: Surface Area = (Base Area) + 4 (Triangle Area). Show students how to calculate the area of the square base and the triangular faces using their measurements. Work through examples step by step, plugging in the numbers and solving for the surface area. Encourage students to practice with different sets of numbers, reinforcing their understanding of how the formula works. Emphasize how the symbols and numbers represent the parts of the pyramid they've been working with.

Vocabulary exercise: collect and display
English language learner support

Begin the lesson by introducing key vocabulary related to square-based pyramids, such as "base," "apex," "slant height," "perpendicular height," "net," and "surface area." Provide students with physical models or diagrams of square-based pyramids and their nets to examine.

As students discuss these terms in pairs or small groups, listen for the language they use to describe the shapes and their components. For example, students might say:
- Base: "the bottom square," or "the flat part at the bottom"
- Apex: "the point at the top," or "where all the triangles meet," or "the tip of the pyramid"
- Slant height: "the length of the triangle side," or " the side from the base up to the top," or "the slanted edge of the pyramid"
- Perpendicular height: "the straight up and down height on the inside," or "the shortest distance from the base to the top," or "how tall the pyramid is"

- Net: "the shape when it's unfolded," or "the flat layout of all the faces," or "what it looks like if you open up the pyramid"
- Surface area: "the total area of all the sides," or "how much it takes to cover the outside," or "adding up all the areas"

Collect their words and phrases, and display them prominently in the classroom on a chart or word wall next to the formal definitions and labeled diagrams or illustrations. For instance, you might create a poster with the term "Apex" at the top, followed by students' descriptions and a picture showing the apex on a pyramid.

Encourage students to refer to this visual display during the lesson as they work on calculating surface areas. By stabilizing and visually reinforcing both the mathematical vocabulary and the students' own expressions, you support English language learners in connecting concepts with the language needed to understand and communicate their ideas effectively.

Exploration

Students: Page 414

> **▶ Interactive exploration**
> Explore online to answer the questions
>
> **⊕ mathspace.co**

Use the interactive exploration in 7.01 to answer these questions:

1. What is the area of the base?
2. What is the area of one of the triangular sides?
3. What is the total surface area?
4. Write a formula for surface area.
5. Write another formula using the variables B for base area and p for perimeter of the base.

Suggested student grouping: Small groups

Students will manipulate the dimensions of a square-based pyramid using sliders and then unravel it to create a net. They will then calculate the area of the base, the area of one of the triangular sides, and the total surface area. They will also be asked to write a general formula for the surface area of a square-based pyramid using variables for the base area and perimeter.

Ideal student responses

These ideal responses may differ from other correct student responses. Less formal responses can be connected with the more precise mathematical language presented here.

1. **What is the area of the base?**

   The area of the base is square of the side length.

2. **What is the area of one of the triangular sides?**

   The area of one of the triangular sides is $\frac{1}{2}$ times the base times the height.

3. **What is the total surface area?**

   The total surface area is the sum of the area of the base and the areas of all the triangular sides.

4. **Write a formula for surface area.**

The formula for the surface area of a square-based pyramid is:

$$SA = \frac{1}{2}ls + \frac{1}{2}ls + \frac{1}{2}ls + \frac{1}{2}ls + lw$$

5. **Write another formula using the variables B for base area and p for perimeter of the base.**

The formula using the variables B for base area and p for perimeter of the base is:

$$B + \frac{1}{2}lp$$

Purposeful questions

- How does changing the side length affect the surface area of the pyramid?
- How does changing the height affect the surface area of the pyramid?
- Why do we multiply the base perimeter by $\frac{1}{2}$ and the slant height in the formula?

Possible misunderstandings

- Students may not understand why the formula for the area of a triangle is used in the calculation of the surface area of a pyramid.
- Students may confuse the height of the pyramid with the slant height when calculating the area of the triangular faces.

Examples

Students: Page 416

Example 1

A square-based pyramid has a base side length of 30 meters and a slant height of 25 meters.

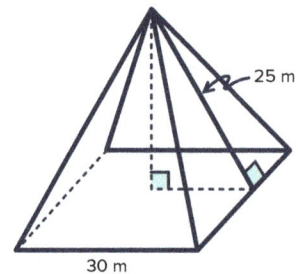

25 m

30 m

a   Create the net of the pyramid.

Create a strategy

To create the net of a pyramid, draw the base square and attach four triangles along each side of the square.

Apply the idea

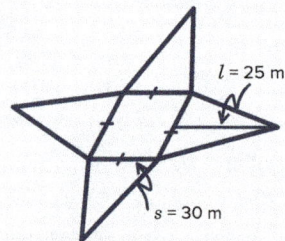

$l = 25$ m

$s = 30$ m

Purpose

Show students how to represent a 3D shape through a 2D net and understand the relationship between them.

b   Use the net to find the surface area.

Create a strategy

We need to find the area of one triangle and then multiply that by 4 because there are 4 congruent, triangular faces. Then add that to the area of the base.

Apply the idea

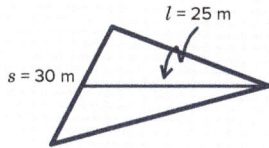

$l = 25$ m

$s = 30$ m

$s = 30$ m

$l = 25$ m

$s = 30$ m

Area of one triangle:

$$\text{Area} = \left(\frac{1}{2}ls\right)$$

$$= \left(\frac{1}{2} \cdot 25 \cdot 30\right)$$

$$= 375 \text{ m}^2$$

Area of all 4 triangles $= 4 \cdot 375 = 1500 \text{ m}^2$

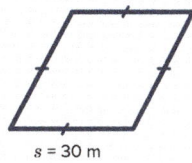

Square base:

$$\text{Area} = s^2$$

$$= 30^2$$

$$= 900 \text{ m}^2$$

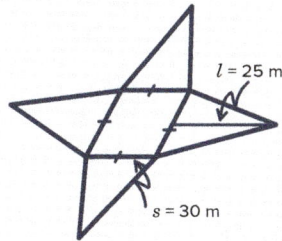

Surface area of the square based pyramid:

$$\text{Surface area} = 1500 + 900$$

$$= 2400 \text{ m}^2$$

Purpose

Challenge students to apply their understanding of area calculation for different shapes and combine them to find the surface area of a 3D shape.

Spatial scaffolding with color-coding

use with Example 1

Student with disabilities support

When teaching students how to create and interpret the net of a square-based pyramid, use color-coding to highlight key components of both the three-dimensional shape and its two-dimensional net. Begin by coloring the base of the pyramid in one color, such as blue, and each of the triangular faces in different colors like red, green, yellow, and orange. When unfolding the pyramid into its net, maintain the same colors for the corresponding parts.

This visual consistency helps students easily see how each face of the 3D shape relates to the 2D net. Encourage students to use the same color-coding in their own drawings and notes to reinforce these connections. Providing templates or diagrams with the color-coding already applied can further support students who struggle with visual-spatial processing.

Example 2

A square-based pyramid has a base side length of 9 centimeters and a height of 3 centimeters. Calculate the surface area of the pyramid.

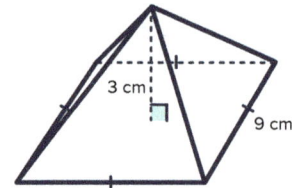

Create a strategy

Calculate the area of the square base using the formula $A_{base} = s^2$. For the triangular sides, use the formula $A_{triangle} = \frac{1}{2} \cdot$ base $\cdot$ slant height. The slant height can be found with the Pythagorean theorem: $a^2 + b^2 = c^2$.

Apply the idea

The area of the square base:

$$A_{base} = 9^2 \qquad \text{Substitute } s = 9$$
$$= 81 \text{ cm}^2 \qquad \text{Evaluate}$$

The slant height can be found with the Pythagorean theorem $a^2 + b^2 = c^2$. First, we need to find the length of the base of the triangle. We can do this by dividing the length of the square side by 2. The base of the triangle is $\frac{9}{2} = 4.5$

$$3^2 + 4.5^2 = c^2 \qquad \text{Substitute } a = 3, b = 4.5$$
$$9 + 20.25 = c^2 \qquad \text{Evaluate the exponents}$$
$$29.25 = c^2 \qquad \text{Evaluate the addition}$$
$$\sqrt{29.25} = \sqrt{c^2} \qquad \text{Take the square root of each side}$$
$$\approx 5.41 \text{ cm}^2 \qquad \text{Evaluate the radical}$$

The area of the four triangles:

$$A_{triangle} = 4 \cdot \frac{1}{2} \cdot 9 \cdot 5.41 \qquad \text{Substitute base} = 9, \text{slant height} = 5.41$$
$$= 97.38 \text{ cm}^2 \qquad \text{Evaluate}$$

To find the surface area of the square based pyramid, we need to add the area of the square base and the area of the triangles.

$$\text{Surface Area} = 81 + 97.38 \qquad \text{Add the } A_{base} \text{ and } A_{triangle}$$
$$= 178.38 \text{ cm}^2 \qquad \text{Evaluate}$$

Reflect and check

We could have also used the formula for the surface area of a pyramid: $SA = lp + B$

First we will find the perimeter of the square base. In this pyramid, each side length is 9, so the perimeter of the square is 36.

Now we can substitute the values into the formula.

$$SA = \frac{1}{2}(5.41)(36) + 81 \qquad \text{Substitute the values}$$
$$= 91.38 + 81 \qquad \text{Evaluate the multiplication}$$
$$= 178.38 \qquad \text{Evaluate the addition}$$

The surface area of the pyramid is 178.38 cm^2

Purpose

Show students how to calculate the surface area of a square-based pyramid using the formula for the area of a square and a triangle and also the Pythagorean theorem.

Expected mistakes

Students may substitute 3 as the slant height rather than solving for the missing value. Remind students that the formula requires slant height and must be found using the Pythagorean theorem.

Example 3

A square pyramid has a surface area of 96 ft^2. Each triangular face has an area of 15 ft^2.

Find the side length of the base of the pyramid, correct to the nearest foot.

Create a strategy

The area of the base can be found by subtracting the surface area by the area of the triangles. There are four congruent triangular faces in a square pyramid.

Once we have the area of the square base, we can solve for the side length of the square.

Apply the idea

There are four triangular faces in a square pyramid. The total area of the triangular faces is given by:

$$4 \cdot 15 \text{ ft}^2 = 60 \text{ ft}^2$$

Area of the base = Surface area — Area of the triangles

$= 96 - 60$	Substitute the values
$= 36$	Evaluate

Area of the square base $= s^2$, where s is the side length of the square base.

$s^2 = s \cdot s$	Rewrite in expanded form
$s \cdot s = 36$	Consider what number times itself is equal to 36
$s = 6$	Evaluate

The side length is 6 ft.

Purpose

Show students how to derive and solve an equation to find the side length of the base in a square pyramid.

Example 4

Joan visited Pyramid of Giza, which is a square pyramid. At the gift shop, she bought a miniature replica of the pyramid that has a slant height of 6 inches and a base length of 4 inches. What is the surface area of the replica?

Create a strategy

We can use the formula $\frac{1}{2}lp + B$ to find the surface area.

Apply the idea

First we will find the area of the base, we have:

$B = (\text{side length})^2$	Area of a square
$= 4^2$	Substitute the side length
$= 16$	Simplify

We can now will calculate the perimeter of the base:

$p = (\text{side length}) \cdot 4$	Perimeter of a square
$= 4 \cdot 4$	Substitute known values
$= 16$	Evaluate the multiplication

Now we can substitute those values into the formula for surface area of a square pyramid: $SA = \frac{1}{2}lp + B$

$SA = \frac{1}{2}lp + B$	Surface area of a square pyramid
$= \frac{1}{2}(6)(16) + (16)$	Substitute known values
$= 48 + 16$	Evaluate the multiplication
$= 64$	Evaluate the addition

The surface area of the replica is 64 square inches.

Purpose

To ensure students can calculate surface area of a pyramid in context.

Reflecting with students

Encourage students to use appropriate units throughout the task to enhance their mathematical precision. In this example, prompt them to include units at every step, such as writing $B = (4 \text{ inches})^2$ and $B = 16 \text{ in}^2$ when calculating the area of the base. Emphasize the importance of carrying units through all calculations, like stating $p = 4 \times 4 \text{ inches} = 16 \text{ inches}$ for the perimeter. When substituting values into the surface area formula, guide students to write $SA = \frac{1}{2} \times 6 \text{ inches} \times 16 \text{ inches} + 16 \text{ in}^2$. By consistently including units, students not only ensure the correctness of their solutions but also deepen their understanding of the measurements involved. This practice helps prevent errors and promotes precision in their mathematical communication.

Students: Page 419

> 💡 **Idea summary**
>
> A square pyramid is a polyhedron with a square base and four faces that are congruent triangles with a common vertex.
>
> To find the surface area of a square pyramid, we can use the equation:
>
> $$\text{Surface area of a square pyramid} = \frac{1}{2}lp + B$$
>
> If we are given the perpendicular height, we can find the slant height by using the Pythagorean theorem: $a^2 + b^2 = c^2$.

Practice

Students: Page 420–422

What do you remember?

1 Consider this regular square pyramid.

   a How many identical triangular faces does it have?

   b Find the area of one triangular face.

   c Find the total surface area of the triangular faces of the pyramid.

   d Find the total surface area of the pyramid.

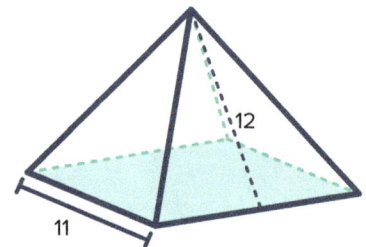

2 Use the net to find the surface area of these square pyramids, round your answer to two decimal places.

   a 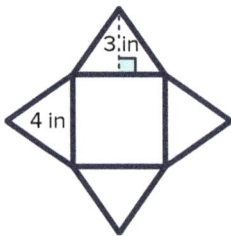   **b** 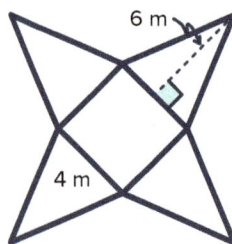   **c** 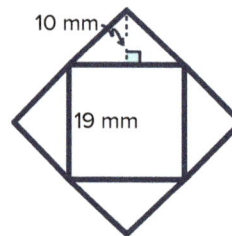   **d**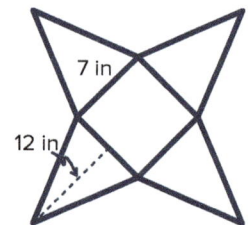

3 Which net matches this square-based pyramid? Justify your answer.

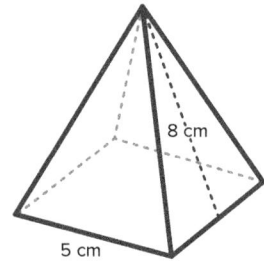

A

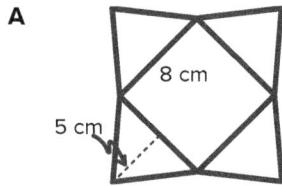

8 cm

5 cm

B

5 cm

8 cm

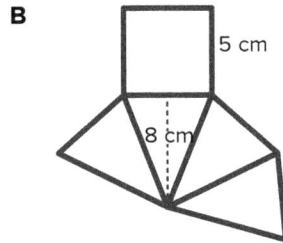

C

8 cm

5 cm

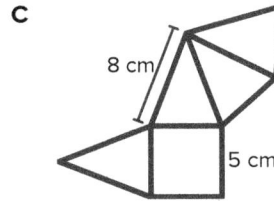

D

5 cm

8 cm

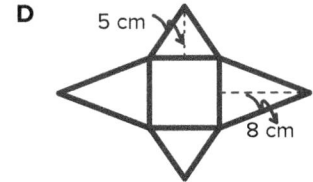

4 Consider this square-based pyramid:
  a Sketch the net of the pyramid.
  b Find the surface area.

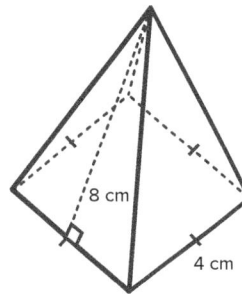

8 cm

4 cm

5 Consider this regular square pyramid:
  a Find the perimeter of the base.
  b Find the area of the base.
  c Identify the slant height.
  d Find the total surface area of the pyramid, rounded to two decimal places.

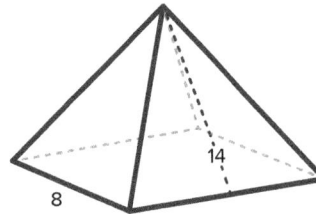

14

8

Let's practice

6 Find the surface area:

a

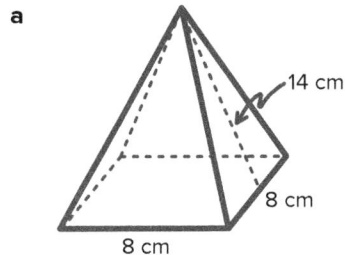

14 cm

8 cm

8 cm

b

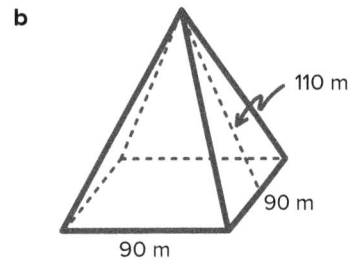

110 m

90 m

90 m

7 Find the surface area of the following square-based pyramids with dimensions:
  a Base side length of 6 ft and a slant height of 5 ft.
  b Base side length of 12 units and a slant height of 15 units.
  c Base side length of 10 cm and a slant height of 15 cm.

8 Calculate the surface area of each square-based pyramid.

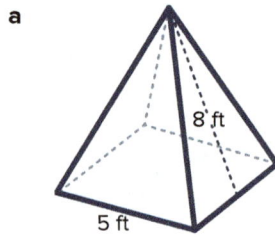

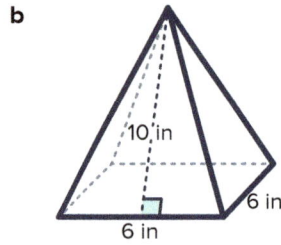

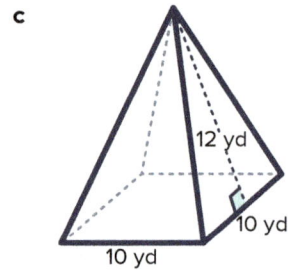

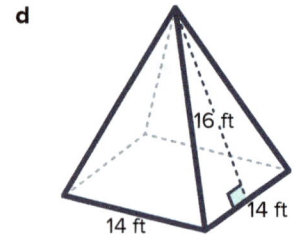

a

8 ft

5 ft

b

10 in

6 in

6 in

c

12 yd

10 yd

10 yd

d

16 ft

14 ft

14 ft

9 Raymond needs to cover the entire surface of this square-based pyramid with paper.

What is the amount of paper he will need?

A 408 square feet **B** 276 square feet

C 672 square feet **D** 528 square feet

11 ft

12 ft

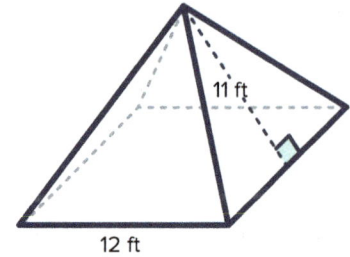

10 As part of an art project, Billy has to paint the surface area of a square-based pyramid.
The pyramid has the dimensions shown.

What is the total surface area of the pyramid?

A 175 square inches **B** 252 square inches

C 301 square inches **D** 441 square inches

9 in

7 in

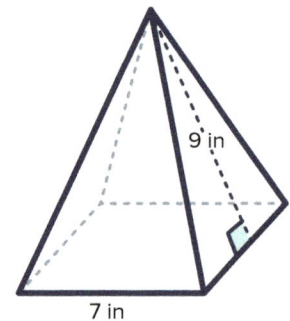

11 Compare the surface area of a square-based pyramid with a base side of 5 cm and a slant height of 10 cm to another with a base side of 10 cm and a slant height of 5 cm.

12 Consider this square-based pyramid:

a Find the surface area of the pyramid. Use the formula:

SA = Area of base + Area of the triangles

b Using the formula

$$SA = \frac{1}{2} \cdot l \cdot p + B$$

where l is the slant height, p is the perimeter of the base, and B is the area of the base.

Calculate the surface area of the pyramid.

c Compare your answers. What do you notice?

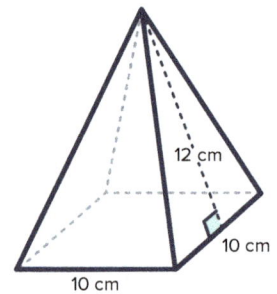

12 cm

10 cm

10 cm

13 The Louvre pyramid is a large glass and metal pyramid which serves as the entrance to the Louvre museum in Paris. It is a square pyramid with a slant height of about 92 ft and a base length of about 112 ft.

If the surface of the pyramid (excluding the base) is entirely covered in glass, how many square meters of glass make up the structure?

14 Some very famous right square pyramids are the Egyptian pyramids located at the Giza Pyramid Complex.

The largest of the pyramids at the site is the Great Pyramid, which has a base of 230 m and each of its triangular faces have a height of 216 m.

Find the surface area of the Great Pyramid, not including the base.

15 A spire is an architectural feature which originated in the 12th century in Germany and can be seen throughout Europe in Gothic and Baroque architecture.

A building has two spires with square bases. The width of the base is 4 ft and the slant height of each side of the spire is 35 ft.

How much paint is required to cover both spires?

Let's extend our thinking

16 A square pyramid has a surface area of 225 ft^2. Each triangular face has an area of 50 ft^2.

Find the side length of the base of the pyramid, rounded to the nearest foot.

17 A square-based pyramid has a surface area of 600 square units and the perimeter of its base is 40 units. Find the slant height of the pyramid.

18 What is the slant height of the square pyramid if its surface area, excluding the base, is 252 square inches and the side length of the base is 9 in?

19 Given this image, find the length of an edge that connects a corner of the square to the apex.

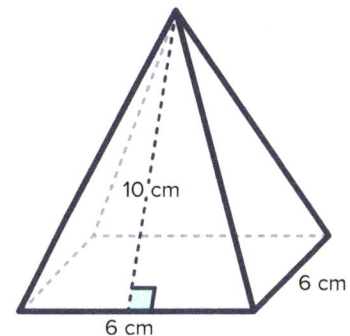

10 cm

6 cm

6 cm

20 The roof of a large public building is in the shape of a square-based pyramid, as shown below:

  a Find the slant height, round your answer to two decimal places.

  b Calculate the surface area of the roof, excluding the base of the pyramid. Round your answer to the nearest square meter.

  c Each tile used on the roof has an area of 600 cm^2. How many tiles are used to cover the roof?

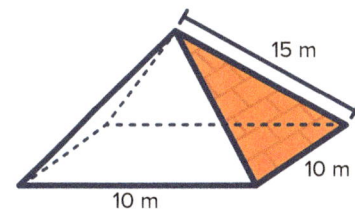

15 m

10 m

10 m

Answers

7.01 Surface area of square-based pyramids

What do you remember?

1 a 4 b 66 units^2 c 264 units^2 d 385 units^2

2 a 40 in^2 b 64 m^2 c 741 mm^2 d 217 in^2

3 Net B is the net that matches the square-based pyramid, as it has the correct dimensions and shape to form the pyramid when folded. The slant height of the triangles matches the 8 cm measurement required to reach from the base to the apex of the pyramid, and the base matches the 5 cm given side length.

4 a

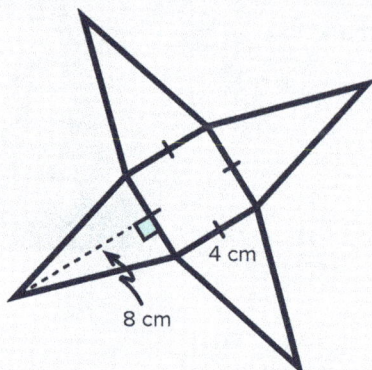

    b 80 cm^2

5 a 32 units b 64 square units
    c 14 units d 296.96 square units

Let's practice

6 a 288 cm^2 b 27 900 m^2

7 a 96 ft^2 b 504 square units
    c 400 cm^2

8 a 105 ft^2 b 156 in^2
    c 340 yd^2 d 644 ft^2

9 **A**

10 **A**

11 For the first pyramid with a base side of 5 cm and a slant height of 10 cm , the surface area is calculated as:

$$5^2 + \frac{1}{2} \cdot 5 \cdot 10 \cdot 4 = 125 \text{ cm}^2$$

For the first pyramid with a base side of 10 cm and a slant height of 5 cm , the surface area is calculated as:

$$10^2 + \frac{1}{2} \cdot 10 \cdot 5 \cdot 4 = 200 \text{ cm}^2$$

Therefore, comparing the surface areas, the second pyramid with a base of 10 cm and a slant height of 5 cm has a larger surface area (200 cm^2) compared to the first pyramid with a base side of 5 cm and a slant height of 10 cm (125 cm^2)

12 a 340 cm^2 b 340 cm^2
    c The surface areas are the same, regardless of which formula is used.

13 20 608 ft^2

14 99 360 m^2

15 560 ft^2 of paint

Let's extend our thinking

16 5 ft

17 $h = 25$ units

18 14 in

19 10.44 cm

20 a 14.14 m b 283 m^2 c 4717

7.02 Volume of square-based pyramids

Subtopic overview

Lesson narrative

In this lesson, students will learn how to calculate the volume of square-based pyramids. A square-based pyramid is a polyhedron with a square base and four triangular faces converging at a single apex. The volume of such a pyramid can be determined using the formula: $(V = \frac{1}{3}Bh)$ where (B) is the area of the base and (h) is the height from the base to the apex. Students will start by understanding the relationship between the volume of pyramids and rectangular prisms. The volume of a pyramid is one-third the volume of a prism with the same base area and height. This concept will be reinforced through examples such as calculating the volume of the Pyramid of Giza by first finding the area of its square base and then using the given height. The exploration involves solving problems where students calculate the volume of pyramids and practice subtracting volumes to find the remaining volume of a modified pyramid. For example, students will determine the volume of a larger pyramid and subtract the volume of a smaller pyramid removed from its top to find the volume of the remaining solid. By the end of the lesson, students should be able to apply the volume formula for square-based pyramids, understand the geometric properties involved, and solve problems involving composite shapes by calculating and subtracting volumes accurately.

Learning objectives

Students: Page 423

After this lesson, you will be able to...
- use concrete objects, nets, diagrams, and formulas to find the volume of square-based pyramids.
- solve real-world problems involving the volume of square-based pyramids.
- explain the relationship between the volume of rectangular prisms and square based pyramids.

Key vocabulary

- volume

Essential understanding

The volume of a square-based pyramid is related to the volume of the cube whose dimensions are the same as the square base of the pyramid.

Standards

This subtopic addresses the following Virginia 2023 Mathematics Standards of Learning standards.

Mathematical Process Goals

MPG1 — Mathematical Problem Solving

Teachers can incorporate this goal into their lessons by guiding students in using the formula for finding the volume of a square-based pyramid. They can provide students with different problem situations that require the application of this formula and support them in identifying the appropriate strategies for solving these problems. Teachers can also encourage students to create their own problems based on real-world scenarios involving square-based pyramids.

MPG4 — Mathematical Connections

Teachers can help students make mathematical connections by linking the concept of volume of a square-based pyramid to the volume of rectangular prisms. They can guide students to visualize and understand how these two shapes are related and show how the formulas for finding their volumes are connected. Additionally, they can present real-world scenarios where both concepts are applicable, such as in architecture or packaging design.

MPG5 — Mathematical Representations

Teachers can promote this goal by providing students with various representations of square-based pyramids, such as concrete models, diagrams, and symbolic notation. They can guide students in understanding how these different representations relate to the formula for finding the volume. Moreover, they can encourage students to create their own representations and explain how they can use these to solve problems involving the volume of square-based pyramids.

Content standards

8.MG.2 — The student will investigate and determine the surface area of square-based pyramids and the volume of cones and square-based pyramids.

8.MG.2b — Determine the volume of cones and square-based pyramids, using concrete objects, diagrams, and formulas.

8.MG.2c — Examine and explain the relationship between the volume of cones and cylinders, and the volume of rectangular prisms and square based pyramids.

8.MG.2d — Solve problems in context involving volume of cones and square-based pyramids and the surface area of square-based pyramids.

Prior connections

7.MG.1 — The student will investigate and determine the volume formulas for right cylinders and the surface area formulas for rectangular prisms and right cylinders and apply the formulas in context.

5.MG.2 — The student will use multiple representations to solve problems, including those in context, involving perimeter, area, and volume.

Future connections

G.DF.1 — The student will create models and solve problems, including those in context, involving surface area and volume of rectangular and triangular prisms, cylinders, cones, pyramids, and spheres.

Teacher introduction

Suggested review

Depending on your students' level of prior knowledge, consider revisiting the following lessons:

Grade 8 — 7.01 Surface area of square-based pyramids

Student lesson & teacher guide

Volume of square-based pyramids

In this lesson, we'll explore how to calculate the volume of pyramids and understand the relationship between their volume and that of rectangular prisms. Starting with the basics, we define volume as the amount of space an object occupies. We then transition to the geometric properties of a rectangular prism and the formula $V = Bh$, where B represents the area of the base and h the height.

▌ Exploration

Students: Page 423

> **▶ Interactive exploration**
> **Explore online to answer the questions**
>
> ⊕ **mathspace.co**

Use the interactive exploration in 7.02 to answer these questions:

1. How many pyramids fit into the prism?

2. Knowing the formula of volume for a prism, how do you think we can find the volume of a pyramid?

Suggested student grouping: Small groups

This exploration will help us derive the formula for the volume of a pyramid, $V = \frac{1}{3}Bh$, emphasizing that the volume of a pyramid is precisely one-third the volume of a prism with the same base area and height.

Ideal student responses

These ideal responses may differ from other correct student responses. Less formal responses can be connected with the more precise mathematical language presented here.

1. **How many pyramids fit into the prism?**
   The image suggests a demonstration where multiple pyramids fit into a single prism. Given the typical geometrical setup where the pyramids perfectly complement the space within the prism without overlap or voids, it is commonly demonstrated that three pyramids can fit into one rectangular prism of the same base and height. This assumes the pyramids are right pyramids with the apex directly above the center of the base.

2. **Knowing the formula of volume for a prism, how do you think we can find the volume of a pyramid?**
   Knowing that the volume V of a prism is given by $V = Bh$ (where B is the base area and h is the height), and from the exploration setup that three pyramids fit into a prism, we can infer that each pyramid occupies one-third of the volume of the prism. Therefore, the volume V of a pyramid can be found using the formula $V = \frac{1}{3}Bh$ This stems from the geometric arrangement where the pyramid occupies a third of the volume of the encompassing prism, assuming the base area and height remain constant.

Purposeful questions

- How does the volume of the pyrmaid relate to the volume of the prism?
- How does the formula you developed relate to the volume of other geometric shapes you know? Why was $\frac{1}{3}$ used?

Possible misunderstandings

- Students may think they need to use the slant height when calculating volume because of its use in the surface area formula. Remind students that the measure they will use for calulating volume is the distance from the center of the base to the apex of the pyramid.

Think algorithmically
Targeted instructional strategies

Encourage students to develop an algorithmic approach to calculating the volume of square-based pyramids, accounting for different given measurements and partial pyramids. Guide them to create a step-by-step procedure that handles cases where the side length of the base is provided instead of the area, and for pyramids that are truncated. Have them practice this algorithm with various examples to test and refine their procedure. Here is an exemplar set of steps they might develop:

1. Write out the given measurements
   - If the area of the base is given, use as is
   - If the side length of the base is given, determine area using: $A = 4s$

2. Write out the formula
   - If a whole pyramid is not given, then rewrite the volume formula by subtracting the volume of the missing section.

3. Substitute all the known values into the formula.

4. Calculate the volume by evaluating the expression.

5. State the final volume, including appropriate units.

Stronger and clearer each time
English language learner support

Begin by asking students to individually write an explanation of why the volume of a square-based pyramid is one-third the volume of a rectangular prism with the same base area and height. Then, have students pair up and share their explanations with a partner. Encourage them to ask each other questions and provide feedback to clarify and expand their ideas.

After the first discussion, ask students to revise their explanations, incorporating any new insights or vocabulary. Repeat this process with a new partner to give students another opportunity to refine their thoughts. Finally, have students write a final version of their explanation, which should show improved clarity and use of mathematical language. This activity helps students deepen their understanding while practicing communicating mathematical concepts in English.

Pattern recognition in formulas
Student with disabilities support

Students can often struggle to remember various formulas, especially those that look similar such as the volume formulas for prisms and pyramids. Encourage students to look for patterns in these formulas to help them remember. For example, the volume formula for a prism is $V = Bh$ and for a pyramid is $V = \frac{1}{3}Bh$. The only difference is the $\frac{1}{3}$ multiplier in the pyramid formula. This reflects the fact that three pyramids can fit inside a prism with the same base and height. This understanding can help students remember which formula to use.

Examples
Students: Page 424

Example 1

The Pyramid of Giza is a square pyramid, that is 280 Egyptian royal cubits high and has a base length of 440 Egyptian royal cubits. What is the volume of the Pyramid of Giza?

Create a strategy

First, we find the area of the base, then we can use that to find the volume. Since this solid is a square pyramid, the base is a square.

Apply the idea

Finding the area of the base, we have:

$$B = \text{side length}^2 \qquad \text{Area of a square}$$
$$= 440^2 \qquad \text{Substitute the side length}$$
$$= 193600 \qquad \text{Simplify}$$

We can now use this to calculate the volume:

$$V = \frac{1}{3}Bh \qquad \text{Volume of a pyramid}$$
$$= \frac{1}{3} \cdot 193600 \cdot 280 \qquad \text{Substitute known values}$$
$$= 18069333\frac{1}{3} \qquad \text{Simplify}$$

So, the volume of the Pyramid of Giza is $18069333\frac{1}{3}$ cubic Egyptian royal cubits.

Purpose

Demonstrate to students how to calculate the volume of a pyramid using real world examples, reinforcing the concepts of area and volume calculations.

🎓 **Advanced learners: Linking volume to scaling and similarity**   use with Example 1
Targeted instructional strategies

Encourage students to explore how scaling the dimensions of a pyramid affects its volume, connecting this problem to the broader concepts of similarity and proportional reasoning in geometry. Prompt students to consider what happens if the dimensions of the Pyramid of Giza are doubled or tripled. Guide them to discover that while the linear dimensions scale by a factor of k, the volume scales by a factor of k^3.

This investigation reinforces their understanding of how changes in one-dimensional measurements impact two-dimensional areas and three-dimensional volumes. By linking the specific calculation of the pyramid's volume to these larger mathematical ideas, students deepen their comprehension of geometric relationships and scaling principles. Use visual representations or scale models to illustrate how the pyramid's volume changes with different scaling factors, making the abstract concept more tangible.

Examples

Students: Page 424–425

Example 2

A small square pyramid of height 4 cm was removed from the top of a large square pyramid of height 8 cm forming the solid shown. Find the exact volume of the solid.

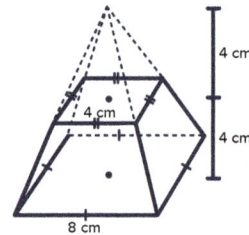

Create a strategy

Subtract the volume of the smaller pyramid from the volume of the larger pyramid.

Apply the idea

Start by calculating the volume of the larger pyramid. We have

$$V = \frac{1}{3}Bh$$ Volume of a pyramid

$$= \frac{1}{3}(64)(8)$$ Substitute $B = 64$ for the area of the base and $h = 8$ for the height of the larger pyramid

$$= \frac{512}{3}$$ Evaluate the multiplication

The volume of the larger pyramid is exactly $\frac{512}{3}$ cm^3.

Now, we can calculate the volume of the smaller pyramid. We have

$$V = \frac{1}{3}Bh$$ Volume of a pyramid

$$= \frac{1}{3}(16)(4)$$ Substitute $B = 16$ for the area of the base and $h = 4$ for the height of the smaller pyramid

$$= \frac{64}{3}$$ Evaluate the multiplication

The volume of the smaller pyramid is exactly $\frac{64}{3}$ cm^3.

The volume of the solid after removing the smaller pyramid is $\frac{512}{3}$ cm^3 $-$ $\frac{64}{3}$ cm^3 $=$ $\frac{448}{3}$ cm^3.

Reflect and check

We could choose to leave the solution in exact form, since the directions do not specify rounding requirements, or we could simplify the solution and round to a measure we find appropriate.

Purpose

Show students how to calculate the exact volume of a complex solid by subtracting the volume of one shape from another.

Expected mistakes

Students might mistakenly believe that if the height of the small pyramid is half the height of the large pyramid, then the base area of the small pyramid is also half the base area of the large pyramid. They may not realize that in similar pyramids, while linear dimensions scale by a factor, areas scale by the square of that factor.

Think-aloud strategy modeling
Student with disabilities support

use with Example 2

Model your thinking process step-by-step as you solve the problem in front of the class. Read the problem aloud and expressing your thoughts:

- "We need to find the exact volume of the solid formed by removing a smaller pyramid from a larger one."
- "First, I'll calculate the volume of the larger pyramid using the formula $V = \frac{1}{3}Bh$. Since the base area B is..."
- "I'm substituting $B = 64$ because the base side length is... Wait, let's make sure that's correct."
- By narrating your thought process, you help students understand how to approach complex problems and maintain focus. Encourage students to practice this strategy by working through similar problems aloud, either individually or with a partner, to reinforce their understanding and attention to each step.

Students: Page 425

Idea summary

The volume of a pyramid can be found by taking one-third the volume of a prism with the same base area and height. The formula for the volume of a pyramid is given by:

$$V = \frac{1}{3}Bh$$

B area of the base

h perpendicular height of the apex from the base

Practice

Students: Page 425–428

What do you remember?

1   Find the volume of the rectangular prism.

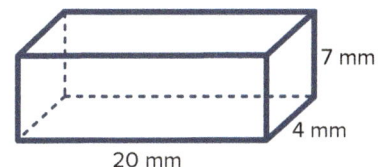

7 mm

4 mm

20 mm

2   Find the volume of a cube with the following side lengths:

   a   10 in

   b $\frac{5}{4}$ ft

3 Consider the image.

    a Identify the:

       i Side length of the base

       ii Height

       iii Area of the base

    b The volume of the square-based pyramid is represented by:

$$V = \frac{1}{3} \cdot (\square) \cdot (\square)$$

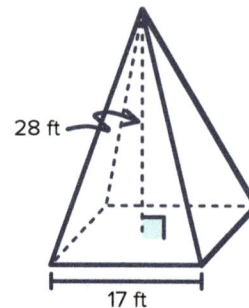

28 ft

17 ft

4 What will be the volume of a square-based pyramid with base sides 10 cm and a height of 18 cm?

    A 600 cm   **B** 600 cm^2 **C** 600 cm^3 **D** 600 cm^4

Let's practice

5 Calculate the volume of the pyramid.

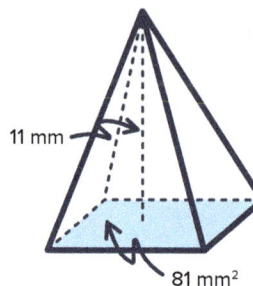

11 mm

81 mm^2

(SOL) 6 What is the volume of the square-based pyramid shown?

    A 196 ft^3 **B** 588 ft^3

    C 4116 ft^3 **D** 12 348 ft^3

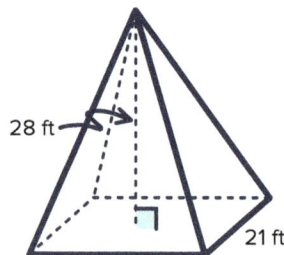

28 ft

21 ft

7 Find the volume of the pyramids shown. Round your answer to two decimal places where necessary.

    a   **b**   **c**   **d**

 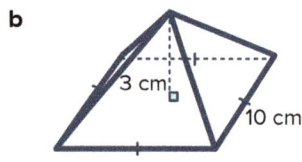 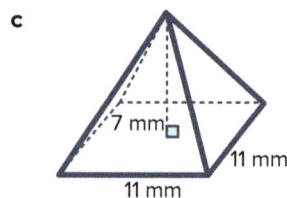 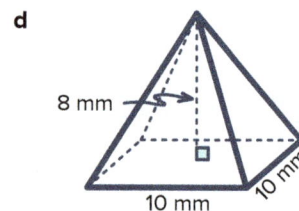

6 cm   10 cm   3 cm   10 cm   7 mm   11 mm   11 mm   8 mm   10 mm   10 mm

(SOL) 8 What is the volume of the square-based pyramid shown?

    A 27 in^3 **B** 32 in^3

    C 80 in^3 **D** 160 in^3

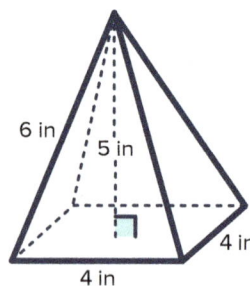

6 in

5 in

4 in

4 in

9 Calculate the volume of the following square-based pyramids. Round your answer to two decimal places where necessary.

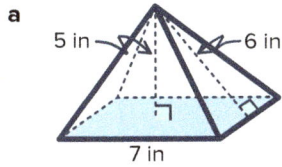

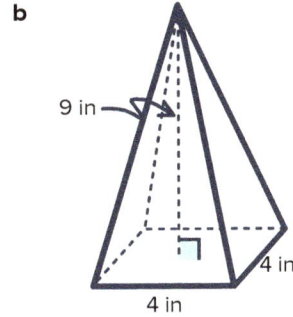

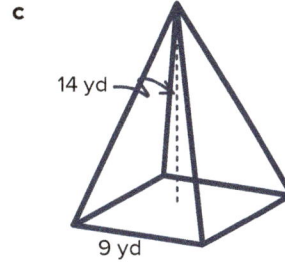

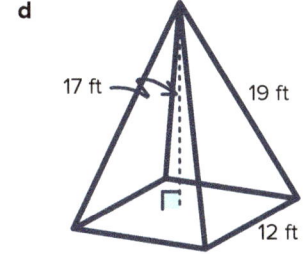

a

5 in 6 in

7 in

b

9 in

4 in

4 in

c

14 yd

9 yd

d

17 ft 19 ft

12 ft

10 Complete the sentence using >, <, or =

The volume of Pyramid 1 is ☐ the volume of Pyramid 2.

Pyramid 1

9 yd

11 yd

11 yd

Pyramid 2

8 yd

12 yd

12 yd

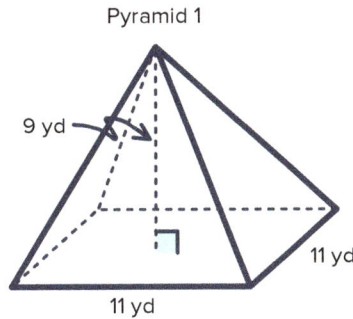

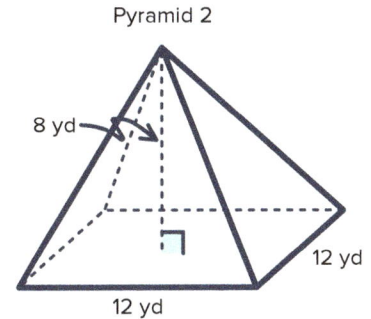

11 A cube has a side length of 12 cm, and a square-based pyramid has the same base side length and height as the cube.

a Calculate the volume of the cube.

b Calculate the volume of the square-based pyramid.

c If multiple square-based pyramids are melted down, how many are needed to completely fill the cube without leaving any space? Justify your answer.

12 A rectangular prism and a square-based pyramid have identical base areas and heights.

a Find the volume of each shape.

b Find the ratio of the volumes. Explain the meaning of the ratio in this context.

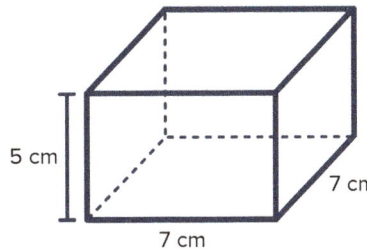

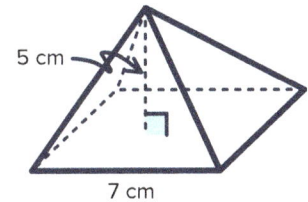

5 cm

7 cm

7 cm

5 cm

7 cm

13 A right square-based pyramid has a height of 24 in and a volume 2592 in³. Find the base length of the pyramid.

14 A right square-based pyramid has a base of length 12 in and a volume of 1248 in³. Find the height of the pyramid.

15 The volume of a pyramid is 80 yd³ with a height of 6 yd.

a Find the area of the pyramid's base.

b Explain how the volume of the pyramid relates to the volume of a rectangular prism with the same height and base.

16 A water tank is in the shape of a rectangular prism with a length and width of 12 m and a height of 7 m. A decorative fountain in the shape of a square-based pyramid has the same base dimensions and height.

If the water tank is full, how much water is needed to completely fill the fountain?

17 Isolda needs to restock on shampoo and found two different bottles as shown.

Bottle A
$12.00

Bottle B
$13.50

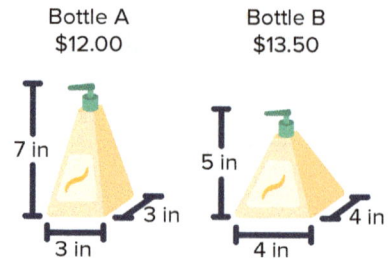

    a The volume of shampoo in bottle B is about how many times the volume in bottle A?

    b Which is a better buy?

18 A square-based pyramid made of iron with base edges of 8.7 ft and height 11.9 ft is melted down and recast in the shape of a cube.

    a Find the volume of the pyramid, correct to two decimal places.

    b Find the side length of the cube, correct to one decimal place.

19 Consider a square-based pyramid and a rectangular prism that both have a base side length of 8 ft. The height of the pyramid is 6 ft and the height of the rectangular prism is 18 ft.

    a Calculate the volume of both shapes.

    b Explain how tripling the height of the pyramid will affect its volume, and compare the new volume to the rectangular prism's volume.

20 A square-based pyramid is designed to fit perfectly inside a rectangular prism so that the pyramid's base and one face of the prism have the same dimensions, and the pyramid's height is equal to the height of the prism.

    a If the rectangular prism has a length and width of 14 cm and a height of 11 cm, calculate the volume of the unfilled space between the pyramid and the prism.

    b Explain how understanding the relationship between their volumes helps in solving this problem.

21 A cube with a pyramid is shown. The pyramid's apex is at the center of the cube, and 6 pyramids fit inside the cube. Find the total volume.

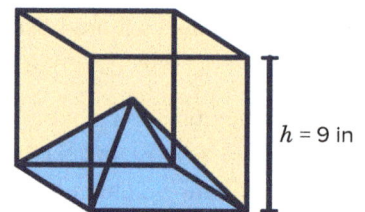

$h = 9$ in

Answers

7.02 Volume of square-based pyramids

What do you remember?

1 560 mm^3

2 **a** 1000 in^3 **b** $\dfrac{125}{64}$ ft^3

3 **a** **i** 17 ft **ii** 28 ft **iii** 289 ft^2

  b $V = \dfrac{1}{3} \cdot (289) \cdot (28)$

4 C

Let's practice

5 297 mm^3

6 C

7 **a** 200 cm^3 **b** 100 cm^3

  c 282.33 mm^3 **d** 266.67 mm^3

8 **A**

9 **a** 81.67 in^3 **b** 48 in^3

  c 378 yd^3 **d** 816 ft^3

10 The volume of pyramid 1 is < the volume of pyramid 2.

11 **a** 1728 cm^3

  b 576 cm^3

  c 3 square-based pyramids can fill the cube without leaving any space. The volume of the cube is 3 times that of the square-based pyramid, so 576 · 3 = 1728

12 **a** Volume of rectangular prism = 245 cm^3

  Volume of square-based pyramid = $81\dfrac{2}{3}$ cm^3

  b The ratio of the volume is 3 : 1. This ratio shows that that the volume of the square-based pyramid is exactly one-third of the volume of a cube with the same side length and height. This illustrates how the formula for the volume of the pyramid is derived from the volume formula of a cube. It also demonstrates that when we calculate the volume of a pyramid, we're essentially finding a fraction of the volume of a corresponding cube.

13 18 in

14 26 in

15 **a** 40 yd^2

  b The volume of the pyramid is one-third of the volume of a rectangular prism with the same base and height.

  Volume of rectangular prism: 40 · 6 = 240 yd^3

  Volume of pyramid: 80 yd^3

  $\dfrac{80}{240} = \dfrac{1}{3}$

16 336 m^3

Let's extend our thinking

17 **a** The volume of shampoo in bottle B is 27 in^3, which is 1.3 times the volume in bottle A of 21 in^3.

  b Bottle B is a better buy.

18 **a** 300.24 ft^3 **b** 6.7 ft

19 **a** Volume of square-based pyramid = 128 ft^3

  Volume of rectangular prism = 1152 ft^3

  b Tripling the height of the pyramid will increase its volume to 384 ft^3, which is equal to $\dfrac{1}{3}$ of the rectangular prism's volume $\left(\dfrac{384}{1152} = \dfrac{1}{3} \right)$.

20 **a** $1437\dfrac{1}{3}$ cm^3

  b Understanding that the volume of a square-based pyramid is equal to one-third of the volume of a rectangular prism when their bases' dimensions and height are equal, suggests that two-thirds of the prism's volume is the unfilled space.

21 729 in^3

7.03 Volume of cones

Subtopic overview

Lesson narrative

In this lesson, students will learn how to calculate the volume of cones. They will begin by recalling the volume formula for cylinders and understanding that the volume of a cone is one-third that of a cylinder with the same base and height. The formula for the volume of a cone is $(V = \frac{1}{3}\pi r^2 h)$, where (r) is the radius of the base and is (h) the height. Students will practice applying this formula through various examples. For instance, given the radius and height, they will use the formula to find the volume of a cone. Additionally, they will solve problems where the volume and one other dimension are provided, and they need to find the missing dimension. The exploration includes problems such as finding the volume of a cone with given dimensions and calculating the height of a cone when the volume and radius are known. By substituting the known values into the formula and solving for the unknown variable, students will enhance their problem-solving skills and understanding of geometric relationships. By the end of the lesson, students should be able to confidently use the volume formula for cones, understand the relationship between the volumes of cones and cylinders, and solve practical problems involving the dimensions and volumes of cones.

Learning objectives

Students: Page 429

🎓 **After this lesson, you will be able to...**

- use concrete objects, nets, diagrams, and formulas to find the volume of cones.
- solve real-world problems involving the volume of cones.
- explain the relationship between the volume of cones and cylinders.

Key vocabulary

- cone
- cylinder

Essential understanding

The volume of a cone is related to the volume of the cylinder with the same height and radius.

Standards

This subtopic addresses the following Virginia 2023 Mathematics Standards of Learning standards.

Mathematical process goals

MPG1 — Mathematical Problem Solving

Teachers can integrate this goal by presenting students with various problems to solve using the formula for the volume of cones. These problems could be based on real-life situations, such as finding the volume of an ice cream cone or a traffic cone. The complexity of the problems can be varied, and students can be encouraged to apply different strategies to solve them.

MPG4 — Mathematical Connections

Teachers can make connections to students' prior knowledge and other disciplines. For example, they can remind students about the volume formula for right cylinders they learned in previous lessons and relate it to the concept of cones. They can also connect the concept of the volume of cones to various fields such as architecture and engineering, demonstrating how mathematics is applied in different disciplines.

MPG5 — Mathematical Representations

Teachers can integrate this goal by encouraging students to represent the concept of cones and their volumes using various methods. They can use concrete models, drawings, and symbols to represent the cones. They can also guide students to make connections between these different representations and understand that they all convey the same mathematical idea. For instance, students can be asked to draw a cone, label its dimensions, calculate its volume using the formula, and then represent this information in a table.

Content standards

8.MG.2 — The student will investigate and determine the surface area of square-based pyramids and the volume of cones and square-based pyramids.

8.MG.2b — Determine the volume of cones and square-based pyramids, using concrete objects, diagrams, and formulas.

8.MG.2c — Examine and explain the relationship between the volume of cones and cylinders, and the volume of rectangular prisms and square based pyramids.

8.MG.2d — Solve problems in context involving volume of cones and square-based pyramids and the surface area of square-based pyramids.

Prior connections

7.MG.1 — The student will investigate and determine the volume formulas for right cylinders and the surface area formulas for rectangular prisms and right cylinders and apply the formulas in context.

5.MG.2 — The student will use multiple representations to solve problems, including those in context, involving perimeter, area, and volume.

6.MG.1 — The student will identify the characteristics of circles and solve problems, including those in context, involving circumference and area.

Future connections

G.DF.1 — The student will create models and solve problems, including those in context, involving surface area and volume of rectangular and triangular prisms, cylinders, cones, pyramids, and spheres.

Lesson Preparation

Suggested review

Depending on your students' level of prior knowledge, consider revisiting the following lessons:

Grade 7 — 7.04 Volume of right cylinders

Tools

You may find these tools helpful:
- Dynamic geometry software
- Physical models of cones
- Printed nets

Student lesson & teacher guide

Volume of cones

Students will apply what they've learned about the formula for the volume of a cylinder to problems involving finding the volume of a cone. They will explore the relationship between a the volume of a cylinder and the volume of a cone.

🎓 Use a Concrete-Representational-Abstract (CRA) approach
Targeted instructional strategies

Concrete: Begin by giving students physical models of cones and cylinders that share the same radius and height. Use materials like plastic cups shaped as cones and cylinders, or have students create them using nets and cardstock. Engage them in an activity where they fill the cone with sand or rice and then pour it into the cylinder. They will observe that it takes exactly three cone-fulls to fill the cylinder. This hands-on experience helps them understand that the volume of a cone is one-third that of a cylinder with the same dimensions. Encourage students to repeat the experiment to reinforce the concept.

Representational: Transition to the representational stage by having students draw diagrams of cones and cylinders on graph paper. Guide them to accurately sketch each shape and label the radius and height. Use shading or coloring to represent the volume inside the shapes. Illustrate how the volume of the cone relates to the cylinder by dividing the cylinder's diagram into three equal sections, showing that one section corresponds to the cone's volume. Encourage students to annotate their diagrams with notes about the one-third volume relationship.

Abstract: Move on to the symbolic representation by introducing the formulas for volume. Write the formula for the volume of a cylinder $V_{cylinder} = \pi r^2 h$ and the volume of a cone $V_{cone} = \frac{1}{3}\pi r^2 h$ on the board. Explain each component of the formulas, ensuring students understand what the symbols represent. Work through example problems together, substituting given values into the formulas to calculate the volume.

Help students connect their hands-on experiences to the drawings and then to the formulas. Ask questions like, "How does filling the cone three times to fill the cylinder relate to the $\frac{1}{3}$ in the cone's volume formula?"

Encourage them to discuss and write about these connections in their own words. Prompt students to consider which representation—models, diagrams, or formulas—helps them understand the concept best and why.

Stronger and clearer each time
English language learner support

Ask students to individually write an explanation of why the volume of a cone is one-third the volume of a cylinder with the same base and height. Encourage them to include diagrams or examples to support their explanations. Next, have students pair up and share their explanations with a partner, asking each other questions to clarify and deepen their understanding. Then, instruct students to find a new partner and repeat the process, refining their explanations further based on the new discussion. Finally, have students revise their original written explanations, incorporating new vocabulary and ideas they gathered from their conversations. This activity will help students refine their understanding and use of mathematical language related to the volumes of cones and cylinders.

Adjusting manipulatives for accessibility
Student with disabilities support

Provide large, easy-to-handle models of cones and cylinders to help all students explore the relationship between their volumes. For students with physical or motor challenges, use sturdy containers like big plastic cones and cylinders that are easier to grasp and less likely to tip over. Instead of small materials like sand or rice, consider using water or large, lightweight balls to demonstrate how the cone fills one-third of the cylinder with the same base and height. Offer measuring tools with enlarged numbers and markings, such as oversized rulers or measuring tapes, to assist in accurately determining dimensions.

When working with nets and diagrams, provide pre-drawn templates or stencils so students who have difficulty with fine motor skills can participate fully. By adjusting these tools and materials, you ensure that all students can engage meaningfully in calculating the volume of cones and understanding their relationship to cylinders.

Exploration

Students: Page 429

> ## Interactive exploration
> **Explore online to answer the questions**
>
> ⊕ **mathspace.co**

Use the interactive exploration in 7.03 to answer these questions:

1. How many cones of water did it take to fill the cylinder?
2. What fraction of the cylinder does one cone fill?
3. Write a formula for the volume of a cone.

Suggested student grouping: Small groups

This exploration uses a Geogebra applet to visually demonstrate the relationship between the volumes of a cone and a cylinder that share the same radius and height. By simulating the pouring of water from a cone into a cylinder, the applet illustrates how the volume of the cone compares to the volume of the cylinder.

Ideal student responses

These ideal responses may differ from other correct student responses. Less formal responses can be connected with the more precise mathematical language presented here.

1. **How many cones of water did it take to fill the cylinder**

   It takes three cones of water to fill the cylinder. This is because the volume of a cone is one-third that of a cylinder with the same height and base area.

2. **What fraction of the cylinder does one cone fill?**

   One cone fills one-third $\left(\dfrac{1}{3}\right)$ of the cylinder.

3. **Write a formula for the volume of a cone.**

   The formula for the volume of a cone is given by:

   $$V = \frac{1}{3}\pi r^2 h$$

   where V is the volume, r is the radius of the base, and h is the height of the cone. This formula is derived from the fact that the cone's volume is a third of the volume of a cylinder with the same base and height.

Purposeful questions

- How do the volumes of the cylinder and the cone relate?
- If the heights of both the cylinder and the cone were increased by the same amount, would the relationship between their volumes change?

Possible misunderstandings

- Students may thing that the volume of the cone is directly proportional to the volume of the cylinder. They may believe that if you double the height or the radius of the cone, its volume will increase in the same proportion as the volume of the cylinder.

Examples

Students: Page 430

Example 1

Find the volume of the cone shown. Round your answer to two decimal places.

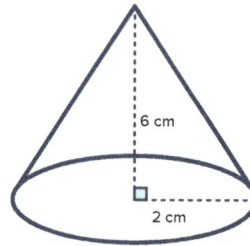

6 cm

2 cm

Create a strategy

Use the formula of the volume of cone.

Apply the idea

We have been given values for radius $r = 2$ and height $h = 6$.

$$V = \frac{1}{3}\pi r^2 h \qquad \text{Write the formula}$$

$$= \frac{1}{3} \cdot \pi \cdot 2^2 \cdot 6 \qquad \text{Substitute } r = 2 \text{ and } h = 6$$

$$= 25.13 \text{ cm}^3 \qquad \text{Evaluate}$$

Purpose

Check that students know the formula for the volume of a cone and can use it to solve for the volume.

Expected mistakes

Students may use the formula for a cylinder instead of a cone when calculating.

Remind students that a cone's volume should always be $\left(\dfrac{1}{3}\right)$ of the volume of a cylinder. So if they memorize the formula for a cylinder, then should have an easier time remembering the formula for a cone.

Encourage students to explore how changes in the cone's dimensions affect its volume by manipulating the formula $V = \frac{1}{3}\pi r^2 h$. Ask them to analyze scenarios such as doubling the radius while keeping the height constant, and predict how the volume will change. Similarly, have them consider the effects of tripling the height or scaling both the radius and height by different factors. This investigation deepens their understanding of proportional relationships and geometric scaling.

You might have students create graphs or tables to represent how varying one dimension impacts the volume, which can highlight patterns and relationships. Including visual aids, like diagrams of cones with different dimensions or even 3D models, can help them visualize these concepts. This approach not only reinforces their grasp of the volume formula but also connects to broader mathematical ideas of scaling and proportion.

Students: Page 430–431

Example 2

An ice cream cone has a volume of 6.28 in^3 and a radius of 1 in. Find the height of the cone.

Create a strategy

The volume of a cone is found by using the formula $\frac{1}{3}\pi r^2 h$.

We can substitute the volume and radius that was given in the problem into the formula to solve for the missing variable h or the height of the cone. Use 3.14 to approximate π

Apply the idea

We are given that the volume is 6.28 in^3 and the radius is 1 in. We can substitute those values for V and r.

$$V = \frac{1}{3}\pi r^2 h \qquad \text{Begin with the formula for volume of a cylinder.}$$

$$6.28 = \frac{1}{3}\cdot 3.14(1)^2 \cdot h \qquad \text{Substitute the given values and 3.14 for } \pi$$

$$= \frac{1}{3}\cdot 3.14(1)\cdot h \qquad \text{Evaluate the exponent}$$

$$= \frac{1}{3}\cdot 3.14 \cdot h \qquad \text{Evaluate the multiplication}$$

$$18.84 = 3.14 \cdot h \qquad \text{Multiply both sides by 3}$$

$$6 = h \qquad \text{Divide both sides by 3.14}$$

The height of the cone is 6 in.

Reflect and check

Our answer would have varied slightly if we had used π instead of 3.14.

$$V = \frac{1}{3}\pi r^2 h \qquad \text{Begin with the formula for volume of a cylinder.}$$

$$6.28 = \frac{1}{3}\pi(1)^2 h \qquad \text{Substitute the given values and 3.14 for } \pi$$

$$= \frac{1}{3}\pi(1)h \qquad \text{Evaluate the exponent}$$

$$= \frac{1}{3}\pi h \qquad \text{Evaluate the multiplication}$$

$$18.84 = \pi h \qquad \text{Multiply both sides by 3}$$

$$6 = h \qquad \text{Divide both sides by } \pi$$

The height of the cone is 5.996 in.

Purpose

Show students how to use the formula for the volume of a cone to solve for a missing dimension, and highlight the importance of precision when using constants like π in calculations.

Idea summary

The volume of a cone is exactly one-third the volume of a cylinder formed from the same base with the same perpendicular height.

To find the volume of a cone, we can use the formula:

$$V_{cone} = \frac{1}{3}\pi r^2 h$$

V_{cone}	Volume of the cone
r	Radius of the cone
h	Height of the cone

Practice

What do you remember?

1   Find the volume of this solid, rounding your answer to one decimal place.

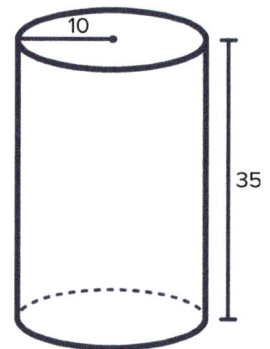

2   Find the volume of a cylinder with radius 7 ft and height 15 ft, rounded to two decimal places.

3   Label the following attributes on the cone.
   - Base
   - Height
   - Radius
   - Diameter
   - Apex

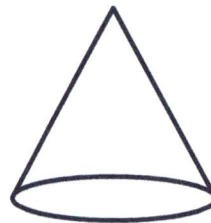

4   Consider the cone:

   a   Write the formula for the volume of a cone in terms of the radius, r, and height, h.

   b   Write the measures of the radius and height of the given cone.

   c   Calculate the volume of the cone, rounded to two decimal places.

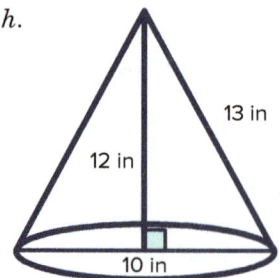

Let's practice

5 Find the volume of the cones. Round your answer to two decimal places.

a
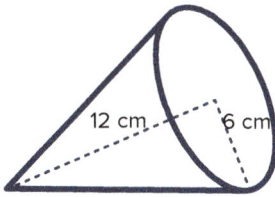
12 cm   6 cm

b
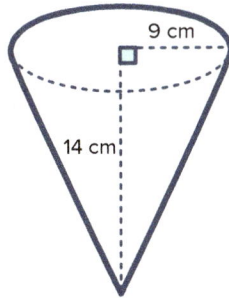
9 cm
14 cm

c
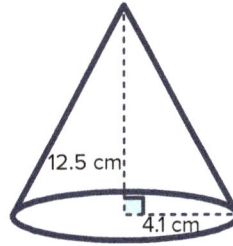
12.5 cm
4.1 cm

d
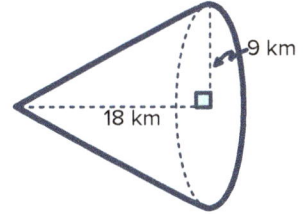
9 km
18 km

6 A cone has a diameter of 8 cm and a vertical length of 6 cm. Calculate the volume of the cone in terms of π.

A $10\pi \text{ cm}^3$ **B** $32\pi \text{ cm}^3$

C $50\pi \text{ cm}^3$ **D** $108\pi \text{ cm}^3$

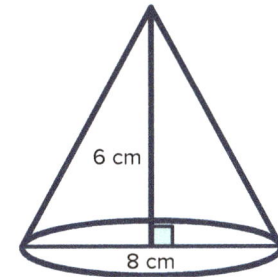
6 cm
8 cm

7 Find the volume of the cones in terms of π.

a Radius = 10 ft, height = 3 ft
b Radius = 12 m, height = 15 m
c Radius = 5 yd, height = 4 yd
d Radius = 10.8 mm, height = 7 mm

8 Find the volume of the cones. Round your answers to two decimal places.

a A cone with a radius of 8 ft and vertical height of 13 ft.
b A cone with a radius of 60 yd and a vertical height of 27.54 yd.
c A cone with a diameter of 6 in and vertical height of 8 in.
d A cone with a diameter of 8 cm and vertical height of 6 cm.

9 Which is the closest to the volume of the cone shown?

A 79.9 cm^3 **B** 357.6 cm^3

C 1972.6 cm^3 **D** 527.5 cm^3

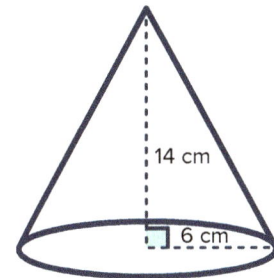
14 cm
6 cm

10 Find the volume of the cone to the nearest cubic meter.

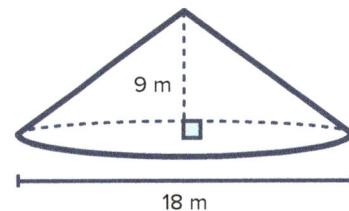
9 m
18 m

11 Match the dimension of each cone to its corresponding volume.

    a Diameter: 12 in   **i** $V = 85.58 \text{ in}^3$
        Height: 6 in

    b Radius: 5 in   **ii** $V = \dfrac{100}{3}\pi \text{ in}^3$
        Height: 4 in

    c Base area: 16 in^2 **iii** $V = 72\pi \text{ in}^3$
        Height: 16 in

12 A sample perfume bottle is shaped like a cone, as shown in the figure. It has a height and radius of 8 cm and 3 cm, respectively.

How much perfume can fit in the cone if its completely filled? Round your answer to one decimal place.

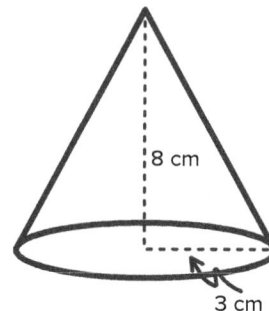

8 cm

3 cm

13 A paper cup shaped like a cone is completely filled with juice. The top of the cup has a radius of 5 cm and a height of 11.7 cm. The juice is then poured into an empty cylindrical glass of the same height and base radius.

    a Find the amount of juice the paper cup holds when full in terms of π.

    b Find the amount of juice the glass holds when full in terms of π.

    c How many paper cups of juice are needed to completely fill the glass? Justify your answer.

14 A pool in the shape of a cylinder has a radius of 7 ft and a height of 5 ft. A water tank in the shape of a cone has the same radius and height. If the pool is full, how much water would be needed to completely fill the water tank?

15 Sam and Jean have opened a small ice cream stand. They have two different sized cones they sell.

    • A small is 4 in in diameter and 4 in in height
    • A large is 5 in in diameter and 6 in in height

Sam wants to charge twice as much for the larger cone because he says that the large cone has twice the volume. Is he correct? Justify your answer.

16 The volume of a cone-shaped hat is 725 cm^3, and its radius is 7 cm as shown in the figure.

If someone is wearing the cone hat on their head, what is the distance from their head to the top of the cone?

In other words, what is the height of the cone? Round your answer to two decimal places.

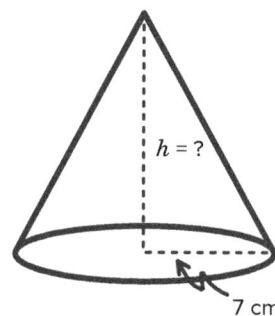

$h = ?$

7 cm

17 Find the radius of the cones:

    a A cone with a volume of 196 ft^3 and the vertical height and radius are equal. Round your answer to two decimal places.

    b A cone with a volume of 12 441.02 in^3 and a vertical height of 30 in. Round your answer to one decimal place.

18 Find the vertical height of the cones:

    a A cone with volume of 26 640.97 in^3 and a radius of 31.9 in. Round your answer to the nearest whole inch.

    b A cone with a volume of 106 532.52 yd^3 and a radius of 54.7 yd. Round your answer to the nearest yard.

19 The radius and the height of a cone are equal, and the volume of the cone is $72\pi \text{ cm}^3$.

    a Find the radius of the cone.

    b Explain how the volume of a cone relates to the volume of a cylinder with the same dimensions.

Let's extend our thinking

20 A vendor at the Sumter County Fair sells kettlecorn in two containers.

Option 1 sells for $3.50:

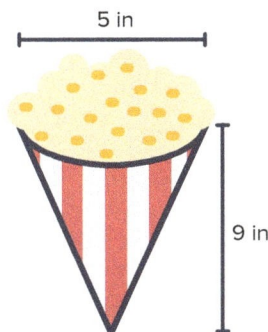

5 in

9 in

Option 2 sells for $10.50:

5 in

9 in

    If you have $12.00 to spend, determine which container option you would purchase. Explain your reasoning.

21 Imagine you have a cylinder that you want to melt down and reshape as a cone. If you want the cone to have the same base and height as the original cylinder, how would you need to cut the original cylinder? Explain your reasoning.

22 Cone X has the same height but twice the radius of Cone Y. How many times larger is the volume of Cone X than the volume of Cone Y? Explain.

23 Which is closest to the height of a cone that has a slant height of 16 in and a radius of 6 in?

    A 20 in   **B** 14.8 in

    C 17.1 in   **D** 10 in

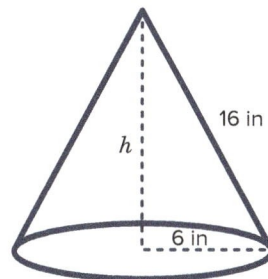

16 in

h

6 in

24 Consider the net of a cone:

    a Which line segment has a length of 18 cm?

    b Which line segment has a length of 30 cm?

    c Find the vertical height, h, of the cone.

    d Find the volume to the nearest cubic centimeter.

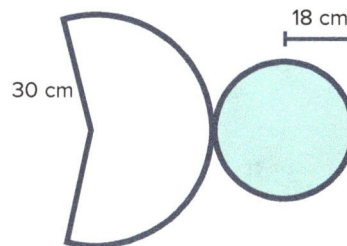

18 cm

30 cm

25 Consider the cone:

    a Calculate the vertical height.

    b Now find the volume of the cone. Round your answer to two decimal places.

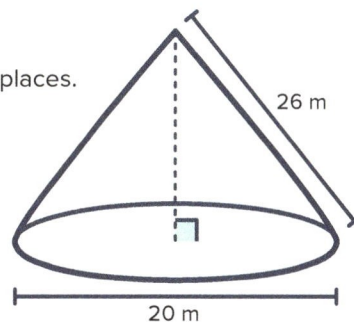

26 m

20 m

Answers

7.03 Volume of cones

What do you remember?

1 10 995.6 units^3

2 2309.07 ft^3

3

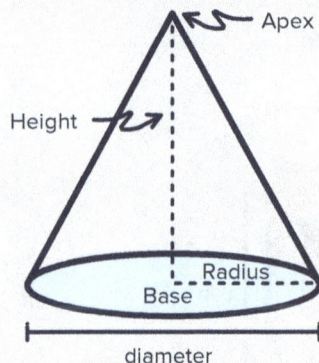

4 a $V = \frac{1}{3}\pi r^2 h$ **b** $r = 5$ in, $h = 12$ in

   c 314.26 in^3

Let's practice

5 a 452.39 cm^3 **b** 1187.52 cm^3

   c 220.04 cm^3 **d** 1526.81 km^3

6 B

7 a 100π ft^3 **b** 720π m^3

   c $\frac{100}{3}\pi$ cm^3 **d** 272.16π mm^3

8 a 871.27 ft^3 **c** 75.40 in^3

   c 103 823.35 yd^3 **d** 100.53 cm^3

9 D

10 763 m^3

11 a iii   **b** ii   **c** i

12 75.4 cm^3

13 a 97.5π cm^3

   b 292.5π cm^3

   c 3 paper cups: $97.5\pi \cdot 3 = 292.5\pi$ cm^3

The juice will fill the cylindrical glass up to one-third of its height. This is because the volume of the cone, which was completely filled, is one-third the volume of the cylinder when they have the same base radius and height.

14 256.56 ft^3

15 The volume of the small cone (with a diameter and height of 4 inches) is approximately 16.7552 cubic inches, and the volume of the large cone (with a diameter of 5 inches and a height of 6 inches) is approximately 39.2699 cubic inches.

The volume of the large cone is not exactly twice the volume of the small cone. It is actually more than twice. To be precise, the volume of the large cone is approximately $\frac{39.2699}{16.7552} \approx 2.34$ times the volume of the small cone.

Therefore, Sam's idea of charging twice as much for the larger cone because he believes it has twice the volume is not accurate. The large cone has more than twice the volume of the small cone, so if pricing were based strictly on volume, the large cone should cost more than twice the price of the small cone.

16 a 14.1 cm

17 a 5.72 ft   **b** 19.9 in

18 a 25 in   **b** 34 yd

19 a $r = 6$ cm

   b The volume of the cylinder is calculated by using the product of the area of its base by its height. The volume of a cone is exactly one third the volume of a cylinder formed from the same base with the same perpendicular height. Since the cone and the cylinder have the same diameter and height, the volume of three of the cones is equal to the volume of one cylinder.

Let's extend our thinking

20 You can order either 3 of Option 1 or 1 of Option 2 to maximize the amount of kettlecorn you buy. Since the cone and the cylinder have the same diameter and height, three of the cones is equal to the volume of one cylinder. Both orders will cost $10.50 total.

21 The volume of a cone is one-third the volume of a cylinder with the same base and height. To make a cone that has the same base and height as the original cylinder, we can only use one-third of the cylinder. Therefore, we would need to cut the cylinder into three equal-sized pieces and only melt down one of those pieces to use for the cone.

22 Let r and h represent the radius and vertical height of Cone Y. The volume of Cone Y is

$$V_Y = \frac{1}{3}\pi r^2 h$$

Since Cone X has the same height but twice the radius of Cone Y, Cone X has the radius $2r$ and height, h. The volume of Cone X is

$$V_X = \frac{1}{3}\pi(2r)^2 h = \frac{4}{3}\pi r^2 h$$

Thus, volume of Cone X is 4 times larger than the volume of Cone Y.

23 B

24 a The radius   **b** The slant height

   c $h = 24$ cm   **d** 8143 cm^3

25 a 24 m   **b** 2513.27 m^3

7.04 Problem solving with volume and surface area

Subtopic overview

Lesson narrative

In this lesson, students will apply their knowledge of volume and surface area to solve real-world problems involving three-dimensional shapes. They will revisit key concepts: volume measures the space inside a shape, while surface area measures the total area of all the faces of the shape. Students will be guided to identify when to use volume or surface area by recognizing contextual clues such as "contain," "fill," "capacity" for volume, and "cover," "wrap," "exterior" for surface area. The exploration includes practical problems such as calculating the volume of a podium formed by cutting off the top of a cone. Students will subtract the volume of the smaller cone from the larger cone to find the remaining volume, using the Pythagorean theorem to determine necessary heights. Additionally, students will solve problems like determining the surface area of a pyramid-shaped terrarium. They will use the formula $(S.A. = \frac{1}{2}lp + B)$, where (l) is the slant height, (p) is the perimeter of the base, and (B) is the area of the base. This includes calculating the surface area of the square base and the triangular faces. By the end of the lesson, students should be able to apply the concepts of volume and surface area to solve complex, real-world problems involving various three-dimensional shapes, understanding the importance of each measure in different contexts.

Learning objectives

Students: Page 436

After this lesson, you will be able to...
- solve real-world problems involving volume of cones and square-based pyramids and the surface area of square-based pyramids.

Key vocabulary

- surface area
- volume

Essential understanding

Volume and surface area of 3D figures can be used to solve real-world problems.

Standards

This subtopic addresses the following Virginia 2023 Mathematics Standards of Learning standards.

Mathematical process goals

MPG1 — Mathematical Problem Solving

Teachers can integrate this goal by presenting students with real-life problems that require the application of volume and surface area formulas. For instance, a problem could involve calculating the amount of paint needed for a pyramid-shaped roof. Teachers can guide students through the problem-solving process, encouraging them to use diagrams and concrete objects to visualize the problem and support their calculations.

MPG3 — Mathematical Reasoning

Teachers can foster mathematical reasoning by emphasizing the importance of understanding the context of a problem and interpreting the results logically. For instance, when comparing the volumes of different cylinders, students should be encouraged to use reasoning to explain their findings and conclusions. Teachers can also promote reasoning by discussing various problemsolving strategies, such as estimating and checking for reasonableness.

MPG4 — Mathematical Representations

Teachers can make mathematical connections by integrating the concepts of volume and surface area with real-world situations. For example, they could present a scenario where students are asked to calculate the volume of water a cylindrical tank can hold, linking mathematical principles to practical applications. Additionally, teachers can connect these concepts to different subjects such as architecture and engineering, highlighting the interdisciplinary nature of mathematics. Furthermore, teachers can emphasize how the current lessons on volume and surface area build upon prior knowledge like the understanding of basic geometric shapes and their properties.

Content standards

8.MG.2 — The student will investigate and determine the surface area of square-based pyramids and the volume of cones and square-based pyramids.

8.MG.2d — Solve problems in context involving volume of cones and square-based pyramids and the surface area of square-based pyramids.

Prior connections

6.MG.2 — The student will reason mathematically to solve problems, including those in context, that involve the area and perimeter of triangles, and parallelograms.

Future connections

G.DF.1 — The student will create models and solve problems, including those in context, involving surface area and volume of rectangular and triangular prisms, cylinders, cones, pyramids, and spheres.

Lesson Preparation

Suggested review

Depending on your students' level of prior knowledge, consider revisiting the following lessons:

Grade 7 — 7.05 Solve problems with volume and surface area

Student lesson & teacher guide

Problem solving with volume and surface area

Students will use their knowledge of the volume formulas for cones to solve real-world problems.

Solve a problem with the STEAM cycle
Targeted instructional strategies

Pique student interest by asking them about the different types of liquid soap dispensers they have encountered. Discuss how the shapes and sizes of these dispensers affect their usability, cost, and environmental impact. Show images or bring in actual examples of various soap dispensers to stimulate interest and conversation.

Use the STEAM cycle to explore this problem using what they have learned about volume and surface area of different solids.

Ask: As a class or in groups, have students ask a particular question like "How can we design a soap dispenser that holds 14oz, but uses the least amount of material to minimize cost and reduce environmental waste?"

Imagine: Facilitate a brainstorming session where students generate ideas for various dispenser designs, encouraging them to sketch their concepts or create simple models. Prompt them to think about how different shapes like cones, square-based pyramids, or for advanced learners, even composite solids. Have them consider how different shapes could affect the volume and surface area.

Plan: Guide students to develop a plan for calculating the surface area and volume of their proposed designs. Assist them in deciding which mathematical formulas to use, determining the necessary measurements, and selecting tools such as rulers or design software to aid in accurate calculations.

Create and test: Support students as they construct prototypes of their dispensers using materials like cardstock, clay, or digital modeling tools. Help them measure dimensions precisely and calculate the surface areas and volumes to test whether their designs meet the specified criteria while using minimal material. The creation of physical models is important, as otherwise students may not recognize that in reality, additional surface area is required to include a pump for the dispenser.

Improve: Encourage students to analyze the results of their tests, identify areas where material usage can be reduced further, and collaborate on refining their designs. Facilitate discussions on how slight adjustments in dimensions or shapes can maintain the desired volume while decreasing surface area, leading to more efficient and cost-effective packaging.

Guided diagram annotation
Student with disabilities support

Provide students with diagrams of the cones and square-based pyramids discussed in the lesson, with key dimensions indicated but not fully labeled. Encourage students to add their own annotations as they solve the problems, such as labeling heights, slant heights, and radii, and noting where they apply the Pythagorean theorem.

For example, give students an outline of a truncated cone (the podium) with the larger and smaller base radii marked, and have them annotate the diagram by calculating and labeling the heights needed to find the volumes before subtracting. Similarly, provide a diagram of a pyramid-shaped terrarium with the base edges labeled, and ask students to add the slant heights and any calculations for the surface areas of the triangular faces. By actively engaging with the diagrams, students enhance their conceptual understanding and support their visual-spatial processing.

Examples

Students: Page 437–438

Example 1

A podium is formed by sawing off the top of a cone. How much space is inside the podium? Round your answer to two decimal places.

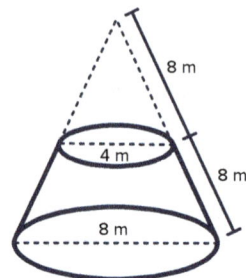

Create a strategy

Because the question is asking about the space inside the podium, we must calculate the volume of the figure. The volume of podium is the difference of the volume of the large cone and the volume of the small cone at the top.

Apply the idea

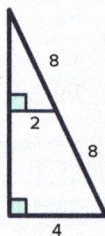

Based on the diagram, the radius of the small cone is $r_1 = 2$ m and the radius of the large cone is $r_2 = 4$ m.

The slant height of the large cone is 16 cm and the slant height of the smaller cone is 8 cm.

To find the perpendicular heights of the cones we can use the Pythagorean theorem.

Let's work with the smaller cone first. Let a be the perpendicular height:

$a^2 + b^2 = c^2$	The Pythagorean theorem
$a^2 + 2^2 = 8^2$	Substitute $b = 2$ and $c = 8$
$a^2 + 4 = 64$	Evaluate the exponents
$a^2 = 60$	Subtract 4 from both sides
$a = \sqrt{60}$	Take the square root of both sides

We will leave $a = \sqrt{60}$ in exact form for now because the longer we wait to round, the more precise the answer will be.

Now that we know the perpendicular height of the smaller cone, we can calculate its volume.

$V_{\text{small}} = \frac{1}{3}\pi r^2 h$	Formula for volume of a cone
$= \frac{1}{3}\pi \cdot 2^2 \cdot \sqrt{60}$	Substitute $r = 2$ and $h = \sqrt{60}$
$= \frac{1}{3}\pi \cdot 4 \cdot \sqrt{60}$	Evaluate the exponent
$= 32.45$ m^3	Use a calculator to evaluate to two decimal places

Now, let's look at the large cone. Let b be the perpendicular height:

$a^2 + b^2 = c^2$	The Pythagorean theorem
$4^2 + b^2 = 16^2$	Substitute $a = 4$ and $c = 16$
$16 + b^2 = 256$	Evaluate the exponents
$b^2 = 240$	Subtract 16 from both sides
$b = \sqrt{240}$	Take the square root of both sides

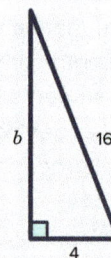

We will also leave $b = \sqrt{240}$ in exact form.

Now that we know the perpendicular height of the larger cone, we can calculate its volume.

$$V_{large} = \frac{1}{3}\pi r^2 h$$ Formula for volume of a cone

$$= \frac{1}{3}\pi \cdot 4^2 \cdot \sqrt{240}$$ Substitute $r = 4$ and $h = \sqrt{240}$

$$= \frac{1}{3}\pi \cdot 16 \cdot \sqrt{240}$$ Evaluate the exponent

$$= 259.57 \text{ m}^3$$ Use a calculator to evaluate to two decimal places

We can now calculate the volume of the podium.

Volume of podium $= 259.57 \text{ m}^3 - 32.45 \text{ m}^3$ Subtract the volume of the small cone from the large cone

$$= 227.12 \text{ m}^3$$ Evaluate

Purpose

Check that students can connect what they know about shapes, finding volume and the Pythagorean theorem to find the volume of a three dimensional shape.

Expected mistakes

Students may have trouble finding a strategy for finding the volume of a shape that is not a cone. Remind students that when given unfamiliar shapes, try to figure out ways to draw or use familar shapes to solve for unfamiliar ones.

🎓 Advanced learners: Derive formula for a frustum
use with Example 1
Targeted instructional strategies

Encourage students to derive the formula for the volume of a frustum themselves rather than providing it upfront. Begin by asking them to recall the formula for the volume of a cone and consider how a frustum relates to a full cone. Guide students to visualize a frustum as a large cone with a smaller cone removed from the top. Prompt them to express the volumes of both cones in terms of their radii and heights.

Then, lead them to subtract the volume of the smaller cone from the larger one to derive the formula for the frustum's volume. Provide a detailed diagram showing the large cone and the smaller cone removed, with all relevant dimensions labeled—radii r_1 and r_2 heights h_1 and h_2, and the slant heights if needed. This process deepens their understanding of geometric relationships and engages them in algebraic manipulation and spatial reasoning.

Students: Page 438–439

Example 2

You are making pyramid-shaped terrarium displays for an art exhibit. Each terrarium has a square base with a width of 9 inches and the slant height of the terrarium is 14 inches. How many square inches of glass do you need to buy to make a terrarium?

Create a strategy

The problem is asking to find the total area of the flat faces making up the pyramid, so the surface area formula $S.A. = \frac{1}{2}lp + B$ will be used to find the surface area of the square-based pyramid.

Apply the idea

The slant height, l, is given as 14 inches, and a side of the square base being 9 inches.

The perimeter, p, of the square base is $4 \cdot 9$, or 36.

The area of the base, B, is 9^2, or 81.

The area of the base, B, is 9^2, or 81.

With all values, we can substitute for the variables in the formula and simplify.

$$\frac{1}{2}lp + B = \frac{1}{2}(14)(36) + (81)$$ Substitute $l = 14$, $p = 36$, and $B = 81$

$$= 252 + 81$$ Evaluate the multiplication

$$= 333$$ Evaluate the addition

The terrarium needs 333 in^2 of glass.

Reflect and check

The net of the pyramid-shaped terrarium with a square base would look like this:

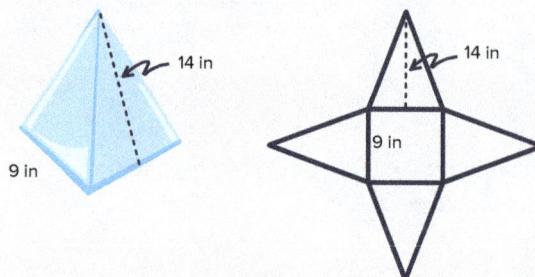

We can use the net to determine that the area of the base would be 81 in^2.

We can use the formula for area of a triangle, $\frac{1}{2}bh$ to determine the area of each of the triangular sides to be 63 in^2. There are four triangular bases, so the area of all of the faces would be 252 in^2.

If we add together the area of the square base and the area of the four triangular bases, we get that the surface area is 252 in^2 + 81 in^2 = 333 in^2. So the surface area of the terrarium is 333 in^2.

Purpose

Show students how to calculate the surface area of a square-based pyramid using given measurements and standard formulas.

Reading comprehension: Three reads
English language learner support

use with Example 2

Apply the "three reads" strategy to support students' understanding of the problem. On the first read, have students focus on grasping the overall context without worrying about details. Ask them, "What is the situation about?" Encourage responses like, "We are making pyramid-shaped terrariums for an art exhibit."

On the second read, guide students to identify and highlight important quantities and vocabulary. Prompt them with questions such as, "What measurements are given?" and "What shapes are involved?" Students should note that the base is a square with a width of 9 inches and the slant height is 14 inches.

For the third read, direct students to determine what the problem is asking them to find. Ask, "What do we need to calculate?" and "Which mathematical concepts or formulas might we use?" Lead them to recognize that they need to find the surface area of the pyramid to determine how much glass is required.

By engaging in these three focused readings, you help students unpack the language of the problem, clarify mathematical terms, and identify the steps needed to solve it. This approach reduces language barriers and enables students to concentrate on the mathematical process.

This strategy utilizes the three reads routine to enhance students' comprehension and supports them in navigating complex language in word problems.

> **💡 Idea summary**
>
> We can use the concept of surface area and volume in real–world problems.
>
> The **surface area** is the sum of the areas of all surfaces of a figure.
>
> The **volume** of a three dimensional shape is the amount of space that is taken up by the shape, or the amount it can hold.

Practice

What do you remember?

1 Find the surface area of the square-based pyramid:

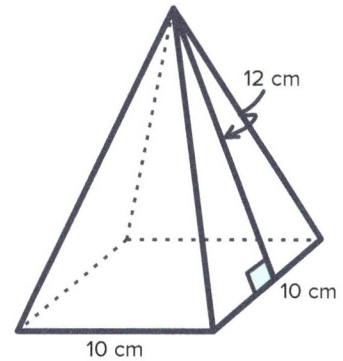

2 Find the volume of the square-based pyramid:

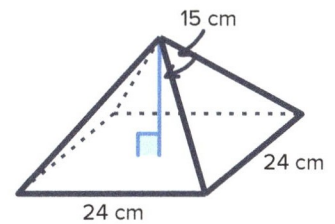

3 Find the volume of the cone, rounded to two decimal places.

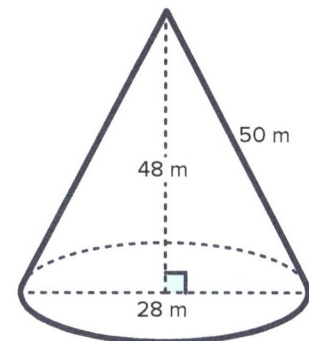

4 For each scenario, determine whether a volume or surface area calculation would be more appropriate.

    a The amount of paint needed to cover a new shed

    b The amount of soil needed to fill a garden bed

c   The water capacity of a swimming pool

d   The heating system for a greenhouse based on its internal air space

e   The amount of metal needed to construct a cylindrical can

f   The amount of chocolate coating needed for a candy bar

g   The space available inside a storage box

Let's practice

5   A city plans to build a monument in the form of a square-based pyramid. The base side length is 50 meters and the monument is to be 120 meters tall.

    a   What will be the volume of the monument?

    b   Explain what the volume represents in this context.

6   A company decides to package their luxury chocolates in a gift box shaped like a square-based pyramid for the holiday season. The box has a base side length of 10 cm and a slant height of 15 cm.

    a   Calculate the total surface area of the gift box, including its base.

    b   Explain what the surface area represents in this context.

7   The Great Pyramid at Giza has a square base with side length 755 ft and the height of the pyramid is 481 ft.

    a   Determine its surface area to the nearest square foot.

    b   Explain what the surface area represents in this context.

8   Find the amount of ice cream the given cone can hold. Round your answer to one decimal place.

9 cm
13 cm

9   A gift shop has pyramid-shaped gift boxes that have a base side of 15 cm, a vertical height of 13.2 cm, and a slant height of 20 cm.

    a   If the box is hollow, how much material is needed to make one gift box?

    b   If you plan to fill the box with chocolate, how much can one box hold?

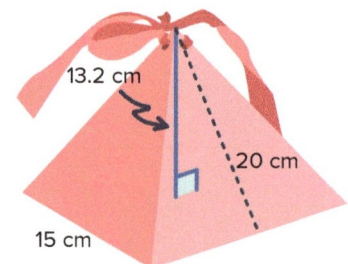
13.2 cm
20 cm
15 cm

10   The height of a birthday cap is 21 cm, and its radius is 3 cm as shown in the figure.

If party favors are made of these hats filled with candy, how much candy can each one hold if they are completely filled?

Round to the nearest cubic centimeter.

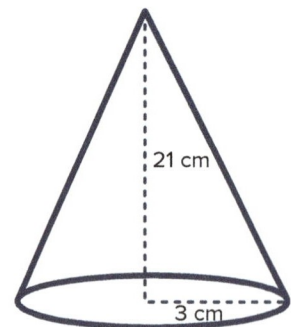
21 cm
3 cm

11   A water tank in the shape of an inverted square-based pyramid needs to be filled. The tank has a base side length of 10 m and a depth of 15 m. How much water is required to fill the tank?

12 A gardening equipment company designs a conical water funnel with a diameter of 24 cm and a height of 30 cm. What is the maximum amount of water that can be in the funnel at any given time?

13 Which pyramid requires more glass to enclose its space?
- The Louvre art museum in Paris, France, has a glass square pyramid at its entrance. The side length of the base of the right pyramid is 35.0 m and its height is 20.6 m. ·
- The Muttart Conservatory in Edmonton, Alberta, has four right square pyramids also with glass faces. One of the largest pyramids has a base side length of 25.7 m and a height of 24.0 m.

14 A famous right square pyramid is the Great Pyramid in Egypt. Its vertical height is approximately 139 m, its base length is 230 m, and its four triangular faces each have a height of 216 m.

How much of the Great Pyramid is exposed to wind erosion?

15 A square pyramid has a base with side length 12 mm and a volume of 1248 mm^3. Find the height of the pyramid.

16 A construction cone has a volume of 157 cubic inches. If its height is 6, what is its radius to the nearest inch?

17 You're tasked with designing a new packaging in the shape of a square-based pyramid for a product. The package must have a volume of 350 cm^3, and you decide the height of the pyramid should be 15 cm.
a Calculate the necessary base side length of the pyramid
b Calculate for the total surface area of the packaging for material cost estimations.

18 A company wants to design soap containers in the shape of square-based pyramids. Each container should hold 250 cm^3 of soap. If the container is to be 10 cm tall, what should be the dimensions of the square base?

19 A paperweight is in the shape of a square-based pyramid with dimensions as shown. The paperweight is made of solid glass.

Find the volume of glass needed to make 3000 paperweights.

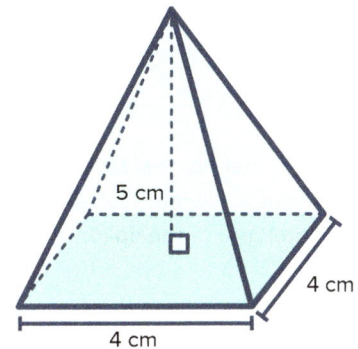

5 cm

4 cm

4 cm

20 Felice bought the square-based pyramid shaped candle shown for her friend's birthday. The package says that the candle burns 20 cm^3 of wax each hour. How many hours will it take the entire candle to burn?

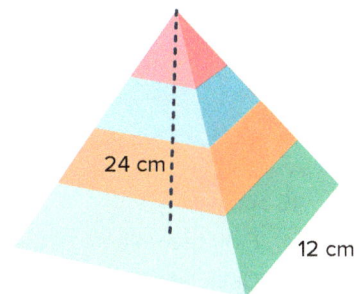

24 cm

12 cm

21 Before 1980, Mount St. Helens was a volcano approximately in the shape of a cone, as shown the first image:
a What was the volume of the mountain, in cubic kilometers? Round your answer to two decimal places.

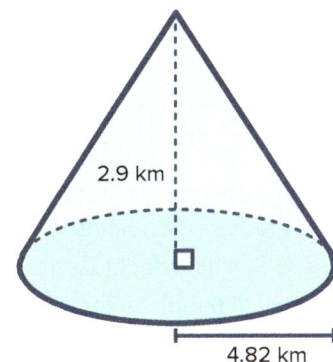

2.9 km

4.82 km

b The tip of the mountain was in the shape of the cone shown in the second image:

Find the volume of the tip in cubic kilometers. Round your answer to two decimal places.

c In 1980, Mount St. Helens erupted and the tip was destroyed.

Find the volume of the remaining mountain, in cubic kilometers. Round your answer to two decimal places.

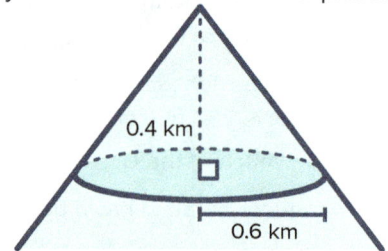

0.4 km

0.6 km

Let's extend our thinking

22 The podium was formed by sawing off the top of a cone. Find the volume of the podium, correct to two decimal places.

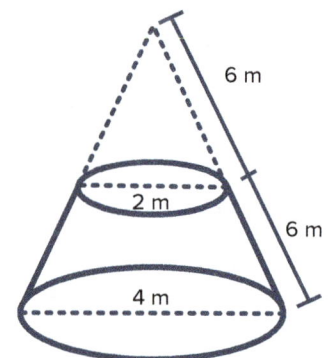

6 m

2 m

6 m

4 m

23 A small square pyramid of height 4 cm was removed from the top of a large square pyramid of height 8 cm forming the solid shown. Find the exact volume of the solid.

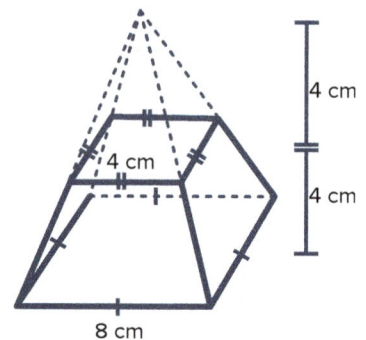

4 cm

4 cm

4 cm

8 cm

24 A weight is constructed by removing the top 38 cm from a 57 cm tall pyramid with a square base of side length 45 cm.

a Find the volume of the original pyramid.

b Calculate the side length of the square on top of the weight.

c Calculate the volume of the removed part of the pyramid.

d Calculate the volume of the weight.

45 cm

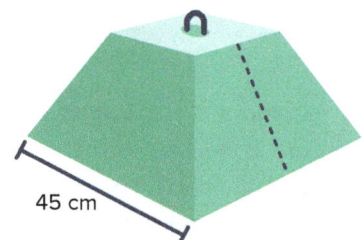

25 A popular ice cream shop wants to design a new cone that can hold 20% more ice cream than their current model, which holds 150 cm^3.

a What should be the volume of the new cone?

b If the diameter remains the same, how much should the height change?

Answers

7.04 Problem solving with volume and surface area

What do you remember?

1. 340 cm^2

2. 2880 cm^3

3. 9852.03 cm^3

4. a Surface Area   b Volume
   c Volume   d Volume
   e Surface Area   f Surface Area
   g Volume

Let's practice

5. a $100\ 000 \text{ m}^3$
   b The volume of $100\ 000 \text{ m}^3$ represents the amount of space the monument occupies, or the amount of material required to build it if the inside of the pyramid is completely filled.

6. a 400 cm^2
   b The total surface area of 400 cm^2 is the space around the outside of the gift box, or the amount of material required to enclose the gift box.

7. a $1, 493, 310 \text{ ft}^2$
   b The surface area of approximately $1, 493, 310 \text{ ft}^2$ represents the area of the outside faces of the pyramid.

8. 1102.7 mL

9. a 825 cm^2 b 990 cm^3

10. 198 cm^3

11. 500 m^3

12. 4523.89 cm^3

13. The Louvre art museum's glass pyramid requires approximately 1892.09 m^2 of glass to enclose its space, while the Muttart Conservatory's largest pyramid requires approximately 1399.29 m^2 of glass.

    Therefore, the Louvre pyramid requires more glass to enclose its space compared to the Muttart Conservatory's pyramid.

14. $99\ 360 \text{ m}^2$

15. 26 mm

16. 5 inches

17. a 8.37 cm b 330.58 cm^2

18. $8.66 \text{ cm} \times 8.66 \text{ cm}$

19. $80\ 000 \text{ cm}^3$

20. 57.6 hours

21. a 70.55 km^3 b 0.15 km^3 c 70.40 km^3

Let's extend our thinking

22. 43.37 m^3

23. $\dfrac{448}{3} \text{ cm}^3$

24. a $38\ 475 \text{ cm}^3$ b 30 cm
    c $11, 400 \text{ cm}^3$ d $27, 075 \text{ cm}^3$

25. a 180 cm^3
    b Since the volume is directly proportional to the height when the radius is constant, the height needs to increase by 20% as well as to achieve 20% increase in volume, assuming the shape of the cone remains similar (i.e., the ratio of height to radius remains constant).

Topic 7 Assessment: 3D Geometry

1 Find the volume of each square-based pyramid or cone, rounding to the nearest hundredth as necessary. Where necessary, use 3.14 for π.

a

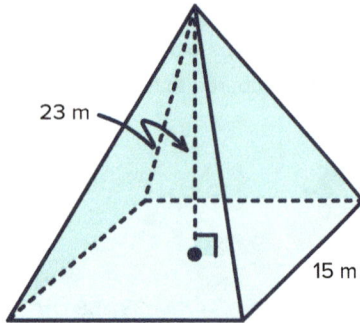
23 m
15 m

b

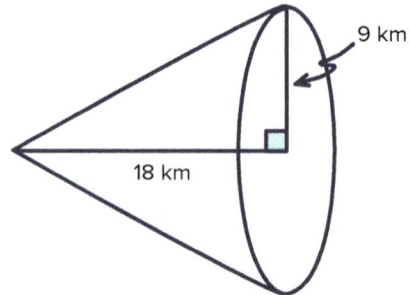
9 km
18 km

c

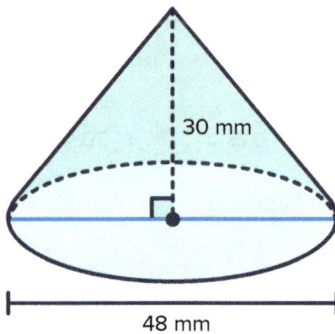
30 mm
48 mm

d

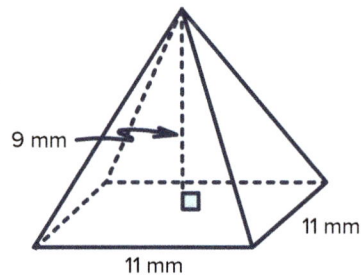
9 mm
11 mm
11 mm

2 Find the surface area of each square-based pyramid.

a

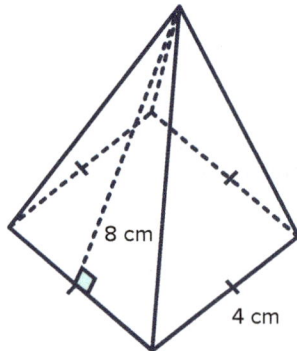
8 cm
4 cm

b

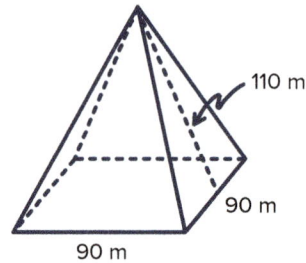
110 m
90 m
90 m

3 Which figure has a greater capacity? Justify your answer. Where needed, round to the nearest tenth and use 3.14 for π.

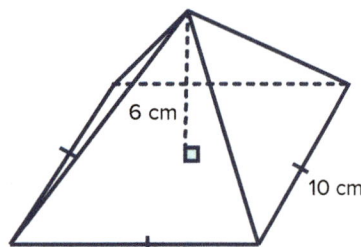
6 cm
10 cm

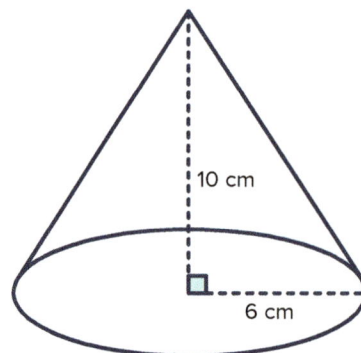
10 cm
6 cm

4 A new board game is going to come out with solid cone-shaped pieces to represent towers. The dimensions of each piece are shown:

   a Solve for the height of each piece to the nearest hundredth.

   b Since the pieces will be made of solid metal, find the number of cubic centimeters needed to make each piece. Round to the nearest whole centimeter and use 3.14 for π.

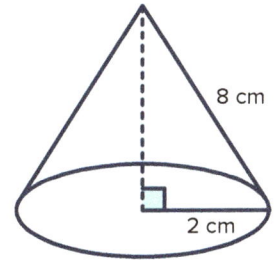

8 cm
2 cm

5 The volume of a pyramid is 60 m^3 with a height of 4 m.

   a Find the area of the base.

   b Explain how the volume of a pyramid relates to the volume of a rectangular prism with the same height and base.

6 Carlos is planning to build a model of a square-based pyramid out of construction paper. His plans shows how the net is folded into the pyramid for a given slant height, l, and base length, b.

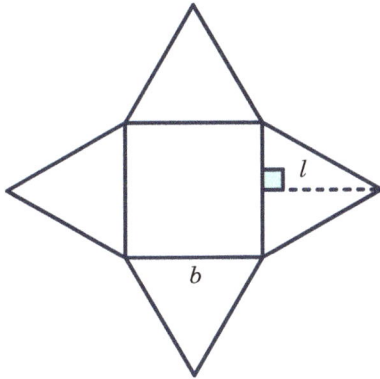

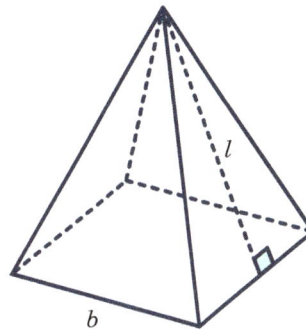

   a Carlos knows that his four triangular faces of the pyramid he wants to build have a total area of 60 in^2, and the height of the triangle, l, is 6 inches. What is the length of the base of each triangle?

   b Carlos has 80 in^2 of paper he can use to build his model. Will it be enough? Justify your answer.

7 A vendor at the Sumter County Fair sells kettlecorn in two containers.

   Option 1 sells for $3.50:

   5 in

   9 in

   Option 2 sells for $10.50:

   5 in

   9 in

   a If you have $12.00 to spend, determine which container option you would purchase. Explain your reasoning.

   b Will this relationship hold true between containers of different sizes? If so, what would need to be true about the containers?

8   A new ice cream shop is very strict about the expectations for its ice cream cones. Each scoop must resemble a perfect sphere placed on the cone. The ice cream sphere should not be wider than the cone. They are hoping as the ice cream melts, it will fill the cone and not overflow.

The dimensions of their cones are as follows:

- Diameter of cone: 3.25 in
- Diameter of ice cream sphere: 3.25 in
- Height of cone: 4.5 in

a   With these dimensions, would the melted ice cream fit inside the cone or overflow? Justify your answer.

b   The ice cream shop is thinking about offering different sizes of cones for different prices. In general, what must be the relationship between the radius and the height for the ice cream not to overflow? Justify your thinking.

Answers

1 **a** 1725 m^3 **b** 1526.04 km^3

    c 18 086.4 mm^3 **d** 363 mm^3

    8.MG.2b

2 **a** 80 cm^2 **b** 27 900 m^2

    8.MG.2a

3 The cone has a bigger capacity. The volume of the pyramid is 200 cm^3, while the volume of the cone is 376.8 cm^3.

    8.MG.2b

4 **a** 7.75 cm   **b** 32 cm^3

    8.MG.2d

5 **a** 45 m^2

    b Similarly to the volume of a cylinder and a cone, the volume of a pyramid is also calculated by taking one-third the volume of a prism that has the same height and base area.

    8.MG.2c

6 **a** 5 in.

    b It will not be enough to build his model. With a base of 5 inches and a slant height of 6 inches, the total surface area of the pyramid will be 85 in^2. This is more than the 80 inches of paper he has available.

    8.MG.2d

7 **a** You can order either 3 of Option 1 or 1 of Option 2 to maximize the amount of kettlecorn you buy. Since the cone and the cylinder have the same diameter and height, three of the cones is equal to the volume of one cylinder. Both orders will cost $10.50 total.

    b The relationship of the cylinder's volume being 3 times the volume of the cone will hold true as long as the base and height are the same size.

    8.MG.2c, 8.MG.2d

Performance task

8 **a** The formula for the volume of the sphere is $V = \frac{4}{3}\pi r^3$.

    So, the volume of ice cream is $\frac{4}{3}\pi(1.625)^3 = 17.97$ in^3.

    The formula for the volume of the cone is $V = \frac{1}{3}\pi r^2 h$.

    So, the volume of ice cream is $\frac{1}{3}\pi(1.625)^2 (4.5) = 12.44$ in^3.

    The volume of the ice cream exceeds the volume of the cone (17.97 in^3 > 12.44 in^3), so the melted ice cream will overflow.

    b In general, the volume of the scoop must be less than that of the cone. Since the formula for the volume of the sphere is $V = \frac{4}{3}\pi r^3$ and the formula for the volume of the cone is $V = \frac{1}{3}\pi r^2 h$, so $\frac{4}{3}\pi r^3 \leq \frac{1}{3}\pi r^2 h$.

    Simplifying this, we get $4r \leq h$. So, the height of the cone must be at least 4 times the radius of the scoop in order not to overflow.

    8.MG.2b, 8.MG.2d, MP1, MP4

www.ingramcontent.com/pod-product-compliance
Lightning Source LLC
Chambersburg PA
CBHW061923250225
22490CB00002B/8